Water Optics and Water Colour Remote Sensing

Special Issue Editors

Yunlin Zhang
Claudia Giardino
Linhai Li

MDPI • Basel • Beijing • Wuhan • Barcelona • Belgrade

MDPI

Special Issue Editors
Yunlin Zhang
Chinese Academy of Sciences
China

Claudia Giardino
National Research Council of Italy
Italy

Linhai Li
University of California San Diego
USA

Editorial Office
MDPI AG
St. Alban-Anlage 66
Basel, Switzerland

This edition is a reprint of the Special Issue published online in the open access journal *Remote Sensing* (ISSN 2072-4292) from 2016–2017 (available at: http://www.mdpi.com/journal/remotesensing/special_issues/watercolorRS).

For citation purposes, cite each article independently as indicated on the article page online and as indicated below:

Author 1; Author 2. Article title. *Journal Name* **Year**, *Article number*, page range.

First Edition 2017

ISBN 978-3-03842-508-3 (Pbk)
ISBN 978-3-03842-509-0 (PDF)

Table of Contents

About the Special Issue Editors

Yunlin Zhang Professor, a physical geographer and limnologist interested in lake optics and water colour remote sensing applications, chromophoric dissolved organic matter (CDOM) biogeochemistry cycle, and lake eutrophication control. Yunlin Zhang received the B.S. degree in geography from Hunan Normal University, Changsha, China, and the Ph.D. degree in physical geography from Nanjing Institute of Geography and Limnology (NIGLAS), Chinese Academy of Sciences, Nanjing, China, in 1999 and 2005, respectively. Since 2005, he has been working in NIGLAS, Nanjing, China, in the general field of water optics and remote sensing of environment. At NIGLAS, he is PI on a number of projects related to lake optics and water colour remote sensing including the Distinguished Young Scholar programme of the National Natural Science Foundation of China (NSFC). He has supervised three PhD and 10 master students. In 2017 he received the first prize of Science and Technology Award of Jiangsu Province from Jiangsu Provincial People's Government for his outstanding contribution in research and application of lake optics and water colour remote sensing.

Claudia Giardino, with a PhD in remote sensing, is a research scientist at CNR in Milano with more than 20 years of experience. She is specialized in quantitative aspects of remote sensing, including radiative transfer modelling and biophysical parameters retrieval. Her main interests include retrieval algorithm development in optically complex water systems, harmful algal blooms, submerged vegetation recognition, satellite data processing and exploitation. She also works in the field of Imaging Spectrometry and performs fieldwork activities for calibration/validation purposes. She has been responsible for CNR research activity in more than 20 national and international projects (mostly funded by the Italian Space Agency, ESA and EU) related to earth observation of inland and coastal waters.

Linhai Li is a postdoctoral researcher at Scripps Institution of Oceanography, University of California San Diego. He is interested in hydrologic optics, optical remote sensing of inland and ocean waters, remote sensing inversion algorithms, radiative transfer of ocean, and inelastic processes in the ocean. He received his B.S. degree in GIS from Wuhan University in 2009, his M.S. degree in Geology from Indiana University in 2011, and his Ph.D. degree in Oceanography from University of California San Diego in 2016. Dr. Li's publication record spans a wide range of topics in hydrologic optics, from fundamental research of ocean optics to remote sensing applications for inland and oceanic waters.

-

Preface to "Water Optics and Water Colour Remote Sensing"

Available water resources, including rivers, reservoirs, lakes, coastal waters, and oceans, are emerging as a limiting factor—not only in quantity, but also in quality—for human development and ecological stability. Declining water quality has become a global issue of significant concern as anthropogenic activities expand and climate change threatens to cause major alterations to the hydrological cycle. Thus, monitoring the physical, chemical, and biological status of those waters is of great importance. Remote sensing has the potential to provide an invaluable complementary source of data at local to global scales. However, accurate, cost-effective, frequent, and synoptic retrieval algorithms of in-water optical and biogeochemical parameters, as well as information on the biophysical properties of the monitored waters, face several challenges. This book is intended to expand our understanding of these relationships and processes.

Along with the journal's announcement of an open call for papers, invitations were issued to experts working in this field. The robust response and strong interest in this Special Issue is evidenced by a total of 50 submissions and the excellent work being undertaken. The resulting 21 papers, plus our editorial paper, were subject to peer review by at least two reviewers. We encourage you to read the papers, which, as explained in our editorial paper, we have grouped into five categories: (1) bio-optical properties; (2) atmospheric correction and data uncertainties; (3) remote sensing estimation of chlorophyll-a; (4) remote sensing estimation of suspended matter and chromophoric dissolved organic matter (CDOM); and (5) water quality and water ecology remote sensing. This Special Issue presents much of the recent progress in this rapidly growing research area including a variety of applications at the global scale (with case studies in Europe, Asia, South and North America, and the Antarctic), achieved with different remote sensing instruments, such as hyperspectral field and airborne sensors, ocean colour radiometry, geostationary platforms, and the multispectral Landsat and Sentinel-2 satellites.

We sincerely thank the excellent authors and peer reviewers, as well as the journal's staff, for their contributions to this Special Issue.

Yunlin Zhang, Claudia Giardino and Linhai Li
Special Issue Editors

![remote sensing logo] *remote sensing*

MDPI

Editorial

Water Optics and Water Colour Remote Sensing

Yunlin Zhang [1,*], Claudia Giardino [2] and Linhai Li [3]

[1] Taihu Laboratory for Lake Ecosystem Research, State Key Laboratory of Lake Science and Environment, Nanjing Institute of Geography and Limnology, Chinese Academy of Sciences, Nanjing 210008, China
[2] Institute for Electromagnetic Sensing of the Environment, National Research Council of Italy, via Bassini 15, 20133 Milan, Italy; giardino.c@irea.cnr.it
[3] Marine Physical Laboratory, Scripps Institution of Oceanography, University of California San Diego, La Jolla, CA 92093, USA; lil032@ucsd.edu
* Correspondence: ylzhang@niglas.ac.cn; Tel.: +86-25-8688-2198; Fax: +86-25-5771-4759

Academic Editor: Prasad S. Thenkabail
Received: 4 August 2017; Accepted: 7 August 2017; Published: 9 August 2017

Abstract: The editorial paper aims to highlight the main topics investigated in the Special Issue (SI) "Water Optics and Water Colour Remote Sensing". The outcomes of the 21 papers published in the SI are presented, along with a bibliometric analysis in the same field, namely, water optics and water colour remote sensing. This editorial summarises how the research articles of the SI approach the study of bio-optical properties of aquatic systems, the development of remote sensing algorithms, and the application of time-series satellite data for assessing long-term and temporal-spatial dynamics in inland, coastal, and oceanic waters. The SI shows the progress with a focus on: (1) bio-optical properties (three papers); (2) atmospheric correction and data uncertainties (five papers); (3) remote sensing estimation of chlorophyll-a (Chl-a) (eight papers); (4) remote sensing estimation of suspended matter and chromophoric dissolved organic matter (CDOM) (four papers); and (5) water quality and water ecology remote sensing (four papers). Overall, the SI presents a variety of applications at the global scale (with case studies in Europe, Asia, South and North America, and the Antarctic), achieved with different remote sensing instruments, such as hyperspectral field and airborne sensors, ocean colour radiometry, geostationary platforms, and the multispectral Landsat and Sentinel-2 satellites. The bibliometric analysis, carried out to include research articles published from 1900 to 2016, indicates that "chlorophyll-a", "ocean colour", "phytoplankton", "SeaWiFS" (Sea-Viewing Wide Field-of-View Sensor), and "chromophoric dissolved organic matter" were the five most frequently used keywords in the field. The SI contents, along with the bibliometric analysis, clearly suggest that remote sensing of Chl-a is one of the topmost investigated subjects in the field.

Keywords: water optics; water colour remote sensing; bibliometric analysis; popular study topics; chlorophyll-a

---

1. Water Optics and Water Colour Remote Sensing from a Bibliometrics Perspective

Bibliometrics, which was first introduced by Pritchard [1], has been widely used to quantitatively and qualitatively analyse scientific production and research trends at the decade to century scale with a large amount of data [2]. Using the Science Citation Index-Expanded (SCI-Expanded, where SCI is the multidisciplinary database of the Institute for Scientific Information, Philadelphia, PA, USA) database of the Web of Science and "bibliometric analysis" as the topic, 1502 papers could be searched by the end of 2016. The bibliography of all articles related to water optics and water colour remote sensing from 1900 to 2016, was compiled by searching, in the online version of SCI-Expanded, the following keywords: TS (Topic) = ("inherent optical propert *" or "apparent optical propert *" or "bio-optical propert *" or "remote sensing reflectance *" or "absorption coefficient *" or "scattering coefficient *″ or CDOM) or (("suspended matter *″ or "particle matter *″ or CDOM or chlorophyll or phycocyanin or "water colour *″ or bloom * or macrophyte *) and "remote sensing *″)) and TS = (lake * or reservoir * or ocean * or coastal * or estuary * or bay *). Keywords contain the most critical information in most papers whose frequency was used to discover study hotspots. Notably, only author keywords, not KeyWords Plus, were used for this analysis [3,4].

The total publication number was 4809 when the aforementioned search strategy was used. We presented the 20 most frequently used keywords of water optics and water colour remote sensing research (Figure 1). "chlorophyll-*a*" (Chl-*a*), "ocean colour", "phytoplankton", "SeaWiFS" (Sea-Viewing Wide Field-of-View Sensor), and "chromophoric dissolved organic matter (CDOM)" were the five most frequently used keywords of the study subject after excluding, for obvious reasons, the keyword "remote sensing". In particular, the frequency of "Chl-*a*" was as high as 532, which was much higher than the second frequently used keyword "ocean colour" (86). The result of SeaWiFS indicated the role played in the past from this sensor, while for recent years sensors such as Moderate Resolution Imaging Spectroradiometer (MODIS), Medium-Resolution Imaging Spectrometer (MERIS), and OLCI might be considered as improved successors. This result indicated that Chl-*a*-related study was the popular study topics in the field of water optics and water colour remote sensing.

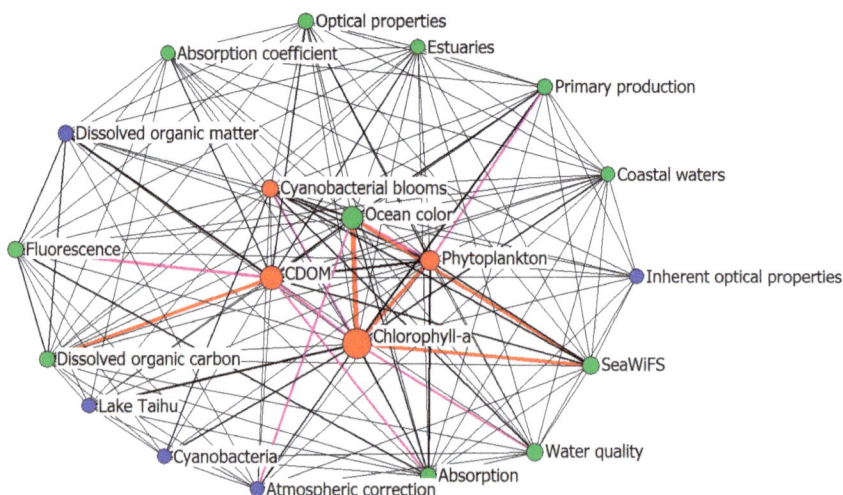

Figure 1. Co-word network of high-frequency keywords in water optics and water colour remote sensing research. The size of lines represents the number of co-occurrences. The colour of lines represents the group of the co-occurrences number: red ≥ 50; 25 ≤ magenta < 50; black ≤ 25. The size of nodes represents the keyword frequency; the colour of nodes represents the group of the number of connections to other nodes: red = 20; 15 ≤ green < 20; 10 ≤ blue < 15.

2. Overview and Scope of the Special Issue

Available water resources, including rivers, reservoirs, lakes, coastal waters, and oceans, are emerging as a limiting factor, not only in quantity, but also in quality, for human development and ecological stability. Declining water quality has become a global issue of significant concern as anthropogenic activities expand and climate change threatens to cause major alterations to the hydrological cycle. Thus, monitoring the physical, chemical, and biological status of those waters is immensely important. Remote sensing has the potential to provide an invaluable complementary source of data at local to global scales. However, accurate, cost-effective, frequent, and synoptic retrieval algorithms of in-water optical and biogeochemical parameters, as well as information on the biophysical properties of the monitored waters, have several challenges.

This Special Issue (SI) focusing on "Water Optics and Water Colour Remote Sensing" is specifically aimed at addressing: (1) issues on water optics including the characterisation of optical properties among rivers, reservoirs, lakes, coastal waters, and open sea, modeling the relationships between apparent optical properties (AOPs) and inherent optical properties (IOPs); and (2) challenges on retrieval algorithm developments, validation, and applications of remote sensing of rivers, reservoirs, lakes, coastal waters, and open ocean. This SI presents many of the recent progresses in this rapidly growing research area.

3. Highlights of Research Articles

The main topics of this SI concentrate on: (1) bio-optical properties (three papers); (2) atmospheric correction and data uncertainties (five papers); (3) remote sensing estimation of Chl-*a* (eight papers); (4) remote sensing estimation of suspended matter and CDOM (four papers); and (5) remote sensing estimation of other water qualities and water ecological parameters (four papers). Overall, Chl-*a* estimation-related study is a major topic in this SI, which is consistent to the previously described bibliometrics analysis.

There are three papers on bio-optical properties in this issue. Ma et al. [5] reported the spatial distribution of the diffuse attenuation coefficient of photosynthetic active radiation (PAR) (K_d(PAR)) and its main regulating factors in 26 lakes and reservoirs in Northeast China. The results showed that total suspended matter was the dominant factor in determining K_d(PAR) values and best correlated with K_d(PAR). Zhou et al. [6] presented a model to describe the polarisation patterns of celestial light, when refracted by wavy water surfaces. Scattering skylight dominates the polarisation patterns, while direct solar light is the dominant source of the intensity of the underwater light field. Wind speed has an influence on disturbing the patterns under water. To optimise atmospheric correction and bio-optical algorithms for the wide variety of lake optical conditions, Eleveld et al. [7] proposed two water type classification schemes through a cluster analysis of in situ hyperspectral remote sensing reflectance spectra collected around the world.

There are five papers on atmospheric correction and data uncertainties in this issue. In each of the three atmospheric correction papers, one or more atmospheric correction strategies were evaluated. The commercial ATCOR4 code was applied to airborne imagery [8], a series three of methods (i.e., ACOLITE NIR, ACOLITE SWIR, and MACCS) based on the use of short-wave infrared (SWIR) spectral bands was applied to satellite data of turbid coastal waters [9], and again three atmospheric correction methods (ACOLITE, 6SV, and Sen2Cor) were applied to Sentinel-2 images of lakes [10]. Two data uncertainties papers discussed the impact of the Signal-to-Noise Ratio of sensor design on the Chl-*a* and total suspended matter algorithms in four lakes located at Mamirauá Sustainable Development Reserve (Amazonia, Brazil) [11], and sensors selection, near-coincident data determination, and potential limitations of remote sensing measurements for surface and near-surface conditions [12].

There are eight papers on the remote sensing estimation of Chl-*a* in this issue. These papers presented Chl-*a* remote sensing estimation models including the height of the 810 peak of remote sensing reflectance [13], the comparison of four empirical models (Blue-Green Ratio Model, Two-Band NIR-Red Ratio Model, Three-Band NIR-Red Model, and Four-Band NIR-Red Model) [14], a coupled atmosphere-hydro-optical model [15], the combination of AisaFENIX sensor and ATCOR4 in image-driven parametrisation [8], iterative stepwise elimination partial least squares regression based on field hyperspectral [16], and an improved baseline fluorescence approach [17]. In addition, long-term spatial and temporal variabilities of Chl-*a* from MODIS and SeaWiFS observations in the Bohai Sea (2000–2012) [18] as well as from MERIS images in Lake Erie (2004–2012) [19] were elucidated.

There are five papers on the remote sensing estimation of suspended matter and CDOM in this issue. Kutser et al. [13] found that the height of the 810 peak was in good correlation with suspended matter and assessed the application of this finding on Landsat 8 and Sentinel-2 images. Wang et al. [20] developed an algorithm using the slope of R_{rs}(490) and R_{rs}(555) to estimate particle cross-sectional area concentration from the Geostationary Ocean Colour Imager (GOCI). An iterative stepwise elimination partial least squares regression based on field hyperspectral was used to estimate water total suspended matter concentration in irrigation ponds in Japan [16]. Campanelli et al. [21] developed an empirical band ratio model using $R_{rs}(667)/R_{rs}(488)$ to estimate CDOM using MODIS ocean colour sensor images and data collected on the North-Central Western Adriatic Sea (Mediterranean Sea).

Lastly, there are four papers on the remote sensing estimation of other water qualities and water ecological parameters in this issue. Turbidity climatology over an eight-year period (2004–2011) in Apalachicola Bay was investigated based on remote sensing estimation from Landsat 5 TM and Landsat 8 OLI imagery using a single-band empirical relationship of band 3 [22]. Deng et al. [23] tested the applicability of the Vertically Generalised Production Model (VGPM) to estimate phytoplankton primary production by comparing the model-derived and the in situ results, investigated the long-term temporal-spatial variations in primary production using MODIS data, and further discussed the potential affecting factors in Lake Taihu. Liang et al. [24] developed a new cyanobacteria and macrophytes index (CMI) based on a blue, a green, and a shortwave infrared band to separate waters with cyanobacterial scums from those dominated by aquatic macrophytes. Deng et al. [25] employed 1375 seasonally continuous Landsat TM/ETM+/OLI data scenes to evaluate the lake water area changes from 1987 to 2015, and found a loss of 241.39 km^2 (10.67%) from 1987–2002 to 2005–2015 in Wuhan Urban Agglomeration, China.

4. Conclusions

This SI presents the state-of-the-art and represents the progress in the field of "Water Optics and Water Colour Remote Sensing". The 21 papers published in this SI present a variety of studies mainly related to bio-optical properties measurement and modeling, atmospheric correction and data uncertainties, remote sensing algorithms for Chl-*a*, suspended matter and CDOM, as well as water quality and water ecology applications. The SI covers a variety of applications at the global scale (with case studies in Europe, Asia, South and North America, and even the Antarctic), with different remote sensing instruments, ranging from hyperspectral field and airborne sensors, to ocean colour radiometry, geostationary platforms, and multispectral sensors such as Landsat and Sentinel-2. In particular, both the bibliometric analysis and a major number of articles in the SI suggest a popular and important research topic in the field—remote sensing of Chl-*a*, a parameter with a global relevance (phytoplankton is a prolific primary producer occurring in oceans and the world's freshwater sources) and one of primary interest for water quality authorities.

Acknowledgments: This study was jointly funded by the National Natural Science Foundation of China (grant 41325001), the Key Program of the Chinese Academy of Sciences (ZDRW-ZS-2017-3-4), and the Key Research Program of Frontier Sciences of the Chinese Academy of Sciences (QYZDB-SSW-DQC016). We would like to thank the staff in the editorial office and all the authors who contribute to the special issue.

Conflicts of Interest: The authors declare no conflict of interest.

References

1. Pritchard, A. Statistical bibliography or bibliometrics? *J. Doc.* **1969**, *25*, 348–349.
2. Zhi, W.; Yuan, L.; Ji, G.; Liu, Y.; Cai, Z.; Chen, X. A bibliometric review on carbon cycling research during 1993–2013. *Environ. Earth Sci.* **2015**, *74*, 6065–6075. [CrossRef]
3. Ho, Y.S.; Siu, E.; Chuang, K.Y. A bibliometric analysis of dengue-related publications in the science citation index expanded. *Future Virol.* **2016**, *11*, 631–648. [CrossRef]
4. Zhang, Y.; Yao, X.; Qin, B. A critical review of the development, current hotspots, and future directions of lake taihu research from the bibliometrics perspective. *Environ. Sci. Pollut. Res.* **2016**, *23*, 12811–12821. [CrossRef] [PubMed]
5. Ma, J.; Song, K.; Wen, Z.; Zhao, Y.; Shang, Y.; Fang, C.; Du, J. Spatial distribution of diffuse attenuation of photosynthetic active radiation and its main regulating factors in inland waters of northeast China. *Remote Sens.* **2016**, *8*, 964. [CrossRef]

6. Zhou, G.; Wang, J.; Xu, W.; Zhang, K.; Ma, Z. Polarization patterns of transmitted celestial light under wavy water surfaces. *Remote Sens.* **2017**, *9*, 324. [CrossRef]

7. Eleveld, M.A.; Ruescas, A.B.; Hommersom, A.; Moore, T.S.; Peters, S.W.M.; Brockmann, C. An optical classification tool for global lake waters. *Remote Sens.* **2017**, *9*, 420. [CrossRef]

8. Markelin, L.; Simis, S.; Hunter, P.; Spyrakos, E.; Tyler, A.; Clewley, D.; Groom, S. Atmospheric correction performance of hyperspectral airborne imagery over a small eutrophic lake under changing cloud cover. *Remote Sens.* **2017**, *9*, 2. [CrossRef]

9. Novoa, S. Atmospheric corrections and multi-conditional algorithm for multi-sensor remote sensing of suspended particulate matter in low-to-high turbidity levels coastal waters. *Remote Sens.* **2017**, *9*, 61. [CrossRef]

10. Martins, V.S.; Barbosa, C.C.F.; de Carvalho, L.A.S.; Jorge, D.S.F.; Lobo, F.d.L.; Novo, E.M.L.d.M. Assessment of atmospheric correction methods for sentinel-2 msi images applied to amazon floodplain lakes. *Remote Sens.* **2017**, *9*, 322. [CrossRef]

11. Jorge, D.; Barbosa, C.; Carvalho, L.S.D.; Affonso, A.; Lobo, F.; Novo, E. Snr (signal-to-noise ratio) impact on water constituent retrieval from simulated images of optically complex amazon lakes. *Remote Sens.* **2017**, *9*, 644. [CrossRef]

12. Hansen, C.H.; Burian, S.J.; Dennison, P.E.; Williams, G.P. Spatiotemporal variability of lake water quality in the context of remote sensing models. *Remote Sens.* **2017**, *9*, 409. [CrossRef]

13. Kutser, T.; Paavel, B.; Verpoorter, C.; Ligi, M.; Soomets, T.; Toming, K.; Casal, G. Remote sensing of black lakes and using 810 nm reflectance peak for retrieving water quality parameters of optically complex waters. *Remote Sens.* **2016**, *8*, 497. [CrossRef]

14. Lins, R.C.; Martinez, J.M.; Marques, D.D.M.; Cirilo, J.A.; Fragoso, C.R., Jr. Assessment of chlorophyll-*a* remote sensing algorithms in a productive tropical estuarine-lagoon system. *Remote Sens.* **2017**, *9*, 516. [CrossRef]

15. Arabi, B.; Salama, M.S.; Wernand, M.R.; Verhoef, W. Mod2sea: A coupled atmosphere-hydro-optical model for the retrieval of chlorophyll-*a* from remote sensing observations in complex turbid waters. *Remote Sens.* **2016**, *8*, 722. [CrossRef]

16. Wang, Z.; Kawamura, K.; Sakuno, Y.; Fan, X.; Gong, Z.; Lim, J. Retrieval of chlorophyll-*a* and total suspended solids using iterative stepwise elimination partial least squares (ise-pls) regression based on field hyperspectral measurements in irrigation ponds in higashihiroshima, Japan. *Remote Sens.* **2017**, *9*, 264. [CrossRef]

17. Zeng, C.; Zeng, T.; Fischer, A.; Xu, H. Fluorescence-based approach to estimate the chlorophyll-*a* concentration of a phytoplankton bloom in ardley cove (Antarctica). *Remote Sens.* **2017**, *9*, 210. [CrossRef]

18. Zhang, H.; Qiu, Z.; Sun, D.; Wang, S.; He, Y. Seasonal and interannual variability of satellite-derived chlorophyll-*a* (2000–2012) in the bohai sea, china. *Remote Sens.* **2017**, *9*, 582. [CrossRef]

19. Zolfaghari, K.; Duguay, C. Estimation of water quality parameters in lake erie from meris using linear mixed effect models. *Remote Sens.* **2016**, *8*, 473. [CrossRef]

20. Wang, S.; Yu, H.; Qiu, Z.; Sun, D.; Zhang, H.; Zheng, L.; Xiao, C. Remote sensing of particle cross-sectional area in the bohai sea and yellow sea: Algorithm development and application implications. *Remote Sens.* **2016**, *8*, 841. [CrossRef]

21. Campanelli, A.; Pascucci, S.; Betti, M.; Grilli, F.; Marini, M.; Pignatti, S.; Guicciardi, S. An empirical ocean colour algorithm for estimating the contribution of coloured dissolved organic matter in north-central western adriatic sea. *Remote Sens.* **2017**, *9*, 180. [CrossRef]

22. Joshi, I.D.; D'Sa, E.J.; Osburn, C.L.; Bianchi, T.S. Turbidity in apalachicola bay, florida from landsat 5 TM and field data: Seasonal patterns and response to extreme events. *Remote Sens.* **2017**, *9*, 367. [CrossRef]

23. Deng, Y.; Zhang, Y.; Li, D.; Shi, K.; Zhang, Y. Temporal and spatial dynamics of phytoplankton primary production in lake taihu derived from modis data. *Remote Sens.* **2017**, *9*, 195. [CrossRef]

24. Liang, Q.; Zhang, Y.; Ma, R.; Loiselle, S.; Li, J.; Hu, M. A modis-based novel method to distinguish surface cyanobacterial scums and aquatic macrophytes in lake taihu. *Remote Sens.* **2017**, *9*, 133. [CrossRef]

25. Deng, Y.; Jiang, W.; Tang, Z.; Li, J.; Lv, J.; Chen, Z.; Jia, K. Spatio-temporal change of lake water extent in wuhan urban agglomeration based on landsat images from 1987 to 2015. *Remote Sens.* **2017**, *9*, 270. [CrossRef]

remote sensing

MDPI

Article

Spatial Distribution of Diffuse Attenuation of Photosynthetic Active Radiation and Its Main Regulating Factors in Inland Waters of Northeast China

Jianhang Ma [1,2], Kaishan Song [1,*], Zhidan Wen [1], Ying Zhao [1,2], Yingxin Shang [1,2], Chong Fang [1,2] and Jia Du [1]

[1]   Northeast Institute of Geography and Agroecology, Chinese Academy of Sciences (CAS), Changchun 130102, China; mmjjhh105@sina.com (J.M.); wenzhidan@iga.ac.cn (Z.W.); zhaoying477@163.com (Y.Z.); goodlucksyx27@163.com (Y.S.); fangchong1991@gmail.com (C.F.); jiaqidu@iga.ac.cn (J.D.)
[2]   University of Chinese Academy of Sciences, Beijing 100049, China
*   Correspondence: songks@iga.ac.cn; Tel.: +86-431-8554-2364

Academic Editors: Linhai Li, Claudia Giardino, Yunlin Zhang, Deepak R. Mishra, Xiaofeng Li and Prasad S. Thenkabail
Received: 15 August 2016; Accepted: 16 November 2016; Published: 21 November 2016

Abstract: Light availability in lakes or reservoirs is affected by optically active components (OACs) in the water. Light plays a key role in the distribution of phytoplankton and hydrophytes, thus, is a good indicator of the trophic state of an aquatic system. Diffuse attenuation of photosynthetic active radiation (PAR) (K_d(PAR)) is commonly used to quantitatively assess the light availability. The PAR and the concentration of OACs were measured at 206 sites, which covered 26 lakes and reservoirs in Northeast China. The spatial distribution of K_d(PAR) was depicted and its association with the OACs was assessed by grey incidences(GIs) and linear regression analysis. K_d(PAR) varied from 0.45 to 15.04 m^{-1}. This investigation revealed that reservoirs in the east part of Northeast China were clear with small K_d(PAR) values, while lakes located in plain areas, where the source of total suspended matter (TSM) varied, displayed high K_d(PAR) values. The GIs and linear regression analysis indicated that the TSM was the dominant factor in determining K_d(PAR) values and best correlated with K_d(PAR) ($R^2 = 0.906$, RMSE = 0.709). Most importantly, we have demonstrated that the TSM concentration is a reliable measurement for the estimation of the K_d(PAR) as 74% of the data produced a relative error (RE) of less than 0.4 in a leave-one-out cross validation (LOO-CV) analysis. Spatial transferability assessment of the model also revealed that TSM performed well as a determining factor of the K_d(PAR) for the majority of the lakes. However, a few exceptions were identified where the optically regulating dominant factors were chlorophyll-*a* (Chl-*a*) and/or the chromophroic dissolved organic matter (CDOM). These extreme cases represent lakes with exceptionally clear waters.

Keywords: light attenuation coefficient; optically active constituents; Northeast China; total suspended matter; water transparency

---

1. Introduction

Diffuse attenuation of photosynthetic active radiation (PAR), expressed as K_d(PAR), indicates the ability of solar radiation to penetrate a water column. The distribution of algae and hydrophytes, which contribute greatly to the lake's primary production, is mainly influenced by the availability of light as well as other factors, for example, temperature and nutrition [1,2]. Euphotic zone depth (z_{eu}), an important input parameter for ecological models that estimate primary production of inland waters [3,4], defined as the depth where 1% of the PAR just beneath the water surface remains [5], can be calculated using the equation $z_{eu} = 4.6/K_d(PAR)$ [5] provided that a water column is homogeneous. The $K_d(PAR)$ is also an important variable to research the heat transfer of lakes [6]. Global climate change and anthropogenic impacts have strong effects on a lakes' ecosystem and this may be depicted by changes in $K_d(PAR)$. Therefore, research of $K_d(PAR)$ and further understanding of its effects can significantly contribute to the development of new approaches for the management and protection of lake environments [7]. Although water transparency, measured with Secchi disk depth (SDD) by human eyes, can also represent the light properties of lakes [8], $K_d(PAR)$ provides a more objective depiction as it is measured with advanced electro-optical instruments [9].

$K_d(PAR)$ can be obtained by fitting the profile of PAR values measured at different depths of the water versus the corresponding depths, according to Lambert–Beer's law [10–12]. Although portable instruments, such as the Li-cor 191, are capable for accurate measurements to estimate $K_d(PAR)$, there are limitations in their application on large scale regions [12] as frequent in situ sampling is costly, labor-intensive and time-consuming. Optical remote sensing imagery is a cost-efficient method to obtain $K_d(PAR)$ values at large regional scale due to the correlation between $K_d(PAR)$ and water leaving radiance, and also due to its spatial and temporal resolution. Numerous remote sensing data such as Landsat/TM/OLI [13], Sea-viewing Wide Field-of-view Sensor (SeaWiFS) [14], the Moderate Resolution Imaging Spectroradiometer (MODIS) [15], the Medium Resolution Imaging Spectrometer (MERIS) [12,16,17] and the Geostationary Ocean Color Imager (GOCI) [18] have been applied to retrieve $K_d(PAR)$ or $K_d(490)$ (which is often used as an agent of $K_d(PAR)$).

Semi-analytical models [15,19–22] for $K_d(PAR)$ or $K_d(490)$ inversion emphasized the importance of inherent optical properties (IOPs) and improved the accuracy of the $K_d(PAR)$ estimation in both open ocean waters (case-I) and coastal or inland waters (case-II). However, large uncertainties still existed in this type of algorithms for turbid and optically complex inland waters [15,19,21]. Though the applications may be confined to specified regions or seasons, empirical models have been widely used to derive $K_d(PAR)$ from remote sensing data for both case-I and case-II waters [12,23–25]. The models were built by calibrating in situ $K_d(PAR)$ with remote sensing reflectance at blue-green [14,24], red [12] or Near-infrared (NIR) [26] bands so the $K_d(PAR)$ can be directly mapped from remote sensing images. $K_d(PAR)$ is governed by the properties of natural water that include both dissolved and particulate organic as well as the inorganic material [2,5,27]. Therefore, the $K_d(PAR)$ can be expressed as a function of the dominant one or some optically active components (OACs) whose concentrations can be estimated from remote sensing data empirically or semi-analytically. Thus, $K_d(PAR)$ can be mapped indirectly from remote sensing images. Chl-*a* plays a significant role in optical property of case-I water so it is rational to estimate $K_d(PAR)$ by the concentration of Chl-*a* that derived from satellite images [9,28]. Dominant factor of case-II water's optical property varied dramatically from Chl-*a* [27] to TSM [10,29] or CDOM [1,30]. Sometimes, there were multiple dominant factors [31] and they changed by seasons [32]. Thereby, comprehensive analysis of the relationships between $K_d(PAR)$ and OACs with in situ data was necessary before indirectly deriving $K_d(PAR)$ from remote sensing data for inland waters. Models to estimate $K_d(PAR)$ from OACs can be built with in situ data [33–35]. Factors that play a significant role in the variance of $K_d(PAR)$ can be identified through the analysis [2] and further used as a guidance for the policy making in the protection of limnology environments.

Northeast China lake zone, with 882 lakes whose area was greater than 1 km^2 in 2010 [36], is one of the five lake zones of China [37]. Some of the lakes and reservoirs in Northeast China are featured with high TSM and CDOM due to the strong wind and shallow water depth, combined with rich soil organic matter that supply much terrigenous dissolved organic matter (DOM) into waters [38–41]. Thus, the optical property and their influence on $K_d(PAR)$ may be different from other lake zones of China. Remote sensing may be the most suitable method to monitor environmental parameters for the widely distributed lakes. Investigations on $K_d(PAR)$ in lakes of East China zone have been carried out [2,7,12,26,31]. Comparatively, much less research on $K_d(PAR)$ in Northeast China lakes have been conducted. Thus, the analysis of dominate OACs of $K_d(PAR)$ could provide a guidance for future investigating of K_d(PAR) by remote sensing. The objectives of this paper are: (1) to depict the spatial distribution pattern of K_d(PAR) of lakes and reservoirs, which represents the trophic state of waters in Northeast China well; (2) to determine the dominant OACs of K_d(PAR) in waters of Northeast China using data collected from 26 lakes and reservoirs with three types of grey incidences (GIs); and (3) to build a relationship between K_d(PAR) and OACs that is meaningful for indirectly mapping K_d(PAR) from remotely sensed imagery.

2. Materials and Methodologies

2.1. Study Area

Northeast China is an important base of agriculture, forestry, energy and heavy industry. The region which extends from 38°N to 54.0°N and 116°E to 136°E, includes all of the Heilongjiang, Jilin and Liaoning Provinces, and parts of Inner Mongolia Province. Its topography is characterized by mountains to the east, north and west that surround the Sanjiang, Songnen, and Liaohe Plains (Figure 1a). The region has a temperate continental monsoon climate which is controlled by the East Asian monsoons. This climate is characterized for its cool and short summers and for its cold and long winters with the lakes being frozen in winters. The annual average temperature ranges between -4 and 12 °C with a gradual increase from north to south. Precipitation is generally higher in the summer and autumn seasons and decreases from 1100 mm in the southeast to 250 mm in the west [42].

Forests are predominantly distributed in mountainous areas where the soil erosion is weak. Grasslands are mainly distributed in the Inner Mongolia plateau where the precipitation is generally less than 400 mm (Figure 1b). A large amount of lakes and reservoirs with various features are distributed in Northeast China. Reservoirs are mainly located in mountainous areas and are very deep with long-narrow shapes. Unlike reservoirs, most of the lakes are generally shallow and are located in plain areas, particularly in Western Songnen Plain. Due to the character of environmental factors like unevenly distributed precipitation, high evaporation and geomorphology of terminal-flow areas, many fresh and saline water bodies are distributed in the Songnen Plain [39].

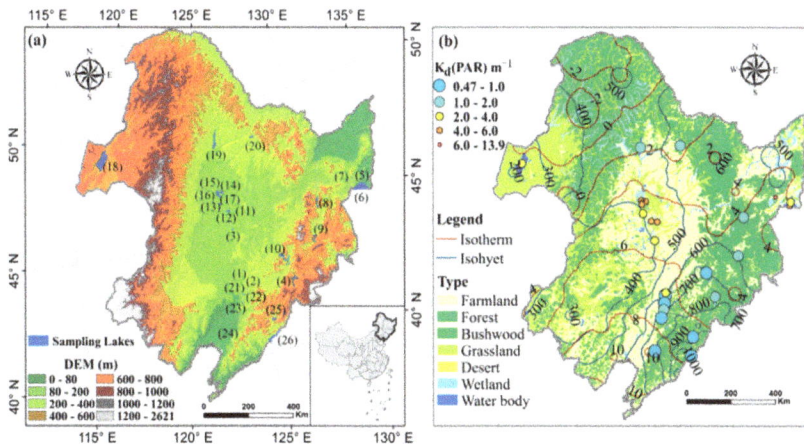

Figure 1. Study area map: (**a**) Digital elevation model (DEM) and sampling lakes with corresponding lake ID; and (**b**) type of vegetation and spatial distribution of the K_d(PAR) derived from in situ measurements, combined with isotherm and isohyets in Northeast China.

2.2. In Situ Data Collection and Analysis

In total, 206 stations with measurement of PAR collected in six field experiments were used. The stations covered 26 lakes and reservoirs in Northeast China. Details about the distribution of the sample points are shown in Figure S1 (Supplementary Materials). There are only two lakes with areas less than 10 km^2. The water approximately 0.5 m below the surface was collected in the acid-washed HDPE bottles for laboratory analysis. The location of each station was recorded with a UniStrong G3 GPS uint. Water transparency (Secchi disk depth, SDD) was measured with a 30 cm diameter black and white quadrant (Secchi) disk. PAR was measured by the Li-Cor 193SA underwater spherical quantum sensor on the sunny side of the boat to avoid any shadow effects. After posing the sensor at one depth in the water, PAR value was continuously recorded for 15 s and output an averaged value by the data logger. This value was regarded as the PAR value at this depth. The PAR measurements were taken at no less than five point's depth for each station. K_d(PAR) was determined by applying the exponential regression model which utilizes Equation (1) [33], provided that the water column was homogeneous. The results were accepted only if the coefficient of determination (R^2) was no less than 0.95 [7] and the number of depth points was no less than 4. In Equation (1), the PAR(Z) represents the PAR value at depth Z and PAR(0^-) represents the PAR value just beneath the surface of the water.

$$\text{PAR}(Z) = \text{PAR}(0^-) \times \exp\left[-K_d(\text{PAR}) \times Z\right] \tag{1}$$

2.3. Water Quality Parameters

To determine the water quality, we calculated the concentrations of TSM and Chl-*a*, and the absorption of CDOM, phytoplankton and non-algal particles (NAP), as follows.

TSM concentration: For all samples, the concentration of TSM was determined gravimetrically. Whatman GF/F glass fiber filters (47 mm in diameter, 0.7 μm in average pore size) were initially combusted at 400 °C for 4 h to remove any organic matters on the filters. After cooling, they were weighed before proceeding to filtration. The volumes of water samples to be filtered were determined by their turbidity. The used filters were stored at 4 °C and re-weighed after drying for 4 h at 105 °C. The concentration of the TSM was calculated by dividing the difference of weight by the volume of the corresponding water sample.

Chl-*a* concentration: Chl-a was determined spectrophotometrically [43,44]. A certain volume (V) of water sample was filtered through GF/F cellulose acetate membrane filters with 47 mm in diameter and 0.47 μm in pore size. The filters were frozen at −20 °C and stored under dark conditions until further analysis. Pigments were extracted by soaking the mashed filters in 10 mL of 90% acetone solution for 24 h under dark conditions. The supernatant was collected after centrifugation (5000 r/min, 20 min) and its absorbance at 630, 647, 664 and 750 nm was measured by the Shimadzu UV-2600 PC spectrophotometer. Concentration of Chl-*a* was calculated by Equation (2) [45], where OD(630), OD(647), OD(664) and OD(750) represented the absorbance at 630, 647, 664 and 750 nm, respectively. The number 10 is the volume of the acetone solution. V is the volume of water sample in liter. L is the cuvette path length in cm. The cuvette with path length of 1 cm was used in this study.

$$\begin{aligned}\text{Chl} - a = \{&11.85 \times [\text{OD}(664) - \text{OD}(750)] - 1.54 \times [\text{OD}(647) - \text{OD}(750)] \\ &- 0.08 \times [\text{OD}(630) - \text{OD}(750)]\} \times 10/(V \times L)\end{aligned} \tag{2}$$

CDOM absorption: The generally high concentration of particles in inland waters results in the difficulties in collecting enough filtrate for measurement by solely filtering water samples through 0.22 μm filters as the particles block the pore easily. Thus, water samples were initially filtered through 0.7 μm (pore size) Whatman GF/F glass fiber filters (pre-combusted at 400 °C for 4 h in a Muffle furnace) and then through 0.22 μm (pore size) nuclepore filters (Whatman) [2]. Then spectrophotometer (Shimadzu UV-2600) was used to measure the CDOM absorbance spectra (OD(λ)) between 200 and 800 nm at 1 nm intervals with the filtrate in 1 cm quartz cuvette and Milli-Q water as reference. The absorption spectrum ($a_{CDOM}(\lambda)$) was calculated from the absorbance using Equation (3) [46], where L is the cuvette path length (0.01 m) and 2.303 is the conversion factor. Some fine particles may have remained in the filtrate so backscattering caused by them should be corrected with absorption of CDOM was assumed to be zero at λ_0, where λ_0 equals to 700 nm [2].

$$a'_{CDOM}(\lambda) = \frac{2.303}{L} \times [OD(\lambda) - OD(\lambda_0) \times (\lambda/\lambda_0)] \tag{3}$$

The absorption coefficient at 355 nm ($a_{CDOM}(355)$) was selected to represent the CDOM concentration [47,48].

Absorption coefficients of phytoplankton (a_{ph}) and NAP (a_{NAP}) were measured by the quantitative filter technique (QFT). A certain volume of each water sample was filtered through Whatman GF/F filter with a nominal pore size of 0.7 μm. Absorption of total particles (a_p) on the filter was measured by spectrophotometer (Shimadzu UV-2600), and then a_{NAP} was measured after the filter was bleached by sodium hypochlorite solution to remove the pigment. As for phytoplankton absorption, a_{ph} was calculated by subtracting a_{NAP} from a_p. The details can be found in [40,49]. Due to the artificial factors during the experiment, only a_p of each sample point were measured for Baishan Reservoir (BSR number 4, Table 1).

Table 1. Summaries of sample points and average values of optical parameters of the sampling lakes. N: counts of sample points; SDD: Secchi disk depth; TSM: total suspended matter; $a_{CDOM}(355)$: the absorption coefficient of chromophroic dissolved organic matter (CDOM) at 355 nm; Chl-*a*: chlorophyll-*a* concentration.

Water Name [a]	Abbreviation (Number)	Area (km²)	Date	N	K_d(PAR) m^{-1}	SDD (m)	TSM (mg/L)	a_{CDOM}(355) m^{-1}	Chl-*a* (μg/L)
Shanmen R.	SMR(1)	1.5	21 April 2015	4	0.77	1.543	3.17	7.86	5.88
Xiasantai R.	XSTR(2)	1.3	21 April 2015	4	2.73	0.703	18.75	10.56	32.84
Xinmiaopao	XMP(3)	26.6	24 April 2015	8	2.53	0.506	26.01	3.51	7.66
Baishan R.	BSR(4)	90.0	4 May 2015	24	1.11	1.283	6.29	5.08	28.02
Xiaoxingkai L.	XXKL(5)	162.1	15 August 2015	13	3.65	0.362	34.77	4.43	6.21
Daxingkai L.	DXKL(6)	1062.4	17 August 2015	8	5.43	0.226	55.17	1.85	6.58
Qingnian R.	QNR(7)	41.1	18 August 2015	4	13.93	0.113	174.50	5.14	4.31
Lianhua L.	LHL(8)	111.7	19 August 2015	14	1.80	1.284	8.12	4.30	16.89
Jingbo L.	JBL(9)	88.8	20 August 2015	11	1.22	1.520	4.61	4.84	23.00
Songhua L.	SHL(10)	216.2	21 August 2015	7	0.68	2.599	1.48	2.41	5.41
Kulipao	KLP(11)	11.6	6 September 2015	4	4.32	0.328	22.10	10.56	3.74
Nanyin R.	NYR(12)	96.4	7 September 2015	3	5.35	0.253	34.17	3.22	40.04
Lamasipao	LMSP(13)	47.9	8 September 2015	7	2.81	0.440	16.10	6.03	59.80
Longhupao	LHP(14)	126.9	9 September 2015	10	4.11	0.280	36.33	4.69	12.27
Talahongpao	TLHP(15)	67.5	10 September 2015	2	5.63	0.240	56.00	6.49	20.62
Xihulupao	XHLP(16)	57.9	10 September 2015	7	6.23	0.254	50.00	4.71	21.25
Huoshaolipao	HSLP(17)	64.3	10 September 2015	8	4.00	0.371	27.18	4.53	9.02
Hulun L.	HLL(18)	2050.2	14 September 2015	28	3.54	0.402	28.38	5.23	7.37
Nierji R.	NEJR(19)	429.6	16 September 2015	16	1.56	1.678	4.24	6.53	4.46
Shankou R.	SKR(20)	64.9	17 September 2015	3	1.27	1.847	1.99	7.59	6.88
Nanchengzi R.	NCZR(21)	10.7	14 April 2016	4	0.85	1.80	1.98	2.45	2.97
Qinghe R.	QHR(22)	17.3	15 April 2016	3	1.49	1.24	8.61	1.98	5.03
Chaihe R.	CHR(23)	12.1	15 April 2016	3	1.00	1.17	4.00	1.83	8.12
Tanghe R.	THR(24)	18.1	18 April 2016	4	0.92	1.85	4.47	1.07	3.25
Huanren R.	HRR(25)	69.6	19 April 2016	2	0.47	2.74	2.63	1.92	3.58
Shuifeng R.	SFR(26)	165.7	20 April 2016	5	0.97	1.45	8.85	1.98	3.94

[a] denotes that L. = Lake; R. = Reservoir; pao means lakes in Northeast China; LHL, JBL and SHL are actually reservoirs but their Chinese names are called lakes.

2.4. Grey Incidences (GIs) Analysis

The grey system theory is a method of processing and analyzing systems with incomplete information [50]. It has been used in remote sensing applications [51]. GIs provide a quantitative description of the system and dominant factor which influences the system's development can be determined with GIs. Given a system contains m factors and one output, for k times of tests of the system, it generates a data sequence with k values of the output as Equation (4). This data sequence is defined as systematic behavior. The corresponding factors compose m data sequences that each contains k values (Equation (5)). This data sequences are defined as factor's behavior.

$$X_0 = \{x_0(k), k = 1, 2 \ldots n\} \tag{4}$$

$$X_i = \{x_i(k), k = 1, 2 \ldots n\} \ (i = 1, 2, 3 \ldots m) \tag{5}$$

In Equations (4) and (5), X_0 and X_i represent the data sequences, $x_0(k)$ is the system output at kth test, and $x_i(k)$ is system factor value of ith factor at kth test. GIs describe the developmental trends of behaviors and factors in a system by analyzing the similarity of geometric patterns between systematic behavior and factor's behavior data sequences [50]. By calculating and comparing GIs between X_0 and X_i, the most influential factors could be identified.

In this research, $K_d(PAR)$ of the sample points was regarded as the systematic behavior meanwhile corresponding Chl-*a*, TSM, and CDOM were regarded as factor's behavior. Three types of GI were calculated. The first one (GI1) was proposed by Deng [52] and the details about calculation can be found in [53]. GI1 was the first model proposed in grey system theory. It measured the trend of system behaviors and factors by the distance between corresponding points of the data sequences. The second one (GI2) was the improved generalized absolute grey incidence model proposed by Cao and the details about its calculation can be found in [50]. GI2 is the modification of the GI proposed by Liu [53]. It measured the trend of system behaviors and factors by the area of the region that surrounded by the curves of the data sequences. Finally, the third one (GI3) was the absolute degree of grey incidence proposed by Mei [54]. According to GI3 model, data sequences X_0 and X_i were converted to $Y_0 = \{y_0(k), k = 1, 2, \ldots, n - 1\}$ and $Y_i = \{y_i(k), k = 1, 2, \ldots, n - 1\}$, respectively, using Equation (6). This conversion calculates the slopes of the adjacent data points in each data sequence.

$$y(k) = x(k + 1) - x(k) \ k = 1, 2, 3, \ldots, n - 1 \tag{6}$$

$$GI(X_0, X_i) = \frac{1}{n-1} \sum_{i=1}^{n-1} \frac{1}{1 + |y_0(k) - y_i(k)|} \tag{7}$$

The GI of two data sequences ($GI(X_0, X_i)$) was calculated from Equation (7). As shown above, the GI3 measured the trend of system behaviors and factors by the slope of the data sequences.

In order to eliminate influence of dimension, the data sequences were standardized to values ranging from 0 to 1 according to Equation (8) before calculating the three kinds of GIs.

$$x_i'(k) = \frac{x_i(k) - \min(X)}{\max(X) - \min(X)} \tag{8}$$

2.5. Linear Regression between $K_d(PAR)$ and OACs

GIs provided the means to identify which of the OACs acts as the dominant factor in determining $K_d(PAR)$, however it does not provide a relationship to quantitatively derive $K_d(PAR)$ from the factor. Therefore, a linear regression analysis was performed to establish the quantitative relationship between $K_d(PAR)$ and the concentration of the OACs. In some researches the regression analysis was also used to identify the dominant factor of $K_d(PAR)$ [2,7] so it can also be used to validate the result of GIs analysis in this paper. In order to further evaluate the applicability of the predicting model, we first used the leave-one-out cross validation (LOO-CV) method and then accessed the model's spatial transferability. For the LOO-CV analysis, $n - 1$ samples were used to calibrate $K_d(PAR)$ and the sample left out was used to validate the model. Similarly, leave one lake out cross validation was used for the assessment of spatial transferability, sample points from $n - 1$ lakes were used to calibrate the $K_d(PAR)$ and the sample points of the lake that was left out were used for validation. The prediction error sum of squares (PRESS) was used to derive the root-mean-square error of cross-validation (RMSECV) of the LOO-CV. Moreover, the relative error (RE), the mean relative error (MRE), and the root-mean-square error (RMSE) were used to assess the accuracy of the model. The details on how to calculate these indicators can be found in [12,44].

2.6. Relationship between K_d(PAR) and SDD

Water transparency is an easily measured indicator of water quality and is used to calculate the trophic state index [55]. SDD is correlated with K_d(PAR) as they are both influenced by the absorption and scatter characteristics of the OACs; in effect, both measurements represent the penetration of light in the water [56]. The difference between them is the varied contributions of the extent of spectral bands as SDD is related to the visible domain (410–665 nm) [57,58] and K_d(PAR) is 400–700 nm [5,33]. K_d(PAR) and SDD are inversely correlated since higher K_d(PAR) values indicate lower water clarity. The relationship of SDD and K_d(PAR) as defined by Holmes [8] is:

$$K_d(PAR) = f \times SDD^{-1} \tag{9}$$

where f has the value of 1.44 for turbid coastal waters [8]. However, other studies have reported different values of f, ranging from 1.7 to 2.3 [59,60]. In this paper, f was determined by linear regression with a fixed intercept at 0. With the relationship K_d(PAR) can be estimated from SDD.

3. Results and Discussion

3.1. Water Quality Characteristics

Due to the different geographical environments, the sampled area included a large diversity of inland waters with varying concentration of OACs (Table 1). The concentration of Chl-*a* (average: 15.16 ± 15.78 µg/L) varied from 1.20 (sample point number 10 of NEJR, number 19, Table 1) to 67.15 µg/L (sample point number 1 of LMSP, number 13, Table 1). BSR (number 4, Table 1) exhibited the largest variation of Chl-*a* ranging from 6.36 to 60.94 µg/L (average: 28.02 ± 20.35 µg/L) attributed to its long-narrow shape. Upstream regions contained high concentration of Chl-*a*. As the water depth becomes deep and flow velocity becomes slow from upstream to downstream regions, the suspended components sunken so concentration of Chl-*a* gradually decreased in the up layer of the water. The TSM concentration ranged from 0.83 (sample point number 3 of SHL, number 10, Table 1) to 184 mg/L (sample point number 4 of QNR, number 7, Table 1), with an average of 4.55 ± 28.25 mg/L. The highest concentration appeared in QNR (average: 174.5 ± 8.85 mg/L) and the lowest was in SHL (average: 1.48 ± 0.97 mg/L). XMP (number 3, Table 1) had the largest variation in the concentration of TSM (range: 10.75 to 64 mg/L, average: 26.01 ± 17.32 mg/L). This variation resulted from one particular sample point with very high TSM concentration (64 mg/L). It was collected in a site with high water turbulence as it located at the junction of a river and the lake. The absorption coefficient of CDOM at 355 nm was high both in XSTR (10.56 ± 0.25 m^{-1}, number 2, Table 1) and KLP (10.56 ± 0.44 m^{-1}, number 11, Table 1) due to their grayish yellow water color.

The SDD ranged from 0.1 to 4.32 m (Table 1). QNR (number 7, Table 1) had the lowest water clarity (0.11 ± 0.01 m), while HRR (number 25, Table 1) was the clearest (2.74 ± 0.34 m). The coefficient f fitted from all sample points was equal to 1.38 (R^2 = 0.97) (Figure 2a). This result was closer to the one reported by Holmes [8] indicating that the equation proposed therein can be used to predict the K_d(PAR) from the SDD measurements.

3.2. Spatial Distribution of K_d(PAR)

The mean K_d(PAR) of each lake and their corresponding geographical distribution are shown in Figure 1b. In general, the reservoirs displayed low K_d(PAR) values. The reason might be due to the fact that they are located in mountainous areas, which are covered by forests that prevent soil erosion, hence, resulting in low concentrations of TSM. Lakes distributed in the Songnen Plain overall displayed high K_d(PAR) values. The soil erosion around these lakes was strong, the lakes were shallow, and the re-suspension caused by strong winds in the area also contributed to high turbidity [39]. These findings were in agreement with the research conducted by Olmanson [61], which revealed that lakes of Minnesota that were located in forest regions were clearer than those located in plain regions.

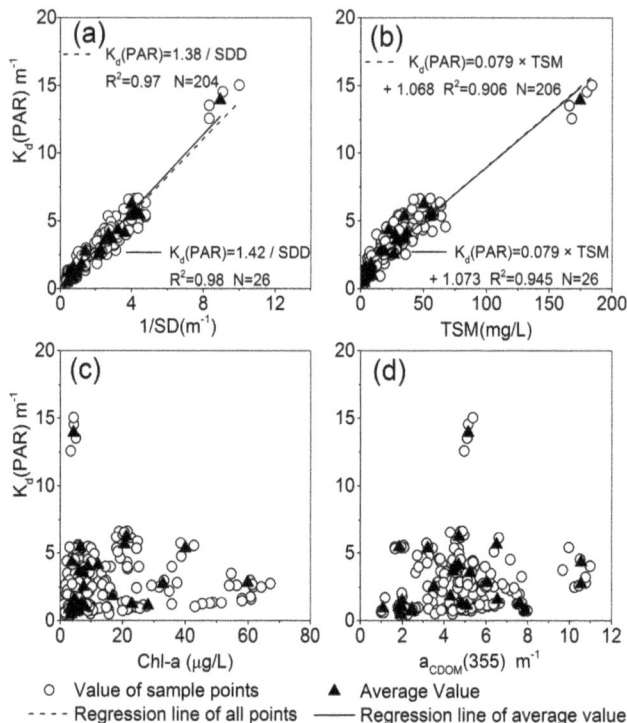

Figure 2. Correlation between K_d(PAR) and the concentration of the water's quality parameters: (**a**) water transparency (SDD); (**b**) total suspended matter (TSM); (**c**) chlorophyll-*a* (Chl-*a*); and (**d**) absorption coefficient of chromophroic dissolved organic matter (CDOM) at 355 nm (a_{CDOM}(355)).

3.3. Grey Incidences between K_d(PAR) and OACs

According to previous studies [48], absorption coefficient at 355 nm of CDOM (a_{CDOM}(355)) was regarded as a representative measure of the concentration of CDOM. GIs were calculated with TSM, Chl-*a* and a_{CDOM}(355) as the factor's behavior data sequences while K_d(PAR) as the systematic behavior data sequence. The factor with higher GIs score indicates a relative bigger influence on K_d(PAR). The GIs of all the sample points were calculated and the results (Table 2) showed that the highest GIs were obtained by comparing between TSM and K_d(PAR). These results indicated that TSM had a higher contribution to K_d(PAR) than Chl-*a* and CDOM, which is consistent with previous studies in Lake Taihu in China [7,12] and UK marine waters [10]. Furthermore, in order to assess the difference of the relationships across various waters, lakes with more than 10 sampling points were also selected to calculate GIs individually. The results (Table 2) indicated that the dominant factors of K_d(PAR) varied according to lakes. For example, K_d(PAR) of BSR may largely influenced by TSM and Chl-*a* as the GIs of two materials did not vary significantly.

Table 2. GIs between the OACs and K_d(PAR) of all the sample points and of lakes whose number of sampling points were greater than 10. The bigger the GI the larger the influence of OACs on K_d(PAR).

Name	OACs	GI 1	GI 2	GI 3	Name	OACs	GI 1	GI 2	GI 3
	TSM	0.82	0.81	0.97		TSM	0.73	0.73	0.94
All points	a_{CDOM}(355)	0.51	0.69	0.86	Jingbo L.	a_{CDOM}(355)	0.63	0.87	0.92
	Chl-a	0.63	0.66	0.85		Chl-a	0.58	0.81	0.83
	TSM	0.77	0.91	0.92		TSM	0.73	0.65	0.86
Baishan R.	a_{CDOM}(355)	0.47	0.78	0.82	Longhupao	a_{CDOM}(355)	0.66	0.45	0.63
	Chl-a	0.73	0.87	0.91		Chl-a	0.71	0.53	0.82
	TSM	0.83	0.77	0.88		TSM	0.73	0.54	0.78
Xiaoxingkai L.	a_{CDOM}(355)	0.57	0.47	0.76	Hulun L.	a_{CDOM}(355)	0.68	0.45	0.76
	Chl-a	0.73	0.74	0.74		Chl-a	0.65	0.30	0.75
	TSM	0.86	0.93	0.92		TSM	0.72	0.71	0.89
Lianhua L.	a_{CDOM}(355)	0.55	0.71	0.89	Nierji R.	a_{CDOM}(355)	0.62	0.56	0.81
	Chl-a	0.66	0.72	0.77		Chl-a	0.57	0.46	0.87

L. = Lake; R. = Reservoir.

3.4. K_d(PAR) Model Calibration and Validation

In this study, TSM was revealed to be the major impact factor for the determination of K_d(PAR). The linear regression analysis provided the means to predict K_d(PAR) by utilizing the OACs data. The results indicated that K_d(PAR) was strongly correlated to TSM, while there was no obvious relationship between K_d(PAR) and Chl-a or CDOM (Figure 2). The mean value of K_d(PAR) and OACs of each lake were determined by averaging all the sample points collected from the corresponding lakes. The coefficient of determination (R^2) was improved when averaged value of each lake, rather than all the single point values, was fitted in the linear model. This is because experimental error may exist in a single sample point, which leads to the dispersion of the data. The average of the data can eliminate the error to some extent and result in a stable performance. The slope of the linear model between K_d(PAR) and TSM in the current study was slightly higher than results in [7] (slope = 0.0626, intercept = 1.6068) and [12] (slope = 0.0563, intercept = 1.52). Considered that our research covered a larger number of lakes than those earlier studies, such a difference was acceptable. The results of linear regression also indicated that the dominant factor of K_d(PAR) for lakes of Northeastern China is TSM as the GIs indicated. This result provides a guidance in band selection to derive K_d(PAR) from remote sensing image that the bands well correlate to TSM may perform well in deriving K_d(PAR) [12]. The model estimating K_d(PAR) from TSM may be applied in mapping K_d(PAR) indirectly from remote sensing image. The dominant factor of K_d(PAR) for lakes in Northeastern China was same to Lake Taihu, a large shallow lake in Eastern China [2,7,12]. However, the coefficients of the linear relationship indicated there was difference in optical properties between Lake Taihu and lakes in Northeastern China.

A multivariate linear regression analysis was also performed with K_d(PAR) as dependent variable and TSM, Chl-a and a_{CDOM}(355) as independent variables. A slight better fit was obtained as the R^2 was 0.916 for all sample point values and 0.960 for lake specified mean values. The result was similar to those of Zhang [2] and Devlin [11], demonstrating that slight improved models were achieved when fit K_d(PAR) with all three explanatory variables rather than solely TSM for waters that K_d(PAR) was mainly influenced by TSM.

The simple linear model may be suitable for predicting K_d(PAR) from TSM as it produced low error rates (RMSE: 0.689 for lake averaged values, and 0.709 for single point values). The RMSECV of the LOO-CV analysis was 0.709 and the MAPE was 0.315. The RE which was used for validation ranged from 0.0016 to 2.2 (Figure 3). The relative error of 61% of the samples was below 0.3 and of 74% of the samples was below 0.4. Altogether our results indicated that the TSM performed well in estimating K_d(PAR). However, care should be taken when applying this model to some cases as TSM might not always represent the major determining factor of K_d(PAR). This could be the case of eight sample points in our study which displayed RE values larger than 1.0 in LOO-CV.

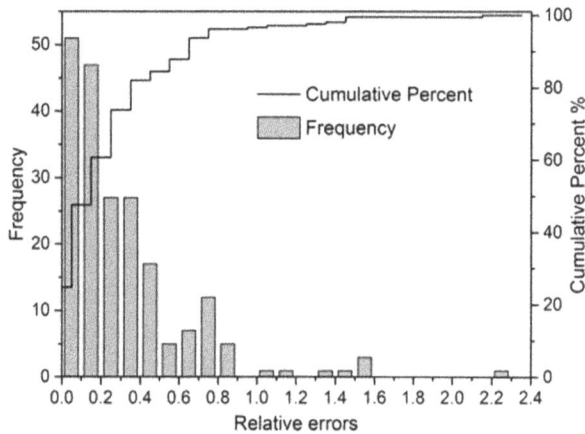

Figure 3. Frequency distribution and cumulative percentage of the relative errors (RE) derived from the leave-out-one cross validation (LOO-CV) of the regression model that had been used to calibrate the K_d(PAR) from the concentration of TSM.

To investigate this further, 32 points, whose RE of LOO-CV was greater than 0.6, were selected to analyze the relationship between K_d(PAR) and OACs. For the 32 sample points, the averaged K_d(PAR) was 0.727 ± 0.150 m^{-1} and the large RE value was mainly caused by the overestimation of the model (see Figure S2). The linear regression analysis (see Figure S3) indicated that K_d(PAR) had a better correlation with the concentration of TSM ($R^2 = 0.60$) than with Chl-*a* and CDOM ($R^2 < 0.1$). However, when compared to the model built by all sample points, the slope and intercept decreased to 0.06 and 0.522, respectively. This explained why the model built with all sample points lead to overestimation of K_d(PAR) for the 32 sample points. However, the GIs revealed that the correlation between the OACs (TSM, Chl-*a*, CDOM) and K_d(PAR) did not vary significantly (GI1: 0.70, 0.65, 0.68), (GI2: 0.55, 0.67, 0.72) and (GI3: 0.91, 0.80, 0.80),which indicated that the dominant factors of K_d(PAR) may not solely TSM for this 32 sample points. The R^2 of multivariate linear regression analysis was 0.64, which implied an improved fit with TSM, Chl-*a* and CDOM.

The integration of the absorption coefficient curves between 400–700 nm were used to calculate the relative contribution of phytoplankton, NAP and CDOM to the total absorption of the samples (Figure 4). Due to the absorption spectra of phytoplankton and NAP in the BSR were not measured, therefore, six data points from BSR were not analyzed further. Nonetheless, the absorption spectra of the total particulate matter showed an obvious peak at 675 nm (see Figure S4), which indicated that the contribution of the phytoplankton to the total absorption is strong. The total absorption of HRR was dominated by CDOM and phytoplankton which might cause the large error observed by solely estimating K_d(PAR) from TSM. Though NAP contributed more than phytoplankton and CDOM to the total absorption of QHR1, CHR1, THR and SFR, high RE were observed when predicting K_d(PAR) solely from the TSM. This might be attributed to the extremely clear water which makes the model unsuitable for this type of environment; therefore, a separate analysis should be done for this lake. For JBL and SHL, the CDOM was the main contributor to the total absorption, thus, the K_d(PAR) may have a good correlation with it, as discussed in Section 3.5.

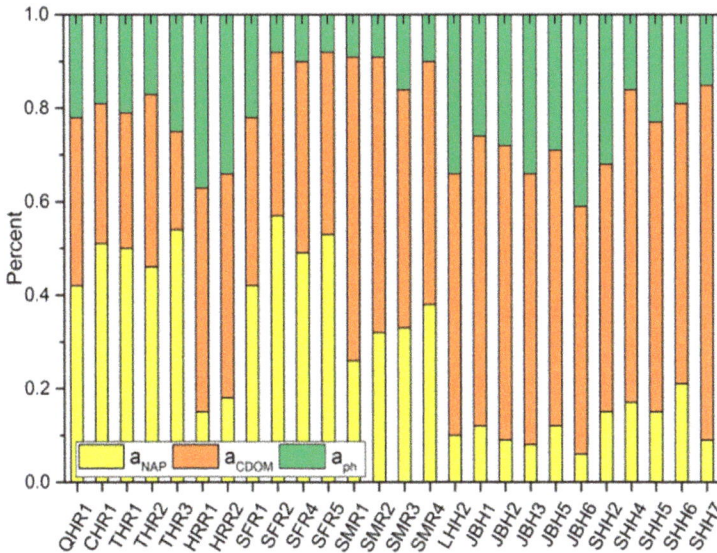

Figure 4. Relative contribution of phytoplankton (a_{ph}), NAP (a_{NAP}) and CDOM (a_{CDOM}) to the total absorption of the sample points with relative errors (RE) greater than 0.6 (based on LOO-CV analysis). Sample points are represented on the x-axis with the abbreviations of the location's name.

3.5. Spatial Transferability

The MRE and RMSE of each lake were calculated in the spatial transferability assessment (Figure 5). QNR (LakeID: 7), KLP (LakeID: 11), NYR (LakeID: 12), and XHLP (LakeID: 16) all had high RMSE values but relatively low MRE values due to their high K_d(PAR). Unlike the RMSE values, which were low for HHR (LakeID: 25), SFR (LakeID: 26), SHL (LakeID: 10) SMR (LakeID: 1), THR (LakeID: 24), JBL (LakeID: 9) and BSR (LakeID: 4) (Figure 5), the MRE (MRE: 1.794, 0.862, 0.860, 0.726, 0.694, 0.641 and 0.532, respectively) indicated that there were large errors in predicting K_d(PAR) in this lakes. These lakes commonly had low levels of TSM, so the relative contributions of Chl-*a* and CDOM could not be eliminated from the analysis as they may also have large influence on K_d(PAR). Otherwise, it may cause the high MRE values in predicting K_d(PAR) solely with TSM.

In order to identify the differential contribution of OACs to the estimation of K_d(PAR) across different lakes, a separate analysis was undertaken for lakes with more than 10 sample points. Although only six sample points were obtained from SHL, it was still selected because it was the second clearest lake in this study, and it had RE greater than 0.4 in the LOO-CV analysis. The GIs for the lakes were calculated respectively and linear regression between OACs and K_d(PAR) was performed. The lake specified averaged absorption spectra were used to calculate the relative contribution of phytoplankton, NAP and CDOM to the total absorption of each lake. The R^2 of linear regression and the relative contribution were shown in Figure 6.

GIs of the lakes are shown in Table 2. For SHL, the GIs between the OACs (TSM, Chl-*a*, CDOM) and the K_d(PAR) are (GI1: 0.62, 0.81, 0.67), (GI2: 0.77, 0.90, 0.80) and (GI3: 0.77, 0.89, 0.83). The GIs of SHL showed that Chl-*a* and CDOM displayed a higher correlation than TSM with the K_d(PAR), and this was consistent with the results obtained by linear regression analysis. The contribution of CDOM and phytoplankton to the total absorption was larger than NAP and this could explain why K_d(PAR) was mainly influenced by Chl-*a* and CDOM. The R^2 of the correlation analysis revealed that the K_d(PAR) of BSR was highly related to both TSM and Chl-*a*, and that there was only a minor variation in the GIs of TSM and Chl-*a*. The K_d(PAR) of NEJR was greatly correlated to both TSM and CDOM, and CDOM and NAP contributed more to the total absorption. This was similar to the analyzing results of XXKH. Overall, the count of dominant factors for determining K_d(PAR) were two rather one in the cases of the lakes/reservoirs described above. This results is consistent with the findings in Danjiangkou Reservoir (averaged K_d(PAR) was 0.726 m^{-1}) in Hubei Province, China [32]. That study revealed that TSM and Chl-*a* dominated K_d(PAR) during the wet season while Chl-*a* and CDOM were the major determinants of K_d(PAR) in the dry season in the reservoir.

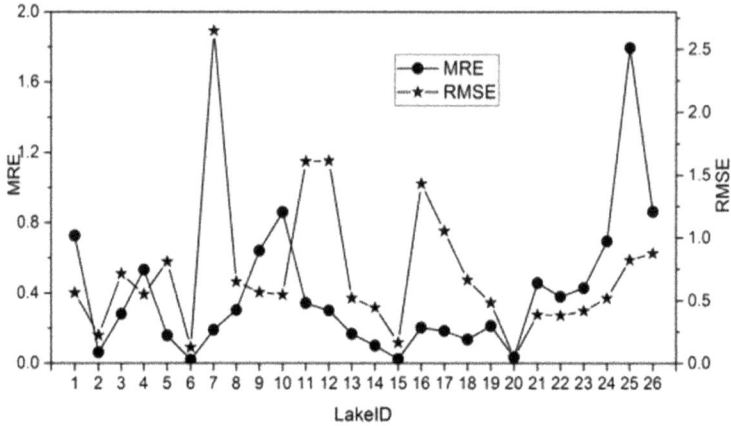

Figure 5. The mean relative error (MRE) and root-mean-square error (RMSE) of the K_d(PAR) regression models for each lake for the evaluation of the spatial transferability of the model. LakeID was corresponding to Table 1 and was represented on the x-axis.

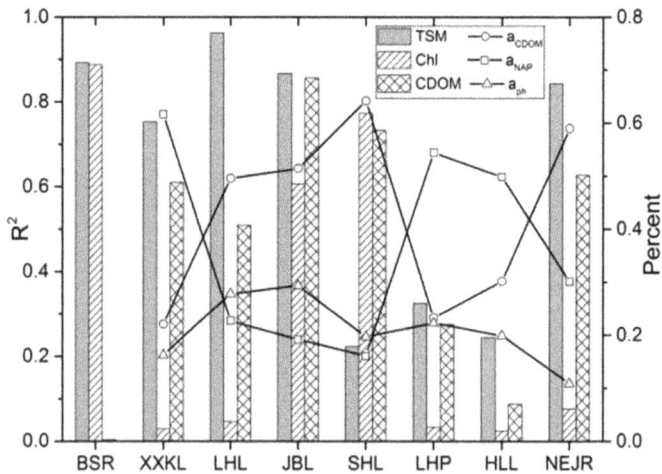

Figure 6. The bars represent the determination coefficients (R^2) of the linear regression analysis between the OACs and the K_d(PAR). The lines represent the relative contribution of phytoplankton (a_{ph}), non-algal particles (a_{NAP}) and CDOM (a_{CDOM}) to the total absorption. Abbreviations of the lakes correspond to those in Table 1 and are represented on the x-axis.

The TSM and CDOM concentrations were well correlated to K_d(PAR) for JBL. Similarly, Chl-*a* also had a modest correlation with K_d(PAR). In contrast to JBL which displayed similar contributions of each constituent to the total absorption, the K_d(PAR) of LHL was mainly correlated to the TSM.

The reason may be that the TSM was more varied in LHL (range: 2.83–22.6 mg/L, StDev: 6.33) than in JBL (range: 2.14–10.2 mg/L, StDev: 2.3), while the Chl-*a* was more varied in JBL (range: 9.94–61.67 µg/L, StDev: 14.61) than in LHL (range: 6.81–43.11 µg/L, StDev: 10.43). Though NAP contributed the most to the total absorption in LHP and HLH, K_d(PAR) had a weak correlation with the TSM. Taking into account that these two lakes are vast in shape and relative shallow in depth (5.92 m for HLH and 2.7 m for LHP) [37], it is possible that the OACs are uniformly distributed in the water causing a little variation in the calculation of K_d(PAR) across different sample points. This might have produced low standard deviations in the values of the dependent and the independent variables, in effect, lowering the significance of the linear correlation. In addition, the measurement of PAR was easily influenced by waves in these lakes which could induce large errors to the calculation of K_d(PAR). These errors could contribute to the reduction of the correlation between the K_d(PAR) and the OACs in HLH and LHP. There was an inconformity that a factor showed a high contribution to the total absorption might not be best correlated to the K_d(PAR). This inconformity was also observed by Shi [29] in Bosten Lake in Xinjiang Province, China. In his study, both CDOM and phytoplankton contributed more than NAP to the total absorption, however, the K_d(PAR) was best correlated to the TSM concentration.

4. Conclusions

Based on the data collected in Northeast China, the relationships between K_d(PAR) and OACs were studied. The optical properties of water bodies were diverse and complex. K_d(PAR) was significantly correlated to the water transparency (SDD) and the coefficient was in agreement with previous studies [8]. According to GIs and linear regression analysis of the data of all the sampled points, it was found that TSM was the most dominant component of the OACs in regulating K_d(PAR). As a systemic analysis method, GIs were effective in identifying the dominant factors of K_d(PAR). The TSM accounted for the most variation of the K_d(PAR) as the R^2 of the linear model only increased 0.01 when TSM, Chl-a, and CDOM were the explanatory variables rather than solely TSM was the explanatory variable. The investigation of the simple linear model demonstrated that it was an effective tool to predict K_d(PAR) from TSM data as it produced small RE in a LOO-CV analysis. However, the dominant regulating factors of K_d(PAR) varied among some of the sample points and lakes, especially for some clear waters, in which the dominant factors may be a combination of two or three kinds of OACs. This variation can be attributed to their distinct geographical environments. Lakes that are open in shape and shallow in depth and distributed in plain areas, where soil erosion is relatively strong, show high K_d(PAR) values. Therefore, the major determining factor of K_d(PAR) in these types of lakes is the concentration of TSM. However, for very clear water bodies, for example in deep and narrow reservoirs, which are distributed in mountainous areas where soil erosion is weak, the concentration of the TSM is low. In such cases, the contribution of the concentrations of Chl-*a* and/or CDOM to the estimation of K_d(PAR) is also significant. There was a discrepancy in the results derived from correlation analysis and from absorption spectra analysis. The material which contributed the most to the total absorption may not be well correlated to the K_d(PAR) due to the scattering of phytoplankton and TSM also affect K_d(PAR) [29]. In general, reservoirs in the east part of Northeast China had low K_d(PAR) values, while lakes located in plain areas showed high K_d(PAR) values. These differences in the K_d(PAR) could reflect variations in the composition of TSM in these distinct geographical locations.

Supplementary Materials: The following are available online at www.mdpi.com/2072-4292/8/11/964/s1, Figure S1: Distribution of sample points of each lakes or reservoirs. The color represents the average K_d(PAR), the number and abbreviation were listed in Table 1; Figure S2: The estimated K_d(PAR) by TSM and in situ K_d(PAR) for sample points that had relative error (RE) greater than 0.6 in LOO-CV. Eight sample points had RE greater than 1 and they were HRR1 (RE: 2.20), JDI I1 (RE: 1.59), JBH3(RE: 1.54), SHH7(RE: 1.50), SFR5(RE: 1.49), HRR2(RE: 1.36), SHH5(RE: 1.15) and QHR1(RE: 1.06); Figure S3: Linear regression analysis between OACs and K_d(PAR) for sample points that had relative error (RE) greater than 0.6 in LOO-CV. (a) total suspended matter (TSM); (b) chlorophyll-a(Chl-a); (c) absorption coefficient of chromophroic dissolved organic matter (CDOM) at

355 nm ($a_{CDOM}(355)$); Figure S4: Absorption coefficient of total particulate materials of BSR. Only the absorption coefficients of total particulate materials were measured during the experiment but the phytoplankton and non-algal particles were not measured. The obvious absorption peak at 675 nm indicated the pigment absorption was dominant in the total particulate absorption.

Acknowledgments: The research was jointly supported by the "One Hundred Talents" program from Chinese Academy of Sciences granted to Kaishan Song, the National Natural Science Foundation of China (No. 41471290, No. 41501387) and the Foundation of China's High Resolution Earth Observation (No. 41-Y20A31-9003-15/17). The authors thank Jia Du, Ming Wang, and Junbin Hou for their persistent assistance with field sampling and in situ measurements. The authors also thank the editors and three anonymous reviewers for their valuable and instructive comments that have strengthened the manuscript greatly.

Author Contributions: Kaishan Song conceived the framework and designed the fieldwork; Zhidan Wen, Ying Zhao, Yingxin Shang, Chong Fang, Jia Du, and Jianhang Ma performed the fieldwork and experiments; Jianhang Ma analyzed the data and wrote the paper; and Jianhang Ma and Kaishan Song revised the paper.

Conflicts of Interest: The authors declare no conflict of interest.

References

1. Lund-Hansen, L.C. Diffuse attenuation coefficients K_d(PAR) at the estuarine North Sea-Baltic Sea transition: Time-series, partitioning, absorption, and scattering. *Estuar. Coast. Shelf Sci.* **2004**, *61*, 251–259. [CrossRef]
2. Zhang, Y.L.; Zhang, B.; Ma, R.H.; Feng, S.; Le, C.F. Optically active substances and their contributions to the underwater light climate in Lake Taihu, a large shallow lake in China. *Fundam. Appl. Limnol. Arch. Hydrobiol.* **2007**, *170*, 11–19. [CrossRef]
3. Behrenfeld, M.J.; Falkowski, P.G. A consumer's guide to phytoplankton primary productivity models. *Limnol. Oceanogr.* **1997**, *42*, 1479–1491. [CrossRef]
4. Behrenfeld, M.J.; Falkowski, P.G. Photosynthetic rates derived from satellite-based chlorophyll concentration. *Limnol. Oceanogr.* **1997**, *42*, 1–20. [CrossRef]
5. Kirk, J.T.O. *Light and Photosynthesis in Aquatic Ecosystems*; Cambridge University Press: New York, NY, USA, 1994.
6. Chang, G.C.; Dickey, T.D. Coastal ocean optical influences on solar transmission and radiant heating rate. *J. Geophys. Res. Oceans* **2004**, *109*. [CrossRef]
7. Zhang, Y.L.; Qin, B.Q.; Hu, W.P.; Wang, S.M.; Chen, Y.W.; Chen, W.M. Temporal-spatial variations of euphotic depth of typical lake regions in Lake Taihu and its ecological environmental significance. *Sci. China Ser. D* **2006**, *49*, 431–442. [CrossRef]
8. Holmes, R.W. The Secchi disk in turbid coastal zones. *Limnol. Oceanogr.* **1970**, *15*, 688–694. [CrossRef]
9. Shang, S.L.; Lee, Z.P.; Wei, G.M. Characterization of MODIS-derived euphotic zone depth: Results for the China Sea. *Remote Sens. Environ.* **2011**, *115*, 180–186. [CrossRef]
10. Devlin, M.J.; Barry, J.; Mills, D.K.; Gowen, R.J.; Foden, J.; Sivyer, D.; Tett, P. Relationships between suspended particulate material, light attenuation and Secchi depth in UK marine waters. *Estuar. Coast. Shelf Sci.* **2008**, *79*, 429–439. [CrossRef]
11. Devlin, M.J.; Barry, J.; Mills, D.K.; Gowen, R.J.; Foden, J.; Sivyer, D.; Greenwood, N.; Pearce, D.; Tett, P. Estimating the diffuse attenuation coefficient from optically active constituents in UK marine waters. *Estuar. Coast. Shelf Sci.* **2009**, *82*, 73–83. [CrossRef]
12. Shi, K.; Zhang, Y.L.; Liu, X.H.; Wang, M.Z.; Qin, B.Q. Remote sensing of diffuse attenuation coefficient of photosynthetically active radiation in Lake Taihu using MERIS data. *Remote Sens. Environ.* **2014**, *140*, 365–377. [CrossRef]
13. Zhang, Y.B.; Zhang, Y.L.; Zha, Y.; Shi, K.; Zhou, Y.Q.; Li, Y.L. Estimation of diffuse attenuation coefficient of photosynthetically active radiation in Xin'anjiang Reservoir based on Landsat 8 data. *Chin. J. Environ. Sci.* **2015**, *36*, 4420–4429, (In Chinese with English Abstract).
14. Mueller, J.L. SeaWiFS algorithm for the diffuse attenuation coefficient, K_d(490), using water-leaving radiances at 490 and 555 nm. SeaWiFS Postlaunch calibration and validation analyses, part. *Med. Today* **2000**, *5*, 16–29.
15. Chen, J.; Cui, T.W.; Tang, J.W.; Song, Q.J. Remote sensing of diffuse attenuation coefficient using MODIS imagery of turbid coastal waters: A case study in Bohai Sea. *Remote Sens. Environ.* **2014**, *140*, 78–93. [CrossRef]

16. Majozi, N.P.; Salama, M.S.; Bernard, S.; Harper, D.M.; Habte, M.G. Remote sensing of euphotic depth in shallow tropical inland waters of Lake Naivasha using MERIS data. *Remote Sens. Environ.* **2014**, *148*, 178–189. [CrossRef]

17. Ramon, D.; Jolivet, D.; Tan, J.; Frouin, R. Estimating photosynthetically available radiation at the ocean surface for primary production (3P Project): Modeling, evaluation, and application to global MERIS imagery. *Proc. SPIE* **2016**. [CrossRef]

18. Frouin, R.; McPherson, J. Estimating photosynthetically available radiation at the ocean surface from GOCI data. *Ocean Sci. J.* **2012**, *47*, 313–321. [CrossRef]

19. Doron, M.; Babin, M.; Mangin, A.; Hembise, O. Estimation of light penetration, and horizontal and vertical visibility in oceanic and coastal waters from surface reflectance. *J. Geophys. Res. Oceans* **2007**, *112*. [CrossRef]

20. Lee, Z.P.; Weidemann, A.; Kindle, J.; Arnone, R.; Carder, K.L.; Davis, C. Euphotic zone depth: Its derivation and implication to ocean color remote sensing. *J. Geophys. Res. Oceans* **2007**, *112*. [CrossRef]

21. Son, S.H.; Wang, M.H. Diffuse attenuation coefficient of the photosynthetically available radiation Kd (PAR) for global open ocean and coastal waters. *Remote Sens. Environ.* **2015**, *159*, 250–258. [CrossRef]

22. Alikas, K.; Kratzer, S.; Reinart, A.; Kauer, T.; Paavel, B. Robust remote sensing algorithms to derive the diffuse attenuation coefficient for lakes and coastal waters. *Limnol. Oceanogr. Methods* **2015**, *13*, 402–415. [CrossRef]

23. Austin, R.W.; Petzold, T.J. *The Determination of the Diffuse Attenuation Coefficient of Sea Water Using the Coastal Zone Color Scanner*; Springer: New York, NY, USA, 1981.

24. Mueller, J.L.; Trees, C.C. Revised SeaWiFS prelaunch algorithm for diffuse attenuation coefficient K (490). *Case Stud. SeaWiFS Calibration Valid.* **1997**, *41*, 18–21.

25. Zhang, T.L.; Fell, F. An empirical algorithm for determining the diffuse attenuation coefficient K_d in clear and turbid waters from spectral remote sensing reflectance. *Limnol. Oceanogr. Methods* **2007**, *5*, 457–462. [CrossRef]

26. Zhang, Y.L.; Liu, X.H.; Yin, Y.; Wang, M.Z.; Qin, B.Q. A simple optical model to estimate diffuse attenuation coefficient of photosynthetically active radiation in an extremely turbid lake from surface reflectance. *Opt. Express* **2012**, *20*, 20482–20493. [CrossRef] [PubMed]

27. Phlips, E.J.; Aldridge, F.J.; Schelske, C.L.; Crisman, T.L. Relationships between light availability, chlorophyll a, and tripton in a large, shallow subtropical lake. *Limnol. Oceanogr.* **1995**, *40*, 416–421. [CrossRef]

28. Morel, A.; Huot, Y.; Gentili, B.; Werdell, P.J.; Hooker, S.B.; Franz, B.A. Examining the consistency of products derived from various ocean color sensors in open ocean (Case 1) waters in the perspective of a multi-sensor approach. *Remote Sens. Environ.* **2007**, *111*, 69–88. [CrossRef]

29. Shi, Z.Q.; Zhang, Y.L.; Yin, Y.; Liu, X.H. Characterization of the underwater light field and the affecting factors in Bosten Lake in summer. *Acta Sci. Circumst.* **2012**, *32*, 2969–2977, (In Chinese with English Abstract).

30. Kratzer, S.; Hakansson, B.; Sahlin, C. Assessing Secchi and photic zone depth in the Baltic Sea from satellite data. *AMBIO J. Hum. Environ.* **2003**, *32*, 577–585. [CrossRef]

31. Li, Y.L.; Zhang, Y.L.; Liu, M.L. Calculation and retrieval of euphotic depth of Lake Taihu by remote sensing. *J. Lake Sci.* **2009**, *21*, 165–172, (In Chinese with English Abstract).

32. Qu, M.Y.; Cai, Q.H.; Shen, H.L.; Li, B. Variation and influencing factors of euphotic depth in Danjiangkou reservoir in different hydrological periods. *Resour. Environ. Yangtze Basin* **2014**, *23*, 53–59. (In Chinese with English Abstract)

33. Pierson, D.C.; Kratzer, S.; Strömbeck, N.; Håkansson, B. Relationship between the attenuation of downwelling irradiance at 490 nm with the attenuation of PAR (400 nm–700 nm) in the Baltic Sea. *Remote Sens. Environ.* **2008**, *112*, 668–680. [CrossRef]

34. Tan, J.; Cherkauer, K.A.; Chaubey, I. Using hyperspectral data to quantify water-quality parameters in the Wabash River and its tributaries, Indiana. *Int. J. Remote Sens.* **2015**, *36*, 5466–5484. [CrossRef]

35. Tan, J.; Cherkauer, K.A.; Chaubey, I. Developing a comprehensive spectral-biogeochemical database of midwestern rivers for water quality retrieval using remote sensing data: A case study of the Wabash River and its tributary, Indiana. *Remote Sens.* **2016**, *8*, 517. [CrossRef]

36. Li, N.; Liu, J.P.; Wang, Z.M. Dynamics and driving force of lake changes in northeast China during 2000–2010. *J. Lake Sci.* **2014**, *26*, 545–551. (In Chinese with English Abstract)

37. Wang, S.M.; Dou, H.S. (Eds.) *Chinese Lake Catalogue*; Science Press: Beijing, China, 1998. (In Chinese)

38. Duan, H.T.; Ma, R.H.; Zhang, Y.L.; Zhang, B. Remote-sensing assessment of regional inland lake water clarity in northeast China. *Limnology* **2009**, *10*, 135–141. [CrossRef]
39. Song, K.S.; Zang, S.Y.; Zhao, Y.; Li, L.; Du, J.; Zhang, N.N.; Wang, X.D.; Shao, T.T.; Guan, Y.; Liu, L. Spatiotemporal characterization of dissolved carbon for inland waters in semi-humid/semi-arid region, China. *Hydrol. Earth Syst. Sci.* **2013**, *17*, 4269–4281. [CrossRef]
40. Wen, Z.D.; Song, K.S.; Zhao, Y.; Du, J.; Ma, J.H. Influence of environmental factors on spectral characteristic of chromophoric dissolved organic matter (CDOM) in Inner Mongolia Plateau, China. *Hydrol. Earth Syst. Sci. Discuss.* **2015**, *12*, 5895–5929. [CrossRef]
41. Zhao, Y.; Song, K.; Wen, Z.D.; Li, L.; Zang, S.Y.; Shao, T.T.; Li, S.J.; Du, J. Seasonal characterization of CDOM for lakes in semiarid regions of Northeast China using excitation–emission matrix fluorescence and parallel factor analysis (EEM–PARAFAC). *Biogeosciences* **2016**, *13*, 1635–1645. [CrossRef]
42. Mao, D.H.; Wang, Z.M.; Wu, C.; Song, K.S.; Ren, C.Y. Examining forest net primary productivity dynamics and driving forces in northeastern China during 1982–2010. *Chin. Geogr. Sci.* **2014**, *24*, 631–646. [CrossRef]
43. Song, K.S.; Wang, Z.M.; Blackwell, J.; Zhang, B.; Li, F.; Zhang, Y.Z.; Jiang, G.J. Water quality monitoring using Landsat Themate Mapper data with empirical algorithms in Chagan Lake, China. *J. Appl. Remote Sens.* **2011**, *5*, 053506. [CrossRef]
44. Song, K.S.; Li, L.; Tedesco, L.P.; Li, S.; Duan, H.T.; Liu, D.W.; Hall, B.E.; Du, J.; Li, Z.C.; Shi, K.; et al. Remote estimation of chlorophyll—A in turbid inland waters: Three-band model versus GA-PLS model. *Remote Sens. Environ.* **2013**, *136*, 342–357. [CrossRef]
45. Jeffrey, S.W.; Humphrey, G.F. New spectrophotometric equations for determining chlorophylls a, b, c1 and c2 in higher plants, algae and natural phytoplankton. *Biochem. Physiol. Pflanz.* **1975**, *167*, 191–194.
46. Bricaud, A.; Morel, A.; Prieur, L. Absorption by dissolved organic matter of the sea (yellow substance) in the UV and visible domains. *Limnol. Oceanogr.* **1981**, *26*, 43–53. [CrossRef]
47. Miller, R.L.; Belz, M.; Del Castillo, C.; Trzaska, R. Determining CDOM absorption spectra in diverse coastal environments using a multiple pathlength, liquid core waveguide system. *Cont. Shelf Res.* **2002**, *22*, 1301–1310. [CrossRef]
48. Shao, T.T.; Song, K.S.; Du, J.; Zhao, Y.; Liu, Z.M.; Zhang, B. Seasonal Variations of CDOM Optical Properties in Rivers across the Liaohe Delta. *Wetlands* **2016**, *36*, 181–192. [CrossRef]
49. Xu, J.P.; Li, F.; Zhang, B.; Song, K.S.; Wang, Z.M.; Liu, D.W.; Zhang, G.X. Estimation of chlorophyll—A concentration using field spectral data: A case study in inland Case-II waters, North China. *Environ. Monit. Assess.* **2009**, *158*, 105–116. [CrossRef] [PubMed]
50. Cao, M.X.; Dang, Y.G.; Mi, C.M. An improvement on calculation of absolute degree of grey incidence. In Proceedings of the 2006 IEEE International Conference on Systems, Man and Cybernetics, Taipei, Taiwan, 8–11 October 2006.
51. Jin, X.L.; Du, J.; Liu, H.J.; Wang, Z.M.; Song, K.S. Remote estimation of soil organic matter content in the Sanjiang Plain, Northest China: The optimal band algorithm versus the GRA-ANN model. *Agric. For. Meteorol.* **2016**, *218*, 250–260. [CrossRef]
52. Deng, J.L. *Basic Methods of Gray System*; Science and Technology University of Central China Press: Wuhan, China, 1987. (In Chinese)
53. Liu, S.F.; Cai, H.; Cao, Y.; Yang, Y.J. Advance in grey incidence analysis modelling. In Proceedings of the 2011 IEEE International Conference on Systems, Man, and Cybernetics (SMC), Anchorage, Alaska, USA, 9–12 October 2011; pp. 1886–1890.
54. Mei, Z.G. The Concept and Computation Method of Grey Absolute Correlation Degree. *Syst. Eng.* **1992**, *10*, 43–44. (In Chinese with English Abstract)
55. Carlson, R.E. A trophic state index for lakes. *Limnol. Oceanogr.* **1977**, *22*, 361–369. [CrossRef]
56. Aas, E.; Høkedal, J.; Sørensen, K. Secchi depth in the Oslofjord-Skagerrak area: Theory, experiments and relationships to other quantities. *Ocean Sci.* **2014**, *10*, 177–199. [CrossRef]
57. Lee, Z.P.; Shang, S.L.; Hu, C.M.; Du, K.P.; Weidemann, A.; Hou, W.L.; Lin, J.F.; Lin, G. Secchi disk depth: A new theory and mechanistic model for underwater visibility. *Remote Sens. Environ.* **2015**, *169*, 139–149. [CrossRef]
58. Lee, Z.P.; Shang, S.L.; Qi, L.; Yang, J.; Lin, G. A semi-analytical scheme to estimate Secchi-disk depth from Landsat-8 measurements. *Remote Sens. Environ.* **2016**, *177*, 101–106. [CrossRef]
59. Raymont, J.E.G. *Plankton and Productivity in the Oceans*; Pergamon Press: Oxford, UK, 1967.

60. Aertebjerg, G.; Bresta, A.M. *Guidelines for the Measurement of Phytoplankton Primary Production*, 2nd ed.; Baltic Marine Biologists Publication: Charlottenlund, Denmark, 1984.

61. Olmanson, L.G.; Bauer, M.E.; Brezonik, P.L. A 20-year Landsat water clarity census of Minnesota's 10,000 lakes. *Remote Sens. Environ.* **2008**, *112*, 4086–4097. [CrossRef]

remote sensing

MDPI

Article

Polarization Patterns of Transmitted Celestial Light under Wavy Water Surfaces

Guanhua Zhou [1,2,3], Jiwen Wang [1], Wujian Xu [1,*], Kai Zhang [4,*] and Zhongqi Ma [1]

1   School of Instrumentation Science and Opto-Electronics Engineering, Beihang University,
    No. 37 Xueyuan Rd., Haidian, Beijing 100191, China; zhouguanhua@buaa.edu.cn (G.Z.);
    wang_jwen@163.com (J.W.); mzq11171079@163.com (Z.M.)
2   Key Laboratory of Space Ocean Remote Sensing and Application, State Oceanic Administration,
    No. 8 Da Huisi Rd., Haidian, Beijing 100081, China
3   State Key Laboratory of Remote Sensing Science, Institute of Remote Sensing and Digital Earth,
    Chinese Academy of Sciences, Beijing 100101, China
4   State Key Laboratory of Environmental Criteria and Risk Assessment,
    Chinese Research Academy of Environmental Sciences, Beijing 100012, China
*   Correspondence: xuwujian@buaa.edu.cn (W.X.); zhangkai@craes.org.cn (K.Z.)

Academic Editors: Linhai Li, Yunlin Zhang, Claudia Giardino, Deepak R. Mishra and Prasad S. Thenkabail
Received: 3 January 2017; Accepted: 27 March 2017; Published: 29 March 2017

Abstract: This paper presents a model to describe the polarization patterns of celestial light, which includes sunlight and skylight, when refracted by wavy water surfaces. The polarization patterns and intensity distribution of refracted light through the wave water surface were calculated. The model was validated by underwater experimental measurements. The experimental and theoretical values agree well qualitatively. This work provides a quantitative description of the repolarization and transmittance of celestial light transmitted through wave water surfaces. The effects of wind speed and incident sources on the underwater refraction polarization patterns are discussed. Scattering skylight dominates the polarization patterns while direct solar light is the dominant source of the intensity of the underwater light field. Wind speed has an influence on disturbing the patterns under water.

Keywords: wavy water; polarized refraction; skylight polarization; oceanic optics

1. Introduction

Polarization is one of the fundamental properties of underwater light fields. Its study can provide a deeper understanding of the nature of hydrologic optics [1], water color remote sensing [2] and other relative applications [3]. However, polarization is typically ignored in previous studies on underwater environments partly because of difficulties in field measurements and the lack of scientific driving. Many of the hydro-ecological processes are assumed to be independent of the polarization of the light field [4]. As the polarization in nature is realized that it can be utilized by many animals to determine geographic direction [5,6], which gives great inspiration to explore underwater polarization and the polarization response of aquatic animals [7]. The occurrence and pattern of light polarization in aquatic environments becomes a subject of wide potential relevance to various problems in oceanography and limnology. Especially in recent years, a tremendous amount of progress has been made in the exploitation of polarization of light in the water column and exiting the sea surface to improve our capacities of observing and monitoring coastal and oceanic environments [8–12]. Kattawar et al. [13] have made a thorough review of polarized light scattering in the atmosphere and ocean. Harmel [14] comprehensively reviewed the recent developments in the use of light polarization for marine environment monitoring from space and discussed the potentialities of polarimetric remote sensing of marine biogeochemical parameters. One of the interesting potentialities is that chlorophyll fluorescence signal could be retrieved from the polarization discrimination technique because the elastically scattered component is partially polarized while the fluorescence signal is totally unpolarized [15]. The polarization-based techniques could play their own part in furthering remote sensing of the marine and inland water environment [14]. Since the Nobel Prize winner Frisch discovered that bees can orient themselves by means of the polarization pattern of skylight [16], which drew great attention of scholars to polarization vision [17] and promoted the development of the research in polarized skylight, reflected off the water surface and underwater fields [18–21]. Refraction of light is associated with polarization according to the Fresnel formulae. Unpolarized sunlight becomes partially linearly polarized after it penetrates through the water surface and the partially polarized skylight will change its state of polarization after transmits from air to water. Research on polarized transmitted light under water is mainly focused on measurement and numerical simulation. The pioneering measurements can trace back to Waterman [5]. He outlined the major characteristics of submarine polarization and discovered that submarine light is substantially polarized in all directions, most linearly but with some ellipticity just beyond the edge of Snell's window [22]. This was followed by Ivanoff and Waterman, who further analyzed the factors that influence the degree of polarization of light under water [23]. The application of Stokes vectors and Mueller matrix makes it easy to describe the polarized light. Voss group contribute greatly to promote the further deep understanding the polarization of underwater by innovative instrumentation development and modeling [24]. Voss and Fry obtained the unified expression of Mueller matrix of seawater [25]. Bhandari and Voss built an imaging system with fisheye lens and measured patterns of full Stokes parameters and polarized light under water accurately [26]. You et al. [27] measured the polarized light field in coastal waters using a hyperspectral and multiangular instrument and discussed the impact of atmospheric conditions and water compositions on underwater degree of linear polarization. Underwater polarization properties have drawn a great attention in the experimental biology community. For example, Cronin investigated the polarized light field in natural marine waters, sampling the spectrum of partially linearly polarized light throughout the celestial light field throughout the day [28]. In a theoretical study, Kattawar numerically simulated the polarization of transmitted light under water with Rayleigh scattering approximation [29]. Horvath computed the polarization pattern of the Rayleigh scattering skylight refracted from a flat water surface without taking the wave water and the intensity distribution of skylight into account [7]. Horvath and his colleagues confined their investigation to the flat water surface; their calculation might be instructive as a first order approximation of the real refracted polarization pattern. Sabbah synthetically described the characteristics of polarized light and explained the polarization phenomenon under water in detail [30]. He also predicted polarization patterns of skylight transmitted through the Snell's window and validated this with experimental results [31]. In the following years, there was a fruitful of series of discoveries on the nature of underwater polarization. For example, Mishchenko and Travis [32] developed a vector radiative model for the atmosphere–ocean system to theoretically simulate several types of satellite retrievals over the ocean with no contributions due to the scattering from within the ocean body. Recently, a 3-D Monte Carlo method has been applied in the simulation of radiative transfer in a dynamic air–water system, which promotes the study of the polarization pattern of light transmitted through the Snell's window [27,33,34]. Mobley [35] proposed a state of the art sea surface model to deal with the polarized reflection and transmission. Wave variance spectral and Fourier transform was used to generate the random sea surface and Monte Carlo polarized ray tracing method was used to compute effective Mueller matrices for reflection and transmission of polarized radiance across the air–water surface. Hieronymmi [36] used polarized ray tracing to investigate air-incident and whitecap-free reflectance and transmittance distributions with high angular resolution subject to sea-characterizing parameters. Though much progress has made on the polarized light propagating through the atmosphere–ocean system, most of the relevant models are not public available. Fortunately, Chami et al. [37] developed a powerful vector radiative transfer model of coupled atmosphere–ocean system for a rough sea surface (OSOAA model) with a friendly graphical user interface shared for the community. More recently, Foster and Gilerson [38] developed a hybrid approach combining vector radiative transfer simulations and the Monte Carlo method is used to determine the transfer functions of polarized light for wind-driven ocean surfaces.

The polarization of water surfaces depends not only on illumination but also to a high degree upon the presence or absence of waves [24]. The existing models in the early stage did not take the polarization characteristics of illumination and the effect of waves into account simultaneously, which depart from the real states. In many cases, the water is assumed to be flat. Realistically, however, water surfaces virtually always have dynamic waves which causes the edges of Snell's Window to be ragged.

The patterns of intensity and polarization under a wave surface are quite complex, and in addition, the measurement of polarization distribution of transmitted light is highly difficulty [4,39]. The existing limited measurements have been explored in specific locations, time and depth [40]. At present, the overall polarization characteristics of the underwater light field are still poorly understood. Thus, it is necessary to develop an effective model to describe the actual distribution of linear polarization underwater. Simulating and extensive modeling of polarization distribution of transmitted light under the wave surface in theory is of great significance to reveal the general rules of transmitted light under wavy water surface in view of the near future launch of polarimetric Earth-observing satellite missions.

The objective of this work is to develop a relatively simple and efficient model to describe the polarization patterns of refracted light under wavy water surfaces which considers the polarization of incident sunlight and skylight and explore the impacts of illumination and ocean state. We confine our investigation to the downwelling celestial light in the vicinity below the wave water surface in clear water. We focus on the dynamic light field as close as about half a meter depth below the water surface. Thus, the scattering effect of water components and reflection from the water bottom and other parameters are omitted in this paper since it is so complex to quantitatively consider all possible influences in a single paper. First (Section 2), a quantitative description of the physics of repolarization of skylight transmitted through the water surface is given. Then the scattering model to simulate the polarization pattern of the skylight is introduced. Then, in Section 3, the polarization model for transmitted light is introduced. The simulation of polarization pattern under various incident conditions and wind speed is presented in Section 4. Finally, the model is validated by the measurements and the feasibility and the limitation of the model is also described.

2. Materials and Methods

2.1. Polarization Characteristics of Refracted Light

Reflection and refraction occur when light passes through the interface between two different mediums, which not only influence the output light intensity, but also change the vibration direction of the E-vector (i.e., plane of polarization). In general, the E-vector of a beam of incident light can be divided into E_p and E_s, which are parallel and vertical to the incident plane, respectively, and the incident plane is determined by incident direction and normal direction of the interface, as demonstrated in Figure 1.

The variations in E-vector of reflected light and refracted light can be expressed by reflectance and transmittance. The reflection coefficients r_s and r_p, transmission coefficients t_s and t_p are all related to the incident angle θ_1, and the vertical components and parallel components are not equal in water medium [41]. When considering the solar incidence, the amplitudes of the electromagnetic waves distribute evenly in all directions perpendicular to the propagation way, so the solar irradiance is unpolarized. The degree of polarization (DOP) of reflected light and transmitted light can be expressed as Equation (1), and the transmittance is shown in Equation (2), where R is the reflectance of the water surface.

$$\begin{cases} Dop_r = \left(r_s{}^2 - r_p{}^2\right) / \left(r_p{}^2 + r_s{}^2\right) \\ Dop_t = \left(t_p{}^2 - t_s{}^2\right) / \left(t_p{}^2 + t_s{}^2\right) \end{cases} \tag{1}$$

$$T = 1 - R = 1 - \left(r_p^2 + r_s^2 \right)/2 \qquad (2)$$

Figure 2a illustrates that how the reflectance and transmittance change with the incident angle, where the refractive index of air n_a equals 1 and the index of water n_w equals 1.33. The intensity of reflected light increases with the increasing incident angle, and the amplitude of E-vector vertical component is bigger than that of the parallel component, consequently the E-vector's direction of vibration turns to vertical when reflection occurs. On the contrary, with the increasing of incident angle, both the vertical and parallel components of the transmitted light E-vectors decrease and then the intensity weakens too. Moreover, the attenuation of the vertical component is higher than that of the parallel component, so that the refracted E-vector's direction of vibration tiles to the horizontal. All of the above accords with energy conservation.

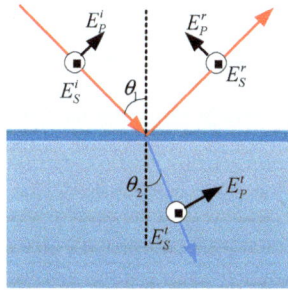

Figure 1. E-vector of an incident beam is reflected and refracted by the interface, E^r is the reflected component, E^t is the refracted component, and E_s and E_p are vertical and parallel component, respectively.

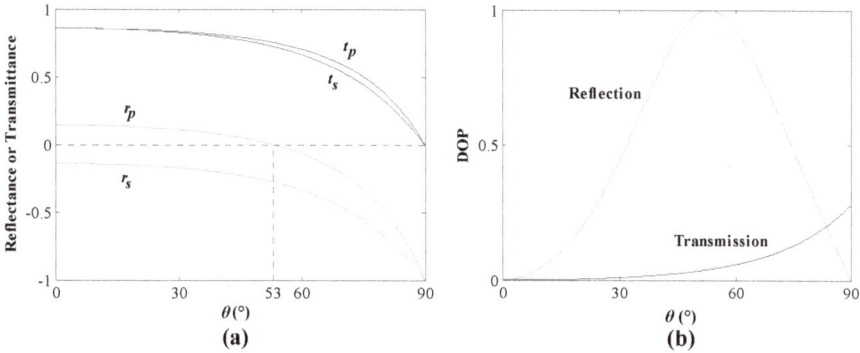

Figure 2. (a) The relation curves between reflection coefficients r_s, r_p, transmission coefficients t_s, t_p and incident angle θ; and (b) the relation curves between the DOP of reflected and transmitted light and the incident angle.

In Figure 2b, we can find that the curve between DOP of reflected light and incident angle has an extreme point. When the incidence zenith angle is equal to the Brewster angle, the DOP is maximal, and the reflected beam only has the vertical component (Figure 2a). Thus, the reflected beam is linearly polarized at the Brewster incident angle, and the polarization direction is perpendicular to the incident plane. However, the DOP of transmitted light rises as the incident angle increases constantly, and the two E-vector components decrease simultaneously. For the direct light incidence, the refracted light beams in all directions under water are partially polarized, and the DOP is much lower than that of reflectance especially in the Snell's window. The DOP of transmitted light steadily increases with the increasing of the incident angle and the maximum is no more than 0.3 for water surface.

To quantitatively describe the polarization of partially polarized incident light refracted by the water surface and simplify the model, we use the Stokes parameters to describe the light beam.

$$\mathbf{S} = (\mathbf{I}, \mathbf{Q}, \mathbf{U}, \mathbf{V})^T, \tag{3}$$

The DOP and the angle of polarization (AOP) can be represented by Stokes parameters as Equations (4) and (5).

$$\text{DOP} = \sqrt{Q^2 + U^2 + V^2}/I \tag{4}$$

$$\text{AOP} = \frac{1}{2}\tan^{-1}(U/Q) \tag{5}$$

Then, we can describe the variation of the polarization when light passes through the air–water interface using the Mueller matrix, and the Mueller matrix of refraction at an interface can be derived using Fresnel's formulas and is shown in Equation (6) [7].

$$\mathbf{M} = \frac{1}{2}\frac{n_2 \cos\theta_2}{n_1 \cos\theta_1}\begin{pmatrix} t_s^2 + t_p^2 & t_s^2 - t_p^2 & 0 & 0 \\ t_s^2 - t_p^2 & t_s^2 + t_p^2 & 0 & 0 \\ 0 & 0 & 2t_s t_p & 0 \\ 0 & 0 & 0 & 2t_s t_p \end{pmatrix} \tag{6}$$

where n_1 and n_2 stands for the refractive indexes of incident and refractive medium. Thus, the Stokes parameters of refracted light can be expressed, as below.

$$\mathbf{S}_t = \mathbf{M} \cdot \mathbf{S}_i \tag{7}$$

where $\mathbf{S}_t$ and $\mathbf{S}_i$ are the Stokes vectors of the refracted light and incident light, correspondingly. The reference direction of the matrix is parallel to the incident plane. Thus, the polarization of the refracted light under the calm surface can be computed directly using Equation (7).

2.2. Intensity and Polarization of Skylight

The skylight transmitted into the ocean is the source of all subsequent radiation which redistributed by the scattering processes occurring from both the fluctuation scattering and the scattering from the hydrosols [3]. The Rayleigh atmosphere model is the simple and effective method to describe clear skylight. According to the Raleigh scattering law, the intensity of the scattering light changes inversely with the fourth power of the wavelength. However, the intensity is not only related to wavelength, but also the scattering angle. To properly compute the distribution of skylight intensity in different directions, this paper adopts a clear skylight radiance distribution model built by Harrison and Coombes [42], the expression is shown in Equation (8).

$$N(\gamma, \theta_s, \theta_v) = (A + Be^{-m\gamma} + C\cos^2\gamma\cos\theta_s)(1 - e^{-\rho\sec\theta_v})(1 - e^{-\tau\sec\theta_s}) \tag{8}$$

where $N(\gamma, \theta_s, \theta)$ is skylight radiance, θ_s is the solar zenith angle (SZA), θ_v is the incidence zenith angle, γ is scattering angle, ρ is a regression coefficient, and τ and m are the optical thickness and mass of atmosphere, respectively. The data are also given by Harrison and Coombes: $A = 1.63$, $B = 53.7$, $C = 2.04$, $m = 5.49$, $\rho = 1.90$, $\tau = 0.53$.

The polarization characteristics of skylight are determined by the scattering of the atmosphere molecule and aerosol. Coulson indicated that the single Rayleigh scattering model can effectively describe the polarized skylight in theory [43]. However, considering the atmospheric disturbances such as multiple scattering, ground reflection, aerosol and the anisotropic of molecules, we compute the skylight polarization patterns using the semi-empirical Rayleigh scattering model [7,43], which is expressed as below.

$$DOP = DOP_{max} \frac{\sin^2 \gamma}{1 + \cos^2 \gamma} \tag{9}$$

where DOP_{max} is an empirical number which is related to the SZA and it equals to 100% in Rayleigh scattering model. When the solar zenith angle θ_s is 0°, 30°, 60° and 90°, the DOP_{max} equals to 56%, 63%, 70% and 77%, respectively. Based on geometrical relationship, then we can compute the scattering angle.

$$\cos \gamma = \cos \theta_s \cos \theta_v + \sin \theta_s \sin \theta_v \cos \varphi \tag{10}$$

where θ_v is the view zenith angle, and φ is the relative azimuth angle.

The skylight in any direction is all partially polarized except for the overcast sky and the four neutral points in the vicinity of the sun and anti-sun, positioned along the solar and anti-solar meridian. The polarization characteristics of skylight were described by the semi-empirical Rayleigh model, which can be considered a good approximation [7,21,43].

2.3. Polarization of Refracted Light under Wavy Water Surface

The Snell's window is the angular extent of the sky above the water's surface after refraction at the air–water interface, which is illustrated in Figure 3a. When observing upward in the water, one can see light rays from a full 180° field-of-view above the water surface. However, those rays are bent and compressed into a field-of-view that is about 97° due to refraction. The shrunken celestial hemisphere seen by submerged observers is called Snell's window [44]. However, the edge of the Snell's window becomes ragged due to the surface waves.

The model to describe the wind-driven ocean waves surface we chose is the classical Cox-Munk model [45], which has been extensively used for ocean color remote sensing studies. In the model, wave surface is considered as collections of wave facets, on which the Snell's refraction law is followed strictly. The wave slope distribution is the function about wind speed and wind direction [45,46]. For every incident sampling point in the celestial hemisphere, the research calculates its Stokes parameters after it is refracted by wavy water surface based on the wave slope distribution, and then gets a weighted average of each Stokes parameter of transmitted beams in all directions.

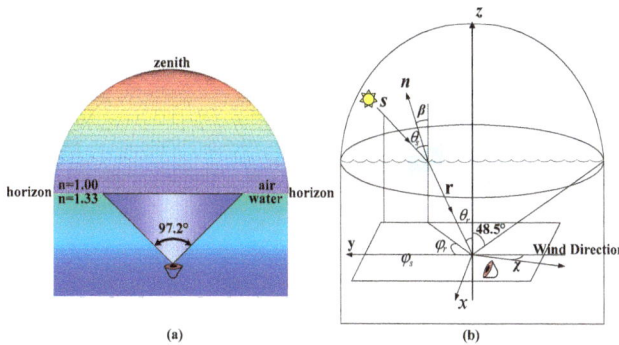

Figure 3. (a) Snell's window for flat water; and (b) the coordinate sketch of the refracted process under the wavy water surface.

Figure 3b illustrates the geometrical process of refraction under the wavy water surface, where θ_s is the angle between the incident light and zenith, φ_s is the incident azimuth angle, and θ_r and φ_r denote the refracted zenith angle and azimuth angle, respectively. The vectors s and r represent the incident and outgoing direction, n is normal to the wave facet. β is the angle between n and zenith.

The wind direction is opposite to the y-axis. The edge of the Snell's window is 48.5° to the zenith. Thus, the vectors s and r can be derived with the space geometrical relationship, and as follows.

$$\begin{cases} s = (\sin\theta_s \sin\varphi_s, \sin\theta_s \cos\varphi_s, \cos\theta_s) \\ r = (\sin\theta_r \sin\varphi_r, \sin\theta_r \cos\varphi_r, \cos\theta_r) \end{cases} \tag{11}$$

For the refraction, s, r and n satisfy the following equation.

$$\begin{cases} n_a s - n_w r = cn \\ c = \sqrt{n_w^2 - 2n_a n_w \cos\alpha + n_a^2} \end{cases} \tag{12}$$

According to the Cox–Munk model, the probability of wave facet with certain slope components can be expressed as Equation (13) [45,46].

$$p(z'_x, z'_y) = \frac{1}{2\pi\sigma_u\sigma_c} e^{-\frac{\xi^2+\eta^2}{2}} \cdot \begin{bmatrix} 1 - \frac{1}{2}C_{21}\eta(\xi^2 - 1) - \frac{1}{6}C_{03}(\eta^3 - 3\eta) + \\ \frac{1}{24}C_{40}(\xi^4 - 6\xi^2 + 3) + \\ \frac{1}{4}C_{22}(\xi^2 - 1)(\eta^2 - 1) + \\ \frac{1}{24}C_{04}(\eta^4 - 6\eta^2 + 3) \end{bmatrix} \tag{13}$$

In this equation, $\xi = z'_x/\sigma_c$ and $\eta = z'_y/\sigma_u$, where $\sigma^2 = \sigma_c^2 + \sigma_u^2$ and σ represents the root mean square (RMS) which is given as below [47].

$$\begin{cases} \sigma_u = \sqrt{0.0053 + 6.71 \times 10^{-4} W} \\ \sigma_c = \sqrt{0.0048 + 1.52 \times 10^{-4} W} \end{cases} \tag{14}$$

In addition, the wave facet can be divided into two parts on x- and y-axes, as below.

$$\begin{cases} z_x = \partial z/\partial x = \frac{\sin\theta_s \sin\varphi_s + \sin\theta_r \sin\varphi_r}{\cos\theta_s + \cos\theta_r} \\ z_y = \partial z/\partial y = \frac{\sin\theta_s \cos\varphi_s + \sin\theta_r \cos\varphi_r}{\cos\theta_s + \cos\theta_r} \end{cases} \tag{15}$$

To consider the effect of wind direction, rotating coordinates to fit with wind direction, which is shown in Equation (16).

$$\begin{cases} z'_x = \cos\chi \cdot z_x + \sin\chi \cdot z_y \\ z'_y = -\sin\chi \cdot z_x + \cos\chi \cdot z_y \end{cases} \tag{16}$$

χ is the angle between wind direction and y-axis. In the results of this paper, χ equals 0. Furthermore, the shadowing effects should be considered because some of the surface elements will be blocked by other elements if viewing from the propagation direction of light [33]. Thus, the shadowing factor $S(\theta_s, \theta_r, \sigma^2)$ is taken into account [48,49], and as follows.

$$\begin{cases} S(\theta_s, \theta_r, \sigma^2) = \frac{1}{1+\Lambda(\cot(\theta_s))+\Lambda(\cot(\theta_r))} \\ \Lambda(x) = \frac{1}{2}\left[\sqrt{\frac{2}{\pi}}\frac{\sigma}{x}\exp\left(-\frac{x^2}{2\sigma^2}\right) - \text{erfc}\left(\frac{x}{\sqrt{2}\sigma}\right)\right] \end{cases} \tag{17}$$

where erfc is the complementary error function. Therefore, the transmission of wave surface can be written as Equation (18).

$$\tau = \frac{\pi \cdot n_w^2 \cdot t(\omega) \cdot \cos\omega \cdot \cos\omega_t}{c^2 \cdot \cos^4\beta \cdot \cos\theta_s \cdot \cos\theta_r} p(z_x, z_y) \cdot S(\theta_s, \theta_r, \sigma^2) \tag{18}$$

To unify the reference plane of *s* and *r*, there is a rotation matrix *C(i)* which is shown as below.

$$\mathbf{C}(i) = \begin{bmatrix} 1 & 0 & 0 & 0 \\ 0 & \cos 2i & \sin 2i & 0 \\ 0 & -\sin 2i & \cos 2i & 0 \\ 0 & 0 & 0 & 1 \end{bmatrix} \tag{19}$$

Finally, we can get the transmitted Mueller matrix of wavy water surface and it can be expressed as Equation (20).

$$\mathbf{T}(\theta_s, \theta_r, \Delta\varphi) = \frac{\pi \cdot n_w^2 \cdot t(\omega) \cdot \cos\omega \cdot \cos\omega_t}{c^2 \cdot \cos^4\beta \cdot \cos\theta_s \cdot \cos\theta_r} p(z_x', z_y') S(\theta_s, \theta_r, \sigma^2) \mathbf{C}(\pi - i_2) \mathbf{M}_r(w) \mathbf{C}(i_1) \tag{20}$$

where $\mathbf{M}_r$ is the Mueller matrix of calm water surface. ω and ω_t are the incident and outgoing angle, respectively. i_1 and i_2 are defined as the rotation angles.

3. Results

Under clear skies, the position of sun dominates the distribution of intensity and polarization of skylight, and then dominantly affects the polarization patterns under the water surface. According to the analysis in Section 2, this paper focuses on the DOP, AOP and transmission intensity distribution of transmitted light through the Snell's window under calm and wavy water surfaces with different solar zenith angles and different wind speeds. Finally, we analyze the polarization patterns of transmitted light under various incident zenith angles, as well as comparing the polarization patterns and intensity of the transmitted light with those of reflected light under the same conditions.

3.1. Polarization and Transmission Patterns of Skylight Transmitted through a Calm Water Surface

When exploring the transmission light polarization, we start with the polarization and transmission patterns of skylight transmitted through a calm water surface. In this case, the transmission Mueller matrix is related to incident direction only, and the direct solar light has no effect on the polarization distribution of the underwater light field because of their unified incident direction. Thus, we only consider the incident skylight when the surface is flat. The entire celestial hemisphere condensed into Snell's window with an angular extent of 97°. Outside the Snell's window, the light from deeper layers is totally reflected and it is dim (under the condition that the substrates are not bright and the scattering constituents of water are normal). Two-dimensional patterns of degree and direction of polarization of refracted skylight are presented for various zenith angles of the sun. The three-dimensional celestial hemisphere (Figure 4) is represented in two dimensions in a polar-coordinate system. The zenith angle and azimuth angle from the solar meridian are measured radially and tangentially. The zenith is at the origin and the horizon corresponds to the outermost circle in this two-dimensional coordinate system.

As shown in Figure 4, the boundary of Snell's window is sharp when the water surface is calm [7]. There is a strong contrast between the bright scene above and the darker reflections from deep water. The polarization patterns are limited within the Snell's window. The pattern of DOP is concentric around the sun position, making it symmetrical about the solar principal plane [22,28]. When the sun is at the zenith, the maximum DOP of transmitted light spreads over the edge of Snell's window. With the increasing of SZA, the DOP of refracted light increases on the whole. When SZA approaches 90°, there is a band of maximum DOP located in the vertical direction of the sun's meridian. The max DOP of about 70% occurs at sunrise and sunset near the underwater surface. However, according to the Fresnel law as shown in Figure 2, the maximum refraction light is only 28% for the unpolarized sunlight incident at the horizon. The polarization state of incident skylight significantly enhances the DOP of downwelling refracted light. Thus, the polarization state of skylight plays a very important role in determining the DOP of downwelling refracted light.

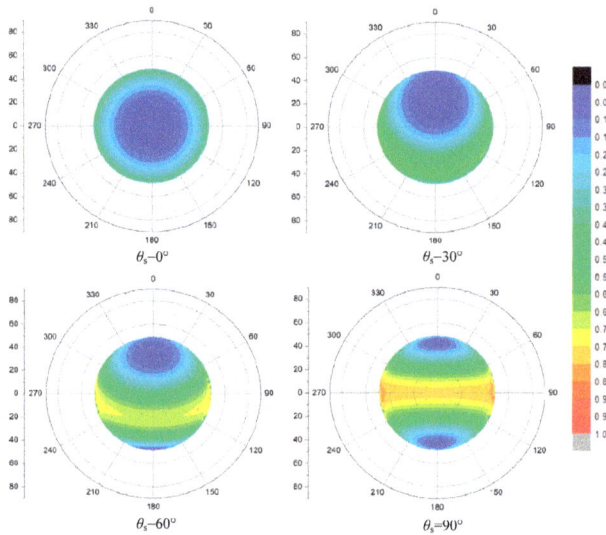

Figure 4. The DOP (degree of polarization) patterns of skylight transmitted through a calm water surface within the Snell's window at different solar zenith angle θ_s = 0°, 30°, 60°, 90°, the DOP ranges from 0 to 1.

As shown in Figure 5, when the sun reaches the zenith, all the incident light E-vectors are horizontal and remain in a horizontal direction after being refracted by a calm surface except for the incident beam in the zenith direction. This is because the attenuation of parallel E-vector component is lower than that of vertical E-vector component and this difference has a lower effect on the incident light polarization. Thus, most of transmitted light is horizontal, only the AOP of incident light near the sun changes after refraction due to the very low DOP. For the same reason, there are two split singular points near the apparent sun position in the AOP pattern of refracted light [7]. In addition, in addition to the normal incidence, the main vibration direction of refracted light is not horizontal for any other angles of incidence, but is parallel to the scattering plane for all incident angles, so that the AOP changes with the movement of the sun position [50].

The distributions of transmitted light under the calm surface at different SZAs are shown in Figure 6. We notice that most energy of incident light can pass through the water surface, so the transmitted intensity varies from 0.9 to 1 as a whole, and rises even further with the increasing of SZA. When the SZA is larger (such as 90°) and enhances the resolution of the color bar, we can find a strip of high intensity close to 1 in the direction perpendicular to the sun's meridian with two bright spots emerging near the Brewster angle. This complies with the two excessively bright areas when observing the whole sky underwater. This is primarily due to the vertical polarized incident skylight with high DOP in that direction when the sun is on the horizon, so the low reflectance leads to the high transmittance. Moreover, the reflectance of incident light near the Brewster angle is lower, so that there are two bright spots [7]. In addition, when light passes through the air–water surface, the n^2 law for radiance is applicable and shown as below [4,51].

$$L_1/L_2 = n_1^2/n_2^2 \tag{21}$$

where n_1 is the refractive index for first medium and n_2 is the refractive index for the second medium. L_1 is the radiance in the first medium and L_2 is the radiance in the second medium.

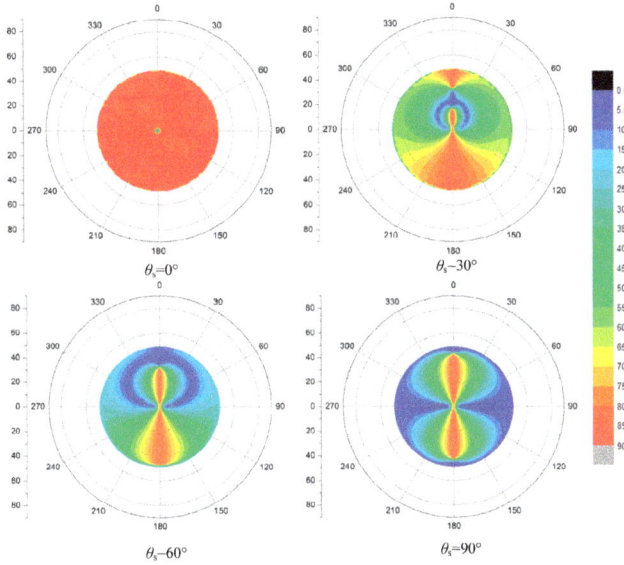

Figure 5. The AOP (angle of polarization) patterns of skylight transmitted through a calm water surface within the Snell's window at different solar zenith angle $\theta_s = 0°, 30°, 60°, 90°$, the AOP ranges from 0 to 1, and the reference plane is the meridian of each observation direction.

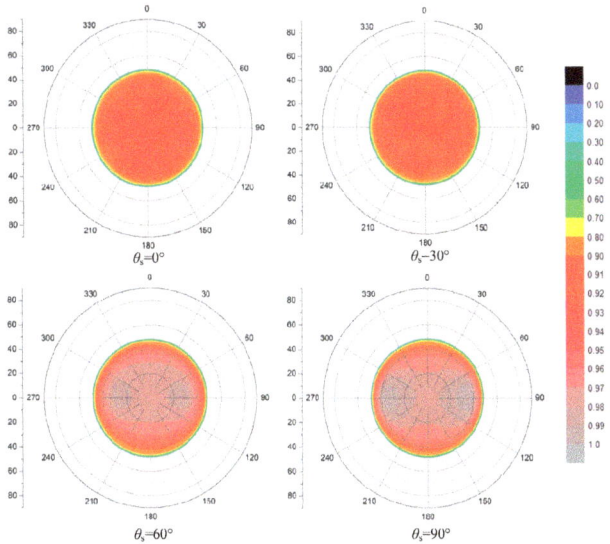

Figure 6. The intensity distributions of skylight transmitted through a calm water surface within the Snell's window at different solar zenith angle $\theta_s = 0°, 30°, 60°, 90°$, the intensity range from 0 to 1.

We can know that the intensity of transmitted light under water is 1.78 times stronger than that of the incident light. Due to the smaller field of view, the radiation intensity within the unit field increases according to the law of energy conservation. In conclusion, the transmitted intensity in the Snell's window is generally high.

3.2. Polarization and Transmittance Patterns of Full Incident Light Passed through a Wavy Water Surface

Undoubtedly, wavy surfaces affect the underwater light field and not only changes the intensity distribution, but also change the polarization patterns, especially in shallow depth water observations. This is mainly induced by the change of the height field of the wave surface and the wave focusing caused by the variation of the real incident angle [52]. The polarization pattern spreads out of the Snell's window owing to the wave surface. Thus, there are two different underwater polarization patterns, one inside and the other outside Snell's window [7].

The polarization and transmission patterns of transmitted light under a wavy water surface at different SZAs are presented in this section, and the effects of various wind speed and incident light sources on the patterns are analyzed.

3.2.1. The Polarization Patterns of Transmitted Light under a Wavy Water Surface

Unlike the calm surface conditions, which only consider the skylight incidence, we consider the whole celestial light, which includes both the contributions of the direct solar light and diffused skylight, to simulate the polarization patterns of transmitted light under a wavy water surface. Just like the calm water, the polarization pattern under wavy water, including DOP and AOP, is symmetrical about the solar meridian plane (see Figures 7 and 8). The ever-changing position of the sun in the sky exerts a major influence on the pattern of polarization underwater. The DOP and AOP patterns under various SZAs are shown in Figures 7 and 8, where the wind speed is 5 m/s, wind direction is 0°. Unlike the calm water, there is no sharp border at the margins of Snell's window which coincides with the measurement by Cronin [28]. The results indicate that the celestial polarization pattern is present within Snell's window, but it is modified due to refraction and repolarization of skylight at the air–water interface. Horvath called the polarization pattern outside Snell's window a bulk transmission-polarization pattern which is created by interaction between water and transmitted light [7]. Both of the inside pattern and outside patterns vary obviously with the sun position under a wavy water surface. Similar to the patterns under a flat surface (Figure 4), when the sun is at the zenith, the maximum DOP also distributes on the edge of the Snell's window. Additionally, the area of maximum DOP lies in the refracted direction normal to the sun's meridian when the sun is located in any other position. We can also find that the variation of polarization patterns of the underwater light field is obvious owing to the waves, which not only make the patterns non-homogeneous, but also distort them along the direction of wind. When the sun was near the zenith, the electric vector of the polarized light was horizontal and the same at all azimuths, which coincides with the measurement by Waterman [5]. Likewise, the maximum AOP distributes along the sun's meridian with the singular points emerging (Figure 8), which are not clear except the point on the zenith due to wave fluctuations. Using the statistical probability of the wave slope distribution model to describe the wave surface, the values of distribution patterns in this paper are weighted averages, not instantaneous results, which are comparable to the measurements. Therefore, the patterns will appear outside the Snell's window and continue with the patterns inside the window.

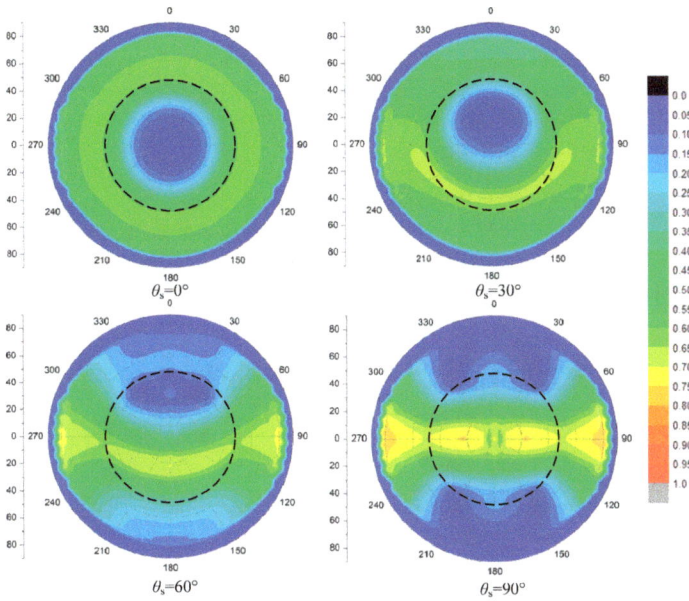

Figure 7. The DOP patterns of full incident light transmitted through a wavy water surface under different SZAs, where wind speed is 5 m/s, wind direction is 0°.

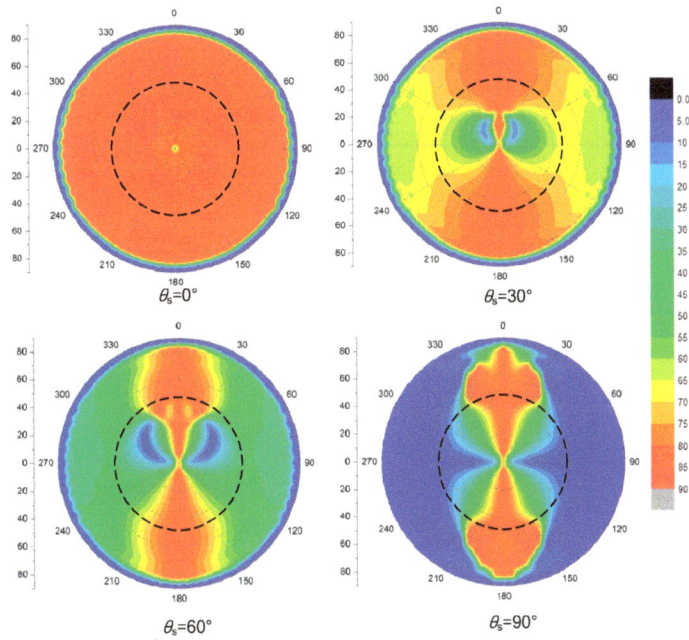

Figure 8. The AOP patterns of full incident light transmitted through a wavy water surface under different SZAs, where wind speed is 5 m/s, wind direction is 0°.

3.2.2. The Transmittance Distribution of Transmitted Light under a Wavy Water Surface

In addition to its degree of polarization and direction of polarization, the intensity of downwelling transmitted light was discovered to also vary systematically with the relative position of the sun and the points observed underwater. Figure 9 illustrates the relative intensity distribution of transmitted light under a wavy water surface under different SZAs. We can see that the intensity distributions focus in the Snell's window, which demonstrates that the Snell's window is the real passageway to form the underwater light field. The direct solar incidence leads to the maximum intensity region near the apparent sun position under water, and there will appear a bright ring on the edge of the window with a greater solar zenith angle. In addition, when the sun is available, the maximum intensity region is where the brightness expands as the SZA increases.

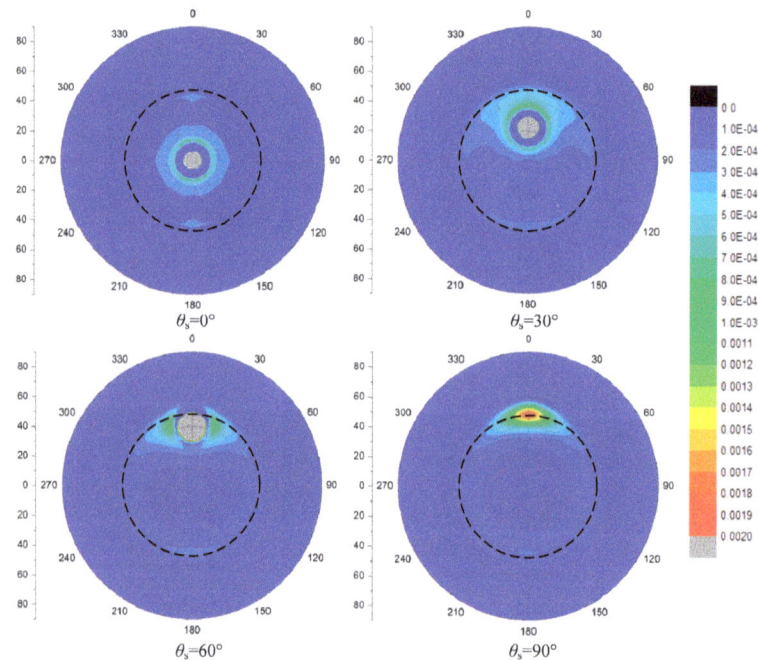

Figure 9. The intensity distributions of transmitted light under a wavy water surface with different SZAs, where wind speed is 5 m/s, wind direction is 0°.

3.2.3. The Effects of Other Factors on Transmitted Light Polarization Patterns

(1) The effect of wave fluctuation

Waves are formed when wind velocity increases; the stronger the wind, the higher the waves. They are called wind-waves. When the wind velocity decreases, there are still waves on the sea for a considerable time, but the length of the waves grows larger and larger. They are so called swell. The swell may have a direction different from that of the wind, which may again rise from another direction and create new wind-waves [22]. In this study, we only take the wind-waves into account and omit the gravity wave for the sake of adoption of the Cox–Munk wave model.

Figure 10 shows the effect of wind speed on the DOP, AOP and intensity patterns of transmitted light under water. We can find that the influence of wind speed (e.g., 1, 5, and 10 m/s) on the patterns of DOP is obvious. The DOP patterns spread slightly to the direction of the wind and the maximum DOP band narrows with the increasing wind speed. The polarization under water also decreases when the surface of the water is undulating, as the sunlight can be refracted in many directions by the waves. Then, there are no more parallel sunbeams underwater, hence there is diminishing polarization of scattered light under the surface. Just as shown in the first row in Figure 10, the DOP decline with the wind speed increasing, as the direct underwater light tends to diffuse due to the wave fluctuation. When the water surface is undulating, the sunlight can be refracted in many directions by the waves. Then, there are no more parallel sunbeams underwater, hence the diminishing polarization of scattered light under the surface. The pattern of AOP is also symmetrical to the sun's meridian plane just that of AOP. Compared with DOP, however, the AOP patterns are more stable. All these could provide stable orientation information sources for polarization sensing aquatic animals for navigation. However, the effect of wind speed on the intensity distribution under water is significant: the bright ring appears on the edge at low wind speed, but disappears with the increasing wind speed, and the bright circle near the apparent sun position expands. In addition, the edge of the Snell's window becomes unclear.

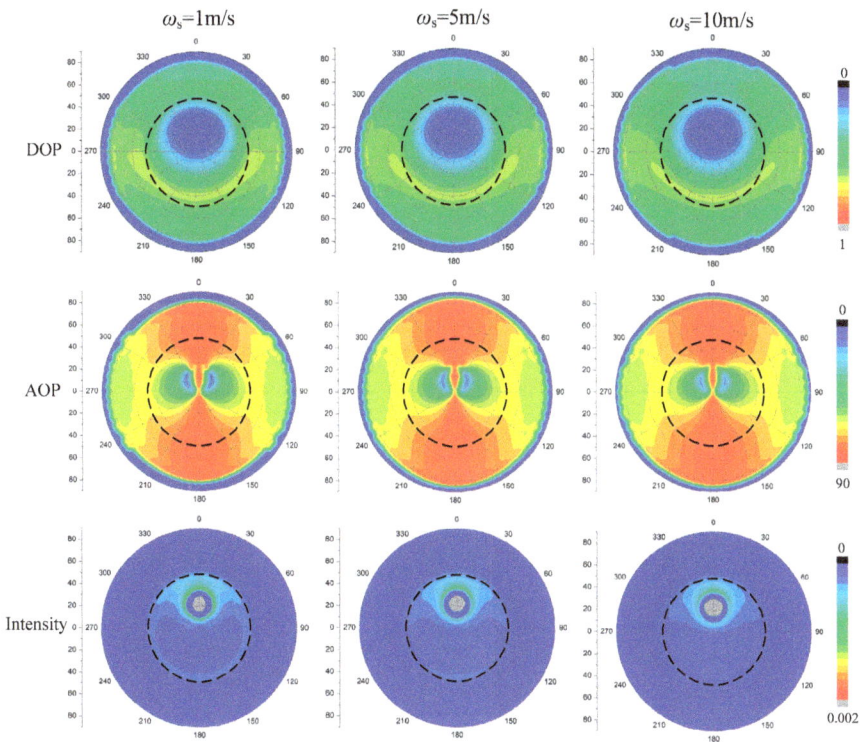

Figure 10. The DOP, AOP and intensity patterns of transmitted light under a wavy water surface with different wind speeds, where the SZA is 30°, wind direction is 0°.

(2) The effect of different incident sources

The refracted polarization patterns of wave water surface depend not only on the presence or absence of waves but also to a high degree upon illumination [22]. In this study, the source of the incident light can be conceived as being from a direct light source (sunlight) or diffuse light source (skylight or scattered light). The former is partially polarized and latter is unpolarized. Unpolarized sunlight becomes slightly partially linear polarized as it refracts through the water surface, but the primary production of the significant linear polarization is caused by the skylight.

Figure 11 indicates that the direct solar incident light only contributes to the polarization and intensity patterns at the apparent sun position under water, while the scattering incident skylight dominates the polarization patterns of the underwater light field. As mentioned above, the maximum DOP of transmission is no more than 0.28 according to the Fresnel's equations. However, the DOP of transmission can exceed 0.6. Therefore, it is obvious that the refracted polarization information underwater mainly come from the skylight rather than from the refraction process. From the three AOP patterns, we can find that the direct light may promote the formation of the singular points except the point on the zenith. In addition, the direct light leads to the brightest part of the transmitted light field under water, and the scattering skylight slightly affects the whole intensity distribution instead. Thus, the direct solar light is the dominant source of the brightness of the underwater light field.

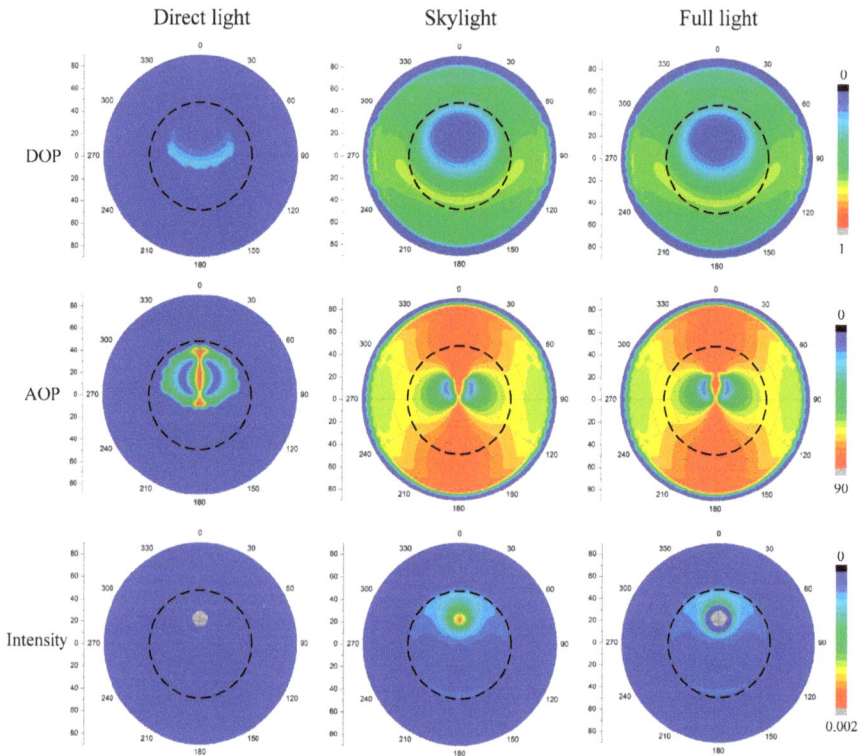

Figure 11. The DOP, AOP and intensity patterns of transmitted light under a wavy water surface with different incident sources (the direct solar light, scattering skylight and combination of above), where the SZA is 30°, wind speed is 5 m/s, and wind direction is 0°.

3.3. Polarization and Transmittance Patterns of Full Incident Light Passed through a Wavy Water Surface

When a beam transfers into the water surface, both reflection and refraction occur. The two processes not only change the intensity, but also the polarization of incident light. This section demonstrates the polarization patterns of the skylight, reflected light and refracted light under various SZAs (30°, 60° and 90°) and then analyzes the differences between the polarization and intensity patterns of reflection and refraction and the relationships between them.

The DOP patterns of skylight at different SZAs are shown in Figure 12. According to the semi-empirical Rayleigh scattering model, the DOP of scattering light under clear skies is directly related to the scattering angle. As the SZA increases, the maximum DOP region whose scattering angle is 90° moves to the zenith where the SZA is 90°. While the DOP of the skylight near the sun is the lowest. That is to say, the further from the sun, the stronger is the DOP, during the sunrise and sunset. Underwater, the apparent sun looks like the sun in the sky, though the wave distorts the DOP patterns, the patterns are still similar to the patterns of the skylight (Figure 12A,C). As shown in Figure 12B,C, with the increasing of SZA, the DOP of reflected light decreases while that of transmitted light rises as a whole. The maximum DOP of reflected light is always distributed near the Brewster angle, and that of transmitted light lies in the refracted direction instead.

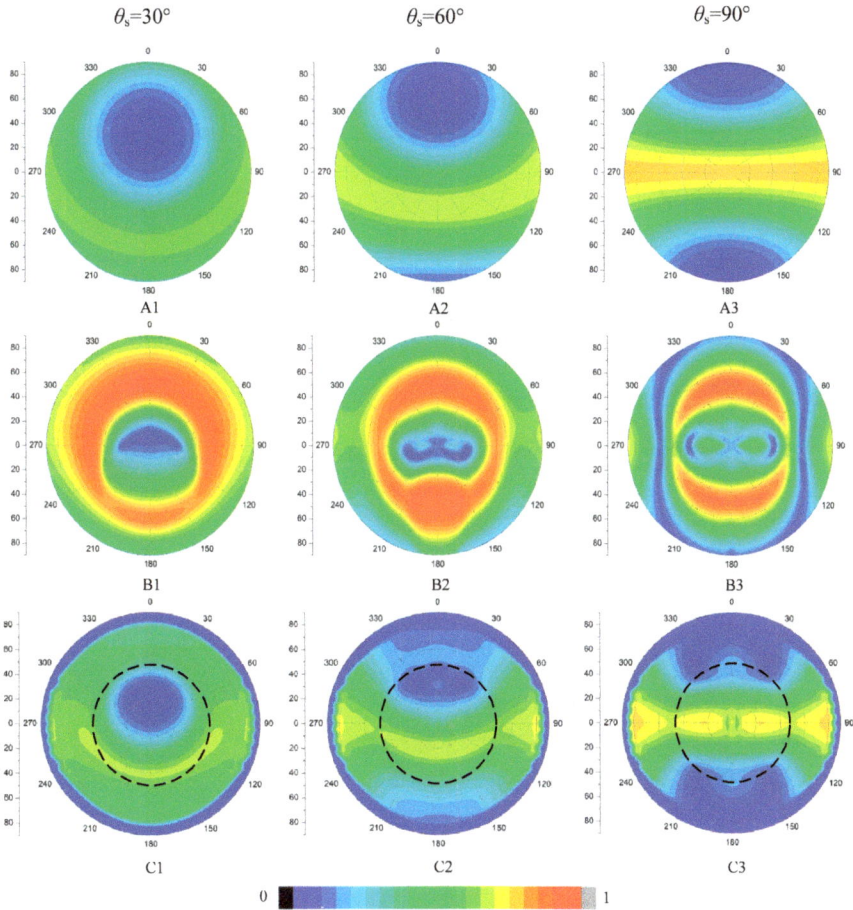

Figure 12. The DOP patterns of the: skylight (**A**); reflected light (**B**); and refracted light (**C**) under different SZAs (30°, 60° and 90°), where the wind speed is 5 m/s, and wind direction is 0°.

In Figure 13A, we can see that the singular spots of the AOP patterns of skylight are always at the zenith and the sun position no matter where the sun is. The light from the sun's meridian is always horizontal polarized. The AOP patterns of transmitted light under water resemble that of incident skylight from Figure 13A,C. This is likely that E-vector of the light does not change its phase when retracted by water surface. However, when light transfers from an optically thinner medium to a denser medium, the reflected light comes up with a phase change of π based on the Fresnel' law. Thus, the reflection changes not only the intensity, but also the phase of incident light, and then the polarization patterns of reflected light are entirely different from that of skylight, as shown in Figures 12B and 13B. We can conclude the polarization characteristics of incident light are reserved in the Snell's window under water when refraction occurs, especially under calm surface conditions.

The pattern of downwelling polarized light under the wavy water surface is superimposed by the pattern of skylight polarization and that alter by refraction and the disturbance of the wave movement.

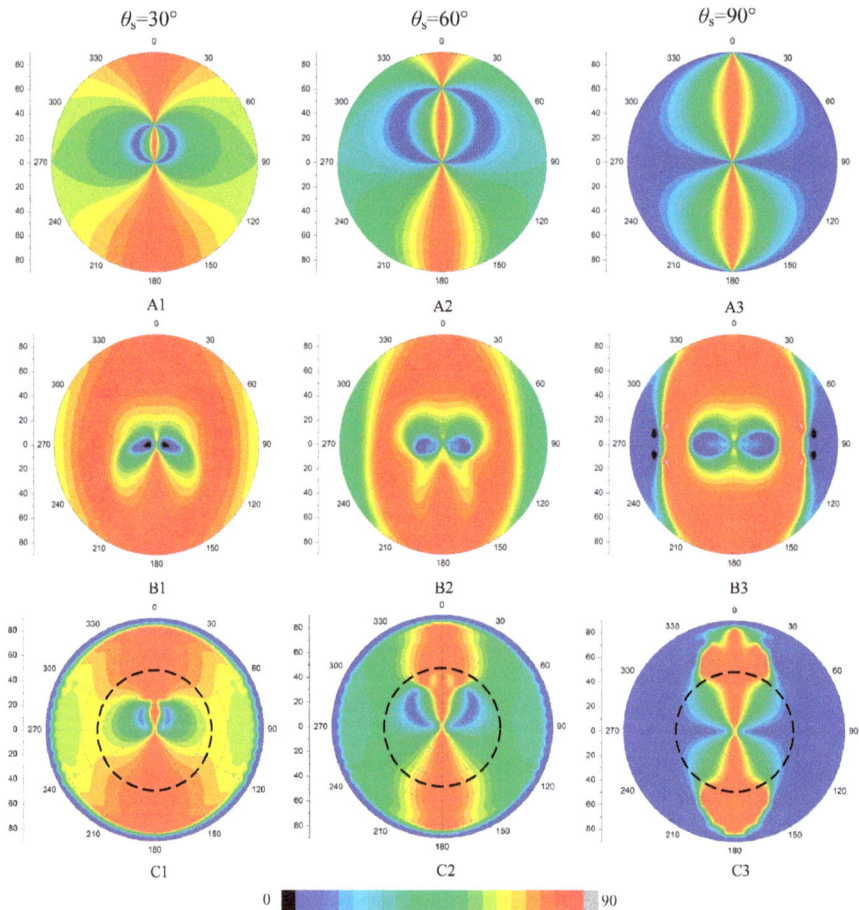

Figure 13. The AOP patterns of the: skylight (**A**); reflected light (**B**); and refracted light (**C**) under different SZAs (30°, 60° and 90°), where the wind speed is 5 m/s, and wind direction is 0°.

4. Discussion

Waterman first measured underwater polarization; subsequently, there are many measurements of linear-polarized light transmitted into water surface [28,31,52–57].

To evaluate above simulation results, we have made a comparison with the data measured by Bhandari and his colleagues in 2009 at the R/P Flip, off Hawaii [24]. The results are shown in Figures 14–16. The measurements are carried out at one meter below the water surface and the water is clear. In that case the scattering effects in water can be omitted. The wind speed is 6 m/s.

The data in Figure 14A were collected at 520 nm in very clear water on 7 September 2009. The data in Figure 14B are the simulated data. In Figure 14, we can find that the majority of downwelling radiance is limited in the Snell's cone. Part of the radiance exceeds the edge of Snell's cone due to the fluctuation of the water surface. The simulated data correspond well with the measured data, not only in the major distribution but also in some details. For example, the radiance in the central part of the Snell's cone is lower than the surrounding circle. The direct solar disk is extended at the edge of Snell's cone. There are also some differences especially at the edge of Snell's cone. The main reason is that our calculation uses the Cox–Munk model, which gives the average results of the transmittance at a certain angle. However, the photos taken by Bhandari reveal the real state of water surface.

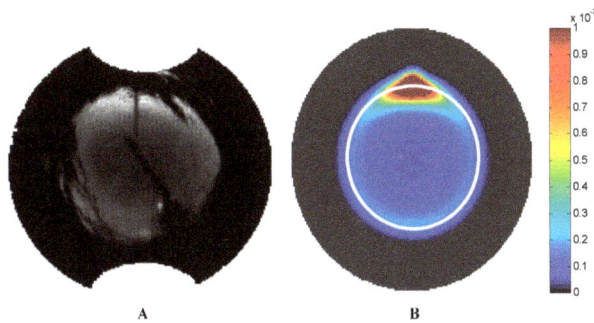

Figure 14. Radiance of downwelling light field. In this and the following figures, the zenith angle for the data increases linearly with the radius from the center. The sun is towards the top of the image and the solar zenith angle is 90°. (**A**) Data adopted from Bhandari's work [24]; and (**B**) simulated results using our model. The white circle indicates the Snell's cone.

Figure 15 shows the comparison between the measured data and our simulation data on Q/I and U/I. In the Snell's cone, the shapes of the two sets of graphs agree well with each other. The correctness of our calculation model is proven in a way. The main differences appear on the edge and outside of the Snell's cone. The graphs near the edge seem to be disordered for the measured data which are probably due to the fluctuation of the water surface. Another reason comes from the measurement error, since the Stokes parameters of downwelling radiance were measured with four lenses. As for the outside circle of the Snell's cone, it is noisy, because the energy of radiance is too low at this part. The simulated data cannot reveal the fluctuation effects and will not be influenced by the low radiance.

It is obvious that the features of DOP and AOP are similar between the measured data and our simulated results. The differences also appear on the edge and outside circle of the Snell's cone. The reason has been explained in the former paragraph. In the Snell's cone, the DOP and AOP patterns are mainly decided by the skylight polarization patterns. The comparison between the measured data and our simulated results shows great similarity, which proves the correctness of our model. The influence of fluctuated water surface mainly appears at the edge of the Snell's cone. The noise outside the Snell's cone comes from the low radiance energy of the downwelling light. Of course, both of the measured data and the simulated results are collected under the situation of shallow depth and clear water.

The research on the polarization characteristics of transmitted light under natural water is an important subject in polarization vision field. However, the polarization pattern of the underwater light field is very complex, it is not only influenced by the sun position, wave fluctuation and incident sources, but also the factors of water depth, turbidity caused by suspended particulates and multiple scattering in the water. The model described in this paper worked well and provides relatively satisfactory results, however there are some points we should emphasize here.

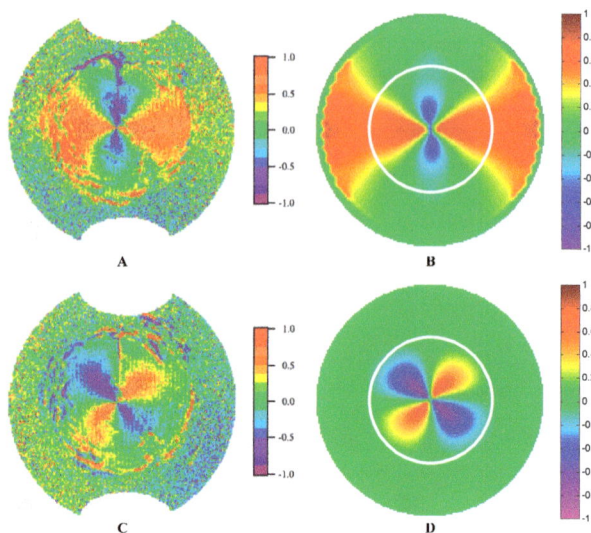

Figure 15. Q/I (**A**); and U/I (**B**) graphs measured by Bhandari [24]. Simulated: Q/I (**C**); and U/I (**D**) graphs.

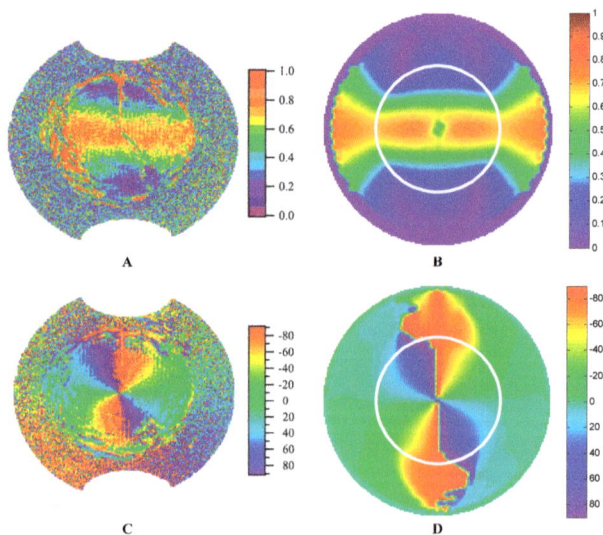

Figure 16. DOP (**A**); and AOP (**B**) graphs measured by Bhandari [24]. Simulated: DOP (**C**); and AOP (**D**) graphs.

In this paper, we focus on the light transmitted downward passing through the wave water surface. The effect of waves and the incident skylight on the polarization pattern of downward light in shallow depth water are emphasized. The contribution of scattering in water and reflection at the water bottom to the polarization of the underwater light field was not considered in this study. Therefore, this model may not validate the situation of upward light or deepwater environments (deeper than about 0.5 m) where the multiple scattering dominates the underwater light field. At great depths (many tens of meters), the illumination is dominated by diffuse light caused by the submarine particle scattering. The polarization is further weakened by the underwater scattering [53]. Moreover, in optical shallow waters, bottom reflection will affect the polarization patterns [23].

Besides linearly polarized light, circularly polarized light can also occur under water. The latter is formed indirectly by total reflection of the existing underwater linear polarized light against the lower side of the water surface [5,22]. We focus on the downwelling refraction of celestial light rather than involve the upwelling refraction or the total refraction process in this study. Thus, the circularly polarized light was also neglected.

We omitted the spectral character of skylight and underwater since the refractive indices of air and water vary slightly with the wavelength of light. Furthermore, according to the field measurements [23,28], the polarization in water is relatively insensitive to wavelength. Patterns of both the overall e-vector orientation and degree of polarization are similar from 360 to 550 nm.

The polarization pattern of skylight is very complex which is influenced by many factors. These include the position of the sun, the coverage of the cloud, the earth's albedo, atmospheric turbidity, multiple scattering and depolarization owing to anisotropy of air molecules. The semi-empirical skylight polarization model adopted in this study is shown to be effective to describe the polarization pattern of the skylight. However, the semi-empirical parameter is difficult to parameterize such factors. Precision modeling can resort to the vector transfer model such as 6SV, RT3, etc. However, coupling such a model needs much more computation time. Furthermore, both of these vector radiative transfer models did not take cloud cover into account. All of these factors above need to be taken into consideration in future research.

5. Conclusions

This work may be briefly summarized as follows. In this paper, we present a model to describe the polarization patterns of celestial light when refracted by the wavy water surface. The polarization characteristics and the radiance distribution pattern of the skylight are both taken into consideration. The polarization patterns and intensity distribution of refracted light off wave water surface were calculated. The model was validated by the underwater experimental measurements. The experimental and theoretical values agree well qualitatively. This work could provide a quantitative description of the repolarization and transmittance of celestial light transmitted through the wave water surface.

The dependence of the submarine light field polarization on the illumination and surface wave conditions, solar zenith and azimuth angles, and the viewing angles are discussed in detail. The polarization properties of transmitted celestial light under shallow depth water are mainly dependent on the incident polarized skylight.

Though the wave surface distorts the polarization patterns under water and leads them to exist out of the Snell's window, the polarization patterns within the Snell's window are similar to that of the skylight. Wind speed slightly affects the underwater polarization distribution, which expands a little along the wind direction as the wind speed rises. The direct solar light dominants the intensity distribution in the light field under water, while the scattering light from the sky is the dominant effect on determining the polarization patterns of transmitted light. We can conclude that the refraction slightly influences the polarization characteristic of incident light, in addition to the stable polarization field under water, which are beneficial to aquatic animals that orientate themselves with polarized sensing.

Comparing the polarization patterns of reflected light to that of transmitted light, we can find that reflection process significantly change the original polarization patterns of incident light, while the patterns of refraction have the basic characteristics of that of the incident light. The maximum DOP of reflected light is always distributed near the Brewster angle, and that of transmitted light lies in the refracted direction of Brewster angle instead.

Underwater polarization has drawn great attention to many fascinating areas of research yet to be explored. Combination of achievement in radiative transfer simulation and novel developed detector technology and data inversion algorithms will certainly provide many new and exciting research topics in ocean optics and ocean color remote sensing [3,14].

Acknowledgments: We are grateful to Chunyue Niu and Victoria Cheung for their constructive help and suggestions. We are also grateful to the anonymous reviewers for their valuable comments and suggestions. This work was supported by the flowing funds: National Natural Science Foundation of China (Grant No. 40901168); Beijing Natural Science Foundation (Grant No. 8162028); Open Fund of Key Laboratory of Space Ocean Remote Sensing and Application (Grant No.201602021); Open Fund of State Key Laboratory of Remote Sensing Science (Grant No. OFSLRSS201506); National Program on Key Basic Research Project (Grant No. 2014CB744204); and Research on data processing theory and methods of the auxiliary lines selection based on satellite remote sensing image (Grant No. GCB17201600036).

Author Contributions: Guanhua Zhou and Wujian Xu conceive the model and wrote the paper. Jiwen Wang and Kai Zhang carried out the calculation and analyzed the data. Zhongqi Ma made the validation and analysis.

Conflicts of Interest: The authors declare no conflict of interest. The founding sponsors had no role in the design of the study; in the collection, analyses, or interpretation of data; in the writing of the manuscript, and in the decision to publish the results.

References

1. Amir, I.; Alexander, G.; Tristan, H.; Alberto, T.; Jacek, C.; Samir, A. The relationship between upwelling underwater polarization and attenuation/absorption ratio. *Opt. Express* **2012**, *20*, 25662–25680.
2. Alberto, T.; Alex, G.; Tristan, H.; Amir, I.; Jacek, C.; Barry, G.; Fred, M.; Sam, A. Estimating particle composition and size distribution from polarized water-leaving radiance. *Appl. Opt.* **2011**, *50*, 5047–5058.
3. Kattawar, G.W. Genesis and evolution of polarization of light in the ocean. *Appl. Opt.* **2013**, *52*, 940–948. [CrossRef] [PubMed]
4. Mobley, C.D. *Light and Water: Radiative Transfer in Natural Waters*; Academic Press: San Diego, CA, USA, 1994.
5. Waterman, T.H. Polarization patterns in submarine illumination. *Science* **1954**, *120*, 927–932. [CrossRef] [PubMed]
6. Horváth, G.; Varjú, D. *Polarized Light in Animal Vision: Polarization Patterns in Nature*; Springer: Heidelberg, Germany, 2004.
7. Horváth, G.; Varjú, D. Underwater refraction–polarization patterns of skylight perceived by aquatic animals through Snell's window of the flat water surface. *Vis. Res.* **1995**, *35*, 1651–1666. [CrossRef]
8. Harmel, T.; Chami, M. Invariance of polarized reflectance measured at the top of atmosphere by PARASOL satellite instrument in the visible range with marine constituents in open ocean waters. *Opt. Express* **2008**, *16*, 6064–6080. [CrossRef] [PubMed]
9. Tonizzo, E.; Gilerson, A.; Harmel, T.; Ibrahim, A.; Chowdhary, J.; Gross, B.; Moshary, F.; Ahmed, S. Estimating particle composition and size distribution from polarized water-leaving radiance. *Appl. Opt.* **2011**, *50*, 5047–5058.
10. Harmel, T.; Alexander, G.; Alberto, T.; Jacek, C.; Alan, W.; Robert, A.; Sam, A. Polarization impacts on the water-leaving radiance retrieval from above-water radiometric measurements. *Appl. Opt.* **2012**, *51*, 8324–8340. [CrossRef] [PubMed]
11. Chowdhary, J.; Cairns, B.; Waquet, F.; Knobelspiesse, K.; Ottaviani, M.; Redemann, J.; Travis, L.; Mishchenko, M. Sensitivity of multiangle, multispectral polarimetric remote sensing over open oceans to water-leaving radiance: Analyses of RSP data acquired during the MILAGRO campaign. *Remote Sens. Environ.* **2012**, *118*, 284–308. [CrossRef]
12. Ibrahim, A.; Gilerson, A.; Harmel, T.; Tonizzo, A.; Chowdhary, J.; Ahmed, S. The relationship between upwelling underwater polarization and attenuation/absorption ratio. *Opt. Express* **2012**, *20*, 25662–25680. [CrossRef] [PubMed]
13. Kattawar, G.W.; Yang, P.; You, Y.; Bi, L.; Xie, Y.; Huang, X.; Hioki, S. Polarization of light in the atmosphere and ocean. In *Light Scattering Reviews 10 Light Scattering and Radiative Transfer*; Kokhanovsky, A.A., Ed.; Springer: Heidelberg, Germany, 2016; pp. 3–40.
14. Harmel, T. Recent developments in the use of light polarization for marine environment monitoring from space. In *Light Scattering Reviews*; Kokhanovsky, A.A., Ed.; Springer: Heidelberg, Germany, 2016; Volume 10, pp. 41–84.

15. Gilerson, A.; Zhou, J.; Oo, M.; Chowdhary, J.; Gross, B.M.; Moshary, F.; Ahmed, S.A. Retrieval of chlorophyll fluorescence from reflectance spectra through polarization discrimination: Modeling and experiments. *Appl. Opt.* **2006**, *45*, 5568–5581. [CrossRef] [PubMed]

16. Frisch, K.V. Die Polarisation des Himmelslichtes als orientierender Faktor bei den Tanzen der Bienen. *Cell. Mol. Life Sci.* **1949**, *5*, 142–148. [CrossRef]

17. Wehner, R. Polarization vision—A uniform sensory capacity. *J. Exp. Biol.* **2001**, *204*, 2589–2596. [PubMed]

18. Brines, M.L.; Gould, J.L. Skylight polarization patterns and animal orientation. *J. Exp. Biol.* **1982**, *96*, 69–91.

19. Horváth, G.; Varjú, D. Reflection–polarization patterns at flat water surfaces and their relevance for insect polarization vision. *J. Theor. Biol.* **1995**, *175*, 27–37. [CrossRef] [PubMed]

20. Waterman, T.H. Polarization of marine light fields and animal orientation. *Proc. SPIE* **1988**, *0925*, 431–437.

21. Zhou, G.; Xu, W.; Niu, C.; Zhao, H. The polarization patterns of skylight reflected off wave water surface. *Opt. Express* **2013**, *21*, 32549–32565. [CrossRef] [PubMed]

22. Können, G.P. *Polarized Light in Nature*; Cambridge University Press: New York, NY, USA, 1985.

23. Ivanoff, A.; Waterman, T.H. Factors, mainly depth and wavelength, affecting the degree of underwater light polarization. *J. Mar. Res.* **1958**, *16*, 283–307.

24. Voss, K.J.; Gleason, A.C.R.; Gordon, H.R.; Kattawar, G.W.; You, Y. Observation of non-principal plane neutral points in the in-water upwelling polarized light field. *Opt. Express* **2011**, *19*, 5942–5952. [CrossRef] [PubMed]

25. Voss, K.J.; Fry, E.S. Measurement of the Mueller matrix for ocean water. *Appl. Opt.* **1984**, *23*, 4427–4439. [CrossRef] [PubMed]

26. Bhandari, P.; Voss, K.J.; Logan, L. An instrument to measure the downwelling polarized radiance distribution in the ocean. *Opt. Express* **2011**, *19*, 17609–17620. [CrossRef] [PubMed]

27. You, Y.; Kattawar, G.W.; Voss, K.J.; Bhandari, P.; Wei, J.; Lewis, M.; Zappa, C.J.; Schultz, H. Polarized light field under dynamic ocean surfaces: Numerical modeling compared with measurements. *J. Geophys. Res.* **2011**, *116*, 1978–2012. [CrossRef]

28. Cronin, T.W.; Shashar, N. The linearly polarized light field in clear, tropical marine waters-spatial and temporal variation of light intensity, degree of polarization and e-vector angle. *J. Exp. Biol.* **2001**, *204*, 2461–2467. [PubMed]

29. Kattawar, G.W.; Adams, C.N. Stokes vector calculations of the submarine light field in an atmosphere–ocean with scattering according to a Rayleigh phase matrix: Effect of interface refractive index on radiance and polarization. *Limnol. Oceanogr.* **1989**, *34*, 1453–1472. [CrossRef]

30. Sabbah, S.; Lerner, A.; Erlick, C.; Shashar, N. Under water polarization vision-a physical examination. *Recent Res. Dev. Exp. Theor. Biol.* **2005**, *1*, 123–176.

31. Sabbah, S.; Barta, A.; Gál, J.; Horváth, G.; Shashar, N. Experimental and theoretical study of skylight polarization transmitted through Snell's window of a flat water surface. *J. Opt. Soc. Am. A* **2006**, *23*, 1978–1988.

32. Mishchenko, M.I.; Travis, L.D. Satellite retrieval of aerosol properties over the ocean using polarization as well as intensity of reflected sunlight. *J. Geophys. Res.* **1997**, *102*. [CrossRef]

33. Zhai, P.; Hu, Y.; Chowdhary, J.; Trepte, C.R.; Lucker, P.L.; Josset, D.B. A vector radiative transfer model for coupled atmosphere and ocean systems with a rough interface. *J. Quant. Spectrosc. Radiat.* **2010**, *111*, 1025–1040. [CrossRef]

34. Xu, Z.; Yue, D.K.P.; Shen, L.; Voss, K.J. Patterns and statistics of in-water polarization under conditions of linear and nonlinear ocean surface waves. *J. Geophys. Res.* **2011**, *116*, 1978–2012. [CrossRef]

35. Mobley, C.D. Polarized reflectance and transmittance properties of windblown sea surfaces. *Appl. Opt.* **2015**, *54*, 4828–4849. [CrossRef] [PubMed]

36. Hieronymi, M. Polarized reflectance and transmittance distribution functions of the ocean surface. *Opt. Express* **2016**, *24*, A1045–A1068. [PubMed]

37. Chami, M.; Lafrance, B.; Fougnie, B.; Chowdhary, J.; Harmel, T.; Waquet, F. OSOAA: A vector radiative transfer model of coupled atmosphere–ocean system for a rough sea surface application to the estimates of the directional variations of the water leaving reflectance to better process multi-angular satellite sensors data over the ocean. *Opt. Express* **2015**, *23*, 27829–27852. [PubMed]

38. Foster, R.; Gilerson, A. Polarized transfer functions of the ocean surface for above-surface determination of the vector submarine light field. *Appl. Opt.* **2016**, *55*, 9476–9494. [CrossRef] [PubMed]

39. Dickey, T.D.; Kattawar, G.W.; Voss, K.J. Shedding new ligdht on light in the ocean. *Phys. Today* **2011**, *64*, 44–49. [CrossRef]

40. Voss, K.J.; Souaidia, N. POLRADS: Polarization radiance distribution measurement system. *Opt. Express* **2010**, *18*, 19672–19680. [CrossRef] [PubMed]
41. Guenther, R.D. *Modern Optics*; Wiley: New York, NY, USA, 1990.
42. Harrison, A.W.; Coombes, C.A. Angular distribution of clear sky short wavelength radiance. *Sol. Energy* **1988**, *40*, 57–63. [CrossRef]
43. Coulson, K.L. *Polarization and Intensity of Light in the Atmosphere*; A. Deepak Publishing: Hampton, VA, USA, 1988.
44. Lynch, D.K. Snell's window in wavy water. *Appl. Opt.* **2015**, *54*, B8–B11. [CrossRef] [PubMed]
45. Cox, C.; Munk, W. Measurement of the roughness of the sea surface from photographs of the sun's glitter. *J. Opt. Soc. Am.* **1954**, *44*, 838–850. [CrossRef]
46. Cox, C.; Munk, W. *Slopes of the Sea Surface Deduced from Photographs of Sun Glitter*; Bulletin of the Scripps Institution of Oceanography of the University of California, La Jolla; University of California Press: Oakland, CA, USA, 1956.
47. Ebuchi, N.; Kizu, S. Probability distribution of surface wave slope derived using sun glitter images from geostationary meteorological satellite and surface vector winds from scatterometers. *J. Oceanogr.* **2002**, *58*, 477–486. [CrossRef]
48. Saunders, P.M. Shadowing on the Ocean and the existence of the horizon. *J. Geophys. Res.* **1967**, *18*, 4643–4649. [CrossRef]
49. Ottaviani, M.; Spurr, R.; Stamnes, K.; Li, W.; Su, W.; Wiscombe, W. Improving the description of sunglint for accurate prediction of remotely sensed radiances. *Quant. Spectr. Radat. Trans.* **2008**, *109*, 2364–2375. [CrossRef]
50. Waterman, T.H. Reviving a neglected celestial underwater polarization compass for aquatic animals. *Biol. Rev.* **2006**, *81*, 111–115. [CrossRef] [PubMed]
51. Wyatt, C.L. Radiometric calibration: Theory and methods. *J. Membr. Biol.* **1978**, *129*, 99–107.
52. Sabbah, S.; Shashar, N. Underwater light polarization and radiance fluctuations induced by surface waves. *Appl. Opt.* **2006**, *45*, 4726–4739. [CrossRef] [PubMed]
53. Cronin, T.W.; Marshall, J. Patterns and properties of polarized light in air and water. *Philos. Trans. R. Soc. B* **2011**, *366*, 619–626. [CrossRef] [PubMed]
54. Shashar, N.; Sabbah, S.; Cronin, T.W. Transmission of linearly polarized light in seawater-implications for polarization signaling. *J. Exp. Biol.* **2004**, *207*, 3619–3628. [CrossRef] [PubMed]
55. Sabbah, S.; Shashar, N. Light polarization under water near sunrise. *J. Opt. Soc. Am. A* **2007**, *24*, 2049–2055. [CrossRef]
56. Bhandari, P. The Design of a Polarmeter and its Use for the Study of the Variation of Downwelling Polarized Radiance Distribution with Depth in the Ocean. Ph.D. Thesis, University of Miami, Coral Gables, FL, USA, 2011.
57. Bhandari, P.; Voss, K.J.; Logan, L. The variation of the polarized downwelling radiance distribution with depth in the coastal and clear ocean. *J. Geophys. Res.* **2011**, *116*. [CrossRef]

remote sensing

MDPI

Article

An Optical Classification Tool for Global Lake Waters

Marieke A. Eleveld [1,2,*,†], Ana B. Ruescas [3,4,†], Annelies Hommersom [5,†], Timothy S. Moore [6,†], Steef W. M. Peters [5,†] and Carsten Brockmann [3,†]

[1]   Deltares, P.O. Box 177, 2600 MH Delft, The Netherlands
[2]   Vrije Universiteit Amsterdam, Institute for Environmental Studies (VU-IVM), De Boelelaan 1087, 1081 HV Amsterdam, The Netherlands
[3]   Brockmann Consult GmbH, Max-Planck-Str.2, 21502 Geesthacht, Germany; ana.ruescas@brockmann-consult.de (A.B.R.); carsten.brockmann@brockmann-consult.de (C.B.)
[4]   Image Processing Laboratory, University of Valencia, P.O. Box 22085, E-46071 Valencia, Spain; ana.b.ruescas@uv.es
[5]   Water Insight, Marijkeweg 22, 6709 PG Wageningen, The Netherlands; hommersom@waterinsight.nl (A.H.); peters@waterinsight.nl (S.W.M.P.)
[6]   University of New Hampshire, 8 College Road, OPAL/Morse Hall, Durham, NH 03824, USA; timothy.moore@unh.edu
*   Correspondence: marieke.eleveld@deltares.nl
†   These authors contributed equally to this work.

Academic Editors: Yunlin Zhang, Claudia Giardino, Linhai Li and Prasad S. Thenkabail
Received: 28 February 2017; Accepted: 23 April 2017; Published: 29 April 2017

Abstract: Shallow and deep lakes receive and recycle organic and inorganic substances from within the confines of these lakes, their watershed and beyond. Hence, a large range in absorption and scattering and extreme differences in optical variability can be found between and within global lakes. This poses a challenge for atmospheric correction and bio-optical algorithms applied to optical remote sensing for water quality monitoring applications. To optimize these applications for the wide variety of lake optical conditions, we adapted a spectral classification scheme based on the concept of optical water types. The optical water types were defined through a cluster analysis of in situ hyperspectral remote sensing reflectance spectra collected by partners and advisors of the European Union 7th Framework Programme (FP7) Global Lakes Sentinel Services (GLaSS) project. The method has been integrated in the Envisat-BEAM software and the Sentinel Application Platform (SNAP) and generates maps of water types from image data. Two variations of water type classification are provided: one based on area-normalized spectral reflectance focusing on spectral shape (6CN, six-class normalized) and one that retains magnitude with no modification to the reflectance signal (6C). This resulted in a protocol, or processing scheme, that can also be applied or adapted for Sentinel-3 Ocean and Land Colour Imager (OLCI) datasets. We apply both treatments to MERIS imagery of a variety of European lakes to demonstrate its applicability. The studied target lakes cover a range of biophysical types, from shallow turbid to deep and clear, as well as eutrophic and dark absorbing waters, rich in colored dissolved organic matter (CDOM). In shallow, high-reflecting Dutch and Estonian lakes with high sediment load, 6C performed better, while in deep, low-reflecting clear Italian and Swedish lakes, 6CN performed better. The 6CN classification of in situ data is promising for very dark, high CDOM, absorbing lakes, but we show that our atmospheric correction of the imagery was insufficient to corroborate this. We anticipate that the application of the protocol to other lakes with unknown in-water characterization, but with comparable biophysical properties will suggest similar atmospheric correction (AC) and in-water retrieval algorithms for global lakes.

Keywords: lakes; reflectance; classification; OWT; atmospheric correction; MERIS; OLCI; water quality

1. Introduction

Freshwater lakes, reservoirs and rivers are an essential resource for human and animal survival. Population increase coupled with change in land use, hydrologic regimes and climate are stressing these systems worldwide, threatening their function as sources for drinking water, socio-economic activities and ecological environments. Over the last decade, there has been an increase in the capacity and availability of remote sensing imagery from satellites for lake systems worldwide, promoting the usage and creating new demands for reliable remotely-sensed datasets. These new capabilities stem in part from newly-launched satellites, such as the MultiSpectral Imager (MSI) on board the European Space Agency's (ESA) Sentinel-2 satellite and the Ocean Land Colour Imager (OLCI) on board ESA's Sentinel-3 satellite. The OLCI sensor is similar in spectral capabilities as the Medium Resolution Imaging Spectrometer (MERIS) sensor (2002–2012), containing spectral channels well suited to derive bio-optical parameters over the large range of optical conditions exhibited in lakes [1,2]. Sentinel-3A was launched in February 2016, and its twin Sentinel-3B is expected to be launched in 2017. The tandem missions of Sentinel-3A/B and follow ups will provide unprecedented monitoring capabilities for lake water quality because of the favorable band settings, high signal/noise ratios, full spatial resolution (300 m) and high overpass frequency.

A prototype infrastructure for handling of bio-optical algorithms and data products specific to freshwater lakes was prepared within the EU Global Lakes Sentinel Services (GLaSS) project (www.glass-project.eu). GLaSS aimed to develop generic methods and tools for Sentinel-2 and Sentinel-3 data, using legacy datasets, and in support of water quality management for any lake worldwide. One of the GLaSS products developed for lake image analysis is a classification tool based on the spectral matching method of Moore et al. [3] as an expression of optical water types (OWTs). The OWT tool operates on atmospherically-corrected and quality-checked images prior to the application of bio-optical algorithms and provides users with a powerful data analysis technique to visualize and discover the (variability of) optical conditions across image scenes.

Classification schemes are more common to terrestrial imagery, but are gaining traction in aquatic applications and share basic similarities [4,5]. In both cases, the classification systems are based on features (i.e., spectral channels) in a spectral signal related to underlying types with ecological meaning. The features stem from the spectral reflectance shape and magnitude and are ultimately limited by the spectral resolution of the sensors when utilized for image classification. For aquatic uses, water types are analogous to land cover types, representing an optical condition, and hence, are referred to as optical water types or OWTs. This notion of optical type has origins in [6,7], where water types were defined by the diffuse attenuation coefficient of downwelling light. These Jerlov types are still used in marine applications [8], and were used in a recent modeling study to generate Inherent Optical Properties (IOPs) for each type [9], directly utilizing type-specific parameters.

More recent water type schemes have been introduced over the last 20 years using a variety of methods based on in situ and/or satellite reflectance data. Regardless of the method, OWTs provide information on the spatial distribution of optical states across image scenes when applied to satellite data. These mapped products function as weighting factors for optimizing bio-optical algorithms and product uncertainties for image scenes [3,10–12]. In these cases, they are intermediary products that are not needed themselves for analysis and are invisible to users. However, OWTs are depictions of optical states, providing information on underlying water conditions that in and of themselves have intrinsic ecological value. They have been used directly for interpretive analysis for ecological diversity [13] and ecological patterns [14] that may not be obvious from other bio-optical products, such as chlorophyll concentration, which may be hard to retrieve in complex lake waters, because of the complex atmospheric and in-water optical properties. In some cases, OWTs have been linked to distinct optical phenomena that relate to specific phytoplankton [15]. These studies collectively illustrate the varying roles and uses for water types, whether freshwater or marine, when applied to remote sensing data.

The GLaSS optical water types are a follow up of [3] that presented OWTs derived from lake and coastal waters. The GLaSS dataset comprises lake data only, encompassing a larger dataset that includes more diverse lakes from across the globe. Within this paper, we introduce this classification method (called GLaSS-OWT or GLaSS optical water type method). The water types were derived from a cluster analysis. The classification system that we present has two main implementation options: a set of optical water types for un-modified reflectance data and a set of water types for normalized reflectance data, an aspect not presented in [3]. The method is designed to be applicable to any lake system, covering a large range of biophysical types from shallow turbid to clear and deep, as well as eutrophic and dark absorbing colored dissolved organic matter (CDOM)-rich waters.

We describe the development of the classification method and demonstrate its application to a variety of lake systems processed with different atmospheric correction schemes. The variations in OWT image products are discussed in the context of atmospheric correction. We also examine the strengths and differences of the different OWT schemes and how they may be appropriate for different global lakes with unknown optical properties.

2. Materials and Methods

2.1. In Situ Data Sources

Conceptually, the OWTs represent optical states that can be determined by the spectral remote sensing reflectance or $R_{rs}(\lambda)$. This term refers to the above-water quantity unless otherwise noted. In practice, they are derived from averaging grouped $R_{rs}(\lambda)$ spectra that share characteristics (e.g., spectral shape), where each individual spectrum is an instance along an optical continuum bound by the outer ranges of the environmental and optical conditions of all water systems. The goal of the GLaSS lake classification is a meaningful partitioning of the full multi-dimensional $R_{rs}(\lambda)$ space into a set of optical water types. This water type-specific approach is intrinsically independent of location and time and therefore designed for global application. Within a water type, there is a range of optical conditions that is represented, and thus, the environmental representation of a water type is that of an average condition.

The GLaSS OWT implementation is based on that of Moore et al. [3], but includes a larger variety of lakes. A motivation for the GLaSS OWT implementation was to develop a lake-specific classification tool for all lakes and conditions. To achieve this, we assembled a dataset of in situ hyperspectral $R_{rs}(\lambda)$ with co-measured Chl-a and Total Suspended Matter (TSM) concentrations and absorption of CDOM at 443 nm (aCDOM) from multiple sources covering a wide dynamic range in optical and environmental conditions.

This dataset includes the 'lake only' dataset portion ($N = 320$) from [3], which consists of measurements from the northeast U.S., the Great Salt Lake [16] and across Spain [17]. We refer the readers to these references for further information on the data collection protocols. These data were combined with the GLaSS in situ dataset (Table 1), which consists of $R_{rs}(\lambda)$ with co-measured Chl-a, TSM and aCDOM from different countries. This dataset contains a large range of Chl-a, TSM and CDOM concentrations that are covered, including the high concentrations (Chl-a > 900 (mgm^{-3}), TSM > 200 (mg m^{-3}), CDOM > 30 (443 m^{-1}), representing a large variety of optical conditions.

The GLaSS $R_{rs}(\lambda)$ measurements were collected above water and processed according to standard protocols [18]. The measurements consisted of: (1) light (radiance) emerging from water (L_w) measured at a 40–45 degree elevation angle from nadir and about a 135 degree azimuth angle from the Sun; (2) radiance from the sky (L_{sky}) measured at the same viewing angles; and (3) downwelling irradiance measurement (E_d). The remote sensing reflectance, R_{rs} (in sr^{-1}) is then computed with:

$$R_{rs}(0,+) = \frac{L_w - \rho L_{sky}}{E_d} \tag{1}$$

where the air-sea interface reflectance factor was fixed at 0.028 at a zenith angle of 42 degrees [19].

Table 1. In situ data from various Global Lakes Sentinel Services (GLaSS) partners and advisory board member Yunlin Zhang. (VIS/NIR, in the visible and near-infrared range; CDOM, colored dissolved organic matter; TSM, total suspended matter; Chl, chlorophyll-a.)

Area	Spectrometers	Spectral Range (nm)	Spectral Resolution in VIS/NIR	Spectral Resolution Interpolated	# Spectra	Range CDOM (443 m^{-1})	Range TSM (gm^{-3})	Range Chl_a (mgm^{-3})
Estonia	TriOS RAMSES	400–800	7 nm	2.5 nm	34	1.7–4.2	1.8–18.7	2.7–45.3
Finland	ASD FieldSpec	350–2500	3 nm	1 nm	16	0.5–10	0.8–3.4	1.7–11
The Netherlands	WI WISP-3	400–800	3.9 nm for E_d 4.9 nm for L_w and L_{sky}	1 nm	177	0.5–1.5	1.3–30	10–50
	Photo Research PR650	380–748	4 nm	4 nm	5 (L.IJsselmeer) 3 (L.Markermeer)		13–26 29.7–39.2	33.4–87.3 36.6–42.6
Italy	ASD FieldSpec Full Range Pro	350–2500	3 nm	1 nm	90	0.04–1.25	0.1–1.5	0.1–10
	SpectraScan Colorimeter PR650	380–780	8 nm	4 nm	3			
	WISP-3	400–800	3.9 nm for E_d 4.9 nm for L_w and L_{sky}	1 nm	13			0
China	ASD	350–1000	3 nm	1 nm	243	0.3–2.4	10–286	5–940
TOTAL					584	0.1–10	0.1–290	1.7–940

A dataset from Lake Erie measured in 2013 ($N = 16$) was also added during the development of the tool. These data included hyperspectral $R_{rs}(\lambda)$ taken with a Field Spec Pro™ VNIR-NIR1 portable spectrometer system from Analytical Spectral Devices (Boulder, Colorado). The protocol for deriving $R_{rs}(\lambda)$ was similar to that of the GLaSS data for Steps 1 and 2, although downwelling irradiance measurement (E_d) was determined from a grey card plaque.

All hyperspectral $R_{rs}(\lambda)$ data were band averaged to 3-nm resolution in the merged dataset ($N = 926$), quality controlled and reduced to $N = 871$ (Figure 1). Quality control measures consisted of visual inspection on every spectral observation and the application of the ocean chlorophyll (OC4) algorithm and MERIS three-band Chl-a algorithms for consistency checking. Observations with noisy or negative spectra were rejected, as were spectra with abnormal Chl-a retrievals. It should be noted that $R_{rs}(\lambda)$ associated with floating algal mats were removed (i.e., high NIR values). We believe this to be a special water type case that will be added in the future. The current dataset contained too few samples for this type to be characterized at present.

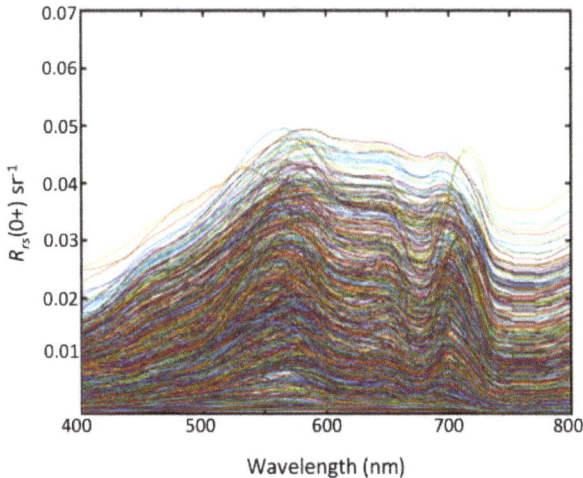

Figure 1. Total remote sensing reflectance, $R_{rs}(\lambda)$ data after quality control.

2.2. Development of the GLaSS Optical Water Types

To create the OWTs, a cluster analysis was applied to the merged, quality controlled $R_{rs}(\lambda)$ data. The goal of the clustering is simply to serve as a mechanism to sort data and to produce a partitioning of meaningful sub-groups. The effectiveness of cluster partitioning depends on the features, in our case R_{rs} as specific wavelengths, represented as a vector, that contribute to separability. In many cases, feature dimensionality can be reduced from the original dataset. This is often necessary to minimize processing time and cluster instability from redundant features or bands that highly covary [20], which is the case with hyperspectral data. Prior to clustering, feature selection and extraction were conducted on $R_{rs}(\lambda)$. The wavelengths chosen were those that matched the MERIS (and several Sentinel 3) visible and NIR band centers—412, 443, 490, 510, 560, 620, 665, 681, 709 and 753 nm—and reduced the dimension of each $R_{rs}(\lambda)$ spectra from 134 down to 10. Note, that this sole purpose of feature reduction is for identifying clusters, not for reducing the spectral dimensionality of the overall dataset.

We applied the fuzzy c-mean (FCM) algorithm [21] to the reduced $R_{rs}(\lambda)$ data. Following [3], these data were transformed to sub-surface values (Equation (2)) following [22]. It should be noted that the clustering and ensuing membership functions use the below-water quantity, but we will retain referencing any spectra as $R_{rs}(\lambda)$ for simplicity.

$$R_{rs}(0,-) = \frac{R_{rs}(0,+)}{0.52 + 1.7 * R_{rs}(0,+)} \qquad (2)$$

The FCM algorithm partitions the input data into a specified number of clusters. The function operates by minimizing the distance between the data points and the prototype cluster centers (means), which are iteratively adjusted until optimization criteria are met. Since the number of clusters is not known beforehand, FCM was applied to the dataset over a range of clusters set from 2–20. Cluster validity functions were used to assess the effectiveness of the cluster performance for each outcome. These functions measure various aspects of the entire cluster partitioning and were used to guide the ultimate choice for the number of optimal clusters [3].

The clusters define the GLaSS OWTs through their means and covariance matrices. While only a subset of bands was used to determine the cluster partitioning, the OWTs were created with the full hyperspectral data allowing for the construction of a membership function (the main component of the classification tool that produces the image classification) to operate on any band configuration within the range of hyperspectral data (400–800 nm) and, thus, on any satellite sensor. It is important to note that the clustering process was applied to the spectrally-reduced $R_{rs}(\lambda)$ data, resulting in a partitioning of the data. This partitioning was simply a means for sorting, and once sorted, the membership functions could be produced from the hyperdimensional $R_{rs}(\lambda)$ data.

There are two different forms of $R_{rs}(\lambda)$ used in classification schemes for depicting OWTs: area-normalized $R_{rs}(\lambda)$, e.g., [13,23] and un-modified or non-normalized $R_{rs}(\lambda)$, e.g., [3]. The rationale behind normalizing is to remove the influence of magnitude on clustering and stressing the spectral shape. The work in [23] showed that coastal turbid waters are susceptible to magnitude shifts based on the concentration of particles of the same type, which are sorted into the same cluster when normalized. Absorption characteristics have more impact on clustering.

The GLaSS OWTs are represented through both approaches, resulting in two different water type sets: a normalized set and a non-normalized set. For the normalized set, we applied a trapezoidal numerical integration over a wavelength range from 400–750 nm (Photosynthetically Active Radiation), hereafter called PAR-normalized, for each spectrum. Each dataset was analyzed separately for cluster analysis, cluster validity and the development of optical water types through the means and covariance matrices. For the non-normalized and the normalized data, the optimal number of clusters (and associated optical water types) was six for each based on validity functions and a priori user knowledge. These are denoted as 6C and 6CN, respectively (Figure 2).

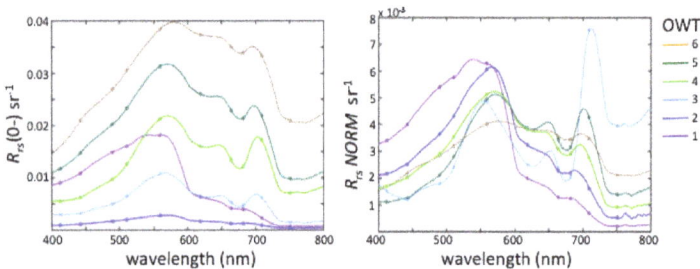

Figure 2. OWT mean spectra for the non-normalized (left) and PAR-normalized (right) clusters. (Normalization as explained in text.) Open circles indicate bands used in the clustering.

2.3. BEAM/SNAP Implementation and the Membership Function

The classification system has been implemented as a processing tool in Brockmann Consult's BEAM software and its successor SNAP and is available for application to satellite imagery (http://www.brockmann-consult.de/cms/web/beam/project, http://step.esa.int/main/toolboxes/snap/). The tool produces class memberships to OWTs (for either configuration) using membership functions, which produce fuzzy partitions for the OWT set.

Membership functions are formed from the mean and covariance matrix for each cluster, and class (OWT) membership values ranging from 0–1 are assigned to observations (pixels) using a two-step fuzzy process. For the first step, the Mahalanobis distance is computed between the observation and the OWT as:

$$Z^2 = (\vec{R}_{rs} - \vec{\mu}_j)^t \Sigma_j^{-1} (\vec{R}_{rs} - \vec{\mu}_j) \tag{3}$$

where $\vec{R}_{rs}$ is the observed remote sensing reflectance vector, $\vec{\mu}_j$ is the mean reflectance vector of the j-th OWT and Σ_j^{-1} is the covariance matrix for the j-th OWT. The Mahalanobis distance is the multivariate equivalent of the standardized random variable $Z = (X - M)/S$, which is the distance of the univariate random variable X from its mean M normalized by the standard deviation S. In other words, the Mahalanobis distance is a weighted form of the Euclidean and is preferable because it incorporates the shape of the distribution of points around the cluster center (i.e., the geometric shape of the point cloud expressed in terms of variance). For the second step, the membership function converts the Mahalanobis distance into a fuzzy membership using a chi-square probability function. In mathematical terms, if the probability distribution of points belonging to the cluster centered at $\vec{\mu}_j$ is normal and $\vec{R}_{rs}$ is a member of that population, then Z^2 as defined by Equation (3) has a chi-squared distribution with n degrees of freedom where n is the dimensionality of $\vec{V}_{rs}$. The likelihood that $\vec{R}_{rs}$ is drawn from the j-th population can be defined as:

$$f_j = 1 - F_n(Z^2) \tag{4}$$

where $F_n(Z^2)$ is the cumulative chi-square distribution function with n degrees of freedom. The fuzzy membership ranges from 0–1 and depicts the degree to which a measured reflectance vector belongs to a given OWT. The value is one if the measured vector is identical to the mean vector of that OWT, and its value diminishes to zero as the Mahalanobis distance increases. This allows for an observation to have memberships to multiple OWTs, although in practice, one or two are typically expressed as present.

2.4. Characteristics of Remote Sensing Data

For inland waters, high backscatter and absorption in both the atmosphere (by land aerosols) and the water (due to high concentrations of optically-active substances) can confuse the coupled atmospheric correction and in-water retrieval software [24]. Furthermore, nearby vegetated land can cause over-radiation of water pixels in the near-infrared (NIR) wavelengths that are used for atmospheric correction. Therefore, we started with radiometrically-corrected MERIS Level-1 TOA radiances. These base datasets were processed with different atmospheric correction algorithms, and the output reflectances (with confidence flags) can subsequently be used in the OWT classification system. The confidence flags are quite strict and will, e.g., indicate extreme reflectances caused by sun glint or vision of the lake bed in optically-shallow waters. The satellite images were processed with and without correction for stray light from adjacent land pixels, using the Improve Contrast over Ocean and Land (ICOL) processor [25]. The images were atmospherically corrected using several processors: Case 2 Regional (C2R, [26]), CoastColour with C2R (CC2R, [27]) and the Modular Inversion and Processing scheme (MIP) [28–30].

This is a subset from the atmospheric correction (AC) methods tested in the GLaSS project [31], because not all AC output was suitable as input for the OWT tool. SCAPE-M (Self-Contained

Atmospheric Parameters Estimation from MERIS data, [32]) is not included in the classification analysis, because of known problems with MERIS Band 2, which would have a large influence on the produced classes. Due to missing spectral bands, the output of the Freie Universität Berlin (FUB/WeW) Water Processor [33] cannot be fed into the OWT tool. The standard MERIS Ground Segment (MEGS) Processor is not included because of the extremely low number of valid pixels it produced in atmospheric correction tests in GLaSS (0–14%, depending on the lake [31]). The output of the 6S (Second Simulation of a Satellite Signal in the Solar Spectrum, [34,35] and the ATCOR [36] processors also did not perform well compared to other atmospheric correction processors for any of the selected lakes, likely because we did not have sufficient information to optimize their parameterization, and they were also not included here.

3. Results

3.1. Properties of the GLaSS OWTs

The cluster analysis for each treatment of $R_{rs}(\lambda)$ resulted in the creation of the OWTs (Figures 2 and 3). The number of optimal clusters was six for each case, which were not directly linked and were coincidental. Tables 2 and 3 show the distributions of class assignments from the cluster analysis for individual in situ lake datasets for each partition. For referencing OWTs within each scheme, we adopt a nomenclature convention of the scheme followed by the OWT. For example, OWT 1 of the non-normalized scheme will be referenced as 6C-1, and OWT 1 of the PAR-normalized scheme will be referenced as 6CN-1, and so forth. For the non-normalized data (Table 2), the distributions across type vary by slightly more than a factor of two maximum (84 points to 6C-6 and 199 to 6C-4).

Table 2. Cluster distribution using the six-class (6C) classification scheme.

Non-Normalized: 6 Classes OWT Type Source	1	2	3	4	5	6	Total
Finnish lakes	0	15	1	0	0	0	16
Taihu	0	0	1	41	108	84	234
Peipsi	0	0	10	21	3	0	34
IJsselmeer	0	0	50	8	0	0	58
Markemeer	0	0	16	57	0	0	73
Italian lakes	93	3	4	3	2	0	105
Betuwe	0	8	8	0	0	0	16
New Hampshire (NH) lakes	32	29	77	39	2	0	179
Spanish lakes	28	72	19	20	1	0	140
Lake Erie	2	0	4	10	0	0	16
Total	155	127	190	199	116	84	871

Table 3. Cluster distribution using the six-class normalized (6CN) classification scheme.

Normalized: 6 Classes OWT Type Source	1	2	3	4	5	6	Total
Finnish lakes	0	11	0	1	4	0	16
Taihu	0	0	8	79	22	125	234
Peipsi	0	0	0	9	25	0	34
IJsselmeer	0	6	0	12	40	0	58
Markemeer	0	8	0	30	35	0	73
Italian lakes	91	12	0	1	1	0	105
Betuwe	0	8	0	6	2	0	16
NH Lakes	27	60	26	32	33	1	179
Spanish lakes	25	52	1	37	25	0	140
Lake Erie	2	7	0	3	4	0	16
Total	145	164	35	210	191	126	871

Individual lake datasets typically group into two or three clusters. For example, the Finnish lakes are mostly grouped into 6C-2, while the Italian lakes are spread across five different OWTs, but mostly are grouped into 6C-1. The normalized $R_{rs}(\lambda)$ cluster distributions change somewhat (Table 3). In some lakes, the data are spread out across more OWTs (e.g., Spanish, New Hamphire (NH) and Finnish lakes), whereas in the case of Lake Peipsi, the data become more concentrated into a single OWT. Still, most of the data sources show just a few dominant types.

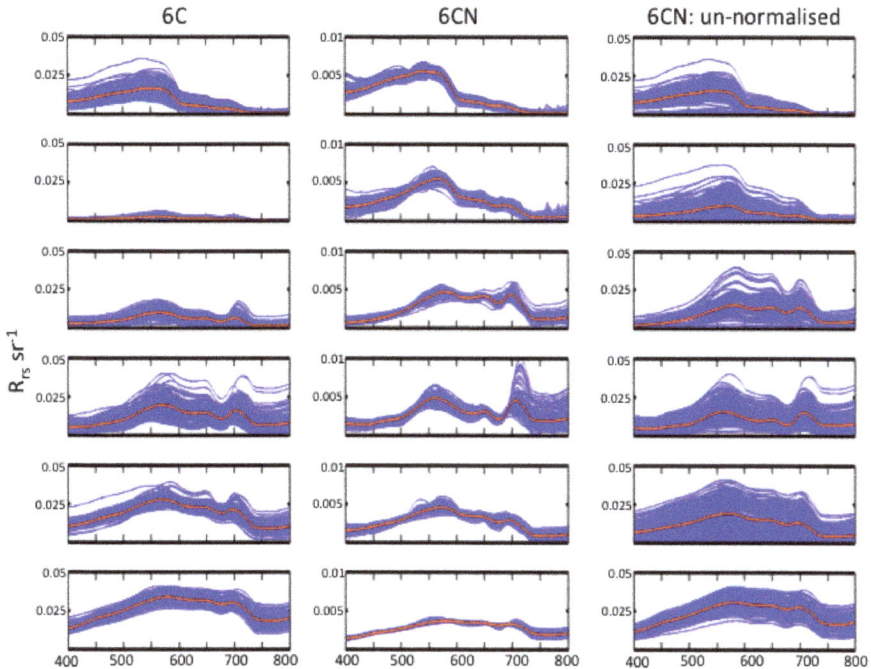

Figure 3. Distribution of individual $R_{rs}(\lambda)$ across the clusters (OWT 1, top; OWT-6, bottom) for the non-normalized (left column) and PAR-normalized (middle column) schemes. The right column shows the same data for the middle column, but not normalized.

The two OWT schemes differ in small, but important ways in how the $R_{rs}(\lambda)$ are distributed and in resulting OWT means. The PAR-normalized treatment effectively removed magnitude effects. For example, 6C-3–6C-6 appear similar in shape and inflection characteristics (e.g., variations on peaks at 550 and 710 nm, depressions at 620 and chlorophyll absorption between 665 and 680 nm), but with different magnitudes and with a general flattening of spectra towards 6C-6, as seen in the mean spectra (Figure 2). For the PAR-normalized system, the spectra belonging to a given cluster cover a wide range of magnitudes, as seen in the $R_{rs}(\lambda)$ when viewed in their non-normalized condition (Figure 3, right column). The same spectra are distributed over several OWTs in the 6C scheme. Conversely, 6C-2 contains low $R_{rs}(\lambda)$ typically associated with high absorption, and these $R_{rs}(\lambda)$ are distributed over several OWTs in 6CN, offering new potential for discrimination within dark or high absorbing waters.

An underlying assumption and early motivation for OWT approaches in the context of bio-optical algorithms is that data assigned to the same cluster share IOP characteristics [12]. Without a full set of co-measured IOP data, it is not possible to verify whether or not $R_{rs}(\lambda)$ associated with the same OWT share similar IOP characteristics. However, the distributions of co-measured Chl-a (all stations), CDOM and TSM concentrations (available for 376 of stations) provide insight into spectral drivers behind the water types (Tables 4 and 5).

Table 4. In-water characteristics for non-normalized OWTs, 6C.

OWT	Chl min	Chl median	Chl max	CDOM min	CDOM median	CDOM max	TSM min	TSM median	TSM max
1	0.1	1.6	12.3	0.04	0.17	1.03	0.15	1.34	14.70
2	0.8	7.2	69.6	0.9	4.8	20.43	0.87	27.18	52.28
3	1.3	24.0	33.0	0.05	2.6	8.0	0.28	16.76	208.9
4	0.9	107.0	705.0	0.27	4.2	18.67	1.70	37.65	190.07
5	0.8	27.0	86.1	0.2	1.17	17.0	3.10	54.03	285.6
6	7.5	22.5	450.0	0.32	0.76	1.03	1.4	67.27	250.36

Table 5. In-water characteristics for normalized OWTs, 6CN.

OWT	Chl min	Chl median	Chl max	CDOM min	CDOM median	CDOM max	TSM min	TSM median	TSM max
1	0.1	1.4	5.8	0.04	0.17	1.03	0.28	1.27	14.70
2	0.3	8.1	69.0	0.17	1.3	2.82	0.15	16.7	52.28
3	1.6	20.5	70.0	3.33	11.4	20.43	10.32	29.95	137.0
4	2.7	120.8	705.0	0.56	0.96	1.52	2.03	47.05	212.6
5	1.7	20.7	450.0	0.27	1.12	12.1	1.7	54.32	227.6
6	7.5	22.5	82.0	0.32	0.85	1.83	1.4	68.3	285.6

The OWT distributions for Chl-a, CDOM and TSM are shown in Figure 4. The trends for mean Chl-a for the 6C scheme show an increase from OWT 1–OWT 4, while TSM increases across all six OWTs, indicating that C6-5 and C6-6 have major inorganic particle contributions. High mean CDOM values are in 6C-2–6C-4, with 6C-2 having the highest value and consistent with the lowest overall mean $R_{rs}(\lambda)$. These combinations are broadly consistent with progressively elevated mean $R_{rs}(\lambda)$, tempered with suppressed spectra with high Chl-a and CDOM OWTs. For the 6CN scheme, mean Chl-a follows that of the 6C scheme. A notable difference in the distribution for CDOM is evident, with 6CN-3 having the highest mean. TSM also follows the trend for the 6C scheme, with the highest TSM distributions associated with 6CN-6 and consistent with the shape of the mean $R_{rs}(\lambda)$.

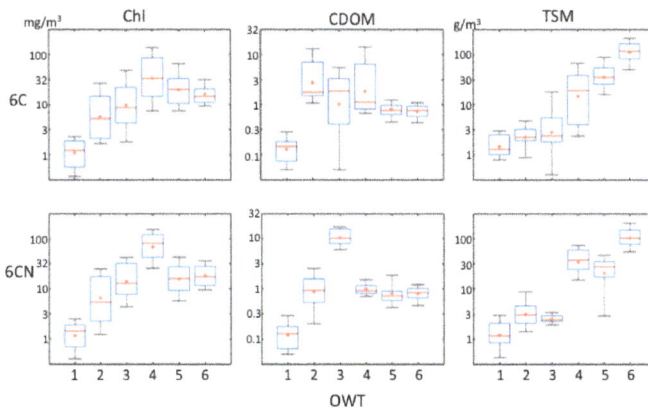

Figure 4. Boxplots for Chl-a, CDOM and TSM across the OWTs for 6C (top row) and 6CN (bottom row).

The general relations between the OWT optical and in-water properties for the two schemes can be summarized as follows: the 6C mean $R_{rs}(\lambda)$ retain absolute shape and are influenced by absorption and scattering properties, ranging from a relatively clear type (6C-1) to turbid, highly scattering waters (6C-6). A very dark water type indicative of high absorption (CDOM-dominated) is also represented (6C-2), and 6C-3–6C-5 generally are associated with increasing levels of phytoplankton biomass in eutrophic waters. In 6CN, peaks and valleys in the red and NIR region are the most differentiating aspect of shape, with 6CN-1 and 6CN-2 relatively flat in this region, and varying levels of shape and magnitude for 6CN-3 through 6CN-6. The largest peak amplitude in the red/NIR region is exhibited by 6CN-3, consistent with the highest Chl-a levels. For the 6CN, there is no 'dark' water OWT, as in the case of the 6C scheme.

3.2. The GLaSS Lakes Case Studies

The GLaSS OWT tool utilizes these schemes with the membership functions to produce mapped products (Figure 5). Mapped products show (1) the fuzzy memberships to each OWT, (2) the dominant water type (determined from the water type with the highest membership) and (3) the membership sum (i.e., the sum of memberships from all water types). Also included are the normalized memberships (not to be confused with the normalized $R_{rs}(\lambda)$). The normalized memberships are constrained to sum to one for every pixel. For these quantities, each membership is divided by the membership sum for that pixel.

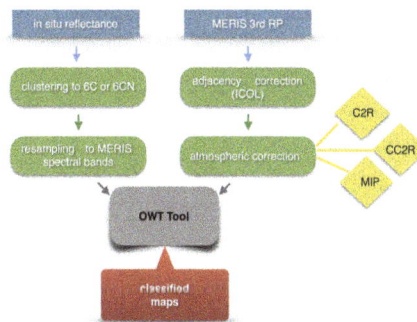

Figure 5. Processing chain for MERIS scenes.

We tested the tool and the two schemes on a selection of MERIS images from GLaSS target lakes, which included lakes in Estonia, Finland, Italy, The Netherlands and Sweden (Table 6). One of our goals was to assess how OWTs are impacted by and can inform us of how to improve the application of atmospheric correction schemes to local imagery. To test this, the satellite images were processed with and without correction for stray light from adjacent land pixels (ICOL [25]) and atmospherically corrected using several AC methods available in GLaSS (see Table 7). The mapped distribution of the dominant OWT for each classification scheme was evaluated on a qualitative basis in consultation with local GLaSS lake experts, since we lack match-up validation data. Atmospheric correction with CC2R gave best results for most lakes, and maps based in these results are discussed in the next sections.

Table 6. List of GLaSS lakes for test case application by country.

Country	Lakes
Estonia	Lake Peipsi, Lake Võrtsjärv
Finland	Lake Päijänne, lake Pääjärvi, Lake Vesijärvi
Italy	Lake Garda, Lake Maggiore, Lake Lugano, lake Idro
The Netherlands	Lake IJsselmeer, lake Markermeer
Sweden	Lake Vättern, Lake Vänern

Table 7. Atmospheric correction method used in each image tested.

Lake	Date yyyy-mm-dd	Atmospheric Correction Method
Estonia	2005-07-18	CC2R
Estonia	2011-07-27	MIP
Finland	2004-08-05	C2R and CC2R
Finland	2006-05-09	C2R and CC2R
Finland	2007-06-01	CC2R
Finland	2007-08-23	MIP
Italy	2009-09-11	CC2R
Italy	2008-05-06	MIP
The Netherlands	2011-04-15	C2R
The Netherlands	2011-04-23	CC2R
The Netherlands	2011-09-28	MIP
Sweden	2003-08-29	CC2R
Sweden	2009-06-26	MIP

Note: CC2R: CoastColour; MIP: Modular Inversion and Processing Scheme; C2R: Case 2 Regional.

3.2.1. Italian Lakes: Deep and Clear

Located in the southern Perialpine region, Lake Garda is the largest Italian lake, typically with meso-oligotrophic conditions. The lake can be divided in two sub-basins: a larger area extending with a N-SW orientation with a deep bottom; and a shallower SE basin. Lake Maggiore is the second largest by surface and volume. It is a very narrow elongated lake with a N-S orientation. The deepest basins (max depth 373 m) are situated in the central and northern parts, with shallower bottoms in the south. Lake Maggiore has experienced eutrophication since the 1960s, but since the 1980s, it has stabilized and cleared, and today, it is classified as oligotrophic. For the Italian lakes, following Tables 2 and 3, we expect 6C-1 and 6CN-1 to occur most of the time, in combination with 6CN-2 [37–39]. Seasonal and daily variation can induce some deviations. Figure 6 shows the classified maps for these lakes after ICOL corrections. The invalid or suspect flags were not applied as masks for the maps, in order not to loose much of the data. Without ICOL, large parts of Garda and all of the other lakes in the area are flagged as 'L2R (level-2 reflectance) invalid' or 'L2R suspect', and the waters are classified as 6C-3 and 6C-4, which is clearly not correct. With ICOL processing, still many of the pixels in the Italian lakes (except for larger Lake Garda) are flagged as 'L2R invalid', but the resulting water types 6C-1, 6C-2 and 6C-3 could be correct for these lakes. However, the percentage presence of 6C-3 is higher than expected, and 6C-4 is assigned to parts of Lake Lugano and Lake Idro, which is not appropriate. The 6CN classifier assigns OWT 6CN-1 to all of the Italian lakes, except of Lake Lugano (6CN-2). This agrees with the known optics of the lakes and with the distribution shown in Table 3. We believe the OWT normalization is appropriate and accurate here.

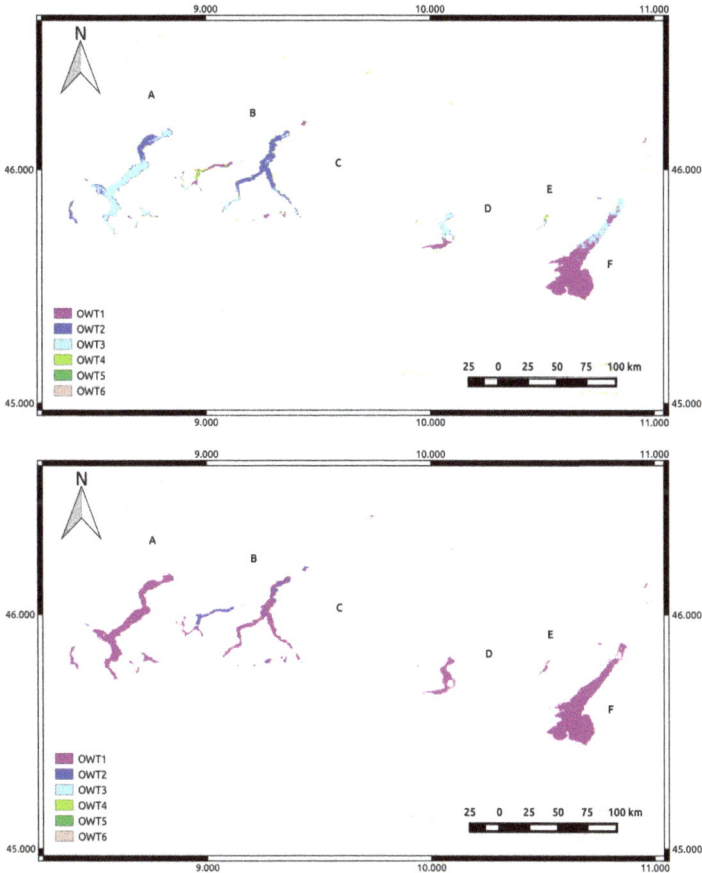

Figure 6. Classification of Italian lakes: Lake Maggiore (**A**), Lake Lugano (**B**), Lake Como (**C**), Lake Iseo (**D**), Lake Idro (**E**) and Lake Garda (**F**). MERIS 2009-09-11 (yyyy-mm-dd) , ICOL + CC2R. No flagging applied. Top: 6C; bottom: 6CN.

3.2.2. The Estonian and Dutch Lakes: Shallow-Turbid and Shallow Phytoplankton-Dominated Lakes

Using Tables 2 and 3 as a guide, we expected Lake Peipsi in Estonia to be classified mostly as 6C-3 and 6C-4 and partly as 6C-5. Lake Võrtsjärv has higher sediment and CDOM loads, and class 6C-5 could therefore be expected. In Figure 7, the results of the 6C (left) and 6CN (right) classifications for Lake Peipsi (east on the map) and Lake Võrtsjärv (west) are shown. 6C-2 and 6C-3 are assigned to the northern part of Lake Peipsi; these are lower OWTs than reported in Table 4. However, the three 6C water types that are found in Lake Peipsi have $R_{rs}(\lambda)$ spectra that are similar to field measurements, and the spatial distribution of the OWT classes seems credible: the northern part with lower classes than the southern part. The southern part of Lake Peipsi (Lake Pihkva) is richer in sediments than the northern part, and Lake Pihkva is very similar to Lake Võrtsjärv, which is confirmed by the classifications [40,41]. At the time of image acquisition (18 July 2005), there was a large phytoplankton bloom in the northern part of Lake Peipsi. In the beginning of July 2005, the measured Chl-a varied between 14 and 74 mg m^{-3} with lower values close to shore and higher values in the center, and in August, the bloom was even more intense. This range of Chl-a concentrations complies somewhat with 6C-2 (Table 4). Importantly, in situ measurements of Lake Peipsi from 2008–2011 [42] show average CDOM absorption at 440 nm of 3.1 m^{-1}. This combination of Chl-a and CDOM concentrations (cf. Table 4) explains that 6C-2 was assigned to this image. OWT 6C-3 indicates the presence of somewhat lower Chl-a concentrations for the areas adjacent to the blooms in Lake Peipsi.

OWTs 6CN-4 and 6CN-5 were expected from the normalized classification of the in situ spectra (Table 3). However, 6CN-2 and 6CN-5 were found in the MERIS image for the northern part of Lake Peipsi. Still, this is a reasonable distribution for the period with during a phytoplankton bloom (elevated Chl) and CDOM concentrations of around 3 m^{-1} (Table 4). The smaller southern part, Lake Võrtsjärv, is assigned class OWT 6C-4 or 6CN-6. The non-normalized classification appears to work best here, as Võrtsjärv has high Chl-a, TSM and CDOM, which agree with OWT 6C-4, but not with OWT 6CN-6 (which had a lower CDOM range in the training set; Table 4).

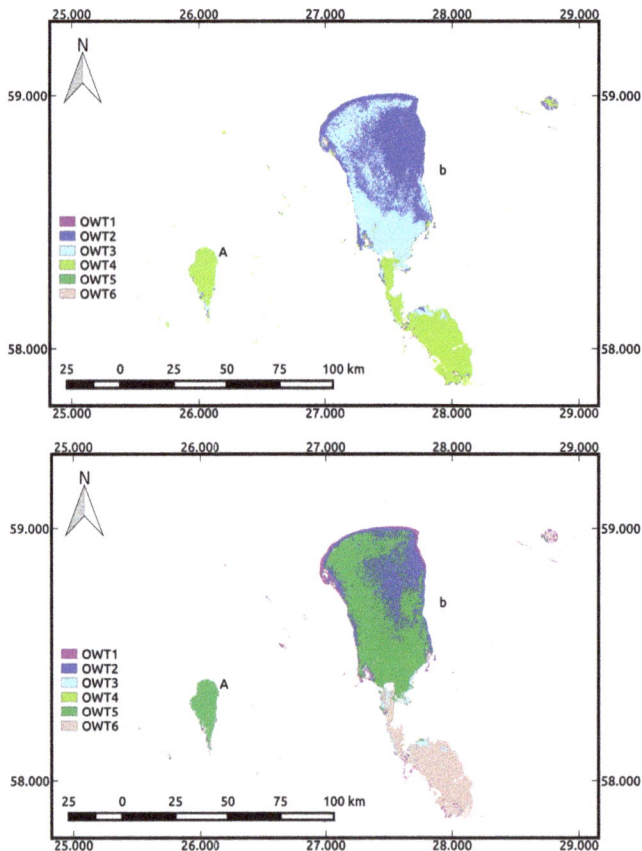

Figure 7. Estonian Lake Peipsi (**A**) and Lake Võrtsjärv (**B**). MERIS 18 July 2005 ICOL + CC2R. Flagged data in black with L2R (Level-2 reflectances) suspect. Left: 6C; right: 6CN.

The Dutch Lake IJsselmeer and its split-off Lake Markermeer have quite distinct optical properties. Markermeer is shallow (average depth of 3.6 m), and bottom sediments are characterized by fine, easily resuspendable sediments with frequently high surface TSM concentrations [24]. River IJssel discharged higher nutrient loads into Lake IJsselmeer, in the past. Lake IJsselmeer is still optically dominated by phytoplankton and cyanobacterial blooms. As expected from Table 2, Markermeer is classified with a combination of OWT 6C-3 and 6C-4, while the majority of IJsselmeer is assigned to OWT 6C-3 (Figure 8). Near the outflow of River IJssel, IJsselmeer also contains OWT 6C-2, which could indicate the presence of CDOM in an otherwise relatively clear region, where mussels filter the water. In both lakes, some OWT 6C-1 pixels are found along the shorelines. This is not correct and is not explained by the adjacency effect, which would lead to a higher (not lower) OWT number. However, all of these 6C-1 pixels are indeed flagged as 'L2R invalid' or 'L2R suspect'. With the normalized classifier, 6CN-2 is dominantly assigned to both lakes IJsselmeer and Markermeer, and 6CN-1 occurs, as well. The latter could be incorrect because the associated concentrations are low (Table 5). In that case, an incorrect atmospheric correction would explain the difference between the results in Table 5 and the MERIS-based maps.

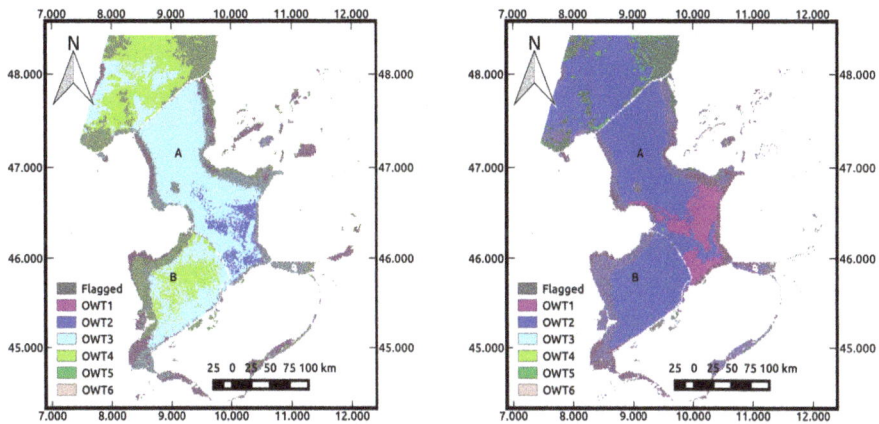

Figure 8. Dutch lakes IJsselmeer (**A**) (north of the dike) and Markermeer (**B**) (south of the dam). MERIS 23 April 2011, ICOL + CC2R. Left: 6C; right: 6CN.

3.2.3. The Finnish and Swedish Lakes: High Absorbing, Low Reflecting Waters

Figure 9 shows a classified MERIS image from 9 September 2006, containing several Finnish lakes. Based on Table 2, the Finnish lakes are almost always classified as 6C-2, because of their overall low $R_{rs}(\lambda)$. After normalization, more differentiation in shape leads to several assigned normalized OWTs (6CN-2, 6CN-4 and one instance of 6CN-5). With the 6C application, most of Päijänne and the central parts of Pääjärvi were classified as 6C-2, which is according to expectation since these lakes have generally low $R_{rs}(\lambda)$ attributable to high CDOM absorption for Pääjärvi and low TSM and particle scattering in Päijänne. Although Lake Vesijärvi is predominantly classified as 6C-3, this is viewed as accurate, because this lake has low CDOM and typically higher TSM and Chl-a and, thus, higher $R_{rs}(\lambda)$. The 'L2R suspect' flag was raised at the shores of Lake Päijänne, which indicates that the OWT 6C-4 pixels might require masking due to excessively high $R_{rs}(\lambda)$ values. The 6CN classifier assigns 6CN-1 to Lake Päijänne and 6CN-2 to Lake Pääjärvi. Other surrounding lakes are classified as 6CN-1, as well. 6CN-1 was not expected according to Table 3, and Lake Vesijärvi and the small surrounding lakes could be classified into several classes. The question is whether the normalization and classification did not work well here, or if something else is disturbing the results. Because Tables 2 and 3 do represent the differences between Finish lakes well, the expectation is that the atmospheric correction might not have been suitable for these lakes [43].

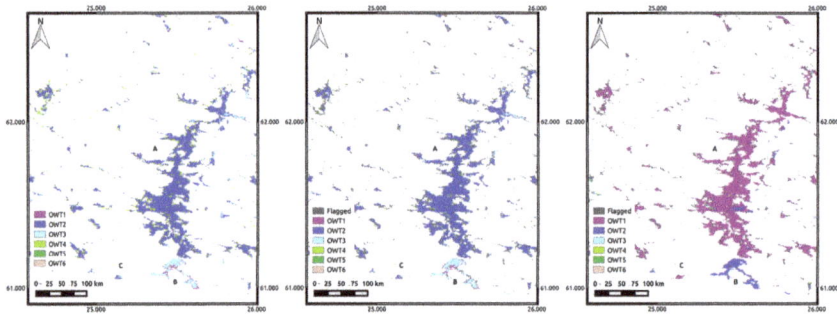

Figure 9. Finnish Lakes: Päijänne (**A**), Pääjärvi, (**B**) and Vesijärvi (**C**). MERIS 2006-05-09, ICOL + CC2R. Left: 6C no flagging; center: 6C 'L2R invalid' flagged out; right: 6CN with flagging.

The two largest Swedish lakes are Lake Vänern and Lake Vättern. Lake Vättern is very clear, with very low concentrations of Chl-a, TSM and CDOM. Lake Vänern typically has low concentrated chlorophyll blooms and relatively high CDOM absorption of around $1\,m^{-1}$ (at 440 nm) [40,44]. Both lakes are mainly classified as OWT 6C-2, Lake Vänern also partly as OWT 6C-3 (Figure 10). The small Bay Dättern, in the south of the eastern basin of Lake Vänern, is very turbid, with high concentrations of TSM ($>30\,g\,m^{-3}$), Chl-a ($>30\,mg\,m^{-3}$ in summer) and very high CDOM concentrations ($3–10\,m^{-1}$). Bay Dättern is classified as OWT 6C-4, and probably due to the high CDOM absorption, it does not fall into OWT 6C-6. The occurrence of OWT 6C-5 was also expected for this bay, but that class was not found. ICOL processing makes a difference in Lake Vättern, but not in a positive sense: after ICOL processing, Vättern is classified as OWT 6C-3, while OWT 6C-1 would have been more appropriate. Bay Dättern continued to be flagged as 'L2R invalid' after ICOL processing. The result without ICOL processing is therefore preferred. With the normalized classifier, Vättern is actually classified as OWT 6CN-1 and seems the most appropriate for this lake. After normalization, Vänern is assigned OWT 6CN-2 and Dättern OWT 6CN-3, which is correct. For the Swedish lakes, the normalization seems an improvement over the non-normalized classification.

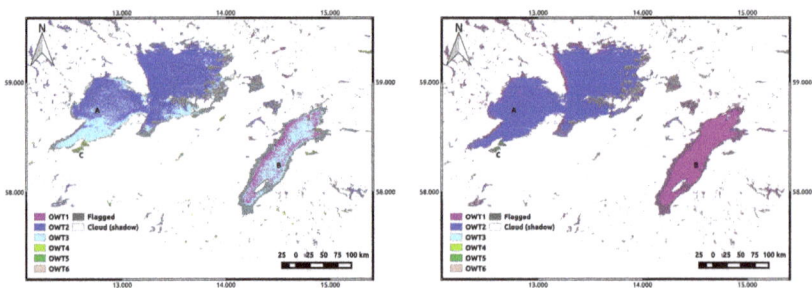

Figure 10. Classification of Swedish Lakes: Lake Vänern (**A**), Lake Vättern (**B**) and the bay of Dättern (**C**). MERIS 25 June 2009, ICOL + CC2R. Only cloud (white), cloud shadow (partly transparent white) and "L2R suspect" flagging applied. Left: 6C; right: 6CN.

4. Discussion

The GLaSS lake OWTs were developed by extending the OWTs derived in [3]. In that earlier study, seven OWTs were identified, but represented coastal marine waters, as well as inland freshwater. The number of OWTs we found in a larger dataset but exclusive to freshwater was six. The impact of adding more data did not significantly alter the partitioning of reflectance space into clusters. We purposefully omitted spectra associated with floating algae because of too few instances to derive stable statistics, but we believe this is a water type that exists and should be incorporated in future renditions.

In addition to the development of the six OWTs on un-modified spectra (6C), we developed a parallel set using normalized spectra (6CN), accentuating absorption features by removing scaled magnitude effects solely attributable to concentration levels [45]. The GLaSS tool contains both options for OWT processing. It is yet to be determined which choice of the classification scheme is best for a given lake (i.e., 6C or 6CN). This will depend on how the classification maps might be used and the nature of the lake system. In the Dutch and Estonian lakes (the lakes with a higher sediment load), the 6C classification performs better, while for the Italian and Swedish lakes (mainly for the clear Lake Vättern), the 6CN classification provided the best results based on the current analysis. This is consistent with our expectations: the normalized method discriminates 'low reflecting' lakes (either clear blue lakes or brown/yellow CDOM lakes) that would otherwise end up in the same 'low' reflectance OWT using non-normalized classification. For the Finnish lakes, however, the results from both classification schemes seem not very convincing: there are large contrasts between the classified in situ reflectance values (Tables 4 and 5) and the image results. We believe this is caused by atmospheric correction problems over dark waters. For these dark absorbing lakes, such as the ones in Finland and Sweden, it is known that the FUB processor performs best. However, due to missing spectral bands, the output of this processor cannot be fed into the OWT tool.

These AC test results highlight a new role for the OWT classification in identifying atmospheric correction problems, as an overall aim of the GLaSS OWT tool is to improve water quality products generated from satellite image processing for any lake system. A general problem with image processing over lakes is that certain AC and bio-optical algorithm retrieval schemes are more suitable for some optical conditions, while other schemes work better for other conditions (e.g., clear versus turbid waters). Selecting the most suitable AC and retrieval algorithm schemes is the most critical decision for producing accurate and meaningful water quality products. The OWT classification provides a mechanism to assist, for example by indicating whether a dark-pixel correction is possible (non-turbid) or not.

Using OWTs for improving the results of atmospheric correction for imagery over lakes would be a new application for these products. Currently, one iteration of atmospheric correction combined with the application of the OWT tool shows the distribution of OWTs over the whole lake. In cases that lakes contain water types that have a better performance with different AC schemes, one could imagine an iterative system where standard AC processing is executed and OWTs are computed, and then, if certain water types are found (assuming error in the turbid areas), a re-application of AC over the scene could be applied with a scheme more suitable to turbid conditions for those pixels assigned to the OWT that is connected to a different AC scheme. This approach is conceptually similar to the switching scheme between the NIR and SWIR AC models originally suggested by [46] and further tested by [47,48] for MODIS imagery over various coastal locations, as well as for MERIS Case 1 and Case 2 atmospheric correction [49–52]. In the present case, multiple AC schemes could be available for selection, with images re-combined similar to the algorithm blending method for in-water retrievals, e.g., [3]. This approach would require further testing of different AC schemes with different image scenes containing a variety of OWTs, but offers an avenue for blending AC schemes within a single image.

One important issue to mention concerns the use of the flags derived from the pre-processing of the scenes and the atmospheric correction method used. In the scenes analyzed, the 'L2R suspect' and the 'L2R invalid' flags removed some misclassified pixels along the lake shores, which could be caused by the remaining adjacency effect or by mixed land-water pixels. Those flags can be used to mask out these problematic pixels, but they can also mask large areas of valid pixels. Therefore, to determine to which class the main part of the lake belongs, it is advised not to use additional flagging besides the land- and cloud-related flags. It is also wise to ignore the much higher classes that are found in the 1–2 pixels along the shores of the lakes when they appear.

The classification maps produced from the GLaSS tool also serve as stand-alone products that provide spatial information for understanding the distributions and long-term trends of optical states that have environmental and ecological linkages. ESA's Diversity II Project (http://www.diversity2.info/) used the dominant OWT class as a monthly inland waters product from MERIS imagery for a variety of globally-distributed lakes. Time series of these classification maps provide indications of how a given system may be trending or changing as expressed OWT changes. Frequency maps of OWTs can be generated and are useful for understanding the distributions of the dominant water types for a given lake, leading to a first order indication of the types of AC schemes and retrieval algorithms that may be needed for processing, e.g., [3]. There is much interest in using remote sensing to support reporting for the European Water Framework Directive and U.S. Clean Water Act [53,54]. Frequent OWT maps can provide an insight into the state and seasonal patterns that occur in lakes. Longer term or unexpected changes can be a reason to perform a full processing and taking additional samples for detailed analysis.

The 6C and 6CN classification methods analyzed here are similar in their implementation, but represent different approaches in classification and interpretation. The non-normalized approach (6C) is based on absolute $R_{rs}(\lambda)$ values and thus can differentiate 'dark' from 'bright' waters more effectively than the normalized scheme (6CN), which essentially removes magnitude effects attributable to particle scattering. The normalized class partitioning is driven by spectral shape effects, largely from spectral-varying absorption properties. These considerations may be relevant to how the classification tool is ultimately used for a lake, such as to determine the most suitable tuning of a bio-optical algorithm or for general optical assessment.

Water classification is a somewhat recent and evolving discipline. Classification schemes exist that use normalized and non-normalized R_{rs} spectra, but there has been no attempt anywhere to connect the two approaches. Currently, each approach has been treated separately without the other, and each has advantages and disadvantages. However, it is possible to unify the two schemes. Figure 11 shows a view of the combined system as a matrix with the normalized and non-normalized normalized input remote sensing reflectance (Rrs) spectra separated into their respective clusters using the six-class scheme for both, resulting in 36 potential variations. Based on our results, 20 of the 36 possibilities are encountered. One approach for integrating the different schemes would be to use one classification system subsidiary to the other. Under this approach, fuzzy memberships would be derived for one scheme as the master factor, and a second sorting could take place according to the dominant OWT of the subsidiary scheme. This approach avoids intermingling fuzzy memberships, which is not yet feasible, but does add a new layer of classification by further discriminating shapes within a class. As an example, the new integrated scheme would use the fuzzy memberships for non-normalized classes as the main fuzzy value for pixel weighting if serving that function, and the dominant class of the normalized data as a subset variation of the non-normalized class. Theoretically, each non-normalized class has six normalized potential assignments when combining 6C and 6CN. This approach has not been tested, but could be a way to take full advantage of the classification tool. It is beyond the scope of this study to develop these concepts further and remains a gap that future work should address.

The selection of algorithms optimized for local conditions can also be facilitated by using the OWT approach to direct the algorithm selection and output blending. As has been demonstrated by [3], the best performing algorithm to a particular water type can be determined through algorithm analysis a priori. During operational classification, the class memberships can then be used to weigh retrievals from multiple algorithms into a blended product. This assumes that algorithm performance for specific OWTs are globally representative. This assumption should be checked for a given lake system though. Variations in local optical drivers or specific Inherent Optical Properties (sIOPs) may deviate from global behavior or conditions.

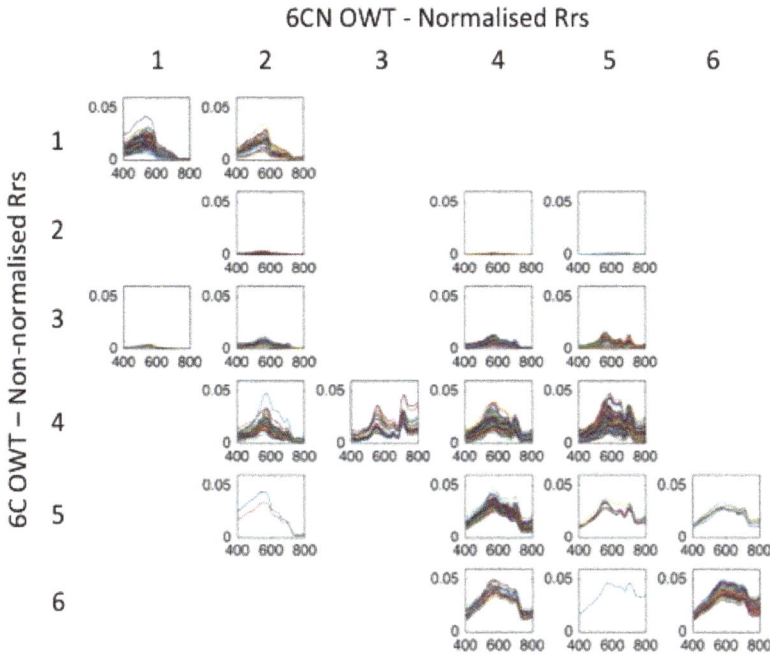

Figure 11. Matrix of non-normalized (GLaSS6C) (rows) and normalized (GLaSS6CN) (columns) clusters with remote sensing reflectance (Rrs) spectra sorted into respective OWT.

5. Conclusions

Optical water type classification is a developing research topic in aquatic remote sensing. It has evolved from Jerlov water types as descriptors of marine waters, to a variety of marine and freshwater schemes designed for use with remote sensing image applications. We have developed a new tool specific for classifying lake remote sensing images, now available in the BEAM software. The development of the tool is an outgrowth of the method presented by [3], differentiated by new data and new scheme configurations. Optical data from different lakes across Europe, the U.S. and China were merged, covering a wide range of environmental conditions, including dark lake waters, turbid waters and highly eutrophic waters comprising cyanobacteria blooms. We have re-developed lake optical water types with both a spectral-normalized and non-normalized treatment, resulting in two separate, but linked schemes. The resulting water types in each scheme were described by in-water concentrations of chlorophyll-a, CDOM and total suspended matter. While each scheme differs at a fundamental level, they serve the same roles for downstream applications, which in the past have included using them as intermediary products for optimizing bio-optical algorithm selection and as stand-alone products for supporting biogeochemical and biodiversity system analysis. We have focused our research on the development of the schemes within the tool and its use with remote sensing imagery from the MERIS sensor for a variety of European lakes as case studies with different optical conditions. We found that each scheme had merits for generating mapped water type products, depending on the lake. While the images used are a small subset of conditions likely to be found

Remote Sens. **2017**, *9*, 420

globally, the analysis is useful as a means for contrasting the different approaches for different lake conditions. The best scheme for any system requires a fundamental a priori knowledge of optical drivers, necessary for interpreting the images. For example, Lake Vättern in Sweden showed better results for the normalized scheme, while other lakes such as Lake Võrtsjärv and Lake Markermeer and IJseelmeer in the Netherlands showed better performance with the non-normalized approach. These approaches were assessed by how accurately we believed the classification maps depicted the underlying optics.

Although a prime use of classification maps is for bio-optical algorithm application, we did not set out to test or develop the tool with algorithms, as was done in [3]. There is a wide variety in algorithms and intended purposes of algorithms, and this type of evaluation with the two-scheme approach was beyond the scope of this work. However, we presented a new use for classification maps as related to guiding atmospheric corrections schemes. As the tool operates directly on the spectral R_{rs}, the atmospheric correction scheme will impact the effectiveness of the classification tool. We tested several different atmospheric correction schemes with each image, producing different classification maps for each test image. The classification results provided feedback on the performance of the atmospheric correction scheme, and we believe that the classification map interpretations are useful in assessing the performance of atmospheric correction when in situ match-up data are not available, which is generally the case. Thus, another use of classification maps is for atmospheric correction assessment and possibly selecting and blending, as well, although we speculate on how this may be done.

These two scheme variations presented here—spectral-normalized and non-normalized—represent the current options available to developers and users of images produced from optical water type classification, regardless of origin. We have shown how these schemes differ spectrally and in use, as well as in in-water characterizations. We have also shown how they are linked through a classification matrix and speculate on the potential to unify the two schemes, which could provide a way to maximize the advantages of each scheme together. This is an evolving area of research, and the guidelines and uses of optical water type schemes are still being explored and discovered.

Acknowledgments: We are very grateful to our GLaSS colleagues Anu Reinart, Krista Alikas (Tartu Observatory), Kerstin Stelzer and Daniel Odermatt (Brockmann Consult), Claudia Giardino, Mariano Bresciani (Consiglio Nazionale delle Ricerche), Karin Schenk, Thomas Heege (EOMAP), Koponen Sampsa, Kari Kalio (Suomen Ymparistokeskus, SYKE) and Petra Philipson (Brockmann Geomatics), as well as the advisory board Arnold G. Dekker (Commonwealth Scientific and Industrial Research Organisation, CSIRO), Yunlin Zhang (Nanjing Institute of Geography and Limnology (NIGLAS), Chinese Academy of Sciences) and Stephen Greb (Wisconsin Department of Natural Resources) for the collaboration, the contribution of bio-optical in situ optical datasets that were used to construct the optical water types, verification of the results and meaningful conversations on this work. We also included in situ optical datasets from Shane Bradt (Department of Biological Sciences, University of New Hampshire) and Antonio Ruiz-Verdu (Image Processing Laboratory, University of Valencia) and would like to thank them for these. We thank ESA for the MERIS data that were obtained through the Brockmann Consult Calvalus system, and Marco Peters (Brockmann Consult) for implementation of the optical water types in BEAM. Much of the research was co-funded by GLaSS (EC FP7 Grant Agreement 313256), and Marieke Eleveld was also supported by Deltares Spearhead Multi-Resolution Modelling and Theme Ecosystems and Environmental Quality; Ana Ruescas was also supported by Brockmann Consult and by the EU under the European Research Council ERC consolidator Grant SEDAL-647423; Tim Moore was supported by NASA Grant NNX15AL43G.

Author Contributions: M.A.E., A.B.R., A.H., T.S.M., C.B. and S.W.M.P. designed the study. A.H. led and compiled the in situ data collection. T.S.M. analyzed the data and performed the statistical modeling for generation of the optical water types. A.B.R. and M.A.E. performed intermediate image processing, and A.H. led and performed the final image processing. M.A.E., A.H., T.S.M. and A.B.R. interpreted the results. T.S.M., M.A.E., A.B.R. and A.H. wrote the paper. S.W.M.P. discussed, commented on and edited the paper.

Conflicts of Interest: The authors declare no conflict of interest.

References

1. Donlon, C.; Berruti, B.; Buongiorno, A.; Ferreira, M.H.; Féménias, P.; Frerick, J.; Goryl, P.; Klein, U.; Laur, H.; Mavrocordatos, C.; et al. The Global Monitoring for Environment and Security (GMES) Sentinel-3 mission. *Remote Sens. Environ.* **2012**, *120*, 37–57.

2. Odermatt, D.; Gitelson, A.; Brando, V.E.; Schaepman, M. Review of constituent retrieval in optically deep and complex waters from satellite imagery. *Remote Sens. Environ.* **2012**, *118*, 116–126.

3. Moore, T.S.; Dowell, M.D.; Bradt, S.; Ruiz-Verdú, A. An optical water type framework for selecting and blending retrievals from bio-optical algorithms in lakes and coastal waters. *Remote Sens. Environ.* **2014**, *143*, 97–111.

4. Hommersom, A.; Wernand, M.R.; Peters, S.; Eleveld, M.A.; van der Woerd, H.J.; de Boer, J. Spectra of a shallow sea-unmixing for class identification and monitoring of coastal waters. *Ocean Dyn.* **2011**, *61*, 463–480.

5. Wernand, M.R.; Hommersom, A.; van der Woerd, H.J. MERIS-based ocean colour classification with the discrete Forel-Ule scale. *Ocean Sci.* **2013**, *9*, 477–487.

6. Jerlov, N.G. *Marine Optics*; Elsevier: Amsterdam, The Netherlands, 1976.

7. Jerlov, N.G. *Optical Studies of Ocean Waters*; Elanders boktr.: Gothenburg, Sweden, 1957.

8. Aas, E.; Hojerslev, N.; Hokedal, J.; Sorensen, K. Optical water types of the Nordic Seas and adjacent areas. *Oceanologia* **2013**, *55*, 471–482.

9. Solonenko, M.; Mobley, C. Inherent optical properties of Jerlov water types. *Appl. Opt.* **2015**, *54*, 5392–5401.

10. Brewin, R.; Ciavatta, S.; Sathyendrenath, S.; Jackson, T.; Tilstone, G.; Curran, K.; Airs, R.; Cummings, D.; Brotas, V.; Organelli, E.; et al. Uncertainty in ocean-colour estimates of chlorophyll for phytoplankton groups. *Front. Mar. Sci.* **2017**, *4*, 104.

11. Le, C.; Li, Y.; Zha, Y.; Sun, D.; Huang, C.; Zhang, H. Remote estimation of chlorophyll a in optically complex waters based on optical classification. *Remote Sens. Environ.* **2011**, *115*, 725–737.

12. Moore, T.S.; Campbell, J.W.; Dowell, M.D. A class-based approach to characterizing and mapping the uncertainty of the MODIS ocean chlorophyll product. *Remote Sens. Environ.* **2009**, *113*, 2424–2430.

13. Mélin, F.; Vantrepotte, V. How optically diverse is the coastal ocean? *Remote Sens. Environ.* **2015**, *160*, 235–251.

14. Trochta, J.; Mouw, C.; Moore, T. Remote sensing of physical cycles in Lake Superior using a spatio-temporal analysis of optical water typologies. *Remote Sens. Environ.* **2015**, *171*, 149–161.

15. Moore, T.; Dowell, M.; Franz, B. Detection of coccolithophore blooms in ocean color satellite imagery: A generalized approach for use with multiple sensors. *Remote Sens. Environ.* **2012**, *117*, 249–263.

16. Bradt, S.R. Development of Bio-Optical Algorithms to Estimate Chlorophyll in the Great Salt Lake and New England Lakes Using In Situ Hyperspectral Measurements. Ph.D. Thesis, The University of New Hampshire, Durham, NH, USA, 2012.

17. Ruiz-Verdú, A.; Simis, S.G.; de Hoyos, C.; Gons, H.J.; Peña-Martinez, R. An evaluation of algorithms for the remote sensing of cyanobacterial biomass. *Remote Sens. Environ.* **2008**, *112*, 3996–4008.

18. Mueller, J.L.; Morel, A.; Frouin, R.; Davis, C.; Arnone, R.; Carder, K.; Lee, Z.P.; Steward, R.G.; Hooker, S.; Mobley, C.D. *Ocean Optics Protocols for Satellite Ocean Color Sensor Validation, Revision 4, Radiometric Measurements and Data Analysis Protocols*; Tech. Memo 2003-21621; Goddard Space Flight Center: Greenbelt, MD, USA, 2003.

19. Mobley, C.D. Estimation of the remote-sensing reflectance from above-surface measurements. *Appl. Opt.* **1999**, *38*, 7442–7455.

20. Webb, A. *Statistical Pattern Recognition*; John Wiley and Sons, Ltd.: Hoboken, NJ, USA, 2002.

21. Bezdek, J. *Patter Recognition with Fuzzy Objective Function Algorithms*; Springer: New York, NY, USA, 1981.

22. Lee, Z.; Carder, K.L.; Arnone, R.A. Deriving inherent optical properties from water color: A multiband quasi-analytical algorithm for optically deep waters. *Appl. Opt.* **2002**, *41*, 5755–5772.

23. Vantrepotte, V.; Loisel, H.; Mélin, F.; Desailly, D.; Duforêt-Gaurier, L. Global particulate matter pool temporal variability over the SeaWiFS period (1997–2007). *Geophys. Res. Lett.* **2011**, *38*, doi:10.1029/2010GL046167.

24. Eleveld, M.A. Wind-induced resuspension in a shallow lake from Medium Resolution Imaging Spectrometer (MERIS) full-resolution reflectances. *Water Resour. Res.* **2012**, doi:10.1029/2011WR011121.

25. Vidot, J.; Santer, R. Atmospheric correction for inland waters application to SeaWiFS. *Int. J. Remote Sens.* **2005**, *26*, 3663–3682.

26. Doerffer, R.; Schiller, H. The MERIS Case 2 water algorithm. *Int. J. Remote Sens.* **2007**, *28*, 517–535.

27. Doerffer, R.; Brockmann, C. *Consensus Case 2 Regional Algorithm Protocols*; Technical Report; Brockmann Consult: Geesthacht, Germany, 2014.

28. Heege, T.; Kiselev, V.; Wettle, M.; Hung, N.N. Operational multi-sensor monitoring of turbidity for the entire Mekong Delta. *Int. J. Remote Sens.* **2014**, *35*, 2910–2926.

29. Heege, T.; Fischer, J. Mapping of water constituents in Lake Constance using multispectral airborne scanner data and a physically based processing scheme. *Can. J. Remote Sens.* **2004**, *30*, 77–86.

30. Heege, T.; Häse, C.; Bogner, A.; Pinnel, N. Airborne Multi-spectral Sensing in Shallow and Deep Waters. *Backscatter* **2003**, *14*, 17–19.

31. GLaSS Deliverable D3.2. Global Lakes Sentinel Services, D3.2: Harmonized Atmospheric Correction Method. 2014. Available online: http://www.glass-project.eu/downloads (accessed on 28 February 2017).

32. Guanter, L.; Ruiz-Verdú, A.; Odermatt, D.; Giardino, C.; Simis, S.; Estellés, V.; Heege, T.; Domínguez-Gómez, J.A.; Moreno, J. Atmospheric correction of ENVISAT/MERIS data over inland waters: Validation for European lakes. *Remote Sens. Environ.* **2010**, *114*, 467–480.

33. Schroeder, T.; Schaale, M.; Fischer, J. Retrieval of atmospheric and oceanic properties from MERIS measurements: A new Case 2 water processor for BEAM. *Int. J. Remote Sens.* **2007**, *28*, 5627–5632.

34. Kotchenova, S.Y.; Vermote, E.F.; Matarrese, R.; Frank, J.; Klemm, J. Validation of a vector version of the 6S radiative transfer code for atmospheric correction of satellite data. Part I: Path radiance. *Appl. Opt.* **2006**, *45*, 6762–6774.

35. Kotchenova, S.Y.; Vermote, E.F. Validation of a vector version of the 6S radiative transfer code for atmospheric correction of satellite data. Part II. Homogeneous Lambertian and anisotropic surfaces. *Appl. Opt.* **2007**, *46*, 4455–4464.

36. Berk, A.; Bernstein, L.S.; Anderson, G.P.; Acharya, P.K.; Robertson, D.C.; Chetwynd, J.H.; Adler-Golden, S.M. MODTRAN cloud and multiple scattering upgrades with application to AVIRIS. *Remote Sens. Environ.* **1998**, *65*, 367–375.

37. Bresciani, M.; Stroppiana, D.; Odermatt, D.; Morabito, G.; Giardino, C. Assessing remotely sensed chlorophyll-a for the implementation of the Water Framework Directive in European perialpine lakes. *Sci. Total Environ.* **2011**, *409*, 3083–3091.

38. Bresciani, M.; Bolpagni, R.; Laini, A.; Matta, E.; Bartoli, M.; Giardino, C. Multitemporal analysis of algal blooms with MERIS images in deep meromictic lake. *Eur. J. Remote Sens.* **2013**, *46*, 445–458.

39. Giardino, C.; Bresciani, M.; Stroppiana, D.; Oggioni, A.; Morabito, G. Optical remote sensing of lakes: An overview on Lake Maggiore. *J. Limnol.* **2014**, *73*, doi:10.4081/jlimnol.2014.817.

40. Alikas, K.; Reinart, A. Validation of the MERIS products on large european lakes: Peipsi, Vanern and Vattern. *Hydrobiologia* **2008**, *599*, 161–168.

41. Alikas, K.; Kratzer, S.; Reinart, A.; Kauer, T.; Paavel, B. Robust remote sensing algorithms to derive the diffuse attenuation coefficient for lakes and coastal waters. *Limnol. Oceanogr. Methods* **2015**, *13*, 402–415.

42. Asuküll, E. Measuring Dissolved Organic Matter From Satellites. Master's Thesis, Tartu University, Tartu, Estonia, 2013.

43. Kallio, K.; Koponen, S.; Ylöstalo, P.; Kervinen, M.; Pyhälahti, T.; Attila, J. Validation of {MERIS} spectral inversion processors using reflectance, {IOP} and water quality measurements in boreal lakes. *Remote Sens. Environ.* **2015**, *157*, 147–157.

44. Philipson, P.; Kratzer, S.; Ben Mustapha, S.; Strombeck, N.; Stelzer, K. Satellite-based water quality monitoring in Lake Vanern, Sweden. *Int. J. Remote Sens.* **2016**, *37*, 3938–3960.

45. Vantrepotte, V.; Loisel, H.; Dessailly, D.; Mauriaux, X. Optical classification of contrasted coastal waters. *Remote Sens. Environ.* **2012**, *123*, 306–323.

46. Wang, M.; Shi, W. The NIR-SWIR Combined Atmospheric Correction Approach for MODIS Ocean Color Data Processing. *Opt. Express* **2007**, *15*, 15722–15733.

47. Shi, W.; Wang, M. An assessment of the black ocean pixel assumption for MODIS SWIR bands. *Remote Sens. Environ.* **2009**, *113*, 1587–1597.

48. Werdell, J.; Franz, B.; Bailey, S. Evaluation of shortwave infrared atmospheric correction for ocean color remote sensing of Chesapeake Bay. *Remote Sens. Environ.* **2010**, *114*, 2238–2247.

49. Antoine, D.; Morel, A. A multiple scattering algorithm for atmospheric correction of remotely sensed ocean colour (MERIS instrument): Principle and implementation for atmospheres carrying various aerosols including absorbing ones. *Int. J. Remote Sens.* **1999**, *20*, 1875–1916.

50. Moore, G.; Lavender, S. *Algorithm Identification: Case II. S Bright Pixel Atmospheric Correction*; Technical Report; ESA: Paris, France, 2011.

51. Moore, G.F.; Aiken, J.; Lavender, S.J. The atmospheric correction of water colour and the quantitative retrieval of suspended particulate matter in Case II waters: Application to MERIS. *Int. J. Remote Sens.* **1999**, *20*, 1713–1733.

52. Antoine, D.; Morel, A. *Atmospheric Correction of the MERIS Observations over Ocean Case 1 Waters*; Technical Report; Laboratoire d'Oceanographie de Villefranche: Villefranche-sur-Mer, France, 2011.

53. GLaSS Deliverable D5.7. Global Lakes Sentinel Services, D5.7: WFD Reporting Case Study Results. 2015. Available online: http://www.glass-project.eu/assets/Deliverables/GLaSS-D5-7.pdf (accessed on 28 February 2017).

54. Schaeffer, B.A.; Schaeffer, K.G.; Keith, D.; Lunetta, R.S.; Conmy, R.; Gould, R.W. Barriers to adopting satellite remote sensing for water quality management. *Int. J. Remote Sens.* **2013**, *34*, 7534–7544.

remote sensing

MDPI

Article

Atmospheric Correction Performance of Hyperspectral Airborne Imagery over a Small Eutrophic Lake under Changing Cloud Cover

Lauri Markelin [1,2,*]**, Stefan G. H. Simis** [1]**, Peter D. Hunter** [3]**, Evangelos Spyrakos** [3]**, Andrew N. Tyler** [3]**, Daniel Clewley** [1] **and Steve Groom** [1]

[1]   Plymouth Marine Laboratory (PML), Prospect Place, The Hoe, Plymouth PL1 3DH, UK; stsi@pml.ac.uk (S.G.H.S.); dac@pml.ac.uk (D.C.); sbg@pml.ac.uk (S.G.)
[2]   Finnish Geospatial Research Institute (FGI), Geodeetinrinne 2, 02430 Masala, Finland
[3]   Department of Biological and Environmental Sciences, University of Stirling, Stirling FK9 4LA, UK; p.d.hunter@stir.ac.uk (P.D.H.); evangelos.spyrakos@stir.ac.uk (E.S.); a.n.tyler@stir.ac.uk (A.N.T.)
*   Correspondence: lauri.markelin@nls.fi; Tel.: +358-29-530-1100

Academic Editors: Yunlin Zhang, Claudia Giardino, Linhai Li, Xiaofeng Li and Prasad S. Thenkabail
Received: 24 August 2016; Accepted: 19 December 2016; Published: 23 December 2016

Abstract: Atmospheric correction of remotely sensed imagery of inland water bodies is essential to interpret water-leaving radiance signals and for the accurate retrieval of water quality variables. Atmospheric correction is particularly challenging over inhomogeneous water bodies surrounded by comparatively bright land surface. We present results of AisaFENIX airborne hyperspectral imagery collected over a small inland water body under changing cloud cover, presenting challenging but common conditions for atmospheric correction. This is the first evaluation of the performance of the FENIX sensor over water bodies. ATCOR4, which is not specifically designed for atmospheric correction over water and does not make any assumptions on water type, was used to obtain atmospherically corrected reflectance values, which were compared to in situ water-leaving reflectance collected at six stations. Three different atmospheric correction strategies in ATCOR4 was tested. The strategy using fully image-derived and spatially varying atmospheric parameters produced a reflectance accuracy of ±0.002, i.e., a difference of less than 15% compared to the in situ reference reflectance. Amplitude and shape of the remotely sensed reflectance spectra were in general accordance with the in situ data. The spectral angle was better than 4.1° for the best cases, in the spectral range of 450–750 nm. The retrieval of chlorophyll-a (Chl-a) concentration using a popular semi-analytical band ratio algorithm for turbid inland waters gave an accuracy of ~16% or 4.4 mg/m^3 compared to retrieval of Chl-a from reflectance measured in situ. Using fixed ATCOR4 processing parameters for whole images improved Chl-a retrieval results from ~6 mg/m^3 difference to reference to approximately 2 mg/m^3. We conclude that the AisaFENIX sensor, in combination with ATCOR4 in image-driven parametrization, can be successfully used for inland water quality observations. This implies that the need for in situ reference measurements is not as strict as has been assumed and a high degree of automation in processing is possible.

Keywords: hyperspectral; airborne; atmospheric correction; ATCOR4; inland waters; water quality; in situ measurements; chlorophyll-a

1. Introduction

Coastal and inland water bodies can receive agricultural, domestic and industrial pollutants and are subject to recreational pressures from leisure, fishing and aquaculture industries. Remote sensing is widely considered as a cost-efficient strategy to complement traditional monitoring methods, in order to meet growing monitoring requirements set out by international environmental legislation [1–4]. Satellite remote sensing is an effective platform for frequent global ocean monitoring, and used increasingly to observe optically-complex coastal waters and inland water bodies of suitable size. The complexity and variability of optically active water constituents as well as the size of many inland water bodies ideally requires a satellite sensor with global coverage, high spatial and temporal resolution and high radiometric sensitivity applied to a set of narrow wavebands. Future hyperspectral satellite missions may well meet this demand [4,5]. Presently, airborne remote sensing is a mature but relatively costly method compared to satellite remote sensing for small water bodies. Studies using airborne hyperspectral sensors have demonstrated that accurate retrieval of optically-active substances in coastal and inland water bodies is possible [4,6–10]. Since the launch of the Medium Resolution Imaging Spectrometer (MERIS), a satellite mission with global coverage and an adequate band set for several inland water types, attention to airborne sensors has, however, waned. Airborne platforms equipped with narrow band multi- or hyperspectral sensorsare, however, still the only remote sensing platforms suitable for observing the majority of inland water bodies, as these are too small to observe with sensors such as the Moderate Resolution Imaging Spectrometer (MODIS), MERIS, and its follow on, Ocean and Land Colour Instrument (OLCI, aboard Sentinel-3).

Representative and accurate in situ reference observations over optically complex inland waters are a key requirement to progress the development of remote sensing of water quality. Ideally, in situ observations are used to validate the whole processing chain of remotely sensed data. This includes both the correction for absorption and scattering in the atmosphere by comparing against in situ measurements of water-leaving reflectance, and the retrieval of in-water optically active components by sampling the concentrations of coloured dissolved organic matter (CDOM), chlorophyll-a (Chl-a) as a proxy for phytoplankton biomass, and suspended particulate matter (SPM). In situ observations may comprise optical or biological point samples with typically high accuracy, but which are limited in their spatiotemporal coverage and relatively labour- and cost-intensive. To be representative of the same conditions, in situ and remote observations need to take place within a narrow time window, especially in dynamic and complex environments such as lakes. There is a relative scarcity in contemporaneous in situ and remote sensing data sets in lakes.

Currently, there is a clear gap between the spatial coverage of in situ and space-borne remote measurements of inland waters, where airborne hyperspectral sensors can be very useful tools for monitoring specific inland, estuarine and coastal waters, either on a regular basis [4], or as a development platform. The high spatial and spectral resolution obtained with airborne hyperspectral instruments makes it possible to develop water quality retrieval algorithms suitable for optically complex waters, which may subsequently be used with current and future satellite sensors [10]. Airborne observations can be used to validate hydrodynamic lake models, and identify spatial dynamics even in small water bodies or systems with complex coastlines [11,12]. We may expect that the use of unmanned airborne vehicles (UAVs) will bring new cost-efficient and agile methods for monitoring inland waters in the future [5,13].

Water bodies typically have low reflectance compared to land, necessitating high radiometric requirements for passive optical sensors and for the accuracy of atmospheric correction. Even though some applications can work directly with the at-sensor radiance data recorded by the airborne sensor [11,14,15], producing water-leaving reflectance spectra through accurate atmospheric correction is a crucial step for most water quality applications over optically complex waters, because in the visible domain only 2%–25% of the total radiance received by the sensor interacted with the water column. Atmospheric correction remains one of the biggest challenges for remote sensing, particularly over coastal and inland waters [16,17].

Although several atmospheric correction models have been developed for satellite observations over coastal and inland waters [18], there is no preferred approach for correcting airborne data collected over water. Most airborne remote sensing applications concern land surfaces and so atmospheric correction over water typically uses land-surface models with or without adaptations to water [16]. The difficulty with using general land-oriented atmospheric correction methods over water is that they typically treat the surface-reflected sky radiance, or Fresnel reflectance, as part of the target reflectance. This is a valid approach for land surface applications, but invalid for water-related applications where only water-leaving radiance is of interest [19,20]. Ideally, generically applicable atmospheric correction methods do not require a priori knowledge of the water-leaving radiance spectrum. Examples of such generic approaches include empirical/semi-empirical methods such as dark pixel subtraction [21,22] and Quick Atmospheric Correction (QUAC) [23,24], or more advanced physics-based radiative transfer methods such as FLAASH (Fast Line-of-sight Atmospheric Analysis of Spectral Hypercubes) [24–26], ACORN (Atmospheric CORrection Now) [27] and ATCOR4 (Atmospheric and Topographic CORrection) [28,29]. The challenge with radiative transfer based methods is that they assume prior knowledge of key atmospheric parameters (aerosol type, horizontal visibility or aerosol optical thickness, water vapour) during the campaign [24,30]. When the hyperspectral sensor has a sufficiently wide spectral range, including visible, near-infrared and shortwave infrared (VIS-NIR-SWIR), it becomes increasingly feasible to derive these parameters from the image data directly [28,31].

The hyperspectral sensors used in airborne water quality studies include various versions of AISA from Specim Ltd. [6,9,20,24,32], CASI [7,22], HyMap [7], APEX [12,32,33] and MIVIS [12,29]. It is essential that sensor performance and subsequent processing chains are validated over a range of water bodies with variable optical characteristics. In a recent study, Moses et al. [24] compared FLAASH and QUAC atmospheric corrections for chlorophyll-a estimation in turbid productive waters using NIR-red algorithms. They concluded that the image-driven QUAC produced more robust and reliable results with a multi-temporal dataset compared to FLAASH. They could not use the automatic aerosol retrieval in FLAASH as the AisaEAGLE sensor lacked the SWIR spectral channels required for the algorithm; Hunter et al. [22] faced this same problem with CASI imagery. Challenges encountered with early AISA sensors have included high instrument noise, especially in the NIR region, and possible radiometric calibration errors [20,24,34]. Giardino et al. [29] used airborne MIVIS imagery to retrieve concentrations of SPM, Chl-a and CDOM in lake Trasimento, Italy. They concluded that for shallow inland water applications high spatial and spectral resolution is needed. Knaeps et al. [35] used APEX imagery to show that in extremely turbid waters one cannot assume that the water reflectance is zero in the wavelength range of 1020–1240 nm and channels in that SWIR range can be used to estimate SPM concentrations. The latest airborne hyperspectral sensor from Specim is AisaFENIX, introduced in 2013. It has wavelength range of 380–2500 nm and up to 622 channels. However, to date, only scientific agricultural applications using the FENIX sensor have been published [36,37] and reports of using AisaFENIX over water bodies are still lacking.

It is inevitable that airborne remote sensing will continue to use a wide range of sensors and processing methods, but the most useful method should be able to produce accurate results without in situ measurements of atmospheric properties and of the target. The hypothesis of this study was that water-leaving reflectance spectra can be acquired even under challenging illumination conditions with a high quality airborne sensor in combination with fully image-driven atmospheric correction, and that the quality of image spectra is enough for realistic water quality parameter retrieval. In this study, we have evaluated three atmospheric correction strategies with ATCOR4, using both in situ and fully image-driven atmospheric parameters, with the airborne hyperspectral imagery collected over a small eutrophic water body with the AisaFENIX sensor. The performance of this sensor over water bodies has not been previously evaluated. The atmospheric correction results are evaluated against in situ hyperspectral measurements of water-leaving reflectance collected with a set of TriOS RAMSES spectrometers from a vessel. An added, but very common, difficulty is introduced by intermitted cloud-cover during the campaign. The validation is based on both qualitative and quantitative comparisons using root-mean-square difference, spectral angle and Chi-square metrics. To assess to what extent the performance of the system is suitable for routine monitoring applications, a semi-analytical band ratio based chlorophyll-a retrieval algorithm was applied to both airborne and in situ radiometric results.

2. Materials and Methods

2.1. Study Area

Loch Leven is located in the Perth and Kinross council area of central Scotland, United Kingdom ($56°12'$N, $3°22'$W, Figure 1). The lake is approximately 6 km long and has a surface area of approximately 13.3 km^2, with mean and maximum depths of 3.9 m and 25.5 m, respectively [38]. It lies at an altitude of 107 m. Loch Leven is a national nature reserve, as well as a Site of Special Scientific Interest and a Special Protection Area. Loch Leven is vital to the local economy, for which nature and wildlife tourism is highly important. The national nature reserve attracts 230,000 visitors per year. Phytoplankton growth in Loch Leven is primarily phosphorus-limited and has a long and well documented history of eutrophication and recovery [39]. In the 40-year period from 1970 to 2010, the total phosphorus concentration has decreased from 100 mg/m^3 to <40 mg/m^3, chlorophyll-a concentration decreased from an annual mean of over 100 mg/m^3 to approximately 40 mg/m^3, and water clarity improved from a Secchi disk depth of approximately 1.0 m to 1.7 m [26,40]. The high phosphorus load of the lake leads regularly to spring and late summer phytoplankton blooms, particularly with frequent sediment resuspension caused by winds, which cause deep mixing in shallow water [26,39,41]. Also, toxin-producing cyanobacteria are an important part of the Loch Leven phytoplankton community, with cyanobacterial blooms common in late summer and early autumn [41].

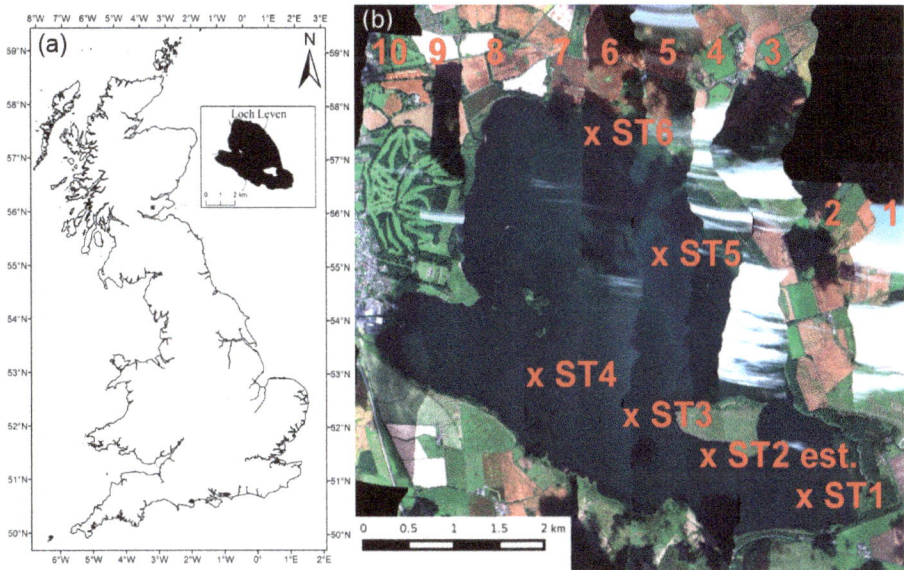

Figure 1. (**a**) map showing the location of Loch Leven within the UK; (**b**) Image mosaic from 10 FENIX flight lines collected during the campaign (colours are illustrative only). Locations of the in situ station measurements are marked with red crosses; the location of station ST2 is approximate.

2.2. Remote Sensing Data

2.2.1. Airborne Data Collection

Airborne hyperspectral data were acquired on 7 August 2014 between 11:20 and 12:27 (local time) with an AisaFENIX sensor (Specim, Spectral Imaging Ltd, http://www.specim.fi/) on board a NERC Airborne Research Facility (National Environment Research Council Airborne Research Facility http://www.bas.ac.uk/nerc-arf) aircraft. In total, 10 flight lines were flown from south to north (Figure 1, Table 1). The flying height was approximately 1500 m, resulting in a ground sampling distance of 2 m. The campaign was carried out in challenging illumination conditions, as clouds and cloud shadows were clearly visible on several FENIX images (Figure 1). The FENIX data used for this publication are available to download from the NERC Earth Observation Data Centre (NEODC; http://neodc.nerc.ac.uk/).

Table 1. Flight lines, in situ reference measurements and atmospheric parameters during the campaign. FL2-9: flight line number, ST1-6: in situ measurement station number. Times are local time (BST). Wind: average wind speed during station measurements. QC pass/All: number of individual spectrum that passed the automatic quality control during station measurements. In situ water vapour and AOT values used for atmospheric correction strategy AC3 are given in italics. wv: columnar water vapour, AOT: aerosol optical thickness, Vis: horizontal visibility, A.: parameter derived by ATCOR (wv and AOT with standard deviations, MT: Microtops sun photometer used for in situ AOT and water vapour measurements.

Line/Station	ST1	FL2	ST2	FL3	FL4	FL5	FL6	FL7	FL8	ST3	FL9	ST4	ST5	ST6
Start time	10:53	11:27	11:29	11:33	11:40	11:47	11:55	12:02	12:10	12:11	12:18	13:07	13:57	15:54
Stop time	10:59	11:28	11:34	11:34	11:41	11:48	11:57	12:03	12:12	12:15	12:20	13:12	14:02	15:59
Wind (m/s)	3.14		1.4							5.03		6.02	3.6	3.1
QC pass/All	5/31		2/30							2/30		1/30	5/30	9/30
MT wv (cm)		1.14		1.13	*1.15*	*1.19*					1.17			
A. wv (cm)		1.11		1.06	1.09	1.23	1.40	1.17	1.15		1.12			
A. wv std		0.03		0.12	0.07	0.07	0.13	0.03	0.03		0.04			
MT AOT		0.193		0.186	*0.150*	*0.135*					0.158			
A. AOT		0.162		0.166	0.176	0.207	0.202	0.203	0.185		0.168			
A. AOT std		0.016		0.020	0.014	0.027	0.023	0.015	0.010		0.013			
A. Vis. (km)		51.6		49.5	45.6	38.1	39.4	36.8	42.3		47.5			

FENIX is a pushbroom sensor with a wavelength range 380–2500 nm and 384 spatial pixels. FENIX has two separate physical detectors to record the whole wavelength range: a CMOS (Complementary Metal Oxide Semiconductor) detector for the VNIR range (380–970 nm) and an MCT (Mercury Cadmium Telluride) detector for the SWIR range (970–2500 nm). The sensor was operated in spectral binning mode, resulting in 448 bands (174 bands in VNIR, 4x binning; 274 bands in SWIR, no binning) with average spectral sampling interval 3.4 nm on VNIR range and 5.7 nm on SWIR range. The spectral resolution is 3.5 nm on VNIR and 12 nm in SWIR range. The peak signal-to-noise ratio (SNR) of the sensor is 500–1000.

Sensor radiometric calibration data from laboratory measurements were applied to the images at NERC-ARF-DAN (NERC Airborne Research Facility—Data Analysis Node, https://nerc-arf-dan.pml.ac.uk) to convert the raw image data to at-sensor radiance data in the original sensor geometry [42].

2.2.2. SNR Estimation

To estimate the FENIX signal-to-noise ratio, at-sensor radiance image data were used and 31 regions of interest (ROI) containing 100 pixels over water (10×10 pixels, approximately 20×20 m) were defined from 7 flight lines. The location of each ROI was manually selected over the most uniform areas of the water body so that the standard deviation of the ROI radiance would be a close representation of sensor noise. The measured mean at-sensor radiance of each ROI was divided by the standard deviation of the ROI to calculate per-band SNR = mean/stdev [43]. ROIs where the standard deviation was lowest (<0.0002 (W/m^2/sr/nm)) and uniform between 450 and 900 nm were considered least affected by in-water variability of optically active constituents, and selected for final SNR calculations. The final estimate of sensor SNR was then calculated as the per-band average of the 17 best quality ROI SNR measurements. This image-based SNR evaluation takes into account the whole sensor–atmosphere–target system and gives estimation of the sensor SNR in real operational conditions.

2.2.3. Atmospheric Correction with ATCOR4

ATCOR4 version 7.0.0 [28], which is based on the radiative transfer model MODTRAN5 [44], was used for the atmospheric correction of the hyperspectral data. ATCOR4 is highly configurable, and includes a number of built-in algorithms to automatically select a number of input parameters, including water vapour, horizontal visibility and aerosol model, from the imagery supplied. These algorithms require narrow bands in the VNIR-SWIR range. If the sensor does not have the required bands and/or there are in situ atmospheric data available, these parameters can also be set manually. The aim of the current analysis was to validate the use of the built-in algorithms for operational processing environments, rather than optimize each input parameter manually and separately for each image.

The following ATCOR4 parameters for the first atmospheric correction (AC1) were used with all flight lines: aerosol model: rural (detected by ATCOR4), spatially variable water vapour (detected by ATCOR4), variable horizontal visibility (detected by ATCOR4). AOT (aerosol optical thickness) over water was set to be interpolated from land values. Water vapour detection was based on bands in the 940 nm and 1130 nm regions. The mean AOT and water vapour values for each flight line, with standard deviation values illustrating the variability of the parameter, are shown in Table 1. The end result of the atmospheric correction was a set of reflectance images with normalized water-leaving reflectance $R_w(\lambda)$ (dimensionless, range [0, 1]) [45] values for each image pixel and band. If the correction would result in negative reflectance values, ATCOR4 will convert them to a constant reflectance value of 0.0025.

Two other built-in ATCOR4 atmospheric correction strategies were also tested with flight lines FL4 and FL5. In the second strategy (named AC2), horizontal visibility and water vapour were detected by ATCOR4, but all of were fixed for the whole image. In the third strategy (AC3), parameters were set to be fixed for the whole image based on in situ measurements (Table 1). Otherwise, all options were kept the same as in AC1. In situ AOT values were converted to horizontal visibility using the Koschmieder equation [28]. The ATCOR4-recommended horizontal visibility values for AC2 was 100 km for both flight lines, FL4 and FL5. The fixed water vapour values in ATCOR4 can only be set by choosing from predefined values, so the closest value in the list, 1.0 cm, was used for all strategies. Both strategies, AC1 and AC2, are fully image driven.

2.3. In Situ Reflectance Measurements

A boat-based sampling campaign was carried out simultaneously with the airborne campaign, between 10:53 and 15:59 BST, to measure remote sensing reflectance R_{rs} [45] with a system based on three TriOS RAMSES spectrometers (http://www.trios.de/). The system recorded spectral water-leaving radiance $L_w(\lambda)$ and sky-radiance $L_{sky}(\lambda)$, as well as downwelling irradiance $E_d(\lambda)$ in the $\lambda = 320$–950.3 nm wavelength range in 192 channels. With a 3.3 nm sampling interval. Six sampling locations (ST1-6, Figure 1) were visited, where the boat was kept stationary and R_{rs} data were recorded for 5–10 min resulting in 30 individual spectra (Table 1). Also, AOT measurements with a Microtops II sunphotometer were performed during the sampling campaign (Table 1, Figure 2).

The target for viewing geometry of the in situ radiance measurements from the boat was for a 40 degrees oblique angle from zenith and 135 degrees from solar azimuth, corresponding to a minimum of sun-glint [45] and avoiding shadows and reflections from the sampling platform. Prevailing weather conditions, with wind speeds 1.4–6.2 m/s, and use of a small boat will have caused occasional deviations from these optimal angles. While cruising between stations and simultaneously recording transect measurements, the GPS heading is sufficiently accurate to derive the azimuth angle of the sensors, and measurements with inappropriate viewing geometry were thus removed from the data set. While cruising between stations ST1 and ST3, GPS information was not recorded, so for these station ST2 the location was estimated by spatiotemporal interpolation between start and destination locations. The station locations are marked in Figure 1, with uncertain location information marked separately.

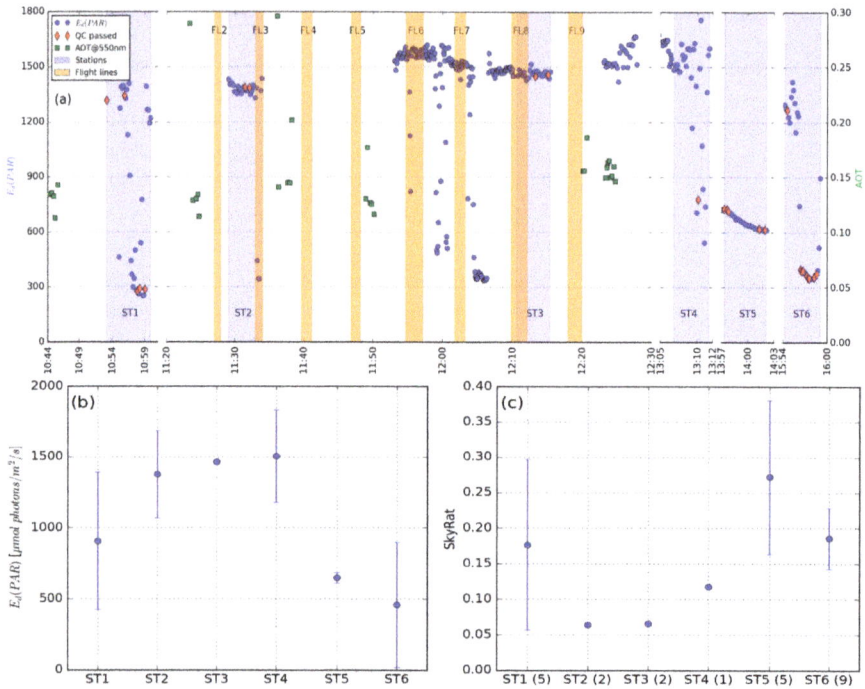

Figure 2. (a) In situ E_d(PAR) and AOT measurements taken during the campaign. In situ observations used for station reference spectra are marked with red diamonds. Orange shading indicates the times of airborne flight lines, and blue shading the times of station measurements; (b) Median and standard deviation of E_d(PAR) from all 30 measurements recorded at each station; (c) SkyRat based on QC passed measurements for calculation of R_{rs} (number of valid observations are shown in horizontal axis label). The SkyRat error bars for stations ST2-ST4 are not shown, as there were <3 valid measurements at these stations.

Post-processing of the radiance data to calculate R_{rs} (units sr^{-1}) followed the protocol of Simis and Olsson [46], which also flags suspect measurements. Only R_{rs} spectra that passed the quality control procedure were used (QC passed). Median and standard deviation R_{rs} spectra were calculated for each station where at least two valid R_{rs} spectra were thus obtained. The median R_{rs} spectrum of each station was used as reference for the airborne observations.

Two additional indicators of illumination conditions during the in situ R_{rs} measurements were calculated during post processing. First, E_d integrated over the spectrum of photosynthetically active radiation (E_d(PAR), 400–700 nm), is used to express the intensity of solar irradiance during a measurement. Second, the ratio $\pi L_{sky}(400)/E_d(400)$ (SkyRat) is indicative of the cloudiness during a measurement. SkyRat < 0.2 are generally indicative of clear sky conditions, whereas values close to unity suggest overcast conditions [46].

Weather Conditions

Weather conditions, in particular cloud cover, varied significantly during the campaign. In situ above-water R_{rs} can be measured under clear, fully overcast, and sometimes even under partially clouded conditions. In general, there is a larger margin for error in relatively turbid waters where water-leaving radiance is high compared to skylight reflected on the water surface. Variable illumination conditions (high standard deviation of SkyRat and E_d(PAR)) are least favourable, because the measured spectrum of sky radiance is less likely to represent the sky radiance reflected on the water surface. Figure 2 shows the variations of in situ-measured E_d(PAR) and AOT during the campaign, with shading indicating the timing of airborne observations and in situ station measurements. As an indication of illumination conditions during in situ R_{rs} station measurements, SkyRat and E_d(PAR) values, with standard deviation error bars for each station, are shown in Figure 2. E_d(PAR) values are based on all 30 observations performed during each station measurement, whereas SkyRat values are based only on spectra for which R_{rs} could be derived, as these are obtained using suitable viewing geometry for the radiance measurements. Stations ST2 and ST3 had the lowest SkyRat values and highest E_d(PAR) radiances, indicating clear sky; still, only two individual spectra passed the quality control for these stations. When stations had more QC-passed measurements (ST1, ST5, ST6), also the standard deviations of these measurements increased. From the E_d(PAR) values in Figure 2 (top) and SkyRat values in Figure 2 (bottom right), it can be seen that illumination conditions varied highly during stations ST1 and ST4, were relatively stable and good during stations ST2 and ST3, were stable and cloudy during station ST5 and were variable but mainly cloudy during station ST6. The measured in situ AOT values varied between 0.12 and 0.3 during the flight campaign. As the weather got cloudier during and after station ST4 measurements, it was not possible to acquire additional in situ AOT measurements.

2.4. Data Analysis

To allow quantitative comparison of image and in situ spectra, in situ R_{rs} spectra were converted to normalized water-leaving reflectance R_w (where $R_w = \pi \times R_{rs}$). This assumes that the upwelling radiance is fully diffuse, whereas its angularity is not strictly known. Subsequently, in situ R_w spectra were interpolated to match the FENIX wavelength grid in the range between 383 nm and 948 nm, where the two sensor systems overlap. To reduce the effect of sensor noise, image R_w data were spatially averaged over 3×3 pixels, approximately 6×6 m, to obtain the final image spectrum for each station. When comparing exclusively the spectral shapes of in situ and airborne R_w, both were standardized between their minimum and maximum values to the range [0, 1].

The dataset allowed ten comparisons between ATCOR4-derived and in situ R_w spectra. There were seven spatial matches between images and in situ station measurements. Due to the image overlap between adjacent flight lines, station ST3 was visible on both flight lines FL5 and FL6. Three additional comparisons were done between stations ST1, ST4 and ST6 and nearly matching flight lines FL3 (45 m distance), FL8 (72 m distance) and FL7 (90 m distance), respectively. The time difference between flight lines and in situ station measurements varied from 9 min to four hours; station ST1 was measured before the airborne observations started, stations ST2 and ST3 during the campaign and stations ST4-6 following the airborne observations. Tables 1 and 2 give the details of the comparisons and their spatial and temporal differences.

Differences between in situ and image R_w spectra were compared using the following numerical evaluations: evaluation of spectral accuracy by using spectral difference, spectral ratio and root mean square difference (RMS), and calculation of Spectral Angle (SA) [47,48] and Chi-square (X^2) metrics to evaluate the similarities in spectral shapes.

Analysis of the reflectance accuracy followed the method described in Markelin et al. [49]. In short, reflectance accuracy was evaluated by first considering the absolute difference (ΔR_w) between measured R_w (R_{w_data}) and the in situ R_w (R_{w_ref}). This difference was then expressed as the relative difference ($\Delta R_w\% = 100\% \ \Delta R_w / R_{w\_ref}$). ΔR_w and $\Delta R_w\%$ were calculated for all comparisons in the matching wavelength range of 383–948 nm. From these differences, root mean square difference values (RMS and RMS%) were calculated using all 10 comparisons and 5 qualitatively best ones as follows:

$$RMS\% = \sqrt{\frac{\Sigma(\Delta R_w\%)^2}{n}}, \tag{1}$$

where n is the number of observations. RMS is calculated using Equation (1) by replacing $\Delta R_w\%$ with ΔR_w.

The spectral angle is a metric comparing spectral shapes, insensitive to spectral amplitude, and is calculated as:

$$\alpha = \cos^{-1}\left(\frac{\sum_{i=1}^{nb} a_i t_i}{\sqrt{\sum_{i=1}^{nb} a_i^2}\sqrt{\sum_{i=1}^{nb} t_i^2}}\right), \tag{2}$$

where a denotes ATCOR4-derived spectra, t are in situ spectra and nb is the number of channels/bands in a spectrum. In practice, the spectral angle is the angle between two vectors (spectra) and is not sensitive to differences in amplitude that are consistent over the whole spectrum. The range for the spectral angle is [0, 180] (or 0–π radians), where values close to 0 indicate high similarity.

Chi-square takes both the shape and the amplitude of the spectra into account, where the sum of all bands is considered for each spectrum, and is calculated as:

$$\chi^2 = \sum_{i=1}^{nb}\left(\frac{(a_i - t_i)^2}{t_i}\right), \tag{3}$$

where a and t are as in (2) and nb is the number of channels/bands in a spectrum.

Additionally, visual comparison of in situ and image R_w spectrum plots was also performed to address specific anomalies, such as areas in the spectrum with consistent error patterns or high noise, the reproduction of key spectral features and to evaluate the homogeneity of the lake.

Table 2. Chi-squared (χ^2) and spectral angle (SA) metrics and Chl-a retrieval difference for all 10 matchups between in situ and image R_w. T.diff: time difference between in situ measurement and aircraft overpass; negative time means that in situ measurement is done before, and positive after, the airborne measurement. Dist.dif: distance between in situ and image measurement locations. Full: full wavelength range 383–948 nm, cen: centre part of the spectra 450–750 nm. Chl-a diff: Chl-a concentration difference between image and in situ-based retrieval in (mg/m^3). Range for the SA is [0, 180]. In both metrics, values closer to 0 indicate more similar spectra. Results from five best matchups are in bold. Measurement location on FL6 was covered with cloud shadow (results in italics). χ^2 and Chl-a diff values for comparison ST5-FL5 full were not relevant as the in situ spectra had negative values above 750 nm, which are not allowed in the calculation of the metric.

In Situ Station		ST1		ST2	ST3		ST4		ST5	ST6	
Flight line		FL2	FL3	FL4	FL5	FL6	FL7	FL8	FL5	*FL6*	FL7
T.dif. (h:mm)		−0:32	−0:38	−0:09	0:26	0:17	1:08	0:59	2:13	*3:59*	3:53
Dist.dif. (m)		-	45	-	-	-	-	72	-	-	90
χ^2	full	0.521	0.294	**0.019**	**0.200**	4.042	**0.219**	**0.147**		*0.244*	0.057
	cen	0.108	0.041	**0.007**	**0.068**	1.025	**0.019**	**0.013**	5.329	*0.233*	0.013
SA°	full	21.9	17.1	**4.2**	**6.0**	16.9	**9.1**	**4.6**	23.2	*14.0*	5.3
	cen	10.4	8.3	**3.0**	**2.9**	8.3	**4.1**	**2.4**	9.6	*7.7*	2.9
Chl-a diff		0.62	1.26	6.87	5.48	3.00	5.96	3.73		*8.10*	10.31

2.5. Chlorophyll-a Retrieval

To evaluate the applicability of the atmospherically corrected imagery in water quality monitoring, we selected semi-analytical band ratio algorithm suitable for turbid inland waters to retrieve Chl-a concentrations from in situ and image R_w spectra. The algorithm presented by Gons et al. [50] was designed for MERIS, and widely used for coastal and inland waters with Chl-a concentrations >10 mg/m^3, such as Loch Leven [26]. First, the backscattering coefficient b_b is calculated as:

$$b_b = \frac{1.61 R_w(778)}{0.082 - 0.6 R_w(778)},$$ (4)

where $R_w(778)$ is water-leaving reflectance of the MERIS equivalent band 12 centered at 778 nm. Then, the Chl-a concentration is calculated as:

$$Chla = \left(\frac{R_w(708)}{R_w(665)}(0.7 - b_b) - 0.4 - b_b^{1.06} \right)/0.016,$$ (5)

where $R_w(708)$ and $R_w(665)$ are water-leaving reflectance for MERIS bands 9 and 6, centered at 708 nm and 665 nm, respectively. As MERIS bands 6 and 9 are 10 nm wide and band 12 is 15 nm wide, the average of three FENIX 3.3 nm wide and TriOS 3.5 nm wide bands were used in calculations. Equation (5) was used to calculate Chl-a concentration in (mg/m^3) from both image-derived and in situ R_w spectra. Image- and in situ reflectance-based concentrations were compared by calculating Chl-a differences and RMS both in mg/m^3 and in percent using Equation (1).

3. Results

3.1. Variability in Water-Leaving Reflectance

From 50 randomly selected R_w spectra representing the whole lake (Figure 3a), it becomes evident that the amplitude of the spectra varies significantly more than their shape. The water is therefore understood to be relatively homogenous in terms of the relative composition of optically active substances. Spectral features caused by chlorophyll-*a* are clearly visible from a reflectance trough near the red absorption peak of the pigment at 650 and 700 nm. The amplitude of spectra is low but non-zero at wavelengths above 950 nm, where the absorption by pure water dominates the optical properties of the water.

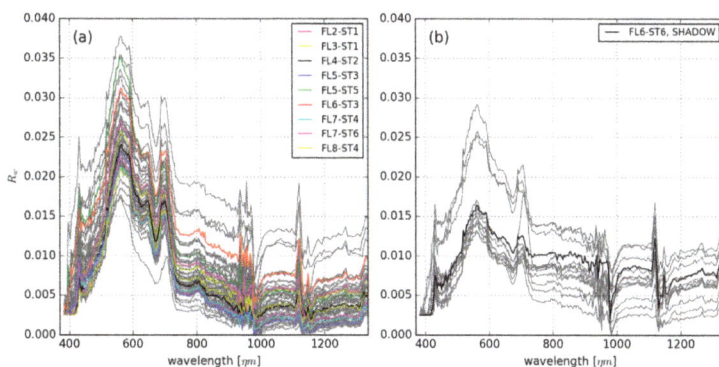

Figure 3. (**a**) R_w spectra measured from atmospherically corrected FENIX images. Spectra measured at station locations are shown in colour, grey spectra are an additional 50 measurements representing all lake areas to indicate the homogeneity of the lake; (**b**) Image-based R_w spectra of water measured from 21 locations under cloud shadows, the black line indicates station ST6 on image FL6. Wavelength range in the x-axis goes up to 1300 nm to indicate the non-zero reflectance in the NIR-SWIR range.

Atmospheric absorption caused by water vapour is seen in at 820, 940 and 1130 nm wavelength regions, where the shape of the spectra is disturbed and the atmospheric correction cannot succeed. Several spectral features likely resulting from suboptimal sensor calibration are also visible in the R_w spectra. The sharp peak visible at 425 nm cannot be associated with in-water optically active substances. Numerous narrow variations in spectral shape, e.g., between 560 and 700 nm, are either sensor or atmospheric correction issues, since water constituents do not show sharply featured optical features in the visible domain (see in situ reference spectra, Figure 4). R_w values of 0.0025 in the area around 400 nm suggest that negative reflectance values resulted from ATCOR4 processing, which ATCOR4 then converted to this low constant value.

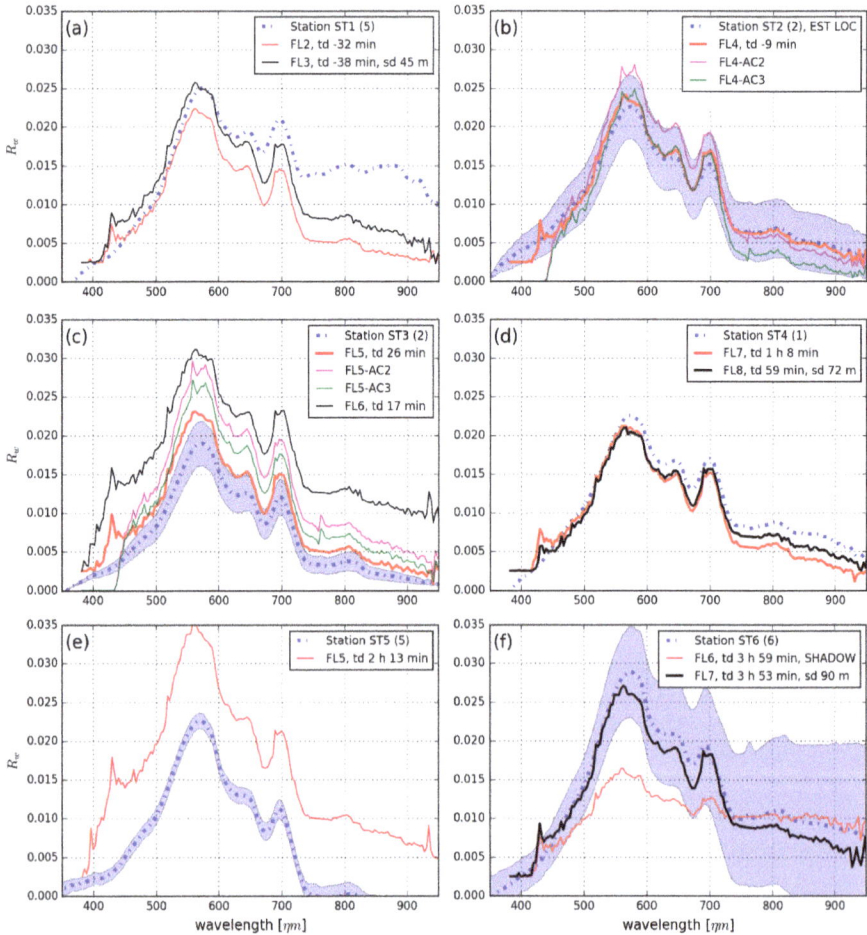

Figure 4. (**a–f**) In situ station and image-derived R_w of each matchup. The best matchups in terms of Chi-square and spectral angle are shown in a thicker line in panels (**b–d,f**); Image R_w from atmospheric correction strategies AC2 and AC3 are included in panels (**b,c**); Bracketed values in the panel legends state the number of spectra used to calculate the in situ spectrum and error bars; td = time difference between in situ measurement and image acquisition, sd = distance between in situ and image matchup measurement, if applicable. Standard deviations for the in situ station measurements are plotted as error bars (blue shading) where available.

An additional 20 spectra were sampled from the images in areas shadowed by clouds (Figure 3b). Here, the R_W spectra were dampened in the 400–750 nm range, but maintained the spectral shape of R_W measured at clear sky locations, suggesting that the interpretation of R_W by ATCOR4 is primarily hampered by an unknown intensity of downwelling irradiance in these shaded areas. At wavelengths beyond 750 nm the signal was comparable to non-shaded locations.

3.2. Evaluation of AisaFENIX SNR

Image-based per-band SNR estimation for FENIX, calculated as the mean of image R_W from 17 locations over water, is shown in Figure 5. Because the SNR is dependent on the radiance recorded at the sensor, the mean at-sensor radiance at these locations is also included. The SNR is approximately proportional to the amplitude of at-sensor radiance, as expected. The edges of linear arrays of the FENIX CMOS (380–410 nm and 960–980 nm) and MCT detectors (960–980 nm) show lower SNR, which is a detector property. The maximum measured SNR ranged from 40–95 in the 450–750 nm wavelength range. Over bright clouds, SNR ranged from 400–550 in the 500–800 nm wavelength range, and approximately 30 elsewhere.

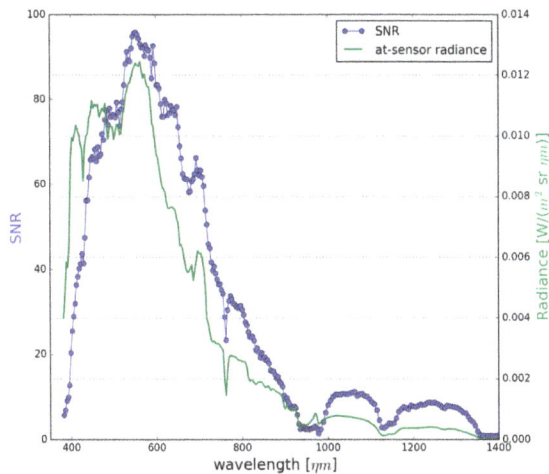

Figure 5. Spectral signal-to-noise ratio (SNR) (blue circles) and at-sensor radiance (green line) of AisaFENIX measured over Loch Leven. Curves are mean of 17 individual measurements from 7 flight lines.

3.3. Effect of Atmospheric Parameters in ATCOR4

Three different atmospheric correction strategies in ATCOR4 were tested to find one suitable for operational use. The image-derived R_W from each strategy used for flight lines FL4 and FL5 are shown in Figure 4b,c. The most accurate results compared to in situ R_W were achieved with strategy AC1, where both water vapour and horizontal visibility were detected by ATCOR4 and those parameters were allowed to vary spatially. When using fixed water vapour and visibility values for whole images (AC2 and AC3), there was a small amplitude difference between image and in situ R_W. The spectral shapes of R_W from AC2 and AC3 were practically identical (Figure 6b,c), meaning that the change in the input visibility caused a fixed shift in amplitude in R_W. As there were only small differences between the three strategies tested, and the primary aim of our work is to derive an operational atmospheric correction approach without dependence on in situ data, the rest of the analysis was performed for image R_W based on atmospheric correction strategy AC1.

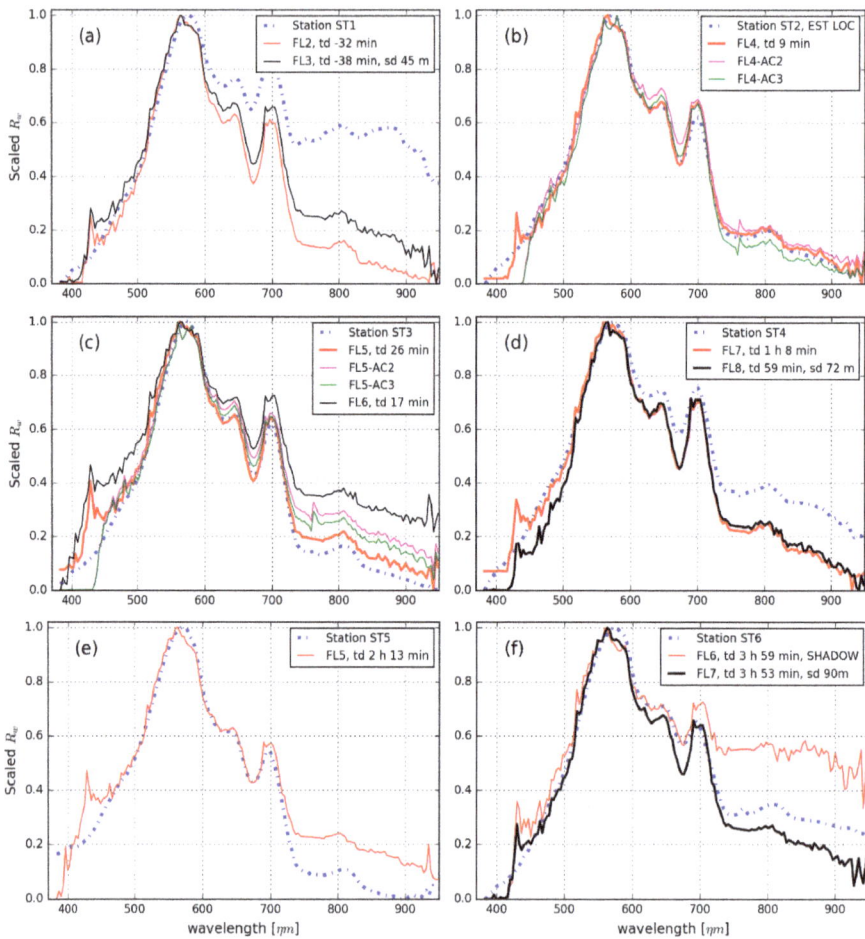

Figure 6. (**a–f**) In situ station and image-derived R_w spectra standardized between 0 and 1. Image spectra corresponding to the best five comparisons are plotted as thicker lines. Image-derived standardized R_w from atmospheric correction strategies AC2 and AC3 are included in panels (**b,c**). In panel legends, td = time difference between in situ measurement and image acquisition, and sd = distance between in situ and image measurement.

3.4. Station-Wise Comparisons of R_w

The campaign resulted in seven spatially matching pairs of airborne and in situ observations. Due to image overlap between adjacent flight lines, station ST3 was visible on flight lines FL5 and FL6. Additionally, three matchups were included from approximately matched station locations (ST1-FL3 at a distance of 45 m, ST4-FL8 at a distance of 72 m and ST6-FL7 at a distance of 90 m). The ten matchups are listed in Table 2. If a station included more than one successful in situ reflectance spectrum, these were used to calculate standard deviation error bars around the median spectrum for that station (Figure 4). Error bars are omitted for station ST1 due to extremely varying illumination conditions. The location of station ST6 on image FL6 was under cloud shadow, but the matchup spectra are included for comparison.

Based on both quantitative (Table 2, Figure 7) and qualitative (Figures 4 and 6) analysis, the best results were obtained for the comparison between station ST2 and image FL4. There, the image and in situ spectrums matched closely over the wavelength range 380–950 nm and the image-derived spectrum was always within the standard deviation of the in situ observations. Other good matches were ST3-FL5, ST4-FL7, ST4-FL8 and ST6-FL7. At these matchups, the spectral shape of image-derived R_W matched well with the in situ spectra, and the difference in amplitude was less than 0.005 in reflectance units. Still, the image-derived R_W included some errors common to all images, most notably the spike at 425 nm. In the other matchups, typical differences between image and in situ spectrums were either a clear difference in amplitude (as in matchups ST3-FL6, ST5-FL5), or shape difference in some part of the spectra (as in matchups ST1-FL2, ST1-FL3, ST6-FL6). The spatial difference between in situ and image measurements did not seem to have any notable effect on the matchups: ST4-FL8 and ST6-FL7, with spatial differences of 72 m and 90 m, respectively, gave good results comparable to matchups with no spatial difference. There was no notable effect of temporal differences between matchups. The time differences for the best five matchups were between −0:09 and 3:59, and between −0:38 and 3:53 for the worst five matchups.

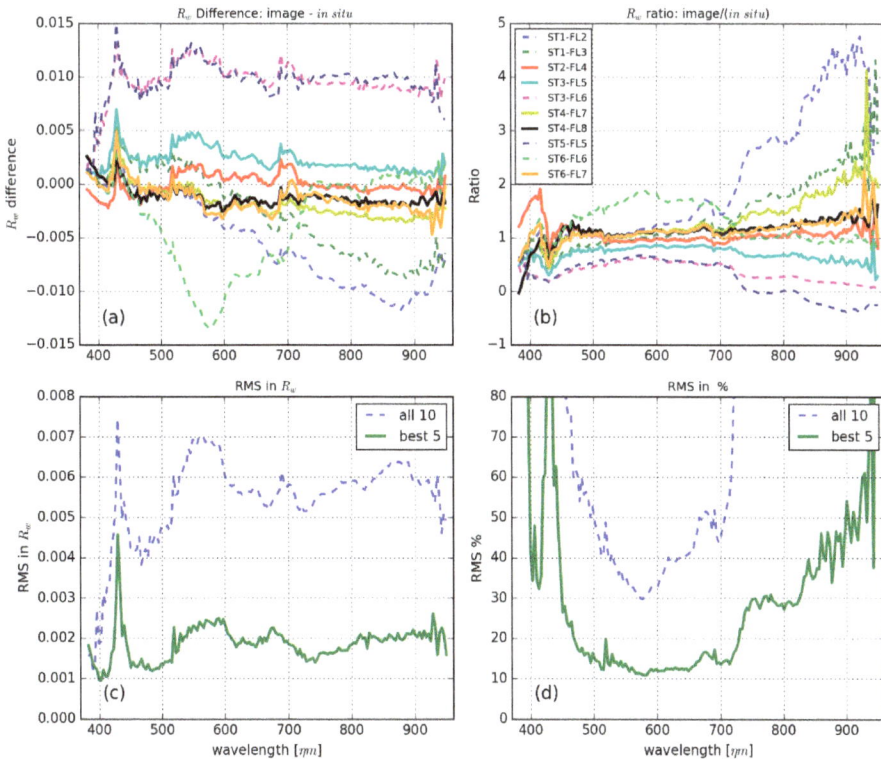

Figure 7. (**a**) Difference between image-derived and in situ R_W; (**b**) ratio of in situ over image-derived R_W; (**c**) RMS of the best five and all matchups, respectively, in units of reflectance; (**d**) Relative RMS. Matchup labels for panes (**a**,**b**) are given in the legend of panel (**b**), and the five best matchups are plotted with a solid line.

Table 2 lists results from Chi-squared and spectral angle metrics used to evaluate spectral differences between in situ and image R_W spectra. Metrics are shown for both the full wavelength range (383–948 nm) and the centre of the spectra (450–750 nm). Both metrics gave similar results in ranking the matchups from best to worst. The best five matchups (ST2-FL4, ST3-FL5, ST4-FL7, ST4-FL8, ST6-FL7) remained the same regardless of using either the full wavelength range or using only the centre part of the spectra. Using only the centre part of the spectra clearly improved the results on both metrics, as both the image and in situ spectrums suffer from low signal in the wavelengths below 450 nm and above 750 nm. The spectral angle gave values of 4.06° or better, and 3.00° on average, for the five best matchups, and 10.4° or better, with 8.85° on average, for the worst five matchups, when considering the centre of the spectra. Chi-square metrics were 0.068 or better, and 0.024 on average, for the five best comparisons, and 5.3 or better, or 1.35 on average, for the remaining five comparisons, when using the centre part of the spectra.

Both the Chi-squared and spectral angle metrics were worse for atmospheric correction strategies AC2 and AC3 than for AC1 for flight lines FL4 and FL5-ST3. The spectral angle gave values of 10.4° and 13.2° for FL4-AC2 and FL4-AC3, and 8.7° and 7.2° for FL5-AC2 and FL5-AC3, respectively, when using the full wavelength range. The respective values when using the centre part of the spectra were 4.2° and 6.0° for FL4-AC2 and AC3, and 3.7° and 2.9° for FL5-AC2 and AC3. Chi-squared values were 0.51 and 0.023 for FL4-AC2 and FL4-AC3, respectively, and 0.401 and 0.204 for FL5-AC2 and FL5-AC3, respectively, when using only the centre part of the spectra.

The RMS and relative RMS difference between in situ and image R_w were calculated for all 10 matchups and for the best five matchups separately, using both Chi-square and spectral angle metrics, with the centre of each spectrum taken into account (Figure 7). The spectral difference plot of image and in situ R_w (Figure 7a), and RMS in R_w (Figure 7c) show that the difference remained relatively stable over the whole wavelength range. This is also visible in the spectral plots of Figure 4. The spectral ratio of image and in situ R_w (Figure 7b) and RMS% plots (Figure 7d) show that, as the water-leaving reflectance signal becomes weaker at wavelengths longer than 700 nm, the uncertainty or error in matchups becomes larger. This effect is also visible in the standardized R_w matchup plots in Figure 6. The best five matchups in terms of RMS had ±0.002 accuracy in reflectance, or better than 15% in terms of RMS%, in the 450–750 nm wavelength range.

Matchup ST6-FL6 was disturbed by cloud shadow in image, and the magnitude of image-derived R_w deviated strongly from in situ R_w in the 450–750 nm wavelength range (Figure 4). When considering only spectral shape in the 450–750 nm wavelength range, the spectral angle value was 7.7°, comparable to other matchups. This is illustrated in the standardized spectra plotted in Figure 6f.

3.5. Comparison in Terms of Retrieved Chlorophyll-a

Chl-a concentrations were calculated from all measured image R_w spectra (Figure 8a), including locations under cloud shadow, and from all in situ station spectra, except ST5, as it included negative reflectance values. Chl-a concentration derived from in situ station spectra varied between 17.8 to 30.9 mg/m^3 and was, on average, 24.9 mg/m^3. Concentrations derived from the image spectra varied between 19.9–43.1 mg/m^3, with an average of 28.7 mg/m^3 and standard deviation of 3.8 mg/m^3. When looking only at the station data, concentrations varied between 25.9 and 35.3 mg/m^3 and the average was 31.0 mg/m^3. Measurements located under cloud shadows produced concentrations between 19.8 and 52.9 mg/m^3, with an average of 30.3 mg/m^3 and standard deviation of 7.2 mg/m^3.

A scatter plot comparing in situ and image-derived Chl-a concentrations is given in Figure 8b, including Chl-a derived from images using atmospheric correction strategies AC2 and AC3. The Chl-a difference values between nine matchups with AC1 are shown in Table 2. It can be seen from Figure 8b that image-derived Chl-a concentrations were consistently higher than concentrations derived from in situ R_w. Also, apart from Chl-a values from ST6 (17.8 mg/m^3), the concentrations are all of the same order, indicating that the water type of Loch Leven is relatively homogenous. The RMS and relative RMS% difference between in situ and image-derived Chl-a concentrations were 5.87 mg/m^3 and 28.3%, respectively, when using all nine matchups. When leaving out the matchups with ST6 that was measured during the most challenging illumination conditions, the RMS and relative RMS% improved to 4.4 mg/m^3 and 16.1%, respectively.

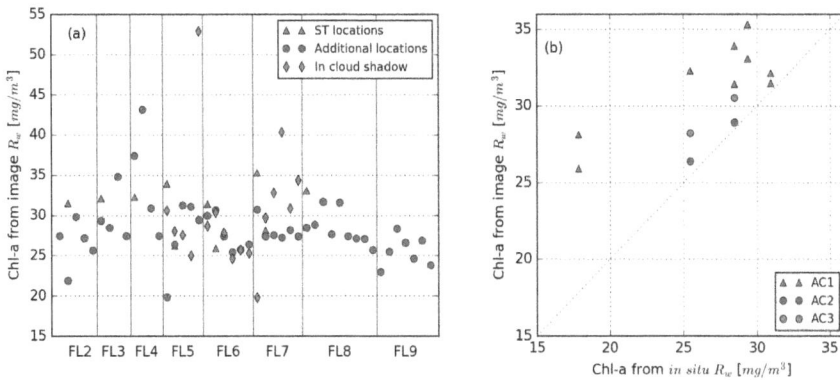

Figure 8. (**a**) Chl-a values in (mg/m^3) derived from all measured image R$_w$; (**b**) scatter plot between Chl-a values derived from in situ R$_w$ and image R$_w$ at station locations.

4. Discussion

4.1. Reflectance Quality of the Best Case Results

AisaFENIX R$_w$ spectra obtained after ATCOR4 atmospheric correction were realistic for Loch Leven and showed good correspondence with concurrent in situ R$_w$, both in terms of spectral shape and magnitude. Different atmospheric correction strategies produced differences, especially in the amplitude of the image R$_w$. The best reflectance accuracy was achieved with the strategy using image-driven retrieval of spatially varying water vapour and horizontal visibility (and subsequently AOT). Five matchups between atmospherically corrected image and in situ R$_w$ data were in excellent agreement in the wavelength range 450–750 nm, preserving features associated with Chl-a absorption. Taking into account only the spectral shape, the Spectral Angle metrics were better than 4.1° for the best five matchups. For comparison, spectral angles between 5.7° and 28.6° (0.1–0.5 rad) have been used as thresholds in Spectral Angle Mapper (SAM) classification, comparing image-derived R$_w$ to reference R$_w$ [51,52]. Errors in water constituent retrieval (e.g., Chl-a) would thus more likely be associated with the presence of systematic errors resulting in an offset of the image-derived spectra than with the correct retrieval of the spectral shape.

The accuracy of ATCOR over inland water bodies using in situ reflectance spectra has not previously been evaluated in literature. In general, the authors of ATCOR4 software estimate that accuracy of retrieved surface reflectance of ±2% for object reflectance <10% can be achieved [53]. Markelin et al. [34] achieved reflectance accuracy better than 10% when using ATCOR4 to correct hyperspectral AisaEAGLE data over land targets. Hunter et al. [26] used FLAASH atmospheric correction with AisaEAGLE imagery collected over Loch Leven and achieved a RMS% range from 26.4% to 77.7% in the 400–800 nm wavelength range, compared to in situ R$_{rs}$. Compared to these results, and taking into account the challenging illumination conditions during the campaign, the reflectance accuracy of better than 15% achieved here can be considered a very promising result.

Acquiring the aerosol properties and their distribution over the campaign area, either by in situ measurements or by image-based algorithms, is crucial for accurate atmospheric correction [20,30]. Differences of 0.01–0.07 between scene-averaged AOT given by ATCOR4 and in situ sunphotometer AOT measurements are in line with results obtained by Pflug et al. [54]. Using satellite imagery, they evaluated the accuracy of the aerosol retrieval in ATCOR and found the mean uncertainty ΔAOT550 nm to be 0.03 ± 0.02 for cloud free conditions and ΔAOT550 nm 0.04 ± 0.03 with inclusion of hazy and cloudy satellite imagery, compared to ground-based sun photometer measurements.

The image-based method used in this study to estimate sensor signal-to-noise ratio gives SNR representative for the whole sensor–atmosphere–dark target system. These values cannot be directly compared to the laboratory-measured SNR values from the sensor manufacturer. The SNR range of 450–800 determined over bright cloud in this study is nevertheless comparable to the laboratory peak SNR of 500–1000 documented by Specim. Measured over water, with average reflectance in the order of 0.02, the FENIX SNR range of 40–95 is similar to the range of 60–120 for the older AisaEAGLE sensor reported by Markelin et al. [34], which was, however, measured over a much brighter target, with average reflectance of 0.26. This indicates that the SNR of FENIX has improved considerably over its predecessor.

4.2. Chlorophyll a Retrieval

The retrieved Chl-a concentration range of 25–35 mg/m^3 is well within the range documented for Loch Leven [26,40]. Concentrations derived from the imagery overestimated Chl-a with respect to in situ R_w. Interestingly, the Chl-a concentrations derived from images corrected with AC2 and AC3 matched similarly or better with in situ R_w, compared to AC1 (Figure 8), even though the spectral match based on Chi-Square and spectral angle metrics was better with AC1. These contrasting results are not well understood and without further validation against Chl-a extracted from water samples we should not consider Chl-a derived from in situ R_w an absolute reference. Although speculative, minor spectral features that are still visible in FENIX-derived spectra and which coincide with wavebands used in the Chl-a retrieval algorithm, such as a small depression in the 700-nm region, may well disturb the results from the NIR-red band ratio algorithm. In addition, variable and at times adverse weather conditions will have added uncertainty to in situ measurements (see Section 4.3). Still, the achieved RMS of 5.87 mg/m^3 and RMS% 28.3% are well in line with Hunter et al. [26]. They evaluated several empirical and semi-analytical algorithms, including the one used in this study, to retrieve Chl-a from hyperspectral AisaEAGLE and CASI imagery collected over Loch Leven and Esthwaite Water, and obtained RMS values from 5.29 to 22.1 mg/m^3 and RMS% from 29.8 to 124%, respectively. Moses et al. [24] obtained similar RMS of 5.54 mg/m^3 when also using semi-analytical three-band Chl-a retrieval algorithm for AisaEAGLE imagery in combination with QUAC atmospheric correction. The Gons et al. [50] semi-analytical NIR-red band ratio algorithm is generally forgiving to spectrally neutral offsets, as long as none of the wavebands used are near-zero. Indeed, Chl-a retrieval from image R_w at locations under cloud shadow gave concentrations comparable to non-shaded locations.

4.3. Uncertainties Related to FENIX, In Situ R_w and Atmospheric Correction

Even though the results achieved in this study are promising, several factors related to the sensor, in situ measurements and atmospheric correction contribute uncertainty to the results. First, SNR values are extremely important for water constituent retrieval, because the very low signal from water causes variations in water quality to be lost in the noise of low SNR systems. The SNR of the AisaFENIX sensor over water is still low compared to dedicated ocean colour satellite instruments such as Sentinel-3 OLCI (SNR 200-1000). The image-based SNR evaluation method is somewhat uncertain in the sense that the standard deviation of each ROI includes small (unknown) contributions from variability in water-leaving radiance. Further, the combination of lower SNR of the FENIX and low signal from the water at wavelengths >750 nm makes comparison of airborne and in situ R_w in these regions challenging.

Low but non-zero image R_w in the order of 0.005, at wavelengths above 1000 nm, can indicate inaccuracies/uncertainties of atmospheric correction, as R_w is expected to be zero even in relatively turbid lakes, due to high absorption by pure water. Natural causes for non-zero R_w in this wavelength region include floating matter, very high turbidity [35], adjacency effects from the land surface, spray, foam (whitecaps), entrained bubbles or thin cloud. Prevailing weather conditions during the campaign, with wind speeds 1.4–6.2 m/s, raises the likelihood of wave effects, which adds uncertainty to both in situ and image R_w spectra.

The differences in the amplitude of the observed R_w spectra can be related to both the inaccuracies of the atmospheric correction and some variations in the relative composition of optically active substances in the water. Different atmospheric correction strategies did result in spectra of different amplitudes (Figure 4b,c) and adherent changes in Chl-a retrieval. One limitation of the used ATCOR4 atmospheric correction method applied to water objects is that it uses the Dark Dense Vegetation (DDV) approach to estimate best horizontal visibility, and subsequently AOT, for an image [28]. Each image must thus include suitable vegetated areas.

In this study, full-width-half-maximum (FWHM) values and centre wavelength for each channel, read from the image headers, and Gaussian spectral shapes, were used to create the FENIX spectral response functions in ATCOR4. This sensor model was then used in ATCOR4 processing. Sensor radiometric laboratory calibration performed by the manufacturer in January 2016 indicated that the used FWHM values may deviate, on average, 0.16 nm from the true FWHM in the visible-NIR range. Even though the error in FWHM can be considered marginal, it may explain some small spectral variation and spikes in the R_w spectra. Two narrow spikes around 700 nm where there should only be a single smooth peak, and a peak around 425 nm, likely result from a sensor calibration or software issue. Indeed, the peak disappeared from newer ATCOR4 runs (AC2 and AC3) performed with an updated sensor model and ATCOR4 version. Unfortunately, it was not possible to rerun all performed ATCOR runs. Thompson et al. [55] concluded that uncertainty in solar irradiance is a likely contributor to fine-scale spectral errors in R_w, especially in the 380–600 nm spectral range. Spectral filtering or polishing, a common operation performed with hyperspectral imagery, could be used to remove these small-scale variations in spectral shape and further improve results [56], if this were shown to consistently improve retrieval accuracy.

The varying weather conditions added considerable uncertainty to in situ R_w spectra. This can be seen in the low number of spectra that passed QC during measurements at several stations (Table 1) and the wide standard deviation of these spectra (Figure 4). The $E_d(PAR)$ and ratio of diffuse sky radiance in Figure 2 show that the illumination conditions were highly variable. Negative reflectance values (not a true target property) of in situ ST5 R_w spectra at wavelengths over 750 nm are a clear indication of these challenging conditions, despite $E_d(PAR)$ being relatively stable at this station. Similarly, the ST1 in situ spectrum (Figure 4a) shows particular signs of disturbance (possibly by waves, ship roll, or glint) in the NIR.

The number of matchup locations is always limited in this type of study, and prevents more thorough analysis of between-image reflectance variations. For example, ATCOR4 offers a sophisticated method called BREFCOR to balance the reflectance differences between neighbouring images [28], which could be included in future studies but ideally requires more ground reference measurements. Relatedly, flight lines and in situ measurements were obtained at different times, varying from 9 min to 4 h difference. Variable winds may cause horizontal differences in sediment and detrital resuspension, although in this study, temporal differences between matchups were not identified.

5. Conclusions

We presented the first results of using the hyperspectral AisaFENIX airborne sensor in water quality applications. The imagery was collected over a small inland water body under changing cloud cover, illustrating highly challenging, but not uncommon, conditions for atmospheric correction. The FENIX sensor showed improved performance in terms of signal-to-noise ratio compared to its predecessor, AisaEAGLE. Atmospheric correction performed with ATCOR4 using fully image-driven atmospheric parameters, and subsequent validation against boat-based in situ hyperspectral measurements of water-leaving reflectance, show reflectance retrieval accuracy of ±0.002, i.e., better than 15% error for the best five matchups between image and in situ R_w spectra. The reflectance accuracy was better when using fully image-driven atmospheric parameters compared to using fixed parameters based on in situ measurements. The spectral angle between image and in situ R_w spectra was better than 4.1° for the best cases and 10° or better for the most challenging

Remote Sens. **2017**, *9*, 2

cases in the spectral range of 450–750 nm. The test with semi-analytical band ratio-based algorithm to retrieve Chl-a concentrations from image R_w spectra gave realistic concentrations and the accuracy was comparable to other studies. The results showed that ATCOR4 can be successfully used for water applications without in situ atmospheric measurements, even in challenging illumination conditions. As ATCOR4 does not a priori assume a water reflectance, it is expected that the presented atmospheric correction method can be applied for images collected over other relatively turbid inland water types as well. The results also indicate that, with accurate atmospheric correction, it could be possible to retrieve reliable Chl-a concentrations at locations covered by cloud shadow. Accurate atmospheric correction is a necessity when airborne (manned or unmanned) optical measurements are performed in challenging illumination conditions, even under cloud cover. We may expect that future airborne monitoring of inland waters will build on advances in unmanned aerial system (UAS) technology and high resolution satellite sensors.

Acknowledgments: Airborne data were acquired by the Natural Environment Research Council Airborne Research Facility (NERC-ARF) as part of project GB12/03. UK Natural Environment Research Council (NERC) GloboLakes projects (Grant Number: NE/J024279/1 www.globolakes.ac.uk) supported the efforts of P. Hunter, V. Spyrakos, A.N. Tyler and S. Simis.

Author Contributions: Lauri Markelin wrote the manuscript with contributions from all authors. Lauri Markelin performed atmospheric correction and data analysis; Lauri Markelin, Stefan G. H. Simis and Steve Groom conceived and designed the analysis; Peter D. Hunter, Evangelos Spyrakos and Andrew N. Tyler planned the airborne campaign and performed the in situ measurements; Daniel Clewley assisted in atmospheric correction and contributed to data processing and analysis.

Conflicts of Interest: The authors declare no conflict of interest.

References

1. Convention on the Protection and Use of Transboundary Watercourses and International Lakes. Available online: http://www.unece.org/env/water/ (accessed on 28 June 2016).

2. The EU Water Framework Directive 2000/60/EC. Available online: http://ec.europa.eu/environment/water/water-framework/index_en.html (accessed on 28 June 2016).

3. United States Federal Water Pollution Control Act. Available online: https://www.epa.gov/laws-regulations/summary-clean-water-act (accessed on 28 June 2016).

4. Mouw, C.B.; Greb, S.; Aurin, D.; DiGiacomo, P.M.; Lee, Z.; Twardowski, M.; Binding, C.; Hu, C.; Ma, R.; Moore, T.; et al. Aquatic color radiometry remote sensing of coastal and inland waters: Challenges and recommendations for future satellite missions. *Remote Sens. Environ.* **2015**, *160*, 15–30. [CrossRef]

5. Malthus, T.J.; Mumby, P.J. Remote sensing of the coastal zone: An overview and priorities for future research. *Int. J. Remote Sens.* **2003**, *24*, 2805–2815. [CrossRef]

6. Kallio, K.; Kutser, T.; Hannonen, T.; Koponen, S.; Pulliainen, J.; Vepsäläinen, J.; Pyhälahti, T. Retrieval of water quality from airborne imaging spectrometry of various lake types in different seasons. *Sci. Total Environ.* **2001**, *268*, 59–77. [CrossRef]

7. Thiemann, S.; Kaufmann, H. Lake water quality monitoring using hyperspectral airborne data—A semiempirical multisensor and multitemporal approach for the Mecklenburg Lake District, Germany. *Remote Sens. Environ.* **2002**, *81*, 228–237. [CrossRef]

8. Hakvoort, H.; de Haan, J.; Jordans, R.; Vos, R.; Peters, S.; Rijkeboer, M. Towards airborne remote sensing of water quality in The Netherlands—Validation and error analysis. *ISPRS J. Photogramm. Remote Sens.* **2002**, *57*, 171–183. [CrossRef]

9. Olmanson, L.G.; Brezonik, P.L.; Bauer, M.E. Airborne hyperspectral remote sensing to assess spatial distribution of water quality characteristics in large rivers: The Mississippi River and its tributaries in Minnesota. *Remote Sens. Environ.* **2013**, *130*, 254–265. [CrossRef]

10. Palmer, S.C.J.; Kutser, T.; Hunter, P.D. Remote sensing of inland waters: Challenges, progress and future directions. *Remote Sens. Environ.* **2015**, *157*, 1–8. [CrossRef]

11. Kallio, K.; Koponen, S.; Pulliainen, J. Feasibility of airborne imaging spectrometry for lake monitoring—A case study of spatial chlorophyll *a* distribution in two meso-eutrophic lakes. *Int. J. Remote Sens.* **2003**, *24*, 3771–3790. [CrossRef]

12. Pinardi, M.; Fenocchi, A.; Giardino, C.; Sibilla, S.; Bartoli, M.; Bresciani, M. Assessing Potential Algal Blooms in a Shallow Fluvial Lake by Combining Hydrodynamic Modelling and Remote-Sensed Images. *Water* **2015**, *7*, 1921–1942. [CrossRef]

13. Anderson, K.; Gaston, K.J. Lightweight unmanned aerial vehicles will revolutionize spatial ecology. *Front. Ecol. Environ.* **2013**, *11*, 138–146. [CrossRef]

14. Pulliainen, J.; Kallio, K.; Eloheimo, K.; Koponen, S.; Servomaa, H.; Hannonen, T.; Tauriainen, S.; Hallikainen, M. A semi-operative approach to lake water quality retrieval from remote sensing data. *Sci. Total Environ.* **2001**, *268*, 79–93. [CrossRef]

15. Koponen, S. Lake water quality classification with airborne hyperspectral spectrometer and simulated MERIS data. *Remote Sens. Environ.* **2002**, *79*, 51–59. [CrossRef]

16. Gao, B.-C.; Montes, M.J.; Davis, C.O.; Goetz, A.F.H. Atmospheric correction algorithms for hyperspectral remote sensing data of land and ocean. *Remote Sens. Environ.* **2009**, *113*, S17–S24. [CrossRef]

17. Amin, R.; Lewis, D.; Gould, R.W.; Hou, W.; Lawson, A.; Ondrusek, M.; Arnone, R. Assessing the Application of Cloud-Shadow Atmospheric Correction Algorithm on HICO. *IEEE Trans. Geosci. Remote Sens.* **2014**, *52*, 2646–2653. [CrossRef]

18. Tyler, A.N.; Hunter, P.D.; Spyrakos, E.; Groom, S.; Constantinescu, A.M.; Kitchen, J. Developments in Earth observation for the assessment and monitoring of inland, transitional, coastal and shelf-sea waters. *Sci. Total Environ.* **2016**, *572*, 1307–1321. [CrossRef] [PubMed]

19. Hu, C.; Carder, K. Atmospheric correction for airborne sensors: Comment on a scheme used for CASI. *Remote Sens. Environ.* **2002**, *79*, 134–137. [CrossRef]

20. Zhang, M.; Hu, C.; English, D.; Carlson, P.; Muller-Karger, F.E.; Toro-Farmer, G.; Herwitz, S.R. Atmospheric Correction of AISA Measurements Over the Florida Keys Optically Shallow Waters: Challenges in Radiometric Calibration and Aerosol Selection. *IEEE J. Sel. Top. Appl. Earth Obs. Remote Sens.* **2015**, *8*, 4189–4196. [CrossRef]

21. Chavez, P.S. An improved dark-object subtraction technique for atmospheric scattering correction of multispectral data. *Remote Sens. Environ.* **1988**, *24*, 459–479. [CrossRef]

22. Hunter, P.D.; Gilvear, D.J.; Tyler, A.N.; Willby, N.J.; Kelly, A. Mapping macrophytic vegetation in shallow lakes using the Compact Airborne Spectrographic Imager (CASI). *Aquat. Conserv. Mar. Freshw. Ecosyst.* **2010**, *20*, 717–727. [CrossRef]

23. Bernstein, L.S. Quick atmospheric correction code: Algorithm description and recent upgrades. *Opt. Eng.* **2012**, *51*, 111719. [CrossRef]

24. Moses, W.J.; Gitelson, A.A.; Perk, R.L.; Gurlin, D.; Rundquist, D.C.; Leavitt, B.C.; Barrow, T.M.; Brakhage, P. Estimation of chlorophyll-a concentration in turbid productive waters using airborne hyperspectral data. *Water Res.* **2012**, *46*, 993–1004. [CrossRef] [PubMed]

25. Adler-Golden, S.M.; Berk, A.; Bernstein, L.S.; Richtsmeier, S.; Acharya, P.K.; Matthew, M.W.; Anderson, G.P.; Allred, C.; Jeong, L.; Chetwynd, J.H., Jr. FLAASH, a MODTRAN4 Atmospheric Correction Package for Hyperspectral Data Retrievals and Simulations. In Proceedings of the 7th Ann. JPL Airborne Earth Science Workshop, Pasadena, CA, USA, 12–16 January 1998; JPL Publication 97-21. pp. 9–14.

26. Hunter, P.D.; Tyler, A.N.; Carvalho, L.; Codd, G.A.; Maberly, S.C. Hyperspectral remote sensing of cyanobacterial pigments as indicators for cell populations and toxins in eutrophic lakes. *Remote Sens. Environ.* **2010**, *114*, 2705–2718. [CrossRef]

27. Miller, C.J. Performance assessment of ACORN atmospheric correction algorithm. *Proc. SPIE* **2002**, *4725*, 438–449.

28. Richter, R.; Schläpfer, D. *Atmospheric/Topographic Correction for Airborne Imagery (ATCOR-4 User Guide, Version 7.0.0, June 2015)*; ATCOR: Langeggweg, Switzerland, 2015.

29. Giardino, C.; Bresciani, M.; Valentini, E.; Gasperini, L.; Bolpagni, R.; Brando, V.E. Airborne hyperspectral data to assess suspended particulate matter and aquatic vegetation in a shallow and turbid lake. *Remote Sens. Environ.* **2015**, *157*, 48–57. [CrossRef]

30. Bassani, C.; Manzo, C.; Braga, F.; Bresciani, M.; Giardino, C.; Alberotanza, L. The impact of the microphysical properties of aerosol on the atmospheric correction of hyperspectral data in coastal waters. *Atmos. Meas. Tech.* **2015**, *8*, 1593–1604. [CrossRef]

31. Schläpfer, D.; Richter, R.; Hueni, A. Recent developments in operational atmospheric and radiometric correction of hyperspectral imagery. In Proceedings of the 6th EARSeL SIG IS Workshop, Tel Aviv, Israel, 16–18 March 2009.

32. Villa, P.; Bresciani, M.; Braga, F.; Bolpagni, R. Mapping aquatic vegetation through remote sensing data: A comparison of vegetation indices performances. In Proceedings of the 6th EARSeL Workshop on Remote Sensing of the Coastal Zone, Matera, Italy, 7 June 2013; pp. 10–15.

33. Knaeps, E.; Ruddick, K.G.; Doxaran, D.; Dogliotti, A.I.; Nechad, B.; Raymaekers, D.; Sterckx, S. A SWIR based algorithm to retrieve total suspended matter in extremely turbid waters. *Remote Sens. Environ.* **2015**, *168*, 66–79. [CrossRef]

34. Markelin, L.; Honkavaara, E.; Takala, T.; Pellikka, P. Calibration and validation of hyperspectral imagery using permanent test field. In Proceedings of the 5th IEEE Workshop on Hyperspectral Image and Signal Processing: Evolution in Remote Sensing (WHISPERS), Gainesville, FL, USA, 2013; pp. 1–4.

35. Knaeps, E.; Dogliotti, A.I.; Raymaekers, D.; Ruddick, K.; Sterckx, S. In situ evidence of non-zero reflectance in the OLCI 1020nm band for a turbid estuary. *Remote Sens. Environ.* **2012**, *120*, 133–144. [CrossRef]

36. Yule, Y.; Pullanagari, R.; Irwin, M.; McVeagh, P.; Kereszturi, G.; White, M.; Manning, M. Mapping nutrient concentration in pasture using hyperspectral imaging. *J. N. Z. Grassl.* **2015**, *77*, 47–50.

37. Pullanagari, R.R.; Kereszturi, G.; Yule, I.J. Mapping of macro and micro nutrients of mixed pastures using airborne AisaFENIX hyperspectral imagery. *ISPRS J. Photogramm. Remote Sens.* **2016**, *117*, 1–10. [CrossRef]

38. Kirby, R.P. The bathymetrical resurvey of Loch Leven, Kinross. *Geogr. J.* **1971**, *137*, 372–378. [CrossRef]

39. May, L.; Defew, L.H.; Bennion, H.; Kirika, A. Historical changes (1905–2005) in external phosphorus loads to Loch Leven, Scotland, UK. *Hydrobiologia* **2012**, *681*, 11–21. [CrossRef]

40. Carvalho, L.; Miller, C.; Spears, B.M.; Gunn, I.D.M.; Bennion, H.; Kirika, A.; May, L. Water quality of Loch Leven: Responses to enrichment, restoration and climate change. *Hydrobiologia* **2012**, *681*, 35–47. [CrossRef]

41. Tyler, A.N.; Hunter, P.D.; Carvalho, L.; Codd, G.A.; Elliott, J.A.; Ferguson, C.A.; Hanley, N.D.; Hopkins, D.W.; Maberly, S.C.; Mearns, K.J.; et al. Strategies for monitoring and managing mass populations of toxic cyanobacteria in recreational waters: A multi-interdisciplinary approach. *Environ. Health* **2009**. [CrossRef] [PubMed]

42. Warren, M.A.; Taylor, B.H.; Grant, M.G.; Shutler, J.D. Data processing of remotely sensed airborne hyperspectral data using the Airborne Processing Library (APL): Geocorrection algorithm descriptions and spatial accuracy assessment. *Comput. Geosci.* **2014**, *64*, 24–34. [CrossRef]

43. Gao, B.-C. An operational method for estimating signal to noise ratios from data acquired with imaging spectrometers. *Remote Sens. Environ.* **1993**, *43*, 23–33. [CrossRef]

44. Berk, A.; Anderson, G.P.; Acharya, P.K.; Bernstein, L.S.; Muratov, L.; Lee, J.; Fox, M.; Adler-Golden, S.M.; Chetwynd, J.H., Jr.; Hoke, M.L.; et al. MODTRAN5: 2006 update. *Proc. SPIE* **2006**. [CrossRef]

45. Mobley, C. *Light and Water: Radiative Transfer in Natural Waters*; Academic Press: Cambridge, MA, USA, 1994.

46. Simis, S.G.H.; Olsson, J. Unattended processing of shipborne hyperspectral reflectance measurements. *Remote Sens. Environ.* **2013**, *135*, 202–212. [CrossRef]

47. Kruse, F.A.; Lefkoff, A.B.; Boardman, J.W.; Heidebrecht, K.B.; Shapiro, A.T.; Barloon, P.J.; Goetz, A.F.H. The spectral image processing system (SIPS)—Interactive visualization and analysis of imaging spectrometer data. *Remote Sens. Environ.* **1993**, *44*, 145–163. [CrossRef]

48. Shanmugam, S.; SrinivasaPerumal, P. Spectral matching approaches in hyperspectral image processing. *Int. J. Remote Sens.* **2014**, *35*, 8217–8251. [CrossRef]

49. Markelin, L.; Honkavaara, E.; Schläpfer, D.; Bovet, S.; Korpela, I. Assessment of Radiometric Correction Methods for ADS40 Imagery. *Photogramm. Fernerkund. Geoinf.* **2012**, *2012*, 251–266. [CrossRef] [PubMed]

50. Gons, H.J.; Rijkeboer, M.; Ruddick, K. Effect of a waveband shift on chlorophyll retrieval from MERIS imagery of inland and coastal waters. *J. Plankton Res.* **2005**, *27*, 125–127. [CrossRef]

51. Kutser, T. Quantitative detection of chlorophyll in cyanobacterial blooms by satellite remote sensing. *Limnol. Oceanogr.* **2004**, *49*, 2179–2189. [CrossRef]

52. Kutser, T.; Paavel, B.; Verpoorter, C.; Kauer, T.; Vahtmäe, E. Remote sensing of water quality in optically complex lakes. *ISPRS Int. Arch. Photogramm. Remote Sens. Spat. Inf. Sci.* **2012**, *XXXIX-B8*, 165–169. [CrossRef]

53. Richter, R.; Schläpfer, D. Geo-atmospheric processing of airborne imaging spectrometry data. Part 2: Atmospheric/topographic correction. *Int. J. Remote Sens.* **2002**, *23*, 2631–2649. [CrossRef]

54. Pflug, B.; Main-Knorn, M.; Makarau, A.; Richter, R. Validation of aerosol estimation in atmospheric correction algorithm ATCOR. *ISPRS Int. Arch. Photogramm. Remote Sens. Spat. Inf. Sci.* **2015**, *XL-7/W3*, 677–683. [CrossRef]
55. Thompson, D.R.; Seidel, F.C.; Gao, B.C.; Gierach, M.M.; Green, R.O.; Kudela, R.M.; Mouroulis, P. Optimizing irradiance estimates for coastal and inland water imaging spectroscopy. *Geophys. Res. Lett.* **2015**, *42*, 4116–4123. [CrossRef]
56. Schläpfer, D.; Richter, R. Spectral Polishing of High Resolution Imaging Spectroscopy Data. In Proceedings of the 7th SIG-IS Workshop on Imaging Spectroscopy, Edinburgh, UK, 11–13 April 2011; pp. 1–7.

remote sensing

MDPI

Article

Atmospheric Corrections and Multi-Conditional Algorithm for Multi-Sensor Remote Sensing of Suspended Particulate Matter in Low-to-High Turbidity Levels Coastal Waters

Stéfani Novoa [1], David Doxaran [1,*], Anouck Ody [1], Quinten Vanhellemont [2], Virginie Lafon [3], Bertrand Lubac [4] and Pierre Gernez [5]

[1] Laboratoire d'Océanographie de Villefranche, UMR7093 CNRS/UPMC, 181 Chemin du Lazaret, 06230 Villefranche-sur-Mer, France; snovoa@gmail.com (S.N.); Anouck.Ody@obs-vlfr.fr (A.O.)
[2] Royal Belgian Institute of Natural Sciences, Brussels 1000, Belgium; quinten.vanhellemont@naturalsciences.be
[3] GEO-Transfert, UMR 5805 Environnements et Paléo-environnements Océaniques et Continentaux (EPOC), Université de Bordeaux, Allée Geoffroy Saint-Hilaire, 33615 Pessac, France; virginie.lafon@i-sea.fr
[4] UMR CNRS 5805 EPOC, OASU, Université de Bordeaux, site de Talence, Bâtiment B18, Allée Geoffroy Saint-Hilaire, 33615 Bordeaux Cedex, France; b.lubac@epoc.u-bordeaux1.fr
[5] Mer Molécules Santé (EA 2160 MMS), Université de Nantes, 2 rue de la Houssinière BP 92208, 44322 Nantes Cedex 3, France; Pierre.Gernez@univ-nantes.fr
* Correspondence: doxaran@obs-vlfr.fr; Tel.: +33-4-9376-3724

Academic Editors: Yunlin Zhang, Claudia Giardino, Linhai Li, Xiaofeng Li and Prasad S. Thenkabail
Received: 20 September 2016; Accepted: 3 January 2017; Published: 12 January 2017

Abstract: The accurate measurement of suspended particulate matter (SPM) concentrations in coastal waters is of crucial importance for ecosystem studies, sediment transport monitoring, and assessment of anthropogenic impacts in the coastal ocean. Ocean color remote sensing is an efficient tool to monitor SPM spatio-temporal variability in coastal waters. However, near-shore satellite images are complex to correct for atmospheric effects due to the proximity of land and to the high level of reflectance caused by high SPM concentrations in the visible and near-infrared spectral regions. The water reflectance signal (ρ_w) tends to saturate at short visible wavelengths when the SPM concentration increases. Using a comprehensive dataset of high-resolution satellite imagery and in situ SPM and water reflectance data, this study presents (i) an assessment of existing atmospheric correction (AC) algorithms developed for turbid coastal waters; and (ii) a switching method that automatically selects the most sensitive SPM vs. ρ_w relationship, to avoid saturation effects when computing the SPM concentration. The approach is applied to satellite data acquired by three medium-high spatial resolution sensors (Landsat-8/Operational Land Imager, National Polar-Orbiting Partnership/Visible Infrared Imaging Radiometer Suite and Aqua/Moderate Resolution Imaging Spectrometer) to map the SPM concentration in some of the most turbid areas of the European coastal ocean, namely the Gironde and Loire estuaries as well as Bourgneuf Bay on the French Atlantic coast. For all three sensors, AC methods based on the use of short-wave infrared (SWIR) spectral bands were tested, and the consistency of the retrieved water reflectance was examined along transects from low- to high-turbidity waters. For OLI data, we also compared a SWIR-based AC (ACOLITE) with a method based on multi-temporal analyses of atmospheric constituents (MACCS). For the selected scenes, the ACOLITE-MACCS difference was lower than 7%. Despite some inaccuracies in ρ_w retrieval, we demonstrate that the SPM concentration can be reliably estimated using OLI, MODIS and VIIRS, regardless of their differences in spatial and spectral resolutions. Match-ups between the OLI-derived SPM concentration and autonomous field measurements from the Loire and Gironde estuaries' monitoring networks provided satisfactory results. The multi-sensor approach together with the multi-conditional algorithm presented here can be applied to the latest generation of ocean color sensors (namely Sentinel2/MSI and Sentinel3/OLCI) to study SPM dynamics in the coastal ocean at higher spatial and temporal resolutions.

Keywords: remote sensing; suspended particulate matter; coastal waters; river plumes; multi-conditional algorithm

---

1. Introduction

The quality of coastal and estuarine waters is increasingly under threat by the intensification of anthropogenic activities. For that reason, the European Union Marine Strategy Framework Directive (MSFD, 2008/56/EC) and Water Framework Directive (WFD, 2000/60/EC and amendments) require member states to monitor the quality of the marine environment and to achieve and maintain a good environmental status of all marine waters by 2020. The directives require member states to assess the ecological quality status of water bodies, based on the status of several elements, including water transparency. Rivers serve as the main channel for the delivery of significant amounts of dissolved and particulate materials from terrestrial environments to the ocean. Along with freshwater, they discharge suspended particulate matter (SPM) that modifies the color and transparency of the water. In addition, SPM is associated with metallic contaminants and bacteria that affect water quality. Hence, monitoring the spatio-temporal distribution of SPM in estuarine and coastal waters is of particular importance, not only to assess water transparency, but also to evaluate the impacts of human activities (e.g., transport of pollutants, dams, offshore wind farms, sand extraction, watershed management) and to study sediment transport dynamics.

SPM field measurements are time-consuming, expensive and specific to a time and/or geographical location, and therefore do not always accurately represent the temporal and spatial dynamics of river, estuarine or coastal systems. Ocean color remote sensing onboard satellite platforms can be very useful to complement field measurements and monitor surface SPM transport in natural waters [1–14]. Most satellite-borne sensors provide a spectral resolution covering the visible and near-infrared (NIR) spectral regions required for atmospheric corrections of satellite data, and for the estimation of biogeochemical material such as SPM (e.g., [15,16]). Sensors such as SPOT (Satellite Pour l'Observation de la Terre) and Landsat-8/OLI (Operational Land Imager), designed for land applications, provide high-spatial-resolution imagery, and their potential for mapping the concentration of SPM in highly turbid waters has been demonstrated [17–19]. These high-resolution sensors combined with high-temporal-resolution satellite data [20,21] have proved to provide valuable information regarding SPM dynamics.

An important issue regarding satellite remote sensing in coastal and estuarine areas is atmospheric correction (AC) failures when applying standard algorithms designed for open ocean methods. This is caused by the presence of high water turbidity and also by the proximity to land. To achieve an accurate atmospheric correction, the top-of-the-atmosphere signal recorded by satellite sensors is separated into marine, gaseous and aerosol contributions. Typical open ocean atmospheric correction methods assume the marine signal to be zero in the near-infrared (NIR) bands due to very high light absorption by pure water and very low light backscattering by suspended particles [22]. The signal in the NIR bands is used to determine an aerosol model, which is then used to extrapolate the aerosol reflectance to visible bands. In turbid waters, the contribution of light backscattering by particles is no longer negligible in the NIR region compared to light absorption. This results in a non-negligible water reflectance signal, an overestimation of the aerosol reflectance, and underestimated or negative water reflectance values in visible bands [23]. Studies have focused on two approaches to develop atmospheric corrections over turbid waters: one is to model the marine contribution in the NIR bands [23,24] and the other involves the use of short-wave infrared (SWIR) bands, where the water signal can be assumed to be zero even in turbid coastal waters [25].

For low to moderately turbid waters, a good correlation is found between the SPM concentration and water reflectance (ρ_w) in the green and red parts of the spectrum (refer to [16]). The corresponding wavebands of wide-swath ocean color instruments, such as the Orbview-2/Sea-viewing Wide Field-of-view Sensor (SeaWiFS), the Aqua/Moderate Resolution Spectrometer (MODIS), the ENVISAT/Medium Spectral Resolution imaging spectrometer (MERIS), the Landsat/Enhanced Thematic Mapper Plus (ETM+) and OLI, and the Visible Infrared Imaging Radiometer Suite (VIIRS), have therefore been successfully used to map SPM in coastal waters for concentrations below ~60 g·m^{-3} [6,13,26–28]. In highly turbid waters (SPM higher than ~60 g·m^{-3}), a saturation of the water reflectance in the green and red bands is usually observed, so a NIR band should be considered to establish relationships with SPM [4,18,29,30]. There are three main types of algorithms commonly used to derive SPM concentration from water reflectance: (1) empirical, (2) semi-analytical and (3) analytical algorithms. Empirical single-band and band-ratio models have been commonly used in coastal and estuarine areas [6,31]. These types of models are dependent on SPM and reflectance ranges, and require calibration with regional measurements. Semi-analytical or analytical models are based on the inherent optical properties (IOPs) and provide a more global application [16,32,33]. However, they can be limited by the validity and accuracy of the hypotheses chosen to model the IOPs. Hence, provided the large choice of SPM algorithms, it is difficult to select one model that will provide accurate SPM concentration retrieval from low- to high-turbidity waters, limiting the study of SPM dynamics over large coastal areas. For that reason, some studies have focused on multi-conditional algorithm schemes composed of several SPM models, as they have been shown to provide a more effective and accurate estimation of SPM over a wide range of turbid waters [34–37]. The difficulty resides in the selection of the proxies and the limiting bounds for each model. Some studies have used ranges of SPM concentration as switching thresholds [35] and others have used reflectance values [36], but the bounds are generally selected through trial and error.

The main objective of this study is to determine the boundaries for switching between different SPM models, based on band comparisons from field water reflectance measurements, then apply this switching algorithm to ocean color satellite data to derive SPM across low- to high-turbidity waters. Since atmospheric correction is a major issue in coastal areas, different atmospheric correction methods are tested for several study areas and sensors, and the most appropriate one is selected. To achieve these aims, this study will focus on three objectives.

(1) To compare atmospheric correction algorithms for OLI, VIIRS and MODIS satellite data over two study areas covering low- to high-turbidity waters;
(2) To develop a reliable multi-conditional algorithm to retrieve SPM from satellite imagery over a wide range of turbidity values and apply it to satellite data;
(3) To inter-compare multi-sensor satellite products (ρ_w and SPM) over turbid coastal, estuarine and river waters.

Two case studies are considered: the Loire Estuary, with the adjacent Bourgneuf Bay, and the Gironde Estuary. For both areas, high quality ρ_w and SPM measurements are available. The paper is organized as follows: first, the methodology for the application of different SPM models over the study areas is developed. Second, a comparison between different atmospheric corrections is presented. Finally, the developed multi-conditional algorithm is applied to atmospherically corrected imagery from multiple satellite sensors.

2. Materials and Methods

2.1. Study Areas

This study covers two areas with a wide range of SPM concentration from low to highly turbid waters (Figure 1). The Gironde Estuary, located in South Western France, is one of the largest estuaries in Europe (length of 90 km and width 3–11 km). It is formed by the confluence of the Garonne and Dordogne rivers. These rivers' watersheds represent 57,000 km² and 24,000 km², and they supply respectively 65% and 35% freshwater inputs into the Estuary. The Garonne's freshwater discharge ranges from less than 100 m³·s⁻¹ to more than 4000 m³·s⁻¹, while the Dordogne discharge fluctuates between 200 and 1500 m³·s⁻¹ [38]. The Gironde's flow rate averages 1100 m³·s⁻¹ and its morphology is typical of a macro-tidal estuary (tidal ranges from 2 to 5 m) impacted by waves. It presents a well-developed turbidity maximum zone formed from tidal asymmetry and density residual circulation [39]. The SPM concentration within surface waters range from about 1 to 50 g·m⁻³ in the plume [40] and from 50 to approximately 3000 g·m⁻³ in the estuary [41,42].

The Loire is the largest river in France: it is 1012 km long and has a watershed area of 117,000 km². Its flow rate ranges between 300 m³·s⁻¹ during the summer droughts and 4000 m³·s⁻¹ during winter floods. The Loire Estuary is 100 km long and has a macro-tidal regime, with a 4 m mean tidal amplitude. It is characterized by high SPM concentration variations, ranging from 50 to more than 1000 g·m⁻³ within surface waters. South from the Loire Estuary, Bourgneuf Bay is a macro-tidal bay with a tidal range between 2 and 6 m. The bay has an area of 340 km², of which 100 km² are intertidal area mostly occupied by mudflats. Due to tidal re-suspension, mudflat and adjacent waters are highly turbid. Bourgneuf Bay is an important oyster-farming site, but in some sectors high SPM concentration may have a negative impact on oyster aquaculture [43].

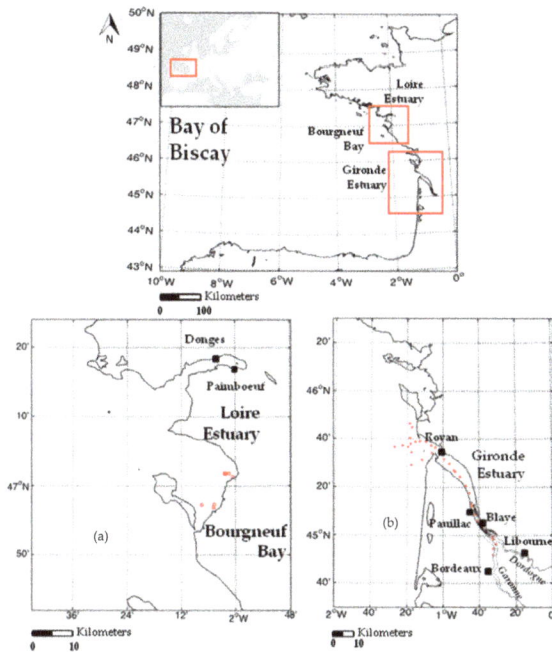

Figure 1. Maps of the study areas: Bourgneuf Bay and Loire Estuary (**a**) and Gironde Estuary and plume area (**b**). Red squares show the location of the in situ measurements performed during the optical cruises. Black squares show the location of the monitoring stations used for match-ups between satellite and in situ data.

2.2. Multi-Conditional SPM Algorithm Development

2.2.1. In Situ SPM and Reflectance Data

Field measurements used for the calibration of SPM models and multi-conditional algorithm were carried out during four bio-optical cruises (from April 2012 to July 2014) in the two selected test sites: the SeaSWIR (2012, 2013) and Rivercolor (2014) surveys in the Gironde Estuary area and the Gigassat (2013) survey in Bourgneuf Bay. Additionally, several measurements conducted in April 2016 in the Bourgneuf-Loire area were used to increase the number of match-ups between satellite and in situ data.

At each station, hyperspectral reflectance measurements were carried out using TriOS-RAMSES radiometers in the same way as the methodology described in [13]. The protocol described in [44] based on the NASA (National Aeronautics and Space Administration) protocols [45] to compute the remote sensing reflectance (R_{rs}, sr^{-1}). R_{rs} spectra of five successive measurements under stable illumination (i.e., downwelling irradiance variations between two measurements lower than 15%) and differing less than 25% from the median of all the spectra, were selected and averaged. A total of 67 R_{rs} spectra were finally selected for the Gironde Estuary and 29 for Bourgneuf Bay.

Simultaneously with the reflectance measurements, water samples were collected with a bucket at about 0.5 m depth. They were directly filtered with pre-weighed Whatman GF/F filters to determine the SPM concentration with the gravimetric method procedure described in [46], based on [47]. Three SPM measurement replicates were conducted per station, and the standard deviation obtained from those measurements were used as the uncertainty for the SPM concentration (error bars in figures). Water turbidity (measured in nephelometric turbidity units, NTU) was measured for most stations using a Hach Portable turbidity meter, following the protocol by [32] SPM and turbidity measurement ranges for each location are shown in Table 1. Three replicate measurements of turbidity were conducted per station to estimate the measurement uncertainties. The SPM vs. turbidity relationship for the Gironde Estuary was established using measurements undertaken during the SeaSWIR surveys at the Pauillac station: SPM (g·m^{-3}) = 0.88 × Turbidity (NTU).

Table 1. The distribution of SPM concentration (g·m^{-3}) and turbidity (NTU) values (field measurements) used for the calibration of the models.

Location	SPM (g·m^{-3})				Turbidity (NTU)			
	Mean	Standard Deviation	Maximum	Minimum	Mean	Standard Deviation	Maximum	Minimum
Gironde	347.1	372.7	1579.1	2.6	310.0	24.19	2045.9	1.5
Bourgneuf	162.4	90.4	340.6	17.8	100.6	78.47	301.3	12.7

2.2.2. SPM Models

The sets of hyperspectral R_{rs} in situ measurements acquired in the Gironde area were convoluted to the relative spectral response function of the green, red and NIR OLI bands (5, 4, 3), VIIRS bands (M4, M5, M7), and MODIS bands (B4, B1, B2) as explained in [16] to derive the band-weighted reflectance values. The same procedure was completed for the in situ measurements collected in Bourgneuf Bay. The resulting R_{rs} values were then expressed as dimensionless water-leaving reflectance ρ_w ($R_{rs} \times \pi$) values, hereinafter referred as ρ.

Figure 2 shows typical in situ measurements of ρ spectra and corresponding SPM concentration. For concentration between 2.6 and 10.5 g·m^{-3}, the ρ between 400 and 600 nm increases rapidly. From the examples shown in Figure 2, it can be observed that red reflectance is more sensitive than green reflectance to concentration changes between 10.5 and 119 g·m^{-3}. For SPM above 119 g·m^{-3}, ρ in the NIR is most sensitive to concentration changes. This implies that models based on the visible bands are not effective in discriminating SPM in highly turbid waters as demonstrated by previous studies [18,31].

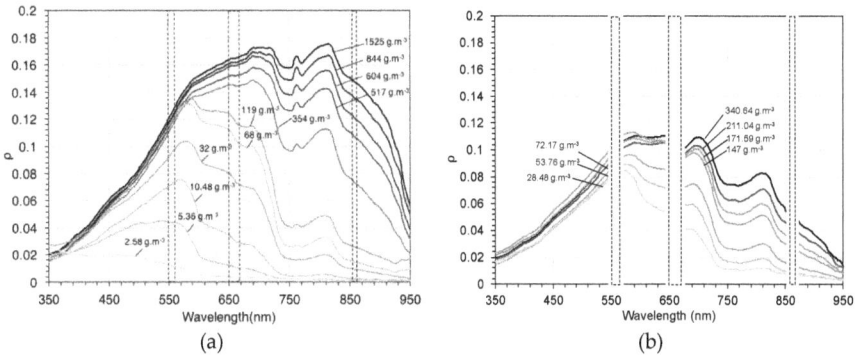

Figure 2. Selected water reflectance spectra ($\rho = R_{rs} \times \pi$) for different SPM concentration (g·m^{-3}) measured in the Gironde Estuary (**a**) and Bourgneuf Bay (**b**). Vertical bars locate the green, red and NIR bands of the considered satellite sensors.

The relationships between ρ (convoluted to OLI bands) and SPM concentration are presented in Figure 3. Analogous relationships were obtained for both MODIS and VIIRS convoluted bands, and are not shown here. Figure 3a shows a linear relationship between ρ in the green band and SPM concentration lower than 10 g·m^{-3}. Above 10 g·m^{-3}, a saturation of ρ is observed in this band. A green band relationship was not established for the Bourgneuf dataset, as low concentration measurements were not available, so the green band relationship obtained for the Gironde was also applied for Bourgneuf Bay.

Water reflectance in the red band is highly sensitive to variations of SPM concentration lower than 50 g·m^{-3} and presents a good linear correlation ($r^2 = 0.89$). Above 50 g·m^{-3}, a saturation of ρ is observed. The NIR band is less sensitive to SPM concentration below 50 g·m^{-3}, however it presents a very good fit above this limit by means of a polynomial regression ($r^2 = 0.97$ see Table 2). Figure 3b presents the sensitivity differences between the three bands (green, red, NIR) to SPM concentration for the ~0–50 g·m^{-3} range. There is a sharper increase in reflectance for the green band compared to the red for concentration below ~10 g·m^{-3}, and a sharper increase in red band reflectance compared to the NIR bands for concentration below 50 g·m^{-3}. This figure also shows the saturation of the green band above ~10 g·m^{-3}, so the r^2 was computed for the values below that concentration.

Several empirical models using single bands and NIR-red band ratios were considered in this study. The semi-analytical model developed by [16] was also re-calibrated with in situ datasets. Table A2 in the appendix shows all the algorithms tested. The performance of each model was assessed using the coefficient of determination (r^2) and the normalized root mean square error (NRMSE, in %) calculated as follows:

$$NRMSE\ (\%) = \frac{\sqrt{\frac{\sum_{i=1}^{N}\left(X_{p,i} - X_{obs,\ i}\right)^2}{N}}}{X_{obs,\ max} - X_{obs,\ min}} \times 100 \tag{1}$$

where x_p and x_{obs} are respectively the model-derived and field-measured SPM concentration, in g·m^{-3}.

In the case of the Gironde Estuary, the dataset ($n = 67$) was divided into two sets, one for calibration ($n = 34$) and one validation ($n = 33$). The NRMSE was computed for the SPM provided by the models with respect to the validation dataset. The best fits and minimum errors were obtained with the empirically-derived polynomial (second order) regression for the NIR band and linear regressions for the red and green bands (Table 2). In the case of Bourgneuf Bay, the best fits were obtained for the NIR band and the semi-analytical equation developed by [16]. Due to the low number of measurements, the Bourgneuf dataset was not separated into calibration and validation sets. Hence, the percent NRMSE (Equation (1)) in this case represents the deviation of the random component within the data. Table 2 summarizes the equations with the best fits for each area and for different SPM ranges. These ranges were established by testing the extent to which the equations predicted accurately the actual value estimated in situ by gravimetry. The equation with the best fits for MODIS and VIIRS bands are shown in the appendix (Table A1).

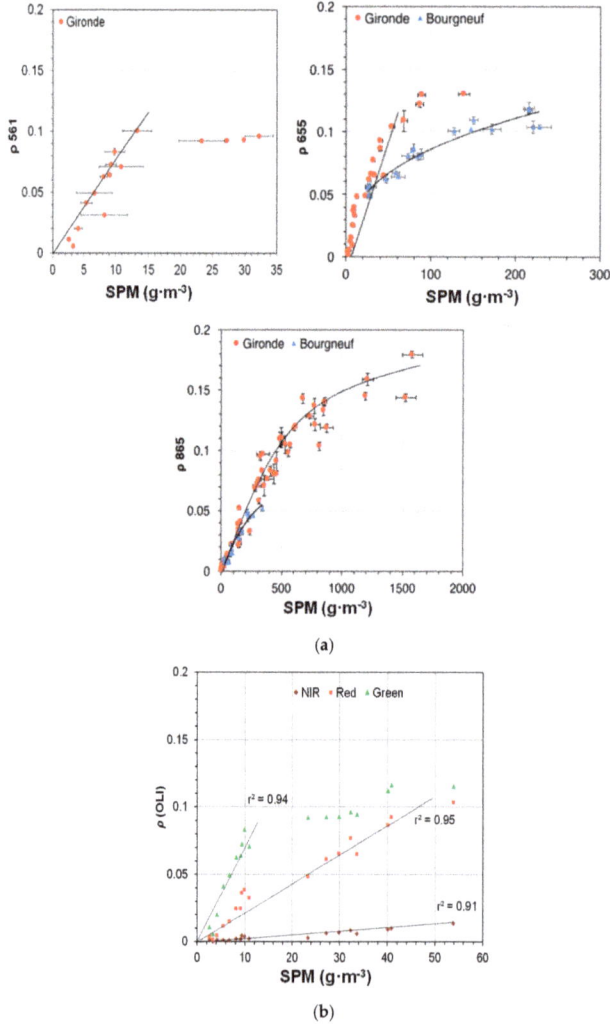

(a)

(b)

Figure 3. Scatter plots showing the comparison between SPM concentration and water reflectance convoluted for OLI bands 561, 655 and 865 nm measured in situ in (**a**) the Gironde Estuary (SeaSWIR 2012–2013, and Rivercolor 2014) and Bourgneuf Bay (2013). (**b**) In situ ρ weighted by sensitivity of red, green and NIR OLI spectral bands vs. in situ SPM concentration for the 0–50 $g \cdot m^{-3}$ range, measured during the two field campaigns in the Gironde Estuary.

Table 2. From the reflectance vs. SPM relationships (Figure 3), summary of the best SPM models for L8/OLI green, red and NIR reflectance bands, for the Gironde and Bourgneuf-Loire areas. The goodness of fit (r^2) and normalized relative root mean square error (NRMSE) are indicated. The most appropriate SPM model (or a combination of two models) is then selected using a radiometric switching criterion (Table 3). Corresponding results for NPP/VIIRS and AQUA/MODIS were calculated but not shown here. These can be requested from the authors; the dataset ($n = 67$) of the Gironde Estuary was divided into two sets, one for calibration ($n = 34$) and one for validation ($n = 33$); the fit and error were computed with respect to the validation set. The Bourgneuf dataset was not separated into calibration and validation sets, so the results represent the deviation of the random component within the data.

Best SPM Model Gironde	Equation	r^2	NRMSE (%)
	Gironde		
Linear green	$130.1 \times \rho\,561$	0.81	16.41
Linear red	$531.5 \times \rho\,655$	0.89	7.23
Polynomial NIR	$37{,}150 \times \rho\,865^2 + 1751 \times \rho\,865$	0.97	9.11
	Bourgneuf-Loire		
Nechad et al. (2010) [16] red (recalibrated)	$\dfrac{477 \times \rho\,655}{1 - \rho\,655/0.1686}$	0.82	18.22
Nechad et al. (2010) [16] NIR (recalibrated)	$\dfrac{4302 \times \rho\,865}{1 - \rho\,865/0.2115}$	0.93	7.81

2.2.3. Algorithm Bounds Selection

The green-to-red and red-to-NIR switching ρ values, S, were selected based on the saturation of the most sensitive bands. The selection was completed by means of band comparison from field water reflectance measurements: ρ (green) vs. ρ (red) and ρ (red) vs. ρ (NIR). The data points were modelled using a logarithmic regression curve. This curve starts as linear for the smaller reflectance values, but bends at the point where the saturation of the most sensitive band starts (see Figure 4). The actual value of this saturation point was computed as the first derivative of the regression curve (i.e., the slope or tangent) is equal to 1, as this is the middle point between a completely horizontal (complete saturation) and a completely vertical line.

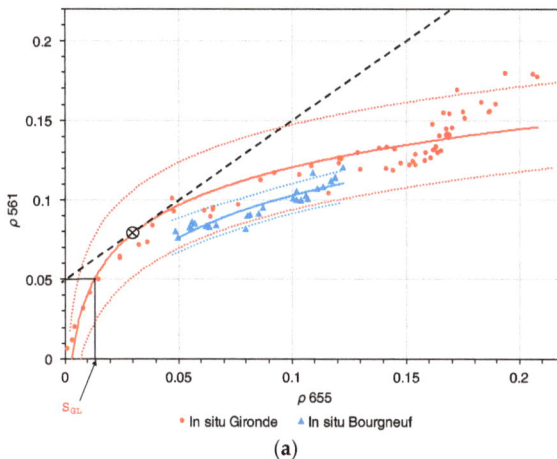

(a)

Figure 4. *Cont.*

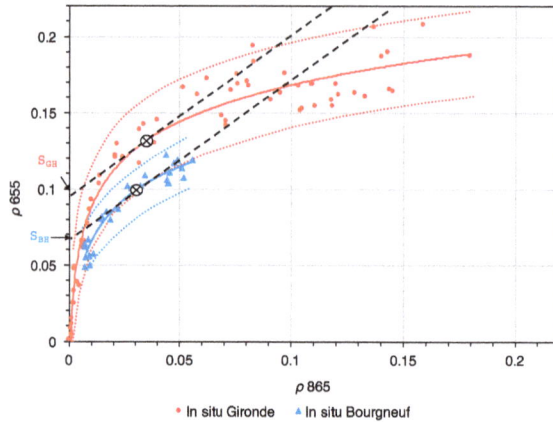

(b)

Figure 4. Scatterplots of reflectance ($\rho = R_{rs} \times \pi$) at 865, 655 and 561 nm for the in situ measurements. (a) Corresponds to the ρ 561-ρ 655 compartive plot and (b) to the ρ 655-ρ 865 comparative plot. The red circles and the blue triangles represent the in situ reflectance values measured in the Gironde Estuary and Bourgneuf Bay, respectively. The solid red and blue lines correspond to the logarithmic regression and the 95% confidence levels for each regression are represented as dashed lines. The circled black crosses ($\otimes$) correspond to the point where the tangent line on the regression curve has a slope = 1 and the black dashed line is the tangent at that point. At the intersection with the y axis, the switching points for each region are indicated, S_{GH} (high switching value for the Gironde area) and S_{BH} (high switching value for the Bourgneuf Bay area). The lower bound switching value is derived from the red-green band regression and expressed as S_{GL}.

Then, the S or switching value was defined as the point of intersection between the tangent line on the regression curve with a slope = 1 (i.e., the saturation point) and the y axis using:

$$\frac{y_0 - y_{sat}}{x_0 - x_{sat}} = 1; \; y_0 = S = y_{sat} - x_{sat} \tag{2}$$

where y_0 corresponds to the switching point S on the y axis (where $x_0 = 0$), and x_{sat} and y_{sat} are the coordinates of the saturation point.

This S value is selected as the transition value to the next SPM vs. ρ equation. Figure 4 shows the regressions between the green-to-red and red-to-NIR bands, based on the in situ reflectance measurements carried out in each region and the switching S values for each case, the S Gironde High ($S_{GH} = 0.13$), the S Bourgneuf High ($S_{BH} = 0.1$) and S Gironde Low ($S_{GL} = 0.03$) (see Table 3 for values). The interval bounds are based on the red ρ, as this is the intermediate band between the green and NIR bands.

The equations were weighted to ensure a smooth transition between the different SPM models for intermediate SPM values. The smoothing bounds ($S_{GL95}{}^-$, $S_{GL95}{}^+$, $S_{GH95}{}^-$, $S_{BH95}{}^+$, $S_{BH95}{}^-$) were derived from the 95% confidence levels (prediction bounds, see dotted red and blue lines on Figure 4) of the regression curve following the same procedure as for the S value calculation. Then, the smooth transition between SPM models using different bands was completed using these smoothing boundary values for the following weighting equations.

Weighted green-red equation:

$$\text{SPM}_{green\text{-}red} = \alpha \times \text{SPM}_{green} + \beta \times \text{SPM}_{red} \tag{3}$$

where

$$\alpha = \ln\left(\frac{S_{GL95+}}{\rho\,655}\right) \div \ln\left(\frac{S_{GL95+}}{S_{GL95-}}\right) \text{ and } \beta = \ln\left(\frac{\rho\,655}{S_{GL95-}}\right) \div \ln\left(\frac{S_{GL95+}}{S_{GL95-}}\right) \tag{4}$$

Weighted red-NIR equation:

$$SPM_{red\text{-}NIR} = \alpha\,SPM\,red + \beta\,SPM\,NIR \tag{5}$$

where

$$\alpha = \ln\left(\frac{S_{GH95+}}{\rho\,655}\right) \div \ln\left(\frac{S_{GH95+}}{S_{GH95-}}\right) \text{ and } \beta = \ln\left(\frac{\rho\,655}{S_{GH95-}}\right) \div \ln\left(\frac{S_{GH95+}}{S_{GH95-}}\right) \tag{6}$$

The transition and smoothing intervals selected for each region and each band are summarized in Table 3. Initially x_{sat} and y_{sat} were selected as switching points (Equation (5)), but the smoothing intervals became too narrow (when selecting the point of the 95% confidence intervals) so the y_0 was selected, and showed to provide good results. For example, if the y_{sat} would have been selected as smoothing bounds (S_{GL95}^-; S_{GL95}^+) the range would have been ρ red = (0.030–0.032) instead of (0.007; 0.016) (see Table 3).

Table 3. Radiometric switching bounds used to select the most appropriate SPM model based on the water reflectance value in the red ($\rho\,655$) for the Gironde and Bourgneuf-Loire. The switching bounds were automatically computed from the green-red and red-NIR reflectance relationships (Figure 4): $S_{GL95} = 0.007$, $S_{GL} = 0.012$; $S_{GL95}^+ = 0.016$; $S_{GH95}^- = 0.08$; $S_{GII} = 0.01$, $S_{GH95}^+ = 0.12$, $S_{BL95}^- = 0.007$, $S_{BL} = 0.012$; $S_{BL95}^+ = 0.016$; $S_{BH95}^- = 0.046$; $S_{GH} = 0.072$, $S_{GH95}^+ = 0.09$, The value of the smoothing coefficients $\alpha_{1,2,3,4}$ and $\beta_{1,2,3,4}$ is given in Equations (3) and (6) below. Parameters for NPP/VIIRS and AQUA/MODIS were calculated but are not shown here. The approximate ranges of SPM concentrations where each model is the most appropriate were computed from the equations given here.

$\rho\,655$ Interval Gironde Model Intevals Interval Values		SPM Model	[SPM] Application (g·m^{-3})
Gironde			
$(0; S_{GL95}^-)$	$(0.007 > \rho\ \text{red})$	Linear green	SPM < 8.5
$(S_{GL95}^-; S_{GL95}^+)\ S_{GL} = 0.012$	$(0.007; 0.016)$	Smoothing interval green-red α_1 Linear green + β_1 Linear red	8.5–9.2
$(S_{GL95}^+; S_{GH95}^-)$	$(0.016; 0.08)$	Linear red	9.2–42.5
$(S_{GH95}^-; S_{GH95}^+)\ S_{GH} = 0.1$	$(0.08; 0.12)$	Smoothing interval red-NIR α_2 Linear red + β_2 Poly NIR	42.5–180
$(S_{GH95}^+<)$	$(0.12 < \rho\ \text{red})$	Poly NIR	SPM > 180
Bourgneuf-Loire			
$(0; S_{BL95}^-)$	$(0.007 > \rho\ \text{red})$	Linear green	SPM < 8.5
$(S_{BL95}^-; S_{BL95}^+)\ S_{GL} = 0.012$	$(0.007; 0.016)$	Smoothing interval green-red α_3 Linear green + β_3 Nechad red	8.5–9.2
$(S_{BL95}^+; S_{BH95}^-)$	$(0.016; 0.046)$	Nechad red	9.2–28.1
$(S_{BH95}^-; S_{BH95}^+)\ S_{BH} = 0.072$	$0.046; 0.09)$	Smoothing interval red-NIR α_4 Nechad red + β_4 Nechad NIR	28.1–180
$(S_{BH95}^+<)$	$0.09 < \rho\ \text{red})$	Nechad NIR	SPM > 180

2.3. Satellite Data and Atmospheric Correction

Satellite images from three sensors were used in this study: Landsat-8/OLI, MODIS, and VIIRS. Detailed information on the characteristics of the three sensors can be found in the literature [19,18 50] (see also Table 4). OLI imagery was initially used for the development of the multi-conditional algorithm due to its high spatial resolution and high quality of the radiometric data in visible, NIR and SWIR spectral bands. Orthorectified and terrain corrected Level 1T OLI data was downloaded from the Landsat-8 portal USGS portal (http://earthexplorer.usgs.gov/) then processed using the ACOLITE software (http://odnature.naturalsciences.be/remsem/acolite-forum/) [19,51] to derive water-leaving reflectance (hereinafter referred as $\rho_w = \pi \times R_{rs}$). ACOLITE establishes a per-tile aerosol type (or epsilon) as the ratio between the Rayleigh corrected reflectance in the aerosol correction bands, for pixels where the marine reflectance can be assumed to be zero (where ρ_w 655 < 0.005, as defined by [19]. The epsilon is then used to extrapolate the observed aerosol reflectance to the visible bands. ACOLITE also provides a choice for aerosol correction using a full tile fixed epsilon, a per pixel variable epsilon or a user defined epsilon. In this study, the first option was selected for the atmospheric correction, as it has been shown to provide good results in highly turbid coastal waters [52]. This software proposes two atmospheric correction (AC) options: the NIR algorithm [51] based on the MUMM approach [23] and using the red (655 nm) and NIR (865 nm) bands, and the SWIR algorithm [19] using the SWIR bands 6 (1609 nm) and 7 (2201 nm). Both atmospheric corrections (NIR and SWIR) were tested in the Gironde area. Four images for the Gironde and four for the Bourgneuf-Loire areas were used for the NIR-SWIR atmospheric correction analysis. Then, the SWIR AC products were compared to the MACCS (Multisensor Atmospheric Correction and Cloud Screening processor) product provided by the Theia Land Data Center (theia.cnes.fr), which was developed by [53]. Its innovation relies on the combination of a multi-spectral assumption that associates the surface reflectance of the red and blue bands of the satellite, with the multi-temporal assumption that observations of a given region on land separated by a few days should yield similar surface reflectance values. They are also corrected for environmental effects. A total of 10 satellite images were used for this ACOLITE-MACCS inter-comparison (Table 5) and to test the SPM multi-conditional algorithm.

Table 4. OLI, MODIS and VIIRS satellite spectral bands and corresponding central wavelengths used in this study.

Sensor/Bands	Landsat 8/OLI (nm)	VIIRS (nm)	MODIS Aqua (nm)
Green	B3—561	M4—551	B4—555
Red	B4—655	M5—671	B1—645
NIR	B5—865	M7—862	B2—859
Atmospheric Correction (SWIR)	B6—1609	M10—1610	B5—1240
	B7—2201	M11—2250	B7—2130
Spatial Resolution	30 m	750 m	250/500 m
Temporal Resolution	1 every 16 days	1 per day	1 per day

Table 5. Date and time of OLI data acquisitions, tidal coefficients and tide times (low and high) at Royan (Gironde Estuary mouth) corresponding to the images used for the comparison between the ACOLITE and MACCS-Theia products.

Date and Time (UTC)	Tidal Range (m)	High Tide Time (UTC)	Low Tide Time (UTC)
10:49 (10 July 2013)	3.4	17:17	11:06
10:49 (30 October 2013)	2.5	12:56	06:32
10:55 (8 December 2013)	3.7	07:59	13:57
10:48 (7 March 2014)	2.8	08:01	14:08
10:47 (22 February 2015)	4.9	06:19	12:26
10:47 (2 September 2015)	4.55	07:20	13:21
10:47 (20 October 2015)	2.4	08:22	14:29
10:48 (7 December 2015)	2.6	13:33	07:06
10:47 (24 January 2016)	4.2	16:32	10:16
10:53 (19 March 2016)	3.0	14:05	07:15

Despite their lower spatial resolution compared to OLI, MODIS and VIIRS imagery were also included in this study to highlight the multi-sensor applicability of the developed algorithm. The VIIRS green (551 nm), red (671 nm) and NIR (862) spectral bands (750 m spatial resolution) were corrected for atmospheric effects using the Gordon and Wang approach in SeaDAS/l2gen (aeropt = −1) with bands M10 (1610 nm) and M11 (2250) as aerosol correction bands. Unfortunately, SeaDAS/l2gen does not allow to process the two VIIRS high spatial resolution bands (I1 and I2, 375 m spatial resolution), so the products presented in this study were generated at a resolution of 250 m by interpolating the 750 m resolution bands (M4, M5, M7). The resulting VIIRS products were compared to OLI products and to in situ measurements carried out during the field campaigns. Then, the multi-conditional SPM algorithm was applied to one image acquired over the Gironde area and another over the Bourgneuf-Loire area. This algorithm is also applied to MODIS (AQUA) images that were atmospherically corrected using the same atmospheric correction as VIIRS images, using the 1240 (B5) and 2130 nm (B7) MODIS SWIR bands. The band 1640 nm was not used due to the presence of faulty detectors on MODIS Aqua. This type of atmospheric correction was selected because it was shown to perform well in highly turbid waters [52]. Reference [54] has shown that VIIRS performance is comparable to MODIS Aqua in corresponding bands in all key performance regions of common spectral coverage, even if there are still some VIIRS calibration issues [55].

Note that to generate satellite products, cloud masking was applied using a reflectance threshold of 0.018 on the 2130 nm (OLI), 2250 nm (VIIRS) and 2130 nm (MODIS) wavebands, which avoids masking turbid waters.

2.4. Multi-Conditional SPM Algorithm Validation

Additional in situ turbidity measurements from the Gironde and Loire Estuary monitoring networks were used to validate the multi-conditional SPM algorithm through match-up with L8/OLI-derived SPM concentration. Due to their larger spatial resolution and the proximity of the in situ stations to the coast, MODIS and VIIRS data were not included in in situ—satellite match-ups.

The Gironde Estuary includes an automated continuous monitoring network, called MAGEST (MArel Gironde ESTuary), [56] comprising four sites (Figure 2): Pauillac in the central Estuary (52 km upstream the mouth); Libourne in the Dordogne tidal river (115 km upstream the mouth), and Bordeaux and Portets in the Garonne river (100 and 140 km upstream the mouth, respectively). The automated stations record dissolved oxygen, temperature, turbidity and salinity every ten minutes at 1 m below the surface. Information on this network can be found at: http://www.magest.u-bordeaux1.fr/. The turbidity sensors (Endress and Hauser, CUS31-W2A) measure values between 0 and 9999 NTU with a precision of 10%. Data from this station were selected within 10 min of the OLI overpasses. The temporal variability (standard deviation) was calculated for measurements conducted at the stations 30 min before and after the satellite overpass. The Loire Estuary also includes the same type of monitoring stations (MAREL), which continuously carry out measurements at different locations. These measurements have been conducted since 2007 in the frame of the SYVEL (Système de veille dans l'estuaire de la Loire) monitoring network operated by the GIPLE (Groupement d'Intérêt Public Loire Estuaire, Nantes, France). Information on the SYVEL network can be found at http://www.loire-estuaire.org/. As in the Gironde Estuary, the sensors are housed inside an instrumented chamber fixed on a pier, where the same type of measurements are recorded. In this study, we used data provided by two of the six stations in the Loire Estuary: the Paimboeuf and Donges stations (see locations on Figure 1). The SYVEL network provides turbidity data, which is then calibrated in SPM concentration using a regional relationship (GIPLE, 2014). In the case of the MAGEST network, turbidity was converted to SPM estimates using the relationship found in the Gironde. Different pixel configurations were selected for the match-ups, for example, using the closest pixel to the MAREL station or an average of several pixels. Only the results obtained with the best method in each area (Gironde or Loire) are presented. The dates to the OLI images used for satellite—in situ match-ups are shown in Table 6, together with tidal ranges and tide times (low and high tides) in the Gironde Estuary (Pauillac) and Loire Estuary (Donges) at those dates.

Table 6. Date, tidal ranges and tide times (low and high tides) in the Gironde Estuary (Pauillac) and Loire Estuary (Donges) corresponding to the OLI images used for satellite—in situ match-ups.

Date and Time (UTC)	Tidal Range (m)	High Tide Time (UTC)	Low Tide Time (UTC)
Gironde Estuary			
10:49 (10 July 2013)	4.85	06:10	13:11
10:49 (11 August 2013)	4.3	7:28	14:18
10:49 (27 August 2013)	3.65	8:45	15:50
10:49 (14 August 2013)	3.60	15:26	10:21
10:49 (30 October 2013)	3.20	14:05	07:59
10:49 (15 November 2013)	4.45	15:38	10:02
10:47 (29 July 2014)	4.25	06:18	13:10
10:47 (22 February 2015)	5.95	07:21	14:25
Loire Estuary			
11:01 (8 July 2013)	4.00	15:37	10:00
10:55 (2 August 2013)	4.65	13:22	06:34
10:54 (8 December 2013)	4.25	08:09	14:05
10:55 (5 March 2014)	5.10	15:37	10:00
10:54 (30 March 2014)	5.75	15:21	09:35
10:53 (15 April 2014)	4.85	15:39	10:03
10:52 (17 May 2014)	5.00	04:52	11:37

3. Results and Discussion

3.1. Atmospheric Corrections

In this section, the ACOLITE NIR and SWIR atmospheric corrections applied to OLI data with fixed scene epsilons are compared. A detailed explanation of these atmospheric corrections can be found in [19,51,57]. Results from ACOLITE are then compared to the MACCS atmospherically corrected products provided by the Theia Data Center. Finally, a comparison is made between OLI and VIIRS water reflectance. This comparison was not conducted for MODIS products as previous studies [19] have already shown a satisfactory correspondence between the OLI and MODIS Aqua atmospheric corrections in turbid coastal waters.

Figure 5 shows the water reflectance values in the OLI green, red and NIR bands obtained applying the ACOLITE-NIR (left column) and ACOLITE-SWIR (right column) atmospheric corrections. The SWIR AC results in water reflectance values at 561 nm 3% (NRMSE) higher than the values obtained applying the NIR AC over the clearest waters (where ρ 561 < 0.001 and ρ 655 < 0.005, as defined by Vanhellemont and Ruddick, 2015) of the transect compared. In the 655 nm band, the difference between the values obtained with the two methods is below a 5% difference for ρ 655 < 0.1. At higher ρ 655 values, found in the Gironde Estuary where the water is more turbid, the NIR AC fails and provides near-zero and even negative values. The spatial variability was calculated for offshore waters (ρ 655 < 0.005), estimating the NRMSE (%) of several 8 × 8 pixel boxes, resulting in an average variability of 5% for the SWIR AC, which could be interpreted as noise. Hence, the mean difference between the NIR AC and SWIR AC for all the bands (areas with low levels of water turbidity in low-turbidity offshore waters) was calculated and proved to be lower than the calculated computed spatial variability. As the NIR AC fails for higher water reflectance (ρ 561 > 0.1, ρ 655 > 0.1, ρ 865 > 0.02), the SWIR AC is the most appropriate correction for the selected study areas. These comparisons are conducted for the transect shown in Figure 5. These results are in accordance with those obtained by [19], who were the first to show the capabilities of Landsat 8/OLI and the SWIR AC to derive accurate ρ values in turbid coastal waters.

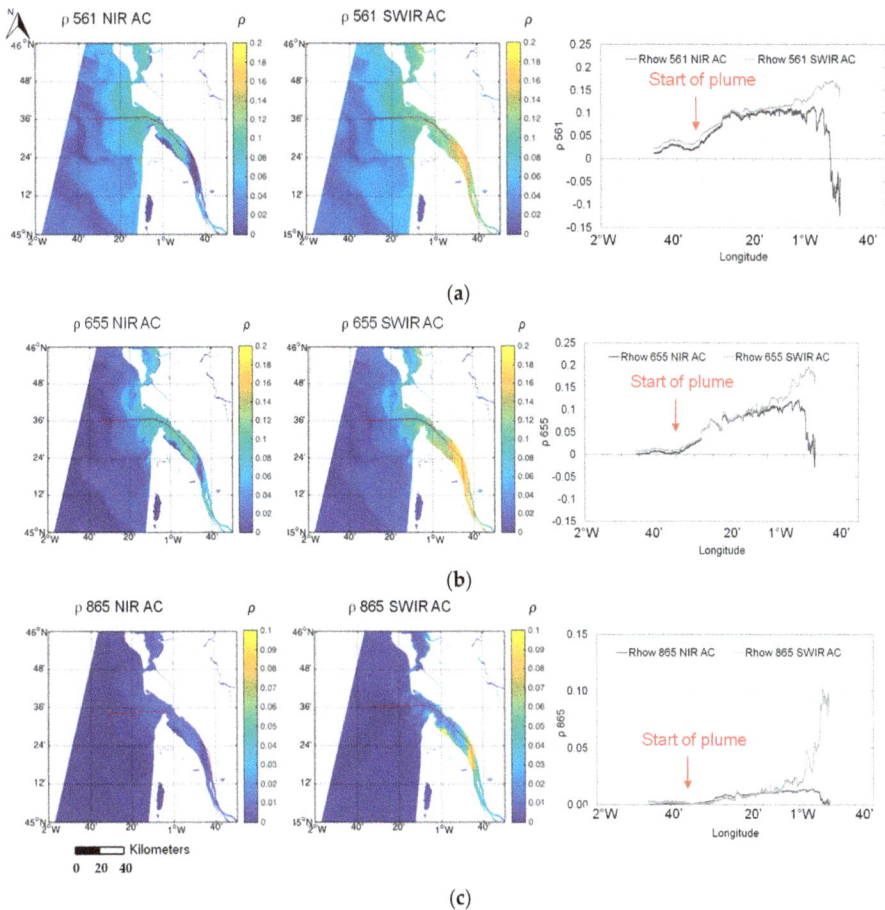

Figure 5. Comparison between NIR (left column) and SWIR (right column) atmospheric corrections applied to OLI data along a transect in the Gironde area on 7 March 2014. OLI bands at (**a**) 561 nm; (**b**) 655 nm; and (**c**) 865 nm (expressed as $\rho = R_{rs} \times \pi$) atmospherically corrected using the NIR and the SWIR options are shown. A transect (red line) over the plume and estuarine waters illustrates the comparison between the reflectance values derived using both atmospheric corrections.

3.1.1. ACOLITE vs. MACCS Products Comparison

Here, the ACOLITE SWIR AC and Theia Data Center MACCS water reflectance products are compared (Figure 6). The highest differences were observed over the less turbid waters, but in general a good agreement exists between both products. The flagged (grey) pixels on MACCS maps over the estuary correspond to the limit of the tile provided by the MACCS-Theia Land Data Center (longitude > 0°45′W). Figure 6b compares ρ values in the green, red and NIR bands obtained by applying both atmospheric corrections to the selected images (Table 5). The selected images were acquired at different periods of the year and for different tidal conditions. The best correlations were obtained for the red and the NIR bands with coefficients of determination of 0.95, a slope close to 1 and a NRMSE around 5%. The maximum differences were observed in the green band (NRMSE = ~7%).

(a)

$y = 0.93x + 0.018$
$r^2 = 0.86$
RMSE = 7.22 %

$y = 1.01x + 0.009$
$r^2 = 0.95$
RMSE = 5.37 %

$y = 1.01x + 0.003$
$r^2 = 0.95$
RMSE= 5.27 %

(b)

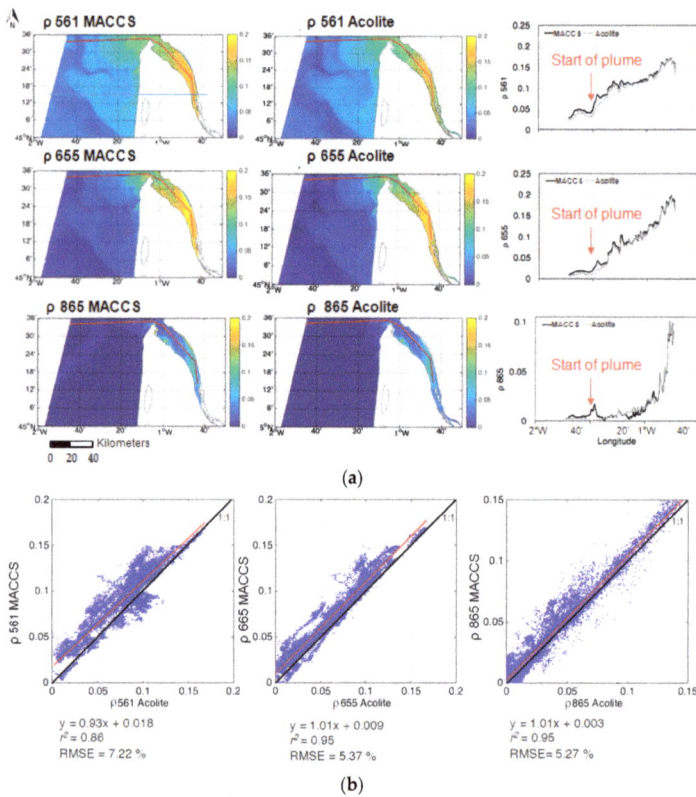

Figure 6. (**a**) Comparison between water reflectance at 561, 655 and 865 nm provided by the ACOLITE-SWIR and MACCS atmospheric corrections in the Gironde area (estuary and plume) for the OLI image acquired on 7 March 2014. The plots on the right represent the reflectance values provided by both products along a transect (red line) over the study area. (**b**) Same results presented as scatter-plots along the transect displayed on Figure 6a for the images dates shown on Table 5. The corresponding best-fitted linear relationships, r^2 coefficients and NRMSE (%) are indicated.

The MACCS algorithm is based on land pixels and estimates the aerosol optical thickness combining a multi-spectral assumption, linking the surface reflectance in the red and blue wavebands, and the assumption that multi-temporal observations of a given area should yield similar surface reflectance when separated by a few days. However, this method is not able to estimate the aerosol model and uses a constant model for a given site, which is a disadvantage for regions where the aerosol model is subject to large spatial variations, such as coastal regions. Instead, the ACOLITE AC method estimates the aerosol type using SWIR bands in clear water pixels for each image. In this study, the aerosol type was assumed to be constant over a single OLI tile (170×185 km^2), but there is an option provided by the ACOLITE software to allow the type to vary spatially. Research [19] demonstrated that products from OLI compare well with those of MODIS Aqua and Terra, using the SWIR bands corrected atmospherically with ACOLITE. For this reason, and since the ACOLITE AC uses an ocean color approach, this method is considered more appropriate for the coastal waters of the study regions, even though the MACCS method is proved to provide satisfactory results over the most turbid waters. Another reason explaining the differences between both atmospheric corrections is that the MACCS AC applies a continental model in this area, while it has been demonstrated that the maritime model is dominant in the Gironde area [58]. The aerosols of maritime origin are almost non-absorbent, so an overestimation of aerosol absorbance is expected when applying a continental model, especially in offshore waters where there is no land aerosol influence. The reason for the green band differences observed is unclear; it could be due to an overestimation caused by the MACCS AC, or an underestimation by the ACOLITE AC caused by a green band overcorrection. However, as the major differences are observed for the lower ρ_w values (Figure 6b), this dissimilarity is probably related to a flawed correction over offshore waters, where low green ρ_w values are usually found. Studies found on the use of the MACCS AC over coastal areas did not provide information that could explain the differences observed [42,58] other than those already mentioned. Since the green band NRMSE percentage remained low enough (<7%) for the purpose of this study, a deeper analysis on the reasons for these differences was considered out of scope.

3.1.2. Validation of Atmospheric Correction

Figure 7a shows the band-to-band ρ value scatter-plots derived from four OLI satellite images of the Gironde Estuary (a and b) and for the Bourgneuf-Loire area (c and d). The same band-to-band ρ value scatter-plots derived from in situ measurements are superimposed. In the case of the Gironde area, the in situ values accurately match the OLI-derived reflectance values, and the logarithmic regressions for both datasets follow the same trend. In the case of the Bourgneuf-Loire area, the trend divergence between in situ and satellite datasets is more significant. This is mainly due to the dispersion observed on the satellite dataset caused by the presence of clouds (Figure 7) on the image acquired on 12 April 2013. The ACOLITE software provides a good mask for clouds, but in some cases cloud shadows or cloud edges are insufficiently masked, introducing significant scatter. Nevertheless, there is a fair overlap between in situ and satellite data for this area as well, taking into account the dataset number difference (in situ stations $n = 29$ vs. satellite pixels $n = \sim20,000$) and the effect of cloud pixels from 12 April 2013. This image was selected because it was acquired on the last day of the field survey carried out in this area, in April 2013.

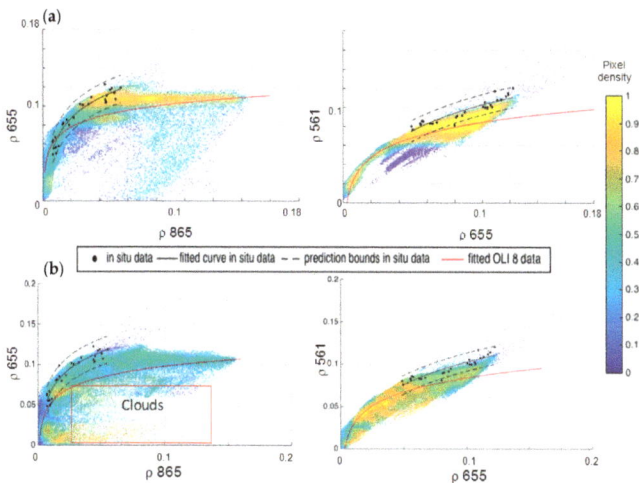

Figure 7. Scatter plots between OLI-derived water reflectance values in bands 865, 655 and 561 nm, extracted from four images 11 August 2013, 7 March 2014, 30 August 2014 and 2 February 2015) acquired over the Gironde Estuary (**a**) and four images acquired over the Bourgneuf-Loire area (**b**) 12 April 2013, 8 December 2013, 2 August 2013, 17 May 2014. The best-fitted logarithmic regression lines are shown in red. Overplot of the water reflectance values measured in the field (black circles) and corresponding regression lines and 95% confidence intervals (solid and dashed black lines, respectively).

Similar ρ band-to-band comparisons were conducted between VIIRS-derived images, and in situ–measured water reflectance at bands centered at 551, 671 and 862 nm (Figure 8). A fair match is obtained between field and satellite datasets for both the Gironde and the Bourgneuf-Loire areas. As can be observed, the in situ (black line) and satellite (red line) regression curves overlap. This demonstrates that the atmospheric correction applied to the VIIRS images, using the SWIR bands and the Seadas (version 7.3), is appropriate for this type of environment.

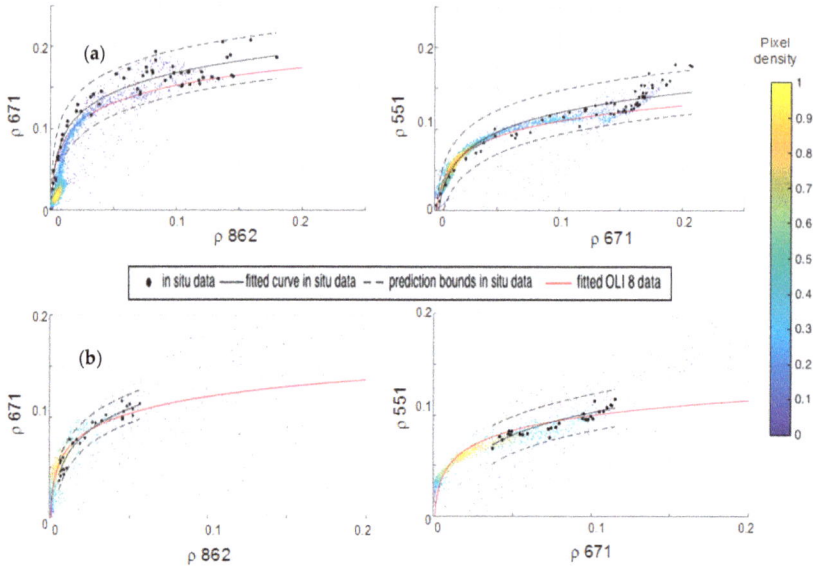

Figure 8. Scatter plots between VIIRS-derived water reflectance values at 862, 671 and 551 (862 vs. 671 and 671 vs. 551) extracted from the image acquired on 7 March 2014 over the Gironde Estuary (**a**) and over Bourgneuf Bay and the Loire Estuary on 12 April 2013 (**b**) and atmospherically corrected using the SWIR bands. The fitted logarithmic regression line of the satellite data is shown in red, the in situ data acquired during the field campaign conducted in 2013 is represented using black dots; the regression line fitted to the in situ data and the 95% confidence intervals are shown, respectively, as solid and dashed black lines.

Figure 9 presents a comparison between in situ–measured and satellite-derived reflectance spectra corresponding to VIIRS and MODIS Aqua images acquired on 12 June 2012, 15 July 2014, 16 July 2014. Satellite data were atmospherically corrected using the SWIR option as well. A good match was obtained for the plume area, but in the estuary, the satellite-derived reflectance was systematically lower than values measured in situ at the Pauillac station. This is due to the size sampling differences (satellite vs. in situ) and the effect of the land reflectance in land/water border pixels. Water pixels located near the shore may be contaminated by the land signal, causing erroneous water reflectance estimates. This underestimation in the case of VIIRS (Figure 9a) is lower than in the case of MODIS, where a particularly sudden decrease is observed for the lower wavelengths. This is caused by the use of the MODIS band B5 (1240 nm), which causes an overcorrection in the shorter wavelength band.

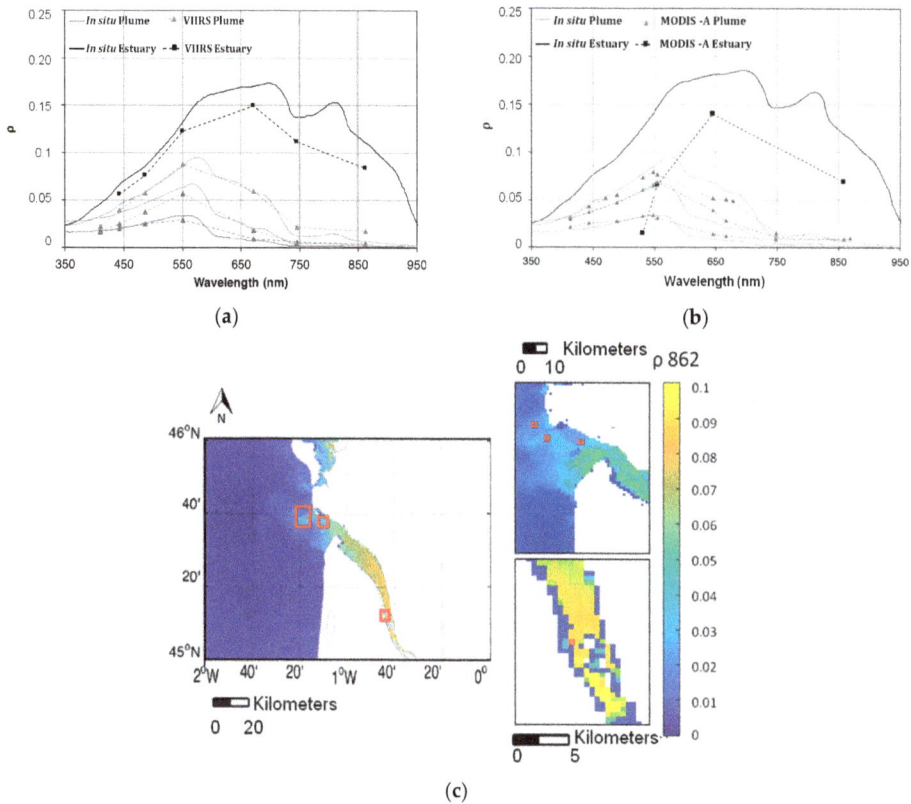

Figure 9. (a) Water reflectance spectra (ρ) measured in the field (12 June 2012, 15 July 2014, 16 July 2014) compared to VIIRS-derived (a) and MODIS-derived (b) water reflectance in the Gironde plume (grey lines) and estuary (black lines). (c) Location of the stations on the ρ 862 VIIRS map. There was a maximum time difference of 20 min between the in situ and the satellite data.

In Figure 10, the reflectance spectra measured in situ on 11 April 2016 at 10:26 (Station 1) and at 11:16 (Station 2) are compared to the OLI-derived values (image acquired at 10:53). There is a closer match between the water reflectance values measured at Station 2 than at Station 1. This could be due to the lower time difference between the image acquisition and the in situ measurement for Station 2 (23 min) than for Station 1 (27 min), together with the significant small-scale variability of the SPM concentration in this specific area given that measurements were conducted during the ebb tide. In general, satellite products appear to underestimate the 'true', i.e., field-measured, water reflectance. Valid match-ups with MODIS and VIIRS were not obtained on the same day.

The comparison between in situ and satellite data products resulted in a good match between the green-red and red-NIR bands, with some discrepancies. In general, the trends observed for the satellite data were lower in the higher reflectance values. This is due to the difference in the amount of data (in situ stations n = 29 vs. satellite pixels n = ~20,000) and to cloud shadow and land effects on coastal pixels. Regarding the OLI red-green bands' (655 vs. 561 nm) comparison, a break was observed between the 0–0.15 and the 0.15–0.2 intervals. If the fit would have been made for values of ρ 655 <0.08, the red and black curves in Figures 7 and 8 would have had a better correspondence for the lower reflectance values, so better results would have been obtained if the comparison was achieved by intervals (e.g., 0–0.08, 0.8–0.2). However, showing the entire data range provides a better understanding of the type of data obtained in situ and from satellite remote sensing measurements, and since the in situ data matches the satellite data, the atmospheric correction was considered to provide realistic reflectance results. The switching bounds were determined using field data, so a better match between the two types of data (in situ vs. satellite) would not affect the switching bound values selected.

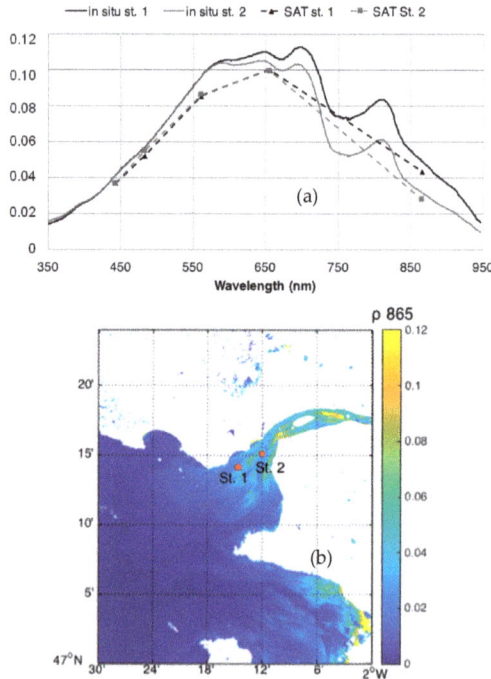

Figure 10. (**a**) Water reflectance spectra (ρ) measured in situ in the Loire Estuary on 11 April 2016 compared to OLI-derived ρ values (SAT); (**b**) Location of the stations on the ρ 865 OLI map. There was a maximum time difference of 20 min between the in situ and the satellite data.

Differences observed between in situ and satellite data, in Figures 9 and 10, could be due to several reasons: (1) an overcorrection of the atmospheric contribution in the SWIR method; (2) the spatial difference between the satellite pixel and field measurements (250/750 m^2 pixel versus ~1 m^2); (3) the near-shore location of the station in the estuary. Option 3 appears to be the most plausible, as the pixel selected for the comparison did not correspond exactly to the location of the Pauillac field station: the next pixel away from the shore was selected to avoid the land effect. Thus, the reflectance values at this location are different to the ones measured in situ at Pauillac. In the case of MODIS (Figure 9b), there is an obvious overcorrection inside the estuary, due most probably to the low pixel resolution for this area combined with the selection of the 1240 SWIR band. This effect was also observed in the highly turbid waters of the La Plata river by [52].

3.2. OLI-VIIRS Comparison

Water reflectance values in the green, red and NIR OLI bands (561, 655 and 865 nm) were re-sampled by neighborhood, averaged to a grid of 750 m resolution and then compared to VIIRS-derived values at 551, 671 and 862 nm bands using a common grid. This comparison was achieved for the image acquired on 12 April 2013 (Table 7). A fair correspondence was found in the

Gironde area for this cloud-free image with a large range of water reflectance values, taking into account the significant overpass time difference between both sensors (~3 h) and the large tidal dynamics occurring in this region, as well as the bandwidth differences between the two satellite sensors. The best correspondence was obtained in the green bands (slope = 0.83, r^2 = 0.71), followed by the red bands (slope = 0.66, r^2 = 0.6), and the NIR bands (slope = 0.29, r^2 = 0.38), even though substantial scatter was present. In the case of the Bourgneuf-Loire area, the best correspondence was obtained between red bands (slope = 0.99, r^2 = 0.42), followed by the NIR bands with a better slope, but with more scatter (slope = 0.82, r^2 = 0.1), and the green bands (slope = 0.64, r^2 = 0.67). Differences between both areas were possibly due firstly to the different optical characteristics and dynamics occurring in both areas, and secondly to a failure of the VIIRS atmospheric correction inside the Gironde Estuary. This exercise was not conducted for MODIS images, because comparisons have already been carried out in similarly turbid waters [19].

Table 7. Slope, offset and determination coefficient derived from the comparison of OLI and VIIRS bands from one image acquired on 12 April 2013 (time difference between OLI and VIIRS data acquisition = 3 h).

Bands Region Compared	Slope	Offset	r^2
green			
Gironde	0.83	0.013	0.71
Bourgneuf-Loire	0.64	0.03	0.67
red			
Gironde	0.66	0.017	0.6
Bourgneuf-Loire	0.99	−0.005	0.42
NIR			
Gironde	0.29	0.008	0.038
Bourgneuf-Loire	0.87	0.008	0.13

Comparisons between OLI and VIIRS products were not found in the literature. However, there is an increasing interest in the exploitation of VIIRS products for coastal studies as this satellite sensor is considered to provide continuity to MODIS. In this study, the VIIRS performance was assessed in highly turbid coastal waters; it showed good results in coastal waters, but the bands' spatial resolution was too coarse to be used inside the estuary. Nevertheless, the results presented here are promising, as this sensor is still being calibrated, its performance is being tested [59] and methods are currently being developed to use the high-resolution bands in coastal waters [60]. The water reflectance values derived from satellite data were slightly lower than the values measured in situ. This is mainly due to the difference in sampling size, as OLI images have a resolution of 30 m, while the water volume sampled in situ was much lower (~10 L). Research [57] showed good agreement between OLI and in situ spectra in low- to high-turbidity waters. This study confirms their findings in study areas with different SPM characteristics with respect to the North Sea [21].

In summary, the SWIR atmospheric correction appears to be the most appropriate for the selected study areas, including low- to high-turbidity waters (Figure 5a). ACOLITE software corrections provided satisfactory water reflectance values for OLI images, in good agreement with MACCS reflectance products (Figure 6a), as already highlighted in previous studies (e.g., [42]). In relation to VIIRS, there is a good correspondence between in situ and satellite-derived water reflectance for both study areas (Figures 8 and 9, Table 7), although additional in situ–satellite match-ups would be necessary to draw further conclusions. The problem with the atmospheric corrections of MODIS data is the use of the 1240 nm band, which can provide reflectance values above zero in highly turbid waters, causing an overcorrection to the visible bands [52,61].

3.3. Multi-Conditional SPM Algorithm

Figure 11 presents the application of the switching SPM model (Table 2) combined with the smoothing procedure (Equations (3)–(6), Table 3) to the OLI image of the Gironde Estuary, acquired on 2 February 2015. The transect along the plume and estuarine area proves that the developed switching and smoothing methods allow a smooth transition between the SPM concentration remotely sensed from the offshore low-turbidity waters to the highly turbid waters inside the estuary. Note that the SPM NIR band values are highly noisy for concentrations lower than 50 g·m^{-3}.

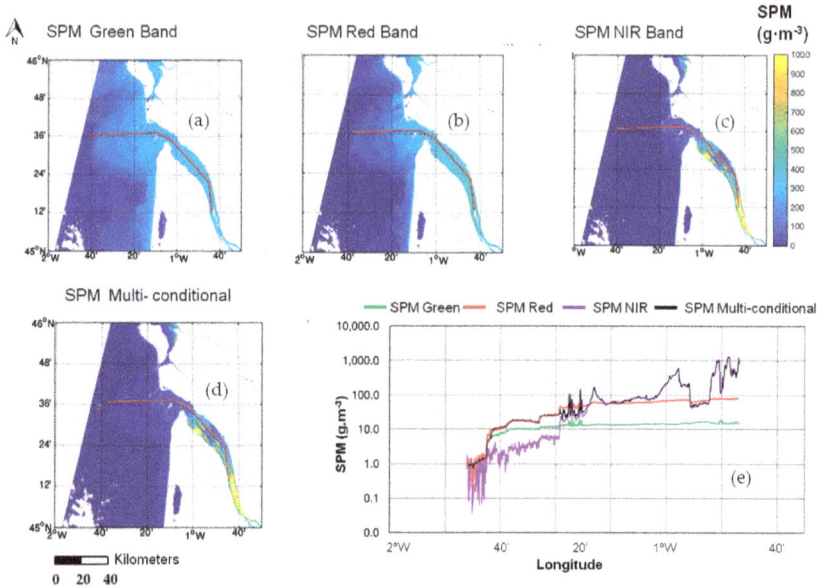

Figure 11. SPM maps derived from the L8/OLI red (**a**), the NIR (**b**), and green bands (**c**) using the equations shown in Table 2, and the resulting map created with the multi-conditional SPM algorithm (**d**) and an OLI image acquired on 2 February 2015. Spatial evolution of SPM concentration provided by the four maps along the transect (**e**).

Figure 12 shows the application of the SPM multi-conditional algorithm to a cloud-free image acquired on 12 April 2016 over the Bourgneuf-Loire area using the procedure presented in Section 2. The smooth transition between the remotely sensed SPM concentration is clearly observed along both the Loire and Bourgneuf transects. As shown in Figure 12b,c, the SPM multi-conditional algorithm (black line) starts estimating SPM using the green band equation in the clear water area (green line) for both the Loire and Bourgneuf transects. Then, as the SPM concentration increases, due to the river plume in the Loire transect and to the mudflat SPM re-suspension in the Bourgneuf area, the multi-conditional algorithm switches to the red band–estimated SPM (red line). Then, as the red band reflectance increases due to the near-shore SPM concentration increase, the algorithm switches to the SPM calculated with the NIR band equation. The figure clearly shows a high variability in SPM estimation using the NIR band equation (purple line) in clear waters (2°30′W–2°20′W), proving the importance of using equations appropriate for each concentration SPM range.

Match-ups between in situ–measured SPM (MAGEST and SYVEL network stations) and OLI-derived SPM concentrations were then analyzed (Figure 13, Table 8) to prove that the multi-conditional algorithm provides accurate SPM concentrations. The dates to the OLI images used for satellite—in situ match-ups are shown in Table 6, together with tidal ranges and tide times (low and high tides) in the

Gironde Estuary (Pauillac) and Loire Estuary (Donges) at those dates. The in situ SPM concentrations for the Gironde Estuary were derived from the turbidity measurements (in FNU) at the stations and the turbidity-SPM relationship was established with in situ measurements taken in the Gironde Estuary. In general, OLI provided good SPM estimates at the Pauillac and Libourne stations (r^2 = 0.8 and 0.95, NMRSE = 16% and 14%, respectively, for Gironde and Loire). Previous match-ups using the MAGEST network measurements and SPOT 5 satellite data provided good results with higher pixel resolutions than OLI, showing the capacity of these stations for satellite imagery validation [42]. Additionally, the SPM concentration calculated using different models (refer to Table A2 in the appendix), including the models of [18,42], were matched to in situ SPM measurements. The resulting match-ups (Table 8) show that the multi-conditional algorithm provided, for both the Gironde and the Loire areas, the best of fit (r^2 = 0.8, 0.67) and low error (16%, 14%) combination when compared to SPM estimated using other models. The combination of several models adapted to each concentration range is the main reason for this result, together with the selection of the best-fitted models. As proved by other studies [35–37], the selection of a specific model combining different bands is an improvement when estimating the SPM concentration in coastal waters.

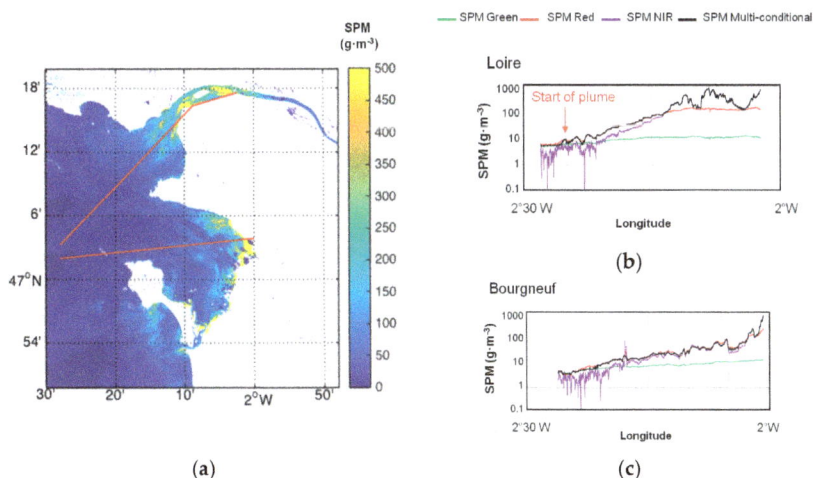

Figure 12. (**a**) SPM map of the Bourgneuf Bay and Loire Estuary derived from OLI satellite image acquired on 12 April 2016 applying the SPM multi-conditional algorithm. The image was atmospherically corrected using the SWIR option of the ACOLITE software. Resulting SPM concentration transects along the Loire Estuary (**b**) and Bourgneuf Bay (**c**) retrieved using the three SPM single-band models (green, red and NIR bands) and with the multi-conditional algorithm.

Table 8 shows the in situ-satellite match-up results using different SPM models. In the case of the Gironde area, the fit, slope, NRMSE (%) and offset when using the band-ratio and single-band exponential and polynomial regressions were similar. Slightly better results were obtained with the exponential NIR-red band ratio models (r^2 = 0.7, slope = 1.1, NRMSE = 22.3%, offset = 16) compared to the exponential NIR band model (r^2 = 0.8, slope = 0.65, NRMSE = 20%, offset = 66), but the polynomial single-band model performed better when using the single NIR band (r^2 = 0.8, slope = 0.8, NRMSE = 16.4%, offset = −3), compared to the band ratio (r^2 = 0.74, slope = 0.9, NRMSE = 19.8%, offset = 33). The SPM concentration estimates did not improve when using equations published in other studies for the same study areas [18,42] providing larger NMRSEs (36.5%, 44.7%) and offsets (−185, −24), compared to the results obtained with the multi-conditional algorithm (slope = 0.8, r^2 = 0.8, NRMSE = 16%, offset = −1.8). In the case of the Bourgneuf-Loire area, the combination of the re-calibrated NIR and red band semi-empirical models from Reference [16] with the multi-conditional algorithm provided the best fit and lowest error (slope = 0.67, r^2 = 0.95, NRMSE = 14%, offset = 121). In this case, the performance of the algorithm will improve with a re-calibration of the models with additional in situ data.

The configuration of the pixels selected for the match-ups is shown in Figure 13. OLI-derived SPM estimates provided good match-up results with the in situ measurements. In the case of the Bourgneuf-Loire area, instead of using one pixel for the match-ups, as in the case of Gironde, an average of four pixels was used, as it provided the best results. Vertical error bars show the standard deviation of the four pixels selected, in the case of the Loire area (see Figure 13b), and of the three nearby pixels in the case of the Gironde area (shown as red rectangles), even if a single pixel was used for the match-ups (red circle) in Figure 13c. Horizontal error bars correspond to the temporal SPM variability ± 10 min from the satellite overpass time. There appears to be an underestimation by the algorithm at SPM concentrations above 100 g·m^{-3} and a slight overestimation below 500 g·m^{-3} (and above 100 g·m^{-3}) in the Bourgneuf-Loire area, while the Gironde algorithm seems to overestimate values above 500 g·m^{-3}. This tendency was also observed by [42] using SPOT 4 data, where the same pixel configuration was used for the Loire match-ups. The imprecision observed for these match-ups is due to several factors, such as the accuracy of the SPM algorithms, the errors and uncertainties in field measurements, the spatial differences between the sampling station and satellite pixel location, as well as uncertainties related to atmospheric corrections. Different SPM models including band ratios, such as those developed by [18,42] for these areas, were applied to the images, but the match-up results did not improve (see Table 8).

Figure 14 compares the SPM concentrations derived from OLI, VIIRS and MODIS (Aqua) satellite data recorded on the same day over the Gironde Estuary. Water reflectances at 655 (OLI), 671 (VIIRS), 645 nm (MODIS aqua) show a similar trend along the transect. VIIRS products provided higher values than OLI in general, up to the most upstream section of the Estuary, where there was a sharp decrease. It corresponded to a failure of the atmospheric correction, due to the low VIIRS spatial resolution for this particular area. This failure was also observed along the MODIS transect, where the most upstream pixels are missing, resulting in a sudden SPM concentration decrease. Generally, the MODIS and OLI transects overlap up to the most upstream section of the estuary, where SPM concentrations are overestimated by MODIS. This is due to the proximity of the land on the last transect pixels, which, as seen in Figure 13d, results in a rapid decrease of the SPM concentration. In this particular region, the reflectance at 859 nm showed a fast increase for MODIS, which was not observed in the OLI 865 band. Hence, particular attention needs to be paid to the near-shore pixels, where the atmospheric correction may provide inaccurate water reflectance estimates, resulting in inaccurate SPM retrievals. For practical reasons, the same switching bounds applied to OLI were used for VIIRS and MODIS Aqua. However, the fine adjustment of the switching bounds to each sensor's spectral bands could provide better results. Despite the fact that the MODIS atmospheric corrections provided underestimated water reflectance values, the transect comparison showed a good match between the three SPM maps.

Overall, the multi-conditional SPM algorithm provides a smooth transition between SPM models for three different sensors over an area that goes from low- to high-turbidity surface waters. The band-switching technique allows keeping the optimal sensitivity of ρ to SPM variations and avoiding the saturation of ρ in (highly) turbid waters. The study areas were characterized by a high amount of cloudy days, and taking into account that OLI images are only available every 16 days, it was difficult to obtain numerous cloud-free images to apply the SPM algorithm. Again, the purpose of this study was not to provide the best SPM models for each area, but to develop a method and test it on selected satellite data recorded for optimal conditions (cloud-free, clear atmosphere) representative of a wide range of SPM concentrations in coastal and estuarine waters. Similar methods have already been developed [35,36], but the procedure presented in the present study (1) automatically selects the model switching bounds based on in situ measurements; (2) fully applies the method to real satellite data provided by three different sensors; and (3) validates the results based on match-ups with field data.

Table 8. Slope, adjusted r^2, percent NRMSE (%) and offset resulting from the match-ups between in situ measurements from the MAGEST (Gironde) and SYVEL (Loire) network stations and the OLI-derived SPM concentration obtained using the different empirical and semi-empirical models shown in Tables 2 and A2 (appendix).

	Exponential 1 Band	Exponential Ratio	Polynomial 1 Band	Polynomial Ratio	Nechad et al. (2010) [16] NIR	Doxaran et al. (2009) [18]	Gernez et al. (2015) [42]	SPM Multi-Conditional
Gironde								
Slope	0.65	1.1	0.8	1.1	0.9	1.4	0.6	0.8
r^2	0.8	0.7	0.8	0.74	0.76	0.6	0.6	0.8
NRMSE (%)	20	22.3	16.4	19.8	17	36.5	44.7	16
offset	66	16	−3	33	−19	−185	−24	−1.8
Loire								
Slope	0.66	0.9	3.53	0.3	0.67	0.3	0.2	0.67
r^2	0.72	0.4	0.76	0.5	0.95	0.4	0.5	0.95
NRMSE (%)	29.7	48	158	40	14	39	46.8	14
offset	58	58	−3	75	−19	46.3	56	121

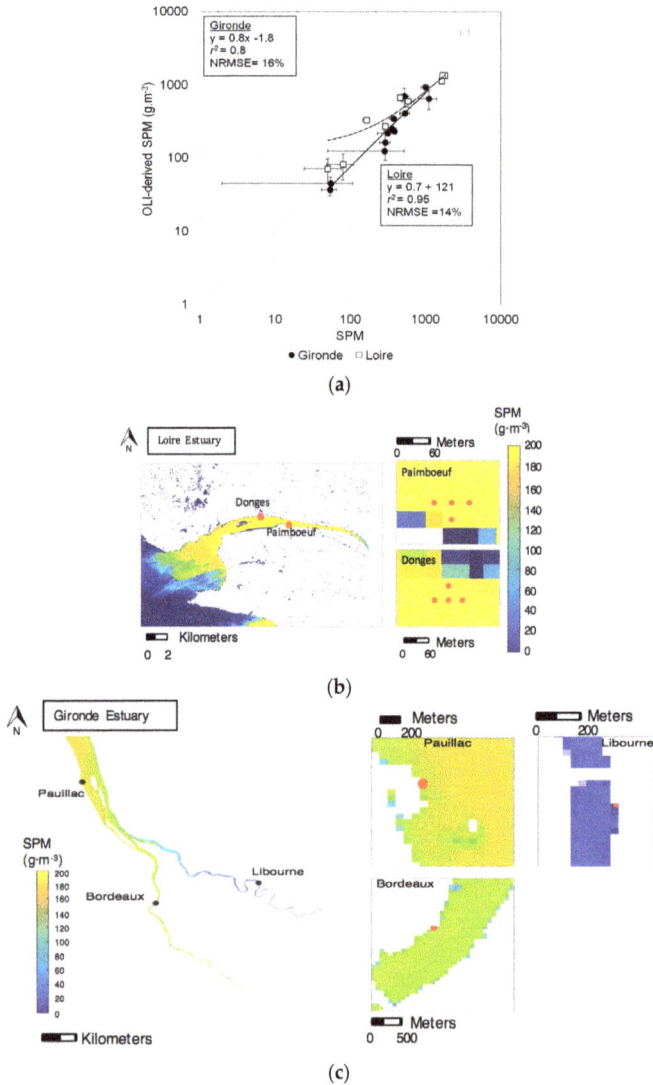

Figure 13. (a) Match-ups between in situ measurements from the MAGEST (Gironde) and SYVEL (Loire) network stations and the OLI-derived SPM concentration obtained using the multi-conditional SPM algorithm. The black solid and dashed black lines show, respectively, the best-fitted linear relationships for the Gironde Estuary stations, the dashed black line shows the correlation for the Loire Estuary stations. Vertical bars show the standard deviation of the four pixels selected for match-up in the case of the Loire area (see (**b**)), and of the three nearby pixels in the case of the Gironde area (shown as red rectangles), even if a single pixel was used for the match-ups (red circle) in (**c**). Horizontal error bars correspond to temporal SPM variability, ±10 min from satellite overpass time. (**b**) Map with the location of the SYVEL network stations and pixels used for the comparison match-ups (average of four pixels = at the Paimboeuf and Donges stations). The map was derived using the OLI image acquired on 8 December 2013 and the SPM multi-conditional algorithm. (**c**) Map with the location of the Pauillac, Bordeaux and Libourne MAGEST network stations: Pauillac, Bordeaux and Libourne. The map was derived from the OLI image acquired on 7 March 2014 applying the SPM multi-conditional algorithm.

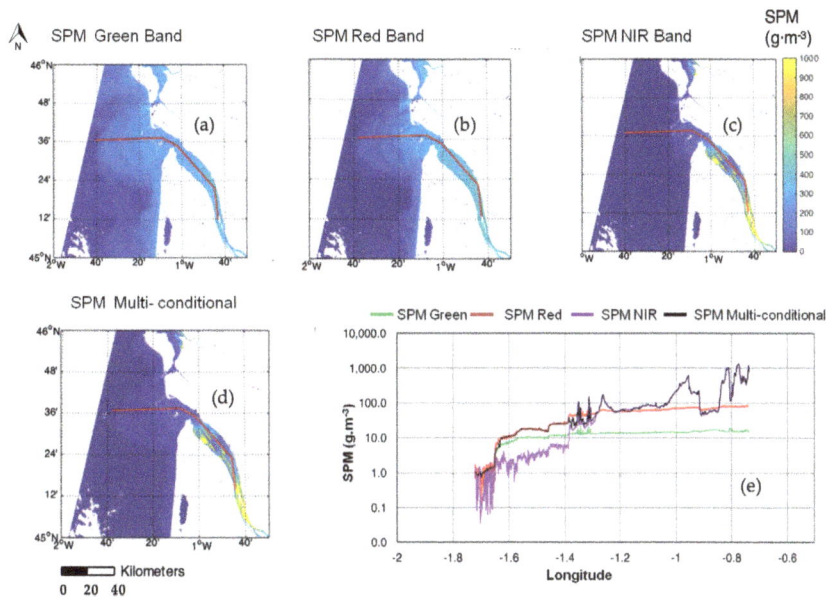

Figure 14. SPM maps derived from OLI, VIIRS and MODIS (Aqua) satellite data applying the SPM multi-conditional algorithm. Resulting ρ 655 (OLI), ρ 671 (VIIRS) and ρ 645 nm (MODIS aqua) (used as switching bands between SPM models) and SPM concentrations (multi-conditional algorithm) along the same transect. The grey sections indicate the switching and smoothing intervals.

4. Conclusions and Perspectives

In this study we have shown that satellite imagery from MODIS, VIIRS and OLI satellites can be used to reliably estimate SPM in coastal waters over a very wide concentration range (1–2000 g·m^{-3}) using a novel multi-conditional algorithm. In situ SPM and water reflectance measurements were used to calibrate switching SPM algorithms based on multiple SPM vs. ρ relationships. For each specific SPM range (<10, 10–50, and >50 g·m^{-3}), the ρ in specific spectral bands (green, red and NIR, respectively) was found to be more sensitive to changes in the SPM concentration, and the corresponding best-fitted SPM vs. ρ relationship was selected. These relationships were established empirically and semi-empirically and calibrated for clear to highly turbid waters in two study site areas: the Gironde and the Bourgneuf-Loire area estuaries. The selected models for each SPM range and each band were chosen based on goodness-of-fit tests (r^2) and NRMSE (%) results from the validation exercises. In the case of the Gironde estuary, the best-performing models were a single-band, second-order polynomial relationship for the NIR band ($r^2 = 0.97$; NRMSE = 9.11%) and a linear relationship for the red ($r^2 = 0.81$; NRMSE = 7.23%) and green bands ($r^2 = 0.81$; NRMSE = 16.41%). In the case of the Bourgneuf-Loire area, the best-performing model was the semi-empirical relationship published by [16], re-calibrated with the in situ dataset, for both the NIR ($r^2 = 0.93$, NRMSE = 7.81%) and red bands ($r^2 = 0.82$, NRMSE = 18.22%).

The bounds for switching between models were based on water reflectance values derived from the saturation points of the most sensitive bands. The bounds were selected by means of band comparisons from field water reflectance measurements: ρ (green) vs. ρ (red) and ρ (red) vs. ρ (NIR). The field data points were modeled using a logarithmic regression curve. The actual value of the saturation point, which is also the switching point, was computed as the point where the first derivative of that regression curve (i.e., the slope or tangent) equals 1, as this is the middle point between a completely horizontal (complete saturation) and a completely vertical line. The switching points selected were based on the red band water reflectance values, being the middle band between the green and the NIR. Then, the models were weighted to ensure a smooth transition between different SPM concentrations. The smoothing bounds were derived from the 95% confidence levels of the regression curve following the same procedure as for the switching value calculation. The switching values for each system are reported in Table 3.

To obtain accurate satellite-derived SPM maps, appropriate atmospheric corrections are required in coastal waters. Several atmospheric correction algorithms were compared, and match-ups between satellite and field measurements showed that SWIR-based atmospheric correction algorithms performed best. Alternative approaches such as the MACCS method initially developed for land applications also provide satisfactory results. Despite some inaccuracies in water reflectance retrieval, the SPM concentration can be reliably estimated using the three sensors (MODIS, VIIRS and OLI) in low (SPM ~1 $g \cdot m^{-3}$) to highly turbid waters (SPM > 2000 $g \cdot m^{-3}$). However, VIIRS and MODIS images fail inside narrow estuaries (here the Gironde) due to low spatial resolution.

The multi-conditional algorithm presented in this study successfully provided a smooth transition between different SPM models, and was then successfully applied to multi-sensor satellite data. It was proved to provide a smooth transition between different SPM models. Results clearly highlighted the need for switching the SPM algorithm in coastal and estuarine waters where (i) the water reflectance in the green and then red spectral regions rapidly saturated with the increasing SPM concentration, while (ii) water reflectance in the NIR is associated with a low signal-to-noise ratio and significantly underestimates the SPM concentration in clear to moderately turbid waters (Figures 9 and 10). A comparison with in situ data showed that the reflectance measurements undertaken in the field corresponded satisfactorily with the satellite-derived water reflectance (Figure 9), except for the MODIS products inside the estuary. Moreover, the match-up exercise using the SYVEL ($r^2 = 0.95$, NRMSE = 14%, slope = 0.7) and MAGEST ($r^2 = 0.8$, NRMSE = 16%, slope = 0.8) autonomous stations provided satisfactory results, proving that the selection of the algorithms was appropriate. Still, additional in situ measurements in these areas would improve the calibration of the models and provide better validation results.

The switching bound selection method presented here can be easily applied to any turbid coastal water area associated with wide turbidity ranges, without the need for in situ measurements. The bounds can be directly selected by comparing the water reflectance in the green, red and NIR wavebands directly derived from satellite data to detect the saturation points and infer the switching values (Figures 7 and 8). This study offers the appropriate methodology to study long-term dynamics and trends using satellite imagery in turbid coastal waters. When applied to multi-sensor satellite data, it can significantly contribute to the understanding of the impact of anthropogenic pressures on coastal environments, monitoring water quality, gaining knowledge of estuarine processes and even studying the impact of recent climate change. The multi-sensor approach presented here can be appropriately applied to the latest generation of ocean color sensors (namely Sentinel2/MSI and Sentinel3/OLCI) to study SPM dynamics in the coastal ocean at higher spatial and temporal resolutions.

Acknowledgments: This work was funded by the European Community's Seventh Framework Programme (FP7/2007-2013) under grant agreement no. 606797 (HIGHROC project). USGS and NASA are acknowledged for the Landsat-8 imagery. NASA is also thanked for the VIIRS and MODIS Aqua data, and for the SeaDAS processing software. We thank the CESBIO and CNES for providing MACCS-corrected Landsat data on the THEIA web portal. In situ reflectance and SPM field data were acquired in the frame of the following projects: SeaSWIR (funding: BELSPO agency PI: Els Knaeps), Rivercolor (funding: CNES; PI: V. Lafon), Gigassat (funding: ANR-12-AGRO-0001-05 and ANR-12-AGRO-0001-08; PI: F. Pernet), and TURBO (funding: CNRS/INSU PNTS-2015-07; PI: P. Gernez). We also acknowledge the MAGEST and SYVEL network teams for the availability of the turbidity time series in the Gironde and Loire estuaries. The MAGEST network is financially supported by the following organizations: AEAG (Agence de l'Eau Adour-Garonne); SMIDDEST (Syndicat MIxte pour le Développement Durable de l'ESTuaire de la Gironde); SMEAG (Syndicat Mixte d'Etudes et d'Aménagement de la Garonne); EPIDOR (Etablissement Public Interdépartemental de la Dordogne); EDF; GPMB (Grand Port Maritime de Bordeaux); Conseil Régional Aquitaine; CG-33 (Conseil Général de Gironde); Ifremer; CNRS; Université Bordeaux 1.

Author Contributions: Stéfani Novoa, David Doxaran, Pierre Gernez and Anouck Ody conceived and designed the experiments, wrote the paper and analyzed the data. Quinten Vanhellemont, Betrand Lubac and Virginie Lafon contributed interpreting results and writing.

Conflicts of Interest: The authors declare no conflict of interest.

Appendix A

Table A1. Selected SPM models for VIIRS and MODIS, NIR, red and green bands. Each model was calibrated and evaluated using in situ measurements acquired in each study region. The goodness of fit (r^2), relative root mean square error percent (%) and appropriate SPM range for each model are displayed. The models for Gironde were evaluated using a calibration and a different validation data set. Due to the low number of measurements, the Bourgneuf-Loire dataset was not separated into calibration and validation sets. The normalized root mean square error percent (%) in this case represents the deviation of the random component within the data as well as the goodness of fit (r^2).

Fit	Equation	r^2	RMSE %	SPM (g·m^{-3})
Gironde				
Polynomial NIR	$32110\,\rho\,862^2 + 2204\,\rho\,862$ (VIIRS)	0.96	4.98	SPM > 50
	$35260\,\rho\,859^2 + 1648\,\rho\,859$ (MODIS)	0.95	9.18	
Linear Red	$575.8 \times \rho\,671$ (VIIRS)	0.88	10.39	50 > SPM
	$511.9 \times \rho\,645$ (MODIS)	0.88	12.18	
Linear Green	$96.6 \times \rho\,551$ (VIIRS)	0.96	18.15	10 > SPM
	$126.86 \times \rho\,555$ (MODIS)	0.91	11.07	
Bourgneuf-Loire				
Nechad et al. (2010) [16] NIR (recalibrated)	$\frac{3734 \times \rho\,862}{1-\rho\,862/0.2114}$ (VIIRS)	0.93	16.82	SPM > 50
	$\frac{3510 \times \rho\,859}{1-\rho\,859/0.2112}$ MODIS	0.88	9.1	
Nechad et al. (2010) [16] Red (recalibrated)	$\frac{571 \times \rho\,671}{1-\rho\,671/0.1751}$ (VIIRS)	0.88	13.31	50 > SPM > 10
	$\frac{441 \times \rho\,645}{1-\rho\,645/0.1641}$ (MODIS)	0.83	11.87	
Linear Green	$96.6 \times \rho\,551$ (VIIRS)	-	-	SPM < 10
	$126.86 \times \rho\,555$ (MODIS)	-	-	

Table A2. Additional equations developed for OLI bands. Each model was calibrated and evaluated using in situ measurements acquired in each study region. The goodness of fit (r^2) and normalized root mean square error percent (%).

Eq.#	Fit	Equation	r^2	RMSE %
		Gironde		
1	Polynomial NIR (1 band)	$37150\,\rho\,865^2 + 1751\,\rho\,865$	0.97	9.11
	Polynomial NIR/Red (Ratio)	$1454 \times \left(\frac{\rho\,865}{\rho\,655}\right)^2 + 28.2 \times \left(\frac{\rho\,865}{\rho\,655}\right)$		
2	Exponential NIR (1 band)	$89.04\,\exp^{(\rho\,865 \times 16)}$		
	Exponential NIR/Red (Ratio)	$60.94\,\exp^{\left(\frac{\rho\,865}{\rho\,655} \times 3.375\right)}$	0.9	7.82
3	Nechad et al. (2010) [16] NIR adjusted	$\frac{2372 \times \rho\,865}{1-\rho\,865/0.2115}$	0.9	9.16
4	Nechad et al. (2010) [16] Red adjusted	$\frac{231.9 \times \rho\,655}{1-\rho\,655/0.1686}$	0.88	5.2
		Bourgneuf-Loire		
1	Exponential (ratio)	$29.12\,exp^{\left(\frac{\rho\,865}{\rho\,655} \times 5.07\right)}$	0.93	25
2	Exponential 1 band	$36.86 \times exp^{(\rho\,865 \times 38)}$	0.74	46
3	Polynomial ratio	$1039.3 \left(\frac{\rho\,865}{\rho\,655}\right)^2 + 12.644 \left(\frac{\rho\,865}{\rho\,655}\right) + 10.828$	0.93	25
4	Polynomial 1 band	$72848\,\rho\,865^2 + 14108\,\rho\,865 - 2.98318$	0.93	22
5	Nechad et al. (2010) [16] NIR	$\frac{2971.3 \times \rho\,865}{1-\rho\,865/0.2115}$	0.93	7.81
6	Nechad et al. (2010) [16] Red	$\frac{289.29\,\rho\,655}{1-\rho\,655/0.1686}$	0.82	18.17

References

1. Tassan, S. Local algorithm using SeaWiFS data for retrieval of phytoplankton pigments, suspneded sediment and yellow substance in coastal waters. *Appl. Opt.* **1994**, *33*, 2369–2378. [CrossRef] [PubMed]
2. Warrick, J.A.; Mertes, L.A.K.; Washburn, L.; Siegel, D.A. Dispersal forcing of southern California river plumes, based on field and remote sensing observations. *Geo-Mar. Lett.* **2004**, *24*, 46–52. [CrossRef]
3. Miller, R.L.; del Castillo, C.E.; McKee, B.A. *Remote Sensing of Coastal Aquatic Environments. Technologies, Techniques and Applications*; Springer: New York, NY, USA, 2005.
4. Doxaran, D.; Froidefond, J.; Castaing, P.; Babin, M. Estuarine, Coastal and Shelf Science. *Science* **2009**, *81*, 321–332.
5. Doxaran, D.; Leymarie, E.; Nechad, B.; Ruddick, K.G.; Dogliotti, A.I.; Knaeps, E. An improved correction method for field measurements of particulate light backscattering in turbid waters State of the Art. *Opt. Express* **2013**, *24*, 3615–3637. [CrossRef] [PubMed]
6. Petus, C.; Chust, G.; Gohin, F.; Doxaran, D.; Froidefond, J.-M.; Sagarminaga, Y. Estimating turbidity and total suspended matter in the Adour River plume (South Bay of Biscay) using MODIS 250-m imagery. *Cont. Shelf Res.* **2010**, *30*, 379–392. [CrossRef]
7. Lorthiois, T.; Doxaran, D.; Chami, M. Daily and seasonal dynamics of suspended particles in the Rhône River plume based on remote sensing and field optical measurements. *Geo-Mar. Lett.* **2012**, *32*, 89–101. [CrossRef]
8. Vanhellemont, Q.; Neukermans, G.; Ruddick, K. Synergy between polar-orbiting and geostationary sensors: Remote sensing of the ocean at high spatial and high temporal resolution. *Remote Sens. Environ.* **2013**, *146*, 49–62. [CrossRef]
9. Eom, J.; Choi, J.-K.; Won, J.-S.; Ryu, J.-H.; Doxaran, D.; Ruddick, K.; Lee, S. Spatiotemporal Variation in Suspended Sediment Concentrations and Related Factors of Coastal Waters Based on Multispatial Satellite Data in Gyeonggi Bay, Korea. *J. Coast. Res.* **2016**, in press. [CrossRef]
10. Hudson, A.S.; Talke, S.A.; Jay, D.A. Using Satellite Observations to Characterize the Response of Estuarine Turbidity Maxima to External Forcing. *Estuaries Coasts* **2016**. [CrossRef]
11. Kubryakov, A.; Stanichny, S.; and Zatsepin, A. River plume dynamics in the Kara Sea from altimetry-based lagrangian model, satellite salinity and chlorophyll data. *Remote Sens. Environ.* **2016**, *176*, 177–187. [CrossRef]
12. Restrepo, J.D.; Park, E.; Aquino, S.; and Latrubesse, E.M. Coral reefs chronically exposed to river sediment plumes in the southwestern Caribbean: Rosario Islands, Colombia. *Sci. Total Environ.* **2016**, *553*, 316–329. [CrossRef] [PubMed]
13. Ody, A.; Doxaran, D.; Vanhellemont, Q.; Nechad, B.; Novoa, S.; Many, G.; Bourrin, F.; Verney, R.; Pairaud, I.; Gentili, B. Potential of high spatial and temporal ocean color satellite data to study the dynamics of suspended particles in a micro-tidal river plume. *Remote Sens.* **2016**, *8*, 245. [CrossRef]
14. Tan, J.; Cherkauer, K.A.; Chaubey, I.; Troy, C.D.; Essig, R. Water quality estimation of River plumes in Southern Lake Michigan using Hyperion. *J. Gt. Lakes Res.* **2016**, *42*, 524–535. [CrossRef]
15. Doxaran, D.; Castaing, P.; Lavender, S.J.; Castaign, P.; Lavender, S.J. Monitoring the maximum turbidity zone and detecting finescale turbidity features in the Gironde estuary using high spatial resolution satellite sensor (SPOT HRV, Landsat ETM) data. *Int. J. Remote Sens.* **2006**, *27*, 2303–2321. [CrossRef]
16. Nechad, B.; Ruddick, K.G.; Park, Y. Calibration and validation of a generic multisensor algorithm for mapping of total suspended matter in turbid waters. *Remote Sens. Environ.* **2010**, *114*, 854–866. [CrossRef]
17. Forget, P.; Ouillon, S. Surface suspended matter off the Rhone river mouth from visible satellite imagery. *Oceanol. Acta* **1998**, *21*, 739–749. [CrossRef]
18. Doxaran, D.; Froidefond, J.; Castaing, P.; Babin, M. Dynamics of the turbidity maximum zone in a macrotidal estuary (the Gironde, France): Observations from field and MODIS satellite data. *Estuar. Coast. Shelf Sci.* **2009**, *81*, 321–332. [CrossRef]
19. Vanhellemont, Q.; Ruddick, K. Advantages of high quality SWIR bands for ocean colour processing: Examples from Landsat-8. *Remote Sens. Environ.* **2015**, *161*, 89–106. [CrossRef]
20. Neukermans, G.; Ruddick, K.; Bernard, E.; Ramon, D.; Nechad, B.; Deschamps, P.Y. Mapping total suspended matter from geostationary satellites: A feasibility study with SEVIRI in the Southern North Sea. *Opt. Express* **2009**, *17*, 14029–14052. [CrossRef] [PubMed]
21. Neukermans, G.; Ruddick, K.G.; Greenwood, N. Diurnal variability of turbidity and light attenuation in the southern North Sea from the SEVIRI geostationary sensor. *Remote Sens. Environ.* **2012**, *124*, 564–580. [CrossRef]

22. Gordon, H.R.; Wang, M. Retrieval of water-leaving radiance and aerosol optical thickness over the oceans with SeaWiFS: A preliminary algorithm. *Appl. Opt.* **1994**, *33*, 443–452. [CrossRef] [PubMed]

23. Ruddick, K.G.; Ovidio, F.; Rijkeboer, M. Atmospheric correction of SeaWiFS imagery for turbid coastal and inland waters: Comment. *Appl. Opt.* **2000**, *39*, 893–895. [CrossRef]

24. Hu, C.; Carder, K.L.; Muller-Karger, F.E. Atmospheric correction of SeaWiFS imagery over turbid coastal waters: A practical method. *Remote Sens. Environ.* **2000**, *74*, 195–206. [CrossRef]

25. Wang, M.; Shi, W. Estimation of ocean contribution at the MODIS near-infrared wavelengths along the east coast of the U.S.: Two case studies. *Geophys. Res. Lett.* **2005**, *32*, L13606. [CrossRef]

26. Gohin, F. Annual cycles of chlorophyll-a, non-algal suspended particulate matter, and turbidity observed from space and in-situ in coastal waters. *Ocean Sci.* **2011**, *7*, 705–732. [CrossRef]

27. Nechad, B.; Alvera-Azcaràte, A.; Ruddick, K.; Greenwood, N. Reconstruction of MODIS total suspended matter time series maps by DINEOF and validation with autonomous platform data. *Ocean Dyn.* **2011**, *61*, 1205–1214. [CrossRef]

28. Van der Woerd, H.; Pasterkamp, R. Mapping of the North Sea turbid coastal waters using SeaWiFS data. *Can. Remote Sens.* **2004**, *30*, 44–53. [CrossRef]

29. Doxaran, D.; Froidefond, J.; Lavender, S.; Castaing, P. Spectral signature of highly turbid waters Application with SPOT data to quantify suspended particulate matter concentrations. *Remote Sens. Environ.* **2002**, *81*, 149–161. [CrossRef]

30. Moore, G.F.; Aiken, J.; Lavender, S.J. The atmospheric correction of water colour and the quantitative retrieval of suspended particulate matter in Case II waters: Application to MERIS. *Int. J. Remote Sens.* **1999**, *20*, 1713–1733. [CrossRef]

31. Doxaran, D.; Froidefond, J.M.; Castaing, P. A reflectance band ratio used to estimate suspended matter concentrations in sediment-dominated coastal waters. *Int. J. Remote Sens.* **2002**, *23*, 5079–5085. [CrossRef]

32. Dogliotti, A.I.; Ruddick, K.G.; Nechad, B.; Doxaran, D.; Knaeps, E. A single algorithm to retrieve turbidity from remotely-sensed data in all coastal and estuarine waters. *Remote Sens. Environ.* **2015**, *156*, 157–168. [CrossRef]

33. Chen, J.; D'Sa, E.; Cui, T.; Zhang, X. A semi-analytical total suspended sediment retrieval model in turbid coastal waters: A case study in Changjiang River Estuary. *Opt. Express* **2013**, *21*, 13018–13031. [CrossRef] [PubMed]

34. Shen, F.; Verhoef, W.; Zhou, Y.; Salama, M.S.; Liu, X. Satellite estimates of wide-range suspended sediment concentrations in Changjiang (Yangtze) estuary using MERIS data. *Estuaries Coasts* **2010**, *33*, 1420–1429. [CrossRef]

35. Feng, L.; Hu, C.; Chen, X.; Song, Q. Influence of the Three Gorges Dam on total suspended matters in the Yangtze Estuary and its adjacent coastal waters: Observations from MODIS. *Remote Sens. Environ.* **2014**, *140*, 779–788. [CrossRef]

36. Han, B.; Loisel, H.; Vantrepotte, V.; Mériaux, X.; Bryère, P.; Ouillon, S.; Dessailly, D.; Xing, Q.; Zhu, J. Development of a semi-analytical algorithm for the retrieval of suspended particulate matter from remote sensing over clear to very turbid waters. *Remote Sens.* **2016**, *8*, 211. [CrossRef]

37. Doxaran, D.; Devred, E.; Babin, M. A 50% increase in the mass of terrestrial particles delivered by the Mackenzie River into the Beaufort Sea (Canadian Arctic Ocean) over the last 10 years. *Biogeosciences* **2015**, *12*, 3551–3565. [CrossRef]

38. Masson, M.; Schäfer, J.; Blanc, G.; Pierre, A. Seasonal variations and annual fluxes of arsenic in the Garonne, Dordogne and Isle Rivers, France. *Sci. Total Environ.* **2007**, *373*, 196–207. [CrossRef] [PubMed]

39. Castaing, P.; Allen, G.P. Mechanisms controlling seaward escape of suspended sediment from the Gironde: A macrotidal estuary in France. *Mar. Geol.* **1981**, *40*, 101–118. [CrossRef]

40. Froidefond, J.M.; Castaing, P.; Mirmand, M.; Ruch, P. Analysis of the turbid plume of the Gironde (France) bases on SPOT radiometric data. *Remote Sens. Environ.* **1991**, *36*, 149–163. [CrossRef]

41. Allen, G.P.; Sauzay, G.; Castaing, P.; Jouanneau, J.M. Transport and deposition of suspended sediment in the Gironde estuary, France. *Estuar. Process.* **1977**, *2*, 63–81.

42. Gernez, P.; Lafon, V.; Lerouxel, A.; Curti, C.; Lubac, B.; Cerisier, S.; Barillé, L. Toward Sentinel-2 high resolution remote sensing of suspended particulate matter in very turbid waters: SPOT4 (Take5) experiment in the loire and gironde estuaries. *Remote Sens.* **2015**, *7*, 9507–9528. [CrossRef]

43. Gernez, P.; Barille, L.; Lerouxel, A.; Mazeran, C.; Lucas, A.; Doxaran, D. Remote sensing of suspended particulate matter in turbid oyster-farming ecosystems. *J. Geophys. Res. C Oceans* **2014**, *119*, 7277–7294. [CrossRef]

44. Ruddick, K.G.; De Cauwer, V.; Park, Y.-J.; Moore, G. Seaborne measurements of near infrared water-leaving reflectance: The similarity spectrum for turbid waters. *Limnol. Oceanogr.* **2006**, *51*, 1167–1179. [CrossRef]

45. Mueller, J.L.; Davis, C.; Amone, R.; Frouin, R.; Carder, K.; Lee, Z.P.; Steward, R.G.; Hooker, S.; Mobley, C.D.; McLean, S. Above-water radiance and remote sensing reflectance measurement and analysis protocols. In *Ocean Optics for Satellite Ocean Color Sensor Validation*; Revision 2; Fargion, G.S., Mueller, J.L., Eds.; NASA: Greenbelt, MD, USA, 2000; pp. 98–107.

46. Tilstone, G.H.; Moore, G.F.; Sorensen, K.; Doerffer, R.; Røttgers, R.; Ruddick, K.G.; Pasterkamp, R.; Jorgensen, P.V. Regional Validation of MERIS Chlorophyll Products in North Sea Coastal Water. Available online: http://citeseerx.ist.psu.edu/viewdoc/download?doi=10.1.1.556.8868&rep=rep1&type=pdf (accessed on 4 January 2017).

47. Van der Linde, D.W. *Protocol for the Determination of Total Suspended Matter in Oceans and Coastal Zones*; Technical Note No. I.98.; CEC-JRC: Ispra, Italy, 1998.

48. Irons, J.R.; Dwyer, J.L.; Barsi, J.A. The next Landsat satellite: The Landsat Data Continuity Mission. *Remote Sens. Environ.* **2012**, *122*, 11–21. [CrossRef]

49. Oudrari, H.; McIntire, J.; Xiong, X.; Butler, J.; Ji, Q.; Schwarting, T.; Lee, S.; Efremova, B. JPSS-1 VIIRS radiometric characterization and calibration based on pre-launch testing. *Remote Sens.* **2016**, *8*, 41. [CrossRef]

50. Xiong, X.; Chiang, K.; Sun, J.; Barnes, W.L.; Guenther, B.; Salomonson, V.V. NASA EOS Terra and Aqua MODIS on-orbit performance. *Adv. Space Res.* **2009**, *43*, 413–422. [CrossRef]

51. Vanhellemont, Q.; Ruddick, K. Turbid wakes associated with offshore wind turbines observed with Landsat 8. *Remote Sens. Environ.* **2014**, *145*, 105–115. [CrossRef]

52. Dogliotti, A.; Ruddick, K. Improving water reflectance retrieval from MODIS imagery in the highly turbid waters of La Plata River. In Proceedings of the VI International Conference Current Problems in Optics of Natural Waters (ONW'2011), St. Petersburg, Russia, 6–9 September 2011; pp. 3–10.

53. Hagolle, O.; Huc, M.; Pascual, D.V.; Dedieu, G. A multi-temporal and multi-spectral method to estimate aerosol optical thickness over land, for the atmospheric correction of FormoSat-2, LandSat, VENμS and Sentinel-2 images. *Remote Sens.* **2015**, *7*, 2668–2691. [CrossRef]

54. Guenther, B.; de Luccia, F.; Mccarthy, J.; Moeller, C.; Xiong, X.; Murphy, R.E. Performance Continuity of the A-Train MODIS Observations: Welcome to the NPP VIIRS. Available online: https://www.star.nesdis.noaa.gov/jpss/documents/meetings/2011/AMS_Seattle_2011/Poster/A-TRAIN%20%20Perf%20Cont%20%20MODIS%20Observa%20-%20Guenther%20-%20WPNB.pdf (accessed on 4 January 2017).

55. Jiang, L.; Wang, M. Improved near-infrared ocean reflectance correction algorithm for satellite ocean color data processing. *Opt. Express* **2014**, *22*, 443–452. [CrossRef] [PubMed]

56. Etcheber, H.; Schmidt, S.; Sottolichio, A.; Maneux, E.; Chabaux, G.; Escalier, J.-M.; Wennekes, H.; Derriennic, H.; Schmeltz, M.; Quéméner, L.; et al. Monitoring water quality in estuarine environments: Lessons from the MAGEST monitoring program in the Gironde fluvial-estuarine system. *Hydrol. Earth Syst. Sci.* **2011**, *15*, 831–840. [CrossRef]

57. Vanhellemont, Q.; Bailey, S.; Franz, B.; Shea, D.; Directorate, O.; Environment, N. Atmospheric Correction of Landsat-8 Imagery Using Seadas. In Proceedings of the Sentinel 2 for Science Workshop, Frascati, Italy, 20–23 May 2014.

58. Bru, D. Corrections Atmosphériques Pour Capteurs À Très Haute Résolution Spatiale En Zone Littorale. Ph.D. Thesis, University of Bordeaux, Bordeaux Cedex, France, 2015.

59. Wang, M.; Liu, X.; Tan, L.; Jiang, L.; Son, S.; Shi, W.; Rausch, K.; Voss, K. Impacts of VIIRS SDR performance on ocean color products. *J. Geophys. Res. Atmos.* **2013**, *118*, 10347–10360. [CrossRef]

60. Vandermeulen, R.A.; Arnone, R.; Ladner, S.; Martinolich, P. Enhanced satellite remote sensing of coastal waters using spatially improved bio-optical products from SNPP-VIIRS. *Remote Sens. Environ.* **2015**, *165*, 53–63. [CrossRef]

61. Knaeps, E.; Ruddick, K.G.; Doxaran, D.; Dogliotti, A.I.; Nechad, B.; Raymaekers, D.; Sterckx, S. A SWIR based algorithm to retrieve total suspended matter in extremely turbid waters. *Remote Sens. Environ.* **2015**, *168*, 66–79. [CrossRef]

remote sensing

MDPI

Article

Assessment of Atmospheric Correction Methods for Sentinel-2 MSI Images Applied to Amazon Floodplain Lakes

Vitor Souza Martins [1,*], Claudio Clemente Faria Barbosa [1], Lino Augusto Sander de Carvalho [1], Daniel Schaffer Ferreira Jorge [1], Felipe de Lucia Lobo [2] and Evlyn Márcia Leão de Moraes Novo [2]

[1]   Image Processing Division, Brazilian Institute for Space Research, São José dos Campos 12227-010, Brazil; claudio.barbosa@inpe.br (C.C.F.B.); lino@dsr.inpe.br (L.A.S.C.); danielsfj@dsr.inpe.br (D.S.F.J.)
[2]   Remote Sensing Division, Brazilian Institute for Space Research, São José dos Campos 12227-010, Brazil; felipe.lobo@inpe.br (F.L.L.); evlyn.novo@inpe.br (E.M.L.M.N.)
*   Correspondence: vitor.martins@inpe.br or vitorstmartins@gmail.com; Tel.: +55-12-3208-6809

Academic Editors: Yunlin Zhang, Claudia Giardino, Linhai Li, Xiaofeng Li and Prasad S. Thenkabail
Received: 16 January 2017; Accepted: 24 March 2017; Published: 29 March 2017

Abstract: Satellite data provide the only viable means for extensive monitoring of remote and large freshwater systems, such as the Amazon floodplain lakes. However, an accurate atmospheric correction is required to retrieve water constituents based on surface water reflectance (R_W). In this paper, we assessed three atmospheric correction methods (Second Simulation of a Satellite Signal in the Solar Spectrum (6SV), ACOLITE and Sen2Cor) applied to an image acquired by the MultiSpectral Instrument (MSI) on-board of the European Space Agency's Sentinel-2A platform using concurrent in-situ measurements over four Amazon floodplain lakes in Brazil. In addition, we evaluated the correction of forest adjacency effects based on the linear spectral unmixing model, and performed a temporal evaluation of atmospheric constituents from Multi-Angle Implementation of Atmospheric Correction (MAIAC) products. The validation of MAIAC aerosol optical depth (AOD) indicated satisfactory retrievals over the Amazon region, with a correlation coefficient (R) of ~0.7 and 0.85 for Terra and Aqua products, respectively. The seasonal distribution of the cloud cover and AOD revealed a contrast between the first and second half of the year in the study area. Furthermore, simulation of top-of-atmosphere (TOA) reflectance showed a critical contribution of atmospheric effects (>50%) to all spectral bands, especially the deep blue (92%–96%) and blue (84%–92%) bands. The atmospheric correction results of the visible bands illustrate the limitation of the methods over dark lakes ($R_W < 1$%), and better match of the R_W shape compared with in-situ measurements over turbid lakes, although the accuracy varied depending on the spectral bands and methods. Particularly above 705 nm, R_W was highly affected by Amazon forest adjacency, and the proposed adjacency effect correction minimized the spectral distortions in R_W (RMSE < 0.006). Finally, an extensive validation of the methods is required for distinct inland water types and atmospheric conditions.

Keywords: Amazon inland water; MAIAC aerosol product; adjacency correction; TOA simulation; MODIS atmospheric product; atmospheric correction

---

1. Introduction

Inland waters are an essential resource for terrestrial life and ecosystem services [1,2]. The Amazon freshwater is an ecosystem bearing one of the highest biodiversities in the world [3]. Amazonian aquatic systems depend on satellite image applications to investigate bio-optical parameters due to the extent and limitations of in-situ measurements [4–7]. Thus, remote sensing images have long been recognized as a potential data source for the continuous modelling and monitoring of the water quality [8].

The new generation of orbital optical sensors, such as Sentinel-2 and Landsat-8, presents a scientific opportunity for inland water research [9]. The Multispectral Imager (MSI) on-board Sentinel-2A delivers images with high spatial (10–30 m), temporal (10 days) and radiometric (12 bits) resolutions [10]. These configurations offer capabilities for the mapping of small and irregular open-water systems, higher sensitivity to bio-optical variables and higher temporal observations enabling the monitoring of changes in the water composition over time. In addition, MSI has been designed with eight spectral bands in the visible and near-infrared (NIR) wavelengths that are feasible for water research of the main optically active components (OACs): chlorophyll-a (Chl-a), total suspended solids (TSS) and coloured dissolved organic matter (CDOM) [11]. In a preliminary assessment of the MSI application, Toming et al. [12] reported reasonable retrievals of Chl-a and CDOM concentrations based on a historical dataset from Estonian lakes. Thus, Sentinel-2 MSI data represent a new perspective for inland and coastal waters [13–15].

Atmospheric correction is a prerequisite to quantify biogeochemical properties based on surface reflectance, once it removes attenuation effects caused by active atmospheric constituents, such as molecular and aerosol scattering and absorption by water vapour, ozone, oxygen and carbon dioxide [16]. In fact, due to the low reflectance, the accurate removal of atmospheric effects is paramount for water surfaces [17]. The surface reflectance quality is highly dependent on the atmospheric correction method, atmospheric-surface characteristics, and sensor design [18].

Several atmospheric correction algorithms are available for multispectral sensors which can be divided in two main categories [19]: (i) image-based approach; and (ii) atmospheric radiative transfer codes (RTCs). In the first category, the atmospheric effects are derived from the image itself and then removed from the TOA signal. For instance, ESA provides a Sentinel toolbox that includes a Sen2Cor processor to generate MSI land products (Level 2A). This processor is a semi-empirical algorithm that integrates image-based retrievals with Look-Up tables (LUTs) from the LibRadtran model to remove atmospheric effects from MSI images [20]. In parallel, Vanhellemont and Ruddick [21] developed an image-based processor, named ACOLITE, for the atmospheric correction of Operational Land Imager (OLI) and MSI images applied to marine and inland water studies. The ACOLITE computes aerosol scattering using Rayleigh-corrected reflectance from NIR bands for clear water and SWIR bands for moderate and turbid water; the water contribution measured in these bands can be negligible [22]. Overall, both ACOLITE and Sen2Cor are image-based approaches available for MSI images and present an advantage in regions without external atmospheric information. In the second category, RTCs compute scattering and absorption of light through the atmosphere to remove them from the signal measured by satellite sensors. The 6S vector version (Second Simulation of a Satellite Signal in the Solar Spectrum) is a well-established RTC that accounts for a wide variety of atmospheric conditions and sensor characteristics [23]. However, the main implication of using RTC is the prior knowledge about atmospheric parameters (e.g., aerosol optical depth (AOD), water vapour and ozone) coinciding with the satellite overpass. In general, this information is available from climatological models [24], sun photometer measurements [25] or satellite atmospheric products [26]. Among the alternatives, recent Moderate Resolution Imaging Spectroradiometer (MODIS) algorithm, named Multi-angle implementation of atmospheric correction (MAIAC), provides a suite of atmospheric products (AOD, cloud mask and water vapour) at fine 1 km resolution [27], which is promising for enhanced quality information in regions with high cloud cover areas such as the Amazon region [28].

In addition to atmospheric correction issues, the contribution of adjacency effects also demands correction [29,30]. In the Amazon context, the presence of dense forest around water bodies contributes to modify the water spectrum measured by orbital sensors. Therefore, remote sensing of the Amazonian water system faces several challenges, such as: (i) dynamic system with optically complex water; (ii) logistical difficulties in collecting water samples and validation data; (iii) seasonal variability of aerosol loading from biomass burning plumes (iv); high cloud cover and cloud cirrus; and (v) forest adjacency effects. In view of these challenges, the critical assessment of atmospheric correction methods applied to new Sentinel 2 MSI image is required, which is still missing for Amazon lakes.

Therefore, our objective is to present an inter-comparison of three atmospheric correction algorithms (6SV based on MAIAC atmospheric product, ACOLITE and Sen2Cor) applied to new a Sentinel-2 MSI image in the case of Amazon floodplain lakes. Regarding atmospheric correction, we conducted a supplementary analysis to understand atmospheric components in the study area, and then, simulated the contribution of atmospheric and surface reflectance to MSI TOA bands. Finally, we developed an adjacency correction based on the Linear Spectral Unmixing (LSU) model for water surfaces, due to the strong forest adjacent effects on the water spectrum. All comparisons were conducted over four Amazon floodplain lakes using in-situ radiometric measurements concurrent to the MSI image overpass.

2. Materials

2.1. Site Description and Field Data

Our study area consists of four Amazon floodplain lakes located in the Mamirauá Sustainable Development Reserve (MSDR), close to the confluence of the Solimões and Japurá Rivers (~25 km) (Figure 1). The MSDR is a complex floodplain ecosystem that remains entirely flooded for 3–6 months due to seasonal water level variation. The annual average amplitude of the water level reaches ~10.6 m [31]. In the MSDR lakes, OACs concentrations change seasonally driven by exchange flow (in- and out-flowing) with large fluvial systems, such as the Japurá and Solimões rivers [32]. Few studies reported the bio-optical properties for this Amazon region (e.g., Affonso et al. [33]), since most efforts concentrated on multidisciplinary reports about the ecological management, fish communities and ecosystem disturbance [34,35]. In general, MSDR integrates a sustainable use of natural resources and preservation practices, where local communities are committed to rational resource exploitation, and biodiversity protection in the reserve [36]. Thus, the MSDR represents an ecological and sustainable model for human–environment relations, and becomes an attractive region for further studies of bio-optical patterns and natural conservation using remote sensing data.

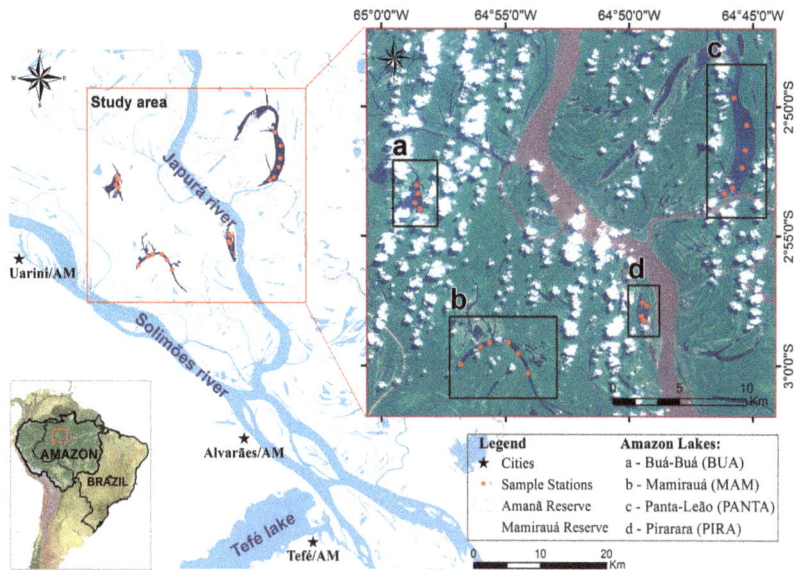

Figure 1. Overview of study area and sample stations over four Amazon floodplain lakes: (**a**) Buá-Buá; (**b**) Mamirauá; (**c**) Panta-leão; and (**d**) Pirarara.

Radiometric measurements were carried out at twenty sample stations during 12–19 August 2016 (Figure 1). Inter-calibrated Trios-RAMSES radiometers were used to measure the above water upwelling radiance (L_w^{+0} [W [watt]$\cdot$m$^{-2}\cdot$sr$^{-1}\cdot$nm^{-1}]), sky radiance (L_{sky}^{+0} [wW [watt]$\cdot$m$^{-2}\cdot$sr$^{-1}\cdot$nm^{-1}]) and above surface downwelling irradiance (E_d^{+0} [W [watt]$\cdot$m$^{-2}\cdot$sr$^{-1}\cdot$nm^{-1}]), within 350–900 wavelengths. The sensors view followed the framed description of Mobley [37], whereas L_w^{+0} has a relative azimuth angle (ϕ_v) within 90°–135° from the sun and a zenith angle (θ_v) of 45° from the nadir, and L_{sky} has a zenith angle (θ'_v) of $\theta_v + 90°$ from nadir. All radiometers operate simultaneously, and measurements were performed within a 3-h interval (10:00 a.m.–13:00 p.m.) to avoid potential impact of specular reflection (glint) at low sun angles. In the pre-processing, all spectroradiometric measurements were interpolated for 1 nm interval (originally ~3.3 nm) and were normalized by sky reference. The remote sensing reflectance (R_{rs}) was calculated at each sampling station according to Mobley [37]:

$$R_{rs}(\theta_v, \phi_v, \lambda) = \left(\frac{L_w^{+0}(\theta_v, \phi_v, \lambda) - r_{sky}(\theta'_v, \phi_v, \theta_0, W) \times L_{sky}^{+0}(\theta'_v, \phi_v, \lambda)}{E_d^{+0}(\lambda)} \right) \tag{1}$$

Afterwards, water reflectance (R_{w^*}) is calculated as:

$$R_{w^*} = \pi \cdot R_{rs} \tag{2}$$

where, r_{sky} is the air-water interface reflection coefficient that minimizes skylight reflection effects, and can be obtained in Mobley [38] as a function of a given view zenith and azimuth angles (θ_v; ϕ_v), sun zenith angle (θ_0) and wind speed W (m/s). In-situ R_{w^*} spectra were weighted by spectral response functions SRF(λ) of MSI bands, thus deriving a multi-spectral data comparable to atmospherically corrected MSI-reflectance from image.

$$R_{w,situ}(\lambda_i) = \frac{\int_k R_{w^*}(\lambda) \times SRF(\lambda) \, d\lambda}{\int_k SRF(\lambda) \, d\lambda} \tag{3}$$

where $R_{w,situ}(\lambda_i)$ is the MSI reflectance simulated from in-situ reflectance, k is bandwidth (nm), λ_i is the central wavelength of spectral band, and *i* is the number of MSI spectral band.

Figure 2 shows the magnitude contrast of the mean and standard deviation of R_{w^*} spectra among lakes. Although all lakes present a typical low spectral reflectance (<2%), Panta-Leão and Pirarara lakes have 2.5 times higher spectral R_{w^*} than those of Buá-Buá and Mamirauá lakes. Indeed, boundary conditions influence bio-optical differences between these lakes, and consequently, contribute to the shape and magnitude contrast of water reflectances. In this context, Mamirauá and Buá-Buá lakes, hereafter called dark lakes, receive a great amount of organic matter content due to the interaction with dense forest reaching heights of up to 40 m (Wittmann et al. 2004). On the other hand, Panta-Leão and Pirarara lakes, hereafter called bright lakes, are directly connected to the Japurá River and exchange a huge volume of water with high sediment loading. We therefore established all discussions based on these two distinct optical conditions; results are referred to as a function of dark and bright lakes.

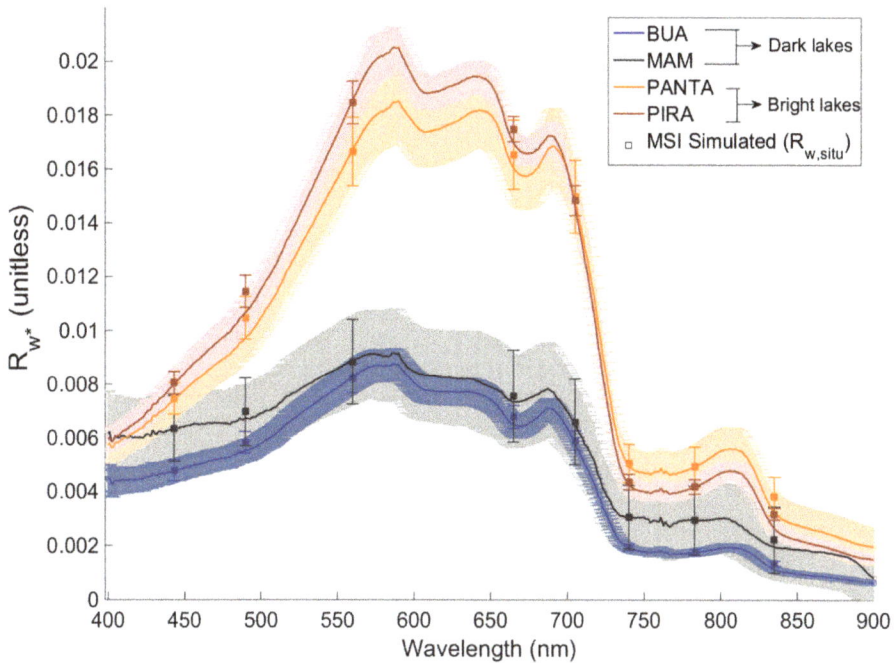

Figure 2. Average (solid line) and standard deviation (shadow-coloured) of water reflectance measured on four Amazon lakes: Buá-Buá (BUA), Mamirauá (MAM), Panta-Leão (PANTA) and Pirarara (PIRA). The square markers and error bars are Multi-Spectral Instrument (MSI) reflectance simulated ($R_{w,situ}$) and their standard deviation, respectively.

2.2. MSI/Sentinel-2 Data

Multi-Spectral Instrument (MSI) on board Sentinel-2 satellite is an optical pushbroom sensor that acquires multi-spectral data for Earth science [39]. The Sentinel-2 mission includes two identical satellites operating in sun-synchronous orbit, with operational Sentinel-2A satellite launched in June 2015, and Sentinel-2B planned for 2017. These twin polar-orbiting satellites allow a high 5-day revisit time of the equator (after the launch of Sentinel-2B), because they are phased at 180° to each other. The high-resolution MSI data include 13 spectral bands from Visible and Near-Infrared (VNIR) to Short Wave InfraRed (SWIR) region, fine spatial resolution (10, 20 and 60 m), and 12 bit quantization (Table 1) [10]. Additionally, the MSI sensor enhances spectral capabilities that include useful bands for land and atmospheric observations, such as the deep blue band (443 nm) for coastal and aerosol retrievals, cirrus detection at 1375 nm and three red-edge bands for vegetation and water studies [40]. Therefore, all those high optical properties configure an attractive sensor for inland water studies, in particular, over high cloud cover regions (e.g., Amazon Basin) due to the high temporal resolution. The standard MSI scene is delivered as Level-1C (L1C) product with radiometric and geometric correction in UTM/WGS84 projection [41].

Simultaneously with the field observations, a cloud-free MSI image was acquired on 12 August 2016 at 14:37 UTC. Our study area is located within MSI granule 20MKB, which was downloaded from the Copernicus Scientific Data Hub website. As a first procedure, all MSI bands were resampled to a 10 m pixel size and the TOA reflectance was divided by a rescaling coefficient of 10,000. The granule has ~2% of cloud cover on granule, and particular cloud-free conditions over our twenty sample stations. Thus, in-situ measurements can be used to compare the MSI surface reflectance derived from atmospheric correction methods.

The time gap between in-situ measurements and the satellite overpass affects the reflectance comparison. Several studies discussed the time gap with respect to reservoir and lake research and the results pointed out that a comparison using measurements with ±3 or up to ±8 days is reasonable when the water and environmental conditions do not present rapid changes [42–45]. In our case, logistical and distances imposed difficulties to access all floodplain lakes concurrently with the satellite overpass. Thus, we started the radiometric measurements on 12 August concurrently with Sentinel-2 overpass, and were sampling new stations every day until all sample stations were completed by 19 August. The number of stations was limited to 2–3 per day to guarantee feasible solar conditions. Note that more than 50% of all sample station data were collected within three days from the satellite overpass, reducing the temporal influence on the radiometric dataset. Although the lakes exchange water with the Japurá River, variation of the water level during this season is a gradual process that relies on channel connections and hydrological periods. Additionally, all lakes presented a depth (>5 m) and low wind speed (~1 m/s) that minimize resuspension and circulation of sediment from the bottom. Therefore, the time gap between MSI image (12 August) and in-situ measurements (12–19 August) was in principle not considered to be an issue for the comparisons.

Table 1. Spectral bands of MSI sensor on-board Sentinel-2 satellite.

MSI Bands (Spatial Resolution)	Central Wavelength (nm)	Bandwidth (nm)	Lref (W·m^{-2}·sr^{-1}·μm^{-1})	SNR at Lref
Band 1 (60 m)	443 (Deep blue)	20	129	129
Band 2 (10 m)	490 (Blue)	65	128	154
Band 3 (10 m)	560 (Green)	35	128	168
Band 4 (10 m)	665 (Red)	30	108	142
Band 5 (20 m)	705 (Red-edge)	15	74.5	117
Band 6 (20 m)	740 (Red-edge)	15	68	89
Band 7 (20 m)	783 (Red-edge)	20	67	105
Band 8 (10 m)	842 (NIR)	115	103	172
Band 8A (20 m)	865 (NIR)	20	52.5	72
Band 9 (60 m)	945 (NIR)	20	9	114
Band 10 (60 m)	1375 (SWIR)	30	6	50
Band 11 (20 m)	1610 (SWIR)	90	4	100
Band 12 (20 m)	2190 (SWIR)	180	1.5	100

2.3. MAIAC Atmospheric Data

Several MODIS algorithms were developed to provide atmospheric products, such as aerosol optical depth, column water vapour and ozone [46–48]. Continuous efforts have been made to enhance the accuracy of atmospheric retrievals from MODIS data. In addition to climate research, these atmospheric products are also used as input in the atmospheric correction of optical images [49]. In this context, the MAIAC algorithm was developed to derive a surface bidirectional reflectance distribution function (BRDF) from MODIS data and a suite of atmospheric products at a high 1 km resolution [27]. These atmospheric products include the cloud and cloud shadow mask, aerosol optical depth at 0.47 and 0.55 μm and column water vapour (U_{H2O}) (see Lyapustin et al. [27,50] for more details). Figure 3 shows the MOD09 surface reflectance product and a comparison of three MODIS aerosol products for the same day of MSI image (12 August 2016): (Figure 3b) MAIAC AOD 1 km; (Figure 3c) MOD04 3 km Collection 6; and (Figure 3d) MOD04 10 km Collection 6. There is a clear difference in the number of valid AOD retrievals between fine resolution MAIAC and MOD4 AOD products. Cloudy conditions limited the wide coverage of MOD04 retrievals at 3 and 10 km resolutions, while the fine-scale MAIAC retrieves AOD information for an individual set of cloud-free pixels. In our case, both Sentinel-2A and Terra satellite have an almost concurrent overpass (~10:30 a.m.), which guarantees fair applications of MODIS atmospheric products as auxiliary information in atmospheric correction. Thus, we selected tile h01v01 of the MAIAC atmospheric product on 12 August 2016. The average of AOD_{550} and U_{H2O} within a 2 km buffer from each lake was used as input for the 6SV model (Table 2). In addition, columnar ozone content was obtained from MODIS global daily product (MOD08_D3-Total_Ozone_mean) on 12 August 2016.

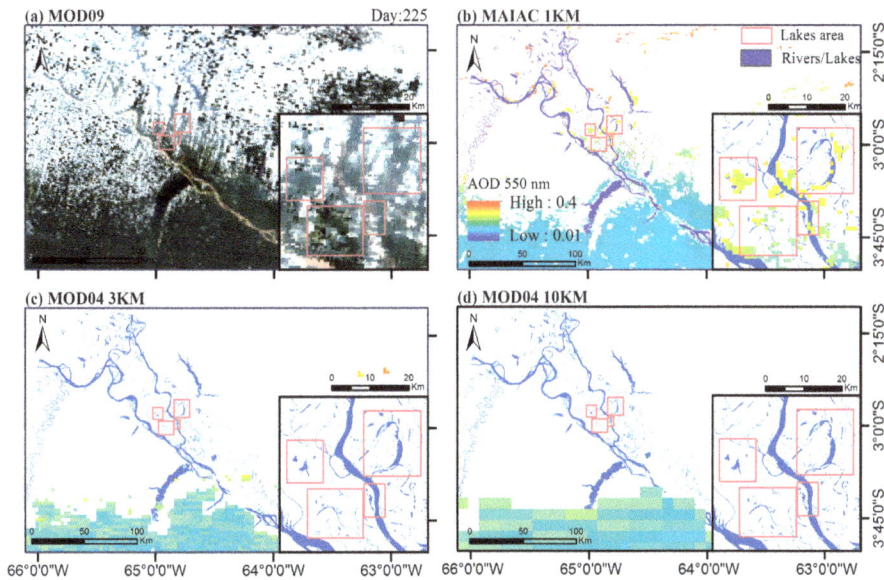

Figure 3. MODIS AOD products concurrently with MSI image on 12 August 2016: (**a**) MOD09 surface reflectance; (**b**) MAIAC AOD$_{550}$ 1 km; (**c**) MOD04 3 km Collection 6; and (**d**) MOD04 10 km Collection 6.

Table 2. Input parameters of 6SV model for Sentinel MSI image.

Parameters	BUA	MAM	PANTA	PIRA
Solar zenith angle (°)	30.96	30.96	30.96	30.96
Solar azimuth angle (°)	53.99	53.99	53.99	53.99
Aerosol Model	Biomass Burning			
AOD at 550 nm [1]	0.3	0.26	0.34	0.3
Ozone (cm-atm)	0.346	0.346	0.346	0.346
Water vapour (g/cm^2)	4.88	4.7	4.06	4.15
Terrain elevation (km)	0.04	0.04	0.04	0.04

[1] AOD adjusted by Terra bias (Section 4.1).

The assessment of MAIAC AOD$_{550}$ is still missing in the Amazon region. Therefore, we performed an evaluation of this satellite AOD product by comparing it with ground measurements. In this context, AERONET program is a global sun-photometer network that provides a multi-spectral sun and sky radiance to derive aerosol optical properties, such as AOD and angstrom exponent information [25]. There are three long-term operational AERONET sites in the Amazon basin: Balbina (1993–2003), Belterra (1996–2005), and Manaus-Embrapa (2011–2016). These sites provide consolidated ground-truth data for the quality evaluation of satellite aerosol products. Therefore, we compared MAIAC AOD retrievals at 550 nm with AERONET observations based on procedures used by Lyapustin et al. [27]. Besides validation of AOD products, we also calculated the monthly average of AOD$_{550}$, columnar water vapour and cloud cover frequency within the Mamirauá region (see red box in Figure 1) using the 15-year MAIAC Terra products (2000–2015). This temporal analysis provides background knowledge of the most variable atmospheric constituents in the study area.

3. Methods

In this section, we present three atmospheric correction methods used for the MSI image and forest adjacency correction: (Section 3.1.) 6SV model based on MAIAC AOD$_{550}$ and U$_{H2O}$ products; (Section 3.2.) ACOLITE algorithm; (Section 3.3) Sen2Cor algorithm; and (Section 3.4) forest adjacency correction based on the LSU model. Based on the MSI-corrected image, we calculated the average of water surface reflectance using a 3 × 3-pixel box centered at each sample station to perform a direct comparison with in-situ measurements.

3.1. 6SV Model + MAIAC Atmospheric Products

The 6SV model is a robust radiative transfer code for the atmospheric correction of different satellite data for a variety of climatological conditions [23]. The atmospheric radiative transfer computes attenuation effects caused by the scattering of molecules and aerosols and gaseous absorption by water vapour (H$_2$O), carbon dioxide (CO$_2$), oxygen (O$_2$), and ozone (O$_3$). Currently, the 6SV model is an operational model used to derive the surface reflectance product from MODIS, ETM+ and OLI images [51,52]. The comparison of 6SV with other complex RTCs, such as SHARM, DISORT and MODTRAN, showed that the vector mode is highly accurate and provides fair agreement results that agree with other RTCs [53]. Due to consistency of multiple sensors and generic features, we used the 6SV model (version 1.1) to evaluate the atmospheric correction of the MSI image. For a given sun-view geometry, sensor characteristics, atmospheric condition and surface reflectance (R$_{sur}$), TOA reflectance can be estimated with the following Equation (4) [23]:

$$R_{TOA}(\lambda, \theta_v, \phi_v, \theta_0, \phi_0) = [R_{R+A}(\lambda, \theta_v, \theta_0, \phi_0) + t_d(\lambda, \theta_0)t_u(\lambda, \theta_v)\frac{R_{sur}(\lambda)}{1-S(\lambda)R_{sur}(\lambda)}]T_g(\lambda, \theta_v, \theta_0,) \tag{4}$$

where T$_g$ refers to gaseous transmission of the principal absorbing constituents (O$_2$, O$_3$, CO$_2$, H$_2$O); R$_{R+A}$ is the molecular and aerosol scattering intrinsic reflectance; t$_d$ and t$_u$ represent the atmospheric transmittance of aerosol and molecular from sun to target and target to sensor, respectively; S is the atmosphere spherical albedo of the atmosphere; and θ$_v$, φ$_v$, θ$_0$, φ$_0$ are the view zenith, view azimuth, solar zenith and solar azimuth angles, respectively. Solving Equation (4) for surface reflectance and simplifying notations of angles, the atmospheric correction proceeds as following Equation (5):

$$R_{sur} = (R_{TOA}/T_g - R_{R+A})/[t_dt_u + S(R_{TOA}/T_g - R_{R+A})] \tag{5}$$

These atmospheric quantities are internally generated when running the model. In this study, the 6SV model was set to MSI bands using a spectral response function. Subsequently, the code was run for each subset of the MAIAC-based atmospheric data according to each lake, applying water vapour (U$_{h_2O}$, measured in g·cm^{-2}), ozone content (U$_{O_3}$, measured in cm·atm^{-1}), aerosol model and AOD at 550 nm described in Table 2. We selected a biomass burning model based on the global aerosol mixture from Taylor et al. [24], which indicated a dominance of biomass burning particles in August in the Mamirauá region (biomass burning: 72%, Sulphate: 22.8%, Maritime: 2.5%, and Dust: 2.7%). The sun angles, date and time of the image acquisition were obtained from the image metadata.

To understand the contribution of the atmosphere and surface to the TOA signal measured by MSI sensor, and to assess the atmospheric effects according to spectral bands, the TOA reflectance was simulated from the above mentioned water reflectance for each lake (Equation (3)). This theoretical TOA reflectance was simulated using the average AOD of August in the lakes (BUA: 0.178; MAM: 0.188; PANTA: 0.181; and PIRA: 0.19), biomass burning model, tropical atmosphere profile, and the average of in-situ reflectances from each lake.

3.2. ACOLITE Algorithm

The Atmospheric Correction for OLI "lite" (ACOLITE) algorithm was developed for the atmospheric correction of OLI/Landsat 8 and MSI/Sentinel 2 images for ocean and inland water studies [21,54,55]. The ACOLITE algorithm removes scattering effects of molecular and aerosol components over clear and turbid water. The Rayleigh scattering was corrected using LUTs from the 6SV model, while aerosol scattering was estimated based on the NIR (842 and 865 nm) bands for clear water and SWIR (1610 and 2130 nm) bands for moderate and turbid water [55]. These bands are very useful to decouple the aerosol reflectance, because the water contribution can be assumed to be negligible. Thus, the aerosol reflectance is retrieved at those bands and extrapolated to VNIR wavelengths based on aerosol type (ε) or on ratio of Rayleigh corrected reflectance in these infrared bands.

The algorithm (version 2016.05.20) allows the user to choose some inputs for the atmospheric correction: (i) derive ε fixe on scene, per pixel or user-defined; (ii) gain factors for radiometric calibration [56,57]; (iii) atmospheric pressure; (iv) smooth window applied to aerosol reflectance values; and (v) cloud mask threshold (default: 0.0215 on the 1610 nm band). In our study, atmospheric correction was performed using the SWIR band approach, as recommended for turbid water [55]; aerosol correction per-pixel; a smooth window of 25 pixels; and cloud dilatation of 16 pixels (default).

3.3. Sen2Cor Algorithm

The Sentinel 2 MSI data are distributed as ortho-image TOA reflectance products. To derive the MSI land products at Level-2A, the Sentinel toolbox provides the Sen2Cor processor for atmospheric correction and scene classification [20]. As a module of the Sen2Cor algorithm, an operational atmospheric correction is applied to the MSI spectral bands to retrieve atmospheric parameters from the image itself, with cirrus correction in a channel at 1375 nm; water vapour retrieval based on the B8A and B9 bands (865, 945 nm) and AOD retrieval [58]. Thus, the algorithm performs a semi-empirical approach that associates image-derived atmospheric properties with the pre-computed Look-up table (LUT) from libRadtran radiative transfer model. The advantage of this image-based approach is that is supports the application in regions without climatological information. The spectral relation of the reflectance of B4 (665 nm) and B2 (490 nm) bands with B12 (2190 nm) band is used for the AOD retrievals in reference areas, such as dense dark vegetation (DDV) surfaces [59]. In the Amazon region, DDV surfaces around lakes benefit AOD retrievals due to the strong reflectance relation between visible and SWIR bands and temporal stability in these preserved areas. When these vegetated areas are not available on scene, the algorithm identifies dark soil and water surfaces, or applies a default visibility of 20 km. To run the Sen2Cor algorithm (version 2.2.1), we chose the rural aerosol, ozone content of 330 D.U., smooth window of 100×100 m^2 box applied to the water vapour map, and adjacency correction within the 1000×1000 m^2 box.

3.4. Adjacency Effect Correction

Adjacency effects are an optical-physical process caused by molecular and aerosol scattering where the target view is affected by radiation reflected from neighbourhood surfaces [60]. These multiple scattering regimes of photons from adjacent areas modify the spectral signal of the target pixel. The magnitude of these effects depends on the: (i) atmospheric turbidity, at a particular aerosol scattering phase function; (ii) spectral contrast of the surface reflectance; and (iii) sensor characteristics [61–63]. The influence of atmospheric scattering on adjacency effects increases as a result of the high optical thickness [64,65]. This impact produces blurring effects that distort the effective spatial resolution, reduce the apparent surface contrast and affect the land cover characterization [62,66]. In case of Sentinel-2 MSI, the actual spatial resolution allows discrimination of small water bodies [67], however, atmospheric scattering might change the effective pixel size and quality of the image according to aerosol microphysical properties (particle-size distribution, composition, and particle shapes) and loading [63]. Moreover, the target spectrum is more susceptible to adjacency effects in heterogeneous areas, especially, at high-spatial resolution [68]. In coastal and inland waters, these effects are more evident due to typical lower reflectance in relation to their neighbourhood surfaces, such as sand close to coastal waters and vegetation around reservoirs and lakes [30].

In our study, the large contrast of reflectance between forest and water surfaces contributes to spectral distortions of water-pixels, particularly, in the NIR signal (Figure 4). For example, the water spectrum of the Mamirauá Lake is highly affected by the forest neighbourhood, due to the typical low reflectance (<1%) and narrow width (200–400 m between margins). Therefore, this phenomenon requires careful evaluation and several studies have proposed solutions (see, for example, Duan et al. [69] and references therein). For example, previous applications derived the contextual information, known as an environmental reflectance function, based on the distance-weighted average reflectance of neighbouring pixels and ratio of direct and diffuse transmittance for adjacency correction [60,70]. However, this environmental reflectance is a critical issue when it is applied to areas with significant spectral contrast between surfaces, such as water and land targets [29]. In addition, different procedures were developed based on atmosphere point spread function (PSF) to quantify surrounding contributions and synthesize the filter correction of the neighbouring reflections [29,69]. Although these methods are routinely used for adjacency correction, the image-based scheme provides a practical sense for remote sensing users and overcomes limitations of atmospheric parameters that are not always available in remote areas such as the Amazon.

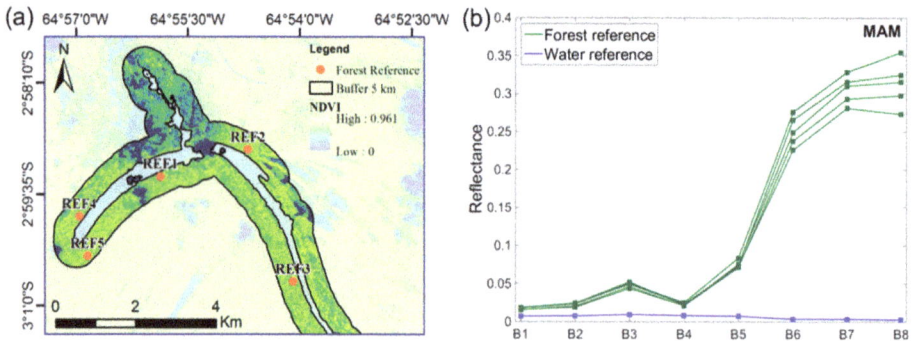

Figure 4. Example of water and forest endmembers selection at Mamirauá Lake. (**a**) Random points in the forest surface near to Mamirauá Lake. (**b**) Water and forest endmembers (Table 3).

Since adjacency effect consists of spectral mixing problems, we performed a simple procedure to decouple water and forest contributions using Linear Spectral Unmixing model (LSU) [71]. The main idea of the LSU model is to decompose the surface reflectance contributions of mixed pixels based on the pure spectrum collection of their surfaces, called endmembers. In this sense, mixed spectrum is given by the linear sum of distinct proportion of each endmember (Equation (6)) and LSU model provides fraction maps of surface contributions per pixel [72].

$$
\begin{aligned}
R_i &= ff_{1,i} \times R_{1,i} + ff_{2,i} \times R_{2,i} + \ldots + ff_{M,i} \times R_{M,i} + w \\
&= \sum_{n=1}^{M} ff_{n,i} \times R_{n,i} + w
\end{aligned}
\tag{6}
$$

where R is the reflectance of pixel, ff is the contribution fraction, R_M is the spectral reflectance of endmember (M), M is the number of endmembers, w is an error term accounting for additive noise (including sensor noise, endmember variability, and other model inadequacies) and $i = [1, 2, \ldots, i_n]$ is the number of spectral bands.

Briefly, this adjacency correction procedure is described in the following steps: (i) selection of water and forest endmembers as an input for the LSU model; (ii) calculation of the adjacency contribution based on a forest fraction map; and (iii) adjacency removal per water pixel of each lake. The selection of the water endmember is critical when all water-pixels are highly affected by adjacency effects. Thus, two main assumptions were made: (i) prior knowledge of the typical water reflectance as input for the LSU model; and (ii) the forest is the only adjacency surface that distorts the water spectrum. Based on these assumptions, we selected the highest spectrum of each sample station to calculate the average of water reflectance per lake (Table 3). For forest reference, five random spectra were selected over a vegetated area with NDVI >0.8 up to 5 km from the lake (example in Figure 4). The average of these vegetation spectra is referred to as forest endmember per lake (Table 3).

After running the LSU model, the forest fraction mapping of the lake was multiplied by forest reference spectrum to estimate adjacency effects for all water-pixels (Equation (7)), which were then removed from the MSI surface reflectance derived from the 6SV model (Equation (8)).

$$R_{adj, i}(x_n, y_n) = R_{fpure, i} \times ff_i(x_n, y_n) \qquad (7)$$

$$R_{cor*, i}(x_n, y_n) = R_{cor, i}(x_n, y_n) - R_{adj, i}(x_n, y_n) \qquad (8)$$

where R_{adj} is the forest adjacency effects, R_{fpure} is the forest reference spectrum (Table 3), ff is the fraction of forest signal affecting the water spectrum, R_{cor} is the MSI-corrected image from 6SV model, R_{cor*} is the adjacency corrected image, $x_n = [x_1, \ldots, x_n]$ is the column of pixel n, $y_n = [y_1, \ldots, y_n]$ is the row of pixel n over water surface, and $i = [1, \ldots, 8]$ is the number of spectral band. Finally, the adjacency corrected MSI image was also evaluated with in-situ measurements.

Table 3. Water and forest endmembers selected as input to LSU per lake.

Lake	Type	B1	B2	B3	B4	B5	B6	B7	B8
BUA	Water	0.005	0.006	0.008	0.007	0.006	0.002	0.002	0.002
	Forest	0.023	0.031	0.062	0.030	0.096	0.289	0.341	0.343
MAM	Water	0.007	0.008	0.009	0.008	0.007	0.004	0.004	0.003
	Forest	0.017	0.022	0.048	0.023	0.075	0.251	0.306	0.313
PANTA	Water	0.008	0.011	0.017	0.017	0.016	0.006	0.006	0.005
	Forest	0.009	0.015	0.042	0.019	0.072	0.249	0.300	0.296
PIRA	Water	0.008	0.012	0.019	0.018	0.015	0.005	0.004	0.003
	Forest	0.017	0.023	0.050	0.025	0.077	0.253	0.312	0.329

4. Results and Discussion

The results will be presented in the following sequence: (i) validation of MAIAC AOD$_{550}$ product in the Amazon region; (ii) seasonal distribution of AOD$_{550}$, U$_{H2O}$ and cloud cover over the study area; (iii) simulation of the TOA reflectance of the MSI spectral bands; (iv) inter-comparison of three atmospheric methods applied to the MSI image from 12 August 2016; and (v) forest adjacency correction based on the LSU model.

4.1. Evaluation of MAIAC AOD$_{550}$

Figure 5 shows a comparison between the average of AERONET measurements within ±30 min of MODIS overpass and the average of MAIAC AOD$_{550}$ within a 25 × 25 km^2 area. The agreement between the satellite and ground measurements was assessed using a linear regression model and the standard expected error (EE) of MODIS atmospheric products (dashed red lines) defined by AOD = ±0.05 ± 0.15 × AOD [46]. EE is the number of MAIAC retrievals falling within the standard expected error, and Remer et al. [46] suggested that EE threshold of 66% represents a satisfactory accuracy of the satellite AOD retrievals. The number of match-ups for Terra (245) was higher than that for Aqua (67), due to the difference of the cloud cover between morning and afternoon periods. In general, our results showed that both Terra and Aqua products agree well with AERONET measurements, with both slope of linear regression and correlation coefficient (R) higher than 0.74. For comparison, the Aqua product presents a slightly better accuracy than that of Terra, with a mean ratio of 1.03 for Aqua and 0.8 for Terra. In this context, forest surfaces represent feasible areas for a strong relation of visible and SWIR bands used for the DDV approach [59] and an increase the sensitivity of aerosol scattering effects in TOA reflectance [73]. Finally, the quality of MAIAC retrievals over the Amazon region are also observed by the number of AOD retrievals falling within standard expected error that presented EE values higher than 66% threshold (EE is 76.7% for Terra and 88.1% for Aqua). To increase the confidence in the AOD product as input for 6SV atmospheric correction, AOD values obtained from MAIAC Terra were corrected (AODcorrected = AOD $\times$ 1/0.803) using a mean ratio of 0.803 (see in Figure 5).

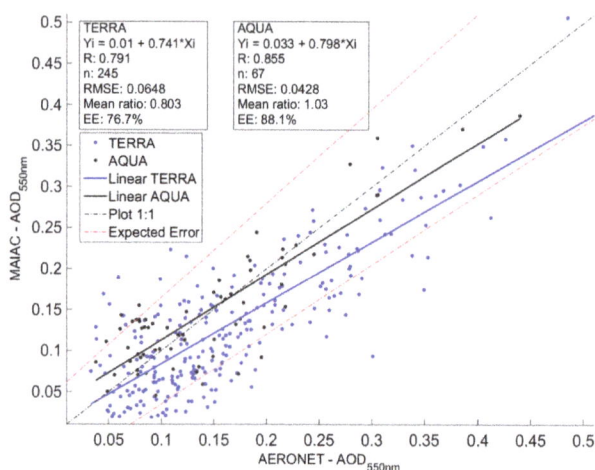

Figure 5. Scatter plot of MAIAC AOD$_{550}$ (y-axis) compared to AERONET AOD$_{550}$ data (x-axis) from three sites in the Amazon region: Balbina, Belterra and Manaus-Embrapa. Solid blue and grey lines are the linear regression fits for Terra and Aqua, respectively. Red dashed lines are the MODIS standard expected error intervals (ΔAOD = $\pm$0.05 $\pm$ 0.15 $\times$ AOD) [74]. Text box: Regression equation, correlation coefficient (R), match-ups (n), root mean square error (RMSE), mean ratio ($\overline{AOD}_{MAIAC} / \overline{AOD}_{AERONET}$), and EE is the number of retrievals falling within standard expected error.

4.2. Background of Atmospheric Constituents

The seasonal variability of atmospheric constituents imposes distinct conditions for atmospheric correction and the background information of these constituents is desirable to understand the effects on remote sensing data [75]. Figure 6 presents the monthly average of AOD$_{550}$, U$_{H_2O}$ and cloud cover frequency using 15-year MAIAC Terra products in the study area (red box in Figure 1). The results showed that atmospheric components have distinct patterns between the first and second half of the year. In the first half of the year, high frequency cloud cover (~90%) was observed, which restricts continuous remote sensing observations of the Mamirauá region in this particular period. However, the probability of cloud-free images increases in July, August, and October, when the cloud cover decreased to 60%–80%. Hilker et al. [76] also reported an increase of cloud-free pixels (40%–60%) in June, July and August based on MAIAC Terra observations from 2007 over the Amazon region and pointed out MAIAC improvements to detect small cloud-free areas compared with the previous MOD09 product. Under these conditions, the temporal resolution of Sentinel 2 MSI images has the potential to increase the probability of cloud-free images over this tropical region thanks to a high five-day revisit scheduled for 2017. In parallel with cloud cover, AOD presents a remarkable seasonal variation in the first and second half of the year. In general, a low AOD (0.02–0.18) is typically observed throughout the year, however, the first half of the year registers the lowest AOD values (0.1–0.14), while the AOD values increase to 0.16–0.18 between August and November. Castro et al. [77] also found typical low AOD values in the Northern Amazon region (see TD1 region), and illustrated the variability between the first and second half of the year caused by local fires. Fires are the cheapest and most effective way for local communities to clear areas in order to introduce pasture and agricultural cultures [78] and can be intentionally used to dissipate insect pests. Therefore, the temporal variation of AOD values imposes particular conditions for atmospheric correction depending on the period, and needs further attention during the second half of the year.

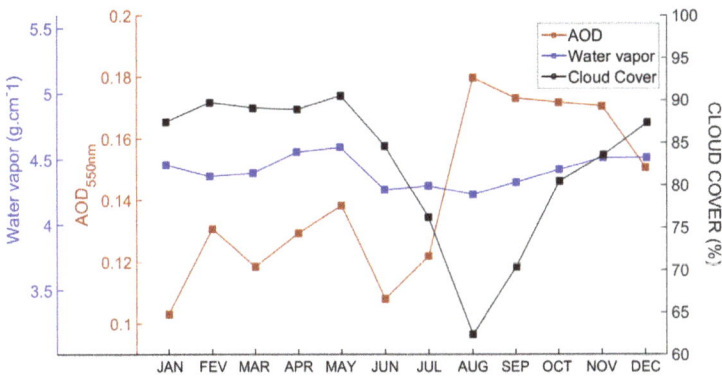

Figure 6. Monthly average of aerosol optical depth (AOD) at 550 nm, water vapour content and cloud cover from MAIAC Terra (2000–2015) in Mamirauá region (see red-box in Figure 1).

The Mamirauá region has high concentrations of water vapour (4.2–4.7 g/cm^2), with quasi-permanent concentration throughout the year. Vermote et al. [70] showed that U_{H2O} content (variation of 0.5–4.1 g/cm^2) implicates on higher atmospheric effects on the reflectance magnitude in the NIR region (3.4%–14.0%) than those for visible bands (0.5%–3.0%) of the Landsat 5 TM bands. Therefore, remote sensing applications using NIR bands should consider the temporal variability of this constituent, for example to vegetation indices. Our field campaign was performed during August, with the lowest cloud cover conditions (~62%) of the year, despite facing the highest AOD and water vapour concentrations. To better understand the atmospheric and surface contributions, we conducted a TOA simulation using the average AOD over the four lakes for the dataset from August, as shown in Section 4.3.

4.3. TOA Simulation Analysis

The contribution of atmosphere and surface reflectance from four Amazonian lakes to the TOA reflectance of MSI spectral bands was simulated using the AOD average of August (Figure 7). The major fraction of theTOA signal is due to atmospheric effects, whose contribution is higher than 75% and 50% over dark and bright lakes, respectively, for all MSI spectral bands. Because atmospheric scattering effects exponentially decay from shorter to longer wavelengths [79], deep blue and blue bands are affected the most by atmospheric signals—representing more than 84% of TOA reflectance. The highest lake surface contribution was observed for red (B4) and red-edge (B5) bands, where the scattering of suspended inorganic particles in the water column increases surface reflectance, which reached ~43% of the TOA reflectance in the Panta-leão Lake. On the other hand, the contribution of dark lakes to TOA reflectance reduced dramatically, mainly in the deep blue band, when compared with bright lakes. For example, the Buá-Buá reflectance had a critical fraction that reached ~4.8% in the deep blue band, which clearly shows the difficulty in accurately removing of atmospheric effects over these low-reflectance surfaces. These results suggest that caution is required when the TOA reflectance band ratio is used to retrieve water quality parameters, even when the algorithms present a better correlation using the TOA reflectance compared with surface reflectance [45]. The atmospheric effects among the bands are quite different, and band ratio therefore does not remove all atmospheric effects.

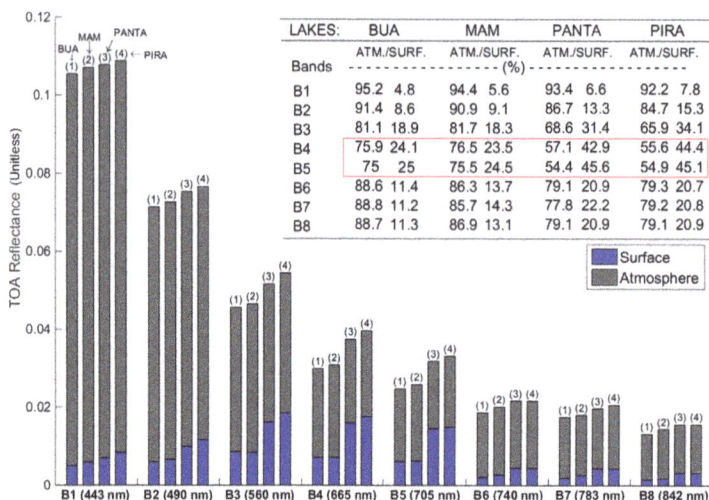

Figure 7. Simulation of TOA reflectance based on month average AOD in August (biomass burning model) and average of water reflectance from four Amazon floodplain lakes. The table at the top right shows the percentage of atmospheric and surface contribution in TOA reflectance simulated for MSI VNIR bands.

In fact, the atmospheric correction is a challenging issue for water quality studies, because inaccuracies of the surface reflectances propagate errors to water quality retrievals. For example, the mapping of CDOM concentration using blue and green bands [80] requires an efficient removal of scattering effects of these (most affected) bands, where atmospheric effects can contribute up to 80% of the TOA signal over dark lakes. On the other hand, red and red-edge bands applied to model TSS concentrations are less affected by atmospheric effects [81]. These examples thus illustrate that OACs modelling also requires a better understanding of atmospheric optical properties to derive accurate surface products.

4.4. Inter-Comparison of Atmospheric Correction Methods

The atmospheric correction performances of three algorithms were assessed using the mean ratio ($\overline{R}_{sur}/\overline{R}_{w,situ}$) and root mean square error (RMSE). The performances are better when the mean ratio is close to unity and RMSE is low. The mean ratio also expresses a relative error, when the values are higher than 2, or relative error is 100%, exceed the maximum bias accepted here. This threshold is important for bio-optical models, because the high inaccuracy of the reflectance affects, for example, the COA retrievals (see Odermatt et al. [82] for distinct bio-models).

Figure 8 presents the comparison of the in-situ measurements acquired at four Amazon floodplain lakes with spectra results from the three atmospheric correction methods, while Figure 9 describes the algorithm performance based on the mean ratio and RMSE. The three methods were capable of removing most of the atmospheric effects, where the reflectance magnitude mainly changed from TOA reflectance of ~12%–14% to surface of ~1%–2% of the deep blue and blue bands (Figure 8). Before atmospheric correction, TOA reflectance was a typical exponential spectrum dominated by

Rayleigh and aerosol scattering effects. After correction, even with caveats, the MSI-corrected spectra (dark blue lines in Figure 8) presented a shape close to that of the in-situ measurements (light blue lines in Figure 8). The results showed that the quality of the MSI-corrected reflectance varies depending on the atmospheric correction method, spectral band, and lake characteristics.

In general, Sen2Cor produced the smallest overall RMSE and the best MSI-corrected shape compared with in-situ data (Figure 9), while the spectral shape of ACOLITE showed a bias varying per sample station. The 6SV method provided a quite similar spectral shape in the visible region for bright lakes (RMSE < 0.006), but the reflectance magnitude was higher than that of the in-situ measurements (Figure 9c). Regarding the comparison among lakes, the MSI-corrected reflectance of bright lakes is closer to that of the in-situ measurements than to that of dark lakes; all methods have a mean ratio of ~0.5–1.5 in the visible bands (Figure 9a–c).

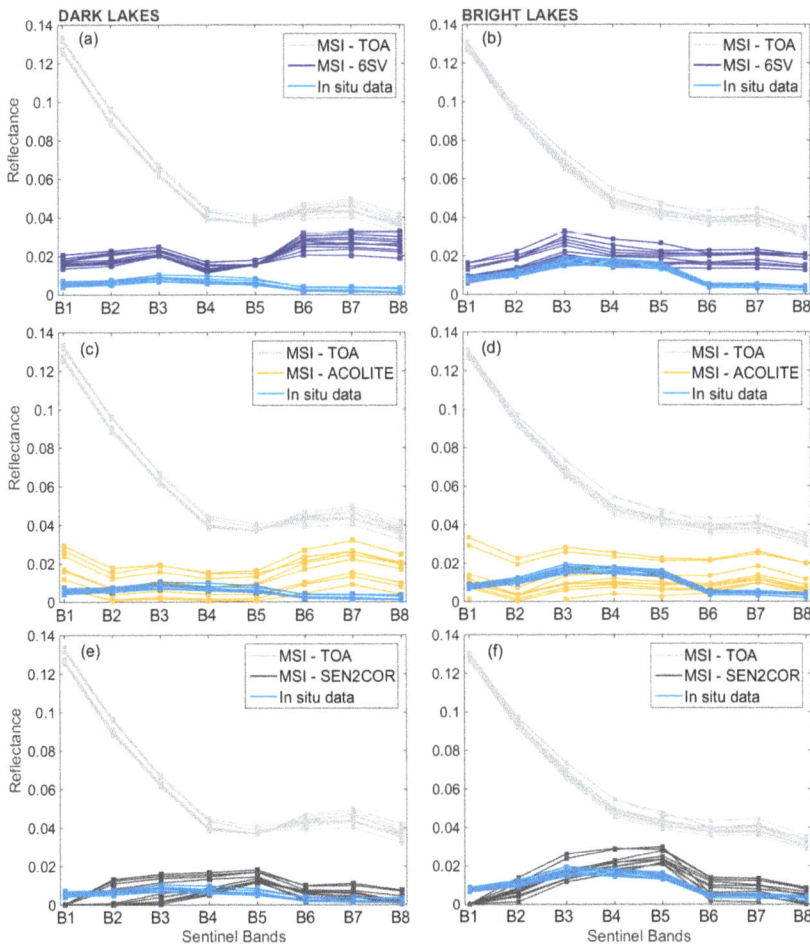

Figure 8. Comparison between MSI reflectance simulated from in-situ measurements, MSI TOA reflectance and MSI-corrected reflectance from three methods: 6SV model based on MAIAC product (**a,b**); ACOLITE (**c,d**); and Sen2Cor (**e,f**). The left column shows reflectance spectra from dark lakes (Buá-Buá and Mamirauá) and, the right column shows reflectance spectra from bright (Pirarara and Panta-Leão) lakes.

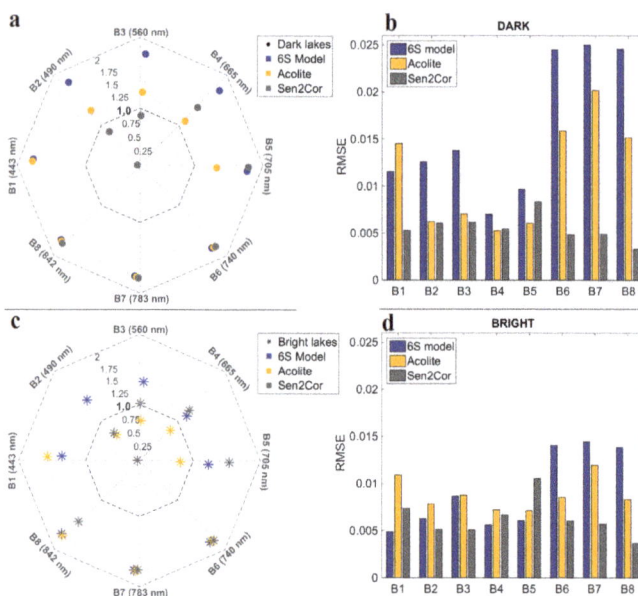

Figure 9. The mean ratio and RMSE for atmospheric correction methods from: dark lakes (**a,b**); and bright lakes (**c,d**). The left column shows the mean ratio ($\overline{R}_{sur}/\overline{R}_{W,situ}$), with better cases being close to unity, and the right column shows the root mean square error (RMSE). Note that all mean ratio values higher than maximum bias accepted here are represented as 2.

In dark lakes, ACOLITE and Sen2Cor showed quite similar RMSE values in the blue, green and red bands (RMSE ~0.006) (Figure 9b), while 6SV was clearly limited under these low reflectance conditions (RMSE ~0.011), with a mean ratio exceeding the maximum threshold (>2) for all spectral bands (Figure 9a). In contrast, the MSI-reflectance corrected by the 6SV model had a lower RMSE (~0.006) and mean ratio of ~1.2–1.50 (overestimation) in the visible bands over bright lakes, compared with the in-situ measurements (Figure 9c). The generic design of 6SV model includes standard aerosol types for land applications, such as biomass burning, urban, continental, desert, and maritime. Bassani et al. [83] highlighted the impact of the reflectance accuracy on coastal water when the standard 6SV aerosol use does not characterize the local aerosol microphysics. In our case, a complex mixture of natural biogenic and biomass burning particles over the central Amazon [78] imposes a constraint for standard aerosol use, which might explain some limitations of the radiative transfer simulation in order to remove atmospheric effects. In addition, satellite aerosol retrievals are susceptive to several uncertainty sources, such as sensor calibration, cloud screen, aerosol models and surface properties [84], which also help to explain the errors derived from the 6SV correction based on MAIAC product. Since most bio-optical models apply water surface reflectance mainly to visible bands (B1–B5) [8], Sen2Cor and 6SV results produce a reasonable spectral shape compared with the in-situ measurements for bright lakes; although they overestimate the reflectance magnitude (Figure 8b–f).

In the visible bands, the ACOLITE method produced a distinct reflectance accuracy among the lakes, where water reflectance was overestimated (mean ratio ~1.1–1.5, except for B1) for dark lakes (Figure 9a) and underestimated (mean bias ~0.63–0.8, except for B1) for bright lakes (Figure 9c). A strong influence of the per-pixel correction was also observed; the random spatial error contributed to biases in the MSI-corrected reflectance for each sample station, while the 6SV produced a spatially correlated error due to the assumption of homogeneous atmospheric effects on lakes. The ACOLITE-SWIR approach requires a high signal-to-noise ratio (SNR) for the accurate quantification of the aerosol effect on TOA reflectance particularly because the effective aerosol scattering is relatively lower at longer wavelengths [85] and accurate aerosol reflectance can be affected by a low SNR. Thus, the propagation of noisy aerosol reflectance from SWIR bands (SNR of 100, see Table 1) to the atmospheric correction of visible bands might explain the spatial variability of error in ACOLITE reflectance retrievals. Therefore, an optional smooth window applied to SWIR bands becomes essential to filter and minimize undesirable effects [21]. Although the ACOLITE produced variable errors in this study, the algorithm has been extensively used for the atmospheric correction of OLI images with satisfactory experiences in coastal and maritime water, for example, sediment plume mapping in the Florida offshore, USA [86], and for turbidity quantification in the Wadden Sea, Germany [87]. Our assessment represents preliminary experience with ACOLITE applied to Sentinel 2 MSI over floodplain water, and further evaluations could contribute to consolidate this image-based approach.

The Sen2Cor algorithm achieves very similar errors for both types of lakes, with RMSE between ~0.003–0.011 for all spectral bands (Figure 9b–d), and better mean ratio for blue, green and red bands (grey marker in Figure 9a–c). In fact, the presence of dense forest close to Amazon floodplain lakes provides feasible areas required for DDV application used by the Sen2Cor algorithm. Although an extensive validation has not been performed yet, first results of the comparison between Sen2Cor AOD retrievals over DDV areas and ground measurements led to an AOD uncertainty of about 0.03 [88], which affects the atmospheric correction accuracy of Sen2Cor algorithm. Toming et al. [12] found the shape similarity of MSI-corrected dataset using the Sen2Cor algorithm compared with a historical dataset (2011–2013), although the MSI reflectance spectra were overestimated by the algorithm. In the Amazon context, the presence of forest surfaces benefits AOD retrievals by the Sen2Cor algorithm, however, this surface neighbour also implies in severe adjacency contamination on the water spectrum measured by orbital sensors.

4.5. Adjacency Effect Correction

Large differences between forest and water reflectance contribute to strong adjacency effects on the water spectrum in the NIR region. For spectral bands above 705 nm, Sen2Cor correction minimizes the adjacency effects over dark lakes leading to a lower RMSE (<0.005) compared with ACOLITE and 6SV (RMSE > 0.013) (Figure 9b–d). Nevertheless, none of the methods presented a suitable performance to remove forest adjacency effects in the NIR region, with a mean ratio exceeding 2 (Figure 9a–c). Dornhofer et al. [11] evaluated the Sen2Cor and ACOLITE performances for bio-optical models in marine water and found an underestimation of the surface reflectance in the visible region for both methods and adjacency effects on the NIR bands due to sand surface of the shoreline. The adjacency correction applied by Sen2Cor is limited over low reflectance surfaces; methods focusing on water surfaces are required for a high spatial resolution [69].

Figure 10 presents the adjacency correction based on LSU model applied to the MSI water reflectance from dark and bright lakes. The results showed that the adjacency correction improves the quality of water reflectance of all spectral bands. In both dark and bright lakes, MSI reflectance exhibits a better agreement with in-situ measurements, especially for MSI bands above 705 nm (B6–B8), where the RMSE decreases from ~0.023 to lower than 0.006 in dark lakes and from ~0.015 to lower than 0.004 in bright lakes (Figure 10). Evidently, the LSU approach solves the unmixing problem of the water spectrum and reduces the discrepancy caused by forest adjacency. However, note that results have limitations to an extensive application, because we assume that only the forest surface causes adjacency effects and the water endmember was based on prior knowledge of radiometric collection over lakes. These assumptions limit the operational correction to reservoir and lakes with historical radiometric collection or to cases where water endmember can be derived from the image itself. Furthermore, adjacency effects typically depend on the surface types in the scene, which makes it reasonable to consider that unmixing problems are caused by other surfaces. Recently, Sterckx et al. [30] proposed a generic-sensor adjacency correction based on the NIR similarity spectrum for water studies, and showed positive or neutral effects on the reflectance accuracy depending on environmental conditions. In general, adjacency correction is a complex procedure that should consider a variety of environmental conditions. Our positive experience using the LSU model is a preliminary assessment to advance towards a more sophisticated image-based approach.

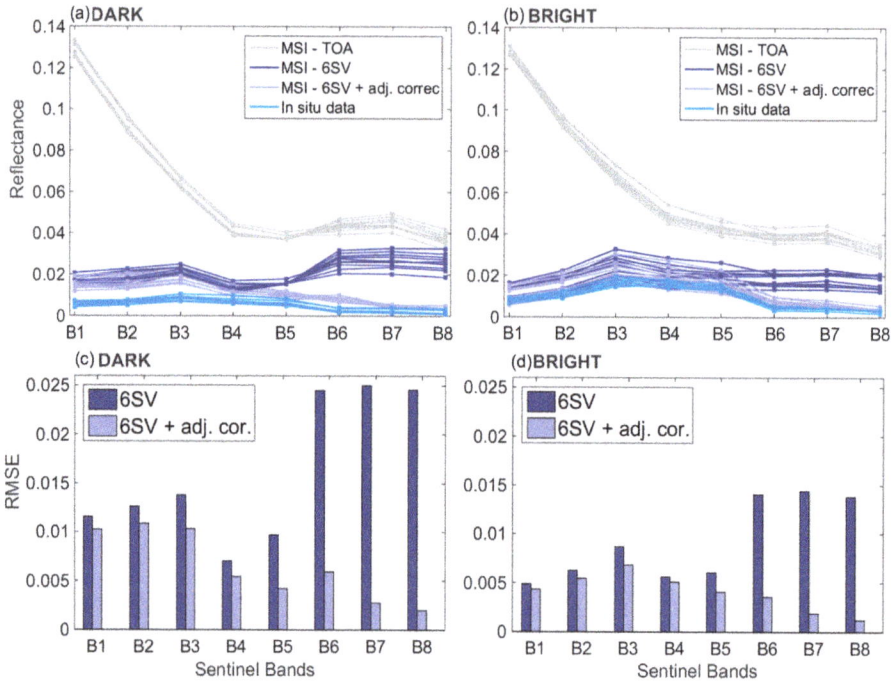

Figure 10. Comparison between MSI TOA reflectance (grey), MSI-corrected reflectance from 6SV model (dark blue), and MSI-adjacency corrected reflectance (purple) and MSI reflectance simulated from in-situ measurements (light blue) from dark (**a,c**) and bright lakes (**b,d**). RMSE bars before (6SV) and after adjacency correction (6SV + adj. cor.) applied to MSI image from 12 August 2016.

5. Conclusions

Sentinel 2 MSI images represent a new opportunity for monitoring small inland aquatic systems. However, to derive an accurate surface reflectance over complex water and consequently produce better water quality products, efforts must focus on the development of efficient atmospheric correction methods. Although our results present a variation in the surface reflectance accuracy along the spectral range, it is important to highlight the complexity of environmental condition in our study area, such as the water reflectance (<2%), narrow and irregular lakes, low spatial variability and strong influence of the adjacency forest.

In terms of the temporal variation of atmospheric constituents (cloud cover, AOD and U_{H_2O}) within the study area, high cloud cover (~90%) limits the cloud-free image during the first half of the year. The cloud cover decreases in the second half of the year (60%–80%), while AOD increases from ~0.12 to ~0.18 and water vapour remains constant ~4.4 g/cm^2 between seasons. The fine resolution of MAIAC AOD product has an acceptable accuracy to support radiative transfer models in the Amazon region. The simulation of the TOA reflectance clearly shows an inherent difficulty faced by inland water studies, because the atmospheric contribution varies from ~50% to ~94% of TOA signal according to spectral band.

The atmospheric correction methods present a notable variation in the MSI-corrected accuracy along spectral bands depending on lake characteristics. For dark lakes, the results indicate method limitations to derive an accurate reflectance spectrum, particularly above 705 nm due to forest adjacency effects. In the visible bands, Sen2Cor and ACOLITE showed quite similar RMSEs for dark lakes whereas the 6SV model exhibited better results for bright lakes. Therefore, the selection of an atmospheric correction method needs to be aligned with the study purposes and user expertise to apply these tools, since our results present advantages and disadvantages of different methods for the reflectance shape and magnitude accuracy.

All methods showed limitations in the accurate retrievals in the NIR region due to forest adjacency effects. The performance of the adjacency correction using the LSU model improved the quality of MSI-corrected reflectance and derived better spectra shape compared with the in-situ measurements. However, there are constraints for operational application, such as selection of the water endmember and other adjacent surfaces, which need to be taken into account when applying the method to different contexts/regions. Therefore, this experience is considered to be a preliminary assessment to advance on image-based approaches to remove adjacency effects.

The assessment of these atmospheric correction methods has an inherent challenge due to the shortage of cloud-free images and the limitations of routine in-situ measurements. However, even for a single Sentinel-2 image, it was possible to explore the available methods applied to complex environmental conditions (low water reflectance, forest adjacency effects and seasonal variability of aerosol atmospheric), such as in the Amazon region. We recommend an extensive validation for distinct water types and atmospheric conditions to further understand the potential and limitations of these atmospheric correction methods.

Acknowledgments: This study was funded by São Paulo Research Foundation (FAPESP) from project no.: 2014/23903-9 and MSA-BNDES from project no.: 1022114003005. V. Martins was funded by the Coordination for the Improvement of Higher Education Personnel (CAPES) Program. E. Novo acknowledges for the research productivity fellowship CNPQ 304568/2014-7. We are very grateful to Renato Ferreira and Adriana Affonso for their assistance and logistical support during field campaign. We would like to thank ESA team for providing free Sentinel-2A data and Sen2Cor processor. We further thank Vanhellemont and RBIN team for the development and distribution of ACOLITE algorithm. We also thank Wilson Robin for providing Py6S (Python interface to 6SV) with full description of package [89]. We thank AERONET PIs, Brent Holben, Paulo Artaxo and their staffs, for establishing and maintaining the AERONET site at Balbina, Belterra and Manaus-Embrapa. Al last, we thank Alexei Lyapustin and Yugie Wang for the updating of MAIAC dataset over Amazon. The MSI response spectral function is available at https://sentinel.esa.int/documents/247904/685211/Sentinel-2A+MSI+Spectral+Responses. We thank the anonymous reviewers for their effort and helpful comments.

Author Contributions: Vitor Martins, Claudio Barbosa, Lino de Carvalho, Daniel Jorge, Felipe Lobo and Evlyn Novo designed the research. Vitor Martins performed the data processing, atmospheric correction and validation, as well as wrote the paper. Claudio Barbosa, Lino de Carvalho and Daniel Jorge were involved in the collection of field data. Claudio Barbosa and Lino de Carvalho supported the processing of in-situ measurements. Felipe Lobo and Evlyn Novo contributed with insights about analysis and methods. All authors equally contributed towards organizing and reviewing the manuscript.

Conflicts of Interest: The authors declare no conflict of interest.

References

1. Dudgeon, D.; Arthington, A.H.; Gessner, M.O.; Kawabata, Z.-I.; Knowler, D.J.; Lévêque, C.; Naiman, R.J.; Prieur-Richard, A.-H.; Soto, D.; Stiassny, M.L.J.; et al. Freshwater biodiversity: Importance, threats, status and conservation challenges. *Biol. Rev.* **2006**, *81*, 163–182. [PubMed]

2. Vörösmarty, C.J.; McIntyre, P.B.; Gessner, M.O.; Dudgeon, D.; Prusevich, A.; Green, P.; Glidden, S.; Bunn, S.E.; Sullivan, C.A.; Liermann, C.R.; et al. Global threats to human water security and river biodiversity. *Nature* **2010**, *467*, 555–561. [CrossRef] [PubMed]

3. Abell, R.; Thieme, M.L.; Revenga, C.; Bryer, M.; Kottelat, M.; Bogutskaya, N.; Coad, B.; Mandrak, N.; Balderas, S.C.; Bussing, W.; et al. Freshwater ecoregions of the world: A new map of biogeographic units for freshwater biodiversity conservation. *Bioscience* **2008**, *58*, 403–414. [CrossRef]

4. Mertes, L.; Smith, M.; Adams, J. Estimating suspended sediment concentrations in surface waters of the Amazon River wetlands from Landsat images. *Remote Sens. Environ.* **1993**, *43*, 281–301. [CrossRef]
5. Park, E.; Latrubesse, E.M. Modeling suspended sediment distribution patterns of the Amazon River using MODIS data. *Remote Sens. Environ.* **2014**, *147*, 232–242. [CrossRef]
6. Lobo, F.L.; Costa, M.P.F.; Novo, E.M.L.M. Time-series analysis of Landsat-MSS/TM/OLI images over Amazonian waters impacted by gold mining activities. *Remote Sens. Environ.* **2014**, *157*, 170–184. [CrossRef]
7. Espinoza Villar, R.; Martinez, J.M.; Guyot, J.L.; Fraizy, P.; Armijos, E.; Crave, A.; Bazán, H.; Vauchel, P.; Lavado, W. The integration of field measurements and satellite observations to determine river solid loads in poorly monitored basins. *J. Hydrol.* **2012**, *444*, 221–228. [CrossRef]
8. Bukata, R.P.; Jerome, J.H.; Kondratyev, A.S.; Pozdnyakov, D.V. *Optical Properties and Remote Sensing of Inland and Coastal Waters*; CRC Press: Boca Raton, FL, USA, 1995.
9. Palmer, S.C.J.; Kutser, T.; Hunter, P.D. Remote sensing of inland waters: Challenges, progress and future directions. *Remote Sens. Environ.* **2015**, *157*, 1–8. [CrossRef]
10. Drusch, M.; Del Bello, U.; Carlier, S.; Colin, O.; Fernandez, V.; Gascon, F.; Hoersch, B.; Isola, C.; Laberinti, P.; Martimort, P.; et al. Sentinel-2: ESA's Optical High-Resolution Mission for GMES Operational Services. *Remote Sens. Environ.* **2012**, *120*, 25–36. [CrossRef]
11. Dörnhöfer, K.; Göritz, A.; Gege, P.; Pflug, B.; Oppelt, N. Water Constituents and Water Depth Retrieval from Sentinel-2A—A First Evaluation in an Oligotrophic Lake. *Remote Sens.* **2016**, *8*, 941. [CrossRef]
12. Toming, K.; Kutser, T.; Laas, A.; Sepp, M.; Paavel, B.; Nõges, T. First Experiences in Mapping Lake Water Quality Parameters with Sentinel-2 MSI Imagery. *Remote Sens.* **2016**, *8*, 640. [CrossRef]
13. Hedley, J.; Roelfsema, C.; Koetz, B.; Phinn, S. Capability of the Sentinel 2 mission for tropical coral reef mapping and coral bleaching detection. *Remote Sens. Environ.* **2012**, *120*, 145–155. [CrossRef]
14. Malthus, T.J.; Hestir, E.L.; Dekker, A.G.; Brando, V.E. The case for a global inland water quality product. In Proceedings of the 2012 IEEE International Geoscience and Remote Sensing Symposium, Munich, Germany, 22–27 July 2012; pp. 5234–5237.
15. Malenovský, Z.; Rott, H.; Cihlar, J.; Schaepman, M.E.; García-Santos, G.; Fernandes, R.; Berger, M. Sentinels for science: Potential of Sentinel-1, -2, and -3 missions for scientific observations of ocean, cryosphere, and land. *Remote Sens. Environ.* **2012**, *120*, 91–101. [CrossRef]
16. Gao, B.C.; Montes, M.J.; Davis, C.O.; Goetz, A.F.H. Atmospheric correction algorithms for hyperspectral remote sensing data of land and ocean. *Remote Sens. Environ.* **2009**, *113*, S17–S24. [CrossRef]
17. International Ocean Colour Coorperating Group (IOCCG). *Atmospheric Correction for Remotely-Sensed Ocean-Colour Products*; International Ocean Colour Coorperating Group (IOCCG): Cape Town, South Africa, 2010.
18. Okin, G.S.; Gu, J. The impact of atmospheric conditions and instrument noise on atmospheric correction and spectral mixture analysis of multispectral imagery. *Remote Sens. Environ.* **2015**, *164*, 130–141. [CrossRef]
19. Hadjimitsis, D.G.; Clayton, C.R.I.; Hope, V.S. An assessment of the effectiveness of atmospheric correction algorithms through the remote sensing of some reservoirs. *Int. J. Remote Sens.* **2004**, *25*, 3651–3674. [CrossRef]
20. Main-Knorn, M.; Pflug, B.; Debaecker, V.; Louis, J. Calibration and validation plan for the L2A processor and products of the Sentinel-2 Mission. *ISPRS Int. Arch. Photogramm. Remote Sens. Spat. Inf. Sci.* **2015**, *40*, 1249–1255. [CrossRef]
21. Vanhellemont, Q.; Ruddick, K. Acolite for Sentinel-2: Aquatic applications of MSI imagery. In Proceedings of the ESA Living Planet Symposium, Pragur, Czech Republic, 9–13 May 2016.
22. Shi, W.; Wang, M. An assessment of the black ocean pixel assumption for MODIS SWIR bands. *Remote Sens. Environ.* **2009**, *113*, 1587–1597. [CrossRef]
23. Vermote, E.F.; Tanré, D.; Deuzé, J.L.; Herman, M.; Morcrette, J.J. Second simulation of the satellite signal in the solar spectrum, 6S: An overview. *IEEE Trans. Geosci. Remote Sens.* **1997**, *35*, 675–686.
24. Taylor, M.; Kazadzis, S.; Amiridis, V.; Kahn, R.A. Global aerosol mixtures and their multiyear and seasonal characteristics. *Atmos. Environ.* **2015**, *116*, 112–129. [CrossRef]
25. Holben, B.N.; Eck, T.F.; Slutsker, I.; Tanré, D.; Buis, J.P.; Setzer, A.; Vermote, E.; Reagan, J.A.; Kaufman, Y.J.; Nakajima, T.; et al. AERONET—A federated instrument network and data archive for aerosol characterization. *Remote Sens. Environ.* **1998**, *66*, 1–16. [CrossRef]

26. King, M.D.; Menzel, W.P.; Kaufman, Y.J.; Tanre, D.; Gao, B.-C.; Platnick, S.; Ackerman, S.A.; Remer, L.A.; Pincus, R.; Hubanks, P.A. Cloud and aerosol properties, precipitable water, and profiles of temperature and water vapor from MODIS. *IEEE Trans. Geosci. Remote Sens.* **2003**, *41*, 442–458. [CrossRef]

27. Lyapustin, A.; Wang, Y.; Laszlo, I.; Kahn, R.; Korkin, S.; Remer, L.; Levy, R.; Reid, J.S. Multiangle implementation of atmospheric correction (MAIAC): 2. Aerosol algorithm. *J. Geophys. Res. Atmos.* **2011**, *116*, 1–15. [CrossRef]

28. Hilker, T.; Lyapustin, A.I.; Tucker, C.J.; Sellers, P.J.; Hall, F.G.; Wang, Y. Remote sensing of tropical ecosystems: Atmospheric correction and cloud masking matter. *Remote Sens. Environ.* **2012**, *127*, 370–384. [CrossRef]

29. Kiselev, V.; Bulgarelli, B.; Heege, T. Sensor independent adjacency correction algorithm for coastal and inland water systems. *Remote Sens. Environ.* **2015**, *157*, 85–95. [CrossRef]

30. Sterckx, S.; Knaeps, S.; Kratzer, S.; Ruddick, K. SIMilarity Environment Correction (SIMEC) applied to MERIS data over inland and coastal waters. *Remote Sens. Environ.* **2015**, *157*, 96–110. [CrossRef]

31. Ramalho, E.E.; Macedo, J.; Vieira, T.M.; Valsecchi, J.; Calvimontes, J.; Marmontel, M.; Queiroz, H.L. Ciclo hidrológico nos ambientes de várzea. *Uakari* **2009**, *5*, 61–87.

32. Affonso, A.G.; Queiroz, H.L.; Novo, E.M.L.M. Abiotic variability among different aquatic systems of the central Amazon floodplain during drought and flood events. *Braz. J. Biol.* **2015**, *75*, 60–69.

33. Affonso, A.G.; Queiroz, H.L.; De Novo, E.M.L.d.M. Limnological characterization of floodplain lakes in Mamirauá Sustainable Development Reserve, Central Amazon (Amazonas State, Brazil). *Acta Limnol. Bras.* **2011**, *23*, 95–108. [CrossRef]

34. Henderson, P.A.; Hamilton, W.D.; Crampton, W.G.R. Evolution and Diversity in Amazonian Floodplain Communities. In *Dynamics of Tropical Communities: 37th Symposium of the British Ecological Society*; Cambridge University Press: Cambridge, UK, 1998; pp. 384–398.

35. Maccord, P.F.L.; Silvano, R.A.M.; Ramires, M.S.; Clauzet, M.; Begossi, A. Dynamics of artisanal fisheries in two Brazilian Amazonian reserves: Implications to co-management. *Hydrobiologia* **2007**, *583*, 365–376. [CrossRef]

36. Castello, L.; Viana, J.P.; Watkins, G.; Pinedo-Vasquez, M.; Luzadis, V.A. Lessons from Integrating Fishers of Arapaima in Small-Scale Fisheries Management at the Mamirauá Reserve, Amazon. *Environ. Manag.* **2009**, *43*, 197–209. [CrossRef] [PubMed]

37. Mobley, C.D. Estimation of the Remote-Sensing Reflectance from Above-Surface Measurements. *Appl. Opt.* **1999**, *38*, 7442. [CrossRef] [PubMed]

38. Mobley, C.D. Polarized reflectance and transmittance properties of windblown sea surfaces. *Appl. Opt.* **2015**, *54*, 4828. [PubMed]

39. Gascon, F.; Cadau, E.; Colin, O.; Hoersch, B.; Isola, C.; López Fernández, B.; Martimort, P. Copernicus Sentinel-2 mission: Products, algorithms and Cal/Val. *Proc. SPIE* **2014**. [CrossRef]

40. Clevers, J.G.P.W.; Gitelson, A.A. Remote estimation of crop and grass chlorophyll and nitrogen content using red-edge bands on Sentinel-2 and -3. *Int. J. Appl. Earth Obs. Geoinf.* **2013**, *23*, 344–351. [CrossRef]

41. Baillarin, S.J.; Meygret, A.; Dechoz, C.; Petrucci, B.; Lacherade, S.; Tremas, T.; Isola, C.; Martimort, P.; Spoto, F. Sentinel-2 level 1 products and image processing performances. In Proceedings of the 2012 IEEE International Geoscience and Remote Sensing Symposium, Munich, Germany, 22–27 July 2012; pp. 7003–7006.

42. Kloiber, S.M.; Brezonik, P.L.; Olmanson, L.G.; Bauer, M.E. A procedure for regional lake water clarity assessment using Landsat multispectral data. *Remote Sens. Environ.* **2002**, *82*, 38–47.

43. Olmanson, L.G.; Bauer, M.E.; Brezonik, P.L. A 20-year Landsat water clarity census of Minnesota's 10,000 lakes. *Remote Sens. Environ.* **2008**, *112*, 4086–4097. [CrossRef]

44. Sriwongsitanon, N.; Surakit, K.; Thianpopirug, S. Influence of atmospheric correction and number of sampling points on the accuracy of water clarity assessment using remote sensing application. *J. Hydrol.* **2011**, *401*, 203–220. [CrossRef]

45. Tebbs, E.J.; Remedios, J.J.; Harper, D.M. Remote sensing of chlorophyll-a as a measure of cyanobacterial biomass in Lake Bogoria, a hypertrophic, saline–alkaline, flamingo lake, using Landsat ETM+. *Remote Sens. Environ.* **2013**, *135*, 92–106. [CrossRef]

46. Remer, L.A.; Kaufman, Y.J.; Tanré, D.; Mattoo, S.; Chu, D.A.; Martins, J.V.; Li, R.-R.; Ichoku, C.; Levy, R.C.; Kleidman, R.G.; et al. The MODIS Aerosol Algorithm, Products, and Validation. *J. Atmos. Sci.* **2005**, *62*, 947–973. [CrossRef]

47. Gao, B.-C.; Kaufman, Y.J. Water vapor retrievals using Moderate Resolution Imaging Spectroradiometer (MODIS) near-infrared channels. *J. Geophys. Res. Atmos.* **2003**. [CrossRef]

48. Hubanks, P.; Platnick, S.; King, M. *MODIS Atmosphere L3 Gridded Product Algorithm Theoretical Basis Document (ATBD)*; Users Guide, ATBDMOD-30; NASA: Greenbelt, MD, USA, 2015.

49. Jiménez-Muñoz, J.C.; Sobrino, J.A.; Mattar, C.; Franch, B. Atmospheric correction of optical imagery from MODIS and Reanalysis atmospheric products. *Remote Sens. Environ.* **2010**, *114*, 2195–2210.

50. Lyapustin, A.; Korkin, S.; Wang, Y.; Quayle, B.; Laszlo, I. Discrimination of biomass burning smoke and clouds in MAIAC algorithm. *Atmos. Chem. Phys.* **2012**, *12*, 9679–9686. [CrossRef]

51. Vermote, E.F.; Kotchenova, S. Atmospheric correction for the monitoring of land surfaces. *J. Geophys. Res.* **2008**. [CrossRef]

52. Vermote, E.; Justice, C.; Claverie, M.; Franch, B. Preliminary analysis of the performance of the Landsat 8/OLI land surface reflectance product. *Remote Sens. Environ.* **2016**, *185*, 46–56. [CrossRef]

53. Kotchenova, S.Y.; Vermote, E.F.; Matarrese, R.; Klemm, F.J., Jr. Validation of a vector version of the 6S radiative transfer code for atmospheric correction of satellite data Part I: Path radiance. *Appl. Opt.* **2007**, *46*, 4455–4464. [CrossRef] [PubMed]

54. Vanhellemont, Q.; Ruddick, K. Turbid wakes associated with offshore wind turbines observed with Landsat 8. *Remote Sens. Environ.* **2014**, *145*, 105–115. [CrossRef]

55. Vanhellemont, Q.; Ruddick, K. Advantages of high quality SWIR bands for ocean colour processing: Examples from Landsat-8. *Remote Sens. Environ.* **2015**, *161*, 89–106. [CrossRef]

56. Pahlevan, N.; Lee, Z.; Wei, J.; Schaaf, C.B.; Schott, J.R.; Berk, A. On-orbit radiometric characterization of OLI (Landsat-8) for applications in aquatic remote sensing. *Remote Sens. Environ.* **2014**, *154*, 272–284. [CrossRef]

57. Franz, B.A.; Bailey, S.W.; Kuring, N.; Werdell, P.J. Ocean Color Measurements from Landsat-8 OLI using SeaDAS. In Proceedings of the Ocean Optics XXII, Portland, MA, USA, 26–31 October 2014; pp. 26–31.

58. Uwe, M.-W.; Jerome, L.; Rudolf, R.; Ferran, G.; Marc, N. Sentinel-2 Level 2a Prototype Processor: Architecture, Algorithms and First Results. In Proceedings of the ESA Living Planet Symposium, Edinburgh, UK, 9–13 September 2013.

59. Kaufman, Y.J.; Wald, A.E.; Remer, L.A.; Gao, B.C.; Li, R.-R.; Flynn, L. The MODIS 2.1-μm channel-correlation with visible reflectance for use in remote sensing of aerosol. *IEEE Trans. Geosci. Remote Sens.* **1997**, *35*, 1286–1298. [CrossRef]

60. Tanre, D.; Herman, M.; Deschamps, P.Y. Influence of the background contribution upon space measurements of ground reflectance. *Appl. Opt.* **1981**, *20*, 3676–3684. [CrossRef] [PubMed]

61. Otterman, J.; Fraser, R.S. Adjacency effects on imaging by surface reflection and atmospheric scattering: cross radiance to zenith. *Appl. Opt.* **1979**, *18*, 2852–2860. [CrossRef] [PubMed]

62. Kaufman, Y.J. Atmospheric effect on spatial resolution of surface imagery. *Appl. Opt.* **1984**, *23*, 4164–4172. [PubMed]

63. Chervet, P.; Lavigne, C.; Roblin, A.; Bruscaglioni, P. Effects of aerosol scattering phase function formulation on point-spread-function calculations. *Appl. Opt.* **2002**, *41*, 6489–6498. [CrossRef] [PubMed]

64. Minomura, M.; Kuze, H.; Takeuchi, N. Adjacency effect in the atmospheric correction of satellite remote sensing data: evaluation of the influence of aerosol extinction profiles. *Opt. Rev.* **2001**, *8*, 133–141. [CrossRef]

65. Dor, B.B.; Devir, A.D.; Shaviv, G.; Bruscaglioni, P.; Donelli, P.; Ismaelli, A. Atmospheric scattering effect on spatial resolution of imaging systems. *Opt. Soc. Am.* **1997**, *14*, 1329–1337.

66. Huang, C.; Townshend, J.R.G.; Liang, S.; Kalluri, S.N.V.; Defries, R.S. Impact of sensor's point spread function on land cover characterization: assessment and deconvolution. *Remote Sens. Environ.* **2002**, *80*, 203–212. [CrossRef]

67. Radoux, J.; Chomé, G.; Jacques, D.C.; Waldner, F.; Bellemans, N.; Matton, N.; Lamarche, C.; D'Andrimont, R.; Defourny, P. Sentinel-2's Potential for Sub-Pixel Landscape Feature Detection. *Remote Sens.* **2016**, *8*, 488. [CrossRef]

68. Sei, A. Efficient correction of adjacency effects for high- resolution imagery: Integral equations, analytic continuation, and Padé approximants. *Appl. Opt.* **2015**, *54*, 3748–3758. [CrossRef]

69. Duan, S.B.; Li, Z.L.; Tang, B.-H.; Wu, H.; Tang, R.; Bi, Y. Atmospheric correction of high-spatial-resolution satellite images with adjacency effects: Application to EO-1 ALI data. *Int. J. Remote Sens.* **2015**, *36*, 5061–5074. [CrossRef]

70. Vermote, E.F.; El Saleous, N.; Justice, C.O.; Kaufman, Y.J.; Privette, J.L.; Remer, L.; Roger, J.C.; Tanré, D. Atmospheric correction of visible to middle-infrared EOS-MODIS data over land surfaces: Background, operational algorithm and validation. *J. Geophys. Res.* **1997**, *102*, 17131–17141. [CrossRef]

71. Burazerovic, D.; Geens, B.; Heylen, R.; Sterckx, S.; Scheunders, P. Unmixing for detection and quantification of adjacency effects. In Proceedings of the 2012 IEEE International Geoscience and Remote Sensing Symposium, Munich, Germany, 22–27 July 2012; pp. 3090–3093.

72. Keshava, N.; Mustard, J.F. Spectral unmixing. *IEEE Signal Process. Mag.* **2002**, *19*, 44–57. [CrossRef]

73. Seidel, F.C.; Popp, C. Critical surface albedo and its implications to aerosol remote sensing. *Atmos. Meas. Tech.* **2012**, *5*, 1653–1665. [CrossRef]

74. Remer, L.A.; Tanre, D.; Kaufman, Y.J.; Levy, R.; Mattoo, S. *Algorithm for Remote Sensing of Tropospheric Aerosol from MODIS: Collection 005*; NASA: Merritt Island, FL, USA, 2006.

75. Kondratyev, K.Y.; Kozoderov, V.V.; Smokty, O.I. *Remote Sensing of the Earth from Space: Atmospheric Correction*; Springer: Berlin, Germany, 2013.

76. Hilker, T.; Lyapustin, A.I.; Hall, F.G.; Myneni, R.; Knyazikhin, Y.; Wang, Y.; Tucker, C.J.; Sellers, P.J. On the measurability of change in Amazon vegetation from MODIS. *Remote Sens. Environ.* **2015**, *166*, 233–242.

77. Videla, F.C.; Barnaba, F.; Angelini, F.; Cremades, P.; Gobbi, G.P. The relative role of amazonian and non-amazonian fires in building up the aerosol optical depth in South America: A five year study (2005–2009). *Atmos. Res.* **2013**, *122*, 298–309. [CrossRef]

78. Artaxo, P.; Rizzo, L.V.; Brito, J.F.; Barbosa, H.M.J.; Arana, A.; Sena, E.T.; Cirino, G.G.; Bastos, W.; Martin, S.T.; Andreae, M.O. Atmospheric aerosols in Amazonia and land use change: From natural biogenic to biomass burning conditions. *Faraday Discuss.* **2013**, *165*, 203–235. [CrossRef] [PubMed]

79. Bodhaine, B.A.; Wood, N.B.; Dutton, E.G.; Slusser, J.R. On Rayleigh optical depth calculations. *J. Atmos. Ocean. Technol.* **1999**, *16*, 1854–1861. [CrossRef]

80. Kutser, T.; Pierson, D.C.; Kallio, K.Y.; Reinart, A.; Sobek, S. Mapping lake CDOM by satellite remote sensing. *Remote Sens. Environ.* **2005**, *94*, 535–540. [CrossRef]

81. Matthews, M.W. A current review of empirical procedures of remote sensing in inland and near-coastal transitional waters. *Int. J. Remote Sens.* **2011**, *32*, 6855–6899. [CrossRef]

82. Odermatt, D.; Gitelson, A.; Brando, V.E.; Schaepman, M. Review of constituent retrieval in optically deep and complex waters from satellite imagery. *Remote Sens. Environ.* **2012**, *118*, 116–126. [CrossRef]

83. Bassani, C.; Manzo, C.; Braga, F.; Bresciani, M.; Giardino, C.; Alberotanza, L. The impact of the microphysical properties of aerosol on the atmospheric correction of hyperspectral data in coastal waters. *Atmos. Meas. Tech.* **2015**, *8*, 1593–1604. [CrossRef]

84. Li, Z.; Zhao, X.; Kahn, R.; Mishchenko, M.; Remer, L.; Lee, K.-H.; Wang, M.; Laszlo, I.; Nakajima, T.; Maring, H. Uncertainties in satellite remote sensing of aerosols and impact on monitoring its long-term trend: A review and perspective. *Ann. Geophys.* **2009**, *27*, 2755–2770. [CrossRef]

85. Eck, T.F.; Holben, B.N.; Reid, J.S.; Dubovik, O.; Smirnov, A.; O'Neill, N.T.; Slutsker, I.; Kinne, S. Wavelength dependence of the optical depth of biomass burning, urban, and desert dust aerosols. *J. Geophys. Res.* **1999**, *104*, 31333–31349.

86. Barnes, B.B.; Hu, C.; Kovach, C.; Silverstein, R.N. Sediment plumes induced by the Port of Miami dredging: Analysis and interpretation using Landsat and MODIS data. *Remote Sens. Environ.* **2015**, *170*, 328–339. [CrossRef]

87. Garaba, S.P.; Zielinski, O. An assessment of water quality monitoring tools in an estuarine system. *Remote Sens. Appl. Soc. Environ.* **2015**, *2*, 1–10. [CrossRef]

88. Louis, J.; Debaecker, V.; Pflug, B.; Main-Knorn, M. Sentinel-2 Sen2Cor: L2A Processor for Users. In Proceedings of the Living Planet Symposium, Prague, Czech Republic, 9–13 May 2016.

89. Wilson, R.T. Py6S: A Python interface to the 6S radiative transfer model. *Comput. Geosci.* **2013**, *51*, 166–171. [CrossRef]

![remote sensing logo] *remote sensing*

MDPI

Article

SNR (Signal-To-Noise Ratio) Impact on Water Constituent Retrieval from Simulated Images of Optically Complex Amazon Lakes

Daniel S. F. Jorge *, Claudio C. F. Barbosa, Lino A. S. De Carvalho, Adriana G. Affonso, Felipe De L. Lobo and Evlyn M. L. De M. Novo

National Institute for Space Research, Avenida dos Astronautas, 1758, São José dos Campos, SP 12227-010, Brazil; claudio.barbosa@inpe.br (C.C.F.B.); lino@dsr.inpe.br (L.A.S.D.C.); affonso@dsr.inpe.br (A.G.A.); felipe.lobo@inpe.br (F.D.L.L.); evlyn.novo@inpe.br (E.M.L.D.M.N.)
* Correspondence: danielsfj@dsr.inpe.br; Tel.: +55-11-94982-6778

Academic Editors: Yunlin Zhang, Claudia Giardino, Linhai Li, Deepak R. Mishra and Prasad S. Thenkabail
Received: 26 February 2017; Accepted: 19 June 2017; Published: 22 June 2017

Abstract: Uncertainties in the estimates of water constituents are among the main issues concerning the orbital remote sensing of inland waters. Those uncertainties result from sensor design, atmosphere correction, model equations, and in situ conditions (cloud cover, lake size/shape, and adjacency effects). In the Amazon floodplain lakes, such uncertainties are amplified due to their seasonal dynamic. Therefore, it is imperative to understand the suitability of a sensor to cope with them and assess their impact on the algorithms for the retrieval of constituents. The objective of this paper is to assess the impact of the SNR on the Chl-a and TSS algorithms in four lakes located at Mamirauá Sustainable Development Reserve (Amazonia, Brazil). Two data sets were simulated (noisy and noiseless spectra) based on in situ measurements and on sensor design (MSI/Sentinel-2, OLCI/Sentinel-3, and OLI/Landsat 8). The dataset was tested using three and four algorithms for TSS and Chl-a, respectively. The results showed that the impact of the SNR on each algorithm displayed similar patterns for both constituents. For additive and single band algorithms, the error amplitude is constant for the entire concentration range. However, for multiplicative algorithms, the error changes according to the model equation and the R_{rs} magnitude. Lastly, for the exponential algorithm, the retrieval amplitude is higher for a low concentration. The OLCI sensor has the best retrieval performance (error of up to 2 µg/L for Chl-a and 3 mg/L for TSS). For MSI, the error of the additive and single band algorithms for TSS and Chl-a are low (up to 5 mg/L and 1 µg/L, respectively); but for the multiplicative algorithm, the errors were above 10 µg/L. The OLI simulation resulted in errors below 3 mg/L for TSS. However, the number and position of OLI bands restrict Chl-a retrieval. Sensor and algorithm selection need a comprehensive analysis of key factors such as sensor design, in situ conditions, water brightness (R_{rs}), and model equations before being applied for inland water studies.

Keywords: signal-to-noise ratio; Remote Sensing Reflectance; bio-optical algorithms; inland waters

---

1. Introduction

Sensor design (spatial, radiometric, and spectral resolution, and signal-to-noise ratio-SNR) is shaped by remote sensing applications (satellite mission). During the last decade, most sensors in orbit were designed for either oceanic water or land applications (e.g., Moderate Resolution Imaging Spectroradiometer (MODIS) Aqua and Terra). Therefore, they were not tuned for inland water applications. Despite numerous studies focusing on inland waters, these sensors are suboptimal, imposing an intense impact on the estimate accuracy [1].

A sensor's SNR is a major issue for the remote sensing community since a large part of the signal comes from atmospheric interference which increases noise. The maximum contribution of the water leaving radiance to the measured signal at the sensor is about 15% [2], whereas the remainder comes from the atmosphere [3–5]. Despite advances in atmospheric correction, residual atmospheric noise remains [6,7]. The SNR of an orbital sensor, measured in the laboratory, can be based on a standard target. A spectrally uniform 5% albedo is commonly used during laboratory calibration with sensors designed for water measurements [8]. However, water leaving radiance is usually lower than that, especially in the longer wavelengths and in waters dominated by organic matter; thus, the actual SNR may fail to reach the prescribed SNR [6,8,9]. The application of orbital sensors to inland waters with low radiance can be highly affected by sensor noise. Degradation of the spatial resolution is usually applied as a tool to reduce noise, thus overcoming SNR limitations. Vanhellemont & Ruddick [10] described the relation between SNR and spatial resolution for Operational Land Imager (OLI) and MODIS images, demonstrating how resampling can reduce the noise in the red band. Regarding the study of small lakes, where only a few pixels are available and resampling is not feasible, the SNR impact is critical. Spectral resolution also impacts the SNR, as the narrower the band width is, the more sensitive it is to the absorption peaks. However, to maintain the SNR requirements for water applications, a sensor's spatial resolution is compromised, since the narrower the bandwidth, the higher the noise. This aspect is observed by comparing the SNR of multispectral and hyperspectral sensors [6,7,11].

In the earliest stages of remote sensing, sensors designed for water color retrieval proved to be very effective for open ocean waters (e.g., IOCCG [12] and Muow et al. [13]). However, monitoring inland waters is more challenging, because of their optical complexity, high spatial frequency of water components, and sensor constraints [14]. New sensors made available in the last few years, such as the Multispectral Instrument (MSI), Ocean and Land Color Instrument (OLCI), Operational Land Imager (OLI), and Hyperspectral imager for the Coastal Ocean (HICO) are potentially useful for inland water studies. The selection of the best sensor, however, is a challenging task because of the limited number of studies regarding the impact of sensor design on inland water color product uncertainty. Gerace et al. [15] compared the quality of four sensors for deriving bio-optical products applied to coastal water studies and concluded that SNR and spatial resolution are the sensor design features with the largest impact on bio-optical product uncertainty. Similarly, Moses et al. [8] focused on the SNR impact on HICO data quality, but only provided information about the average impact of the SNR on the bio-optical products, without assessing how differences in the model equations (additive and/or multiplicative operations and linear or logarithmic fit), magnitude, and shape of each R_{rs} spectrum contributed to the relative impact of SNR on product uncertainty.

The uncertainties in the R_{rs} spectrum related to sensor design can also be further amplified depending on the target characteristics. In the case of the Amazon floodplain lakes, those uncertainties can be even larger because they are usually isolated, surrounded by dense vegetation characterized by very high trees (up to 35 m) and subjected to seasonal variation in size, depth, and optical composition due to the Flood Pulse [16]. In addition to these threshold constraints related to inland waters, atmospheric correction in the Amazon region can be another major source of uncertainty due to the spatial and temporal variation of cloud cover, including cloud shadow and aerosol scattering properties [10]. To successfully obtain water color products in these challenging conditions, it is crucial to assess the intrinsic capability of the available sensors and quantify their impact on the water leaving signal and water color algorithms.

Given this lack of information concerning the uncertainties caused by the interplay of sensor design and target optical features, the objective of this paper is to assess the impact of the SNR on water color products derived from satellite images applied in threshold conditions such as the Amazon floodplain lakes. To accomplish this objective, optical and limnological in situ measurements were collected at Mamirauá Sustainable Development Reserve (RDSM), in Central Amazon, Brazil, and used as the input to simulate three orbital image sensors (MSI/Sentinel-2, OLCI/Sentinel-3, OLCI/Landsat 8).

2. Materials and Methods

2.1. Study Site Description

The selected lakes are located in the RDSM (Figure 1a,b). This is the first and largest Sustainable Development Reserve in Brazil dedicated exclusively to the protection of the Amazonian floodplain, comprising approximately 1,124,000 hectares. This Conservation Unit was created by the State of Amazonas in 1996 and is one of the Brazilian sites of the United Nations Ramsar Convention [17]. It consists of a pristine floodplain inundated by sediment-rich whitewater rivers at the confluence of the Japurá and Solimões rivers, and forms a complex mosaic of seasonally flooded forests, lakes, and channels. Rivers and lakes undergo constant change due to the transport of sediments and organic matter, caused by the annual water level variation of up to 12 m [17–20]. Moreover, the seasonal flooding changes the proportion of suspended and dissolved components in the water by altering its physical-chemical conditions [21]. Consequently, this affects the ecosystem where these waters circulate [22]. The flood pulse starts in May and ends in July, while the drought period lasts from September to November. The rising of the water level begins in January and the water starts receding in September. The flood pulse has a monomodal annual pattern (Figure 1c), and the changes in the water level are due to changes in the snowmelt in the Andes and precipitation in the pre-andean region and in the median Amazon basin [19]. The management plan of the RDSM accounts for more than 5000 lakes in the area, that vary in shape (elongated, circular, and complex), size (from 1.5 ha to 900 ha), and connection to the main rivers and channels, which will influence the lake hydraulic residence and water flow connection during the low water level phase [23].

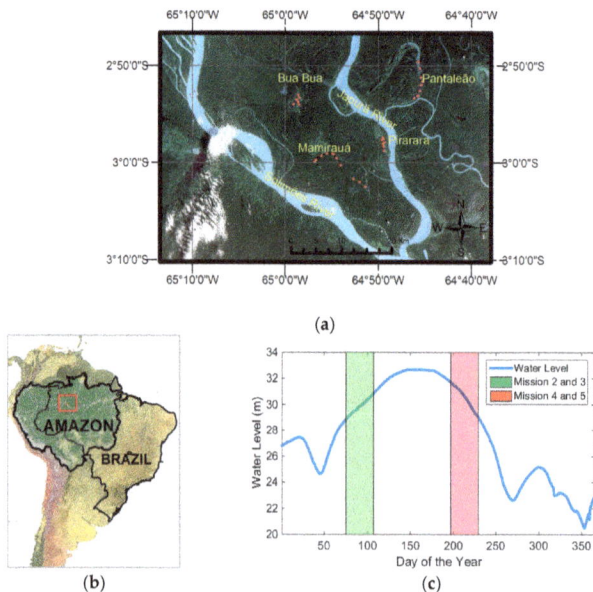

(a)

(b)   (c)

Figure 1. (a) OLI true color image for the study area showing the selected lakes inside the MSDR. Red dots represent the distributions of points at each lake. The image is from December 4th of 2014. (b) Brazil figure with RDSM location evidenced in red (c) Hydrograph for the year 2016, showing water level variation for missions 2, 3 (in green) 4 and 5 (in red). Mission 1 occurred in the same flood phase as missions 4 and 5 in 2015.

The criteria for lake selection included their potential for remote sensing analysis (lake size and shape), as well as accessibility throughout the year. The sampling points were selected to include the main observable changes in the lake water color. Based on the above restrictions, the Bua-Buá (triangular shape, 1 km × 2.1 km), Mamirauá (elongated shape, up to 0.4 km × 4 km), Pirarara (lozengular shape, up to 0.9 km × 2.7 km), and Pantaleão (rectangular shape, up to 1.5 km × 6 km) lakes were selected (Figure 1a).

2.2. In Situ Dataset

Five field missions (subsequently named M1, M2, M3, M4, and M5) were carried out for two years. Missions M2 and M3 were conducted during the rising water period (March and April 2016) and missions M1, M4, and M5 during the receding water period (July 2015, July and August 2016). In each mission, three to six sampling points were visited per lake, resulting in a total of 102 sampling points. At each sampling point, limnological and radiometric data were obtained.

For the limnological measurement, water samples were collected at the subsurface (10 cm) and kept light-free and cooled in ice for a maximum of 3 h, before being filtered. For the Chlorophyll-a concentration (µg/L), water samples were filtered through Whatman GF/F (0.7 µm) filters and for the Total Suspended Solids (TSS) (mg/L) and its Inorganic (TSIS) and Organic (TSOS) fractions, water was filtered through Whatman GF/C (1.2 µm), both of which included 45 mm filters. A maximum of 500 mL was filtered for each sample. Chl-a was analyzed according to Nush [24] and TSS and its fractions according to Wetzel & Likens [25], in replicates.

The Colored Dissolved Organic Matter (CDOM) spectral absorptions $(a_{CDOM}(\lambda))$ (m^{-1}) were determined using a 10 cm quartz cuvette in a single beam mode of the UV-2600 Shimadzu spectrophotometer, scanning from 300 to 800 nm, with 1 nm increments. $a_{CDOM}(\lambda)$ was generated based on the $a_{CDOM}(\lambda)$ measured, following Tilstone et al. [26], and the a_{CDOM} exponential model for each $a_{CDOM}(\lambda)$ measured at 420 nm and the slope of each curve. Table 1 shows the range of magnitude of limnological data, illustrating the optical diversity of the lakes and the changes along the flood pulse.

Table 1. Limnological dataset for each lake and flooding phase. The names refer to the four lakes (Bua-Buá, Mamirauá, Pantaleão, and Pirarara). The mean value is shown followed by the standard deviation in parenthesis. Chl-a is Chlorophyll-a in µg/L, TSS is in mg/L, and a_{CDOM} is $a_{CDOM}(420)$ in m^{-1}.

	Rising Water				Receding Water			
	Bua	**Mam**	**Pant**	**Pira**	**Bua**	**Mam**	**Pant**	**Pira**
Chl-a	14.7	18.1	11	8.3	8.2	7.6	12.1	9.3
	(9.2)	(6.2)	(5.6)	(3.4)	(4.9)	(4.7)	(5.6)	(3.6)
TSS	9.5	9.7	18.5	25.9	5.5	5.2	7.5	6.8
	(3.2)	(2.6)	(4.8)	(6.8)	(2.4)	(1.1)	(1.5)	(1.3)
a_{CDOM}	5.6	6.4	2.1	2.2	2.5	2.6	2.5	2.1
	(0.7)	(1.5)	(0.2)	(0.2)	(0.2)	(0.3)	(0.4)	(0.2)

Radiometric measurements were carried out for all sampling points, using three intercalibrated RAMSES–Trios sensors. The sensors were used to estimate the R_{rs}, above water radiance $(L_w, W \cdot m^{-2} \cdot sr^1 \cdot nm^{-1})$, sky radiance $(L_{SKY}, W \cdot m^{-2} \cdot sr^1 \cdot nm^{-1})$, and downwelling irradiance $(E_D, W \cdot m^{-2} \cdot nm^{-1})$, between 350 and 900 nm. During the measurements, the sensors were positioned with azimuth angles between 90° and 135° in relation to the sun and a Zenith angle of 45° to avoid sun glint effects [27]. The measurement framework followed Mobley [28]. All of the measurements were made between 10:00 and 13:00 and at least 15 samples were obtained for each measured depth. The dataset was processed using MSDA_XE and Matlab. The R_{rs} estimate followed Mobley [28], with sun glint correction based on each sampling point spectrum.

Figure 2 shows the R_{rs} magnitude for all of the sampling points, split according to the water stage (rising water Figure 2a and receding water Figure 2b). The threshold conditions influence the bio-optical properties of the four lakes, with a high input of organic matter throughout the season, especially at Bua-Buá and Mamirauá during the rising water period, whereas Pantaleão and Pirarara's R_{rs} is enhanced by the high sediment loading, particularly during the rising water period. RDSM lakes can be considered dark when compared to sediment loaded inland lakes, making it a remarkable study site to evaluate the impact of the SNR on bio-optical products.

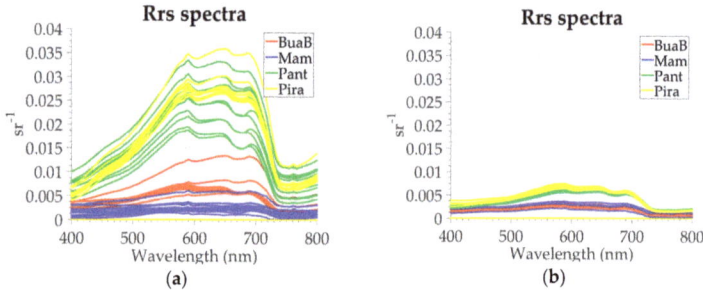

Figure 2. Rrs spectra for each lake in different water phases. (**a**) Mission 2 and 3 (rising water period); (**b**) Mission 1, 4, and 5 (receding water period). Each lake is represented by a specific color; Mamirauá in Blue, Bua-Buá in red, Pantaleão in green, and Pirarara in yellow.

2.3. Data Processing

2.3.1. Dataset Sensor Simulation

The impact of SNR on the R_{rs} spectrum was assessed by simulating two datasets: Noisy spectra and noiseless spectra. The input data for the simulation were in situ measurements and the sensor design specifications. The steps for the simulation are similar for both datasets, except for the noise addition, and are described below. It is important to highlight that the simulation assumes optimum conditions such as perfect atmospheric correction, algorithm calibration, and errorless in situ measurements to isolate the noise impact.

A total of 1000 noisy orbital R_{rs} spectra were simulated for each of the 102 in situ R_{rs} measurements based on the characteristics of MSI/Sentinel-2 [29], OLCI/Sentinel-3 [30], and OLI/Landsat 8 [31], resulting in a total of 306.000 spectra. The simulation workflow (Figure 3) consists of the following five steps: (1) Resampling in situ spectra to sensors that are band-weighted; (2) Computation of a sensor's specific noise; (3) Noise addition to simulated spectra (step 1); (4) Spectra quantization; and (5) Conversion of TOA (Top of atmosphere) irradiance solar spectrum to surface irradiance.

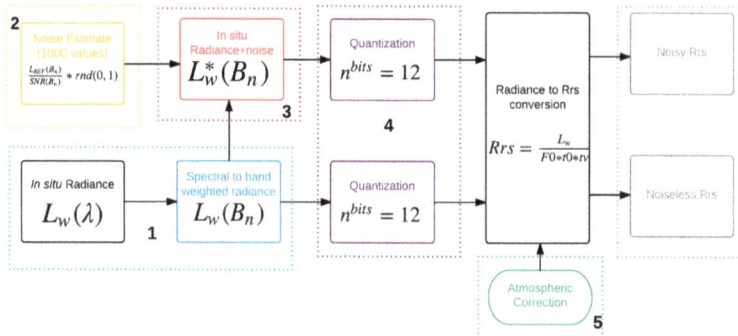

Figure 3. Framework of the proposed simulation for a single spectrum of one orbital sensor.

Image simulation was carried out according to the following steps:

(1) Resampling in situ spectra to sensors that are band-weighted. The conversion of each radiance spectrum to the band-weighted radiance of each sensor was based on the sensor's response function (Tables 2–4), applied according to Equation (1):

$$L_w(B_n) = \frac{\int_{\Delta B_n} L_w(\lambda) \times RF_{B_n}(\lambda)d\lambda}{\int_{\Delta B_n} RF_{B_n}(\lambda)d\lambda} \tag{1}$$

where B is the sensor band; n is the band number, varying from 1 to n, according to sensor design; $L_w(B_n)$ is the water leaving radiance for each band in the unit of (W·m^{-2}·sr^1·nm^{-1}), ΔB_n is the band width; and RF_{B_n} is the response function for each sensor band.

(2) Computation of a sensor's specific noise. The TOA radiance was converted to equivalent noise using the reference radiance at the TOA of each band and sensor (Tables 2–4) and the respective SNR [8] (Equations (2) and (3)).

$$Noise(B_n) = \frac{L_{REF}(B_n)}{SNR(B_n)} \times rnd \tag{2}$$

where L_{REF} is the reference radiance used to generate the specific *SNR*, *SNR* is the Signal to Noise Ratio for each sensor band, and rnd is a random number obtained from a standard normal distribution (mean equal to zero and standard deviation equal to one N (0,1)). For each sensor, a total of 1000 $Noise(B_n)$ spectra were generated.

(3) Noise addition to simulated spectra (step 1). The $Noise(B_n)$ was added to the $L_w(B_n)$ Equation (3). To exempt the impact of error propagation due to atmospheric correction on the noise, no atmospheric uncertainty and "optimum" atmospheric correction was assumed.

$$L_w^*(B_n) = L_w(B_n) \pm Noise(B_n) \tag{3}$$

where $L_w^*(B_n)$ is the noisy water leaving radiance for each band.

(4) Spectra quantization. $L_w^*(B_n)$ quantization was carried out according to Equation (4).

$$L_w^{*b}(B_n) = L_w^*(B_n) / \frac{L_{TOA}max(B_n)}{2^{nbit}} \tag{4}$$

where $L_w^{*b}(B_n)$ is the quantized noisy water leaving radiance for each band, $L_{TOA}max(B_n)$ is the maximum radiance measured by the sensor at B_n, and *nbit* is the number of bits for each sensor (12 bits).

(5) Conversion of TOA (Top of atmosphere) irradiance solar spectrum to surface irradiance. The propagation of the solar spectrum to water level was based on the algorithm described by Vanhellemont & Rudick [10,32], using a standard atmosphere and equations derived from Kaskaoutis & Kambezidis [33], Bird & Riordan [34] and Leckner [35]. The relationship between water leaving radiance (L_w) and reflectance (ρ) is described as:

$$\rho = \frac{\pi \times L_w \times d^2}{F0 \times \cos\theta_0} \tag{5}$$

where d^2 is the Earth-Sun distance in Astronomic Units, θ_0 is the Sun zenith angle, $F0$ is the solar irradiance, and ρ is the reflectance. This paper assumed a value of 1 for d, and $F0$ from Gueymard [36] and Gueymard et al. [37]. $F0$ was propagated to water level as follows [10,32]:

$$F0_{ml} = t_0 \times t_v \times (F0_{TOA}) \tag{6}$$

where t_0 and t_v are the sunwater and sea-sensor diffuse transmittance, $F0_{wl}$ is the solar irradiance at water level, and $F0_{TOA}$ is the solar irradiance at the TOA. For each wavelength, the diffuse transmittance t_0 and t_v was calculated by replacing θ with θ_0 and θ_v in:

$$t = exp\left[-\left(\frac{\tau_r}{2} + \tau_{OZ}\right)/cos\theta\right] \tag{7}$$

where τ_r and τ_{OZ} are the Rayleigh and Ozone optical thickness for a given atmosphere composition, respectively. The impact of water absorption and aerosol absorption on the atmospheric transmittance is ignored in this part of the process.

Table 2. MSI/Sentinel-2 sensor configurations used as the input for simulation. CW is the central wavelength (nm), BW is the band width (nm), SR is the Spatial Resolution (m), L_{REF} (W·m^{-2}·sr^{-1}·μm^{-1}) is the radiance in which the SNR was calculated, Quant is the quantization, and $L_{TOA}max$ (W·m^{-2}·sr^{-1}·μm^{-1}) is the maximum radiance that can be measured by the sensor.

Bands	CW	BW	SR	L_{REF}	SNR	Quant	$L_{TOA}max$
B1	443	20	60	129	129	12	588
B2	490	65	10	128	154	12	615.5
B3	560	35	10	128	168	12	559
B4	665	30	10	108	142	12	484
B5	705	15	20	74.5	117	12	449.5
B6	740	15	20	68	89	12	413
B7	783	20	20	67	105	12	387

Table 3. OLCI/Sentinel-3 sensor configurations used as the input for simulation. CW is the central wavelength (nm), BW is the band width (nm), SR is the Spatial Resolution (m), L_{REF} (W·m^{-2}·sr^{-1}·μm^{-1}) is the radiance in which the SNR was calculated, Quant is the quantization, and $L_{TOA}max$ (W·m^{-2}·sr^{-1}·μm^{-1}) is the maximum radiance that can be measured by the sensor.

Bands	CW	BW	SR	L_{REF}	SNR	Quant	$L_{TOA}max$
B1	400	10	300	63	2188	12	413.5
B2	412	10	300	74	2061	12	501.3
B3	442	10	300	66	1811	12	466.1
B4	490	10	300	51	1541	12	483.3
B5	510	10	300	44	1488	12	449.6
B6	560	10	300	31	1280	12	524.5
B7	620	10	300	21	997	12	397.9
B8	665	10	300	16	855	12	364.9
B9	673	7.5	300	16	707	12	443.1
B10	681	7.5	300	15	745	12	350.3
B11	708	10	300	13	785	12	332.4
B12	753	7.5	300	10	605	12	377.7
B13	778	15	300	9	812	12	277.5
B14	865	20	300	6	666	12	229.5
B15	885	10	300	6	395	12	281

Table 4. OLI/Landsat 8sensor configurations used as the input for simulation. CW is the central wavelength (nm), BW is the band width (nm), SR is the Spatial Resolution (m), L_{REF} (W·m^{-2}·sr^{-1}·μm^{-1}) is the radiance in which the SNR was calculated, Quant is the quantization, and $L_{TOA}max$ (W·m^{-2}·sr^{-1}·μm^{-1}) is the maximum radiance that can be measured by the sensor.

Bands	CW	BW	SR	L_{REF}	SNR	Quant	$L_{TOA}max$
B1	443	20	30	190	232	12	782
B2	482	65	30	190	355	12	800
B3	565	75	30	194	296	12	738
B4	660	50	30	150	222	12	622
B5	867	40	30	150	199	12	381

The Rayleigh optical depth (τ_r) was calculated using the model proposed by Kaskaoutis & Kambezidis [33], with improvements proposed by Leckner [35]:

$$\tau_r \lambda = 0.008735 \left(\frac{P}{P0} \right) \lambda^{-4.08} \tag{8}$$

where λ is the wavelength in micrometers, P is the atmospheric pressure at the site (1014 hPa), and $P0$ is the reference sea level pressure (1013.25 hPa).

The t_v was calculated according to Kaskaoutis & Kambezidis [33], which was based on Bird & Riordan [34] and Leckner [35] (Equations (9) and (10)):

$$\tau_v(\lambda) = \exp(\alpha_0 \times \lambda \times O_3 \times M_0) \tag{9}$$

where α_0 is the ozone absorption coefficient, O_3 is the ozone concentration (atm.·cm^{-1}), and M_0 is the ozone mass. The ozone absorption coefficient was linearly interpolated from Bird & Riordan [34], and the ozone mass was calculated following Leckner [35].

$$M_0 = (1 + h_0/6370) \left(cos^2 Z + 2\, h_0/6370 \right)^{0.5} \tag{10}$$

where h_0 is the height of the maximum ozone concentration, assumed as 22 km, and Z is the zenithal angle. Input parameters (Tables 5 and 6) were used for solar spectrum propagation throughout the atmosphere.

Table 5. Parameters used during the conversion of the TOA irradiance solar spectrum to surface irradiance for all of the sensors.

Parameter	Range or Value
Date	1 January
Time	12 h 00 min (GMT)
Latitude	0
Ground Elevation	40 m
Sensor Zenith Angle	0
Sensor Azimuth Angle	0
Ozone Amount	0.3 atm cm^{-1}
Height of Maximum Ozone Concentration	22 km
Atmopsheric pressure at site	1014

Table 6. Example of the parameters obtained and used during the atmospheric simulation for the MSI sensor. $F0_{TOA}$ is the band-weighted extraterrestrial solar irradiance, τ_r is the Rayleigh optical thickness for a standard atmosphere, and τ_{OZ} is the ozone optical thickness for 300 DU of atmospheric ozone.

Band	$F0_{TOA}$ $\left(Wm^{-2}\, \mu m^{-1} \right)$	τ_r	τ_{OZ}
B1 (443)	1938.2	0.2405	0.0004
B2 (490)	1916.5	0.1543	0.0087
B3 (560)	1845.9	0.0934	0.0309
B4 (665)	1524.7	0.0464	0.0167
B5 (705)	1402.5	0.0366	0.0063
B6 (740)	1290	0.0298	0.0030
B7 (783)	1184.8	0.0238	0.0002

The simulation of the noiseless datasets followed the same steps, except for steps 2 and 3

At the end of the dataset sensor simulation, the quantized noiseless water leaving radiance for each band $\left(L_w^b(B_n)\right)$ and $L_w^{*b}(B_n)$ were converted to $R_{rs}(B_n)$ according to Equation (11), in order to be used as the input for bio-optical algorithms.

$$R_{rs}(B_n) = \frac{L_w^b(B_n)}{F0_{wl}(B_n) \times t_0(B_n) \times t_v(B_n)} \tag{11}$$

where $R_{rs}(B_n)$ is the Remote Sensing Reflectance at B_n, $F0_{wl}$ is the solar irradiance at water level, and t_0 and t_v are the sunwater and sea-sensor diffuse transmittance, respectively.

2.3.2. Impact of Sensors Characteristics on Chl-a and TSS Algorithms

The two simulated datasets (noisy and noiseless) were used to assess the impact of the optical sensor configuration on the Chl-a and TSS algorithms currently in use [38–43]. Seven algorithms were applied for different sensors and study sites [38–43], with few changes in the central wavelengths according to band availability. Algorithms were chosen based on the diversity of bands and mathematical operations involved so as to encompass a range of model equations. For brevity, such models were classified in this paper as additive (subtraction and addition), multiplicative (division and multiplication), and exponential.

Three empirical algorithms were tested for TSS (*TSS_linear*, *TSS_exp*, *TSS_NSSI*) and four algorithms for Chl-a (*CLH*, *2B*, *3B*, *NDCI*) (Table 7). Each algorithm was calibrated using the noiseless dataset and in situ measurements of either TSS or Chl-a. In order to remove the uncertainty of model calibration, the calibrated model was applied to the noiseless dataset, instead of the in situ dataset, and compared with the model results based on the noisy dataset input. This method has two assumptions: (i) the concentration provided by calibrated data is the reference concentration ("ground truth") against which the simulation results are assessed; (ii) the uncertainty between the noisy and noiseless outputs is only due to the changes in SNR. The modeled concentration is compared to the "ground truth", so that changes in magnitude are solely based on two aspects: algorithm constants (e.g., *a*, *b*, *c*, and *d* (Table 7)) and R_{rs}.

Table 7. Chl-a and TSS algorithms. The exact wavelength used changed for each sensor. *CLH* is a chlorophyll line height model, *2B* is a red/NIR band ratio model, *3B* is a red/NIR 3 band model, *NDCI* is a red NIR 2 band model, *TSS_linear* is a linear red band model, *TSS_exp* is an exponential red band model, and *TSS_NSSI* is a red/green exponential band ratio model.

Model Name	Linear Model ($a \times x + b$)	Reference
CLH	$x = R_{rs}(708) - (R_{rs}(665) + R_{rs}(740))/2$	[38]
2B	$x = R_{rs}(665) \times R_{rs}(708)^{-1}$	[39]
3B	$x = (R_{rs}(665)^{-1} - R_{rs}(708)^{-1}) \times R_{rs}(753)$	[39]
NDCI	$x = R_{rs}(\text{red}) - R_{rs}(\text{NIR})/(R_{rs}(\text{red}) + R_{rs}(\text{NIR}))$	[40]
TSS_linear	$x = R_{rs}(\text{red})$	[41]
	Non Linear model	**Reference**
TSS_exp	$TSS = ((a\,R_{rs}(\text{red}))/b)c + d$	[42]
TSS_NSSI	$NSSI = (R_{rs}(\text{green}) - R_{rs}(\text{red}))/(R_{rs}(\text{green}) + R_{rs}(\text{red}))$ $TSS = a\,e^{-b\,NSSI}$	[43]

Most approaches employed to quantify the SNR impact on water algorithms use the Normalized Root-Mean-Square Error (NRMSE) (e.g., Moses et al. [8] and Gerace et al. [15]) as a statistic measurement for assessing model quality. Although accuracy measurements such as NRMSE give an insight regarding the proportional error of the model fitting, they do not remove the intrinsic error due to the choice of algorithm equations, which might lead to an over- or underestimation of SNR error. For this reason, this paper focuses on the relationship between the algorithm equation, R_{rs} magnitude, and the shape of the modeled concentration distribution. Considering the available bands for each sensor, Chl-a algorithms were applied to MSI and OLCI sensors, while TSS algorithms were applied to the three sensors.

3. Results and Discussion

3.1. Dataset Simulation

The simulated R_{rs} values for the three sensors and the relative error are shown in Figure 4. Given the sensor design, OLCI (Figure 4c) presented the highest spectral resolution and number of bands, allowing an accurate portrayal of the water R_{rs} spectrum. On the other hand, the spatial resolution (300 m) limits its application in the study of small and narrow lakes, as opposed to OLI and MSI (up to 30 m).

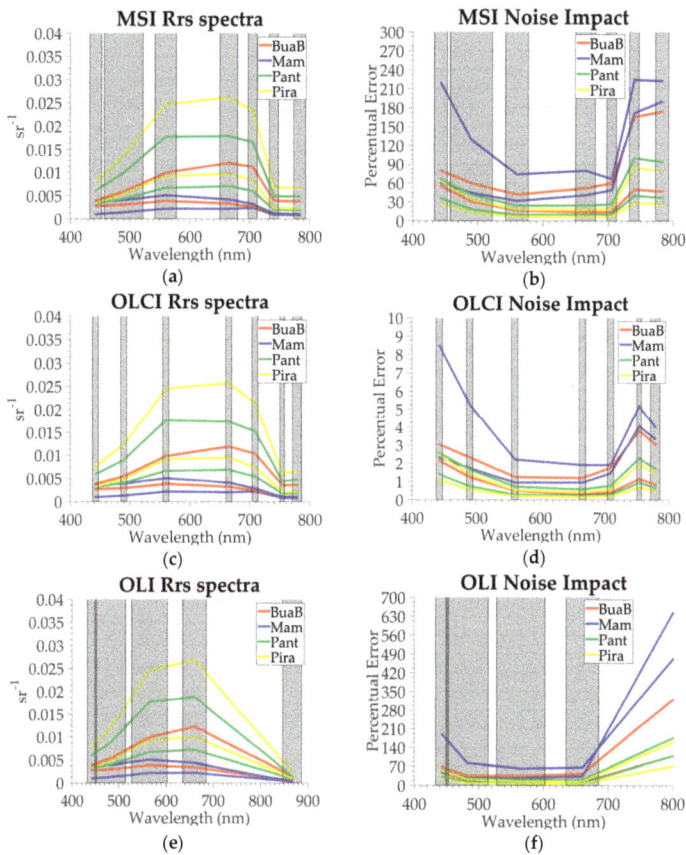

Figure 4. (a–f) Noise band-weighted R_{rs} spectra for each sensor (**a,c,e**) and normalized error due to the optical characteristics of each sensor band (**b,d,f**), with two spectra of each lake. (**a,b**) MSI; (**c,d**) OLCI; (**e,f**) OLI. Each lake is represented by a specific color; Mamirauá in Blue, Bua-Buá in red, Pantaleão in green, and Pirarara in yellow. The noise used was equal to $\frac{L_{TOA}}{SNR} \times .rnd$, in which *rnd* is the standard deviation of a standard normal distribution (equal to 1). The percent error is equal to $100 \times \frac{R_{rs}(noisy) - R_{rs}}{R_{rs}}$.

Disregarding the sensor design, the error percentage due to noise is higher in the blue and NIR bands (Figure 4b,d,f) than that in the remaining bands. The relative high impact in the blue band is due to the higher signal from the atmosphere in relation to longer wavelengths. In the Near Infra-Red (NIR) band, nevertheless, the relative error is due to the small SNR (Tables 2–4). OLI has the highest relative noise impact, reaching 700% in the NIR, whereas MSI's maximum value is 250% and OLCI's maximum value is only 10%. Considering these results, the best algorithms for the retrieval of optical components should include bands between 550 and 700 nm, where relative errors are below 50% in all of the cases. Another important aspect is the relative error for each lake, as a function of water clarity. The highest errors are observed at Mamirauá and Bua-Buá (dark lakes), while the errors at Pirarara and Pantaleão (bright lakes) are below 50% for all bands. The results indicate that before selecting the sensor, it is crucial to consider the R_{rs} amplitude range, as long as the spatial resolution suits the lake area and shape constraints.

3.2. Algorithm Evaluation

3.2.1. Chlorophyll-a

The results show that the relative accuracy of Chl-a retrieval is highly dependent on the model equation, SNR, and R_{rs} magnitude (Figures 5 and 6). In general, the concentration error increases from additive towards multiplicative band operations. Additionally, the highest error for all of the models was observed for the MSI sensor. Although the bandwidth and position are additional sources of uncertainty for algorithms, this method compared the noisy and noiseless datasets of each sensor, assuming that error amplitude is only related to the SNR of each sensor.

In the *CLH* model, the relationship among the bands is additive, so the noise impact is reduced when compared to the multiplicative models (Figures 5 and 6). The concentration error is affected by the algorithm slope, while the intercept contribution is constant for all concentrations. The error magnitude and distribution (Figures 5a and 6a) are the same for all concentrations; so, a higher relative error is expected for low concentrations due to the algorithm intercept uncertainty. For these algorithms, the concentration changes do not depend on the R_{rs} magnitude.

A different pattern is observed for the multiplicative Chl-a algorithms (*2B, 3B, NDCI*–Figures 5 and 6b–d) which apply band ratios. In this case, the noise interference can be either constructive or destructive. When compared to the additive model (*CLH*), a higher error amplitude is expected for all concentrations. The highest relative impact is observed in low concentrations.

The SNR impact on the Chl-a concentration changes according to the model equation (Figures 5 and 6). Most approaches quantifying the SNR impact on water algorithms use RMSE or NRMSE. For example, Moses et al. [8] observed sensor relative errors of up to 40% for the OC4 algorithm and 25% for the two bands red-NIR algorithm for HICO. Gerace et al. [15] used an optimization algorithm for OLI and MERIS and observed errors of 35% for Chl-a. These errors, however, were computed assuming that the R_{rs} magnitude, model equation, and noise are independent. Based on our results and given that OC4 is a fourth-degree polynomial (not tested in this work), one would expect a higher error amplitude in Chl-a estimates due to the SNR.

For the *CLH* model, the error amplitude changed for each sensor, but its distribution shape remained the same. The error amplitude is constant for all concentrations, with a value of around 1 µg/L for MSI and 0.1 µg/L for OLCI. The relative error is higher for low concentrations (up to 11%) and is halved at the max concentration (Figures 5 and 6a).

For the *2B, 3B,* and *NDCI* models, the shape of the noisy data distributions seems to be erratic, with different amplitudes for similar concentrations for the three sensors (Figures 5 and 6b–d). The analysis of the in situ spectrum for model *2B, 3B,* and *NDCI,* showed a higher error amplitude (up to 2 µg/L) for spectra with a low R_{rs} magnitude (<0.005 sr^{-1}) (Figure 7a). On the other hand, for spectra with a higher R_{rs} (>0.01 sr^{-1}) (Figure 7b), the amplitude is similar to that of the *CLH* model. Based on these results, the impact of the SNR on the Chl-a estimation is higher in dark lakes such as Mamirauá and Bua-Buá (errors of up to 2 µg/L) (Figure 7c,d) for OLCI. For the MSI sensor, the three proposed algorithms showed a higher impact than that of OLCI due to the poorer SNR, with errors above 10 µg/L.

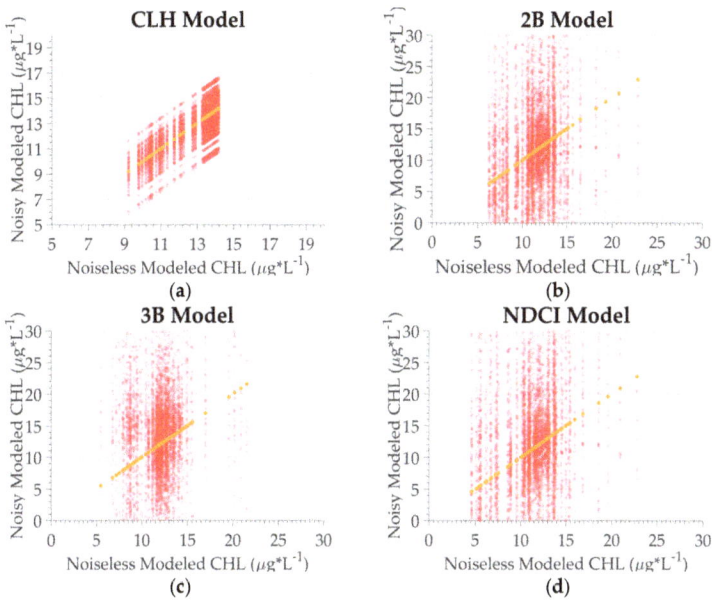

Figure 5. Algorithm performance for the chl-a concentration of the MSI sensor. (**a**) *CLH*; (**b**) *2B*; (**c**) *3B*; (**d**) *NDCI*. Circles in red represent the data with noise, circles in yellow the noiseless data. The transparency is based on the frequency distribution, in which opaque indicates a higher frequency, and transparent indicates a lower frequency.

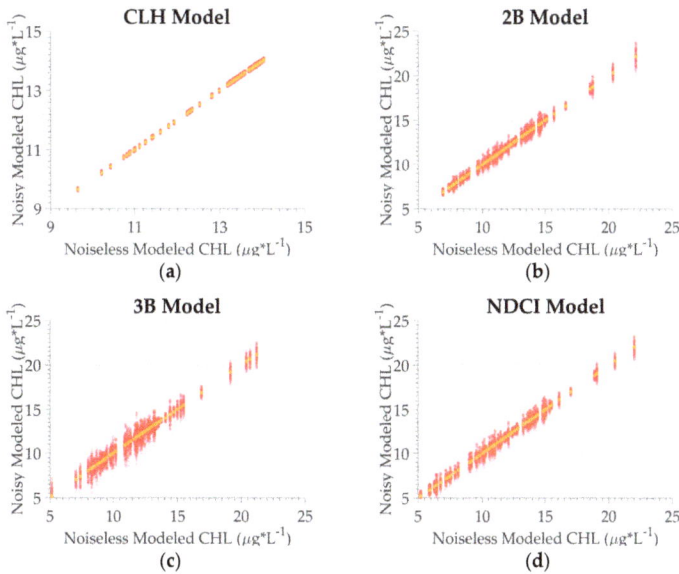

Figure 6. Algorithm performance for the chl-a concentration of the OLCI sensor. (**a**) *CLH*; (**b**) *2B*; (**c**) *3B*; (**d**) *NDCI*. Circles in red represent the data with noise, circles in yellow the noiseless data. The transparency is based on the frequency distribution, in which opaque indicates a higher frequency, and transparent indicates a lower frequency.

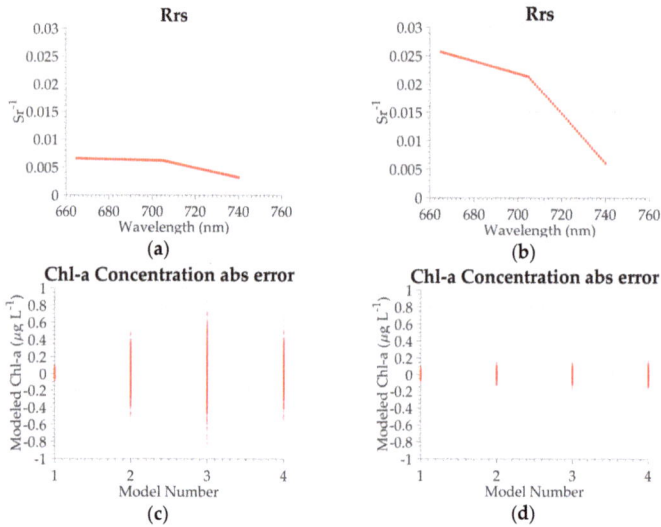

Figure 7. Example of the OLCI R$_{rs}$ magnitude and Chl-a concentration obtained for the four models for two lakes. (**a**) R$_{rs}$ for three bands for one sample station of Bua-Buá, (**b**) R$_{rs}$ for three bands for one sample station of Pirarara, (**c**) Chl-a concentration for the Bua-Buá sampling point, (**d**) Chl-a concentration for the Pirarara sampling point. The numbers 1, 2, 3, and 4 refer to the *CLH*, *2B*, *3B*, and *NDCI* models, respectively.

For spectra with a low R$_{rs}$, the error distribution amplitude was higher than those with a high R$_{rs}$. Therefore, future sensors for imaging inland water applications should require both a higher SNR minimum and minimum spatial resolution to cope with small and narrow lakes. Another aspect to be considered is the model equation. According to Luck [44], higher order algorithms increase the fitness between in situ and modeled data. However, these algorithms propagate the uncertainty in Chl-a estimates due to the SNR under specific conditions (e.g., low R$_{rs}$ magnitude and model equation), decreasing the algorithm accuracy. Thus, the balance of those two aspects should also be considered when applying water quality algorithms, as shown by the results from the four models (Figures 5–7).

As the *CLH* model uses an additive operation, it can be applied to the four lakes using the three sensors, without any preprocessing. For the multiplicative models, however, spatial resampling may be necessary in order to mathematically increase the SNR. For example, with a 2 × 2 and 3 ×3 pixel window, it is possible to increase the SNR by up to two and three times, respectively. These SNR increments can be calculated by the square root of the window size times the original SNR [45]. In the case of MSI, the pixel size of the selected bands is up to 30 m, so resampling may computationally increase the SNR without compromising the results for suitable sized lakes. On the other hand, an OLCI 300 m pixel size is not appropriate for resampling for most of the small lakes due to spectral mixing [46]. Therefore, when comparing the design of sensors for inland water application, it is imperative to assess the sensor suitability for any specific study site, as well as the post processing feasibility (if required).

3.2.2. TSS

Similarly to Chl-a algorithms, the relative accuracy of TSS retrieval depends on the model equation, SNR, and R$_{rs}$ magnitude (Figures 8–10. In general, the TSS concentration error increases from additive to multiplicative equations and from linear to exponential models.

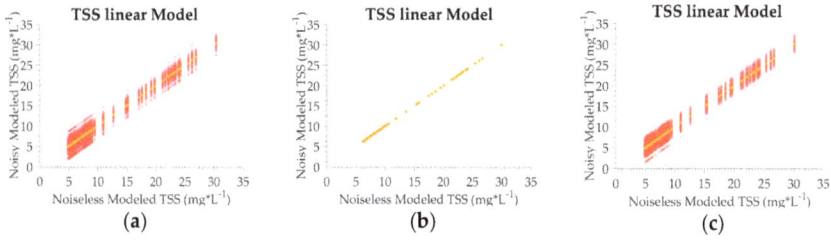

Figure 8. *TSS_linear* model performance for TSS concentration. (**a**) MSI sensor; (**b**) OLCI sensor; (**c**) OLI sensor. Circles in red represent the data with noise, circles in yellow noiseless data. The transparency is based on the frequency distribution, in which opaque indicates a higher frequency, and transparent indicates a lower frequency.

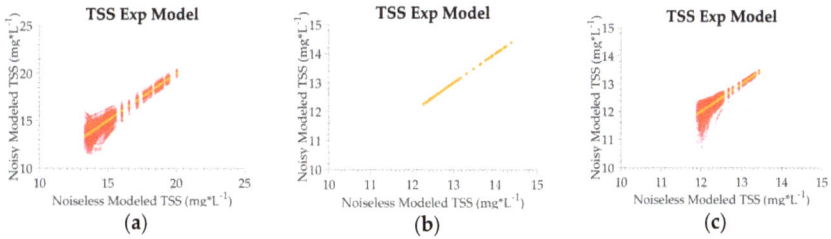

Figure 9. *TSS_exp* algorithm performance for TSS concentration. (**a**) MSI sensor; (**b**) OLCI sensor; (**c**) OLI sensor. Circles in red represent the data with noise, circles in yellow noiseless data. The transparency is based on the frequency distribution, in which opaque indicates a higher frequency, and transparent indicates a lower frequency.

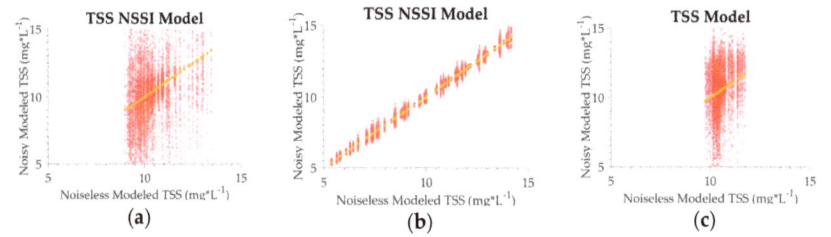

Figure 10. *TSS_NSSI* algorithm performance for TSS concentration. (**a**) MSI sensor; (**b**) OLCI sensor; (**c**) OLI sensor. Circles in red represent the data with noise, circles in yellow noiseless data. The transparency is based on the frequency distribution, in which opaque indicates a higher frequency, and transparent indicates a lower frequency.

For the *TSS_linear* model (Figure 8), the noise impact is linear for all the sensors and throughout the concentration range. The highest error is observed for the MSI sensors (5 mg/L), followed by OLI (3 mg/L) and OLCI (negligible error). The relative contribution of the error is higher with low concentrations. Given that the absolute impact of the noise is constant throughout the concentrations, the relative error is lower for higher concentrations. For the *TSS_exp* model (Figure 9), a distinct pattern is observed; in low TSS concentrations, the error is higher, and the distribution follows an exponential curve towards a higher concentration. The maximum errors for the *TSS_exp* model reach 3 mg/L and 1 mg/L for MSI and OLI, respectively (Figure 9). Lastly, the *TSS_NSSI* model (Figure 10) showed an erratic pattern, similar to that displayed in the Chl-a multiplicative models (Figure 6b–d), suggesting a dependency on the R_{rs} magnitude due to the band ratio approach.

Moses et al. [8] found similar results for HICO, with errors of up to 40% for concentrations below 3 mg/L, and up to 5% for concentrations above 5 mg/L. Although the authors used the optimized error minimization approach, the results displayed similar patterns as observed in this study. Gerace et al. [15] observed mean errors of up to 15% due to the SNR in TSS algorithms.

The uncertainty in the TSS retrieval due to the SNR is highly dependent on the model. The model equation and the concentration range control the need of pixel resampling in order to reduce the noise impact. An MSI and OLI pixel size of 30 m is adequate for most lakes, but the SNR can be a limitation. One band model is usually enough for TSS retrieval, so the OLI spectral resolution is not a restriction, but the SNR, in some cases, needs to be mathematically increased similarly to Chl-a algorithms. In spite of the good congruence of the OLCI estimates, its pixel size of 300 m is a serious constraint for the study of small lakes, as for those in the RDSM.

4. Conclusions

The experiment carried out to assess the impact of the SNR on water color products indicated that, regardless of the estimated parameter (TSS or Chl-a) and sensor design (OLI, OLCI, and MSI), the error pattern is similar for any given algorithm. It is important to highlight that the simulation assumes optimum conditions such as perfect atmospheric correction, algorithm calibration, and errorless in situ measurements to isolate the noise impact. For an actual orbital image, under the described suboptimal conditions, there is an increase in the uncertainty of TSS and Chl-a retrievals.

The amplitude of the retrieved concentration due to the noise is constant for the entire concentration range when using additive and single band algorithms. However, when using multiplicative algorithms, the amplitude changes according to the model equation and to the R_{rs} magnitude. Finally, we observed that the retrieval amplitude is higher for a low concentration regarding the exponential algorithm.

The noise impact on band ratio algorithms applied to Chl-a and TSS retrieval is amplified when using a lower R_{rs}. While this impact is less substantial for a higher R_{rs}, it is similar to that of the additive algorithms. Although the OLCI sensor presents the best performance due to its narrow band width and high SNR (error of up to 2 µg/L for Chl-a and 3 mg/L for TSS), its spatial resolution (300 m) can be restrictive to most remote sensing studies in RDSM.

For the MSI sensor, despite its low SNR, the error magnitudes of the linear single bands algorithm used to retrieve the TSS and additive algorithms for Chl-a are low (up to 5 mg/L and 1 µg/L, respectively). Even though multiplicative algorithms using MSI data to retrieve Chl-a presented an error above 10 µg/L, the sensor could be applied if the lake size and shape enable resampling.

The OLI simulation indicated that its design is slightly better than that of MSI for all TSS algorithms, resulting in errors below 3 mg/L and 5 mg/L, respectively. However, the number and position of OLI bands are clear restrictions for Chl-a retrieval.

The sensor and algorithm selection need a comprehensive analysis before inland water studies are carried out. In this analysis, the sensor design, in situ conditions (cloud cover, lake size/ shape, and adjacency effects), water brightness (R_{rs}), and model equation (mathematical operation and fitting model) are the key factors considered.

The methods developed in this study will be applied in the near future under real conditions, in order to investigate the role of each of those aspects on the uncertainty caused by the real noise on bio-optical products.

Acknowledgments: This study was funded by the São Paulo Research Foundation (FAPESP) from project no.: 597 2014/23903-9, National Council for Scientific and Technological Development (CNPq) from project no.: 461469/2014-6, and MSA-BNDES from project no.: 1022114003005. Jorge was funded by the CNPq. We are very grateful to Vitor Martins, Renato Ferreira, Jean Farhat, Franciele Sarmento, and Waterloo Pereira Filho for their assistance during field missions. We would like to thank the Mamirauá Institute and staff for all the support during those two years.

Remote Sens. **2017**, *9*, 644

Author Contributions: Daniel S. F. Jorge, Claudio C. F. Barbosa, Lino A. S. de Carvalho, Adriana G. Affonso, Felipe de L. Lobo, and Evlyn M. L. M. Novo designed the research. Daniel S. F. Jorge performed the simulation and wrote the paper. Daniel S. F. Jorge, Claudio C. F. Barbosa, Lino A. S. de Carvalho, and Adriana G. Affonso were involved in the collection of field data. Claudio C. F. Barbosa and Lino A. S. de Carvalho supported the development of the Matlab routines used in the proposed methodology. All authors contributed insights for the analysis and methodology, and equally contributed towards organizing and reviewing the manuscript.

Conflicts of Interest: The authors declare no conflict of interest.

References

1. IOCCG. *Minimum Requirements for an Operational Ocean-Colour Sensor for the Open Ocean*; Reports of the International Ocean-Colour Coordinating Group, No. 1; IOCCG: Dartmouth, NS, Canada, 1997.
2. Martins, V.S.; Barbosa, C.C.F.; de Carvalho, L.A.S.; Jorge, D.S.F.; Lobo, F.L.; Novo, E.M.L.M. Assessment of Atmospheric Correction Methods for Sentinel-2 MSI Images Applied to Amazon Floodplain Lakes. *Remote Sens.* **2017**, *9*, 322. [CrossRef]
3. Brando, V.E.; Decker, A.G. Satellite hyperspectral remote sensing for estimating estuarine and coastal water quality. *IEEE Trans. Geosci. Remote Sens.* **2003**, *41*, 1378–1387. [CrossRef]
4. Gao, B.C.; Montes, M.J.; Davis, C.O.; Goetz, A.F.H. Atmospheric correction algorithms for hyperspectral remote sensing data of land and ocean. *Remote Sens. Environ.* **2009**, *113*, S17–S24. [CrossRef]
5. Chen, J.; Lee, Z.; Hu, C.; Wei, J. Improving satellite data products for open oceans with a scheme to correct the residual errors in remote sensing reflectance. *J. Geophys. Res. Oceans* **2016**, *121*, 3866–3886. [CrossRef]
6. Giardino, C.; Bresciani, M.; Cazzaniga, I.; Schenk, K.; Rieger, P.; Braga, F.; Matta, E.; Brando, V.E. Evaluation of Multi-Resolution Satellite Sensors for Assessing Water Quality and Bottom Depth of Lake Garda. *Sensors* **2014**, *14*, 24116–24131. [CrossRef] [PubMed]
7. Braga, F.; Giardino, C.; Bassani, C.; Matta, E.; Candiani, G.; Strömbeck, N.; Adamo, M.; Bresciani, M. Assessing water quality in the northern Adriatic Sea from HICO™ data. *Remote Sens. Lett.* **2013**, *4*, 1028–1037. [CrossRef]
8. Moses, W.J.; Bowles, J.H.; Lucke, R.L.; Corson, M.R. Impact of signal-to-noise ratio in a hyperspectral sensor on accuracy of biophysical parameter estimation in case II waters. *Opt. Express* **2012**, *20*, 4309–4330. [CrossRef] [PubMed]
9. Pahlevan, N.; Lee, Z.; Wei, J.; Schaaf, C.B.; Schott, J.R.; Berk, A. On-orbit radiometric characterization of OLI (Landsat-8) for applications in aquatic remote sensing. *Remote Sens. Environ.* **2014**, *154*, 272–284. [CrossRef]
10. Vanhellemont, Q.; Ruddick, K. Advantages of high quality SWIR bands for ocean color processing: Examples from Landsat-8. *Remote Sens. Environ.* **2015**, *161*, 89–106. [CrossRef]
11. Lobo, F.L.; Novo, E.M.L.M.; Barbosa, C.C.F.; Galvão, L.S. Reference spectra to classify Amazon water types. *Int. J. Remote Sens.* **2012**, *33*, 3422–3442. [CrossRef]
12. IOCCG. *Remote Sensing of Ocean Colour in Coastal, and Other Optically-Complex Waters*; Sathyendranath, S., Ed.; Reports of the International Ocean-Colour Coordinating Group, No. 3; IOCCG: Dartmouth, NS, Canada, 2000.
13. Mouw, C.B.; Greb, S.; Aurin, D.; DiGiacomo, P.M.; Lee, Z.; Twardowski, M.; Binding, C.; Hu, C.; Ma, R.; Moor, T.; et al. Aquatic color radiometry remote sensing of coastal and inland waters: Challenges and recommendations for future satellite missions. *Remote Sens. Environ.* **2015**, *160*, 15–30. [CrossRef]
14. Palmer, S.C.J.; Kutser, T.; Hunter, P.D. Remote sensing of inland waters: Challenges, progress and future directions. *Remote Sens. Environ.* **2015**, *157*, 1–8. [CrossRef]
15. Gerace, A.D.; Schott, J.R.; Nevins, R. Increased potential to monitor water quality in the near-shore environment with Landsat's next-generation satellite. *J. Appl. Remote Sens.* **2013**, *7*, 073558. [CrossRef]
16. Junk, W.J.; Bayley, P.B.; Sparks, R.E. The flood pulse concept in river-floodplain systems. *Can. Spec. Publ. Fish. Aquat. Sci.* **1989**, *106*, 110–127.
17. Queiroz, H.L. A reserva de desenvolvimento sustentável Mamirauá. *Estudos Avançados* **2005**, *19*, 183–203. [CrossRef]
18. Ayres, J.M. *As Matas de Várzea do Mamirauá Médio Rio Solimões*; CNPq-Programa Trópico Úmido e Sociedade Civil Mamirauá: Brasília, Brasil, 1993; p. 123.
19. Queiroz, H.L. Classification of water bodies based on biotic and abiotic parameters at the várzeas of Mamirauá Reserve, Central Amazon. *Uakari* **2007**, *3*, 19–34.

20. Ramalho, E.E.; Macedo, J.; Vieira, T.M.; Valsecchi, J.; Calvimontes, J.; Marmontel, M.; Queiroz, H. Ciclo hidrológico nos ambientes de várzea da Reserva de Desenvolvimento Sustentável Mamirauá: Médio Rio Solimões, período de 1990 a 2008. *Uakari* **2009**, *5*, 61–87.

21. Melack, J.M.; Bruce, R.F. Biogeochemistry of Amazon Floodplain. In *The Biogeochemistry of the Amazon Basin*; Oxford University Press: New York, NY, USA, 2001; p. 235.

22. Forsberg, B.R.; Devol, A.H.; Rickey, J.E.; Martinelli, L.A.; Santos, H. Factors controlling nutrient concentrations in Amazon floodplain lakes. *Limnol. Oceanogr.* **1988**, *33*, 41–56. [CrossRef]

23. Affonso, A.G.; Queiroz, H.; Novo, E.M.M.L. Change in macrophyte coverage may affect pirarucu (*Arapaima gigas*) abundance, fishery and conservation in Amazon floodplain lake. In Proceedings of the 6th World Fisheries Congress, Edinburgh, UK, 7–12 May 2012.

24. Nush, EA. Comparison of different methods for chlorophyll and phaeopigment determination. *Arch. Hydrobiol.* **1980**, *14*, 14–39.

25. Wetzel, R.G.; Likens, G.E. *Limnological Analyses*; Springer: New York, NY, USA, 1991; p. 391.

26. Tilstone, G.H.; Moore, G.F.; Sorensen, K.; Doerffer, R.; Rottgers, R.; Ruddick, K.; Pasterkamp, R.; Jorgensen, P.V. REVAMP Regional validation of MERIS chlorophyll products in North. Sea coastal waters. In Proceedings of the Working meeting on MERIS and AATSR Calibration and Geophysical Validation (ENVISAT MAVT-2003), Frascati, Italy, 20–24 October 2003.

27. Mueller, J.L.; Fargion, G.S. *Ocean Optics Protocols for Satellite Ocean Color Sensor Validation*; Revision 3; NASA TM 2002-210004; NASA Goddard Space Flight Center: Greenbelt, MD, USA, 2002; p. 308.

28. Mobley, C.D. Estimation of the remote-sensing reflectance from above-surface measurements. *Appl. Opt.* **1999**, *38*, 7442–7455. [CrossRef] [PubMed]

29. ESA Sentinel Online. Sentinel-2 MSI Introduction. Available online: https://earth.esa.int/web/sentinel/user-guides/sentinel-2-msi (accessed on 26 February 2017).

30. ESA Sentinel Online. Sentinel-3 OLCI Introduction. Available online: https://earth.esa.int/web/sentinel/user-guides/sentinel-3-olci (accessed on 26 February 2017).

31. OSCAR Observing Systems Capability Analysis and Review Tool. Available online: www.wmo-sat.info/oscar/instruments/view/375 (accessed on 26 February 2017).

32. Vanhellemont, Q.; Ruddick, K. Turbid wakes associated with offshore wind turbines observed with Landsat 8. *Remote Sens. Environ.* **2014**, *145*, 105–115. [CrossRef]

33. Kaskaoutis, D.G.; Kambezidis, H.D. Investigation into the wavelength dependence of the aerosol optical depth in the Athens area. *Q. J. R. Meteorol. Soc.* **2006**, *132*, 2217–2234. [CrossRef]

34. Bird, R.; Riordan, C. Simple Solar Spectral Model for Direct and Diffuse Irradiance on Horizontal and Tilted Planes at the Earth's Surface for Cloudless Atmospheres. *J. Clim. Appl. Meteorol.* **1984**, *25*, 87–97. [CrossRef]

35. Leckner, B. The spectral distribution of solar radiation at the earth's surface elements of a model. *Sol. Energy* **1978**, *20*, 143–150. [CrossRef]

36. Gueymard, C. Parameterized transmittance model for direct beam and circumsolar spectral irradiance. *Sol. Energy* **2001**, *71*, 325–346. [CrossRef]

37. Gueymard, C.; Myers, D.; Emery, K. Proposed reference irradiance spectra for solar energy systems testing. *Sol. Energy* **2002**, *73*, 443–467. [CrossRef]

38. Matthews, M.W.; Bernard, S.; Robertson, L. An algorithm for detecting trophic status (chlorophyll-a), cyanobacteria-dominance, surface scums and floating vegetation in inland and coastal waters. *Remote Sens. Environ.* **2012**, *124*, 637–652. [CrossRef]

39. Moses, W.J.; Gitelson, A.A.; Berdinikov, S.; Povazhnyy, V. Satellite estimation of chlorophyll-a concentration using the red and NIR bands of MERIS-the Azov Sea case study. *IEEE Geosci. Remote Sens. Lett.* **2009**, *6*, 845–849. [CrossRef]

40. Mishra, S.; Mishra, D.R. Normalized difference chlorophyll—A retrieval in turbid, productive estuaries: Chesapeake Bay case study. *Remote Sens. Environ.* **2007**, *109*, 464–472.

41. Binding, C.E.; Bowers, D.G.; Mitchelson-Jacob, E.G. An algorithm for the retrieval of suspended sediment concentrations in the Irish Sea from SeaWiFS ocean colour satellite imagery. *Int. J. Remote Sens.* **2003**, *24*, 3791–3806. [CrossRef]

42. Chen, X.; Han, X.; Feng, L. Towards a practical remote-sensing model of suspended sediment concentrations in turbid waters using MERIS measurements. *Int. J. Remote Sens.* **2015**, *36*, 3875–3889. [CrossRef]

43. De Lucia Lobo, F.; Costa, M.P.; Leao de Moraes Novo, E.M. Time-series analysis of Landsat-MSS/TM/OLI images over Amazonian waters impacted by gold mining activities. *Remote Sens. Environ.* **2015**, *157*, 170–184. [CrossRef]

44. Friedman, J.; Hastie, T.; Tibshirani, R. *The Elements of Statistical Learning*; Series in Statistics; Springer: Berlin, Germany, 2001; Volume 1.

45. Luck, S.J. *An Introduction to the Event-Related Potential Technique*; MIT Press: Cambridge, MA, USA, 2014.

46. Roberts, D.A.; Gardner, M.; Church, R.; Ustin, S.; Scheer, G.; Green, R.O. Mapping chaparral in the Santa Monica Mountains using multiple endmember spectral mixture models. *Remote Sens. Environ.* **1998**, *65*, 267–279. [CrossRef]

remote sensing

MDPI

Article

Spatiotemporal Variability of Lake Water Quality in the Context of Remote Sensing Models

Carly Hyatt Hansen [1,*], Steven J. Burian [1], Philip E. Dennison [2] and Gustavious P. Williams [3]

[1] Department of Civil and Environmental Engineering, University of Utah, Salt Lake City, UT 84112, USA; steve.burian@utah.edu
[2] Department of Geography, University of Utah, Salt Lake City, UT 84112, USA; dennison@geog.utah.edu
[3] Department of Civil and Environmental Engineering, Brigham Young University, Provo, UT 84602, USA; gus.p.williams@byu.edu
* Correspondence: carly.hansen@utah.edu

Academic Editors: Yunlin Zhang, Claudia Giardino, Linhai Li, Deepak R. Mishra and Prasad S. Thenkabail
Received: 25 February 2017; Accepted: 21 April 2017; Published: 26 April 2017

Abstract: This study demonstrates a number of methods for using field sampling and observed lake characteristics and patterns to improve techniques for development of algae remote sensing models and applications. As satellite and airborne sensors improve and their data are more readily available, applications of models to estimate water quality via remote sensing are becoming more practical for local water quality monitoring, particularly of surface algal conditions. Despite the increasing number of applications, there are significant concerns associated with remote sensing model development and application, several of which are addressed in this study. These concerns include: (1) selecting sensors which are suitable for the spatial and temporal variability in the water body; (2) determining appropriate uses of near-coincident data in empirical model calibration; and (3) recognizing potential limitations of remote sensing measurements which are biased toward surface and near-surface conditions. We address these issues in three lakes in the Great Salt Lake surface water system (namely the Great Salt Lake, Farmington Bay, and Utah Lake) through sampling at scales that are representative of commonly used sensors, repeated sampling, and sampling at both near-surface depths and throughout the water column. The variability across distances representative of the spatial resolutions of Landsat, SENTINEL-2 and MODIS sensors suggests that these sensors are appropriate for this lake system. We also use observed temporal variability in the system to evaluate sensors. These relationships proved to be complex, and observed temporal variability indicates the revisit time of Landsat may be problematic for detecting short events in some lakes, while it may be sufficient for other areas of the system with lower short-term variability. Temporal variability patterns in these lakes are also used to assess near-coincident data in empirical model development. Finally, relationships between the surface and water column conditions illustrate potential issues with near-surface remote sensing, particularly when there are events that cause mixing in the water column.

Keywords: spatiotemporal variability; water quality; chlorophyll-a; near-coincident remote sensing

1. Introduction

Over the past decade, remote sensing of water quality has become more widely used and the extent of applications has grown tremendously, especially in non-coastal environments. Notable inland water quality applications of remote sensing include large-scale quality and clarity surveys [1–4] and real-time tracking and forecasting of nuisance algal blooms (NABs) or harmful algal blooms (HABs) [5,6]. The general process of developing an empirical remote sensing model for algal blooms typically involves: downloading and processing of remote sensing imagery (which may include atmospheric correction and conversion from digital numbers to reflectance at the near-surface of the water body), collecting coincident (or near-coincident) field measurements of chlorophyll-a (or other parameters related to biomass or levels of toxins), and using regression or other statistical modeling techniques to develop a relationship between the field-measured concentrations and remotely sensed reflectance from the corresponding pixel or group of pixels. Multiple sensors offer greater coverage with varying overpass frequencies and extents, and band combinations which are more optimal for characterization of water quality conditions. Increased availability of imagery data and processed data products has also facilitated increased use and application. Despite all of these advances, there are a number of issues that remain to be addressed to support more effective and accurate remote sensing model development and application. Many of these issues stem from traditional assumptions associated with the use and application of remote sensing data, and do not consider conditions and processes that are specific to the water bodies of interest.

Water quality conditions, particularly algal growth, in lakes and reservoirs have been shown to change relatively quickly (i.e., seasonally or sub-seasonally) [7–9]. Algal bloom variability in inland waters also occurs on smaller spatial scales than in the open ocean. Spatial and temporal variability in water quality may be caused by a number of processes, such as resuspension of suspended sediments and point-source inflow of nutrients [10]. Increased variability in lake and reservoir water quality requires that in situ data used to develop remote sensing water quality models represent conditions at the time of the imagery acquisition–to the extent possible. Often, the historical records do not provide exact temporal matches between the in situ samples and the satellite overpass, requiring the use of "near-coincident" data, or some relaxation of a definition of a "match." Coastal and lake water clarity and quality remote sensing literature report a wide range of time-windows for considering data to be near-coincident. Reported windows range from ±3 h [11], same day [12], one day [4,13], seven days [2,14], to ±10 days [1] between the satellite image acquisition and the field samples used for calibration. Often, a particular time-window for near-coincident matches is arbitrarily chosen (e.g., using an arbitrary increase in the percentage of samples that match with a satellite image [15]), or the study states that the relaxation of the time-window improved the model fit, without detailing the actual improvement [1].

Another issue that is often overlooked in water quality remote sensing applications is thorough review and evaluation of appropriate sensors in the context of a specific water body (which has unique spatial and temporal characteristics). Sensor characteristics can have large implications for the utility of the resulting model and dataset. Model application determines the sensor choice and could depend on a number of factors: the spatial resolution (which is limited by the size of the water body or multiple waterbodies in a region), the spatial variability within the water body, the desired return time (which is influenced by the temporal variability of the water quality processes), the length of historic record, spectral resolution (which determines the ability of the sensor to discriminate or more accurately determine conditions and which parameters can be estimated), the available processing resources (from the imagery data and data products to the personnel who will perform data processing and analysis), and the scope of the application (both spatial and temporal). For empirical model development, information from the field (e.g., concentration of chlorophyll-a measured at a single point on the water body) is matched to information from the satellite (reflectance averaged over a single pixel or group of pixels). Therefore, the spatial variability of the water body may influence the choice of satellite. For example, if the algae concentrations vary substantially on the order of 20–40 m, then a satellite with a resolution of 30 m will be sufficient, while a satellite with a resolution of 500 or 1000 m would be too coarse to adequately represent the variability of the chlorophyll concentrations. One review suggests different medium spatial resolution satellites (e.g., Landsat) and coarser spatial resolution satellites (e.g., MODIS) for water clarity and quality studies be selected based primarily on the size of the water body [16], however, other characteristics of the lake, namely the ability of different spatial resolutions (e.g., Landsat resolution of 30 m or SENTINEL-2 resolution of 10–60 m compared to MODIS resolution of 250–1000 m) to represent spatial variability within the lake or the ability of more frequent overpasses to address temporal variability (e.g., Landsat every 16 days compared to SENTINEL-2 every 5 days and MODIS every 1–2 days) are not considered.

Finally, remotely sensed data are limited by the optical depth of the water column (the depth at which light is able to penetrate), which means that the estimates are limited to near-surface algae populations. Optical depth is also a function of chlorophyll concentration; as the near-surface algae populations increase, optical depth decreases. However, algae thrive not only at the surface but exist throughout the water column. Algal population characteristics (species, diversity, etc.) may vary with depth, especially when the water column is stratified and there are differences in oxygen or salinity [17,18]. Concerns have been raised about the utility of only sensing and estimating the surface of the lake given these variable conditions throughout the water column. It is therefore important to explore the relationship between surface and water-column algae concentrations and the variability within the water column when evaluating the limitations of remotely-sensed surface estimates.

This study uses field measurements of chlorophyll to evaluate techniques and assumptions that are often used in remote sensing models of algae and surface water quality. While there are many additional considerations for water quality (particularly algae/chlorophyll concentrations) this paper focuses on the three issues outlined above: (1) selecting sensors which are suitable for the spatial and temporal variability in the water body; (2) determining appropriate uses of near-coincident data in empirical model calibration; and (3) recognizing potential limitations of remote sensing measurements which are biased toward surface and near-surface conditions.

Study Area

The study area for this paper is the Utah Lake and Great Salt Lake (GSL) system. This lake system is important for recreation and ecosystem services for the urban areas that are concentrated in the hillsides and valleys to the east of these lakes. During the summer of 2016, Utah Lake and Farmington Bay of the GSL experienced massive cyanobacterial algal blooms. While large algal blooms in these lakes are not particularly rare, the rapid development and magnitude of the recent blooms spurred widespread attention and motivated increased interest in monitoring these waters, particularly through remote sensing because the size of the lakes make them difficult to monitor through field sampling alone. Data were collected with water quality sondes at a number of locations throughout the system (shown on the map in Figure 1) throughout the summer of 2016 to support this research.

Previous studies in the Utah Lake and GSL system have explored variation in algal speciation throughout the growing season and environmental factors which contribute to species diversity [19–22]. Historical sampling campaigns on Utah Lake revealed typical algal succession, with diatoms and then green algae dominating in early summer, and then cyanobacteria dominating during the late summer months, and a general decrease in species diversity throughout the summer [21,22]. In Farmington Bay and the GSL, studies have focused on speciation and presence of toxins in cyanobacteria. These studies have found seasonal trends in algae growth and have observed stark differences between algae types in different regions of the GSL and Farmington Bay [19,20,23,24]. These studies improve understanding of the algae populations in this lake system; however, they lack important information about spatial or temporal variability at scales that are necessary for improving remote sensing model development.

The Great Salt Lake is divided roughly in half by a railroad causeway which runs East-West, separating the much more saline (roughly 28% salinity) North Arm, which includes Gunnison Bay and Bear River/Willard Bay, from the South Arm (Gilbert Bay and Bridger Bay) and Farmington Bay, which is further separated by an automobile causeway. These bays maintain a salinity between 11% and 15% [25] and at the north end of Farmington Bay, salinity is typically around 8% [20]. These lakes are relatively shallow, with an average depth of approximately 4.2 m in Gilbert Bay and an average depth of approximately 1 m in Farmington Bay. Secchi depth (as a measure of transparency) ranges between 2 and 5 m in the South Arm of the GSL, while in Farmington Bay, it is regularly less than 0.3 m [26]. Utah Lake, which flows into the Great Salt Lake through the Jordan River is also a shallow lake (average depth of 2.74 m) and while it is a freshwater lake, it has high dissolved solids, resulting in slightly saline conditions [27]. High rates of suspended sediments result in high turbidity, and prior to the large algal bloom in 2016, the Secchi depth in the middle of Utah Lake was roughly 0.2 m.

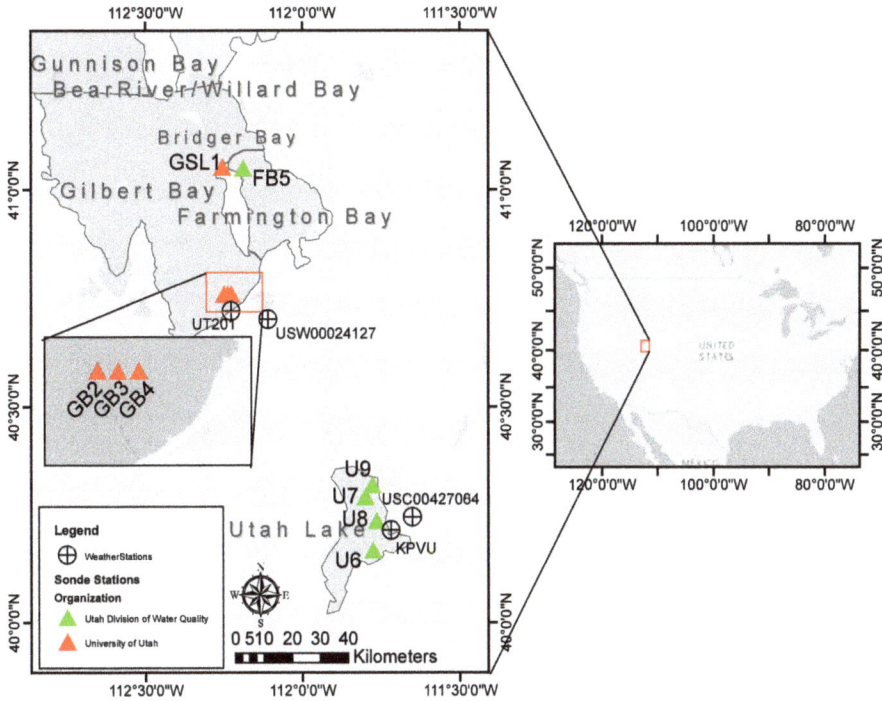

Figure 1. Sampling Locations and Study Area.

2. Materials and Methods

2.1. Data Collection

The collection of water quality samples was designed to provide information about algae biomass (measured as chlorophyll-a) and its: (1) temporal variability (through repeated sampling visits and high-frequency sampling); (2) spatial variability (through multiple sites and/or offsets); and (3) surface–water column relationships. Chlorophyll-a data were collected by researchers at the University of Utah (U of Utah) using a Hydrolab DS5 (OTT Hydromet) multiparameter sonde equipped with a submersible fluorescence Chlorophyll-a sensor (range of 0.03–500 µg/L). Chlorophyll-a data were also provided by the Utah Division of Water Quality (UDWQ) measured using YSI EXO 2 multiparameter sonde (with submersible fluorescence Chlorophyll-a sensor (range of 0–400 µg/L) coupled with a Nexsens CB-450 buoy platform. Sampling locations were chosen based on accessibility. During the study period, low water levels, exposed reef-like bioherms, and deep sediments restricted boat and individual access to many locations in the lakes that may otherwise have been sampled. Details of the sampling at each station are summarized below and in Table 1, including the duration of sampling periods and the types of samples collected. Durations and frequencies of data collection were determined by the availability of equipment and personnel, and local weather conditions. Data collected by the University of Utah are shared under the Creative Commons Attribution CC BYU License [28] and data collected by the UDWQ are available through the iUTAH Time Series Analyst data portal.

Table 1. Summary of Data Collection Periods and Methods.

Lake	Stations	Organization	Sampling Periods (2016)	7.5 m Offsets	Surface (<1 m)	Water Profiles	Approximate Lake Depth During Study Period (m)
Main GSL	GSL1	U of Utah	23–31 July	X	X	-	0.8
	GB2; GB3; GB4	U of Utah	6–16 June; 6–14 July; 12–22 Aug	X	X	X	5.1
Farmington Bay	FB5	UDWQ	8 July–28 July	-	X	-	0.5
Utah Lake	U6; U7; U8	UDWQ	28 Aug–13 Sept	-	X	-	1–1.5
	U9	UDWQ	15 July–8 Aug	-	X	-	1–1.5

2.1.1. UDWQ Data

UDWQ sondes were installed in a variety of locations in Utah Lake and Farmington Bay following the large July 2016 algal blooms. The site names for these sites have been modified to maintain consistency with the naming convention of the University of Utah sites. One temporary fixed sonde was placed approximately 0.75 m below the surface at station U9 (UDWQ Site 4917310) in Utah Lake, providing daily measurements between 15 July and 8 August, 2016. The sondes in stations U6 (UDWQ Site 4917390), U7 (UDWQ Site WVineyard), and U8 (UDWQ Site WProvo) were installed on buoys anchored at the locations shown in Figure 1, and provided daily measurements at approximately 0.3 m below the surface between 28 August and 13 September 2016. Water depths in Utah Lake during this time period were between 1 and 1.5 m. Finally, a fixed sonde in Farmington Bay at station FB5 (UDWQ Site 4895200) provided daily measurements between 8 July and 28 July, 2016 at a depth of approximately 0.3 m below the surface (due to extremely low water levels, which were approximately 0.5 m at this time). The measurements for these sondes (which were reported at a 15-min frequency) were averaged between 11:00–11:30 a.m. in order to maintain consistency in day-to-day comparisons (reducing the effect of diurnal patterns of algae on the chlorophyll measurements which peaks during midday and then drops in the evening). These daily measurements were used in exploring temporal variability.

2.1.2. University of Utah Data

While the fixed UDWQ sondes in Utah Lake and Farmington Bay provide stationary data for exploration of temporal variability, data collection by the University of Utah was designed to explore temporal variability as well as variability on different spatial scales. Data collected by the University of Utah was focused in the main body of the South Arm of the GSL (Gilbert Bay and Bridger Bay). Surface data at the Gilbert Bay sites were consistently collected between 9:00 and 11:30 a.m. (again, to minimize the effects of diurnal patterns of photosynthesis). Data collection took place during three periods: 6, 8, 9, 10 and 13 June; 6, 7, 8, 12 and 14 July; and 12, 15, 16, 17 and 22 August. At these sites (GB2, GB3 and GB4), approximately 20–30 measurements were taken at a 1-s frequency at an average depth of 0.4 m below the surface and averaged. The Gilbert Bay sites (prefixed with GB) which were navigable by boat, were located approximately 1000 m apart, which is the same scale as the coarsest MODIS spatial resolution. At each of these sites, data were also collected at offsets to the site center to represent sub-Landsat and sub-SENTINEL-2 resolution. These offset samples were spaced at approximately 7.5 m increments (i.e., 7.5, 15, 22.5 and 30 m) from the original sites GB2, GB3 and GB4. The offsets were identified with suffixes a, b, c and d, so that the first offset (7.5 m) from GB2 was identified as GB2a, the second offset (15 m) from GB2 was GB2b, etc.) At these sites, lake current and wind patterns differed from one sampling day to the next, resulting in variable drift directions between the GB sites and their offsets, though it was generally consistently in the southwest direction. Nonetheless, relative distances between the original sites and the offsets were maintained. Approximately 20–30 measurements at the GSL1 site were collected at a 1-s frequency approximately 0.3 m below the surface and averaged in a July sampling period (23, 24, 27, 30 and 31 July). Data collection at this site also included sampling at offsets at the same increments (7.5, 15, 22.5 and 30 m) east of the original site.

The data at Bridger Bay were averaged at approximately 0.3 m below the surface (due to low lake levels at this location), and were consistently collected in the afternoon (due to equipment availability and to reduce effect of diurnal patterns).

In addition to the surface data obtained at the Gilbert Bay sites, measurements were collected throughout the water column to examine relationships between chlorophyll measurements at different depths. At sites GB2, GB3, and GB4, data were collected over the water profile, by manually lowering the sonde at approximately 0.3 m/s and recording at a 1-s frequency. Profiles were created by averaging the concentrations over 1 m intervals from 0–6 m) to represent different ranges of the water column.

For the sites reached by boat, we approached the locations from the opposite direction of the lake current and turned off the engine, allowing the boat to drift to the sites and offsets in an effort to reduce the amount of artificial mixing caused by the engine. Despite these efforts, some amount of mixing from the engine may have occurred which would have an effect on the measured concentrations and subsequent variability, particularly near the surface. The FB site and offsets were reached by foot, and mixing may have been caused by stirring up sediments.

2.1.3. Meteorological Data

In order to examine conditions that may contribute to surface mixing in the lakes, meteorological data were collected from MesoWest weather stations located near the Gilbert Bay sampling locations (Site UT201, at 40.72255, −112.22569) and near Provo Bay in Utah Lake (Site KPVU, at 40.21667, −111.71667). Parameters including wind speed (kilometers per hour) and peak wind gust (kilometers per hour) were recorded at 10 min intervals for UT201 and at 5 min intervals for KPVU. Wind speed is averaged over a daily scale and the daily peak wind gust is the maximum peak wind gust. Daily precipitation data totals (mm) and maximum temperatures (degrees Celsius) were obtained from NOAA Stations USW00024127 at 40.7034, −112.109 and USC00427064 at 40.2458, −111.6508. Comparable meteorological data near the Farmington Bay site were not available for study period.

2.2. Statistical and Graphical Analysis

To evaluate the variation over time, we computed the autocorrelation function or estimates of autocovariance [29]. These estimates were calculated for each site with regular daily sampling (all of the UDWQ sites in Utah Lake and Farmington Bay) using the "acf" function, which is built in to the R statistical software [30]. At each of the lags for these sites, we tested for statistically significant autocovariance of surface chlorophyll measurements. The autocorrelation function could not be computed for the main GSL sites (GB and GSL), since these data were not collected at regular intervals, and there were insufficient points for alternative analyses (e.g., constructing a temporal variogram). Instead, for these sites, temporal variation was analyzed graphically by calculating the difference in chlorophyll measurements between subsequent samples (for short-term variation), as well as the mean and standard deviation for each of the sampling periods (for seasonal variation).

We also examined spatial variation of surface chlorophyll concentrations with respect to the spatial resolutions of several commonly-used sensors. As noted, the distances between sites and offsets for the samples are representative of the spatial resolution of Landsat/SENTINEL-2 and MODIS band regions. The observed differences in measurements between the offsets and the sites offer insight into fine-scale variability (<30 m) that would occur at the sub-Landsat and SENTINEL-2 spatial resolutions and coarser-scale variability (1000 m) that corresponds with the spatial resolution of MODIS. To evaluate the differences between offsets, we calculated the difference and percent differences in surface measurements between the sites and their respective offsets using Equations (1) and (2):

$$Difference = Chl_{x,j} - Chl_{y,j} \tag{1}$$

$$Percent\ Difference = \left(\frac{Chl_{x,j} - Chl_{y,j}}{Chl_{x,j}} \right) * 100 \tag{2}$$

where *Chl* is the mean chlorophyll concentration between 0 and 1 m below the surface for the sampling date *j* at site *x* (e.g., GB2) and corresponding offset *y* (e.g., GB2a, GB2b, etc.).

Finally, we used linear regression to evaluate relationships between conditions at the surface and throughout the water column for the GB2, GB3, and GB4 sites for each of the sampling periods. Due to extremely low lake levels in Farmington Bay, Utah Lake, and Bridger Bay, samples at multiple depths were not possible at these locations. The regressions follow the general form of Equation (3):

$$Chl_{x,k} = m \cdot Chl_{x,l} + b \tag{3}$$

where *Chl* is the mean chlorophyll concentration at site *x*, at depth *k* below the surface, and *l* is the depth of 0–1 m below the lake surface. The strength of the relationship is measured through the correlation coefficient, or R^2. For this case, the correlation coefficient translates to the amount of variance at intermediate depths that is explained by the surface measurements.

3. Results

3.1. Temporal Variability

The results of the autocorrelation function are visualized in a correlogram, showing the autocorrelation of surface chlorophyll values versus the lag (days). The correlograms for each of the sites with daily sampling, shown in Figure 2, graphically illustrate how the time series is correlated with itself, or how similar measurements are from one day to measurements from some lagged time period.

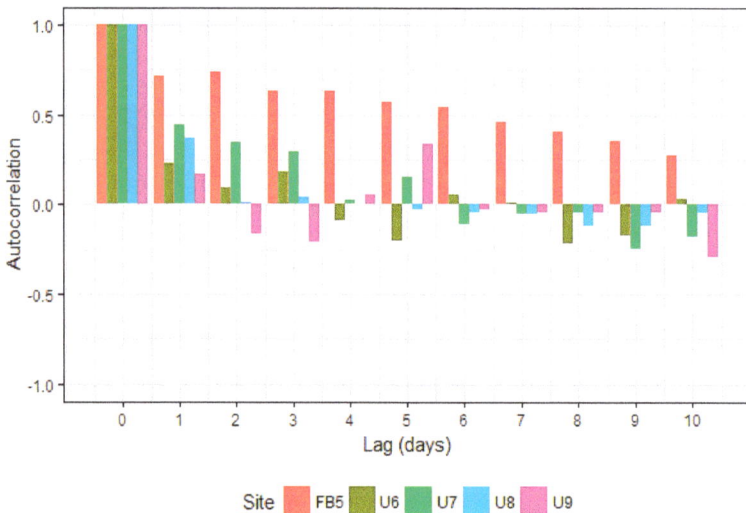

Figure 2. Autocorrelation for Utah Lake (U6, U7, U8 and U9) and Farmington Bay (FB5) Sites.

The null hypothesis, which is tested at each lag, is that there is no autocorrelation between the lagged samples. The different patterns of autocorrelation in Figure 2 show that there are major differences in the temporal autocorrelation in different parts of the lake system. At α = 0.05, there is no statistically significant autocorrelation for all time lags for Utah Lake sites U9 and U6, and near-statistically significant autocorrelation for a lag of one day for U8 and U7. The rapid decrease in autocorrelation for many of the Utah Lake sites is evidence of high short-term variability in this body. In clear contrast with the patterns observed in Utah Lake, there is significant autocorrelation for all lags up to 11 days for the site in Farmington Bay (FB5).

For sites where it was not possible to calculate an autocorrelation function, the differences in chlorophyll measurements between subsequent samples for each of the sampling periods are shown in Figure 3.

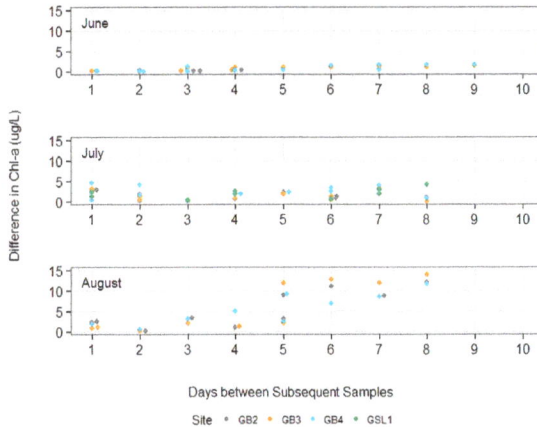

Figure 3. Temporal Variation between Subsequent Samples by Sampling Periods at the GB and GSL Sites.

In the samples from June and July, there is relatively small variation (<2 and 5 µg/L, respectively), even at 8 and 10 days between subsequent samples. In August, however, the data show a clear positive trend of increasing differences between surface chlorophyll measurements, that is, the difference between the subsequent samples increases as time between the samples increases. The data also show the variation in between subsequent measurements increases throughout the summer season. For example, in June, the mean difference at seven days between subsequent samples is 1.02 µg/L, while the mean differences in July and August at seven days are 3.05 µg/L and 9.67 µg/L, respectively. This seasonal increase in variability is also evident in comparisons of the standard deviation of surface measurements during each sampling period, shown in Figure 4. There was also a general positive trend in chlorophyll concentrations throughout the sampling period (meaning that both magnitudes of chlorophyll and variance increased throughout the summer).

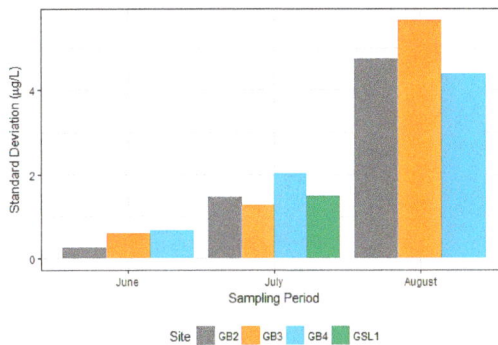

Figure 4. Standard Deviation for Surface Chlorophyll at GB and GSL Sites by Sampling Period.

3.2. Spatial Variability

To illustrate the differences in spatial resolution of several commonly-used sensors, Figure 5 compares the coverage of a portion of the study area (Utah Lake) with resolutions ranging from 30 m (Landsat 8, Band 2, 19 July 2016), to 60 m (SENTINEL-2, Band 1, 22 July 2016) and 500 m (MODIS, Band 3, 20 July 2016).

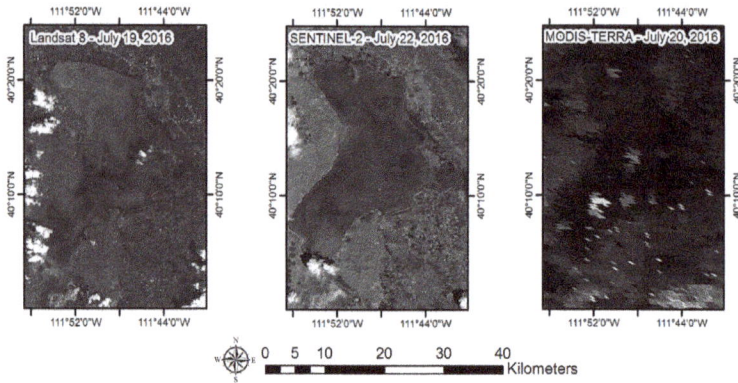

Figure 5. Comparison of Spatial Resolution in Coverage of Utah Lake at 30 m (Landsat 8, Band 2), 60 m (SENTINEL-2, Band 1) and 500 m (MODIS, Band 3).

The resolutions of Landsat and SENTINEL-2 show clear definition between the lake and the shore, and variability in surface conditions (including the extent of the large algal bloom) can be detected at both these scales. On the other hand, the coarse resolution of the MODIS image makes it difficult to delineate the shoreline and while there is some variability between the in-lake pixels, the extent of the bloom is difficult to distinguish. In the GB sites, surface chlorophyll data collected at sites and offsets correspond with the range of spatial scales for these sensors. The differences in surface chlorophyll for fine spatial scales (corresponding with Landsat/SENTINEL-2) and coarse spatial scales (corresponding with the coarsest resolution of MODIS, 1000 m) are shown in Figures 6 and 7.

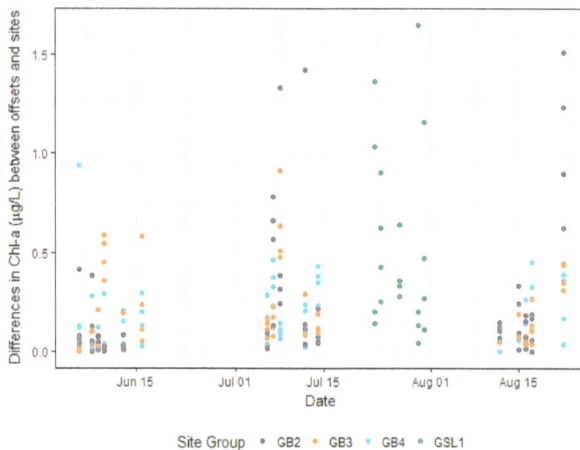

Figure 6. Variability between Sites and Offsets (<30 m distances or Sub-Landsat/Sub-SENTINEL-2 Scales) in the Great Salt Lake (GB and GSL Sites).

For site groups (where each site group includes the site and its offsets) GSL1, GB2, GB3 and GB4, there was generally less than 30 percent difference between the surface measurements at the offsets and those at the site. The plots show that the highest differences between the offsets and the sites occur in the later summer months, while relatively small differences are observed in early summer. Throughout the entire season, the maximum difference in magnitude between a site and its offsets is 1.7 µg/L.

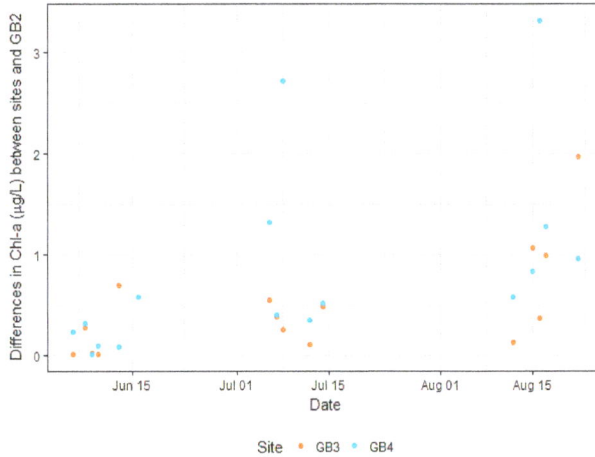

Figure 7. Variability between Sites (Approximately 1000 m distances, or MODIS Scale) in the Great Salt Lake (GB Sites).

This figure shows that even at this larger scale, the differences are still generally small (below 30 percent), though the actual difference in magnitude was higher (with a maximum difference of 3.4 µg/L) than those at the sub-pixel distances on the Landsat/SENTINEL-2 scale. Again, greater differences are observed in later summer months compared to early summer.

3.3. Surface/Water Column Measurements

The linear relationships between average surface measurements (0–1 m below the surface) and various depths (1–2 m, 2–3 m and 3–4 m) from data collected in Gilbert Bay (where water depths allowed for water column measurements) are shown in Figure 8.

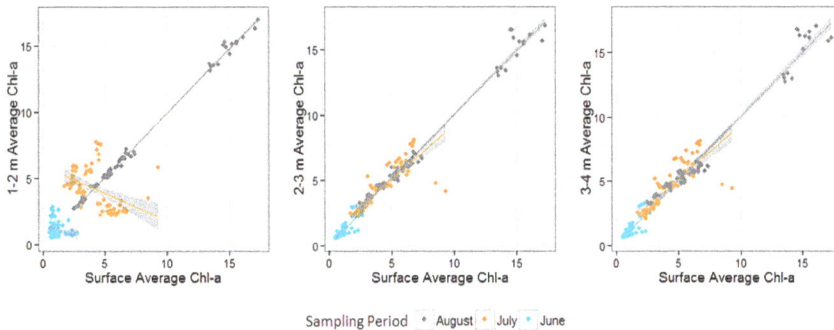

Figure 8. Relationships between Surface and Depths throughout the Water Column for GB Sites.

For 1–2 m, the overall (across all sampling periods) R^2 is 0.79; for 2–3 m it is 0.97; and for 3–4 m it is 0.96. However, the relationship is highly dependent on the sampling period, particularly at depths of 1–2 m. For June and July, there are virtually no relationships between the surface chlorophyll and chlorophyll at 1–2 m below the surface, and the relationships at other depths are weaker for these sampling periods than for the August sampling period.

3.4. Meteorological Record

Short-term weather events such as rainfall and high wind events have the potential to cause surface mixing and subsequently affect the observed temporal and spatial variability patterns, as well as conditions throughout the water column. Records of the daily average values for wind speed, peak daily wind gust, total daily precipitation and maximum temperature are shown for two weather stations near the Great Salt Lake and Utah Lake are shown in Figure 9.

During the periods of data collection for Utah Lake sites, conditions were relatively stable with respect to precipitation and temperature. The extremely shallow lake was likely heavily influenced by the wind, allowing for a great deal of mechanical mixing to occur. This corresponds with the low autocorrelation values in the Utah Lake sites. Other seasonal patterns in variability, such as the general increase in concentrations observed in the GB sites, correspond with the fairly stable and favorable weather conditions (lack of any large precipitation events during the mid-summer months, sustained high temperatures in late July, and a steady cooling through August).

The seasonality of the surface/water column relationship may be partially explained by weather conditions and short-term events, such as the variable temperature in June and July, and the slightly higher wind and precipitation events in the GSL in June. It is important to note that poor correlations between surface and 1–2 m depths may also be influenced by mechanical mixing caused by turbulence from the boat, which could create artificially high variability near the surface.

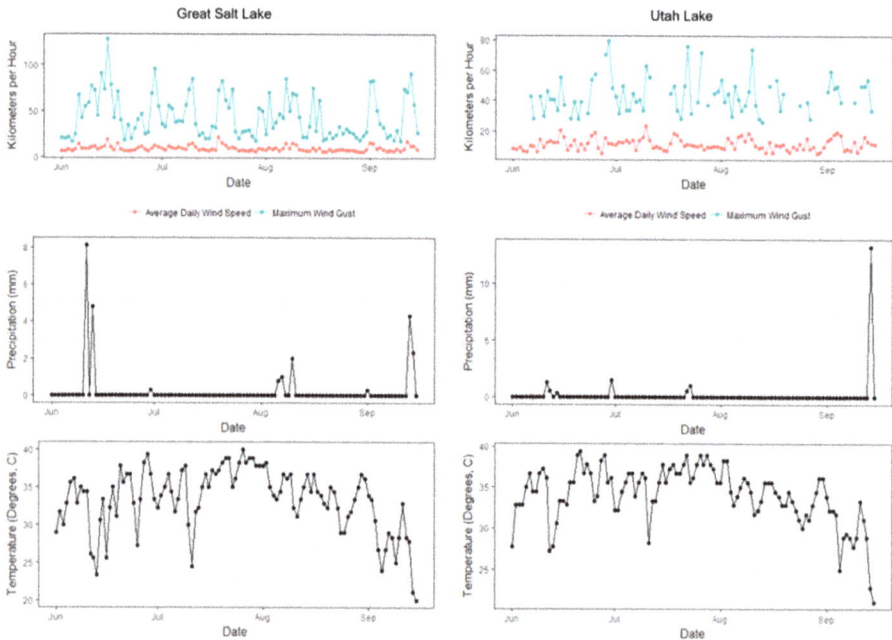

Figure 9. Daily Wind, Precipitation, and Temperature Records near the Great Salt Lake (GSL) and Utah Lake over the Period of Data Collection.

4. Discussion

The measures of variability over time (including autocorrelation, magnitude of differences between subsequent samples, and standard deviation for different sampling periods) suggest that the water bodies in the Great Salt Lake system have distinct temporal characteristics. These characteristics have important implications for remote sensing modeling techniques. The Utah Lake samples showed non-significant autocorrelation after one day, while the Farmington Bay samples showed statistically significant autocorrelation for up to 11 days. This indicates that the Utah Lake conditions are much more variable than those in Farmington Bay, with Utah Lake variation on a daily scale, rather than the near-weekly scale exhibited in Farmington Bay. In a remote sensing context, this means that shorter time windows may be needed for calibrating Utah Lake models, while longer time windows may be justified for Farmington Bay models. In the GB and GSL locations, where sampling frequencies were irregular, there was a general trend of increasing differences in chlorophyll concentrations as the time between samples increased. These differences and the overall variation increased throughout the summer, indicating that the temporal correlation may not be stationary, but decreases throughout the growing season. This increase in variability could justify a shorter time-window for near-coincident data in the later summer months than the earlier summer months.

The observed temporal patterns provide additional information for evaluating suitability of the Landsat, SENTINEL-2, and MODIS sensors for this lake system. For example, events in Utah Lake may be completely missed by the revisit time of Landsat sensors, requiring the use of multiple sensors to adequately capture the rapidly changing conditions and acknowledgment of the limitations of the temporal resolution of this sensor and its ability to describe short-term changes.

The comparisons of surface measurements between the GB and GSL sites and offsets as well as among sites were also useful in evaluating different spatial resolutions of commonly-used sensors. The relatively small variation between sites and offsets indicates that there is low variability over the distances measured by a single pixel for Landsat/SENTINEL-2 or MODIS. This suggests that these platforms, or others with similar spatial resolution, are suitable for monitoring the main body of the GSL. These results also suggest that finer spatial resolution products (such as those obtained by airborne sensors) would not necessarily provide significantly more information for this part of the system.

Finally, the linear models between concentrations at the surface and those at different depths in the water column in the GB sites show that these relationships are both depth and seasonally dependent. This result is interesting because it shows a stronger relationship between the measurements at the surface and greater depths (2–3 and 3–4 m) than between the surface and subsurface (1–2 m) measurements. If the data are analyzed by sampling period, the relationship between the surface data and the 1–2 m data exhibit a relatively strong fit for August, but not in June or July. The data at greater depths, however, exhibit relatively strong relationships during all of the sampling periods. The high variability observed at the surface and near-surface depths indicates that surface-biased estimates may be influenced by short-term weather events or human activity that causes mixture. The strong linear relationships for the other depths and for 1–2 m depths during August suggest that near-surface estimates provided by remote sensing may be strongly correlated with conditions throughout the water column, especially during periods of low surface mixing. In summary, the different relationships between surface and water column conditions highlight that surface conditions do not always reflect the conditions throughout the water column, and that the mechanical mixing processes which are unique to each water body should be taken into account before assuming any relationship between surface and water column conditions.

The spatial and temporal patterns observed in these lakes add to previous observational studies in these lakes which have focused largely on speciation and the diversity of algal populations. As species diversity decreases throughout the summer, the observations in this study also show that overall algae biomass magnitudes and variability in algae biomass increases. This relationship has both positive and negative implications for remote sensing; it provides additional motivation for using remote sensing methods during the late summer months when conditions are highly variable and more likely to be worse than early summer months, but it also highlights potential challenges associated with remote sensing of conditions when there is high species variability (leading to greater potential variability in the spectral signature of the surface waters).

5. Conclusions

The observations and analysis provided valuable insights into the Utah and GSL lake systems; however, it is important to acknowledge that the results may not be representative for all portions of the system. In particular, the surface/water column analyses in the lower portion of the GSL are not representative of the surface water/water column relationship in Utah Lake. Utah Lake is consistently much more turbid than the southern arm of the GSL, in general is shallower, and has far different mixing patterns. We recommend that this kind of analysis should be conducted in areas where unique or localized hydrodynamic disturbances exist (such as elevated exposure to wind and surface mixing, or near outfalls from wastewater treatment plants or streams where there may be increased mixing or stirring up of bottom sediments).

The temporal and spatial analysis presented in this study supports development of specific methods for future remote sensing work in this region. This support includes selecting appropriate sensors and defining appropriate time-windows for using near-coincident data. The seasonal differences in temporal correlation (as inferred by differences between subsequent samples) suggest the use of a shorter time-window for near-coincident data in calibrating empirical models in the late summer season than in the earlier summer months. We recommend that for modeling development in the main body of the GSL, near-coincident matches be limited to ±2 days, though more relaxed time-windows could be used for early summer matches. Based on the autocorrelation of the samples in Utah Lake and Farmington Bay, we recommend limiting the time windows for considering near-coincident matches to ±1 day for Utah Lake, while Farmington Bay may use a more relaxed time window.

Our spatial analysis showed small variations between offsets and sampling sites, indicating that Landsat/SENTINEL-2 resolution and MODIS resolutions would be appropriate for the southern arm of GSL, while finer-scale resolutions may be unnecessary as there is little variation at these smaller scales. As with the surface/water column analysis, this type of sampling in other parts of the lake system would be helpful in determining the most appropriate methods based on their unique spatial variability characteristics. From a temporal standpoint, the Landsat return time of 16 days is offset by the fact that there are multiple sensors which may be used, for example both Landsat 5 and 7 provide data for historical applications, while Landsat 8 and SENTINEL-2 provide data for more recent and ongoing applications (from 2013 and 2015, respectively). These instruments provide imagery on a more frequent basis (assuming no interference from cloud cover). However, our temporal analysis of the sensor data in Utah Lake and the main body of the GSL, shows that lake conditions change on shorter periods, and this revisit frequency may miss important changes in surface algae conditions. This is contrasted by Farmington Bay, where the conditions do not change as drastically over these time scales.

The information about spatiotemporal patterns should be considered along with other factors including: the spectral resolution of the sensors and how well the spectral measurements can describe the measures of algal biomass in certain lake environments [31], data availability (both field samples and imagery), and the historical scope (which may restrict the types of sensors which can be used) in order to meet the needs of the specific region of interest and the application. While focused on the GSL region and its unique characteristics, this study demonstrates a number of sampling and analysis techniques that could be applied in other settings to inform and improve the design of remote sensing studies. Information about the unique spatial and temporal variability patterns in a water body should be incorporated into the process of remote sensing model development, to help guide modeling decisions and assumptions.

Acknowledgments: This article was developed under Assistance Agreement No. 83586-01 awarded by the U.S. Environmental Protection Agency to Michael Barber. It has not been formally reviewed by EPA. The views expressed in this document are solely those of the authors and do not necessarily reflect those of the Agency. EPA does not endorse any products or commercial services mentioned in this publication. Additional funding was provided by the Department of Civil and Environmental Engineering at the University of Utah. The authors would like to acknowledge Marshall Baillie and others at the Utah Division of Water Quality for sharing data.

Author Contributions: S.B., P.D. and G.W. provided input on study concept, advising on data analysis and edits to manuscript drafts. C.H. performed data collection and analysis, and prepared the manuscript.

Conflicts of Interest: The authors declare no conflict of interest.

References

1. Olmanson, L.G.; Bauer, M.E.; Brezonik, P.L. A 20-year Landsat water clarity census of Minnesota's 10,000 lakes. *Remote Sens. Environ.* **2008**, *112*, 4086–4097. [CrossRef]
2. Kloiber, S.M.; Brezonik, P.L.; Olmanson, L.G.; Bauer, M.E. A procedure for regional lake water clarity assessment using Landsat multispectral data. *Remote Sens. Environ.* **2002**, *82*, 38–47. [CrossRef]
3. Sayers, M.; Fahnenstiel, G.L.; Shuchman, R.A.; Whitley, M. Cyanobacteria blooms in three eutrophic basins of the great lakes: A comparative analysis using satellite remote sensing. *Int. J. Remote Sens.* **2016**, *37*, 4148–4171. [CrossRef]
4. Hansen, C.H.; Williams, G.P.; Adjei, Z.; Barlow, A.; Nelson, E.J.; Miller, A.W. Reservoir water quality monitoring using remote sensing with seasonal models: Case study of five central-Utah reservoirs. *Lake Reserv. Manag.* **2015**, *31*, 225–240. [CrossRef]
5. Wynne, T.T.; Stumpf, R.P.; Tomlinson, M.C.; Fahnenstiel, G.L.; Dyble, J.; Schwab, D.J.; Joshi, S.J. Evolution of a cyanobacterial bloom forecast system in western Lake Erie: Development and initial evaluation. *J. Gt. Lakes Res.* **2013**, *39*, 90–99. [CrossRef]
6. Glasgow, H.B.; Burkholder, J.M.; Reed, R.E.; Lewitus, A.J.; Kleinman, J.E. Real-time remote monitoring of water quality: A review of current applications, and advancements in sensor, telemetry, and computing technologies. *J. Exp. Mar. Biol. Ecol.* **2004**, *300*, 409–448. [CrossRef]
7. Goldman, C.R.; Jassby, A.D.; Hackley, S.H. Decadal, interannual, and seasonal variability in enrichment bioassays at Lake Tahoe, California-Nevada, USA. *Can. J. Fish. Aquat. Sci.* **1993**, *50*, 1489–1496. [CrossRef]
8. Jiang, Y.-J.; He, W.; Liu, W.-X.; Qin, N.; Ouyang, H.-L.; Wang, Q.-M.; Kong, X.-Z.; He, Q.-S.; Yang, C.; Yang, B. The seasonal and spatial variations of phytoplankton community and their correlation with environmental factors in a large eutrophic Chinese lake (Lake Chaohu). *Ecol. Indic.* **2014**, *40*, 58–67. [CrossRef]
9. Wynne, T.T.; Stumpf, R.P. Spatial and temporal patterns in the seasonal distribution of toxic cyanobacteria in western Lake Erie from 2002–2014. *Toxins* **2015**, *7*, 1649–1663. [CrossRef] [PubMed]
10. Mouw, C.B.; Greb, S.; Aurin, D.; DiGiacomo, P.M.; Lee, Z.; Twardowski, M.; Binding, C.; Hu, C.; Ma, R.; Moore, T. Aquatic color radiometry remote sensing of coastal and inland waters: Challenges and recommendations for future satellite missions. *Remote Sens. Environ.* **2015**, *160*, 15–30. [CrossRef]
11. Bailey, S.W.; Werdell, P.J. A multi-sensor approach for the on-orbit validation of ocean color satellite data products. *Remote Sens. Environ.* **2006**, *102*, 12–23. [CrossRef]
12. Giardino, C.; Pepe, M.; Brivio, P.A.; Ghezzi, P.; Zilioli, E. Detecting chlorophyll, Secchi Disk Depth and surface temperature in a sub-alpine lake using Landsat imagery. *Sci. Total Environ.* **2001**, *268*, 19–29. [CrossRef]
13. Lesht, B.M.; Barbiero, R.P.; Warren, G.J. A band-ratio algorithm for retrieving open-lake chlorophyll values from satellite observations of the great lakes. *J. Gt. Lakes Res.* **2013**, *39*, 138–152. [CrossRef]
14. McCullough, I.M.; Loftin, C.S.; Sader, S.A. Landsat imagery reveals declining clarity of Maine's lakes during 1995–2010. *Freshw. Sci.* **2013**, *32*, 741–752. [CrossRef]
15. Johnson, R.; Strutton, P.G.; Wright, S.W.; McMinn, A.; Meiners, K.M. Three improved satellite chlorophyll algorithms for the southern ocean. *J. Geophys. Res. Oceans* **2013**, *118*, 3694–3703. [CrossRef]
16. Olmanson, L.G.; Brezonik, P.L.; Bauer, M.E. Evaluation of medium to low resolution satellite imagery for regional lake water quality assessments. *Water Resour. Res.* **2011**, *47*. [CrossRef]
17. Mouser, J.E.; Baxter, B.K.; Spear, J.R.; Peters, J.W.; Posewitz, M.C.; Boyd, E.S. Contrasting patterns of community assembly in the stratified water column of Great Salt Lake, Utah. *Microb. Ecol.* **2013**, *66*, 268–280. [CrossRef] [PubMed]

Remote Sens. **2017**, *9*, 409

18. Klausmeier, C.A.; Litchman, E. Algal games: The vertical distribution of phytoplankton in poorly mixed water columns. *Limnol. Oceanogr.* **2001**, *46*, 1998–2007. [CrossRef]
19. Goel, R.; Myers, L. Evaluation of cyanotoxins in the Farmington Bay, Great Salt Lake, Utah. Available online: http://cdsewer.org/GSLRes/2009_CYANOBACTERIA_PROJECT_REPORT.pdf (accessed on 1 January 2016).
20. Marden, B.; Richards, D. Factors Influencing Cyanobacteria Blooms in Farmington Bay, Great Salt Lake, Utah. Available online: https://www.researchgate.net/profile/David_Richards20/publication/305488678_Factors_Influencing_Cyanobacteria_Blooms_in_Farmington_Bya_Great_Salt_Lake_Utah/links/5790eefe08ae0831552f92ab.pdf (accessed on 12 March 2015).
21. Rushforth, S.R.; Squires, L.E. New records and comprehensive list of the algal taxa of Utah Lake, Utah, USA. *Gt. Basin Nat.* **1985**, *45*, 237–254.
22. Whiting, M.C.; Brotherson, J.D.; Rushforth, S.R. Environmental interaction in summer algal communities of Utah Lake. *Gt. Basin Nat.* **1978**, *38*, 31–41.
23. Wurtsbaugh, W.A.; Marcarelli, A.M.; Boyer, G.L. Eutrophication and Metal Concentrations in Three bays of the Great Salt Lake (USA). Available online: http://digitalcommons.usu.edu/cgi/viewcontent.cgi?article=1548&context=wats_facpub (accessed on 1 July 2012).
24. Naftz, D.; Angeroth, C.; Kenney, T.; Waddell, B.; Darnall, N.; Silva, S.; Perschon, C.; Whitehead, J. Anthropogenic influences on the input and biogeochemical cycling of nutrients and mercury in Great Salt Lake, Utah, USA. *Appl. Geochem.* **2008**, *23*, 1731–1744. [CrossRef]
25. USGS. Great Salt Lake—Salinity and Water Quality. Available online: https://ut.water.usgs.gov/greatsaltlake/salinity/ (accessed on 5 April 2017).
26. Wurtsbaugh, W.; Marcarelli, A. *Eutrophication in Farmington Bay, Great Salt Lake, Utah 2005 Annual Report*; Central Davis Sewer District: Kaysville, UT, USA, 2006.
27. Utah DEQ. *Utah Lake Report*; Utah Department of Environmental Quality: Salt Lake City, UT, USA, 2006.
28. Hansen, C. *Great Salt Lake Water Quality*; Hydroshare: Cambridge, MA, USA, 2017.
29. Cressie, N.; Wikle, C.K. *Statistics for Spatio-Temporal Data*; John Wiley & Sons: Hoboken, NJ, USA, 2015.
30. R Core Team. *R: A Language and Environment for Statistical Computing*; R Foundation for Statistical Computing: Vienna, Austria, 2016.
31. Olmanson, L.G.; Brezonik, P.L.; Finlay, J.C.; Bauer, M.E. Comparison of Landsat 8 and Landsat 7 for regional measurements of CDOM and water clarity in lakes. *Remote Sens. Environ.* **2016**, *185*, 119–128. [CrossRef]

remote sensing

MDPI

Article

Remote Sensing of Black Lakes and Using 810 nm Reflectance Peak for Retrieving Water Quality Parameters of Optically Complex Waters

Tiit Kutser [1,2,*], Birgot Paavel [1], Charles Verpoorter [2,3], Martin Ligi [4], Tuuli Soomets [1], Kaire Toming [1] and Gema Casal [1]

[1] Estonian Marine Institute, University of Tartu, Mäealuse 14, 12618 Tallinn, Estonia; birgot.paavel@ut.ee (B.P.); Tuuli.Soomets@ut.ee (T.S.); Kaire.Toming.001@ut.ee (K.T.); gema.casal@marine.ie (G.C.)

[2] Evolutionary Biology Centre, Limnology, University of Uppsala, Norbyvägen 18D, 75236 Uppsala, Sweden; Charles.Verpoorter@univ-littoral.fr

[3] Laboratoire d'Oceanologie et des Geosciences, Universite de Lille Nord de France, ULCO, 32 Avenue Foch, 62930 Wimereux, France

[4] Tartu Observatory, 61602 Tõravere, Tartumaa, Estonia; Martin.Ligi@to.ee

* Correspondence: Tiit.Kutser@ut.ee; Tel.: +372-6718-947

Academic Editors: Yunlin Zhang, Claudia Giardino, Linhai Li, Deepak R. Mishra and Prasad S. Thenkabail
Received: 11 March 2016; Accepted: 7 June 2016; Published: 14 June 2016

Abstract: Many lakes in boreal and arctic regions have high concentrations of CDOM (coloured dissolved organic matter). Remote sensing of such lakes is complicated due to very low water leaving signals. There are extreme (black) lakes where the water reflectance values are negligible in almost entire visible part of spectrum (400–700 nm) due to the absorption by CDOM. In these lakes, the only water-leaving signal detectable by remote sensing sensors occurs as two peaks—near 710 nm and 810 nm. The first peak has been widely used in remote sensing of eutrophic waters for more than two decades. We show on the example of field radiometry data collected in Estonian and Swedish lakes that the height of the 810 nm peak can also be used in retrieving water constituents from remote sensing data. This is important especially in black lakes where the height of the 710 nm peak is still affected by CDOM. We have shown that the 810 nm peak can be used also in remote sensing of a wide variety of lakes. The 810 nm peak is caused by combined effect of slight decrease in absorption by water molecules and backscattering from particulate material in the water. Phytoplankton was the dominant particulate material in most of the studied lakes. Therefore, the height of the 810 peak was in good correlation with all proxies of phytoplankton biomass—chlorophyll-a ($R^2 = 0.77$), total suspended matter ($R^2 = 0.70$), and suspended particulate organic matter ($R^2 = 0.68$). There was no correlation between the peak height and the suspended particulate inorganic matter. Satellite sensors with sufficient spatial and radiometric resolution for mapping lake water quality (Landsat 8 OLI and Sentinel-2 MSI) were launched recently. In order to test whether these satellites can capture the 810 nm peak we simulated the spectral performance of these two satellites from field radiometry data. Actual satellite imagery from a black lake was also used to study whether these sensors can detect the peak despite their band configuration. Sentinel 2 MSI has a nearly perfectly positioned band at 705 nm to characterize the 700–720 nm peak. We found that the MSI 783 nm band can be used to detect the 810 nm peak despite the location of this band is not in perfect to capture the peak.

Keywords: lakes; CDOM; remote sensing; hyperspectral; Sentinel-2; chlorophyll-a; suspended matter; Landsat 8

1. Introduction

Lakes are an important source of drinking water, they provide different services from fisheries to tourism, support biodiversity, and are an important component in the global carbon cycle [1–3]. Monitoring the water quality and understanding the physical, chemical, and biological status of inland waters is hard to achieve without using remote sensing [4]. However, there are many obstacles in the way to achieve sufficient accuracy of inland water remote sensing products. Some of them are related to optical complexity of the waters, some to the methodology (e.g., atmospheric correction), and some to the technology (radiometric, spatial, and spectral resolution of sensors) used [4].

Only the visible part of electromagnetic radiation can potentially provide us information about the water constituents in most waterbodies as water itself absorbs light strongly at other wavelengths [5–7]. The exceptions here are waters with high concentrations of suspended matter [8,9] or phytoplankton [10,11] where the water reflectance in the near infrared (NIR) part of spectrum can also provide us useful information. However, there may be extreme environments where the water-leaving signal is negligible also in the visible part of the spectrum, automatically preventing the use of most current remote sensing algorithms and methods. For example, in this study we investigated lakes where the CDOM concentrations are so high that the water-leaving signal is practically zero at all visible wavelengths and the above water measured signal consisted predominantly of sun and sky glint. One may assume that the number of such extreme lakes is small as the only reflectance data we were able to find from almost as dark lakes was published only recently [7]. However, the global inventory of lakes [12] shows that majority of lakes on the Earth are between 55N and 75N, meaning boreal and arctic lakes, are the most abundant. Many boreal lakes have high CDOM concentration [7,13–15]. Most of the arctic lakes are actually permafrost thaw ponds that should be rich in dissolved organic carbon, DOC, and its coloured component, CDOM, although Sobek *et al.* [16] have shown that the lake DOC pattern at higher latitudes is quite complicated. Thus, at present there is very fragmented information on the possible abundance of CDOM-rich lakes as most of them are probably in inhabited and hardly accessible regions.

It was shown recently [17,18] that the iron bound to DOC makes lake water absorbance higher and variable iron to carbon ratio makes remote sensing retrieval of CDOM and DOC concentrations complicated. It means that the number of lakes in which remote sensing is challenging due to low water leaving signal, caused by high absorbance, should be relatively high globally. Retrieving the lake CDOM and DOC concentrations is important from both a drinking water perspective [19] and the global carbon cycle studies point of view [2]. The drinking water industry needs this information also in near real time as sudden heavy precipitation may increase the amount of carbon quickly and require modifications in water treatment processes. Consequently, there is a strong need to study optical properties of CDOM-rich lakes and the potential for retrieval of water quality parameters by means of remote sensing in such lakes.

Field radiometers have become remote sensing instruments on their own right rather than being just calibration and validation devices of satellite measurements. For example, routine reflectance measurements carried out from ferries (ferriscope.org) and hand-held devices have been developed for quick monitoring of lake water quality [20]. Many of the black lakes are small. Therefore, in this study, we focused mainly on field radiometry rather than satellite remote sensing.

The main aim of the study was to investigate black lakes with nearly negligible water leaving signal in the visible part of the spectrum and to estimate whether remote sensing retrieval of water constituents in such extreme CDOM-rich lakes is feasible by means of hyperspectral sensors. The next step was applying the results obtained in black lakes on all other lakes for which we had field radiometry data. Satellites with sufficient spatial resolution for small lake studies, like the Landsat 8 and Sentinel-2, became available recently. Therefore, it was reasonable to evaluate are these satellites suitable from their band configuration point of view for remote sensing of CDOM-rich lakes. This evaluation was performed by recalculating hyperspectral field radiometry data into spectral bands of Landsat 8 and Sentinel-2 as well as using actual satellite imagery.

2. Materials and Methods

2.1. Study Sites

We chose three nearly black water lakes in South-Eastern Estonia for our study—Nohipalu Mustjärv, Meelva, and Mustjärv (Figure 1). These lakes were chosen based on our previous knowledge about the optical water properties there. The lakes are small—their area varies between 0.2 km^2 and 0.8 km^2.

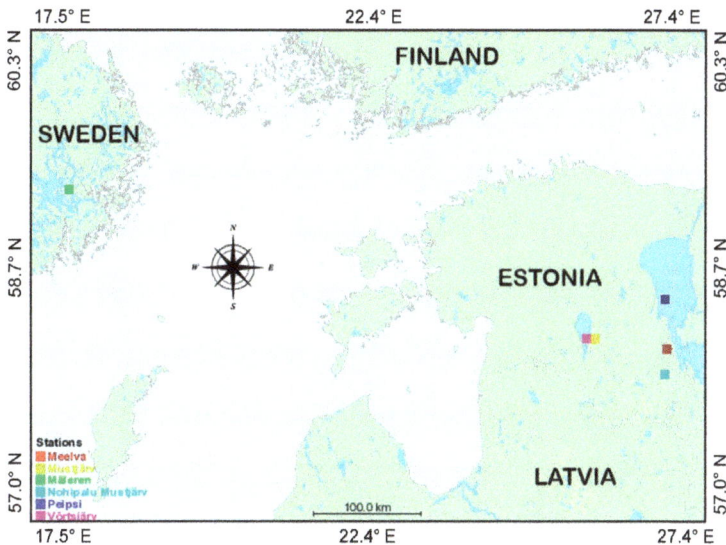

Figure 1. Locations of the study sites (GPS Visualizer).

In order to test how similar/dissimilar from remote sensing point of view are these extreme lakes from typical boreal lakes we chose other four lakes in Sweden and Estonia. Lake Mälaren is the third largest lake in Sweden (1140 km^2). It is a gemorphologically sophisticated lake where different basins are often connected only through narrow straits. Optical properties of these basins vary in a wide range [17,21,22] and can be considered as different water bodies from an optical point of view. We also used data from three Estonian lakes where optical measurements are carried out in semi-regular basis. These lakes are Lake Peipsi, the fourth largest lake in Europe (3555 km^2), Lake Võrtsjärv (270.7 km^2) and Lake Harku (1.64 km^2). All the other lakes, besides Mälaren, are shallow (maximum depth between 2.5 m and 12.9 m, Mälaren 61 m).

2.2. Field Measurements

Field measurements were carried out during three years 2011–2013. The total number of sampling stations was 105. The extreme CDOM-rich lakes Mustjärv and Nohipalu Mustjärv were sampled once, Meelva twice. Data from the other boreal lakes was collected in the frame of different projects. Lake Mälaren in Sweden was sampled once (7 sampling stations) while three Estonian lakes were studied almost on the monthly bases during three ice free seasons. We sampled Lake Võrtsjärv 9 times (19 samples in total), Lake Harku 11 times (one station each time), and Lake Peipsi 11 times (65 samples in total).

Water reflectance measurements were carried out using two Ramses (TriOS) sensors one measuring downwelling irradiance and one measuring upwelling radiance from the nadir. Ramses is sampling with 3.3 nm interval in the wavelength range 350–900 nm. Each reflectance spectrum used in the study is an average of 10 measured spectra.

We knew in advance that significant part of the measured reflectance is actually sun and sky glint, especially in the darkest lakes. Therefore, we carried out reflectance measurements in two different ways. In each sampling station we first measured reflectance as a ratio of upwelling radiance to downwelling irradiance, L_u/E_d, where both Ramses sensors were above the water surface *i.e.*, these were normal reflectance measurements carried out from a boat with zenith and nadir looking sensors. A second set of measurements was carried out putting the 5 cm black plastic tube, surrounding the Ramses radiance sensor head, just under the water surface. This way we can measure water-leaving radiance, Lw, without sun and sky glint. Dividing the subsurface radiance with the downwelling irradiance measured above the water surface we got glint-free reflectance spectra. The methodology was described in more detail in [23].

Optical properties of the water were measured with WetLabs optical package containing a CTD, AC-S absorption and attenuation sensor, ECO-BB3 backscattering sensor and ECO-VSF3 volume scattering sensor. Complementary wavelengths were chosen for the BB3 and VSF3. This allows us to measure backscattering coefficient at six wavelengths as backscattering coefficient can also be calculated from the ECO-VSF3 data.

2.3. Laboratory Analysis

Water samples were collected from the surface layer (0.2 m) directly into 2.5 litre canisters that were then stored in the dark and cold before filtering in the evening of each sampling day (less than 10 h between the collection and filtering). The volume of lake water filtered through Whatman GF/F-filters depended on particle load (0.1–1 litre). Phytoplankton pigments were extracted from the filters with 96% ethanol at 20 °C for 24 h and measured spectrometrically both before and after acidification with dilute hydrochloride acid. Later, optical density values were converted respectively to chlorophyll-a and phaeophytin-a concentrations according to Lorenzen [24] formulas. The concentration of total suspended matter, TSS, was measured gravimetrically after filtration of the same amount of water through pre-weighed and dried (103–105 °C for 1 h) filters. The inorganic fraction of suspended matter, SPIM, was measured after combustion of filters at 550 °C for 30 min. The organic fraction of suspended matter, SPOM, was determined by subtraction of SPIM from TSS [25].

Absorption by colored dissolved organic matter, $a_{CDOM}(\lambda)$, was measured with a spectrophotometer (Hitachi U-3010 UV/VIS, in the range 350–750 nm with 1 nm resolution) in water filtered through Millipore 0.2 µm filter. The measurements were carried out in a 5 cm cuvette against distilled water and corrected for residual scattering according to Davis-Colley & Vant [26]. The water from the three extreme CDOM lakes was diluted 1:1 with distilled water before the measurements as otherwise most of the light beam was absorbed in the 5 cm cuvette.

2.4. Satellite Data

Suitability of Sentinel-2 and Landsat 8 was tested in two ways. First, we simulated theoretical performance of the satellites in the black lakes by recalculating Ramses field reflectance spectra using spectral response functions of the two satellites. Secondly, we used a Landsat 8 image from 8 July 2013 acquired simultaneously with our field campaign in Lake Peipsi and 1–2 days before sampling on the black lakes and Lake Võrtsjärv. Atmospheric correction was performed with four different methods from which ATCOR23 was selected as the best performing method for most of the lakes [27]. More detail about the field campaign and different atmospheric correction methods tested is provided in [27]. Sentinel-2 imagery was collected in 11 and 14 August 2015 and processed with SNAP Sentinel-2 toolbox and Sen2Cor atmospheric correction module provided together with the Sentinel Toolbox. Detailed description of Sentinel-2 processing and using the MSI imagery in mapping of different lake water quality parameters is given in [28].

2.5. Remote Sensing Algorithms

There are a variety of methods for relating the remote sensing signal to optical water properties like chlorophyll-a, CDOM or suspended matter. For example, single band algorithms, band ratios, more sophisticated colour indices or analytical methods for retrieving three or more water properties simultaneously have been used [29–33]. We chose to use peak height algorithms as the measured reflectance had only two peaks and very low signal at other wavelengths. We tested spectral differences using the reflectance values of the peak and at one of the nearby wavelengths where the reflectance is the lowest. However, the best performing algorithms were three band algorithms where the peak height was calculated against the baseline of two wavelengths where the reflectance is low (Equations (1) and (2)).

It is known that the peak near 700–720 nm is moving towards longer wavelengths with the increasing phytoplankton biomass [34]. Therefore, the first peak height was calculated as the difference between the maximum reflectance in the 700–720 nm wavelength range against the 676–770 nm baseline (Equation (1)). The baseline wavelengths were chosen based on the shape of the reflectance spectra (Figure 2).

$$P_1 = R_{max}(700 - 720) - [R(646) + R(770)]/2 \tag{1}$$

where P_1 is the height of the peak near 710 nm, $R_{max}(700–720)$ is maximum reflectance value in the 700–720 nm wavelength range, and the $R(646)$ and $R(770)$ are reflectance values at these two wavelengths.

The second peak height was calculated simply as the difference between the reflectance value at 810 nm and the 770–840 nm baseline as the location of the maximum value was always at 810 nm (Equation (2)).

$$P_2 = R(810) - [R(770) + R(840)]/2 \tag{2}$$

where P_2 is the height of the peak at 810 nm, $R(810)$, $R(770)$, and $R(840)$ are the reflectance values at these three wavelengths. Most of researchers measure reflectance just above the water surface and do not have measurements of glint free spectra. In order to investigate the effect of glint on the retrieval results we calculated the peak heights from both "normal" (L_u/E_d) and glint-free (L_w/E_d) reflectance spectra.

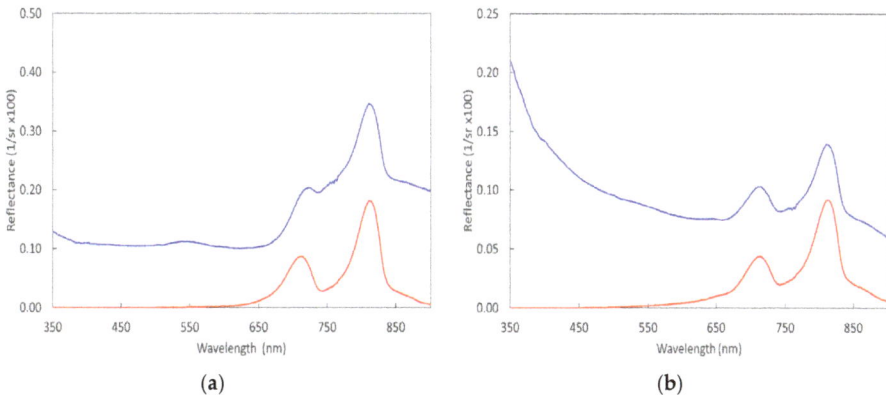

Figure 2. Reflectance spectra of extreme CDOM-rich lakes Nohipalu Mustjärv (**a**) and Meelva (**b**). The reflectance spectra shown with blue line were measured with both L_u and E_d sensor above the water surface. The reflectance spectra shown with red line were measured putting the radiance sensor surrounded with plastic tube a few centimetres below the water surface (reflectance without glint).

3. Results

The optical properties of the studied lakes were variable as seen from the Table 1. For example, chlorophyll-a concentration varied between 2.14 mg·m^{-3} and 203.31 mg·m^{-3} whereas TSS varied between 0.75 mg·L^{-1} and 63.33 mg·L^{-1}. All the lakes were relatively CDOM-rich—the a_{CDOM}(400) varied between 3.23 m^{-1} and 63.05 m^{-1}.

Table 1. Minimum, maximum, and mean concentrations of optically active substances measured in lakes under investigation. Concentration of Chl is in mg·m^{-3}, TSS, SPIM, and SPOM in mg·L^{-1} and a_{CDOM}(400) in m^{-1}. Only maximum values are provided if there is a single measurement from a particular lake.

	Nohipalu Mustjärv	Meelva	Mustjärv	Mäleren	Harku	Võrtsjärv	Peipsi
			Chl, mg·m^{-3}				
Min		11.04		7.13	36.31	15.06	2.14
Mean		14.56		24.28	123.93	33.74	15.10
Max	4.67	18.07	7.34	50.82	203.31	57.83	38.98
			a_{CDOM}(400), m^{-1}				
Min		41.45		3.23	6.12	3.76	3.23
Mean		44.52		5.70	9.77	6.20	6.54
Max	63.05	49.48	47.60	10.04	13.99	11.33	15.11
			TSS, mg·L^{-1}				
Min		9.00		18.89	10.67	3.33	0.75
Mean		9.00		29.0	36.17	14.22	7.90
Max	12	9.00	26.00	43.05	63.33	21.00	23.8
			SPIM, mg·L^{-1}				
Min		5.50		10.05	0.67	0.00	0.00
Mean		5.75		11.47	7.58	3.42	3.41
Max	0.80	6.00	17.00	14.05	22.40	8.67	17.84
			SPOM, mg·L^{-1}				
Min		3.00		7.37	6.33	3.33	0.00
Mean		3.25		17.58	28.58	10.79	4.84
Max	16.80	3.50	9.00	32.00	62.5	15.50	10.67

Reflectance spectra of the extreme CDOM lakes are shown in Figure 2 and reflectance spectra of all studied lakes are shown in Figure 3. It is seen in the Figure 2 that in the extreme lakes water reflectance (red spectrum) is negligible in almost the entire visible part of the spectrum and the only usable signal is in the form of two peaks, which have maxima near 710 nm and 810 nm. It is also seen in the Figure 2 that significant part of the remote sensing signal measured above the black lakes (blue spectrum) is light reflected from the water surface. In the 350–600 nm spectral range the whole signal is glint. The only chance to get information about the water constituents in such lakes is to use these two reflectance peaks. We calculated the height of the two (710 nm and 810 nm) peaks in order to try to estimate concentrations of optically active substances chlorophyll-a, TSS (total suspended solids), SPIM (suspended particulate inorganic matter), SPOM (suspended particulate organic matter), and CDOM. Besides the peak heights themselves, we also used differences, sums, and ratios of these two peak heights.

Reflectance spectra of all studied lakes together are presented in Figure 3. Most of the spectra have been collected in lakes with significant cyanobacterial biomass as there is a phycocyanin absorption feature at 620 nm and a peak at 650 nm (Figure 3) typical to only cyanobacteria [11].

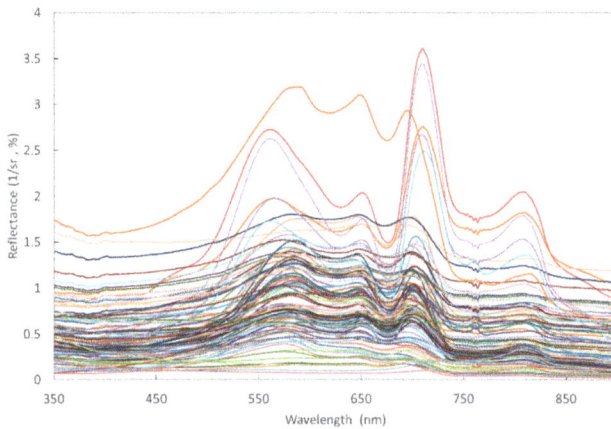

Figure 3. Reflectance spectra of all studied Estonian and Swedish lakes.

The peak near 710 nm, caused by combined effect of absorption by water molecules and very high reflectance of phytoplankton in the infrared part of spectrum, is often used for chlorophyll-a retrieval in many waterbodies [35–37]. Not surprisingly, there was also good correlation between the peak height, P_1, and chlorophyll-a in the lakes studied by us (Figure 4).

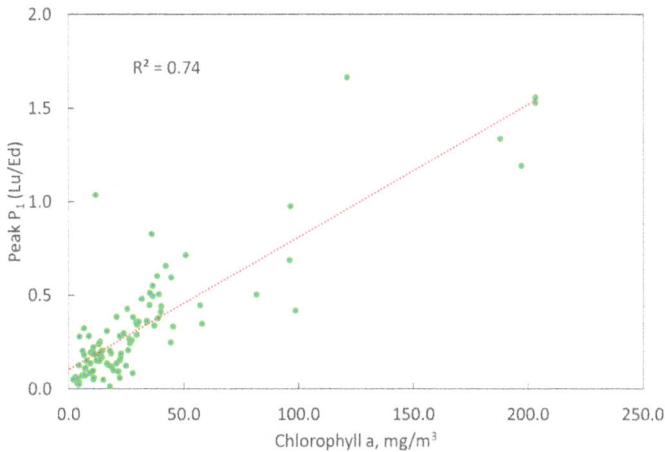

Figure 4. Correlation between the 700–720 nm peak height (P_1, Equation (1)) and chlorophyll-a concentration in all studied Estonian and Swedish lakes.

The height of the 810 nm peak is clearly higher in reflectance spectra of black lakes than the height of the 700–720 nm peak (Figure 2). Therefore, we decided to test whether the height of this peak, P_2, is in correlation with the concentrations of optically active substances. Figure 5 illustrates the correlation between P_2 and chlorophyll-a and Figure 6 the correlation between the P_2 and total suspended matter. The correlation was good for both parameters when data from all Estonian and Swedish lakes was used.

It was not surprising that the height of the 710 nm peak was in good correlation with chlorophyll-a concentration—R^2 was 0.74 and 0.72 for above water (Figure 4) and glint free reflectance respectively. However, it was surprising that the 810 nm peak height correlated even better with the chlorophyll-a concentration—R^2 = 0.77 (Figure 5). This result was obtained for glint-free spectra. R^2 was just 0.37 if the peak height was calculated from the above water reflectance spectra. This stresses the importance of removing glint from aquatic reflectance spectra. There have been studies relating the elevated NIR reflectance values to high mineral suspended matter concentration [8,9]. Therefore, the good correlation between the 810 nm peak and TSS was expected to certain extent.

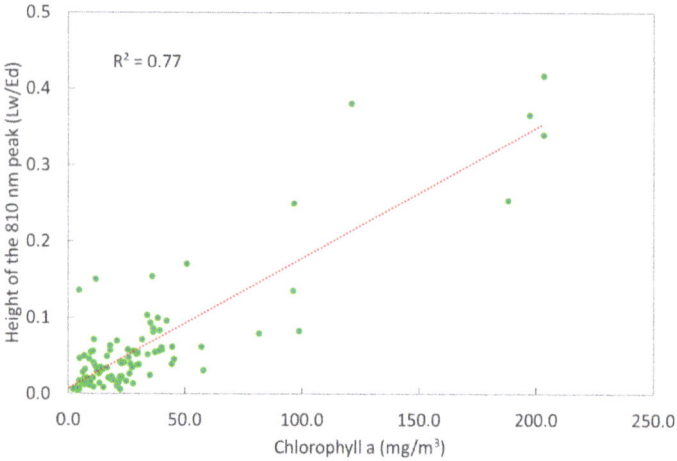

Figure 5. Correlation between the 810 nm peak height (P$_2$, Equation (2)) and chlorophyll-a concentration in all studied Estonian and Swedish lakes.

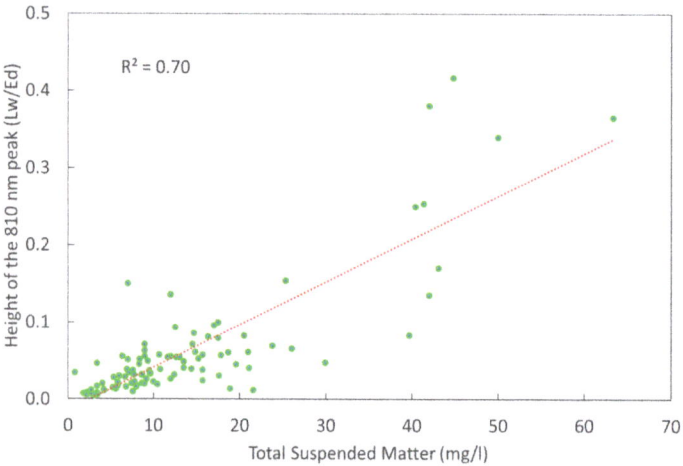

Figure 6. Correlation between the 810 nm peak height (P$_2$, Equation (2)) and total suspended matter concentration in all Estonian and Swedish lakes.

In order to test theoretical performance of Sentinel-2 and Landsat 8 sensors in picking up the two peaks containing information about the water properties in the case of black lakes we took the *in situ* measured glint-free spectrum of Nohipalu Mustjärv (Figure 2a) and recalculated it using spectral response functions of Sentinel-2 MSI and Landsat 8 OLI sensors. The results are given in Figure 7. It is clearly seen that Landsat 8 band configuration does not allow detection of either of the peaks.

Sentinel-2 MSI does not have narrow spectral bands near the 810 nm peak. However, the 783 nm centered band 7 allows to detect the peak to certain extent. Especially, because the bands 6 and 8a are located at wavelengths where the lake reflectance values are low. The 705 nm band 5 of Sentinel-2 is almost perfectly located for detection of elevated biomass in waterbodies as we have also shown in our study focusing on using Sentinel-2 imagery in lake research [28].

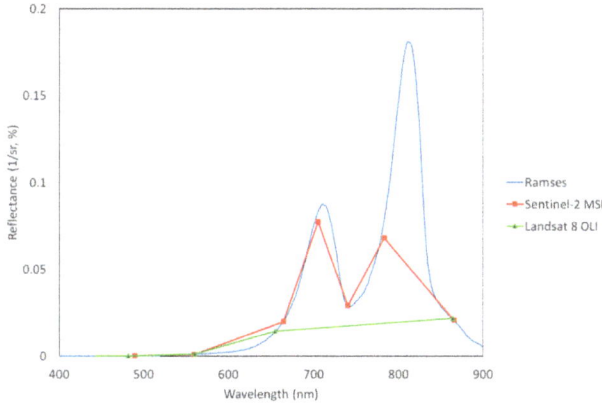

Figure 7. A glint-free reflectance spectrum of black-water Nohipalu Mustjärv measured with field radiometer Ramses (blue spectrum) and reflectance spectra calculated from the same spectrum using Sentinel-2 MSI (red) and Landsat 8 OLI (green) spectral response functions.

The results obtained from actual satellite imagery resemble those obtained from field measurements spectra as can be seen in Figure 8. Both satellite reflectances are slightly elevated (not zero) in the blue to green part of spectrum where the water leaving signal is practically zero as can be seen in Figure 2. This indicates that the satellite signal also contains glint from the water surface which may be significant compared to the water leaving signal as is clearly seen in Figure 2b. Another potential source of the non-negligible reflectance is the adjacency effect as the black lakes are small and water leaving signal very low compared to the potential signal contamination from the adjacent land.

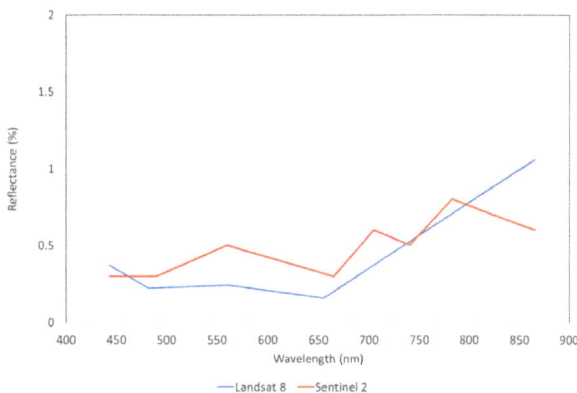

Figure 8. Reflectance spectra of black-water Nohipalu Mustjärv from two different satellites. The Blue line is ATCOR23 corrected Landsat 8 data and red line is Sen2Cor corrected Sentinel 2 data.

4. Discussion

As was mentioned earlier our main aim was to study the extreme CDOM lakes with field radiometers in order to understand is it possible to retrieve water quality parameters of lakes where the water leaving signal is close to zero in visible part of spectrum. The first question that arises is quite subjective—what is a black or extreme CDOM lake? For example, Duan *et al.* [38,39] investigated black water blooms in Lake Taihu where the CDOM absorption at 443 nm reached up to 1.68 m^{-1} (~3.3 m^{-1} at 400 nm). Such waters seem black compared to turbid, highly backscattering, waters of the rest of the lake. We found so low CDOM values only in a few Lake Peipsi stations (minimum value 3.23 m^{-1}). In the lakes we would call black the a$_{CDOM}$(400) varied between 41.45 m^{-1} and 63.05 m^{-1}. Black water lakes have been studied also in the USA. For example, Brezonik *et al.* [7] studied a few lakes where CDOM absorption at 440 nm reached up to 25.1 m^{-1} (~49 m^{-1} at 400 nm). Thus, the terms black, CDOM-rich, extreme CDOM lakes are quit arbitrary and depend on the background CDOM levels nearby rather than absolute absorption values.

The relativeness of water colour is clearly seen also in the Figure 9. The lake Võrtsjärv shown in the left part of the image has nearly twice as high mean CDOM concentration (Table 1) than the black water blooms in Lake Taihu [38,39]. Nevertheless, the Võrtsjärv water looks bright green compared to the Mustjärv in the same scene. There are two reasons for that. First of all the Võrtsjärv water contains relatively high amounts of particulate matter (both organic and inorganic) and is therefore a relatively bright object. On the other hand the visual appearance of all objects in processed satellite imagery depends also on the image stretch and brightness of other object in the scene.

Figure 9. Fragment of a Sentinel-2 image with a small fraction of Lake Võrtsjärv (left half of the image) and an extreme CDOM-rich Lake Mustjärv.

In most lakes the absorption by CDOM is negligible in red and near infrared parts of spectrum. However, in the three black lakes studied by us the absorption of CDOM and water molecules are equal at 700 nm or the CDOM absorption is even higher (Figure 10). Thus, the light backscattered from phytoplankton has to overcome both water and CDOM absorption in order to form detectable signal in reflectance spectra. Absorption by water molecules increases almost exponentially with increasing wavelength after 690 nm [5–7]. Therefore, the elevated signal forms a relatively narrow peak near 700 nm in the case of high biomass (or benthic vegetation in shallow water) and the maximum of the peak is moving towards longer wavelength with increasing biomass. In the lakes where CDOM absorption is still strong near 700 nm it first of all causes the decrease in the height of the peak often used to estimate phytoplankton biomass in water, but it also causes slight shift of the maximum in reflectance spectra towards NIR as the CDOM absorption decreases exponentially with increasing wavelength.

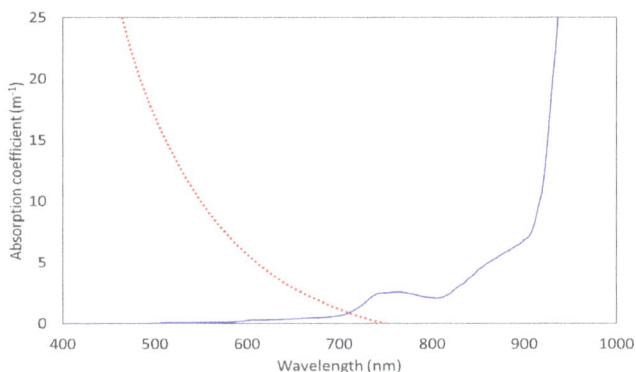

Figure 10. Absorption coefficient of CDOM from lake Nohipalu Mustjärv (red graph) and pure water absorption spectrum (blue line) [5].

The reason why the 810 nm peak occurs in reflectance spectra is a small decrease in water absorption coefficient approximately between 770 nm and 860 nm (see Figure 10) with the lowest absorption coefficient at 810 nm. Presence of the peak is obvious in all the reflectance spectra collected by us in different lakes (Figure 3) not only in the black lakes.

Brezonik *et al.* [7] presented a few reflectance spectra from CDOM-rich lakes in the USA. These spectra were similar to our black lakes—nearly negligible reflectance in the visible part of spectrum and a peak near 710 nm. Unfortunately, the graphs in their paper did not show reflectance beyond 800 nm. Therefore, we do not have reflectance data from very CDOM-rich lakes in other parts of the world to compare with our black lakes. Reflectance spectra with the second near infrared peak (around 810–850 nm) have been published in several papers [10,23]. However, the reason and magnitude of this peak was not discussed or even mentioned in any of these papers.

Doxaran *et al.* [8] attributed the high reflectance values in NIR part of spectrum to high concentration of suspended matter and used different band ratios that included NIR band of different satellites (SPOT 790–890 nm, Landsat 750–900 nm, and SeaWiFS 845–885 nm) for retrieving concentrations of suspended matter. There are several publications [8,9,40] showing that the elevated signals in the NIR part of the spectrum is a good predictor of TSM. All these studies were carried out in waters very rich in mineral particles. Therefore, the elevated NIR signal has been attributed to high mineral particle load in the water, not high phytoplankton concentration (chlorophyll-a).

This may seem contradictory to our results, but it is not. The peak near 810 nm is caused by high amount of scattering particles and local dip in absorption coefficient of water molecules. If the scattering material in the water is of mineral origin, like in previous studies [8,9,32], then the peak height is in good correlation with the concentration of mineral particles. In lakes studied by us (and many other lakes) the dominating scattering material in water is phytoplankton. Consequently, the height of the peak at 810 nm has to be in correlation with all parameters describing phytoplankton abundance in the water like chlorophyll-a, TSS, and—its organic component—SPOM. In the lakes studied by us, there was no correlation ($R^2 = 0.15$) between the concentration of mineral particles, SPIM, and the 810 nm peak height. On the other hand, the analysis of the *in situ* data showed that there was good correlation between chlorophyll-a and SPOM ($R^2 = 0.74$) and TSS and SPOM ($R^2 = 0.87$) indicating that majority of the suspended particles were organic and significant fraction of them were living phytoplankton cells.

Backscattering coefficient values were relatively high in the extreme CDOM-rich lakes. For example, the backscattering coefficient at 595 nm, $b_b(595)$, varied between 0.05 and 0.15 m^{-1} in all studied lakes, whereas it was between 0.08 and 0.15 m^{-1} in the extreme CDOM lakes. Thus, the reflectance in these lakes was low not because of low backscattering, but because the high CDOM absorption masks the backscattering signal.

The concentration of phytoplankton (chlorophyll-a) was relatively high in the extreme CDOM lakes (Table 1) and caused the appearance of the peaks at 710 nm and 810 nm. One may assume that in these lakes the amount of light available for photosynthesis is very low limiting the growth of phytoplankton, but this was not the case. The explanation here is species composition of phytoplankton. Cyanobacteria are the most dominant group in Estonian lakes and Lake Mälaren in Sweden during summer season. Many species of cyanobacteria can regulate their buoyancy and in calm weather conditions can choose the water depth most optimal for their growth. The extreme CDOM-rich lakes studied by us are relatively small and surrounded by forest. This means that the wind speed and water mixing are usually low and cyanobacteria can stay close to the surface where light is available for primary production. This also has an effect on the water reflectance. We have shown [41] that vertical distribution of phytoplankton biomass has significant impact on the remote sensing signal and the biomass close to the surface has quite different reflectance than the same biomass uniformly mixed in the water column. Therefore, it is not surprising that the biomass located in a thin surface layer (as there is no light at depths of a few decimetres) is producing a strong remote sensing signal and spectral features typical to cyanobacteria (Figure 3).

There are studies [42] showing that the 810 nm peak is suitable for mapping water depth in very shallow (less than 1 m deep) waters. This is reasonable as benthic vegetation has high reflectance in NIR part of spectrum [43,44] and there is a decrease in water absorbance at 810 nm making the bottom signal detectable in this spectral region. All reflectance measurements of this study were carried out in optically deep waters with no bottom contribution. Therefore, we are sure that the height of the 810 nm peaks is only due to water constituents and there is no contribution from benthic vegetation.

Our results show that the 810 nm peak height is relatively sensitive to glint. The glint-free reflectance spectra (measured with the radiance sensor just below the water surface) produced better results that the "normal" reflectance spectra measured above the water surface. The glint removal method developed by us [23] performed well in the case of most measurements except for the most extreme Nohipalu Mustjärv. Most probably, the cause of the failure of the glint removal method was cyanobacterial biomass floating on the water surface producing high values of reflectance in the NIR part of spectrum. The glint removal procedure is not applicable when the NIR signal is higher than the UV signal.

Our results show that both the 710 nm and 810 nm peaks are very useful for retrieving chlorophyll-a and total suspended matter concentrations not only in the CDOM-rich lakes, where there is no measurable signal in the visible part of spectrum, but also in a much wider variety of lakes. The 710 nm peak continues to be the most useful spectral feature for retrieving phytoplankton biomass in productive waters. However, our study shows that the 810 nm peak is more useful in the extreme lakes where the CDOM absorption is still strong at 710 nm.

These two peaks can be used in the interpretation of remote sensing data in the cases where hyperspectral instruments are used (airborne and hand held devices). The only spaceborne instrument sufficient for lake studies spectral (10 nm) and spatial resolution (30 m) was Hyperion on board the EO-1. It was an experimental sensor that did not provide global coverage. The launch of Landsat 8 and Sentinel-2 opened great potential for lake remote sensing from a spatial and radiometric resolution point of view. The spectral resolution of Landsat is not very good from a water quality monitoring perspective. For example, it does not have spectral bands near the 700–720 nm peak, which is most often used to estimate chlorophyll-a concentration in coastal and inland waters [32,34,36,37,45]. Landsat series satellites have been used for mapping lake chlorophyll content [45] for several decades. However, it has been done mainly in eutrophic lakes where biomass is high and the total suspended matter (causing the changes in lake reflectance) is mainly phytoplankton. It is clearly seen comparing the Figures 3 and 7 that the band configuration of Landsat 8 is not optimal for lake water quality monitoring.

Sentinel-2 spatial resolution is finer than that of Landsat 8, but more important for lake studies is its spectral resolution and band configuration. The narrow 705 nm band opens great opportunities in lake chlorophyll-a remote sensing studies as we have demonstrated for black lakes in this study and for a wider variety of lakes in our Sentinel-2 lake remote sensing study [28]. We showed that, although the band 7 of Sentinel-2 OLI sensor is not positioned optimally to capture the 810 nm peak, the 783 nm band is still useful for this purpose. The suitability of this band in lake remote sensing has to be tested in the future.

5. Conclusions

We have shown with field reflectance data that, in black lakes, the water leaving signal may be very close to zero in most of the visible part of the spectrum. The measured visible part of spectrum remote sensing reflectance consists mainly of glint in such lakes.

We showed that the height of the 810 nm peak in reflectance spectra is in correlation with the parameters describing phytoplankton biomass (Chlorophyll-a, TSS, SPOM) in a wide variety of lakes. This is especially useful in black lakes where the 700–720 nm peak, normally used in retrieval of chlorophyll-a, is still affected by CDOM absorption.

Previous studies have shown that the NIR peak is caused by large amount of mineral particles in water. Our results show that the 810 nm peak is caused by combined effect of decreased water absorption between 760 nm and 860 nm and scattering by particles in the water column. If the particles in the water are primarily phytoplankton (like in the lakes studied by us), then the height of the 810 nm peak is in good correlation with chlorophyll-a and other parameters describing phytoplankton biomass (SPOM and TSS). If the scattering material in water is mainly of mineral origin (like in previous coastal and river studies) then the 810 peak is in correlation with SPIM and TSS concentration.

Landsat 8 bands are not suitable for detecting the two peaks occurring in reflectance spectra of many lakes. On the other hand, Sentinel-2 band 5 (705 nm) is almost perfectly located for mapping phytoplankton biomass (chlorophyll-a) and the band 7 (783 nm) also allows detection of the 810 nm peak in water reflectance spectra. This is especially useful in the case of black lakes as CDOM absorption may still affect the peak at 705 nm.

Acknowledgments: Fieldwork and data analysis costs were covered by Estonian Science Foundation Grants 8576 and 8654, KESTA program project VeeObs, Estonian Basic Science Research Grant SF0180009s11, and FP7 project eratH2Observe. Lake Mälaren data collection and analysis costs were covered by the FORMAS project "The Colour of water—interplay with climate, and effects on drinking water supply".

Author Contributions: Tiit Kutser wrote the manuscript with input from all co-authors and participated in all field campaigns. Birgot Paavel, Martin Ligi, and Tuuli Kauer participated in the Estonian fieldworks, processed and analyzed the field data, and carried out laboratory analysis. Charles Verpoorter participated in the Swedish field campaign, carried out all the laboratory analysis, and processed and analyzed Lake Mälaren data. Kaire Toming was responsible for processing and analyzing Sentinel-2 imagery while Gema Casal was responsible for processing and analyzing Landsat 8 imagery.

Conflicts of Interest: The authors declare no conflict of interest.

References

1. Dudgeon, D.; Arthington, A.H.; Gessner, M.O.; Kawabata, Z.I.; Knowler, D.J.; Lévêque, C.; Naiman, R.J.; Prieur-Richard, A.H.; Soto, D.; Stiassny, M.L.; *et al.* Freshwater biodiversity: Importance, threats, status and conservation challenges. *Biol. Rev.* **2006**, *81*, 163–182. [CrossRef] [PubMed]

2. Tranvik, L.J.; Downing, J.A.; Cotner, J.B.; Loiselle, S.; Striegl, R.G.; Ballatore, T.J.; Dillon, P.; Finlay, K.; Fortino, K.; Knoll, L.B.; *et al.* Lakes and impoundments as regulators of carbon cycling and climate. *Limnol. Oceanogr.* **2009**, *54*, 2298–2314. [CrossRef]

3. Bastviken, D.; Tranvik, L.J.; Downing, J.A.; Crill, P.M.; M, P.; Enrich-Prast, A. Freshwater methane emissions offset the continental carbon sink. *Science* **2011**, *331*, 50. [CrossRef] [PubMed]

4. Palmer, S.J.; Kutser, T.; Hunter, P.D. Remote sensing of inland waters: Challenges, progress and future directions. *Remote Sens. Environ.* **2015**, *157*, 1–8. [CrossRef]

5. Pope, R.M.; Fry, E.S. Absorption spectrum (380–700 nm) of pure water. II. Integrating cavity measurements. *Appl. Opt.* **1997**, *36*, 8710–8723. [CrossRef] [PubMed]

6. Segelstein, D. The Complex Refractive Index of Water. Master's Thesis, University of Missouri-Kansas City, Kansas City, MO, USA, 1981.

7. Brezonik, P.L.; Olmanson, L.G.; Finlay, J.C.; Bauer, M.E. Factors affecting the measurement of CDOM by remote sensing of optically complex inland waters. *Remote Sens. Environ.* **2015**, *157*, 199–215. [CrossRef]

8. Doxaran, D.; Froidefond, J.M.; Castaing, P. Remote-sensing reflectance of turbid sediment-dominated waters. Reduction of sediment type variations and changing illumination conditions effects by use of reflectance ratios. *Appl. Opt.* **2003**, *42*, 2623–2634. [CrossRef] [PubMed]

9. Doxaran, D.; Cherukuru, R.C.N.; Lavender, S.J. Use of reflectance band ratios to estimate suspended and dissolved matter concentrations in estuarine waters. *Int. J. Remote Sens.* **2004**, *26*, 1753–1769. [CrossRef]

10. Quibell, G. Estimating chlorophyll concentrations using upwelling radiance from different freshwater algal genera. *Int. J. Remote Sens.* **1992**, *14*, 2611–2621. [CrossRef]

11. Kutser, T. Quantitative detection of chlorophyll in cyanobacterial blooms by satellite remote sensing. *Limnol. Oceanogr.* **2004**, *49*, 2179–2189. [CrossRef]

12. Verpoorter, C.; Kutser, T.; Seekell, D.; Tranvik, L.J. A global inventory of lakes based on high-resolution satellite imagery. *Geophys. Res. Lett.* **2014**, *41*, 6396–6402. [CrossRef]

13. Kutser, T.; Pierson, D.C.; Kallio, K.; Reinart, A.; Sobek, S. Mapping lake CDOM by satellite remote sensing. *Remote Sens. Environ.* **2005**, *94*, 535–540. [CrossRef]

14. Kutser, T.; Tranvik, L.J.; Pierson, D.C. Variations in colored dissolved organic matter between boreal lakes studied by satellite remote sensing. *J. Appl. Remote Sens.* **2009**, *3*, 033538.

15. Weyhenmeyer, G.A.; Prairie, Y.T.; Tranvik, L.J. Browning of boreal freshwaters coupled to carbon-iron interactions along the aquatic continuum. *PLoS ONE* **2014**, *9*, e88104. [CrossRef] [PubMed]

16. Sobek, S.; Tranvik, L.J.; Prairie, Y.T.; Kortelainen, P.; Cole, J.J. Patterns and regulation of dissolved organic carbon: An analysis of 7500 widely distributed lakes. *Limnol. Oceanogr.* **2007**, *52*, 1208–1219. [CrossRef]

17. Köhler, S.J.; Kothawala, D.; Futter, M.N.; Liungman, O.; Tranvik, L. In-lake processes offset increased terrestrial inputs of dissolved organic carbon and color to lakes. *PLoS ONE* **2013**, *8*, e70598. [CrossRef] [PubMed]

18. Kutser, T.; Alikas, K.; Kothawala, D.N.; Köhler, S.J. Impact of iron associated to organic matter on remote sensing estimates of lake carbon content. *Remote Sens. Environ.* **2015**, *156*, 109–116. [CrossRef]

19. Eikebrokk, B.; Vogt, R.D.; Liltved, H. NOM increase in Northern European source waters: Discussion of possible causes and impacts on coagulation/contact filtration processes Water Science and Technology. *Water Supply* **2004**, *4*, 47–54.

20. Hommersom, A.; Kratzer, S.; Laanen, M.; Ansko, I.; Ligi, M.; Bresciani, M.; Giardino, C.; Beltrán-Abaunza, J.M.; Moore, G.; Wernand, M.; *et al.* Intercomparison in the field between the new WISP-3 and other radiometers (TriOS Ramses, ASD FieldSpec, and TACCS). *J. Appl. Remote Sens.* **2012**, *6*, 063615. [CrossRef]

21. Kutser, T.; Verpoorter, C.; Paavel, B.; Tranvik, L.J. Estimating lake carbon fractions from remote sensing data. *Remote Sens. Environ.* **2015**, *157*, 136–146. [CrossRef]

22. Kutser, T. The possibility of using the Landsat image archive for monitoring long time trends in coloured dissolved organic matter concentration in lake waters. *Remote Sens. Environ.* **2012**, *123*, 334–338. [CrossRef]

23. Kutser, T.; Vahtmäe, E.; Paavel, B.; Kauer, T. Removing glint effects from field radiometry data measured in optically complex coastal and inland waters. *Remote Sens. Environ.* **2013**, *133*, 85–89. [CrossRef]

24. Lorenzen, C.J. Determination of chlorophyll and phaeopigments; spectrophotometric equations. *Limnol. Oceanogr.* **1967**, *12*, 343–346. [CrossRef]

25. *ESS Method 340.2: Total Suspended Solids, Mass Balance, Volatile Suspended Solids*; Environmental Sciences Section, Inorganic Chemistry Unit, Wisconsin State Lab of Hygiene: Madison, WI, USA, 1993.

26. Davies-Colley, R.J.; Vant, W.N. Absorption of light by yellow substance in freshwater lakes. *Limnol. Oceanogr.* **1987**, *32*, 416–425. [CrossRef]

27. Kutser, T.G.; Pascual, C.; Barbosa, C.; Paavel, B.; Ferreira, R.; Carvalho, L.; Toming, K. Mapping inland water carbon content with Landsat 8 data. *Int. J. Remote Sens.* **2016**, in press.

28. Toming, K.; Kutser, T.; Laas, A.; Sepp, M.; Paavel, B.; Nõges, T. Mapping lake water quality parameters with Sentinel-2 MSI imagery. *Remote Sens.* **2016**. submitted.

29. Kallio, K.; Kutser, T.; Hannonen, T.; Koponen, S.; Pulliainen, J.; Vepsäläinen, J.; Pyhälahti, T. Retrieval of water quality from airborne imaging spectrometry of various lake types in different seasons. *Sci. Total Environ.* **2001**, *268*, 59–77. [CrossRef]

30. Arst, H.; Kutser, T. Data processing and interpretation of sea radiance factor measurements. *Polar Res.* **1994**, *13*, 3–12. [CrossRef]

31. Lee, Z.P.; Carder, K.L.; Mobley, C.D.; Steward, R.G.; Patch, J.S. Hyperspectral remote sensing for shallow waters. 2. Deriving bottom depths and water properties by optimization. *Appl. Opt.* **1999**, *38*, 3831–3843. [CrossRef] [PubMed]

32. Kutser, T. Passive optical remote sensing of cyanobacteria and other intense phytoplankton blooms in coastal and inland waters. *Int. J. Remote Sens.* **2009**, *30*, 4401–4425. [CrossRef]

33. Kutser, T.; Herlevi, A.; Kallio, K.; Arst, H. A hyperspectral model for interpretation of passive optical remote sensing data from turbid lakes. *Sci. Total Environ.* **2001**, *268*, 47–58. [CrossRef]

34. Gitelson, A.A. The peak near 700 nm on radiance spectra of algae and water: Relationships of its magnitude and position with chlorophyll concentration. *Int. J. Remote Sens.* **1992**, *13*, 3367–3373. [CrossRef]

35. Gower, J.F.R.; King, S.; Borstad, G.A.; Brown, L. Detection of intense plankton blooms using the 709 nm band of the MERIS imaging spectrometer. *Int. J. Remote Sens.* **2005**, *26*, 2005–2012. [CrossRef]

36. Gitelson, A.A.; Schalles, J.F.; Hladik, C.M. Remote chlorophyll-a retrieval in turbid, productive estuaries: Chesapeak Bay case study. *Remote Sens. Environ.* **2007**, *109*, 464–472. [CrossRef]

37. Matthews, M.W.; Bernard, S.; Robertson, L. An algorithm for detecting trophic status (chlorophyll-a), cyanobacterial-dominance, surface scums and floating vegetation in inland and coastal waters. *Remote Sens. Environ.* **2012**, *124*, 637–652. [CrossRef]

38. Duan, H.; Ma, R.; Loiselle, S.A.; Shen, Q.; Yin, H.; Zhang, Y. Optical characterization of black water blooms in eutrophic waters. *Sci. Total Environ.* **2014**, *482–483*, 174–183. [CrossRef] [PubMed]

39. Duan, H.; Loiselle, S.A.; Li, Z.; Shen, Q.; Du, Y.; Ma, R. A new insight into black blooms: Synergies between optical and chemical factors. *Estuar. Coast. Shelf Sci.* **2016**, *175*, 118–125. [CrossRef]

40. Knaeps, E.; Dogliotti, A.I.; Raymaekers, D.; Ruddick, K.; Sterckx, S. *In situ* evidence of non-zero reflectance in the OLCI 1020 nm band for a turbid estuary. *Remote Sens. Environ.* **2012**, *120*, 133–144. [CrossRef]

41. Kutser, T.; Metsamaa, L.; Dekker, A.D. Influence of the vertical distribution of cyanobacteria in the water column on the remote sensing signal. *Estuarine Coast. Shelf Sci.* **2008**, *78*, 649–654. [CrossRef]

42. Bachmann, C.M.; Montes, M.J.; Fusina, R.A.; Parrish, C.; Sellars, J.; Weidemann, A.; Goode, W.; Nichols, C.R.; Woodward, P.; McIlhany, K.; *et al.* Bathymetry retrieval from hyperspectral imagery in the very shallow water limit: A case study from the Virginia Coast Reserve (VCR'07) multi-sensor campaign. *Mar. Geodesy* **2010**, *33*, 53–75. [CrossRef]

43. Vahtmäe, E.; Kutser, T.; Martin, G.; Kotta, J. Feasibility of hyperspectral remote sensing for mapping benthic macroalgal cover in turbid coastal waters. *Remote Sens. Environ.* **2006**, *101*, 342–351. [CrossRef]

44. Kotta, J.; Remm, K.; Vahtmäe, E.; Kutser, T.; Orav-Kotta, H. In-air spectral signatures of the Baltic Sea macrophytes and their statistical separability. *J. Appl. Remote Sens.* **2014**, *8*, 083634. [CrossRef]

45. Dekker, A.G.; Peters, S.W.M. The use of the Thematic Mapper for the analysis of eutrophic lakes: A case study in The Netherlands. *Int. J. Remote Sens.* **1993**, *14*, 799–821. [CrossRef]

remote sensing

MDPI

Article

Assessment of Chlorophyll-a Remote Sensing Algorithms in a Productive Tropical Estuarine-Lagoon System

Regina Camara Lins [1], Jean-Michel Martinez [2], David da Motta Marques [3], José Almir Cirilo [1] and Carlos Ruberto Fragoso Jr. [4,*]

[1]   Department of Civil Engineering, Federal University of Pernambuco, 50670-901 Recife, Brazil; reginacamaralins@hotmail.com (R.C.L.); almir.cirilo@gmail.com (J.A.C.)

[2]   Géosciences Environnement Toulouse (GET), Unité Mixte de Recherche 5563, IRD/CNRS/Université Toulouse III, 31400 Toulouse, France; martinez@ird.fr

[3]   Hydraulic Research Institute, Federal University of Rio Grande do Sul, CP 15029 Porto Alegre, Brazil; dmm@iph.ufrgs.br

[4]   Center for Technology, Federal University of Alagoas, 57072-970 Maceió, Brazil

*   Correspondence: ruberto@ctec.ufal.br; Tel.: +55-82-996-328-814

Academic Editors: Yunlin Zhang, Claudia Giardino, Linhai Li, Deepak R. Mishra and Prasad S. Thenkabail
Received: 27 February 2017; Accepted: 19 May 2017; Published: 24 May 2017

Abstract: Remote estimation of chlorophyll-a in turbid and productive estuaries is difficult due to the optical complexity of Case 2 waters. Although recent advances have been obtained with the use of empirical approaches for estimating chlorophyll-a in these environments, the understanding of the relationship between spectral reflectance and chlorophyll-a is based mainly on temperate and subtropical estuarine systems. The potential to apply standard NIR-Red models to productive tropical estuaries remains underexplored. Therefore, the purpose of this study is to evaluate the performance of several approaches based on multispectral data to estimate chlorophyll-a in a productive tropical estuarine-lagoon system, using in situ measurements of remote sensing reflectance, R_{rs}. The possibility of applying algorithms using simulated satellite bands of modern and recent launched sensors was also evaluated. More accurate retrievals of chlorophyll-a ($r^2 > 0.80$) based on field datasets were found using NIR-Red three-band models. In addition, enhanced chlorophyll-a retrievals were found using the two-band algorithm based on bands of recently launched satellites such as Sentinel-2/MSI and Sentinel-3/OLCI, indicating a promising application of these sensors to remotely estimate chlorophyll-a for coming decades in turbid inland waters. Our findings suggest that empirical models based on optical properties involving water constituents have strong potential to estimate chlorophyll-a using multispectral data from satellite, airborne or handheld sensors in productive tropical estuaries.

Keywords: shallow productive estuary; chlorophyll-a; remote sensing; Sentinel

1. Introduction

Estuarine systems, which are transitional complex zones between rivers and oceans, exhibit spatial variability, seasonal cycles and distributions of organisms strongly influenced by various environmental factors, most notably: (a) a spatio-temporal salinity gradient; (b) high variability of temperature and light intensity; (c) inflow and nutrient discharge; and (d) hydrodynamic patterns [1]. These specific features make estuaries among the most productive habitats in the world, as they provide natural resources to maintain the use of local fisheries and aquaculture, allowing significant socio-economic development [2].

Although numerous physical and biological characteristics of these systems are well-known, their susceptibility to anthropic aggression (e.g., effluent discharge, fishing and aquiculture), which affects the trophic structure, has also been recognized [3]. For instance, the continuous increase in both nutrient loading and organic matter has led to estuaries with higher eutrophication levels. This degradation process usually results in an increase in water turbidity due to blooms of cyanobacteria or green algae, which subsequently affect the entire trophic structure. Thus, methodological approaches to quantify eutrophication are essential to improve the understanding of the ecosystem dynamics of estuaries and to develop tools for accurate decision making.

Long-term monitoring of chlorophyll-a (a phytoplankton biomass indicator) is frequently carried out to assess the eutrophic state of an estuary [4–6]. However, this monitoring requires representative (in time and space) field sampling and laboratory measurements to adequately cover the distribution of phytoplankton. Unfortunately, it is not always possible to have technical and financial resources available to carry out such monitoring, mainly in estuaries that exhibit high spatio-temporal heterogeneity [7].

Data from satellite sensors may provide better information on chlorophyll-a variability in comparison to conventional field monitoring because most modern sensors (e.g., Moderate Resolution Imaging Spectroradiometer (MODIS), Medium Resolution Imaging Spectrometer (MERIS) and Sea-viewing Wide Field-of-view Sensor (SeaWiFS)) have improved capabilities with respect to spectral, radiometric, temporal and spatial resolutions [8–10]. Recent advances have resulted in significant progress in the remote assessment of chlorophyll-a in turbid and productive waters [11]. However, some problems, such as atmospheric correction and complexity of optical properties involving water constituents, make it difficult to use a simple and universal empirical algorithm to estimate chlorophyll-a from satellite data, being an additional challenge to inland water remote sensing [12,13]. Therefore, for a geographic and/or seasonal region, it is essential to conduct a thorough investigation of the local relationships between in situ-measured chlorophyll-a and spectral bands of airborne or handheld sensors before using satellite datasets [14].

In general, these empirical algorithms are often based on a relationship between chlorophyll-a and reflectance, $R_{rs}(\lambda)$, which is derived from the bio-optical theory of inherent optical properties (IOPs), such as total absorption (a) and backscattering (b_b) coefficients [15,16]. The underlying principle is that changes in the concentrations and distribution of organic and inorganic particulates and dissolved substances in the water affect the observed reflectance, $R_{rs}(\lambda)$, in different wavelengths, λ, according to Gons [17] and Preisendorfer [18]:

$$R_{rs}(\lambda) = \frac{f(\lambda)}{Q(\lambda)} \frac{b_b(\lambda)}{a(\lambda) + b_b(\lambda)} \tag{1}$$

$$a = a_{Chl-a} + a_{NAP} + a_{CDOM} + a_{water} \tag{2}$$

$$b_b = b_{b,water} + b_{b,particles} \tag{3}$$

where $f(\lambda)$ describes the sensitivity of the reflectance to variations in the solar zenith angle [19], and $Q(\lambda)$ expresses the bidirectional properties of reflectance [20]; a_{Chla}, a_{NAP}, a_{CDOM} and a_{water} are the chlorophyll-a, the non-algal particles (NAP), the colored dissolved organic matter (CDOM) and the pure water absorption coefficients, respectively; and $b_{b,water}$ and $b_{b,particles}$ are backscattering due to water and organic/inorganic particles in suspension, respectively. As values for both a_{water} and $b_{b,water}$ can be assigned as constants [21–23], it is still necessary to identify the contribution of some important optically active compounds (i.e., a_{NAP}, a_{CDOM} and $b_{b,particles}$) in order to estimate chlorophyll-a concentrations from multispectral data.

With respect to oligotrophic and clear waters, Case 1 [24], reflectance in the blue region is dominated by the spectral response of chlorophyll-a, which has an absorption maximum around 440 nm. In these environments, models based on the blue/green ratio have shown the best performance [25–27] since CDOM and NAP show a strong correlation with chlorophyll-a because these constituents are derived from processes related to phytoplankton (i.e., mortality and exudation). However, in turbid and productive waters, i.e., Case 2 waters (IOCCG 2000), blue/green ratio models do not adequately represent the chlorophyll-a variability since CDOM and NAP may originate from additional sources, such as runoff of sediments, nutrients and organic matter and the resuspension of sediments from shallow bottoms [28,29].

Thus, major efforts have been made in the last decade to test and evaluate different algorithms to estimate chlorophyll-a in inland and coastal waters using multispectral data from datasets with different sources, such as handheld, airborne or spaceborne sensors [30–43]. In turbid and productive estuaries, the best empirical approaches for estimating chlorophyll-a have been obtained with the use of NIR-Red models [44,45]. However, the understanding of these relationships has mainly focused on temperate and subtropical estuarine systems using modern satellite sensors, such as MODIS, MERIS and SeaWiFS [14,30,44–48]. Therefore, there is a notable lack of knowledge on bio-optical variability in tropical estuarine systems, which could be a novel branch for scientific investigations to develop new spectral reflectance-based models using in situ measured (i.e., spectral reflectance and chlorophyll-a) or satellite-derived data, especially for recently launched satellite, such as Sentinel-2 and Sentinel-3, which opened a new potential to estimate chl-a in optically complex waters where fine spectral, spatial and temporal resolutions are required [49]. Moreover, this work is one of the first assessing the variability of the water color of a tropical lagoon contributing to expand the knowledge of inland water optical properties.

The purpose of this study is to evaluate the performance of several approaches based on spectral bands to estimate chlorophyll-a in a productive tropical estuarine-lagoon system, using in situ observations. The possibility of applying the algorithm to remote sensing satellite images of modern and recent spaceborne sensors is also discussed in this study. This knowledge is a first step towards obtaining comprehensive and reliable reflectance-based models for tropical turbid and productive waters, which may be further tested with satellite data in order to remotely estimate chlorophyll-a.

2. Materials and Methods

2.1. Study Area

The Mundaú-Manguaba Estuarine-Lagoon System (MMELS) is a shallow (max. depth of 3.5 m) tropical lagoon system located in the state of Alagoas, northeastern Brazil, between 9°35'00"S and 9°46'00"S latitude and 35°34'00"W and 35°58'00"W longitude (Figure 1). The system is composed of three compartments: (a) the Mundaú Lagoon (27 km^2) in the eastern MMELS, which receives freshwater mainly from the Mundaú river basin (annual average discharge of 35 m^3/s); (b) the Manguaba Lagoon (42 km^2) in the western MMELS, which receives an average annual fresh water discharge of 28 m^3/s from the Paraíba do Meio and Sumaúma river basins; and (c) the mangrove-lined narrow channel system (12 km^2), which connects both lagoons via a single 250 m wide tidal inlet to the Atlantic Ocean. MMELS exhibits a tropical semi-humid climate with well-defined dry (from October to December) and wet (from May to July) seasons. The average annual mean temperature is 25 °C, and the winds blow predominantly from a southeasterly direction, governed mainly by trade winds in summer [50]. Manguaba lagoon is generally less saline than Mundaú lagoon. However, both lagoons are characterized by eutrophic conditions and are dominated by phytoplankton. Additional physical characteristics of the Mundaú and Manguaba lagoons can be observed in Table 1.

Table 1. Main physical features of the Mundaú and Manguaba lagoons.

Features	Mundaú	Manguaba
Volume (106 m^3)	43	97.7
Average depth (m)	1.5	2.2
Tidal range (m)	0.2	0.03
Tidal prism (106 m^3)	17.3	6.1
Average freshwater discharge (m^3/s)	35	28
Retention time (days)	16	36

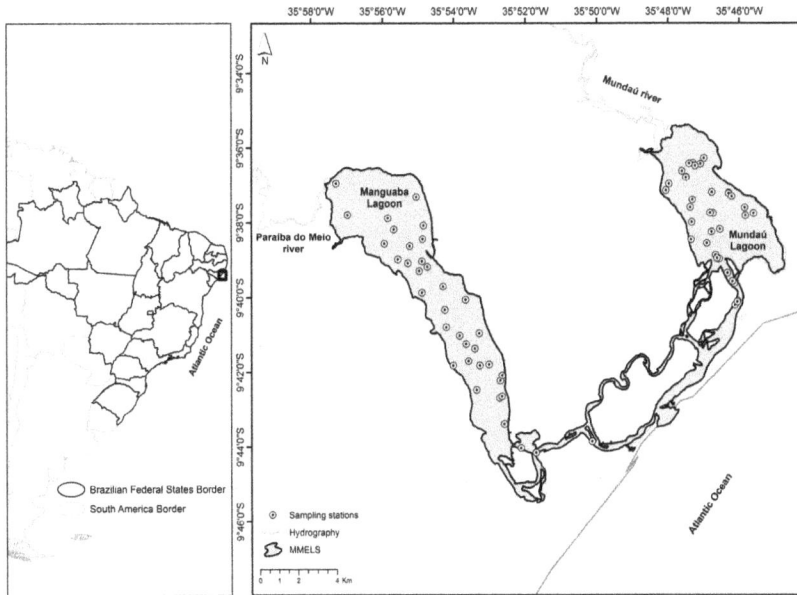

Figure 1. Mundaú-Manguaba Estuarine-Lagoon System (MMELS) study site and spatial distribution of sampling stations, which were used to collect water samples and reflectance measurements.

2.2. Field Measurements

Shipboard data were collected during six major field campaigns conducted between May and September 2015. During each field campaign, a set of 12 sampling stations, well-distributed across MMELS, was established (Figure 1). At each sampling station, above-water optical measurements and water sample collection for laboratory analysis were carried out simultaneously. Surface water samples (each sample = 2.0 L volume) were collected at a depth of 0.2 m below the water surface. The samples were stored in a cooler with ice under dark conditions and were transported to the laboratory to determine the chlorophyll-a (Chl-a) and total suspended solids (TSS) concentrations. Above-water hyperspectral reflectance measurements were performed using TriOs RAMSES radiometers, with a spectral resolution of approximately 3.3 nm, following the ocean optics protocols recommended by NASA (see [51]). All of the radiometers were mounted in an aluminum pole vertically positioned on the top of the boat. An irradiance sensor (operating in the range 350–721 nm) was used to measure downwelling irradiance above the water surface, $E_d(\lambda)$, and two radiance sensors (operating in the range 350–950 nm with a 7° field of view) were used to measure upwelling radiance above the surface water, $L_u(\lambda)$, as well as the sky radiance that was used to correct for the skylight reflection effect at the air-water interface, $L_s(\lambda)$.

The remote sensing reflectance, $R_{rs}(\lambda)$, was calculated as follows:

$$R_{rs}(\lambda) = \frac{L_u(\lambda) - \rho.L_s(\lambda)}{E_d(\lambda)} \tag{4}$$

The above-water upwelling radiance, L_u, is the sum of the upwelling radiance, $L_w(0_+)$, and the sky radiance directly reflected by the air-water interface, L_r. Because only L_u is directly measureable, and $L_w(0_+)$ and L_r are not measured, L_r is assessed as $L_r = \rho \cdot L_s$, where ρ is a proportionality factor. The factor ρ is not an inherent optical property of the surface and is dependent on the sky conditions, wind speed, solar zenith angle, and viewing geometry. Mobley [52] used a radiative transfer code to estimate the variability of ρ as a function of the different forcing factors. These results showed that when L_u was acquired with a viewing direction of 40° from the nadir and 135° from the sun, the variability of ρ was considerably reduced under clear-sky conditions, and a value of 0.028 was considered acceptable at wind speeds less than 5 m·s^{-1}. Finally, $L_w(0_+)$ was calculated by the subtraction of L_u and $\rho \cdot L_s$. To limit the effects of external factors, all radiometric measurements were acquired within the viewing geometry defined by Mobley [52], under low-wind (0–4 m·s^{-1}) and clear-sky conditions, and for sun zenith angles ranging from 0 to 30°.

2.3. Water Sample Analysis

All samples were filtered on land, no later than 12 h after sampling, using Whatman GF/F glass fiber filters (pore size of 0.45 µm). The filters were wrapped in aluminum foil and kept frozen until analysis. Chlorophyll-a from algae concentrated on filters was extracted into 90% ethanol for 18 h in an amber flask and measured using a spectrophotometric trichromatic method [53]. Water samples also were analyzed for total suspended solids (TSS), which were filtered using Whatman GF/F filters; the residue retained was dried to a constant weight at 103 to 105 °C and measured gravimetrically [53].

2.4. Reflectance Spectra Classification

All 72 reflectance spectra were classified into four homogeneous groups, which were sufficient to identify possible spatial or temporal dependencies in the remote sensing reflectance data. The optical classes were divided using a k-means clustering analysis, an unsupervised classification technique that categorizes the data set based on the natural distribution of the data in multivariate space. Each $R_{rs}(\lambda)$ spectrum was previously normalized by its integral, calculated over the entire spectrum [54], in order to enhance the spectral shape of the $R_{rs}(\lambda)$ spectra in the classification. This classification may support the use of multiple retrieval algorithms for each lagoon, homogeneous regions or a global retrieval algorithm for a whole system [55].

2.5. Models to Estimate Chlorophyll-a from $R_{rs}(\lambda)$

Among several models used to estimate chlorophyll-a in inland waters from spectral reflectance data (see a review conducted by Matthews [56]), we evaluated the performance of four algorithms using the available dataset. The following models were chosen due to their wide and successful use in previous studies and because these algorithms are based on physical fundamentals.

2.5.1. The Blue-Green Ratio Model

The first tested model to estimate chlorophyll-a uses a simple ratio between reflectance in the blue region at 440 nm, $R(\lambda_{Blue})$, in which Chl-a and carotenoids strongly absorb light, and reflectance in the green region at 550 nm, $R(\lambda_{Green})$, where reflectance is minimally absorbed by pigments. This model was initially proposed by Morel and Prieur [24], and it is still widely used for ocean color [25,27].

$$Chla \propto R(\lambda_{Blue}) / R(\lambda_{Green}) \tag{5}$$

2.5.2. The Two-Band NIR-Red Ratio Model

The second tested model follows the same principle of the blue-green ratio model, but it considers a ratio between reflectance in the near-infrared, $R(\lambda_{NIR})$, and reflectance in the red region, $R(\lambda_{Red})$, according to:

$$Chla \propto R(\lambda_{NIR})/R(\lambda_{Red}) \tag{6}$$

where λ_{Red} is a wavelength usually located around the point of maximum chlorophyll-a absorption, which is restricted to the range 660 nm < λ_{Red} < 690 nm [33]. One may assume that the absorption by non-algal particles, yellow substances and backscattering can be considered non-significant in comparison to the chlorophyll-a concentration in this wavelength, or $a_{chl-a}(\lambda_1) \gg a_{NAP}(\lambda_1) + a_{CDOM}(\lambda_1)$ and $a_{chl-a}(\lambda_1) \gg b_b(\lambda_1)$. In general, the near-infrared wavelength, λ_{NIR}, may be found at two different positions in the NIR: (a) between 700 nm and 720 nm, known as λ_2 [39,44,57], where absorption of the water constituents is minimal; or (b) beyond 710 nm, known as λ_3 [31,45,58], where absorption is mostly dominated by water (i.e., $a_{chl-a} + a_{NAP} + a_{CDOM} \sim 0$). In general, $b_b(\lambda)$ may be assumed to be approximately equal at both wavelengths.

2.5.3. The Three-Band NIR-Red Model

This model was initially developed to estimate pigments in terrestrial vegetation, but it has recently been used to determine chlorophyll-a in turbid and productive waters [14,31,33,35]. The algorithm has the form

$$Chla \propto \left[R(\lambda_1)^{-1} - R(\lambda_2)^{-1} \right] \times R(\lambda_3) \tag{7}$$

where $R(\lambda_i)$ is the measured reflectance in the spectral band λ_i. Similar to the two-band NIR-Red ratio model, λ_1 is the wavelength at which maximum chlorophyll-a absorption occurs.

This approach is based on the following three assumptions: (a) effect of CDOM and detrital absorption on $R(\lambda_1)$ is significant in the two-band ratio model, which can be minimized by subtraction of $R(\lambda_2)$; (b) the absorption of these constituents must be approximately the same at both wavelengths λ_1 and λ_2 (i.e., $a_{NAP}(\lambda_2) + a_{CDOM}(\lambda_2) \approx a_{NAP}(\lambda_1) + a_{CDOM}(\lambda_1)$); and (c) the chlorophyll-a absorption in λ_2 must be much smaller than that in λ_1 ($a_{chl-a}(\lambda_2) \ll a_{chl-a}(\lambda_1)$). Dall'Olmo, Gitelson and Rundquist [31] suggested λ_2 values between 690 nm and 730 nm. The third band, λ_3, is used to compensate the variability in backscattering between samples, and it is usually located where absorption is dominated by water with the same recommended range of the previous model. At this position, $R(\lambda_3)$ is influenced by backscattering only, and the backscattering, $b_b(\lambda)$, is approximately equal at the three wavelengths.

2.5.4. The Four-Band NIR-Red Model

The four-band model was developed by Le [36] to improve the performance of the three-band model for highly turbid waters. A fourth band, λ_4, located at NIR wavelengths, was included to minimize the impacts of absorption and backscattering of suspended solids in λ_3 [36,38]. Therefore, CDOM and detrital absorptions at λ_4 are similar to those at λ_3 (i.e., $a_{NAP}(\lambda_4) + a_{CDOM}(\lambda_4) \approx a_{NAP}(\lambda_3) + a_{CDOM}(\lambda_3)$), and $b_b(\lambda)$ is approximately equal at all four wavelengths.

$$Chla \propto \left[R(\lambda_1)^{-1} - R(\lambda_2)^{-1} \right] / \left[R(\lambda_4)^{-1} - R(\lambda_3)^{-1} \right] \tag{8}$$

2.6. Algorithm, Model Evaluation and Validation

An algorithm was developed in MATLAB® to identify the optimal wavelength positions in the NIR-Red models based on reflectance measurements. This algorithm consists of testing all possible sets of linear best-fit functions between observed and estimated chlorophyll-a within the spectral ranges recommended in previous studies for each optimal wavelength. Thus, a linear best-fit function was chosen for each model considering the following, in sequential order: the maximum and minimum values of the coefficient of determination (r^2) and the root-mean square error (RMSE) of each chlorophyll-a model. The models were evaluated considering different subsets, such as MMELS, Mundaú lagoon, Manguaba lagoon and optical classes defined in the reflectance spectra classification.

In order to evaluate the accuracy and stability of the best models based on reflectance measurements, the MMELS dataset was divided into calibration and independent validation subsets. The models were calibrated using data collected from 2/3 of the total number of stations, corresponding to the first two field campaigns for each lagoon, and were validated using data collected from 1/3 of the stations, which corresponded to the last field campaign for each lagoon. Validation analysis was not extended to smaller subsets (e.g., Mundaú lagoon, Manguaba lagoon or optical classes) in order to reduce risks associated with obtaining non-representative samples for both calibration and validation sets.

2.7. Retrieval of Chlorophyll-a Using Models Based on Simulated Satellite Bands

Satellite bands were simulated using reflectance measurements in order to check the potential for satellite application in MMELS. For this analysis, we only selected satellite sensors capable of accurately describing the spatial-temporal variability of the optical properties in the MMELS, considering the size and retention time of the lagoons (see Table 1). Such physical features limited the use of most sensors with free satellite imagery, which have spatial and temporal resolutions larger than 500 m and one week, respectively. These criteria resulted in the selection of four satellite sensors (Table 2): (a) the still operational NASA sensors, MODIS-Terra (launched in 1999) and MODIS-Aqua (launched in 2002); (b) the non-operational MERIS sensor associated with the Envisat satellite, which was launched by ESA's Copernicus programme (2002–2013); (c) the MSI (MultiSpectral Imager) sensor related to Sentinel-2A (launched in June 2015) and Sentinel-2B (launched in March 2017); and (d) ESA's OLCI (Ocean and Land Colour Instrument) sensor on-board Sentinel-3A (launched in February 2016), which has visible and short-wave infrared radiances for ocean, inland and coastal waters in order to reach levels of accuracy and precision equivalent to those of the MERIS sensors.

Table 2. Characteristics of sensors and their associated satellites used to retrieve chlorophyll-a information in MMELS.

Sensor	Satellite	Spectral (Bands)	Temporal (Days)	Radiometric (Bit)	Spatial (m)	Central Wavelength (400–900 nm)
MODIS	Terra/Acqua	36	1	12	250 500	645, 858 * 469, 555
MERIS	Envisat	15	3	16	300	412, 443, 490, 510, 560, 620, 665, 681, 709, 754 *, 761 *, 779 *, 865 *, 885 *
MSI	Sentinel-2	13	<5	12	10 20 60	490, 560, 665, 842 * 705, 740, 783, 865 * 443
OLCI	Sentinel-3	21	<2	16	300	400, 412, 442, 490, 510, 560, 620, 665, 674, 681, 709, 754 *, 761 *, 764 *, 767 *, 779 *, 865 *, 885 *

The "Resolution" header spans the Spectral, Temporal, Radiometric, and Spatial columns.

* Bands beyond the spectral range of operation of the R_{rs} measured in-situ.

MODIS and MERIS sensors have been widely used to estimate chlorophyll-a in inland and coastal waters [35,40,46], which can allow time-series reconstitution of the last 20 years. The recently available sensors MSI and OLCI (successor of MERIS) have a good spatial and temporal resolution and new spectral bands, which were positioned at strategic locations to improve chlorophyll-a estimates.

The best NIR-Red models using reflectance measurements with $r^2 > 0.8$ were chosen to test the performance of the band models using satellite-based wavelengths. The optimal positions found in the models based on R_{rs} measured in-situ were tuned to the nearest simulated satellite bands, which were obtained using R_{rs} in-situ data weighted according to the corresponding spectral response function (SRF) of each ocean color sensor.

3. Results

3.1. Constituent Concentrations

Notably, the chlorophyll-a and TSS concentrations in the water samples exhibited high variability (Table 3). The levels of chlorophyll-a differed between the lagoons, ranging between 0.97 and 48.9 mg/m^3 (average value of 12.86 mg/m^3) in Mundaú lagoon, and between 5.99 and 117.54 mg/m^3 in Manguaba lagoon (average value of 42.77 mg/m^3). In contrast to the Chl-a concentrations, the TSS concentrations were higher in Mundaú lagoon (range of 15.2–61.0 mg/L and average of 32.8 mg/L) in comparison with Manguaba lagoon (range of 9.0–44.0 mg/L and average of 22.7 mg/L). In addition, we did not find a strong correlation between the chlorophyll-a and TSS concentrations in the MMELS ($r^2 = 0.08$; data not shown).

Table 3. Statistics of water constituents considering different subsets.

Subset	Chlorophyll-a (mg/m^3)				SST (mg/L)			
	Min	Max	Mean	Stdev	Min	Max	Mean	SD
Mundaú (N = 36)	0.97	48.90	12.86	9.72	15.15	61.00	32.80	11.99
Manguaba (N = 36)	5.99	117.54	42.77	24.22	9.00	44.00	22.86	9.34
MMELS (N = 72)	0.97	117.54	27.81	23.72	9.00	61.00	27.83	11.79

3.2. Reflectance Spectra and Classification

The measured reflectance spectra (range of 400–720 nm) of the surveyed sampling points in the Mundaú and Manguaba lagoons are shown in Figure 2a,b, respectively. The hyperspectral reflectance differed slightly between the two datasets. Nevertheless, in both lagoons, it was possible to identify spectral features similar to reflectance spectra previously observed for turbid, productive waters [32,59], such as a slight depression at 440 nm; a prominent peak around 565 nm; a trough at 625 nm (more intense in Manguaba lagoon) followed by a discrete peak around 660 nm; and a prominent trough at 670 nm followed by a prominent peak close to 705 nm. With respect to Manguaba lagoon, we observed that the reflectance magnitude peak around 705 nm was comparable to that in the green region, except for the 14 July campaign, where we did not observe a well-defined peak in the green region.

Minimum reflectance values were observed around 670 nm, where the maximum absorption of chlorophyll-a occurs in the red region. The minimum reflectance near 670 nm did not show a linear correlation with the chlorophyll-a concentration ($r^2 = 0.0006$; data not shown), in contrast to the difference between peak reflectance around 700 nm and minimum reflectance near 670 nm ($r^2 = 0.69$, Figure 2c). In addition, we found a strong relationship between the peak position in the red region and the chlorophyll-a concentration ($r^2 = 0.8$, Figure 2d) but a very poor correlation between maximum reflectance values near 700 nm and chlorophyll-a ($r^2 = 0.1$; data not shown).

Clustering analyses resulted in four classes of normalized reflectance, which were sufficient to explain the spatial or temporal dependencies in the remote sensing reflectance data over the entire system (Figure 3). The behavior of the mean and standard deviation in the spectra of the four classes can be observed in Figure 4. A comprehensive description of each group is presented below.

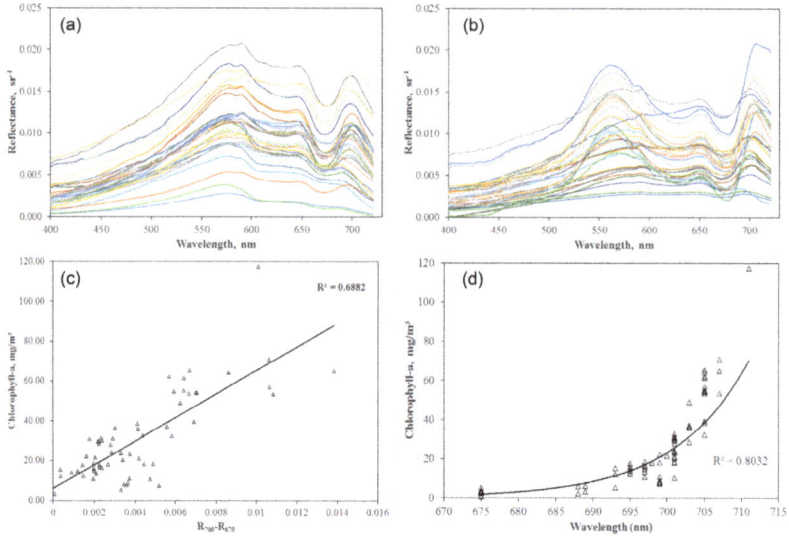

Figure 2. Typical spectral profiles in the MMELS waters measured in 2015: (**a**) Mundaú (5 May, 10 June, and 8 August); and (**b**) Manguaba (14 July, 3 September, and 22 September). Each line represents the spectral reflectance measured at a certain sampling point. The following relationships are also presented: (**c**) chlorophyll-a concentration versus the difference between the peak reflectance around 700 nm and the minimum reflectance near 670 nm; (**d**) and the chlorophyll-a concentration versus the peak position in the red region.

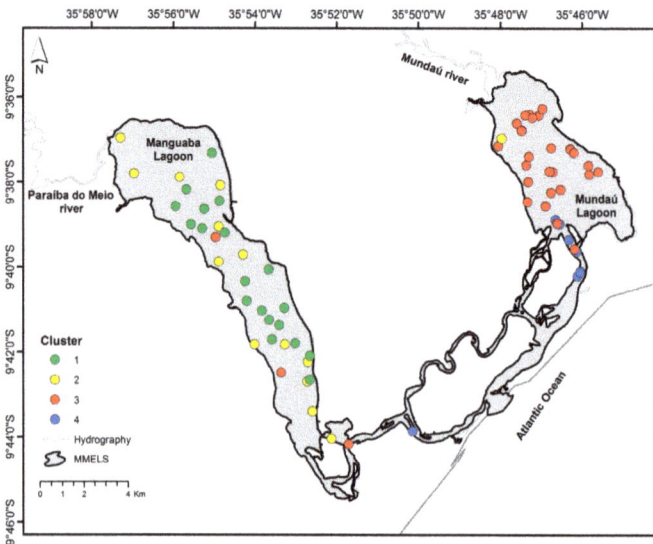

Figure 3. Reflectance spectra k-means clustering classification for normalized data in MMELS.

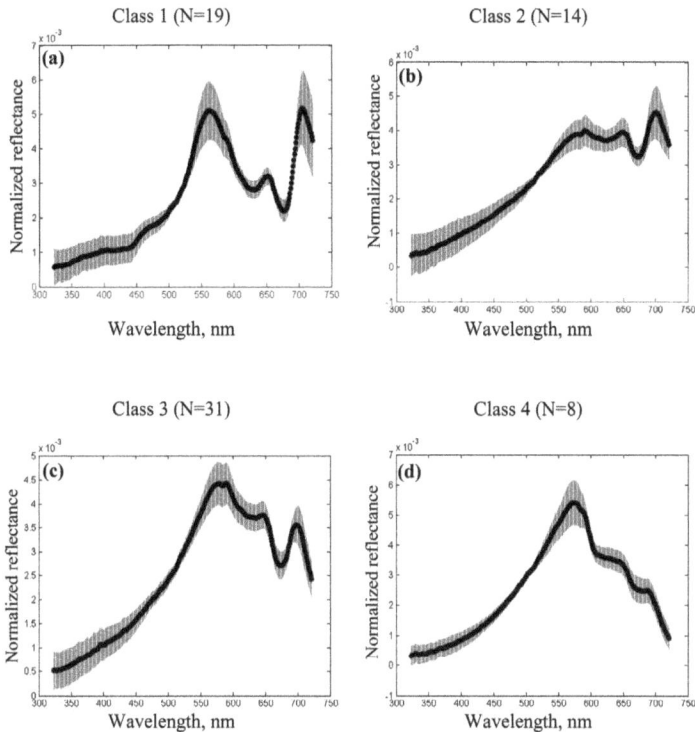

Figure 4. Mean (black dots) and standard deviation (grey ranges) for four classes of the normalized reflectance spectra: (**a**) Class 1; (**b**) Class 2; (**c**) Class 3; and (**d**) Class 4.

Class 1 (Figure 4a) consisted only of sampling stations in the Manguaba lagoon (N = 19). This class was characterized by a slight depression at around 440 nm followed by two well-marked troughs at 625 and 675 nm. Two distinct peaks with comparable magnitudes at around 560 and 710 nm were also well pronounced. This class had the highest chlorophyll-a concentrations and the lowest TSS concentrations with average values of approximately 56.40 mg/m^3 and 21.50 mg/m^3, respectively.

Class 2 (Figure 4b) contained all 12 samples collected during the 14 July campaign in Manguaba lagoon and only one sample located closest to the main river of the Mundaú lagoon (N = 14). This class was characterized by an indistinct peak in the green region, a strong trough at 675 nm and a peak at around 700 nm. This class had moderate values of chlorophyll-a and TSS, with average values of approximately 27.60 mg/m^3 and 24.00 mg/m^3, respectively.

Class 3 (Figure 4c) mainly represented the vast majority of the Mundaú lagoon waters and a few samples that were spatially distributed in the Manguaba lagoon (N = 31). This class was characterized by a clearly defined trough at 675 nm and a distinct peak around 700 nm. The chlorophyll-a concentrations were lower than those observed in Class 2, with average and maximum values of 14.87 mg/m^3 and 32.64 mg/m^3, respectively. This class also had the highest TSS concentrations with average values of approximately 34.00 mg/m^3.

Class 4 (Figure 4d) consisted of points located near the channels connecting the lagoons to the ocean (N = 8). This class showed no specific spectral features and was characterized by the lowest chlorophyll-a concentrations (average of 3.3 mg/m^3) and moderate TSS concentrations (average of 25.90 mg/m^3).

3.3. Assessment of Chl-a Retrieval Models

A summary of the accuracy assessment of the best band models for classified and non-classified waters—considering three distinct datasets (i.e., Mundaú lagoon, Manguaba lagoon and MMELS)—can be observed in Table 4.

Table 4. Slopes (p) and intercepts (q) of the linear best-fit function between observed and retrieved chlorophyll-a with corresponding coefficient of determination (r^2) and root mean square error (RMSE in mg/m^3) for the monitoring datasets. The MMELS dataset (global) was divided into calibration (cal) and independent validation (val) subsets to evaluate the accuracy and stability of the best models based on the reflectance measurements.

	Water	Models	p	q	r^2	RMSE
Non-classified	Mundaú (N = 36)	R_{440}/R_{550}	3.55	11.54	0.00	9.86
		R_{713}/R_{682}	25.71	−10.91	0.54	6.71
		$(R_{690}{}^{-1} - R_{706}{}^{-1}) \cdot R_{721}$	81.96	15.13	0.60	6.23
		$(R_{690}{}^{-1} - R_{695}{}^{-1})/(R_{709}{}^{-1} - R_{702}{}^{-1})$	10.83	7.54	0.74	5.02
	Manguaba (N = 36)	R_{440}/R_{550}	−73.45	65.29	0.26	19.05
		R_{714}/R_{690}	52.84	−21.84	0.81	9.54
		$(R_{690}{}^{-1} - R_{714}{}^{-1}) \cdot R_{721}$	57.87	31.15	**0.82**	**9.51**
		$(R_{689}{}^{-1} - R_{713}{}^{-1})/(R_{721}{}^{-1} - R_{720}{}^{-1})$	0.76	31.45	0.71	11.73
	MMELS (N = 72)	R_{440}/R_{550} (cal, N = 48)			0.06	21.20
		R_{440}/R_{550} (val, N = 24)	−39.32	38.22	0.00	153.52
		R_{440}/R_{550} (global, N = 72)	−64.89	49.74	0.15	20.43
		R_{721}/R_{660} (cal, N = 48)	39.52	−12.90	**0.84**	**8.78**
		R_{721}/R_{660} (val, N = 24)			0.66	12.25
		R_{713}/R_{690} (global, N = 72)	56.21	−29.30	0.83	9.06
		$(R_{690}{}^{-1} - R_{717}{}^{-1}) \cdot R_{721}$ (cal, N = 48)	56.70	27.53	**0.86**	**8.32**
		$(R_{690}{}^{-1} - R_{717}{}^{-1}) \cdot R_{721}$ (val, N = 24)			0.73	11.06
		$(R_{690}{}^{-1} - R_{714}{}^{-1}) \cdot R_{720}$ (global, N = 72)	64.03	26.01	0.84	8.81
		$(R_{660}{}^{-1} - R_{695}{}^{-1})/(R_{721}{}^{-1} - R_{720}{}^{-1})$ (cal, N = 48)	1.22	11.37	**0.87**	**7.91**
		$(R_{660}{}^{-1} - R_{695}{}^{-1})/(R_{721}{}^{-1} - R_{720}{}^{-1})$ (val, N = 24)			0.15	19.42
		$(R_{660}{}^{-1} - R_{713}{}^{-1})/(R_{721}{}^{-1} - R_{720}{}^{-1})$ (global, N = 72)	0.81	21.11	0.72	11.78
Classified	Class 1 (N = 19)	R_{440}/R_{550}	−27.33	63.33	0.04	18.76
		R_{721}/R_{690}	43.36	0.88	0.66	11.26
		$(R_{690}{}^{-1} - R_{721}{}^{-1}) \cdot R_{721}$	43.36	44.25	0.65	11.26
		$(R_{660}{}^{-1} - R_{713}{}^{-1})/(R_{711}{}^{-1} - R_{712}{}^{-1})$	0.14	45.35	**0.75**	**9.52**
	Class 2 (N = 14)	R_{440}/R_{550}	−30.64	40.49	0.19	8.45
		R_{711}/R_{690}	122.20	−94.94	0.80	4.16
		$(R_{690}{}^{-1} - R_{711}{}^{-1}) \cdot R_{721}$	142.90	27.34	0.81	4.09
		$(R_{663}{}^{-1} - R_{703}{}^{-1})/(R_{707}{}^{-1} - R_{706}{}^{-1})$	0.58	11.37	**0.93**	**2.48**
	Class 3 (N = 31)	R_{440}/R_{550}	−18.04	21.82	0.06	7.15
		R_{700}/R_{660}	60.49	−49.12	0.45	5.49
		$(R_{690}{}^{-1} - R_{699}{}^{-1}) \cdot R_{721}$	99.12	10.53	0.47	5.37
		$(R_{660}{}^{-1} - R_{713}{}^{-1})/(R_{721}{}^{-1} - R_{720}{}^{-1})$	8.82	34.62	**0.51**	**5.16**
	Class 4 (N = 8)	R_{440}/R_{550}	6.32	1.33	0.07	1.66
		R_{690}/R_{687}	41.83	−38.11	0.14	1.6
		$(R_{687}{}^{-1} - R_{691}{}^{-1}) \cdot R_{721}$	128.20	3.95	0.17	1.57
		$(R_{690}{}^{-1} - R_{701}{}^{-1})/(R_{698}{}^{-1} - R_{694}{}^{-1})$	28.87	78.65	**0.75**	**0.86**

For non-classified waters, the blue-green ratio model Equation (5) showed a very poor correlation with the measured Chl-a concentration for the MMELS dataset ($r^2 = 0.06$ and 0.00 for calibration and validation subsets, respectively), as well as for each lagoon ($r^2 < 0.27$). The NIR-Red band models and the two-, three- and four-band models had similar performance considering the calibration subset, with strong coefficients of determination for the whole system ($r^2 > 0.84$). These models also showed a good performance for Manguaba lagoon ($r^2 > 0.81$). A more accurate and stable Chl-a retrieval was obtained using a three-band model for MMELS, with optimal spectral band positions for λ_1, λ_2 and λ_3 at 690, 717 and 721 nm, respectively, which was followed by a two-band model for MMELS with wavelengths λ_1 and λ_2 equal to 660 and 721 nm, respectively. Moreover, a remarkable decrease in performance of the four-band model was observed for the validation subset, suggesting an unstable Chl-a retrieval using this model. For Manguaba lagoon, the four-band model did not perform better than the three-band model, in contrast to Mundaú lagoon, where an improvement in performance was observed with an increase in model complexity (i.e., from the two-band to the four-band model).

An improvement in performance with increasing model complexity was also observed for classified waters. In general, the two- and three-band models for classified waters did not improve chlorophyll-a retrieval in comparison with band models for non-classified waters. However, the four-band model for Class 2 showed the best performance ($r^2 = 0.93$, RMSE = 2.48 mg/m^3) among all models for both classified and non-classified waters, with optimal spectral band positions for λ_1, λ_2, λ_3 and λ_4 at 663, 703, 706 and 707 nm, respectively. Strong linear relationships between the four-band model and the measured Chl-a were established for Class 1 ($r^2 = 0.75$, RMSE = 9.52 mg/m^3) and Class 4 ($r^2 = 0.75$, RMSE = 0.86 mg/m^3). For Class 3, the four-band model also showed a reasonable performance ($r^2 = 0.51$, RMSE = 5.16 mg/m^3) but was still lower than the band model for Mundaú lagoon (non-classified waters).

3.4. Retrieval of Chlorophyll-a Using Models Based on Simulated Satellite Bands

The applicability of satellite sensors such as MODIS, MERIS, OLCI and MSI was evaluated considering two- and three-band models for MMELS and the four-band model for Class 2 since these models showed the best performance with respect to chlorophyll-a retrieval using reflectance measurements.

The accuracy assessment of the best band models for the MMELS and Class 2 waters with respect to the four satellite sensors can be observed in Table 5. Notably, the two-band models showed better performance compared with the other models based on simulated satellite bands for MMELS and Manguaba lagoon. Considering the operating range, it was not possible to find a good combination of MODIS data, which resulted in the lower performance of this sensor among the two-band models. For both the MERIS and OLCI sensors, λ_2 was represented by the 9th channel (centered at 709 nm), and λ_1 was repositioned to the 8th channel (centered at 681 nm). However, the best performance of the MERIS and OLCI sensors was observed with greater displacement of λ_1 to the 7th channel (centered at 665 nm). In addition, a good model performance was also obtained using simulated MSI bands (R_{705}/R_{665}). For Class 2 waters, the four-band model could not be applied for Chl-a estimation using satellite sensors, in contrast to the four-band model based on R_{rs} measured in-situ. However, it was possible to accurately retrieve Chl-a using the three-band model for the MERIS and OLCI sensor data.

Table 5. Slopes (p) and intercepts (q) of the linear best-fit function between observed and retrieved chlorophyll-a with corresponding coefficient of determination (r^2) and root mean square error (RMSE in mg/m^3) for various satellite bands.

Water	Models	p	q	r^2	RMSE
Manguaba (N = 36)	MODIS − R_{645}/R_{555}	38.21	−11.32	0.56	14.55
	MERIS − R_{709}/R_{681}	19.08	7.27	0.64	13.29
	OLCI − R_{709}/R_{681}	19.97	6.51	0.65	13.07
	MERIS − R_{709}/R_{665}	29.78	−6.35	**0.72**	**11.76**
	OLCI − R_{709}/R_{665}	28.43	−4.53	**0.71**	**11.87**
	MSI − R_{705}/R_{665}	34.39	−12.52	**0.72**	**11.73**
	MERIS − $(R_{681}^{-1} - R_{709}^{-1}) \cdot R_{665}$	58.77	18.98	**0.71**	**11.84**
	OLCI − $(R_{681}^{-1} - R_{709}^{-1}) \cdot R_{674}$	73.60	18.97	**0.70**	**12.11**
MMELS (N = 72)	MODIS − R_{645}/R_{555}	34.63	−17.27	0.31	18.42
	MERIS − R_{709}/R_{681}	23.98	−6.73	0.70	12.12
	OLCI − R_{709}/R_{681}	24.95	−7.25	0.71	11.91
	MERIS − R_{709}/R_{665}	34.12	−17.29	**0.77**	**10.66**
	OLCI − R_{709}/R_{665}	32.83	−15.74	**0.76**	**10.77**
	MSI − R_{705}/R_{665}	39.07	−23.40	**0.78**	**10.44**
	MERIS − $(R_{681}^{-1} - R_{709}^{-1}) \cdot R_{665}$	39.74	20.92	0.60	14.00
	OLCI − $(R_{681}^{-1} - R_{709}^{-1}) \cdot R_{674}$	42.17	22.74	0.55	14.87
	MERIS − $(R_{665}^{-1} - R_{709}^{-1})/(R_{709}^{-1} - R_{681}^{-1})$	1.80	28.87	0.01	22.03
	OLCI − $(R_{674}^{-1} - R_{709}^{-1})/(R_{709}^{-1} - R_{681}^{-1})$	0.67	27.84	0.00	22.15

Table 5. *Cont.*

Water	Models	p	q	r^2	RMSE
	$MODIS - R_{645}/R_{555}$	18.12	10.48	0.02	9.30
	$MERIS - R_{709}/R_{681}$	44.75	−28.22	0.65	5.58
	$OLCI - R_{709}/R_{681}$	48.42	−31.57	0.67	5.42
Class 2 (N = 14)	$MSI - R_{705}/R_{665}$	45.14	−28.87	0.53	6.39
	$MERIS - (R_{681}^{-1} - R_{709}^{-1}) \cdot R_{665}$	71.65	14.33	**0.70**	**5.15**
	$OLCI - (R_{681}^{-1} - R_{709}^{-1}) \cdot R_{674}$	81.47	14.78	**0.71**	**5.02**
	$MERIS - (R_{665}^{-1} - R_{709}^{-1})/(R_{709}^{-1} - R_{681}^{-1})$	−6.15	22.77	0.28	7.95
	$OLCI - (R_{674}^{-1} - R_{709}^{-1})/(R_{709}^{-1} - R_{681}^{-1})$	−3.69	24.52	0.33	7.69

4. Discussion

MMELS can be characterized as an optically complex environment (Case 2 water) since a low correlation between TSS and Chl-a was found ($r^2 = 0.08$). This result suggests that these constituents do not necessarily covary over space and time, in contrast to oligotrophic and clear waters [24,60]. Furthermore, the high turbidity observed in the lagoons, with Secchi disk values ranging between 0.58 m and 0.95 m (see [61]), suggests a low or nonexistent effect of the bottom sediment on the spectral reflectance since the depth of the photic zone (light penetration zone) is lower than the water depth in MMELS (see Table 1).

The low reflectance values in the blue region (from 400 nm to around 470 nm) indicate the presence of yellow substances, which usually dominate the absorption in this spectral range for waters influenced by river runoff [62], suggesting a significant CDOM contribution from the rivers discharging to both lagoons. The slight depression observed around 440 nm can be explained by absorption peaks of chlorophyll-a in this region.

The first reflectance peak in the green range, around 560 nm in Mundaú lagoon and 570 nm in Manguaba lagoon, indicates minimal absorption by most algal pigments, but this reflectance is still influenced by CDOM absorption [63]. Thus, the backscattering by inorganic suspended matter and phytoplankton predominates in determining the spectral reflectance.

The minimum, observed around 625 nm, is related to phycocyanin absorption due to the presence of cyanobacteria [34,64,65]. In Manguaba lagoon, this feature was more prominent than in Mundaú lagoon, which may be explained by two factors: (a) the backscattering caused by the high TSS concentration in Mundaú lagoon may have masked the effect of phycocyanin absorption; and/or (b) the biomass of cyanobacteria was relatively low in Mundaú lagoon. The maximum absorption peak of chlorophyll-a in the red region, around 670 nm [39], was present in almost all reflectance spectra. It is likely that at this wavelength, the reflectance is still influenced by other optically active components since a weak correlation was observed between minimum reflectance near 670 nm and the chlorophyll-a concentration ($r^2 = 0.0006$).

The second prominent reflectance peak around 700 nm occurred because of minimal absorption of water constituents (Chl-a, NAP, and CDOM) and particulate backscattering, which controls the reflectance variations in this region. Although the peak magnitude near 700 nm vs. the Chl-a concentration indicated a very poor relationship ($r^2 = 0.1$), the increase in the Chl-a concentration caused the displacement of the peak position in the red region ($r^2 = 0.8$, Figure 2d), which is usually observed in turbid and productive waters [33,45,57,66]. Indeed, the peak positions near 700 nm in Manguaba lagoon occurred at wavelengths slightly longer than those in Mundaú lagoon, which exhibited lower concentrations of chlorophyll-a.

The optical classes resulting from the clustering analysis indicated a homogeneous spatial behavior over Manguaba lagoon for two distinct classes related to the wet (Class 2) and dry (Class 1) seasons. Water samples associated with Class 2 showed optical characteristics of waters dominated by chlorophyll-a and suspended sediment, which exhibited higher concentrations due to the increase in sediment loading from river discharge in the wet season. In addition, the combined effect of increasing light availability (i.e., fewer suspended sediments) and higher water temperature can explain the distinct optical behavior associated with waters dominated by chlorophyll-a observed in the dry season.

The water samples from the Mundaú lagoon were mostly classified as Class 3, indicating the homogeneity of the optical properties of these waters. This optical class had intermediate concentrations of chlorophyll-a and higher concentrations of TSS, which can be explained by the following: (a) higher nutrient and organic matter loading in Mundaú lagoon in comparison with Manguaba lagoon, since a considerable amount of sewage from surrounding municipalities, mainly Maceió city, is regularly discharged into Mundaú lagoon; and (b) more intensive anthropogenic activities (i.e., agriculture and dredging along the river-bed) in the Mundaú river basin, which favor soil erosion and sediment transport into Mundaú lagoon.

Regarding the evaluated models for non-classified waters, the blue-green ratio model was not a good algorithm to estimate the chlorophyll-a concentration for MMELS or the individual lagoons. As previously reported, this model is not suitable for turbid and productive waters (Case 2 water body) since reflectance in the blue and green ranges is strongly influenced by NAP and CDOM absorption, as well as by backscattering caused by inorganic and non-living organic suspended matter [63,67].

More accurate retrievals of chlorophyll-a were found using the NIR-Red band models. The two- and three-band NIR-Red models for Manguaba lagoon and MMELS had high coefficients of determination ($r^2 > 0.80$) and similar performance with respect to other chlorophyll-a retrievals reported for turbid and productive estuaries [45]. The inclusion of the third band did not result in a significant improvement in performance relative to the two-band NIR-Red model, possibly because the effect of absorption by CDOM and NAP on reflectance was not relevant for the wavelengths chosen as optimal in Manguaba lagoon ($\lambda_1 = 690$ nm, $\lambda_2 = 714$ nm) and MMELS ($\lambda_1 = 660$ nm, $\lambda_2 = 721$ nm).

Despite the better performance of the three-band NIR-Red model, the optimal position for λ_3 was found in the upper limit of the operating range. At this wavelength, the reflectance was still high (e.g., $R_{721} > R_{670}$ was observed in some reflectance curves for both lagoons). This finding suggests that the absorption caused by pure water was not completely dominant in this region, and hence this reflectance was strongly influenced by the absorption and backscattering of the other water constituents. In general, values greater than 720 nm are usually found for λ_3 for three-band NIR-Red models applied to inland and coastal waters [33,37,44]. Therefore, it is likely that the spectral range of operation may have limited the displacement of λ_3 for higher NIR wavelengths.

Considering the four-band NIR-Red model applied to MMELS and Manguaba lagoon, the addition of the fourth band did not improve model's performance with respect to chlorophyll-a retrieval in comparison to the two- and three-band NIR-Red models. Interestingly, for both of these datasets, the optimal positions for λ_3 and λ_4 were close to the upper limit of the operating range. In addition, the distance between these two wavelengths was only 1 nm, which is considerably smaller than the resolution of the radiometers used in the study. Naturally, such positioning aimed to minimize the effect of absorption by CDOM and NAP, as well as backscattering, in λ_3, but clearly this was not sufficient, especially for Manguaba lagoon water samples with higher concentrations of chlorophyll-a. This resulted in a decrease in performance of the four-band model in relation to the three-band model. In general, optimal values of λ_4 are found around 730 nm [38,44], which are beyond the spectral range of operation of this study.

The same pattern was not observed in Mundaú lagoon, where an increase in model complexity (i.e., from the two to the four-band NIR-Red model) resulted in an improvement of chlorophyll-a retrieval. However, the best model for Mundaú lagoon (four-band NIR-Red model, $r^2 = 0.74$) had a lower performance in relation to the two- and three-band NIR-Red models applied to both MMELS and Manguaba lagoon. The work in [33] reported that in turbid waters, two-band NIR-Red models tend to overestimate chlorophyll-a due to an increase in reflectance caused by backscattering of the huge amount of suspended particles in both the visible and NIR ranges. Thus, a lower ability to estimate chlorophyll-a in Mundaú lagoon was expected since this lagoon can be considered a highly turbid system that is less productive than Manguaba lagoon, as most water samples exhibit chlorophyll-a concentrations less than 25 mg/m^3. In these environments, the absorption coefficient of chlorophyll-a may be comparable, in terms of magnitude, to either the backscattering coefficient of suspended particles or the absorption coefficient of NAP and CDOM, not satisfying the condition assumed for NIR-Red models (i.e., $a_{chl-a}(\lambda_1) >> a_{NAP}(\lambda_1) + a_{CDOM}(\lambda_1)$ and $a_{chl-a}(\lambda_1) >> b_b(\lambda_1)$). Furthermore, the high concentration of TSS in Mundaú lagoon favored the four-band NIR-Red model since this algorithm minimizes the effect of absorption by CDOM and NAP in NIR wavelengths [36]. Nevertheless, it may be possible to improve the performance of the four-band NIR-Red model in Mundaú lagoon by searching for optimal values of λ_4 beyond the upper limit of the spectral range of operation (i.e., for $\lambda_4 > 721$ nm). For instance, it is common to find optimal values of λ_4 around 740 nm [36].

Regarding the evaluated models for classified waters, the inversion algorithms applied to homogeneous groups generally did not improve the chlorophyll-a retrieval in comparison with the band models using the non-classified dataset; this may be associated with the following: (a) the derivation of models from a reduced number of water samples in each class; (b) the classification method chosen (e.g., unsupervised classification clustering, fuzzy logic classification and reflectance shape characteristics calculation approach), which may control the number and optical characteristics of each class [38,54,55]; and (c) a higher uniformity of both optical characteristics and concentration values of the water constituents in each class, which may result in a lower performance of the models, mainly for classes with higher concentrations of suspended matter and lower concentrations of chlorophyll-a. However, the pre-classification of the reflectance spectra dataset into homogeneous groups has shown to be satisfactory to validate the band models using a different handheld dataset without repositioning of optimal wavelength values for each model [38,48]. Therefore, the classified band models obtained for Manguaba lagoon can be considered promising, especially the model for Class 2.

Although estuarine environments have more intensified dynamics in relation to other inland water bodies (i.e., lakes and reservoirs), the high retention times of both the Mundaú (>2 weeks) and Manguaba (>1 month) lagoons suggest slight water quality changes over time. Thus, MMELS demonstrates considerable potential for the application of models using satellite sensors with spatial and temporal resolutions less than 500 m and one week, respectively, similar to the sensors evaluated in this study. In general, the two-band models based on simulated satellite bands showed better performance than the three-band models for MMELS, which is in agreement with other studies [37,44,48]. The results also indicate that the two-band model using the MODIS sensor (R_{645}/R_{555}) failed to estimate chlorophyll-a. This likely occurred because it was not possible to use band 15 (centered at 748 nm), which is widely used in band models based on MODIS spectral bands [32,33,37] but is still beyond the spectral range of operation of this study. Nevertheless, enhanced chlorophyll-a retrievals have been found using a two-band algorithm based on simulated Sentinel-2/MSI, Envisat/MERIS and Sentinel-3/OLCI satellite bands, indicating a promising application of these sensors in MMELS and Manguaba lagoon. Regarding Class 2 waters, a decrease in performance of the four-band model based on simulated MERIS and OLCI bands in comparison with the two- and three-band models can be explained by the inability to identify a good combination of simulated satellite bands near the wavelengths chosen as optimal for the models using R_{rs} measured in-situ. The assessment of models to directly estimate chlorophyll-a using satellite images captured by the MSI and OLCI sensors encourages further future investigation in MMELS, since images from these recently spaceborne sensors were not available during the monitoring period (May to September 2015). Despite the spectral range operational constraints, our findings suggest that empirical models based on optical properties involving water constituents have a strong potential to estimate chlorophyll-a using spectral data from satellite, airborne or handheld sensors in productive tropical estuaries, as was also observed in temperate, tropical and subtropical inland waters [14,30,33,68].

Despite the wide range of measured chlorophyll-a in the dataset (0.97–117.24 mg·m^{-3}), the monitoring period was limited from May to September 2015, which corresponds to an inter-seasonal period between the end of the wet season and the beginning of the dry season. A recent study, using approximately 100 water samples for each season, indicated that a unique model based on spectral reflectance can be used to estimate chlorophyll-a for all four seasons [42], suggesting that the empirical models described in this study can also be applied outside of the monitoring period.

5. Conclusions

In this study, we examined the performance of several algorithms based on spectral bands to estimate chlorophyll-a in a shallow, turbid, productive tropical estuarine-lagoon system using in situ reflectance spectra. We also investigated the potential of model application using satellite sensors. Our results showed accurate retrievals of chlorophyll-a ($r^2 > 0.80$) using the NIR-Red three-band model for MMELS, which exhibits a light penetration zone lower than the water depth due to high turbidity. These accurate models open up a novel branch for scientific investigations searching for new spectral reflectance-based models using in situ reflectance measurements beyond the spectral range of operation in this study. In addition, these models can be used to explore the possibility of extending this approach to other productive tropical estuaries.

We also observed accurate chlorophyll-a retrievals using the two-band algorithm based on the simulated Sentinel-2/MSI, Envisat/MERIS and Sentinel-3/OLCI satellite bands with temporal and spatial resolutions less than one week and 500 m, respectively, indicating a promising application of these sensors to remotely estimate chlorophyll-a in MMELS and Manguaba lagoon. However, further validation to test the application of these sensors must be addressed considering the characteristics of each sensor, in particular, the signal-to-noise ratio and radiometric and spatial resolutions.

Acknowledgments: We are grateful to institutional grants provided by the Agency for Financing Studies and Projects (Finep), research project No. 01.13.0419.00, and The Brazilian National Council for Scientific and Technological Development (CNPq), through the Universal Research Project (476789/2013-3), which funded all equipment and operating expenses of this study. We also thank the Research Support Foundation of the State of Pernambuco (FACEPE), who granted a doctoral scholarship (PBPG-1148-3.01/12) to the first author, and IMA/AL, who provided the boat that was used for collecting the in situ data. The authors would also like to thank the anonymous referees, who helped to significantly improve the quality and the clarity of this manuscript.

Author Contributions: R.C.L., J.-M.M., J.A.C. and C.R.F.Jr. developed the idea for the study, as well as its design. R.C.L., D.d.M.M. and C.R.F.Jr. were responsible for the construction and validation of the dataset. R.C.L., J.-M.M., J.A.C., D.d.M.M. and C.R.F.Jr. participated in the analysis and development of the discussion. R.C.L. and C.R.F.Jr. finalized the manuscript. All authors read and approved the manuscript.

Conflicts of Interest: The authors declare no conflict of interest.

References

1. Mitchell, S.B.; Jennerjahn, T.C.; Vizzini, S.; Zhang, W. Changes to processes in estuaries and coastal waters due to intense multiple pressures—An introduction and synthesis. *Estuar. Coast. Shelf Sci.* **2015**, *156*, 1–6. [CrossRef]
2. Hardisty, J. Introduction to estuarine systems. In *Estuaries: Monitoring and Modeling the Physical System*; Blackwell Publishing: Hoboken, NJ, USA, 2007; pp. 3–22.
3. Cadee, G.C. Book review: Nutrients and eutrophication in estuaries and coastal waters. *Aquat. Ecol.* **2004**, *38*, 616–617. [CrossRef]
4. Boyer, J.N.; Kelble, C.R.; Ortner, P.B.; Rudnick, D.T. Phytoplankton bloom status: Chlorophyll a biomass as an indicator of water quality condition in the southern estuaries of florida, USA. *Ecol. Indic.* **2009**, *9*, S56–S67. [CrossRef]
5. Scanes, P.; Coade, G.; Doherty, M.; Hill, R. Evaluation of the utility of water quality based indicators of estuarine lagoon condition in nsw, australia. *Estuar. Coast. Shelf Sci.* **2007**, *74*, 306–319. [CrossRef]
6. Paerl, H.W.; Valdes-Weaver, L.M.; Joyner, A.R.; Winkelmann, V. Phytoplankton indicators of ecological change in the eutrophying pamlico sound system, north carolina. *Ecol. Appl.* **2007**, *17*, S88–S101. [CrossRef]
7. Navalgund, R.R.; Jayaraman, V.; Roy, P.S. Remote sensing applications: An overview. *Curr. Sci.* **2007**, *93*, 1747–1766.

8. Bukata, R.P. Retrospection and introspection on remote sensing of inland water quality: "Like déjà vu all over again". *J. Great Lakes Res.* **2013**, *39*, 2–5. [CrossRef]

9. Odermatt, D.; Gitelson, A.; Brando, V.E.; Schaepman, M. Review of constituent retrieval in optically deep and complex waters from satellite imagery. *Remote Sens. Environ.* **2012**, *118*, 116–126. [CrossRef]

10. Tyler, A.N.; Hunter, P.D.; Spyrakos, E.; Groom, S.; Constantinescu, A.M.; Kitchen, J. Developments in earth observation for the assessment and monitoring of inland, transitional, coastal and shelf-sea waters. *Sci. Total Environ.* **2016**, *572*, 1307–1321. [CrossRef] [PubMed]

11. Cannizzaro, J.P.; Carder, K.L. Estimating chlorophyll a concentrations from remote-sensing reflectance in optically shallow waters. *Remote Sens. Environ.* **2006**, *101*, 13–24. [CrossRef]

12. Palmer, S.C.J.; Kutser, T.; Hunter, P.D. Remote sensing of inland waters: Challenges, progress and future directions. *Remote Sens. Environ.* **2015**, *157*, 1–8. [CrossRef]

13. Mouw, C.B.; Greb, S.; Aurin, D.; DiGiacomo, P.M.; Lee, Z.; Twardowski, M.; Binding, C.; Hu, C.; Ma, R.; Moore, T.; et al. Aquatic color radiometry remote sensing of coastal and inland waters: Challenges and recommendations for future satellite missions. *Remote Sens. Environ.* **2015**, *160*, 15–30. [CrossRef]

14. Xie, C.-H.; Chang, J.-Y.; Zhang, Y.-Z. A new method for estimating chlorophyll-a concentration in the pearl river estuary. *Optik* **2015**, *126*, 4510–4515. [CrossRef]

15. Le, C.; Hu, C.; English, D.; Cannizzaro, J.; Chen, Z.; Kovach, C.; Anastasiou, C.J.; Zhao, J.; Carder, K.L. Inherent and apparent optical properties of the complex estuarine waters of tampa bay: What controls light? *Estuar. Coast. Shelf Sci.* **2013**, *117*, 54–69. [CrossRef]

16. Tzortziou, M.; Subramaniam, A.; Herman, J.R.; Gallegos, C.L.; Neale, P.J.; Harding, L.W., Jr. Remote sensing reflectance and inherent optical properties in the mid chesapeake bay. *Estuar. Coast. Shelf Sci.* **2007**, *72*, 16–32. [CrossRef]

17. Gons, H.J. Optical teledetection of chlorophyll a in turbid inland water. *Environ. Sci. Technol.* **1999**, *33*, 1127–1132. [CrossRef]

18. Preisendorfer, R.W. *Application of Radiative Transfer Theory to Light Measurements in the Sea, Monograph No. 10*; L'Institut Géographique National: Champigneulles, France, 1961.

19. Morel, A.; Gentili, B. Diffuse reflectance of oceanic waters: Its dependence on sun angle as influenced by the molecular scattering contribution. *Appl. Opt.* **1991**, *30*, 4427–4438. [CrossRef] [PubMed]

20. Morel, A.; Gentili, B. Diffuse reflectance of oceanic waters. II. Bidirectional aspects. *Appl. Opt.* **1993**, *32*, 6864–6879. [CrossRef] [PubMed]

21. Pope, R.M.; Fry, E.S. Absorption spectrum (380–700 nm) of pure water. II. Integrating cavity measurements. *Appl. Opt.* **1997**, *36*, 8710–8723. [CrossRef] [PubMed]

22. Smith, R.C.; Baker, K.S. Optical properties of the clearest natural waters (200–800 nm). *Appl. Opt.* **1981**, *20*, 177–184. [CrossRef] [PubMed]

23. Morel, A. Optical properties of pure water and pure sea water. In *Optical Aspects of Oceanography*; Jerlov, N.G., Nielsen, E.S., Eds.; Academic Press: New York, NY, USA, 1974; pp. 1–24.

24. Morel, A.; Prieur, L. Analysis of variations in ocean color1. *Limnol. Oceanogr.* **1977**, *22*, 709–722. [CrossRef]

25. Gordon, H.R.; Morel, A.Y. In-water algorithms. In *Remote Assessment of Ocean Color for Interpretation of Satellite Visible Imagery*; Springer: Berlin, Germany, 1983; pp. 24–67.

26. Hu, C.; Lee, Z.; Franz, B. Chlorophyll aalgorithms for oligotrophic oceans: A novel approach based on three-band reflectance difference. *J. Geophys. Res. Oceans* **2012**, *117*. [CrossRef]

27. O'Reilly, J.E.; Maritorena, S.; Mitchell, B.G.; Siegel, D.A.; Carder, K.L.; Garver, S.A.; Kahru, M.; McClain, C. Ocean color chlorophyll algorithms for seawifs. *J. Geophys. Res. Oceans* **1998**, *103*, 24937–24953. [CrossRef]

28. Harding, J.L.W.; Magnuson, A.; Mallonee, M.E. Seawifs retrievals of chlorophyll in chesapeake bay and the mid-atlantic bight. *Estuar. Coast. Shelf Sci.* **2005**, *62*, 75–94. [CrossRef]

29. Wang, M.; Shi, W.; Tang, J. Water property monitoring and assessment for china's inland lake taihu from MODIS-aqua measurements. *Remote Sens. Environ.* **2011**, *115*, 841–854. [CrossRef]

30. Le, C.; Hu, C.; English, D.; Cannizzaro, J.; Chen, Z.; Feng, L.; Boler, R.; Kovach, C. Towards a long-term chlorophyll-a data record in a turbid estuary using MODIS observations. *Prog. Oceanogr.* **2013**, *109*, 90–103. [CrossRef]

31. Dall'Olmo, G.; Gitelson, A.A.; Rundquist, D.C. Towards a unified approach for remote estimation of chlorophyll-a in both terrestrial vegetation and turbid productive waters. *Geophys. Res. Lett.* **2003**, *30*. [CrossRef]

32. Dall'Olmo, G.; Gitelson, A.A.; Rundquist, D.C.; Leavitt, B.; Barrow, T.; Holz, J.C. Assessing the potential of seawifs and MODIS for estimating chlorophyll concentration in turbid productive waters using red and near-infrared bands. *Remote Sens. Environ.* **2005**, *96*, 176–187. [CrossRef]

33. Gitelson, A.A.; Dall'Olmo, G.; Moses, W.; Rundquist, D.C.; Barrow, T.; Fisher, T.R.; Gurlin, D.; Holz, J. A simple semi-analytical model for remote estimation of chlorophyll-a in turbid waters: Validation. *Remote Sens. Environ.* **2008**, *112*, 3582–3593. [CrossRef]

34. Randolph, K.; Wilson, J.; Tedesco, L.; Li, L.; Pascual, D.L.; Soyeux, E. Hyperspectral remote sensing of cyanobacteria in turbid productive water using optically active pigments, chlorophyll a and phycocyanin. *Remote Sens. Environ.* **2008**, *112*, 4009–4019. [CrossRef]

35. Chavula, G.; Brezonik, P.; Thenkabail, P.; Johnson, T.; Bauer, M. Estimating chlorophyll concentration in lake malawi from MODIS satellite imagery. *Phys. Chem. Earth Parts A/B/C* **2009**, *34*, 755–760. [CrossRef]

36. Le, C.; Li, Y.; Zha, Y.; Sun, D.; Huang, C.; Lu, H. A four-band semi-analytical model for estimating chlorophyll a in highly turbid lakes: The case of taihu lake, china. *Remote Sens. Environ.* **2009**, *113*, 1175–1182. [CrossRef]

37. Gurlin, D.; Gitelson, A.A.; Moses, W.J. Remote estimation of chl-a concentration in turbid productive waters—Return to a simple two-band nir-red model? *Remote Sens. Environ.* **2011**, *115*, 3479–3490. [CrossRef]

38. Le, C.; Li, Y.; Zha, Y.; Sun, D.; Huang, C.; Zhang, H. Remote estimation of chlorophyll a in optically complex waters based on optical classification. *Remote Sens. Environ.* **2011**, *115*, 725–737. [CrossRef]

39. Yacobi, Y.Z.; Moses, W.J.; Kaganovsky, S.; Sulimani, B.; Leavitt, B.C.; Gitelson, A.A. Nir-red reflectance-based algorithms for chlorophyll—A estimation in mesotrophic inland and coastal waters: Lake kinneret case study. *Water Res.* **2011**, *45*, 2428–2436. [CrossRef] [PubMed]

40. Lyu, H.; Li, X.; Wang, Y.; Jin, Q.; Cao, K.; Wang, Q.; Li, Y. Evaluation of chlorophyll-a retrieval algorithms based on MERIS bands for optically varying eutrophic inland lakes. *Sci. Total Environ.* **2015**, *530–531*, 373–382. [CrossRef] [PubMed]

41. Palmer, S.C.J.; Hunter, P.D.; Lankester, T.; Hubbard, S.; Spyrakos, E.; Tyler, A.N.; Présing, M.; Horváth, H.; Lamb, A.; Balzter, H.; et al. Validation of envisat MERIS algorithms for chlorophyll retrieval in a large, turbid and optically-complex shallow lake. *Remote Sens. Environ.* **2015**, *157*, 158–169. [CrossRef]

42. Wang, L.; Pu, H.; Sun, D.-W. Estimation of chlorophyll-a concentration of different seasons in outdoor ponds using hyperspectral imaging. *Talanta* **2016**, *147*, 422–429. [CrossRef] [PubMed]

43. Loisel, H.; Vantrepotte, V.; Ouillon, S.; Ngoc, D.D.; Herrmann, M.; Tran, V.; Mériaux, X.; Dessailly, D.; Jamet, C.; Duhaut, T.; et al. Assessment and analysis of the chlorophyll-a concentration variability over the vietnamese coastal waters from the MERIS ocean color sensor (2002–2012). *Remote Sens. Environ.* **2017**, *190*, 217–232. [CrossRef]

44. Le, C.; Hu, C.; Cannizzaro, J.; English, D.; Muller-Karger, F.; Lee, Z. Evaluation of chlorophyll-a remote sensing algorithms for an optically complex estuary. *Remote Sens. Environ.* **2013**, *129*, 75–89. [CrossRef]

45. Gitelson, A.A.; Schalles, J.F.; Hladik, C.M. Remote chlorophyll-a retrieval in turbid, productive estuaries: Chesapeake bay case study. *Remote Sens. Environ.* **2007**, *109*, 464–472. [CrossRef]

46. Hu, C.; Chen, Z.; Clayton, T.D.; Swarzenski, P.; Brock, J.C.; Muller-Karger, F.E. Assessment of estuarine water-quality indicators using MODIS medium-resolution bands: Initial results from tampa bay, FL. *Remote Sens. Environ.* **2004**, *93*, 423–441. [CrossRef]

47. Chen, J.; Quan, W. An improved algorithm for retrieving chlorophyll-a from the yellow river estuary using MODIS imagery. *Environ. Monit. Assess.* **2012**, *185*, 2243–2255. [CrossRef] [PubMed]

48. Sun, D.; Hu, C.; Qiu, Z.; Cannizzaro, J.P.; Barnes, B.B. Influence of a red band-based water classification approach on chlorophyll algorithms for optically complex estuaries. *Remote Sens. Environ.* **2014**, *155*, 289–302. [CrossRef]

49. Toming, K.; Kutser, T.; Laas, A.; Sepp, M.; Paavel, B.; Nõges, T. First experiences in mapping lake water quality parameters with Sentinel-2 msi imagery. *Remote Sens.* **2016**, *8*, 640. [CrossRef]

50. Oliveira, A.M.; Kjerfve, B. Regular article: Environmental responses of a tropical coastal lagoon system to hydrological variability: Mundaú-manguaba, brazil. *Estuar. Coast. Shelf Sci.* **1993**, *37*, 575–591. [CrossRef]

51. Mueller, J.; Davis, C.; Arnone, R.; Frouin, R.; Carder, K.; Lee, Z.P.; Steward, R.G.; Hooker, S.; Mobley, C.; McLean, S. *Above-Water Radiance and Remote Sensing Reflectance Measurement and Analysis Protocols*; Goddard Space Flight Space Center: Greenbelt, MD, USA, 2003; pp. 21–31.

52. Mobley, C.D. Estimation of the remote-sensing reflectance from above-surface measurements. *Appl. Opt.* **1999**, *38*, 7442–7455. [CrossRef] [PubMed]

53. Rice, E.W.; Baird, R.B.; Eaton, A.D.; Clesceri, L.S. *Standard Methods for the Examination of Water and Wastewater*; American Public Health Association: Washington, DC, USA, 2005.

54. Lubac, B.; Loisel, H. Variability and classification of remote sensing reflectance spectra in the eastern english channel and southern north sea. *Remote Sens. Environ.* **2007**, *110*, 45–58. [CrossRef]

55. Martinez, J.-M.; Espinoza-Villar, R.; Armijos, E.; Silva Moreira, L. The optical properties of river and floodplain waters in the amazon river basin: Implications for satellite-based measurements of suspended particulate matter. *J. Geophys. Res. Earth Surf.* **2015**, *120*, 1274–1287. [CrossRef]

56. Matthews, M.W. A current review of empirical procedures of remote sensing in inland and near-coastal transitional waters. *Int. J. Remote Sens.* **2011**, *32*, 6855–6899. [CrossRef]

57. Gitelson, A. The peak near 700 nm on radiance spectra of algae and water-relationships of its magnitude and position with chlorophyll concentration. *Int. J. Remote Sens.* **1992**, *13*, 3367–3373. [CrossRef]

58. Stumpf, R.P.; Tyler, M.A. Satellite detection of bloom and pigment distributions in estuaries. *Remote Sens. Environ.* **1988**, *24*, 385–404. [CrossRef]

59. Ruddick, K.G.; De Cauwer, V.; Park, Y.-J.; Moore, G. Seaborne measurements of near infrared water-leaving reflectance: The similarity spectrum for turbid waters. *Limnol. Oceanogr.* **2006**, *51*, 1167–1179. [CrossRef]

60. Mélin, F.; Vantrepotte, V. How optically diverse is the coastal ocean? *Remote Sens. Environ.* **2015**, *160*, 235–251. [CrossRef]

61. Silva, E.D.A.; Nogueira, E.M.S.; Dué, A.; Carnaúba, M.P.; Guedes, E.A.C. Microalgas perifiticas em caiçaras situadas nas lagoas: Mundaú e manguaba do sistema lagunar de maceió. In *Congresso de Ecologia do Brasil*; Sociedade de Ecologia do Brasil (SEB): Caxambu, Brazil, 2005. Available online: www.seb-ecologia.org.br/viiceb/resumos/31a.pdf (accessed on 21 October 2016).

62. Mobley, C.D. *Light and Water: Radiative Transfer in Natural Waters*; Academic Press: Cambridge, MA, USA, 1994; p. 592.

63. Brezonik, P.L.; Olmanson, L.G.; Finlay, J.C.; Bauer, M.E. Factors affecting the measurement of cdom by remote sensing of optically complex inland waters. *Remote Sens. Environ.* **2015**, *157*, 199–215. [CrossRef]

64. Mishra, S.; Mishra, D.R.; Lee, Z.; Tucker, C.S. Quantifying cyanobacterial phycocyanin concentration in turbid productive waters: A quasi-analytical approach. *Remote Sens. Environ.* **2013**, *133*, 141–151. [CrossRef]

65. Lyu, H.; Wang, Q.; Wu, C.; Zhu, L.; Yin, B.; Li, Y.; Huang, J. Retrieval of phycocyanin concentration from remote-sensing reflectance using a semi-analytic model in eutrophic lakes. *Ecol. Inform.* **2013**, *18*, 178–187. [CrossRef]

66. Tao, B.; Mao, Z.; Pan, D.; Shen, Y.; Zhu, Q.; Chen, J. Influence of bio-optical parameter variability on the reflectance peak position in the red band of algal bloom waters. *Ecol. Inform.* **2013**, *16*, 17–24. [CrossRef]

67. Binding, C.E.; Bowers, D.G.; Mitchelson-Jacob, E.G. Estimating suspended sediment concentrations from ocean colour measurements in moderately turbid waters; the impact of variable particle scattering properties. *Remote Sens. Environ.* **2005**, *94*, 373–383. [CrossRef]

68. Augusto-Silva, P.B.; Ogashawara, I.; Barbosa, C.C.F.; De Carvalho, L.A.S.; Jorge, D.S.F.; Fornari, C.I.; Stech, J.L. Analysis of MERIS reflectance algorithms for estimating chlorophyll-α concentration in a brazilian reservoir. *Remote Sens.* **2014**, *6*, 11689–11707. [CrossRef]

remote sensing

MDPI

Article

MOD2SEA: A Coupled Atmosphere-Hydro-Optical Model for the Retrieval of Chlorophyll-a from Remote Sensing Observations in Complex Turbid Waters

Behnaz Arabi [1,*], Mhd. Suhyb Salama [1], Marcel Robert Wernand [2] and Wouter Verhoef [1]

[1] Faculty of Geo-Information Science and Earth Observation (ITC), Department of Water Resources, University of Twente, P.O. Box 217, 7500AE Enschede, The Netherlands; s.salama@utwente.nl (M.S.S.); w.verhoef@utwente.nl (W.V.)

[2] Department of Coastal Systems, Marine Optics and Remote Sensing, Royal Netherlands Institute for Sea Research (NIOZ), P.O. Box 59, 1790AB Den Burg, Texel, The Netherlands; marcel.wernand@nioz.nl

* Correspondence: b.arabi@utwente.nl or arabi.behnaz@gmail.com; Tel.: +31-534-874-288

Academic Editors: Magaly Koch and Prasad S. Thenkabail
Received: 20 June 2016; Accepted: 27 August 2016; Published: 1 September 2016

Abstract: An accurate estimation of the chlorophyll-a (Chla) concentration is crucial for water quality monitoring and is highly desired by various government agencies and environmental groups. However, using satellite observations for Chla estimation remains problematic over coastal waters due to their optical complexity and the critical atmospheric correction. In this study, we coupled an atmospheric and a water optical model for the simultaneous atmospheric correction and retrieval of Chla in the complex waters of the Wadden Sea. This coupled model called MOD2SEA combines simulations from the MODerate resolution atmospheric TRANsmission model (MODTRAN) and the two-stream radiative transfer hydro-optical model 2SeaColor. The accuracy of the coupled MOD2SEA model was validated using a matchup data set of MERIS (MEdium Resolution Imaging SpectRometer) observations and four years of concurrent ground truth measurements (2007–2010) at the NIOZ jetty location in the Dutch part of the Wadden Sea. The results showed that MERIS-derived Chla from MOD2SEA explained the variations of measured Chla with a determination coefficient of $R^2 = 0.88$ and a RMSE of 3.32 mg·m^{-3}, which means a significant improvement in comparison with the standard MERIS Case 2 regional (C2R) processor. The proposed coupled model might be used to generate a time series of reliable Chla maps, which is of profound importance for the assessment of causes and consequences of long-term phenological changes of Chla in the turbid Wadden Sea area.

Keywords: Chlorophyll-a; remote sensing; MERIS; atmospheric correction; MODTRAN; 2Seacolor; C2R; Wadden Sea

1. Introduction

Effective management of water quality in coastal regions and turbid waters requires accurate information about water quality parameter changes on prolonged time scales. Although this may sound simple, it is an extremely challenging task. One of the most important water quality parameters is chlorophyll-a (Chla) concentration, which is an important factor controlling light attenuation in the water column and is used as a measure of the eutrophic state [1]. Chla concentration is a very crucial factor to understanding the planetary carbon cycle [2] and is considered as an important indicator of eutrophication in marine ecosystems that may influence human life [3,4]. Chla abundance can be affected by anthropogenic nutrient supply from industrial and agricultural sources, where simultaneously the aquaculture industries and fisheries are influenced by Chla abundance [5].

Long-term monitoring of Chla concentration using field measurements and laboratory analysis requires conventional cruise surveys with satisfactory temporal and spatial coverage. Unfortunately, this is often not feasible for most coastal regions due to lack of financial resources and technical equipment while it is impossible in practice to collect in situ measurements for the whole regions using cruise measurements. The spatiotemporal coverage provided by remote sensing can considerably overcome some of these deficits in the current in situ monitoring programs for water quality parameters [6]. Satellite ocean color is especially important since it is the only remotely sensed property that directly identifies a biological component of the ecosystem [2]. Regarding the spatial and temporal sampling capabilities of satellite data, remote sensing of ocean color is considered as the principal source of data for investigating long-term changes in Chla concentration and phytoplankton biomass in many coastal areas' estuaries [7].

The maintenance of a good environmental status in European coastal regions and sea has become a crucial concern embodied in European regulations (Marine Strategy Framework Directive, Directive 2008/56/EC of the European Parliament and of the Council, "establishing a framework for community action in the field of marine environmental policy") [8]. One of the most important European coastal zones which has aroused increasing attention from all of Europe is the Wadden Sea. For the assessment of the current role of the Wadden Sea as a source of Chla and organic matter, and for the ongoing discussion on eutrophication problem areas, it is of great interest to obtain more detailed knowledge on the phytoplankton and Chla changes and their regulating factors in this turbid coastal region of the North Sea [9]. In addition, monitoring of this area is mandatory due to its nature reserve status and its July 2009 inclusion on the UNESCO World Heritage List [10]. Recently some research into the analysis of long-term variations and trends in the optically active substances (Chla, Suspended Particulate Matter (SPM), Colored Dissolved Organic Matters (CDOM)) and water color changes using in situ measurements have been conducted over different parts of the Wadden Sea [11–14]. However, using satellite observations for Chla estimation remains problematic in this area due to its optical complexity and the critical application of an accurate atmospheric correction. Recent efforts show that researchers are confronted with two main problems in improving the accuracy of derived water parameter concentration using remote sensing techniques in the Wadden Sea. First, most atmospheric correction methods fail in this region [15,16]. Second, the general water property retrieval models do not work well in this complex turbid water [17,18]. Thus, the main purpose of this research is to tackle these two problems aiming to increase the accuracy of Chla concentration retrieval from earth observation data in this area.

1.1. Atmospheric Correction

Quality of the atmospheric correction is one of the most limiting factors for the accurate retrieval of water constituents from earth observation data in coastal waters [19]. The standard atmospheric correction method by Gordon and Wang [20] assumes a zero water-leaving reflectance due to high absorption by seawater in the near-infra-red (NIR) and can be performed by extrapolating the aerosol optical properties to the visible from the NIR spectral region [21]. This is not always the case when in turbid waters (which often are optically complex) [22], higher concentrations of Chla and SPM can cause a significant water-leaving reflectance in the NIR [23]. Indeed, most of the atmospheric correction methods fail in these areas due to the complexity of the recorded top of atmosphere (TOA) radiance signal at satellite images [24] as these signals are associated with aerosols from continental sources [25]. In addition, in coastal waters, photons from nearby land areas can enter the field-of-view of the sensor (the adjacency effect) and contribute to total NIR backscatter [26], whereas in shallow waters, TOA radiances can also be influenced by the bottom effect [10]. Consequently, the black pixel assumption tends to overestimate the aerosol scattered radiance and thus underestimates the water-leaving radiance in these areas [27]. In recent years, some studies have been conducted to improve the atmospheric correction over turbid waters [28–30]. For example, some efforts were made to improve the atmospheric correction method by assuming a zero water-leaving reflectance in the shortwave infrared, even in the case of highly turbid waters [31,32]. However, in further studies, researchers found that for extremely high turbidities, even in the shortwave infrared region, the water-leaving reflectance was not absolutely equal to zero [33]. In addition, other studies focused on the non-negligible water-leaving reflectance assumption in the NIR [34,35]. For example, Carder et al. [36] investigated the ratio of water-leaving reflectance at two NIR bands. This ratio was either assumed constant [37] or estimated from neighboring pixels of open oceans [38]. Although the assumption of a known relationship between the values of water-leaving reflectance in two NIR bands is necessary, it is not sufficient. Indeed, accurate information about visibility and aerosol type is still needed [34]. Shen et al. [39] used the radiative transfer model MODTRAN to perform atmospheric correction for MERIS images over highly turbid waters. As shown by Verhoef and Bach [40], for an assumed visibility and aerosol type, MODTRAN can be used to extract the necessary atmospheric parameters to remove the scattering and absorption effects of the atmosphere and to obtain calibrated surface reflectance, as well as correcting the adjacency effects. However, this technique assumes a spatially homogeneous atmosphere [41], while in reality not only visibility but also the aerosol type may vary spatially within the extent of satellite images (in the presence of local haze variations). For example, in the case of coastal waters, some aerosol types (e.g., urban or rural) might exist in the regions close to the land and other pixels might have the maritime aerosol type. Consequently, the assumption of a homogeneous atmosphere may lead to wrong establishment of visibility and aerosol model in different parts of the image and may result in overestimation or underestimation of water constituent concentrations from ocean-color observations. The Case-2 regional (C2R) processor provided by ESA for MERIS L1 products in the MERIS regional coastal and case 2 water projects [42], performs atmospheric correction pixel by pixel and contains procedures for determining inherent optical properties that are delivered as MERIS L2 products, including reflectance, inherent optical properties (IOPs), and water quality parameters. However, the C2R processor may be invalid for very chlorophyll-rich waters like some eutrophic lakes [43] and for highly turbid waters [39]. In this paper, by applying radiative transfer modeling for the non-homogeneous atmosphere and comparing the results with the C2R processor, we tried to improve the atmospheric correction technique over this coastal area.

1.2. Hydro-Optical Model

After improving the atmospheric correction technique, water constituent concentration-dependent optical modeling of turbid waters is the next step. Improving the accuracy of water properties retrievals in coastal waters requires generic models that can be applied to these complex water bodies [44]. For open oceans, estimation of Chla from earth observation data is well established [45]. An empirical algorithm is in use that, with slight modifications for the actual band settings, has proven to work well for instruments like SeaWiFS (Sea-Viewing Wide Field-of-View Sensor), MODIS (Moderate Resolution Imaging Spectroradiometer) and MERIS [46–48]. However, satellite estimation of Chla concentration is still difficult for coastal waters, where Chla, SPM and CDOM occur in various mixtures which complicate the derivation of their concentrations from reflectance observations [49].

Therefore, there is a pressing need to develop, implement and validate a self-consistent, generic and operational retrieval model of water quality in turbid waters [49]. In this study, the forward analytical model known as 2SeaColor developed by Salama and Verhoef [50] was applied for the first time to retrieve Chla concentration in the Wadden Sea. The 2SeaColor model is based on the solution of the two-stream radiative transfer equations for incident sunlight and also performs well for turbid waters, while the commonly applied water quality algorithms might suffer from saturation in the presence of a high turbidity [44].

After defining the main problems of remote sensing of coastal waters described above, and motivated by the need for a high-quality, satellite-based long-term Chla retrieval in the turbid waters of the Wadden Sea, this research focused on the following objectives: (1) improving the accuracy of Chla concentration (mg·m^{-3}) retrieval from MERIS data by applying a coupled MODTRAN−2SeaColor model (MOD2SEA) for the Wadden Sea and (2) comparing the accuracy of the coupled MOD2SEA in performing atmospheric correction and retrieving Chla concentration values with the ESA standard C2R processor. The paper is arranged as follows. The case study is described first. Then, the datasets used for C2R and MODTRAN simulations as well as the 2SeaColor model are briefly introduced. Next, we validate the derived Chla concentration and water-leaving reflectance values for both MOD2SEA and C2R processor against the ground truth measurements at the NIOZ jetty station. Then, we evaluate the remote sensing (MOD2SEA and C2R) retrievals and compare the variation of MOD2SEA results with similar in situ studies in the Wadden Sea. Finally, we suggest some recommendations for further remote sensing studies in complex turbid waters like the Wadden Sea and discuss the applicability of this approach to other estuaries and satellite ocean color missions.

2. Materials and Methods

2.1. Study Area

The Dutch Wadden Sea is a coastal area located between the mainland of the Netherlands and the North Sea. The area is located between the Marsdiep near Den Helder in the southwest and the Dollard near Groningen in the northeast and comprises a surface area of 2500 km^2 (Figure 1). This region is a shallow, well-mixed tidal area that consists of several separated tidal basins. Each basin comprises tidal flats, subtidal areas and channels. Basins are connected to the adjacent North Sea by relatively narrow and deep tidal inlets between the barrier islands [51].

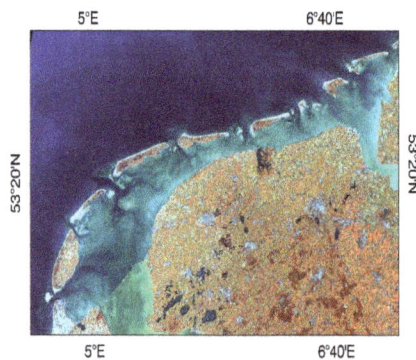

Figure 1. One Landsat-8 OLI image covering the Dutch Wadden Sea and parts of IJsselmeer lake acquired on 20 July 2016 (Color composite of red = band 5, green = band 3 and blue = band 1).

The high near-surface concentrations of water constituents as well as the spatial, tidal and seasonal variations of the optically active substances (Chla, SPM and CDOM) make this region an optically very complex area and a good representative for remote sensing studies in turbid coastal waters [10].

2.2. Ground Truth Dataset

The ground truth data have been extensively used to investigate the accuracy of remote sensing radiometric products (i.e., the remote sensing reflectance) from the recorded TOA radiance in satellite observations like MERIS images [52]. In this study, the ground truth above-water radiometric dataset was provided by the research jetty of the Royal Netherlands Institute for Sea Research (NIOZ) at Texel, located in the Dutch part of the Wadden Sea. Every quarter of an hour, radiometric color measurements of the water, sun and sky (including meteorological conditions), as well as Chla and mineral concentration, were recorded for over a decade [53]. The data were collected at the NIOZ jetty station (53°00′06″N; 4°47′21″E) [54], where the newest generation of hyperspectral radiometers was installed for "autonomous" monitoring of the Wadden Sea from 2001 until the present [53] (Figure 2).

The footprint size of the radiometer is less than a meter, and the viewing direction is not nadir but oblique, so the measurements on the ground are only partially representative of the nadir water reflectance from 300 m pixels as sensed by MERIS.

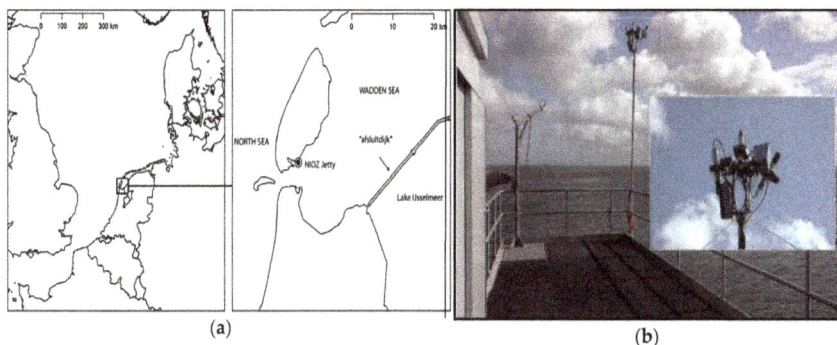

Figure 2. (**a**) The location at the NIOZ jetty sampling station in the western part of the Dutch Wadden Sea [54]; (**b**) The optical system mounted on a pole on the platform of the NIOZ jetty in the Wadden Sea [53].

In addition, specific inherent optical properties (SIOPs) of water constituents in the Wadden Sea were obtained from Hommersom et al. [11], who documented SIOP measurements in 2007 at 37 stations in this area.

2.3. Satellite Observations

The MERIS sensor, operational on board the European environmental satellite ENVISAT between 2002–2012, was primarily intended for ocean, coastal and continental water remote sensing. MERIS was an orbital sensor with 15 bands covering the spectral range from 400 to 950 nm and was succeeded by the Ocean and Land Color Instrument (OLCI) on board Sentinel-3 beyond 2015 [55]. The high sensitivity and large dynamic range of the MERIS sensor has been widely used for ocean and coastal water remote sensing [56–59]. In this study, ocean color data were obtained from ESA archive of MERIS images (full resolution: 300 m) covering the Wadden Sea during 2002–2012 (data provided by European Space Agency). MERIS has a revisit time of three days over the Dutch Wadden Sea at around 10:30 a.m. local time. The MERIS 1b image provides TOA radiance information and some environmental parameters for each pixel. Some of these environmental parameters (such as sun zenith angle (SZA), view zenith angle (VZA), relative azimuth angle (RAA), water vapor (H_2O) and ozone (O_3)) were used as input parameters to perform MODTRAN simulations in this study.

2.4. Ground Truth and Satellite Observation Data Matchups

Validation of ocean color products (i.e., biogeochemical parameters, inherent optical properties (IOPs) and water-leaving radiance), theoretically, should be performed from ground truth measurements acquired simultaneously to the satellite overpass over the same location (the so-called matchup points) [60]. In this study, the following criteria were used to find matchup points between satellite observations and ground truth measurements: (1) all available MERIS images over the Dutch part of Wadden Sea between 2002 and 2012 were checked to select the cloud-free images; (2) a narrow time window of ±1 h was used; (3) five-by-five pixel kernels centered on the ground truth measurement coordinates were then extracted from the MERIS images using BEAM software (version 5.0) (no aggregation method was used to avoid possible spectral contamination); (4) finally, 35 suitable MERIS images were concurrent with ground truth-measured concentrations of Chla at the NIOZ jetty station during 2007–2010.

3. Methodology

The accuracy of the coupled MOD2SEA model in doing atmospheric correction and deriving Chla concentration values was evaluated against ground truth measurements and was compared with C2R results.

3.1. The Coupled MOD2SEA Model

The developed MOD2SEA method combined two lookup tables (LUTs) from 2SeaColor and MODTRAN as schematically shown in Figure 3.

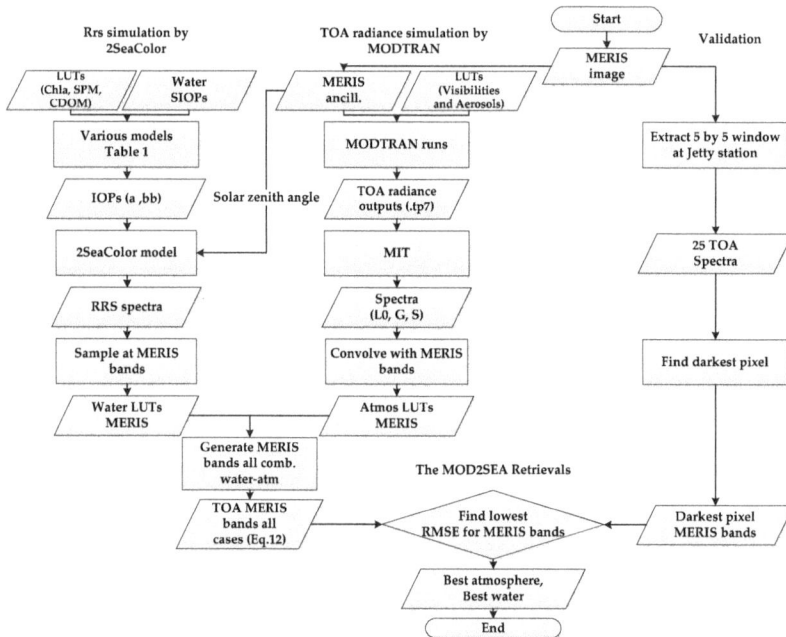

Figure 3. Diagram of the coupled MOD2SEA model (pixel based).

These LUTs were generated by simulating the water-leaving reflectance for varying ranges of the governing biophysical variables (with respect to range of these water quality variables at the NIOZ jetty station (Table 1)) and MODTRAN parameters based on different combinations of visibilities and aerosol models at specific viewing-illumination geometries for every MERIS image separately. Table 1 presents the LUT composition of the 2SeaColor model and the MODTRAN input variables in this assessment.

Table 1. Lookup table composition of MOD2SEA model.

LUT Variables	Range	Increment	Unit
Chla	0–150	5; 0.1	$mg \cdot m^{-3}$
SPM	0–150	5; 0.1	$g \cdot m^{-3}$
CDOM absorption at 440 nm	0–2.5	1; 0.1	m^{-1}
Visibility	5–50	1	km
Aerosol type	Rural, Maritime, Urban	-	-

The details on simulation of R_{RS} by the 2SeaColor model and TOA radiance by the MODTRAN radiative transfer code are described as follows:

3.1.1. Reflectance (R_{RS}) Simulation by 2SeaColor Forward Model

The 2SeaColor model is based on the solution of the two-stream radiative transfer equations including direct sunlight, as described by Duntley (1942, 1963) [61,62]. Both the analytical forward model and the inversion scheme are provided in detail in Salama and Verhoef [50]. The reflectance result predicted by the 2SeaColor model is r_{sd}^{∞}, the directional-hemispherical reflectance of the semi-infinite medium, which is linked to IOPs by Salama and Verhoef [50]:

$$r_{sd}^{\infty} = \frac{\sqrt{1+2x}-1}{\sqrt{1+2x}+2\mu_w} \tag{1}$$

where x is the ratio of backscattering to absorption coefficients ($x = b_b/a$), and μ_w is the cosine of the solar zenith angle beneath the water surface. The reflectance factor r_{sd}^{∞} can be approximated by $Q \times R(0^-)$ under sunny conditions, where $Q = 3.25$ and $R(0^-)$ is the irradiance reflectance beneath the surface [63], which can be converted to above-surface remote sensing reflectance (R_{RS}) by Lee et al. [64].

$$R_{RS} = \frac{0.52 \times R\left(0^-\right)}{1 - 1.7 \times R\left(0^-\right)} \tag{2}$$

Total absorption and backscattering coefficient of water constituents (a and b_b) were calculated using Equations (3) and (4) respectively [27,48].

$$a\left(\lambda\right) = a_w\left(\lambda\right) + a_{chl}\left(\lambda\right) + a_{nap}\left(\lambda\right) + a_{cdom}\left(\lambda\right) \tag{3}$$

$$b_b\left(\lambda\right) = b_{bw}\left(\lambda\right) + b_{b,chl}\left(\lambda\right) + b_{b,nap}\left(\lambda\right) \tag{4}$$

where the subscripts *w*, *chl*, nap and *cdom* stand for water molecules, chlorophyll, non-algae particles and colored dissolved organic matter, respectively. As implemented in Salama and Shen [65], the absorption coefficients of the water constituents (a) are parameterized by (Bricaud et al. [66]; Lee et al. [67]; Lee et al. [68]). Also, the backscattering coefficients of the water constituents (b_b) were parametrized by (Doxaran et al. [69] and Morel et al. [70]).

Table 2. Summary of the used parameterizations.

Variable	Parametrization	Equation	Reference
Chla absorption	$a_{chl}\left(\lambda\right) = [a_0\left(\lambda\right) + a_1\left(\lambda\right) \times \ln a_{chl}\left(443\right)] \times a_{chl}\left(443\right)$ $a_{chl}\left(443\right) = 0.06 \times [Chl]^{0.65}$	(5)	[68]
CDOM absorption	$a_{cdom}\left(\lambda\right) = a_{cdom}\left(440\right) \times \exp\left[-S_{cdom}\left(\lambda - 440\right)\right]$	(6)	[66]
Nap absorption	$a_{nap}\left(\lambda\right) = a_{nap}\left(440\right) \times \exp\left[-S_{nap} \times \left(\lambda - 440\right)\right]$ $a_{nap}\left(440\right) = a_{nap}^*\left(440\right) \times [SPM]$	(7)	[67]
Chla backscattering	$b_{b,chl}\left(\lambda\right) = \left\{0.002 + 0.01 \times [0.5 - 0.25 \times \log_{10}[C_{chl}] \times \left(\frac{\lambda}{550}\right)^n]\right\} \times b_{b,chl}\left(550\right)$ $b_{b,chl}\left(550\right) = 0.416 \times [Chl]^{0.766}$	(8)	[70]
NAP backscattering	$b_{nap}\left(\lambda\right) = b_{nap}\left(550\right) \times \left(\frac{550}{\lambda}\right)^{-\gamma} - [1 - \tanh\left(0.5 \times \gamma^2\right)] \times a_{nap}\left(\lambda\right)$ $b_{nap}\left(550\right) = b_{nap}^*\left(550\right) \times I \times [SPM]$	(9)	[69]
Scattering of water molecules	$b_{bw}\left(\lambda\right)$; Listed values, Table (3.8), page 104.	(10)	[71]
Absorption of water molecules	$a_w(\lambda)$; Listed values	(11)	[72]

In Table 2, [*Chl*], [*SPM*] and a_{cdom} (440) stand for Chla concentration, SPM concentration and the CDOM absorption at 440 nm respectively. The absorption and backscattering coefficients of water molecules (a_w and b_{bw}) were taken from previous studies (Mobley [71]; Pope and Fry [72]) and a_0 and a_1 were given in Lee et al. [67]. The initial values of non-algae particle absorption (a^*_{nap} (440) = 0.036 $m^2 \cdot g^{-1}$), spectral slope of non-algae particles (S_{nap} = 0.011 nm^{-1}), spectral slope of

CDOM (S_{cdom} = 0.013 nm^{-1}) and specific scattering coefficient of non-algae particles (b_{nap}^* (550) = 0.282) were taken from the Hommersom et al. [11] SIOP measurements at 37 stations in the Wadden Sea. Also, the initial values of γ and I (γ = 0.6 and I = 0.019) for the North Sea were taken from Doxaran et al. [69] and Petzold [73], respectively. In this study, we used the 2Seacolor forward model and the various parameterizations described in Table 2 to simulate the water-leaving reflectance (R_{RS} spectra) values for a series of combinations of Chla, SPM and CDOM concentration (Table 1) and for the given SZA associated with every MERIS image separately. The simulated values of R_{RS} spectra for all MERIS bands were stored in a water LUT for the MERIS bands and then used as R_{RS} input parameters for MODTRAN to calculate the TOA radiances in the MERIS bands.

3.1.2. Top of Atmosphere (TOA) Radiance Simulation by MODTRAN

MODTRAN is the successor of the atmospheric radiative transfer model LOWTRAN [74]. It is publicly available from the Air Force Research Laboratory in the USA. The latest version of MODTRAN (5.2.1) contains large spectral databases of the extraterrestrial solar irradiance and the absorption of all relevant atmospheric gases at a high spectral resolution. The accurate calculation of atmospheric multiple scattering makes it a very appropriate tool for reliable simulation and interpretation of remote sensing problems in the optical and thermal spectral regions [75]. To apply MODTRAN simulations, first of all several parameters describing the real atmospheric conditions should be determined as inputs for this model. Table 3 shows the standard definition of MODTRAN inputs with respect to the ranges of average values of atmospheric and geometric variables variation over one image for four years of all available MERIS images between 2007 and 2010 over the Dutch part of the Wadden Sea. In the MERIS image, some of the local atmospheric (O_3, H_2O) and geometric variables (VZA, SZA and RAA) can be used as input for MODTRAN. Note that for every MERIS image a separate input file was created by establishing the local atmospheric (O_3, H_2O, CO_2) and geometric variables (VZA, SZA, RAA) of that specific run to MODTRAN (Figure 3). These parameters could be retrieved from MERIS ancillary data per pixel using Matlab.

Table 3. Input parameters for MODTRAN4 simulations.

Parameter	Range or Value	Unit
Atmospheric profile	Mid Latitude Summer	-
Correlated-k option	Yes	-
DISORT number of streams	8	-
Concentration of CO_2 *	380–390	ppm
H_2O	0.5–4.5	g·cm^{-2}
O_3	250–450	DU
SZA	30–80	degree
VZA	5–30	degree
RAA	0–150	degree
Visibility	5–50 (1 km increment)	km
Aerosol Model	Rural, Maritime, Urban	-
Surface height	0	km
Sensor Height	800	km
Molecular band model resolution	1.0	cm^{-1}
Start, ending wavelength	−1000	nm

* Annual CO_2 concentration level can be in Global Greenhouse Reference Network [76].

In this study, we varied the aerosol type (rural, maritime and urban) and visibility (5 to 50 km with 1 km step) and thus made a total of 135 scenarios for each lookup table and given atmospheric state and angular geometry, which were extracted from the MERIS image ancillary data per image. For each scenario the MODTRAN Interrogation Technique (MIT) was applied by using surface albedos of 0.0, 0.5 and 1.0 (the MIT is explained in detail by Verhoef and Bach [75]). The output .tp7 file of MODTRAN quantified the TOA radiance spectrum for each simulated wavelength from 350 to

1000 nm. Then in the MIT the .tp7 file was used as input to derive three MODTRAN parameters (gain factor (G), path radiance (L_0), and spherical albedo (S)). These parameters are spectral variables depending on various atmospheric conditions [75]. The spectral response functions (SRF) of the MERIS bands were convolved with the MODTRAN parameters to compute L_0, G and S for every MERIS band and these simulations were stored in the atmospheric LUTs (Atmos LUTs MERIS).

3.1.3. The MOD2SEA Retrievals

The simulated TOA radiance of MERIS data in the MODTRAN output file, L_{TOA} ($Wm^{-2} \cdot sr^{-1} \cdot \mu m^{-1}$), Can be expressed in surface reflectance r by the following equation [77]:

$$L_{TOA} = L_0 + \frac{Gr}{1 - Sr} \quad (12)$$

where r is the hemispherical reflectance ($= \pi R_{RS}$) leaving the water surface, L_0 is the total radiance for zero surface albedo ($Wm^{-2} \cdot sr^{-1} \cdot \mu m^{-1}$), S is the spherical albedo of the atmosphere and G is the overall gain factor. In this study, the LUTs of water-leaving reflectance generated by the 2SeaColor model were used as R_{RS} input parameters of Equation (12) to calculate TOA radiance for all combinations of water properties and atmospheric conditions and then organized in a water-atmosphere lookup table (water-atmosphere). The simultaneous retrieval of Chla, SPM, CDOM concentration, aerosol type and visibility was then performed by spectrally fitting the MOD2SEA-simulated TOA radiances (using RMSE) to MERIS TOA radiances for all MERIS bands except the band numbers 1, 2 and 11. Band 11 is located in the O_2-A absorption band and can give erroneous results due to sampling errors of MERIS. Bands 1 and 2 gave systematic deviations in R_{RS} after atmospheric correction. The cause of this problem is presently still unknown. In this retrieval, Chla retrieval using the coupled MOD2SEA model was performed in two steps. First the increments of 5, 5 and 1 were taken for Chla concentration ($mg \cdot m^{-3}$), SPM concentration ($g \cdot m^{-3}$) and CDOM absorption at 440 nm (m^{-1}), respectively, to find an approximate solution. Later, in the refined step, the step size of the LUTs composition was reduced to 0.1 for all water constituents in the identified rough range resulting from the first step. Applying this approach led to speeding up the running of the Matlab code and to obtaining more precise results. Although Figure 3 suggests the storage of a fixed LUT for water R_{RS} for each MERIS image, this LUT was only generated in a loop, and not stored, in order to reduce memory requirements. The best fitting combination of water properties and atmospheric conditions was found during the generation of the water LUT, but this water LUT was never stored as such, contrary to the atmospheric LUT, which was actually stored. This approach also allowed greater flexibility by applying the two-step procedure in finding the best-fitting water properties, by first applying a rough search in the first round with large steps in the three concentrations, and in the next round a refined search with small steps over much smaller ranges. It should be noted that the current procedure applied to a single pixel per matchup date is not suitable to be applied pixel by pixel, and this issue is left for a future study.

3.2. MERIS Case-2 Regional (C2R) Processor

The Case-2 regional processor (C2R) [42], available in the Basis ERS and ENVISAT (A) ATSR and MERIS Toolbox (BEAM) software, has been widely used to derive water quality parameters from MERIS images [78–81]. The C2R processor consists of two procedures, one for atmospheric correction and one for the bio-optical part for retrieving the IOPs of water columns. The Neural Networks (NN) in C2R were trained with Hydrolight [71] simulations and in situ measurements in the German bight and from other cruises in European seas [42]. More details can be found in Doerffer and Schiller [42]. The output of the C2R processor, including IOPs: the absorption coefficient of Chla at wavelength 443 nm (a_{chl} (443)), the absorption coefficient of CDOM (a_{CDOM} (443)), the total absorption (a_{tot} (443)), and the scattering coefficient of SPM (b_{spm} (443)) were then used to define water quality parameters such as Chla and SPM. Equations to relate BEAM processor IOPs to water quality concentrations of Chla and SPM are presented as follows:

$$[Chla] = 21.0 \times a_{chl}(443)^{1.04} \tag{13}$$

$$[SPM] = 1.72 \times b_{spm}(443) \tag{14}$$

where [Chl], [SPM], $a_{chl}(443)$ and $b_{smp}(443)$ stand for Chla concentration, SPM concentration, the Chl absorption at 443 nm and SPM scattering coefficient at 443 nm, respectively.

3.3. Validation

To evaluate the accuracy of the MOD2SEA coupled model and the C2R processor, we applied these two models to the 35 matchup moments of MERIS observations and four years of concurrent Chla measurements (2007–2010) at the NIOZ jetty location, separately. The validation of model simulations were performed in two different levels of atmospheric correction and water retrieval models. Since the NIOZ jetty station is located close to the land, for every image, the darkest pixel from 5 by 5 pixels around the location of this station was extracted first. By selecting the darkest pixel from the 5 × 5 neighborhood centered on the jetty station, we exclude cloudy and land pixels, as well as water pixels close to the shore that are possibly influenced by an adjacency effect due to the near land area. Of course an underlying assumption in our approach is that the water of the darkest pixel has the same composition as found at the location of the jetty station. However, since the water current is mostly strong near the inlet to the Wadden Sea, we are confident that the water is well-mixed, and local gradients in water properties are small.

3.3.1. Atmospheric Correction

The accuracy of atmospheric correction methods using the coupled MOD2SEA model and C2R processor was evaluated against the ground truth water-leaving reflectance for all 35 matchups between 2007 and 2010 at the NIOZ jetty station. Four statistical parameters, the root mean square error (RMSE), the determination coefficient (R^2), the normalized root mean square error (NRMSE) and relative root mean square error (RRMSE) [82] were used to quantify the goodness-of-fit between derived and measured water-leaving reflectance values at the NIOZ jetty data where near-concurrent (±1 h) MERIS measurements were available. To do this, three MERIS bands 3, 5 and 7 were selected. Finally, the accuracy of the proposed MOD2SEA model in doing atmospheric correction was compared against C2R processor products. The results of this assessment are presented in Section 4.2.

3.3.2. Water Model Inversion

The accuracy of retrieved Chla concentration values using the coupled MOD2SEA model and the C2R processor were evaluated against ground truth Chla measurements for all 35 matchup points at the NIOZ jetty station between 2007 and 2010. The results of this evaluation are presented in Section 4.3. It should be mentioned that in view of the main objective of this study (retrieval of Chla concentration) and the availability of ground truth measurements, investigation of changes in other water constituents (SPM and CDOM concentration) was considered to fall outside of the scope of this study, although these were retrieved along with Chla using MOD2SEA. In addition, the visibility and aerosol type were retrieved simultaneously with water quality parameter concentration which were used in the model to simulate water-leaving reflectance values based on the best matching TOA radiance by MOD2SEA coupled model.

4. Results

4.1. Variability of MODTRAN Parameters (L_0, G and S) at Different Atmospheric Conditions

The case of the three aerosol types—rural, maritime and urban—for a visibility of 20 km on 7 October 2007 was used as an example to display the result of applying MODTRAN to the MERIS bands for three atmospheric conditions. We used MIT method [75] to derive L_0, G and S values using surface albedos of 0.0, 0.5 and 1.0 for the mentioned visibilities and aerosol types in these figures.

The atmospheric path radiance L_0 represents the case when the surface reflectance is zero and the radiance at the top of atmosphere comes from atmospheric scattering alone. As Figure 4 shows, L_0 values decrease with wavelength, which means at longer wavelengths the atmosphere scatters less. The S presents the spherical albedo values which are not large and show a similar trend to L_0. The gain factor G contains the product of the extraterrestrial solar irradiance and the total two-way transmittance through the atmosphere, and shows a maximum at about 500 nm. L_0, S and G vary with different combinations of aerosol types and visibilities, while for maritime and rural aerosol types, they have similar values. The urban aerosol model has a stronger absorption and always has lower values when compared to the other two aerosol models. Examples of the MODTRAN path radiance simulations (L_0) from 7 October 2007, for visibilities of 5, 10 and 40 km while water-leaving reflectance is zero as representative for a range of haze conditions and three different aerosol models are presented in Figure 5.

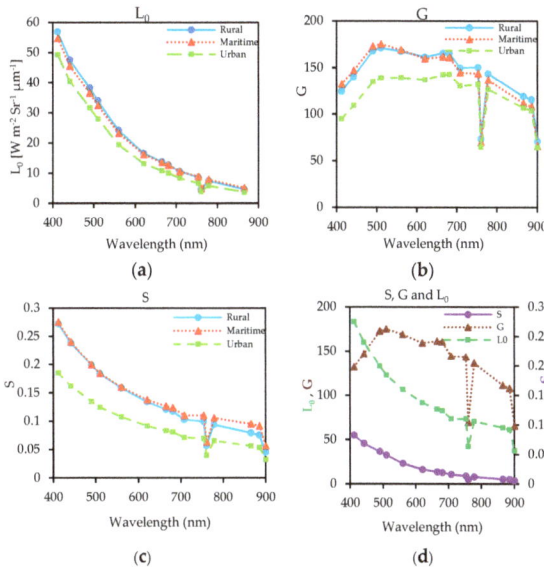

Figure 4. (a–c) L_0, G and S values at the visibility of 20 km and different aerosol types; (d) The atmospheric parameters L_0, S and G for the maritime aerosol type and a visibility of 20 km.

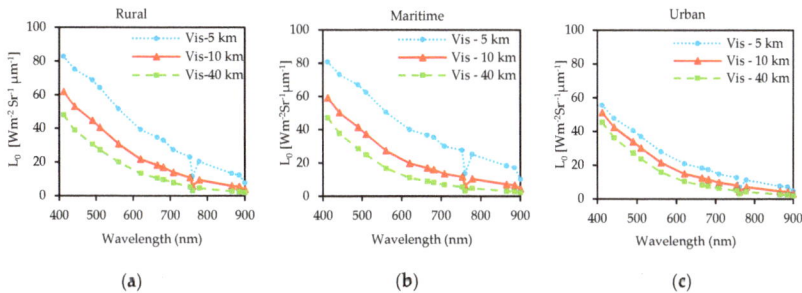

Figure 5. TOA radiance simulated by MODTRAN for (a) rural; (b) maritime and (c) urban aerosol types respectively.

As this figure shows, the calculated TOA radiances for the urban aerosol type show a lower range of variation compared to the maritime and rural cases. All the values of TOA radiance for the urban aerosol type are between 0 and 60 ($Wm^{-2}{\cdot}sr^{-1}{\cdot}\mu m^{-1}$), while these values for maritime and rural ones vary between 0 and 80 ($Wm^{-2}{\cdot}sr^{-1}{\cdot}\mu m^{-1}$). On the other hand, the simulated TOA radiances by MODTRAN differ significantly not only with aerosol type, but also with visibility. Lower visibility gives higher TOA radiances. Consequently, a wrong assumption about visibility or aerosol type leads to a wrong calculation of water-leaving reflectance and as a result the water parameter concentrations may be overestimated or underestimated.

4.2. Atmospheric Correction Validation

The results of performing of the coupled MOD2SEA model and the ESA MERIS standard C2R processor to derive water-leaving reflectances for MERIS bands of 3, 5 and 7 against ground truth measurements are shown in Figure 6. The statistical analysis regarding this assessment are presented in Table 4.

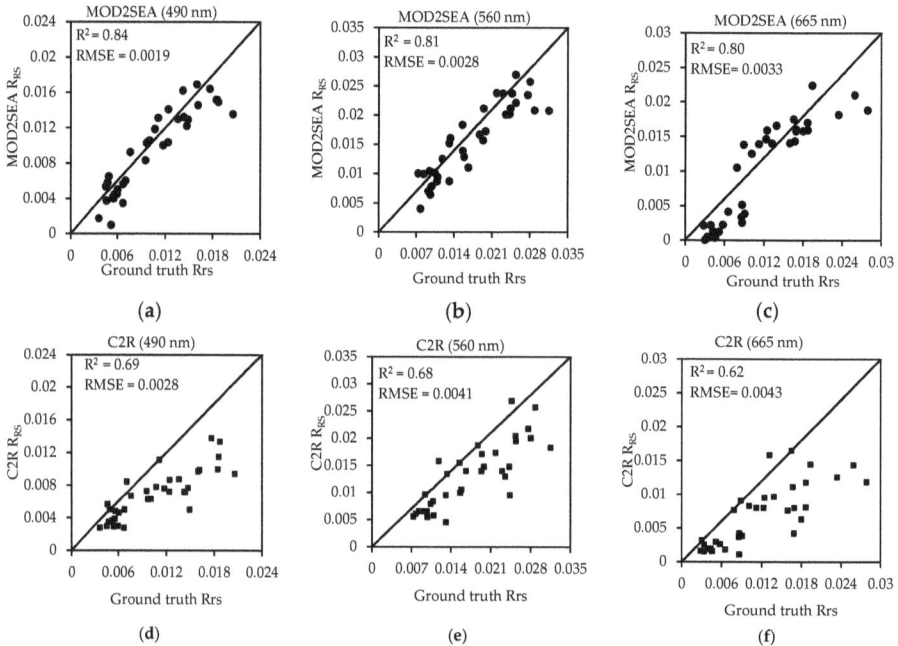

Figure 6. Comparison between MERIS-retrieved values and ground truth measurements for water-leaving reflectance values (Rrs) for 35 matchup in 2007–2010 at NIOZ jetty location. (**a**–**c**) represent the retrieved Rrs using the coupled MOD2SEA model against ground truth measurements for MERIS band of 3, 5 and 7 (band centers: 490, 560 and 665 nm) respectively; (**d**–**f**) represent the retrieved Rrs values using C2R processor against ground truth measurements for MERIS bands centers of 3, 5 and 7 (band centers: 490, 560 and 665 nm) respectively.

As it can be seen from Figure 6, the coupled MOD2SEA model provides significant improvements in the atmospheric correction and the resulting water-leaving reflectance in comparison with C2R processor in all MERIS bands of 3, 5 and 7. More details of this evaluation are presented in Table 4.

As the statistical measures show, performing atmospheric correction by applying the MODTRAN lookup table proposed in the MOD2SEA coupled model resulted in a reasonable accuracy against ground truth above the water radiometric dataset for 35 matchups between 2007–2010 at the NIOZ jetty station for bands 3, 5 and 7 respectively. In addition, the MOD2SEA coupled model shows significant improvement especially in band 3 with $R^2 = 0.84$, RMSE = 0.0022, NRMSE = 13.18% and RRMSE = 21.08% in comparison with C2R. The standard C2R processor also shows higher accuracy for band 3 ($R^2 = 0.69$, RMSE = 0.0047) in comparison with bands 5 ($R^2 = 0.68$, RMSE = 0.0058) and 7 ($R^2 = 0.62$, RMSE = 0.0063), respectively.

Table 4. Models' performance evaluation in atmospheric correction part.

Statistical Measures	R^2		RMSE		NRMSE (%)		RRMSE (%)	
MERIS bands/Model	MOD2SEA	C2R	MODSEA	C2R	MOD2SEA	C2R	MOD2SEA	C2R
3	0.84	0.69	0.0022	0.0047	13.18	28.03	21.08	44.81
5	0.81	0.68	0.0034	0.0058	14.38	24.20	20.07	33.78
7	0.80	0.62	0.0035	0.0063	14.39	25.51	30.94	54.87

4.3. Water Retrieval Validation

The comparisons of C2R and MOD2SEA Chla retrieval against ground truth measurements are shown in Figure 7 and related statistical analysis are presented in Table 5.

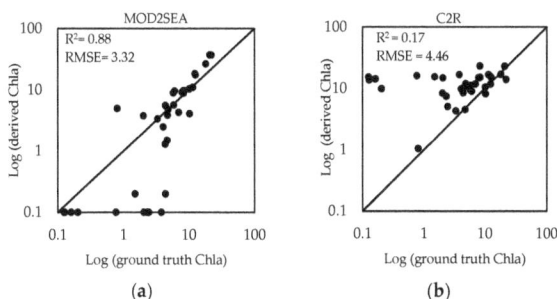

Figure 7. Comparison between MERIS-derived and measured log Chla (mg·m^{-3}) for 35 matchup moments. (**a**) MOD2SEA and (**b**) C2R.

Assessing the model accuracy using R^2 and RMSE shows the reasonable agreement between the measured and retrieved Chla (mg·m^{-3}) for all the matchup points during 2007–2010 at the NIOZ jetty location with a significant regression (Figure 7: $R^2 = 0.88$ and RMSE = 3.32 mg·m^{-3}) during the period of four years. In addition, the comparison of this model with the C2R processor shows significant improvement in retrieval of Chla. The result of this comparison is presented in Table 5.

Table 5. Models performance evaluation Chla retrieval.

Statistical Measures	R^2	RMSE	NRMSE (%)	RRMSE (%)
MOD2SEA	0.88	3.32	15.25	53.31
C2R	0.17	4.42	20.30	70.98

There are several possible reasons for the improvement of MOD2SEA in the retrieval of Chla in comparison with the C2R procedure, but the most obvious one is probably that the SIOPs used in the training of the C2R neural network might be more generic and thus different from the ones used in this study and which are more applicable to the Wadden Sea. In addition, the derived Chla data for 35 matchups between 2007–2010 by the MOD2SEA coupled model was examined to see how well the ground truth values (mg·m^{-3}) agreed with those derived from the MERIS images (mg·m^{-3}) at the NIOZ jetty location (Figure 8). In this figure, the X-axis presents the date while the Y-axis presents the

Chla concentration for ground truth data (in blue), the MOD2SEA coupled model (in red) between 2007 and 2010.

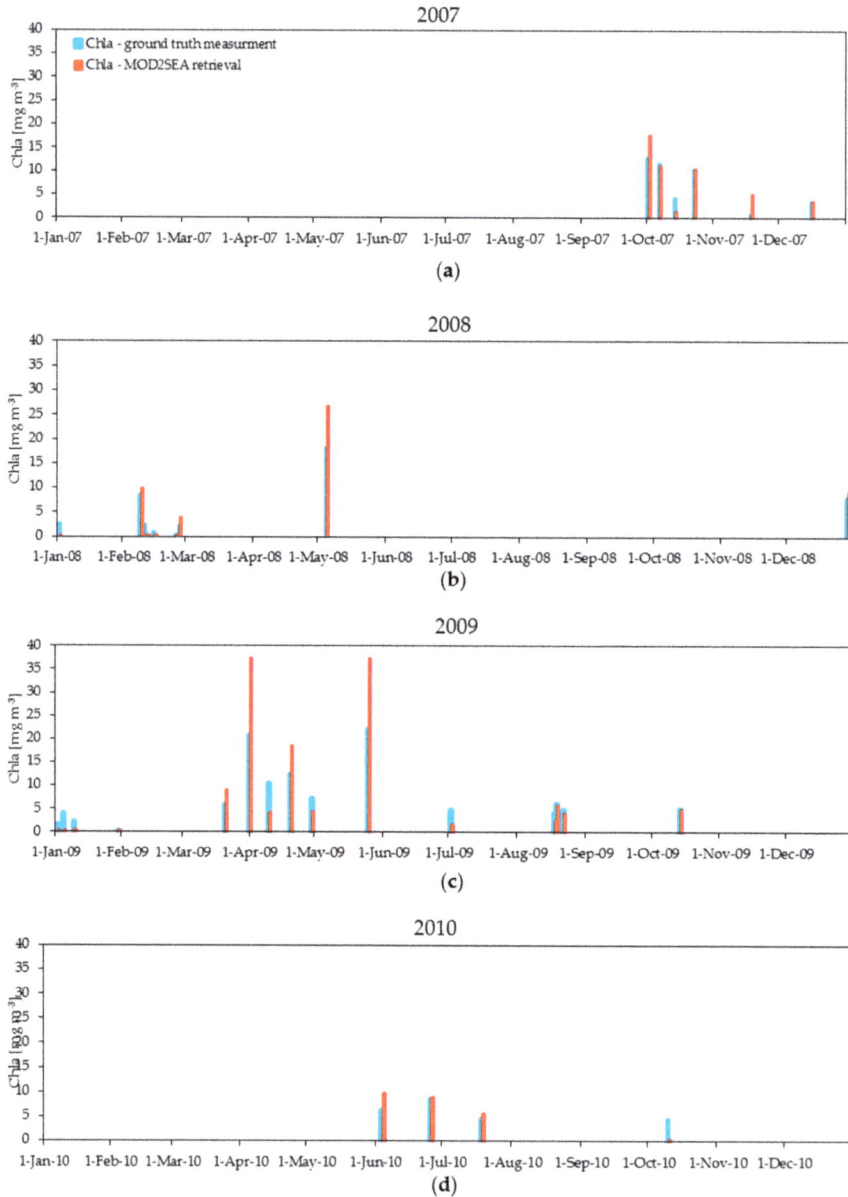

Figure 8. (**a–d**) The four-year comparison of derived Chla values using the coupled MOD2SEA model (red line) and ground truth measurements (blue line) (mg·m^{-3}) from 2007–2010 at matchup moments.

As Figure 8 shows, the derived Chla concentration values using the MOD2SEA model shows reasonable agreement during 2007–2010, with maximum retrieved values of around 40 mg·m^{-3} and minimum values just above zero. However, despite the agreement between MERIS-derived and ground truth Chla in a four-year period, systematic overestimations at high Chla concentration values (during April and May) $mg \cdot m^{-3}$ were also identified. Chla products, particularly during the phytoplankton bloom seasons of spring and summer, require further development. This overestimation might be explained by the Chla parametrization of the Lee et al. [68] model, since it appears that the Chla model calibration based on that model does not fit that well for the Wadden Sea. This Chla overestimation using satellite images was also in agreement with a Chla retrieval overestimation in most of the European seas studies by Zibordi et al. [58].

5. Discussion

Accurate estimation of water-leaving reflectance from satellite sensors is a fundamental goal for ocean color satellite missions [83]. Basically, the commonly applied atmospheric correction methods based on zero water-leaving reflectance in the near-infrared bands fail when applied to turbid waters since the high concentrations of water constituents lead to a detectable water-leaving reflectance in the near-infrared region in satellite image. In this study, we focused on the long-term retrieval of Chla concentration from MERIS images in the Wadden Sea, and the MOD2SEA coupled model is proposed as a tool to improve the retrieval of Chla concentration from earth observation data in this area.

Calculating accurate water-leaving reflectance spectra in order to translate them into Chla concentration under different atmospheric conditions is a crucial part of this study, since the atmosphere, in most cases, contributes more than 90% of the TOA radiance signal [41]. We can attribute the success of the MOD2SEA coupled model to its capability of combining simulations from 2SeaColor with the MODTRAN radiative transfer model for different combinations of aerosol type, visibility and water constituent concentrations for all MERIS bands to simulate TOA radiances, instead of applying routine atmospheric correction and water retrieval algorithms separately. Furthermore, based on a heterogeneous atmosphere assumption of the coupled MOD2SEA model, this technique can help suppress the influence of local haze variations in satellite images. Thus, applying this method results in a considerable improvement of the accuracy of the atmospheric correction, which is the most problematic part of remote sensing data processing for turbid waters like the Wadden Sea.

However, satellite estimation of Chla concentration is still difficult for coastal waters, where Chla, SPM and CDOM occur in various mixtures which complicate the derivation of their concentrations from reflectance observations. The 2SeaColor model performed well while the commonly applied water quality algorithms might fail in water constituent retrieval. Figure 9 shows an example of coupled MOD2SEA model spectral matchings for 2 October 2007.

As this figure shows, good matches are found between modelled and observed TOA radiance as well as modelled and atmospherically corrected water-leaving reflectance with respect to identified visibility and aerosol type by the coupled model. All in all, assessing the MOD2SEA Chla retrievals from MERIS data at one location (NIOZ jetty station) for a period of four years (2007–2010) shows reasonable agreement with ground truth measurements ($R^2 = 0.88$, RMSE = 33.2%). The 33.2% RMSE appears reasonable enough, as compared with the validation of the SeaWiFS Chla data product for global open ocean waters with a relative RMSE of about 58% [1]. In addition, this model shows considerable improvement to retrieve Chla concentration from satellite images in comparison with similar studies for the Wadden Sea [10,18]. The results of retrieved Chla concentration using this coupled model are within the range of measured Chla concentration ($mg \cdot m^{-3}$) on the ground reported by other researchers, while a clear seasonal pattern is observed with the peak values during spring (in May). For example, Hommersom [10] reported Chla concentration range variations in the Wadden Sea during eight surface water sampling campaigns in 2006–2007 in 156 stations while Chla also showed a strong seasonal pattern with the highest values during spring in May. Chang et al. [84] showed the higher Chla concentrations occurred in May. Reuter et al. [85] provided continuous data on Chla concentration of the Wadden Sea at a time series station established in autumn 2002 by the University of Oldenburg. They reported a clear seasonal pattern in Chla concentration between 2007–2008 which has the highest values in May and the lowest values in November. Tillman et al. [86]

showed a large variability of Chla concentration over the year in the Wadden Sea. Winter concentrations were much lower than summer concentrations while in spring a phytoplankton bloom with peak concentrations occurs. Cadée and Hegeman [87] showed that yearly patterns of Chla concentration were similar in the Wadden Sea, although the overall inter-annual variability is large, as well as the maxima measured during spring bloom (in May). All in all, regarding the reasonable agreement of the MOD2SEA results with ground truth measurements and considering the turbid nature and complex heterogeneity in the turbid Wadden Sea, the performance of this coupled model should be regarded as encouraging and satisfactory enough.

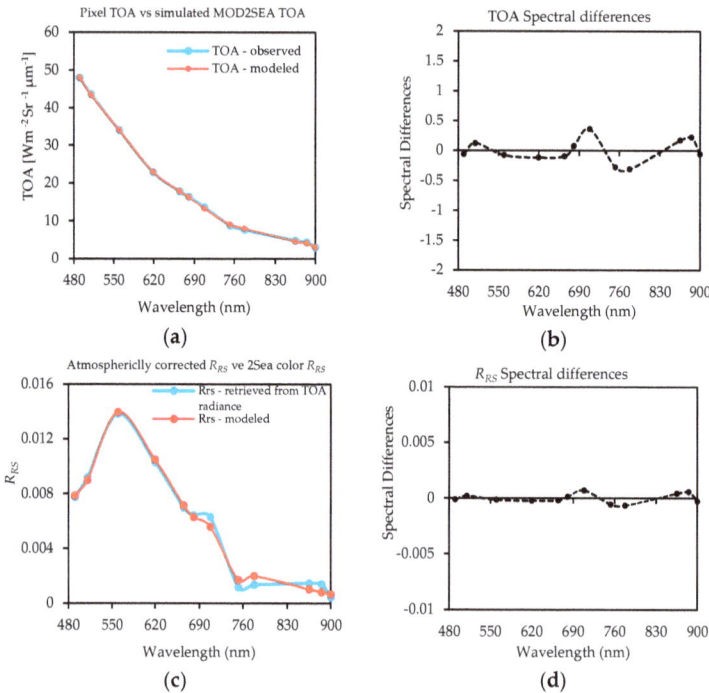

Figure 9. (**a**) The best match identified by the coupled model between simulated TOA radiances vs. pixel TOA radiance; (**b**) Spectral differences between observed and simulated TOA radiance; (**c**) The simulated R_{RS} (extracted from the best TOA radiance match) vs. the simulated 2SeaColor R_{RS}; (**d**) Spectral differences between simulated R_{RS} by 2SeaColor and atmospherically corrected R_{RS} from observed TOA radiance by MOD2SEA.

It is also worth mentioning that in shallow coastal waters like the Wadden Sea the bottom might influence the reflected signal to the sensor. This is not the case for the NIOZ jetty data where, due to the depth of >5 m and the high turbidity of the water (Table 1) near the NIOZ jetty and the surrounding area, the bottom effect on observed reflectance is negligible. This has been confirmed in the quality check of the NIOZ jetty data and the corresponding MERIS pixels. However, in the other shallower parts of the Wadden Sea, the bottom effect might contribute substantially in the visible region of the spectrum. As can be seen in Figure 1, the effect of the bottom is visible in large areas of the Wadden Sea satellite image. Thus, for shallow waters it is recommended in future studies to develop water constituent retrieval algorithms by incorporating sea bottom effects in the hydro-optical model. We speculate that developing a hydro-optical model including the bottom effect may lead to significant improvements in the derived water constituent concentrations from earth observation data in this shallow coastal region. That is why, in the next phase of this research, we are going to include the bottom effect contribution into the TOA radiance calculation to derive and provide Chla concentration maps over the Wadden Sea.

For the Wadden Sea, and many other estuaries, knowledge of local specific inherent optical properties to locally calibrate retrieval algorithms is often lacking, and more research is still needed. For the Wadden Sea, Peters et al. [88] reported a complete set of SIOPs for Chla, SPM and CDOM measurements. However, the data of Peters were all collected at one location (the Marsdiep inlet) and only for two days (in May 2000). After that, the only published set of SIOP measurements in the Wadden Sea was constructed by Hommersom et al. [11]. Using Hommersom's measurements, SIOPs increased the accuracy of the derived Chla concentration significantly in comparison to previous efforts in this region. However, Hommersom's measurements lack seasonal information on the SIOPs while there is currently not much information on the SIOPs to be the basis for a hydro-optical model for the Wadden Sea. Without any doubt, having seasonal SIOPs may lead to an improved accuracy of retrieved Chla using this coupled model. Thus, more in situ data (especially on SIOPs) is still necessary for the model calibration.

Although our current efforts are centered on validating the proposed coupled atmospheric-hydroptical model in the highly turbid Wadden Sea using MERIS satellite images, it is unclear how broadly applicable this coupled model will be and to what extent these findings could be generalized. Thus, we suggest to extend this study to other parts of the world using various ocean color remote sensors. However, to apply this method to other regions, first the availability of valid SIOPs (water quality constituent's absorption and backscattering coefficient), in addition to the accurate ecological and geophysical knowledge of the interest area (i.e., the ranges of water constituent concentrations) are needed. Furthermore, spectral response functions of the desired sensor as well as atmospheric parameters and illumination geometry of the satellite image to run MODTRAN are required. As a consequence, access to accurate in situ water-leaving reflectance and water quality parameter concentration is essential for the assessment of primary data products from satellite ocean color missions [55] using the proposed approach.

The water Framework Directive regulations from the European Union force member states to monitor all their coastal areas (Environment Directorate-General of the European Commission, 2000). Availability of one decade of MERIS images (2002–2012) over the Wadden Sea, gives the opportunity to provide long-term Chla distribution maps using this coupled model with reasonable accuracy and to conduct a one-decade phenological analysis in this area. To provide Chla concentration maps with reasonable accuracy, the proposed method should be applied pixel by pixel for the whole region of interest. To speed up the pixel-based approach, a filter can be introduced to remove those combinations of visibilities and aerosol types from the MODTRAN lookup table which result in negative water-leaving reflectance values in any band, by considering the recorded TOA signal per pixel. On the other hand, other water quality parameters like SPM and CDOM as well as visibility and aerosol model maps can be produced as output of the MOD2SEA code. Of particular interest when analyzing the variability in the MERIS-derived Chla data trend for the Wadden Sea is whether any significant decreasing trend from 2002–2012 would indicate the effect of prior nutrient reduction management actions. This has significant implications for identifying positive anomaly events and may act as an alert for management actions. Clearly, climatic variability needs to be considered carefully when interpreting the long-term data trends and when making management decisions [1]. Furthermore, this established MERIS-based Chla data record may serve as baseline data to continuously monitor the estuary's eutrophic state, and the validated algorithm may extend such observations to the future using various satellite continuity missions. The Ocean Land Color Instrument (OLCI), embedded on the Sentinel-3 platform, is a sensor especially adapted for aquatic remote sensing [89] and succeeded the MERIS sensor in 2015 [90]. The launch of Sentinel-3 and OLCI will secure future consistent operational monitoring by medium resolution data for water quality assessment also of coastal zones and bays [89]. OLCI is designed mainly for global biological and

Remote Sens. **2016**, *8*, 722

biochemical oceanography, which constrains its spatial resolution. On the other hand, the asymmetric view of OLCI will offer sun-glint free images in 21 spectral bands (from ultraviolet to near-infrared wavelengths) with an improved spatial coverage and temporal frequency. OLCI will provide high quality optical ocean observations (e.g., normalized water-leaving radiance, inherent optical properties, spectral attenuation of downwelling irradiance, photosynthetically active radiation, particle size distribution) and allow more accurate retrieval of the ocean color variables (e.g., Chlorophyll, SPM and CDOM concentrations) [91] where the OLCI bands are optimized to measure ocean color over open ocean and coastal zones. Sentinel-3 was successfully launched in February 2016 and will give free access to satellite data of the Wadden Sea. It is expected that the MOD2SEA coupled model will also operate successfully to derive Chla concentration using OLCI images for highly turbid waters and that it will result in an accuracy improvement in atmospheric correction and Chla retrieval aspects in comparison with the MERIS sensor. Thus, applying this method for further studies using OLCI data over the Wadden Sea is recommended.

6. Conclusions

A coupled atmospheric-hydro-optical model (MOD2SEA) has been proposed and validated to derive long-term Chla concentration (mg·m^{-3}), visibility and aerosol type from MERIS observations for the coastal turbid area of the Wadden Sea. At one location, the model validation showed a good agreement between MERIS-derived and measured Chla concentration for a period of four years (2007–2010). We attribute the success of this approach to the simultaneous retrieval of atmosphere and water properties. In addition, we have found that water and atmospheric properties have different effects on TOA radiance spectra and therefore these are separately retrievable from MERIS data if the coupled MOD2SEA model is used. Using this coupled atmospheric-hydro-optical model led to considerable improvement for the simultaneous retrieval of water and atmosphere properties using earth observation data, with significant results in the accuracy in comparison with other algorithms applied to derive Chla in the Wadden Sea.

Acknowledgments: The authors would like to thank three anonymous reviewers for their constructive comments and suggestions to improve the quality of our paper. We also would like to thank the European Space Agency for providing the MERIS data within the framework of the "Integrated Network for Production and Loss Assessment in the Coastal Environment (IN PLACE)" project, which is funded by NWO-ZKO.

Author Contributions: All authors have made major and unique contributions in this research. The satellite data from ESA archive of MERIS images was provided by Suhyb Salama (as a part of IN PLACE project). Ground truth measurements from NIOZ jetty station were provided by Marcel Wernand. The Matlab code was written by Behnaz Arabi and Suhyb Salama and was then modified and extended by Wouter Verhoef. Behnaz Arabi performed in situ R_{RS} (λ), Chla and MERIS imagery categorization, matchup points finding, MODTRAN and model simulations, data analysis as well as the results interpretation. The original manuscript was written by Behnaz Arabi. In addition, Wouter Verhoef and Suhyb Salama supervised the research and discussed the results and provided substantial contribution in revising the manuscript at all stages. All co-authors participated in the improvement of the manuscript. All authors have read and approved the final version of this manuscript.

Conflicts of Interest: The authors declare no conflict of interest.

References

1. Le, C.; Hu, C.; English, D.; Cannizzaro, J.; Chen, Z.; Feng, L.; Boler, R.; Kovach, C. Towards a long-term chlorophyll-a data record in a turbid estuary using MODIS observations. *Prog. Oceanogr.* **2013**, *109*, 90–103. [CrossRef]

2. Casal, G.; Furey, T.; Dabrowski, T.; Nolan, G. Generating a Long-Term Series of Sst and Chlorophyll-a for the Coast of Ireland. *ISPRS Int. Arch. Photogramm. Remote Sens. Spat. Inf. Sci.* **2015**, *XL-7/W3*, 933–940. [CrossRef]

3. Werdell, P.J.; Bailey, S.W.; Franz, B.A.; Harding, L.W.; Feldman, G.C.; McClain, C.R. Regional and seasonal variability of chlorophyll-a in Chesapeake Bay as observed by SeaWiFS and MODIS-Aqua. *Remote Sens. Environ.* **2009**, *113*, 1319–1330. [CrossRef]

4. Moradi, M.; Kabiri, K. Spatio-temporal variability of SST and Chlorophyll-a from MODIS data in the Persian Gulf. *Mar. Pollut. Bull.* **2015**, *98*, 14–25. [CrossRef] [PubMed]

5. Peters, S.W.M.; Woerd, H.; Van der Pasterkamp, R. CHL maps of the Dutch coastal zone: A case study within the REVAMP project. *Structure* **2004**, *2003*, 10–13.

6. Van der Woerd, H.; Pasterkamp, R.; Van der Woerd, H.; Pasterkamp, R. Mapping the North Sea turbid coastal waters using SeaWIFS data. *Can. J. Remote Sens.* **2004**, *30*, 44–53. [CrossRef]

7. Le, C.; Hu, C.; English, D.; Cannizzaro, J.; Kovach, C. Climate-driven chlorophyll-a changes in a turbid estuary: Observations from satellites and implications for management. *Remote Sens. Environ.* **2013**, *130*, 11–24. [CrossRef]

8. Mélin, F.; Vantrepotte, V.; Clerici, M.; D'Alimonte, D.; Zibordi, G.; Berthon, J.-F.; Canuti, E. Multi-sensor satellite time series of optical properties and chlorophyll-a concentration in the Adriatic Sea. *Prog. Oceanogr.* **2011**, *91*, 229–244. [CrossRef]

9. Hommersom, A.; Wernand, M.R.; Peters, S.; De Boer, J. A review on substances and processes relevant for optical remote sensing of extremely turbid marine areas, with a focus on the Wadden Sea. *Helgol. Mar. Res.* **2010**, *64*, 75–92. [CrossRef]

10. Hommersom, A. "Dense water" and "Fluid Sand" Optical Properties and Methods for Remote Sensing of the Estremely Turbid Wadden Sea. Ph.D. Thesis, Vrije University Amsterdam, Amsterdam, The Netherlands, 2010.

11. Hommersom, A.; Peters, S.; Wernand, M.R.; de Boer, J. Spatial and temporal variability in bio-optical properties of the Wadden Sea. *Estuar. Coast. Shelf Sci.* **2009**, *83*, 360–370. [CrossRef]

12. Philippart, C.J.M.; Salama, M.S.; Kromkamp, J.C.; van der Woerd, H.J.; Zuur, A.F.; Cadée, G.C. Four decades of variability in turbidity in the western Wadden Sea as derived from corrected Secchi disk readings. *J. Sea Res.* **2013**, *82*, 67–79. [CrossRef]

13. Poremba, K.; Tillmann, U.; Hesse, K.-J. Distribution patterns of bacterioplankton and chlorophyll-a in the German Wadden Sea. *Helgol. Mar. Res.* **1999**, *53*, 28–35. [CrossRef]

14. Philippart, C.J.M.; Beukema, J.J.; Cadée, G.C.; Dekker, R.; Goedhart, P.W.; Van Iperen, J.M.; Leopold, M.F.; Herman, P.M.J. Impacts of nutrient reduction on coastal communities. *Ecosystems* **2007**, *10*, 95–118. [CrossRef]

15. Gemein, N.; Stanev, E.; Brink-spalink, G.; Wolff, J.-O.; Reuter, R. Patterns of suspended matter in the East Frisian Wadden Sea: Comparison of numerical simulations with MERIS observations. *Eur. Assoc. Remote Sens. Lab. Proc.* **2006**, *5*, 180–198.

16. Bartholdy, J.; Folving, S. Sediment classification and surface type mapping in the Danish Wadden Sea by remote sensing. *Neth. J. Sea Res.* **1986**, *20*, 337–345. [CrossRef]

17. Hommersom, A.; Researcher, I. A case study on the use of hydropt in the wadden sea. In Proceedings of the 2nd MERIS (A) ATSR User Workshop, Frascati, Italy, 22–26 September 2008.

18. Hommersom, A.; Peters, S.; Woerd, H.J.; Van der Eleveld, M.A. Tracking wadden sea water masses with an inverse bio-optical and endmember model satellite data. *EARSeL Proc.* **2010**, *9*, 1–12.

19. Schroeder, T.; Behnert, I.; Schaale, M.; Fischer, J.; Doerffer, R. Atmospheric correction algorithm for MERIS above case-2 waters. *Int. J. Remote Sens.* **2007**, *28*, 1469–1486. [CrossRef]

20. Gordon, H.R.; Wang, M. Retrieval of water-leaving radiance and aerosol optical thickness over the oceans with SeaWiFS: A preliminary algorithm. *Appl. Opt.* **1994**, *33*, 443–452. [CrossRef] [PubMed]

21. Goyens, C.; Jamet, C.; Schroeder, T. Evaluation of four atmospheric correction algorithms for MODIS-Aqua images over contrasted coastal waters. *Remote Sens. Environ.* **2013**, *131*, 63–75. [CrossRef]

22. Jamet, C.; Loisel, H.; Kuchinke, C.P.; Ruddick, K.; Zibordi, G.; Feng, H. Comparison of three SeaWiFS atmospheric correction algorithms for turbid waters using AERONET-OC measurements. *Remote Sens. Environ.* **2011**, *115*, 1955–1965. [CrossRef]

23. Siegel, D.A.; Wang, M.; Maritorena, S.; Robinson, W. Atmospheric correction of satellite ocean color imagery: the black pixel assumption. *Appl. Opt.* **2000**, *39*, 3582–3591. [CrossRef] [PubMed]

24. Carpintero, M.; Polo, M.J.; Salama, M.S. Simultaneous atmospheric correction and quantification of suspended particulate matter in the Guadalquivir estuary from Landsat images. *Proc. Int. Assoc. Hydrol. Sci.* **2015**, *368*, 15–20. [CrossRef]

25. Mélin, F.; Zibordi, G.; Berthon, J.-F. Assessment of satellite ocean color products at a coastal site. *Remote Sens. Environ.* **2007**, *110*, 192–215. [CrossRef]

26. Santer, R.; Schmechtig, C. Adjacency effects on water surfaces. Primary scattering approximation and sensitivity study. *Appl. Opt.* **2000**, *39*, 361–375. [CrossRef] [PubMed]

27. Sathyendranath, S. International ocean-colour coordinating group remote sensing of ocean colour in coastal, and other optically-complex. *Waters* **2000**, *3*, 2–11.

28. Hu, C.; Carder, K.L.; Muller-Karger, F.E. Atmospheric correction of SeaWiFS imagery over turbid coastal waters: A practical method. *Remote Sens. Environ.* **2000**, *74*, 195–206. [CrossRef]
29. Wang, M.; Shi, W. The NIR-SWIR combined atmospheric correction approach for MODIS ocean color data processing. *Opt. Express* **2007**, *15*, 15722–15733. [CrossRef] [PubMed]
30. Ruddick, K.G.; De Cauwer, V.; Park, Y.-J.; Moore, G.F. Seaborne measurements of near infrared water-leaving reflectance: The similarity spectrum for turbid waters. *Limnol. Oceanogr.* **2006**, *51*, 1167–1179. [CrossRef]
31. Wang, M. Remote sensing of the ocean contributions from ultraviolet to near-infrared using the shortwave infrared bands: Simulations. *Appl. Opt.* **2007**, *46*, 1535–1547. [CrossRef] [PubMed]
32. Wang, M.; Shi, W. Estimation of ocean contribution at the MODIS near-infrared wavelengths along the east coast of the U.S.: Two case studies. *Geophys. Res. Lett.* **2005**, *32*, 1–5. [CrossRef]
33. Wang, M.; Shi, W.; Tang, J. Water property monitoring and assessment for China's inland Lake Taihu from MODIS-Aqua measurements. *Remote Sens. Environ.* **2011**, *115*, 841–854. [CrossRef]
34. Salama, M.S.; Shen, F. Simultaneous atmospheric correction and quantification of suspended particulate matters from orbital and geostationary earth observation sensors. *Estuar. Coast. Shelf Sci.* **2010**, *86*, 499–511. [CrossRef]
35. Doxaran, D.; Lamquin, N.; Park, Y.-J.; Mazeran, C.; Ryu, J.-H.; Wang, M.; Poteau, A. Retrieval of the seawater reflectance for suspended solids monitoring in the East China Sea using MODIS, MERIS and GOCI satellite data. *Remote Sens. Environ.* **2014**, *146*, 36–48. [CrossRef]
36. Carder, K.L.; Cattrall, C.; Chen, F.R. MODIS clear water epsilons (ATBD 21). *Ocean. Color. Web.* **2002**, *2002*, 1–4.
37. Gould, R.W.; Arnone, R.A.; Martinolich, P.M. Spectral dependence of the scattering coefficient in case 1 and case 2 waters. *Appl. Opt.* **1999**, *38*, 2377–2383. [CrossRef] [PubMed]
38. Ruddick, K.G.; Ovidio, F.; Rijkeboer, M. Atmospheric correction of SeaWiFS imagery for turbid coastal and inland waters. *Appl. Opt.* **2000**, *39*, 897–912. [CrossRef] [PubMed]
39. Shen, F.; Verhoef, W.; Zhou, Y.X.; Salama, M.S.; Liu, X.L. Satellite estimates of wide-range suspended sediment concentrations in Changjiang (Yangtze) estuary using MERIS data. *Estuar. Coasts* **2010**, *33*, 1420–1429. [CrossRef]
40. Verhoef, W.; Bach, H. Coupled soil-leaf-canopy and atmosphere radiative transfer modeling to simulate hyperspectral multi-angular surface reflectance and TOA radiance data. *Remote Sens. Environ.* **2007**, *109*, 166–182. [CrossRef]
41. Shen, F.; Verhoef, W. Suppression of local haze variations in MERIS images over turbid coastal waters for retrieval of suspended sediment concentration. *Opt. Express* **2010**, *18*, 12653–12662. [CrossRef] [PubMed]
42. Doerffer, R.; Schiller, H. The MERIS Case 2 water algorithm. *Int. J. Remote Sens.* **2007**, *28*, 517–535. [CrossRef]
43. Duan, H.; Ma, R.; Simis, S.G.H.; Zhang, Y. Validation of MERIS case-2 water products in Lake Taihu, China. *GIScience Remote Sens.* **2012**, *49*, 873–894. [CrossRef]
44. Salama, M.S. *Roadmap Waterkwaliteit: Helder Beeld op Troebel Water*; ITC: Boston, MA, USA, 2014.
45. O'Reilly, J.E.; Maritorena, S.; Mitchell, B.G.; Siegel, D.A.; Carder, K.L.; Garver, S.A.; Kahru, M.; McClain, C. Ocean color chlorophyll algorithms for SeaWiFS. *J. Geophys. Res.* **1998**, *103*, 24937. [CrossRef]
46. O'Reilly, J.; Maritorena, S. Ocean color chlorophyll a algorithms for SeaWiFS, OC2, and OC4: Version 4. In *SeaWiFS Postlaunch Calibration and Validation Analyses*; NASA: Washington, DC, USA, 2000; pp. 8–22.
47. Dasgupta, S.; Singh, R.P.; Kafatos, M. Comparison of global chlorophyll concentrations using MODIS data. *Adv. Space Res.* **2009**, *43*, 1090–1100. [CrossRef]
48. Arnone, R.; Babin, M.; Barnard, A.H.; Boss, E.; Cannizzaro, J.P.; Carder, K.L.; Chen, F.R.; Devred, E.; Doerffer, R.; Du, K.; et al. *Remote Sensing of Inherent Optical Properties: Fundamentals, Tests of Algorithms, and Applications*; The International Ocean-Colour Coordinating Group: Dartmouth, NH, Canada, 2006.
49. Salama, M.S.; Radwan, M.; van der Velde, R. A hydro-optical model for deriving water quality variables from satellite images (HydroSat): A case study of the Nile River demonstrating the future Sentinel-2 capabilities. *Phys. Chem. Earth* **2012**, *50–52*, 224–232. [CrossRef]
50. Salama, M.S.; Verhoef, W. Two-stream remote sensing model for water quality mapping: 2SeaColor. *Remote Sens. Environ.* **2014**, *157*, 111–122. [CrossRef]
51. Ridderinkhof, H.; Zimmerman, J.T.F.; Philippart, M.E. Tidal exchange between the North Sea and Dutch Wadden Sea and mixing time scales of the tidal basins. *Neth. J. Sea Res.* **1990**, *25*, 331–350. [CrossRef]

52. Zibordi, G.; Berthon, J.F.; Mélin, F.; D'Alimonte, D. Cross-site consistent in situ measurements for satellite ocean color applications: The BiOMaP radiometric dataset. *Remote Sens. Environ.* **2011**, *115*, 2104–2115. [CrossRef]

53. Wernand, M.R.; Poseidon, S. Paintbox Historical Archives of Ocean Colour in Global-Change Poseidon's Verfdoos. Ph.D. Thesis, Utrecht University, Utrecht, The Netherlands, 2011.

54. Ly, J.; Philippart, C.J.M.; Kromkamp, J.C. Phosphorus limitation during a phytoplankton spring bloom in the western Dutch Wadden Sea. *J. Sea Res.* **2014**, *88*, 109–120. [CrossRef]

55. Zibordi, G.; Berthon, J.F.; Mélin, F.; D'Alimonte, D.; Kaitala, S. Validation of satellite ocean color primary products at optically complex coastal sites: Northern Adriatic Sea, Northern Baltic Proper and Gulf of Finland. *Remote Sens. Environ.* **2009**, *113*, 2574–2591. [CrossRef]

56. Schroeder, T.; Schaale, M.; Fischer, J. Retrieval of atmospheric and oceanic properties from MERIS measurements: A new Case-2 water processor for BEAM. *Int. J. Remote Sens.* **2007**, *28*, 5627–5632. [CrossRef]

57. Peters, S.W.M.; Eleveld, M.; Pasterkamp, R.; Van Der Woerd, H.; Devolder, M.; Jans, S.; Park, Y.; Ruddick, K.; Block, T.; Brockmann, C.; et al. *Atlas of Chlorophyll-a Concentration for the North; Sea Based on MERIS Imagery of 2003*; Vrije Universiteit: Amsterdam, The Netherlands, 2005.

58. Zibordi, G.; Mélin, F.; Berthon, J.F.; Canuti, E. Assessment of MERIS ocean color data products for European seas. *Ocean Sci.* **2013**, *9*, 521–533. [CrossRef]

59. Zibordi, G.; Mélin, F.; Berthon, J.F. Comparison of SeaWiFS, MODIS and MERIS radiometric products at a coastal site. *Geophys. Res. Lett.* **2006**, *33*, 1–4. [CrossRef]

60. Loisel, H.; Vantrepotte, V.; Jamet, C.; Dat, D.N. Challenges and new advances in ocean color remote sensing of coastal waters. In *Oceanography Research*; In Tech: Greer, SC, USA, 2013.

61. Duntley, S.Q. The optical properties of diffusing materials. *J. Opt. Soc. Am.* **1942**, *32*, 61–70. [CrossRef]

62. Duntley, S.Q. Light in the sea. *J. Opt. Soc. Am.* **1963**, *53*, 214–233. [CrossRef]

63. Morel, A.; Gentili, B. Diffuse reflectance of oceanic waters. II Bidirectional aspects. *Appl. Opt.* **1993**, *32*, 6864–6879. [CrossRef] [PubMed]

64. Lee, Z.; Carder, K.L.; Arnone, R.A. Deriving inherent optical properties from water color: A multiband quasi-analytical algorithm for optically deep waters. *Appl. Opt.* **2002**, *41*, 5755–5772. [CrossRef] [PubMed]

65. Salama, M.S.; Shen, F. Stochastic inversion of ocean color data using the cross-entropy method. *Opt. Express* **2010**, *18*, 479–499. [CrossRef] [PubMed]

66. Briucaud, A.; Morel, A.; Prieur, L. Absorption by dissolved organic matter of the sea (yellow substance) in the UV and visible domains. *Limnol. Oceanogr.* **1981**, *26*, 43–53. [CrossRef]

67. Lee, Z.; Carder, K.L.; Mobley, C.D.; Steward, R.G.; Patch, J.S. Hyperspectral remote sensing for shallow waters. I. A semianalytical model. *Appl. Opt.* **1998**, *37*, 6329–6338. [CrossRef] [PubMed]

68. Lee, Z.; Carder, K.L.; Mobley, C.D.; Steward, R.G.; Patch, J.S. Hyperspectral remote sensing for shallow waters. 2. Deriving bottom depths and water properties by optimization. *Appl. Opt.* **1999**, *38*, 3831–3843. [CrossRef] [PubMed]

69. Doxaran, D.; Ruddick, K.; McKee, D.; Gentili, B.; Tailliez, D.; Chami, M.; Babin, M. Spectral variations of light scattering by marine particles in coastal waters, from visible to near infrared. *Limnol. Oceanogr.* **2009**, *54*, 1257–1271. [CrossRef]

70. Morel, A.; Maritorena, S. Bio-optical properties of oceanic waters: A reappraisal. *J. Geophys. Res.* **2001**, *106*, 7163–7180. [CrossRef]

71. Mobley, C.D. *Light and Water: Radiative Transfer in Natural Waters*; Academic Press: Cambridge, MA, USA, 1994.

72. Pope, R.M.; Fry, E.S. Absorption spectrum (380–700 nm) of pure water. II. Integrating cavity measurements. *Appl. Opt.* **1997**, *36*, 8710–8723. [CrossRef] [PubMed]

73. Petzold, T. *Volume Scattering Functions for Selected Ocean Waters*; No. SIO-REF-72-78; Scripps Institution of Oceanography La Jolla Ca Visibility Laboratory: San Diego, CA, USA, 1972.

74. Kneizys, F.X.; Shettle, E.P.; Abreu, L.W.; Chetwynd, J.H.; Anderson, G.P. *Users Guide to LOWTRAN 7*; No. AFGL-TR-88-0177; Air Force Geophysics Lab Hanscom: Bedford, MA, USA, 1988.

75. Verhoef, W.; Bach, H. Simulation of hyperspectral and directional radiance images using coupled biophysical and atmospheric radiative transfer models. *Remote Sens. Environ.* **2003**, *87*, 23–41. [CrossRef]

76. Global Greenhouse Reference Network 2017. Available online: http://www.esrl.noaa.gov/gmd/ccgg/trends/ (accessed on 31 August 2016).

77. Berk, A.; Anderson, G.P.; Acharya, P.K.; Shettle, E.P. *MODTRAN 5.2. 1 User's Manual*; Spectral Sciences Inc.: Burlington, MA, USA, 2011.

78. Attila, J.; Koponen, S.; Kallio, K.; Lindfors, A.; Kaitala, S.; Ylöstalo, P. MERIS Case II water processor comparison on coastal sites of the northern Baltic Sea. *Remote Sens. Environ.* **2013**, *128*, 138–149. [CrossRef]

79. Beltrán-Abaunza, J.M.; Kratzer, S.; Brockmann, C. Evaluation of MERIS products from Baltic Sea coastal waters rich in CDOM. *Ocean Sci.* **2014**, *10*, 377–396. [CrossRef]

80. Smith, M.E.; Bernard, S.; O'Donoghue, S. The assessment of optimal {MERIS} ocean colour products in the shelf waters of the KwaZulu-Natal Bight, South Africa. *Remote Sens. Environ.* **2013**, *137*, 124–138. [CrossRef]

81. Ambarwulan, W.; Verhoef, W.; Mannaerts, C.M.; Salama, M.S. Estimating total suspended matter concentration in tropica coastal waters of the Berau estuary, Indonesia. *Int. J. Remote Sens.* **2012**, *33*, 4919–4936. [CrossRef]

82. Bayat, B.; Van der Tol, C.; Verhoef, W. Remote sensing of grass response to drought stress using spectroscopic techniques and canopy reflectance model inversion. *Remote Sens.* **2016**, *8*. [CrossRef]

83. Zibordi, G.; Ruddick, K.; Ansko, I.; Moore, G.; Kratzer, S.; Icely, J.; Reinart, A. In situ determination of the remote sensing reflectance: An inter-comparison. *Ocean Sci.* **2012**, *8*, 567–586. [CrossRef]

84. Chang, T.S.; Joerdel, O.; Flemming, B.W.; Bartholomä, A. The role of particle aggregation/disaggregation in muddy sediment dynamics and seasonal sediment turnover in a back-barrier tidal basin, East Frisian Wadden Sea, southern North Sea. *Mar. Geol.* **2006**, *235*, 49–61. [CrossRef]

85. Reuter, R.; Badewien, T.H.; BartholomÄ, A.; Braun, A.; Lübben, A.; Rullkötter, J. A hydrographic time series station in the Wadden Sea (southern North Sea). *Ocean Dyn.* **2009**, *59*, 195–211. [CrossRef]

86. Tillmann, U.; Hesse, K.-J.; Colijn, F. Planktonic primary production in the German Wadden Sea. *J. Plankton Res.* **2000**, *22*, 1253–1276. [CrossRef]

87. Cadée, G.C. Accumulation and sedimentation of Phaeocystis globosa in the Dutch Wadden Sea. *J. Sea Res.* **1996**, *36*, 321–327. [CrossRef]

88. Peters, S.W.M. *MERIMON-2000: MERIS for Water Quality Monitoring in the Belgian-Dutch-German Coastal Zone*; Food and Agriculture Organization of the United Nations: Rome, Italy, 2001.

89. Harvey, E.T.; Kratzer, S.; Philipson, P. Satellite-based water quality monitoring for improved spatial and temporal retrieval of chlorophyll-a in coastal waters. *Remote Sens. Environ.* **2014**, *158*, 417–430. [CrossRef]

90. Saulquin, B.; Fablet, R.; Bourg, L.; Mercier, G.; d'Andon, O.F. MEETC2: Ocean color atmospheric corrections in coastal complex waters using a Bayesian latent class model and potential for the incoming sentinel 3-OLCI mission. *Remote Sens. Environ.* **2016**, *172*, 39–49. [CrossRef]

91. Malenovský, Z.; Rott, H.; Cihlar, J.; Schaepman, M.E.; García-Santos, G.; Fernandes, R.; Berger, M. Sentinels for science: Potential of Sentinel-1, -2, and -3 missions for scientific observations of ocean, cryosphere, and land. *Remote Sens. Environ.* **2012**, *120*, 91–101.

remote sensing

MDPI

Article

Retrieval of Chlorophyll-*a* and Total Suspended Solids Using Iterative Stepwise Elimination Partial Least Squares (ISE-PLS) Regression Based on Field Hyperspectral Measurements in Irrigation Ponds in Higashihiroshima, Japan

Zuomin Wang [1], Kensuke Kawamura [2], Yuji Sakuno [1,*], Xinyan Fan [3], Zhe Gong [3] and Jihyun Lim [3]

[1]   Graduate School of Engineering, Hiroshima University, 1-4-1 Kagamiyama, Higashihiroshima, Hiroshima 739-8527, Japan; wangzuomin123@gmail.com
[2]   Social Sciences Division, Japan International Research Center for Agricultural Sciences (JIRCAS), 1-1 Ohwashi, Tsukuba, Ibaraki 305-8686, Japan; kamuken@affrc.go.jp
[3]   Graduate School for International Development and Cooperation (IDEC), Hiroshima University, 1-5-1 Kagamiyama, Higashihiroshima, Hiroshima 739-8529, Japan; xinyanfan5160@gmail.com (X.F.); gongzhe79@gmail.com (Z.G.); limjihyun7@gmail.com (J.L.)
*   Correspondence: sakuno@hiroshima-u.ac.jp; Tel.: +81-82-424-7773

Academic Editors: Yunlin Zhang, Claudia Giardino, Linhai Li, Deepak R. Mishra and Prasad S. Thenkabail
Received: 16 December 2016; Accepted: 9 March 2017; Published: 13 March 2017

Abstract: Concentrations of chlorophyll-*a* (Chl-*a*) and total suspended solids (TSS) are significant parameters used to assess water quality. The objective of this study is to establish a quantitative model for estimating the Chl-*a* and the TSS concentrations in irrigation ponds in Higashihiroshima, Japan, using field hyperspectral measurements and statistical analysis. Field experiments were conducted in six ponds and spectral readings for Chl-*a* and TSS were obtained from six field observations in 2014. For statistical approaches, we used two spectral indices, the ratio spectral index (RSI) and the normalized difference spectral index (NDSI), and a partial least squares (PLS) regression. The predictive abilities were compared using the coefficient of determination (R^2), the root mean squared error of cross validation (RMSECV) and the residual predictive deviation (RPD). Overall, iterative stepwise elimination based on PLS (ISE–PLS), using the first derivative reflectance (FDR), showed the best predictive accuracy, for both Chl-*a* ($R^2 = 0.98$, RMSECV = 6.15, RPD = 7.44) and TSS ($R^2 = 0.97$, RMSECV = 1.91, RPD = 6.64). The important wavebands for estimating Chl-*a* (16.97% of all wavebands) and TSS (8.38% of all wavebands) were selected by ISE–PLS from all 501 wavebands over the 400–900 nm range. These findings suggest that ISE–PLS based on field hyperspectral measurements can be used to estimate water Chl-*a* and TSS concentrations in irrigation ponds.

Keywords: chlorophyll-*a*; hyperspectral; irrigation ponds; partial least squares regression; total suspended solids

---

1. Introduction

Agriculture is by far the greatest water consumer in the world, and consequently, a major cause of water pollution. The primary pollutants from agriculture are excess nutrients and pesticides [1]. In agricultural activity, non-point source pollution, such as irrigation water and surface runoff water containing fertilizer from farmland, contributes to excessive nutrient concentrations [2]. Meanwhile, excess nutrients that cause eutrophication, hypoxia and algal blooms in surface water bodies and coastal areas contribute to the primary global water quality problem [1]. Eutrophication has become a widespread matter of concern during the past 50 years, especially in coastal and inland waters [3].

The chlorophyll-*a* (Chl-*a*) concentration in water is the most widely applied parameter to assess the water quality status of lakes, particularly with respect to their trophic quality [4]. Since Chl-*a* is the primary photosynthetic pigment of all plant life [5], the concentration of Chl-*a* indicates phytoplankton biomass and eutrophication in lakes [6]. The concentration of total suspended solids (TSS) is another commonly used indicator for water quality assessment [7]. TSS consists of organic and inorganic materials suspended in the water [8]. Increased TSS decrease light transmission through the water [9], and therefore affect light availability to phytoplankton, thus resulting in a decrease of phytoplankton primary production [10].

However, traditional water quality monitoring requires in situ measurements and sampling, then returning the samples to the laboratory to measure water quality indicators (e.g., Chl-*a* and TSS), which is costly and time consuming [11]. Remote sensing makes it possible to monitor the state of the globe routinely, and is cost effective and useful, with the benefits of its passive nature and wide spatial coverage [12]. Earlier studies have demonstrated several algorithms developed for satellite sensors to estimate ocean and coastal water quality parameters, such as the Chl-*a* algorithm OC3, created for the moderate resolution imaging spectroradiometer (MODIS) data, and OC4, created for sea-viewing wide field-of-view sensor (SeaWiFS) data [13]. The geostationary ocean color imager (GOCI) also shows good performance, using the linear combination index (LCI) method to monitor Chl-*a* [14]. Further, a three-band semi-analytical reflectance model, originally developed by Gitelson et al. (2003) [15], and a normalized difference chlorophyll index (NDCI) [16], both performed well for assessing Chl-*a* in turbid productive water [16–18]. For estimating TSS concentrations, an algorithm with a single wavelength created for MODIS and medium spectral resolution imaging spectrometer (MERIS) data has been proved to be satisfactory [19].

Unlike ocean and coastal water, inland water usually has a smaller surface area and more complicated spectral features, especially irrigation ponds, which are often impacted by human use such as agriculture activities. Consequently, inland water quality monitoring presents higher requirements for both temporal and spatial resolution of satellite sensor data; hence currently used satellite sensors often have limited practical applicability in assessing relatively smaller inland water bodies. Since there are a limited number of wavebands for Landsat and other multispectral sensors, finding more informative wavebands to improve the performance of water quality estimation is necessary. With respect to in situ measurements, a two-band ratio approach, for example the ratio spectral index (RSI), has performed well for estimating Chl-*a* concentrations in inland waters [18,20,21], especially using the ratio of near-infrared (NIR) regions to red wavebands, such as the reflectance ratio of 705 nm to 670 nm performed by Han et al. (1997) [22]. Normalized difference spectral indices (NDSI) are another type of spectral indices frequently used to select the optimum bands for spectral analysis. As similar studies that have been done before mainly focused on vegetation parameters retrieval [23–25], optimum bands have been calculated from combinations of all available bands in the hyperspectral spectrum, a considerable range for hyperspectral analysis. Water quality parameters retrieval requires a similarly broad approach.

Partial least squares (PLS) regression, which was developed by Wold (1966) [26], is widely used to extract valuable information for spectroscopic analysis. PLS regression uses all available wavebands without multi-collinearity issues. The eigenvectors of the explanatory variables are manipulated such that the corresponding scores (latent variables) not only explain the variance of the explanatory variables (wavebands) themselves, but also are highly correlated with the response variables (Chl-*a* and TSS) [27]. However, PLS is considered limited because it treats each wavelength as independent, which incorporate noise created by non-informative wavelengths [28]. There is increasing evidence to indicate that wavelength selection can affect the performance of PLS analysis [29], since wavelength selection for PLS models is performed to eliminate uninformative variables and choose the variables that contribute the most to the predictive ability of the calibration model [30].

Iterative stepwise elimination PLS (ISE–PLS), developed by Boggia et al. (1997) [31], combines PLS regression and the most useful information from hundreds of wavebands into the first several factors [32]. This method was developed to eliminate useless wavebands in PLS analysis.

The objective of this study is to develop models to estimate Chl-*a* and TSS using in situ spectral reflectance data and statistical approaches. We used several regression analyses including (a) a simple linear regression at each waveband of reflectance and the first derivative reflectance (FDR) to explore informative wavelength regions for Chl-*a* and TSS estimation; (b) all available two-band combination spectral indices (RSI and NDSI); and (c) a PLS regression using original reflectance and FDR datasets. In the PLS analyses, the predictive ability of ISE–PLS was compared with that of a standard full spectrum PLS (FS–PLS) and the spectral indices (RSI and NDSI).

2. Study Area

The study area is located in Higashihiroshima, Japan, as shown in Figure 1. Higashihiroshima is a core city in the central region of Hiroshima Prefecture, with a total area of 635.32 km^2 covering nearly 7.5% of the prefecture's total area. Paddy fields, totalling 36.8 km^2, cover 14.9% of the Hiroshima Prefecture. Consequently, Higashihiroshima has the largest rice production of the 86 cities, towns and villages in Hiroshima Prefecture [33]. The city has an estimated population of 183,834 people, and its population density was 289.36 people per km^2 in 2011. The number of irrigation ponds in Hiroshima Prefecture approaches approximately 21,000. This qualifies as the second largest number in Japan; a quarter of the total irrigation ponds in Japan are in Higashihiroshima, the average beneficiary area is 3.36 ha, and the average number of beneficiary farmhouses is approximately 9 [34]. The monthly mean temperature ranges from 2.2 °C in January to 25.8 °C in August, and the monthly precipitation ranges from 43.3 mm in December to 232.1 mm in July, referring to the minimum and maximum values, respectively. To assess changes in water quality status and environments, six ponds, including both eutrophic ponds and non-eutrophic ponds, were selected for this study. Descriptions of the six ponds are listed in Table 1.

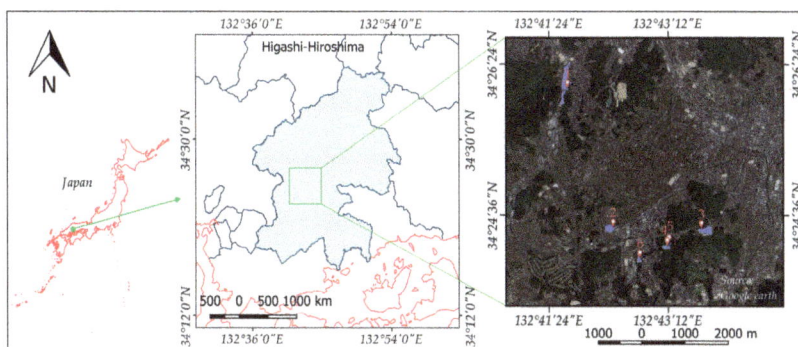

Figure 1. Locations of Higashihiroshima and the six irrigation ponds used in this study.

Table 1. The six irrigation ponds in the study.

No.	Name of Pond	Alt. (m)	Depth (m)	Area (ha)	Coordinate
1	Nanatsu-ike	245	2.3	8.1	34°26′06.46″N 132°41′39.69″E
2	Shitami-Oike	221	1.5	2.5	34°24′28.56″N 132°42′22.09″E
3	Okuda-Oike	228	3.3	2.9	34°24′25.24″N 132°43′43.16″E
4	Yamanaka ike	231	2.6	1.2	34°24′14.15″N 132°43′12.21″E
5	Yamanakaike-kamiike	231	1.1	0.1	34°24′15.29″N 132°43′14.45″E
6	Budou-ike	210	1.6	1	34°24′02.78″N 132°42′45.89″E

3. Materials and Methods

3.1. Measurement of Water Surface Reflectance

Measurements of water surface reflectance were performed using an ASD FieldSpec HandHeld-2 spectrometer (ASD Inc., Boulder, CO, USA) with a spectral range of 350–1050 nm and a probe field angle of 10°. Spectral readings were taken approximately 1 m above the water surface between 10:30 and 13:00 on a day with clear skies. Surveys were conducted six times between 3 January 2014, and 28 June 2014. From these data, a total of 36 datasets were obtained.

With respect to the spectral data, the ranges 325–399 nm and 901–1075 nm from each spectrum were identified as noise and removed. Subsequently, spectral data were smoothed using a moving and normalized Gaussian filter with a sigma (standard deviation) of 2.5. The FDR was also computed and compared with the original reflectance.

3.2. Water Sampling and Chemical Analysis

The water sampling sites were consistent with the spectral reflectance measurements. Immediately after measurement of spectral reflectance, water samples were collected into two 1 L containers. The samples were maintained at constant temperature and protected from light until they were received at the laboratory for analysis.

Chl-*a* and TSS concentrations were determined at the laboratory of the Graduate School for International Development and Cooperation (IDEC), Hiroshima University, Japan. Chl-*a* was extracted using 90% acetone, the absorption of Chl-*a* was measured by a spectrophotometer (UVmini-1240, SHIMADZU Co., Kyoto, Japan) and pigment concentration was calculated using the equations from UNESCO. To measure the TSS, the water sample was filtered using 47 mm diameter GF/F filters. The filters were weighed before and after drying with an oven drier (SANYO Electric Co., Moriguchi, Osaka, Japan) at 105 °C for two hours. The TSS contents were quantified by the difference in the weight of the filter paper before and after filtration.

3.3. Ratio Spectral Index and Normalized Difference Spectral Indices

A combination of spectral indices between all wavebands is performed to select the optimum two-band combination. The aim of spectral indices is to construct a mathematical combination of spectral wavebands to enhance information content with respect to the parameter under study [35]. Moreover, normalization in the NDSI is effective at cancelling atmospheric disturbance or other sources of error, while enhancing and standardizing the spectral response to the observed targets [23].

For this study, two of the most commonly used spectral indices (RSI and NDSI) were calculated using the reflectance dataset. The forms to express them are as follows:

$$\text{RSI}(i,j) = \frac{R_i}{R_j} \tag{1}$$

$$\text{NDSI}(i,j) = \frac{R_i - R_j}{R_i + R_j} \tag{2}$$

where R_i and R_j are the intensity values for bands i and j, respectively. Using both indices, the optimum wavebands from all combinations of two separate wavelengths were obtained.

3.4. Full Spectrum Partial Least Squares Regression

We performed PLS regression to estimate Chl-*a* and TSS concentrations using the reflectance and FDR datasets ($n = 36$). The standard FS–PLS regression equation is as follows:

$$y = \beta_1 x_1 + \beta_2 x_2 + \ldots + \beta_i x_i + \varepsilon \tag{3}$$

where the response variable y is a vector of the water quality parameters (Chl-a and TSS), the predictor variables x_1 to x_i are surface reflectance or FDR values for spectral bands 1 to i (400, 401, ... , 900 nm), respectively, β_1 to β_i are the estimated weighted regression coefficients, and ε is the error vector. The latent variables were introduced to simplify the relationship between response variables and predictor variables. To determine the optimal number of latent variables (NLV), leave-one-out (LOO) cross validation was performed to avoid overfitting of the model, which was based on the minimum value of the root mean squared error (RMSECV). The RMSECV is calculated as follows:

$$RMSECV = \sqrt{\frac{\sum_{i=1}^{n}\left(y_i - y_p\right)^2}{n}} \qquad (4)$$

where y_i and y_p represent the measured and predicted water quality parameters (Chl-a and TSS) for sample i, and n is the number of samples in the dataset (n = 36).

3.5. Iterative Stepwise Elimination Partial Least Squares Regression

The ISE–PLS is a model-wise technique [31], which is based on the wavelengths selection function of the ISE method. To improve the performance of the PLS model, the optimum wavelengths with good predictive ability are selected for model calibration. The wavelengths elimination process depends on the importance of the predictors (z_i), described as follows:

$$z_i = \frac{|\beta_i|s_i}{\sum_{i=1}^{I}|\beta_i|s_i} \qquad (5)$$

where β_i is the regression coefficient and s_i is the standard deviation of predictor, both corresponding to the predictor variable of the waveband i.

Initially, all available wavebands (501 bands, 400–900 nm) are used to develop the PLS regression model. Then variables are ranked from most contributed to least contributed according to the predictor z_i; in other words, the predictor z_i represents the weight of each variable. The least contributed variable is eliminated and the PLS model is recalibrated with the remaining predictor variables [36]. The model building procedure is repeated, and in each cycle the predictor variable with the minimum importance (i.e., the less informative wavelength) is eliminated, until the final variable is eliminated. To determine the optimum number of wavelengths to include in the final model, LOO cross validation is conducted after each calibration. The final model with the maximum predictive ability is calibrated by the minimum value of RMSECV [37].

3.6. Evaluation of Predictive Ability

The coefficient of determination (R^2) and RMSECV were selected as indices to evaluate the FS–PLS and ISE–PLS calibration models' accuracy by using LOO cross validation. High results for R^2 and low RMSECV indicate the best model to predict Chl-a and TSS concentrations. In addition, the residual predictive deviation (RPD) was used to evaluate the predictive ability of the models, which was defined as the ratio of standard deviation (SD) of reference data in prediction to RMSECV [38]. For determining the performance ability of the calibration models, the goal RPD was at least 3 for agriculture applications; RPD values between 2 and 3 indicate a model with good prediction ability, $1.5 < RPD < 2$ is an intermediate model needing some improvement, and an $RPD < 1.5$ indicates that the model has poor prediction ability [39].

All data handling and linear regression analyses were performed using Matlab software ver. 8.6 (MathWorks, Sherborn, MA, USA).

4. Results

4.1. Chl-a and TSS Concentrations in Irrigation Ponds

Descriptive statistics are shown in Table 2, including the sampling data, the number of samples, the minimum (Min), the maximum (Max), the mean, the standard deviation (SD) and the coefficient of variation (CV). In total, 36 samples were collected from six irrigation ponds in six sets of field measurements (3 January, 19 January, 24 March, 9 April, 24 May, and 28 June in 2014). Field samples (n = 36) provided a wide range of both Chl-a (SD = 46.1 µg/L, CV = 2.0) and TSS (SD = 12.8 mg/L, CV = 1.65). In the datasets, Chl-a ranged from 0 to 169.5 µg/L, and TSS ranged from 0.1 to 53 mg/L, which indicates that this study involves various water quality conditions from different ponds.

Table 2. Descriptive statistics for the Chl-a and TSS concentrations.

Date	n	Chl-a (µg/L)					TSS (mg/L)				
		Min	Max	Mean	SD	CV	Min	Max	Mean	SD	CV
3 January 2014	6	0.1	98.7	20.7	39.1	1.9	0.1	16.8	6.1	7.2	1.2
19 January 2014	6	0.1	169.5	36.0	67.5	1.9	0.1	26.5	7.6	11.0	1.5
24 March 2014	6	0	169.1	36.8	67.3	1.8	0.4	38.0	10.2	15.5	1.5
9 April 2014	6	0.5	48.5	8.7	19.5	2.2	0.5	33.5	6.5	13.2	2.0
24 May 2014	6	0.9	37.7	9.2	14.6	1.6	0.2	26.0	5.8	10.0	1.7
28 June 2014	6	1.6	133.9	27.1	52.5	1.9	0.3	53.0	10.4	20.9	2.0
Total	36	0	169.5	23.1	46.1	2.0	0.1	53.0	7.8	12.8	1.65

SD = standard deviation; CV = coefficient of variation; n = number of samples.

4.2. Comparison of Simple Linear Regression Models

In this study, several simple linear regression models were constructed, and the accuracy was compared with that of the PLS method. As shown in Table 3, distinct bands were selected as the optimal bands with respect to accuracy for all models. In the model that used the gaussian smoothed water surface reflectance and FDR, the 730 nm and 705 nm wavebands were selected, based on the linear correlation coefficient shown in Figure 2, to estimate Chl-a concentration (R^2 = 0.14 and 0.54); 722 nm and 704 nm were selected to estimate TSS (R^2 = 0.05 and 0.46). Figure 2 shows the correlation coefficient (r) between reflectance/FDR and Chl-a/TSS with regard to each waveband. It is clear that FDR obviously improved correlation with Chl-a and TSS; moreover, spectra reflectance and absorption features were also enhanced (Figure 2b). An NIR/red algorithm developed by Han et al. (1997) [22] was introduced for comparison of the RSI selected wavebands and accuracy. The NIR/red model showed a higher R^2 and lower RMSE than the single waveband models. However, based on the regression between the reflectance of each waveband and Chl-a and TSS, the RSI model selected the R719/R662 ratio as the best band combination, which enhanced the performance of ratio model, giving the highest R^2 value of 0.72 for Chl-a. The R717/R630 ratio was the best band combination for TSS, with an R^2 of 0.52 (Figure 3a,b). A three-band semi-analytical algorithm for estimating Chl-a concentration was conducted, as a previous study suggested [40], and the optimal wavebands of model were tuned according to the optical properties of the water bodies. Bands 660, 703, and 740 nm were final selected for the three-band model with an R^2 of 0.71 and RMSE of 29.32. For another algorithm introduced in a previous study, the NDCI was evaluated using remote sensing reflectance R_{rs} at an absorption peak of 665 nm (R_{rs665}), which is closely related to absorption by Chl-a pigments and a reflectance peak of 708 nm (R_{rs708}), which was sensitive to variations in Chl-a concentration in water, with a result of an R^2 of 0.60 and an RMSE of 28.82. For the NDSI model, bands 719 and 663 nm were the best combination for estimating Chl-a (R^2 = 0.64), and bands 704 and 698 nm were the best combination for TSS (R^2 = 0.55) (Figure 3c,d). The results showed the lowest RMSECV in the RSI model for Chl-a (24.14) and in the NDSI model for TSS (8.48). Among the models, the RSI or the NDSI showed higher R^2 values and lower RMSECV values than those of the two types of single-band models in the estimation of both Chl-*a* and TSS.

Table 3. Regression models used to estimate Chl-*a* and TSS concentrations with two spectral data types (reflectance and FDR) and two spectral indices (RSI and NDSI).

Parameter	Spectral Index	Model	R^2	RMSE
Chl-*a*	Reflectancce	Chl-*a* = 0.0004 × R_{730} + 0.0396	0.14	51.00
	FDR	Chl-*a* = 1 × 10^{-5} × R_{705} − 0.0004	0.54	51.01
	NIR/red (Han et al. (1997) [22])	Chl-*a* = 94.748 × R_{705}/R_{670} − 88.897	0.60	28.78
	Three-band (Gitelson et al. (2003) [15])	Chl-*a* = 0.0036 × $(R^{-1}_{660} − R^{-1}_{703})$ × R_{740} − 0.0665	0.71	29.32
	NDCI (Mishra et al. (2012) [16])	Chl-*a* = 253.16 × $(R_{rs708} − R_{rs665})/(R_{rs708}+R_{rs665})$ + 36.535	0.60	28.82
	RSI	Chl-*a* = 119.27 × R_{719}/R_{662} − 88.052	0.72	24.14
	NDSI	Chl-*a* = 253.16 × $(R_{719} − R_{663})/(R_{719} + R_{663})$ + 36.535	0.64	27.19
TSS	Reflectancce	TSS = 0.0009 × R_{722} + 0.0501	0.05	14.81
	FDR	TSS = 5 × 10^{-5} × R_{704} − 0.0003	0.46	14.83
	RSI	TSS = 31.419 × R_{717}/R_{630} − 17.913	0.52	8.73
	NDSI	TSS = 300.45 × $(R_{704} − R_{698})/(R_{704} + R_{698})$ + 6.3868	0.55	8.48

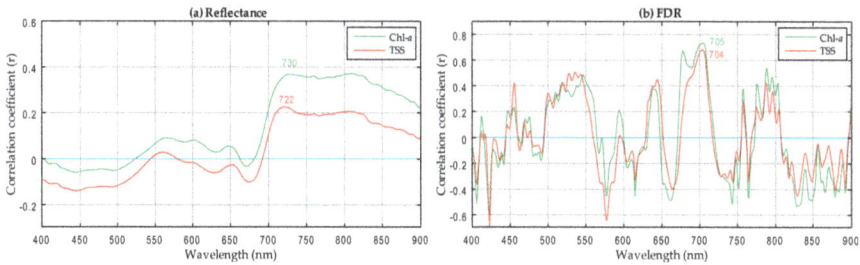

Figure 2. Correlation coefficients (*r*) between water quality parameters (Chl-*a* and TSS) at each wavelength: (**a**) reflectance; (**b**) FDR.

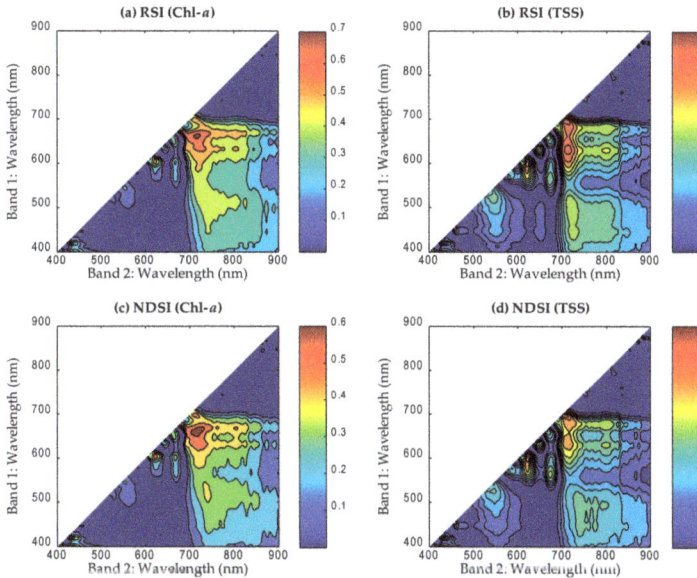

Figure 3. Distributions of R^2 between two wavebands using RSI (**a**) Chl-*a*; (**b**) TSS and NDSI (**c**) Chl-*a*; (**d**) TSS.

4.3. FS–PLS and ISE–PLS Models

Calibration and cross validation results between reflectance/FDR spectra and Chl-*a*/TSS using FS–PLS and ISE–PLS are shown in Table 4. The results showed that the optimum NLV ranged between 4 and 8 in FS–PLS and between 5 and 11 in ISE–PLS, which was determined by the LOO cross validation based on the lowest RMSECV. The RPD ranged between 1.22 and 1.32 (low accuracy) in FS–PLS and between 1.45 and 7.44 (excellent accuracy) in ISE–PLS. In particular, the selected number of wavebands and the percentage to full spectrum (that is, selected wavebands number/all ($n = 501$) × 100%) were calculated to evaluate the informative wavebands for ISE–PLS. Results showed the selected wavebands number ranged between 9 and 85, and the percent ratio ranged between 1.80 and 16.97. Overall, for Chl-*a*, ISE–PLS using FDR showed the highest R^2, highest RPD, and lowest RMSECV ($R^2 = 0.98$, RMSECV = 6.15, RPD = 7.44); NLV = 11, and 85 wavebands were selected. Similarly, with respect to TSS, ISE–PLS using FDR showed the highest R^2, highest RPD and lowest RMSECV ($R^2 = 0.97$, RMSECV = 1.91, RPD = 6.64); NLV = 11, and 42 wavebands were selected.

Table 4. Optimum NLV, R^2 and RMSECV using the LOO method in FS–PLS and in ISE–PLS using the entire dataset ($n = 36$), with the residual predictive deviation, the number of selected wavebands and the percent ratio with respect to the full spectrum ($i = 501$).

Parameter	Spectral Data Type	Regression	Calibration			Cross Validation			Selected Wavebands Number	Selected Wavebands (%)
			NLV	R^2	RMSEC	R^2	RMSECV	RPD		
Chl-*a*	Reflectance	FSPLS	4	0.59	29.26	0.41	35.44	1.28		
	Reflectance	ISEPLS	6	0.70	25.01	0.60	29.27	1.55	9	1.80
	FDR	FSPLS	8	0.99	3.25	0.43	35.15	1.32		
	FDR	ISEPLS	11	1	1.14	0.98	6.15	7.44	85	16.97
TSS	Reflectance	FSPLS	6	0.61	7.87	0.35	10.36	1.22		
	Reflectance	ISEPLS	5	0.62	7.76	0.53	8.73	1.45	13	2.59
	FDR	FSPLS	5	0.93	3.39	0.40	9.98	1.27		
	FDR	ISEPLS	11	1	0.84	0.97	1.91	6.64	42	8.38

FDR = first derivative reflectance; NLV = number of latent variables; RMSEC = root mean square error from calibration; RMSECV = root mean square error from cross validation; RPD = the residual predictive deviation.

The relations between observed and predicted Chl-*a* and TSS are shown in Figure 4. The data in this figure were used to evaluate goodness of fit in the FS–PLS and ISE–PLS models. Comparisons between the FS–PLS and ISE–PLS models were presented in combination with the R^2 and RMSE from the cross validation listed in Table 4. For Chl-*a*, the ISE–PLS using FDR showed a higher R^2 and lower RMSECV. The scatter distribution also showed a better linear relation, which can be judged by the red dots clustered along the 1:1 line in Figure 4b. Similarly, with respect to TSS, the ISE–PLS model using FDR showed better results than the others (Figure 4d, red dot). However, both red and green dots clustered vertically, particularly in Figure 4a,c, showing a large variation in the predicted values and nearly no variation in the observed values, indicating that plenty of observed Chl-*a* and TSS samples had low concentrations. This vertical clustering also indicates the FS–PLS and ISE–PLS using reflectance had lower predictive abilities than using FDR.

The selected wavebands in ISE–PLS using the reflectance and FDR datasets are shown in Figure 5. In the reflectance dataset, the selected wavebands were primarily in the red wavelengths (650–680 nm) for Chl-*a*. For TSS, the selected wavebands were in green wavelengths (560 nm), red wavelengths (620–630 nm) and red-edge wavelengths (720 nm). In the FDR datasets, a cluster of wavebands focus on the red region (670–680 nm, 690–710 nm) for Chl-*a*, and wavebands were also selected from other regions: blue (around 410), green (around 490 nm, 510 nm), red (around 603, 615), and the NIR region between 820 nm and 900 nm. Similarly, more wavebands were selected for TSS using FDR than using reflectance, especially in the red (around 620 nm, 680 nm, and 700 nm) and NIR (around 730 nm) regions.

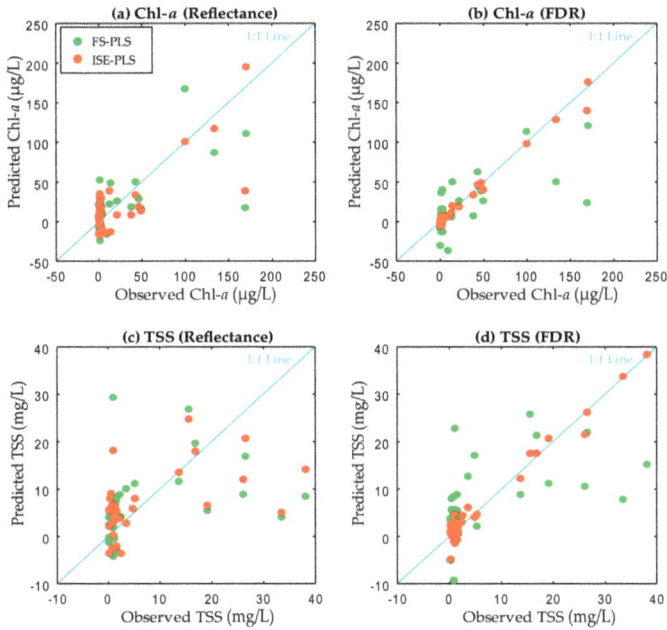

Figure 4. Relations between measured and cross-validated prediction values of Chl-*a* (**a**) Reflectance; (**b**) FDR and TSS; (**c**) Reflectance; (**d**) FDR using FS–PLS and ISE–PLS.

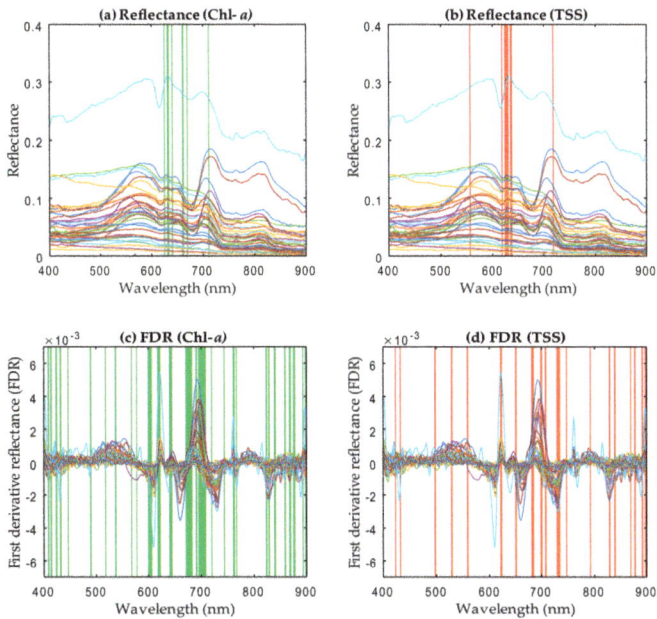

Figure 5. Selected wavebands in ISE–PLS using reflectance or FDR datasets (*n* = 36) to estimate: (**a**) and (**c**) Chl-*a*; (**b**) and (**d**) TSS. Green bars = Chl-*a*; red bars = TSS.

5. Discussion

5.1. Evaluation of the Predictive Abilities of Simple Linear Regression Models

In the present study, models established by single waveband and two waveband combinations were compared using PLS. For single waveband models, FDR showed a better R^2 than smoothed reflectance both for Chl-*a* and TSS, indicating that the accuracy can be improved by enhancing the features of absorption and reflectance from the smoothed reflectance. However, all single waveband models showed poor accuracy for estimating both Chl-*a* and TSS concentrations. According to previous research, single band focus on 670–750 nm is better at determining TSS concentrations [19], especially in turbid water. Single band focus showed no predictive ability in that research, and simple linear regression using two wavebands combinations showed poor accuracy for TSS, which may indicate that TSS is difficult to detect using single-band or two-band combinations in relatively clear water; as shown in our results, most observed TSS values were low. The three-band model was successfully used for Chl-a retrieval in turbid water bodies [18,40]. As for this research, the optimal spectral bands selected from the iterative band tuning are in accord with the previous research [17]; however, even the result shows a considerable R^2, but the relatively high RMSE may indicate a low accuracy model, which may be attributed to different compositions of optically active constituents (Chl-*a*, tripton, CDOM) [40]. The NDCI is a special case of the NDSI: two bands of NDCI are determined by the reflectance peak and spectral absorption peak, and the normalizing of two bands reflectance can eliminate uncertainties in the estimation of R_{rs} [16]. As a comparison, the result of the NDSI has a slight improvement with an R^2 of 0.64 and an RMSE of 27.19 than the NDCI with an R^2 of 0.60 and an RMSE of 28.82, which may indicate that a combination of wavebands at 719 and 663 nm in the NDSI can better reflect the Chl-*a* variations in this research area. Among all tested combinations of the RSI and the NDSI, the best R^2 values were obtained using the NIR waveband (719 nm) and the red region (662 nm for the RSI, 663 nm for the NDSI) to estimate Chl-*a* concentrations, which agrees with the findings of other research. In most available research on the measurement of chlorophyll content in water, the absorption trough is located at near 670 nm, caused by absorption of Chl-*a* [22,41] and the reflectance peak near 710 nm, caused by the fluorescence of Chl-*a* [42–44]. On account of these characteristics, the two waveband models, particularly the NIR/red ratio, have been widely used for Chl-*a* retrieval, and a variety of algorithms have been based mainly on the ratio of reflectance peak (about 710 nm) to reflectance trough (about 670 nm) [22,45]. Similarly, in the present study two wavebands, from the NIR and red regions respectively, were selected by the NDSI, confirming the water body reflection characteristics.

5.2. Evaluation of the Predictive Abilities of FS–PLS and ISE–PLS

As we expected, the PLS models exhibited better predictive abilities than models that use single wavebands or the index-based (RSI and NDSI) approaches, which shows the PLS method is potentially useful in retrieval of inland water quality parameters [46,47]. In our PLS analyses, results using ISE–PLS models with the FDR dataset showed higher R^2 and lower RMSECV values than those of the reflectance dataset. These results are consistent with the research of Han and Rundquitst (1997) [22], who noted that FDR was better correlated with chlorophyll concentration than raw reflectance, and that random noise and the effects of suspended matter could be reduced by FDR [46]. After eliminating outliers and useless predictors, ISE–PLS calibrated more potential models than FS–PLS, both for Chl-*a* and TSS, with the wavelengths relevant to water quality. As a consequence, predictive ability was further enhanced, which is reflected in the results of evaluation indices. PLS-based waveband selection greatly improved predictions for both Chl-*a* (R^2 from 0.43 to 0.98, RMSECV from 35.15 to 6.15, RPD from 1.32 to 7.44) and TSS (R^2 from 0.40 to 0.97, RMSECV from 9.98 to 1.91, RPD from 1.27 to 6.64). The PLS models in combination with wavelength selection had an improved performance also supported by other previous research [29,36,48]. However, the R^2 for Chl-*a* using ISE–PLS reached 0.98, a result that does not rule out the possibility of overfitting; therefore, the solution method for this condition should be the subject of additional research and validation.

5.3. Importance of Selected Wavebands in ISE–PLS

Our results showed 16.97% of all available wavelengths that were selected for predicting Chl-*a* and 8.38% were also selected for predicting TSS by ISE–PLS, which indicates that less than 20% of the waveband information from field hyperspectral data contributes to the prediction for water quality parameters (Chl-*a* and TSS) and over 80% were redundant. In the reflectance dataset, wavebands primarily in the red wavelengths were selected: between 630 and 710 nm for Chl-*a*; for TSS, 560 nm, 620–630 nm, and 720 nm. In the FDR dataset, the selected wavebands for estimating both Chl-*a* and TSS involved more regions than the reflectance dataset. Nevertheless, similar wavelengths in the visible and NIR regions were selected; blue (410 nm), green (approximately 490 nm, 510 nm), and red (approximately 603 nm, 615 nm) for Chl-*a*; and blue (approximately 420 nm), green (approximately 500 nm), red (approximately 620 nm, 680 nm and 700 nm), and NIR (approximately 730 nm) for TSS. Intensive absorption by Chl-*a* resulted in reflectance troughs around 440 and 670nm (Figure 5a) [49]. Low absorption of algal pigments or the scattering of phytoplankton cells and inorganic suspended materials might cause the reflectance peak near 570 nm [41]. The reflectance spectrum peak near 700 nm had a strong correlation with Chl-*a* concentration [42,50,51]. Several previous studies of inland water quality also proved these wavelengths have the potential to predict Chl-*a* and TSS concentrations [52–54]. This study brings obvious evidence that the ISE-PLS model may be considered as a unified approach for remote quantification of constituent concentrations in water quality assessment. Using this method, more informative wavebands can be selected from hundreds of hyperspectral wavebands, which indicates the accuracy and efficiency can be enhanced by ISE-PLS when it comes to using hyperspectral sensors in satellites with a high temporal and spatial resolution to monitor relatively small area inland water quality in the future.

6. Conclusions

The present study develops models for estimating Chl-*a* and TSS concentrations in irrigation ponds using water surface reflectance spectral data. Our results show that PLS regression analysis has high potential for predicting Chl-*a* and TSS based on field hyperspectral measurements, and that ISE wavebands selection in combination with PLS regression analysis can enhance predictive ability. Chl-*a* and TSS concentrations were estimated with high accuracy by using ISE-PLS, which explains 98% of the variance for Chl-*a* and 97% of the variance for TSS. The important wavebands for estimating Chl-*a* and TSS using ISE–PLS represented 16.97% and 8.38%, respectively, of all 501 wavebands over the 400–900 nm range. The selected wavebands approximately match the absorption peaks published by previous researchers. Compared to the estimation of water quality parameters by satellite sensors such as MODIS, ISE–PLS selected more informative wavebands, especially the wavelength at approximately 700 nm. These results provide useful insights for future analyses on the assessment of water quality in irrigation ponds, especially when using satellite imagery.

Acknowledgments: This study was supported by the Environmental Research and Technology Development Fund (S9) of the Ministry of the Environment, Japan, and JSPS KAKENHI (24560623, 15K14041, 16H05631).

Author Contributions: Yuji Sakuno and Kensuke Kawamura designed this study and the fieldwork; Xinyan Fan, Zhe Gong, Jihyun Lim, and Zuomin Wang performed the fieldwork; Zuomin Wang carried out the laboratory analysis, analyzed the data and wrote the manuscript; Zuomin Wang, Kensuke Kawamura and Yuji Sakuno revised the paper.

Conflicts of Interest: The authors declare no conflicts of interest.

References

1. Mateo-Sagasta, J.; Burke, J. *SOLAW Background Thematic Report—TR08*; FAO: Rome, Italy, 2010.
2. Yang, X.E.; Wu, X.; Hao, H.L.; He, Z.L. Mechanisms and assessment of water eutrophication. *J. Zhejiang Univ. Sci. B* **2008**, *9*, 197–209. [CrossRef] [PubMed]

3.   Rönnberg, C.; Bonsdorff, E. Baltic Sea eutrophication: Area-specific ecological consequences. *Hydrobiologia* **2004**, *514*, 227–241. [CrossRef]

4.   World Health Organization (WHO). *Guidelines for Drinking-Water Quality*, 4th ed.; WHO: Geneva, Switzerland, 2011.

5.   Latif, Z.; Tasneem, M.A.; Javed, T.; Butt, S.; Fazil, M.; Ali, M.; Sajjad, M.I. Evaluation of Water-Quality by Chlorophyll and Dissolved Oxygen. *Water Resour. South Present Scenar. Future Prospect.* **2003**, *7*, 123–135.

6.   Lu, F.; Chen, Z.; Liu, W.; Shao, H. Modeling chlorophyll-*a* concentrations using an artificial neural network for precisely eco-restoring lake basin. *Ecol. Eng.* **2016**, *95*, 422–429. [CrossRef]

7.   Sikorska, A.E.; Del Giudice, D.; Banasik, K.; Rieckermann, J. The value of streamflow data in improving TSS predictions—Bayesian multi-objective calibration. *J. Hydrol.* **2015**, *530*, 241–254. [CrossRef]

8.   Fondriest Environmental, Inc. *Turbidity, Total Suspended Solids and Water Clarity*; Fundamentals of Environmental Measurements, 2014. Available online: http://www.fondriest.com/environmental-measurements/parameters/water-quality/turbidity-total-suspended-solids-water-clarity (accessed on 3 November 2016).

9.   Bash, J. *Effects of Turbidity and Suspended Solids on Salmonids*; Center for Streamside Studies, University of Washington: Seattle, WA, USA, 2001; p. 74.

10. Davies-Colley, R.J.; Smith, D.G. Turbidity, suspended sediment, and water clarity: A review. *J. Am. Water Resour. Assoc.* **2001**, *37*, 1085–1101. [CrossRef]

11. Shafique, N.A.; Fulk, F.; Autrey, B.C.; Flotemersch, J. Hyperspectral Remote Sensing of Water Quality Parameters for Large Rivers in the Ohio River Basin. In Proceedings of the First Interagency Conference on Research in the Watersheds, USDA Agricultural Research Service, Washington, DC, USA, 27–30 October 2003.

12. Voutilainen, A.; Pyhälahti, T.; Kallio, K.Y.; Pulliainen, J.; Haario, H.; Kaipio, J.P. A filtering approach for estimating lake water quality from remote sensing data. *Int. J. Appl. Earth Obs. Geoinform.* **2007**, *9*, 50–64. [CrossRef]

13. O'Reilly, J.E.; Maritorena, S.; Mitchell, B.G.; Siegel, D.A.; Carder, K.L.; Garver, S.A.; Kahru, M.; McClain, C. Ocean color chlorophyll algorithms for SeaWiFS. *J. Geophys. Res.* **1998**, *103*, 24937–24953. [CrossRef]

14. Sakuno, Y.; Makio, K.; Koike, K.; Maung-Saw-Htoo-Thaw; Kitahara, S. Chlorophyll-a estimation in Tachibana bay by data Fusion of GOCI and MODIS using linear combination index algorithm. *Adv. Remote Sens.* **2013**, *2*, 292–296. [CrossRef]

15. Gitelson, A.A.; Gritz, U.; Merzlyak, M.N. Relationships between leaf chlorophyll content and spectral reflectance and algorithms for nondestructive chlorophyll assessment in higher plant leaves. *J. Plant Physiol.* **2003**, *160*, 271–282. [CrossRef] [PubMed]

16. Mishra, S.; Mishra, D.R. Normalized difference chlorophyll index: A novel model for remote estimation of chlorophyll-*a* concentration in turbid productive waters. *Remote Sens. Environ.* **2012**, *117*, 394–406. [CrossRef]

17. Dall'Olmo, G.; Gitelson, A.A.; Rundquist, D.C. Towards a unified approach for remote estimation of chlorophyll-a in both terrestrial vegetation and turbid productive waters. *Geophys. Res. Lett.* **2003**, *30*, 1938. [CrossRef]

18. Gitelson, A.A.; Dall'Olmo, G.; Moses, W.; Rundquist, D.C.; Barrow, T.; Fisher, T.R.; Gurlin, D.; Holz, J. A simple semi-analytical model for remote estimation of chlorophyll-*a* in turbid waters: Validation. *Remote Sens. Environ.* **2008**, *112*, 3582–3593. [CrossRef]

19. Nechad, B.; Ruddick, K.G.; Park, Y. Calibration and validation of a generic multisensor algorithm for mapping of total suspended matter in turbid waters. *Remote Sens. Environ.* **2010**, *114*, 854–866. [CrossRef]

20. Pulliainen, J.; Kallio, K.; Eloheimo, K.; Koponen, S.; Servomaa, H.; Hannonen, T.; Tauriainen, S.; Hallikainen, M. A semi-operative approach to lake water quality retrieval from remote sensing data. *Sci. Total Environ.* **2001**, *268*, 79–93. [CrossRef]

21. Dall'Olmo, G.; Gitelson, A.A.; Rundquist, D.C.; Leavitt, B.; Barrow, T.; Holz, J.C. Assessing the potential of SeaWiFS and MODIS for estimating chlorophyll concentration in turbid productive waters using red and near-infrared bands. *Remote Sens. Environ.* **2005**, *96*, 176–187. [CrossRef]

22. Han, L.; Rundquist, D.C. Comparison of NIR/RED ratio and first derivative of reflectance in estimating algal-chlorophyll concentration: A case study in a turbid reservoir. *Remote Sens. Environ.* **1997**, *62*, 253–261. [CrossRef]

23. Inoue, Y.; Peñuelas, J.; Miyata, A.; Mano, M. Normalized difference spectral indices for estimating photosynthetic efficiency and capacity at a canopy scale derived from hyperspectral and CO_2 flux measurements in rice. *Remote Sens. Environ.* **2008**, *112*, 156–172. [CrossRef]
24. Stagakis, S.; Markos, N.; Sykioti, O.; Kyparissis, A. Monitoring canopy biophysical and biochemical parameters in ecosystem scale using satellite hyperspectral imagery: An application on a phlomis fruticosa Mediterranean ecosystem using multiangular CHRIS/PROBA observations. *Remote Sens. Environ.* **2010**, *114*, 977–994. [CrossRef]
25. Inoue, Y.; Sakaiya, E.; Zhu, Y.; Takahashi, W. Diagnostic mapping of canopy nitrogen content in rice based on hyperspectral measurements. *Remote Sens. Environ.* **2012**, *126*, 210–221. [CrossRef]
26. Wold, H. Estimation of Principal Components and Related Models by Iterative Least Squares. In *Multivariate Analysis*; Krishnaiaah, P.R., Ed.; Academic Press: New York, NY, USA, 1966; pp. 391–420.
27. Song, K.; Li, L.; Li, S.; Tedesco, L.; Duan, H.; Li, Z.; Shi, K.; Du, J.; Zhao, Y.; Shao, T. Using partial least squares-artificial neural network for inversion of inland water Chlorophylla. *IEEE Trans. Geosci. Remote Sens.* **2014**, *52*, 1502–1517. [CrossRef]
28. Ghasemi, J.; Niazi, A. Genetic-algorithm-based wavelength selection in multicomponent spectrophotometric determination by PLS: Application on copper and zinc mixture. *Talanta* **2003**, *59*, 311–317. [CrossRef]
29. Kawamura, K.; Watanabe, N.; Sakanoue, S.; Inoue, Y. Estimating forage biomass and quality in a mixed sown pasture based on partial least squares regression with waveband selection. *Grassl. Sci.* **2008**, *54*, 131–145. [CrossRef]
30. Swierenga, H.; Groot, P.J.; Weijer, A.P.; Derksen, M.W.J.; Buydens, L.M.C. Improvement of PLS model transferability by robust wavelength selection. *Chemom. Intell. Lab. Syst.* **1998**, *41*, 237–248. [CrossRef]
31. Boggia, R.; Forina, M.; Fossa, P.; Mosti, L. Chemometric study and validation strategies in the structure-activity relationships of new class of cardiotonic agents. *Quant. Struct Act. Relatsh.* **1997**, *16*, 201–213. [CrossRef]
32. Kawamura, K.; Watanabe, N.; Sakanoue, S.; Lee, H.; Inoue, Y.; Odagawa, S. Testing genetic algorithm as a tool to select relevant wavebands from field hyperspectral data for estimating pasture mass and quality in a mixed sown pasture using partial least squares regression. *Grassl. Sci.* **2010**, *56*, 205–216. [CrossRef]
33. Derbalah, A.S.H.; Nakatani, N.; Sakugawa, H. Distribution, seasonal pattern, flux and contamination source of pesticides and nonylphenol residues in Kurose River water, Higashi–Hiroshima, Japan. *Geochem. J.* **2003**, *37*, 217–232. [CrossRef]
34. Abe, H.; Shinohara, S. A study on irrigation ponds in Higashihiroshima: A statistical approach. *J. Fac. Appl. Biol. Sci. Hiroshima Univ.* **1996**, *35*, 27–34.
35. Stratoulias, V.; Heino, T.I.; Michon, F. Lin-28 regulates oogenesis and muscle formation in Drosophila melanogaster. *PLoS ONE* **2014**, *9*, e101141. [CrossRef] [PubMed]
36. Forina, M.; Lanteri, S.; Oliveros, M.; Millan, C.P. Selection of useful predictors in multivariate calibration. *Anal. Bioanal. Chem.* **2004**, *380*, 397–418. [CrossRef] [PubMed]
37. D'Archivio, A.A.; Maggi, M.A.; Ruggieri, F. Modelling of UPLC behaviour of acylcarnitines by quantitative structure–retention relationships. *J. Pharm. Biomed. Anal.* **2014**, *96*, 224–230. [CrossRef] [PubMed]
38. Williams, P.C. Implementation of Near-Infrared Technology. In *Near-Infrared Technology in the Agricultural and Food Industries*, 2nd ed.; Williams, P.C., Norris, K., Eds.; Association of Cereal Chemists Inc.: Eagan, MN, USA, 2001; pp. 145–169.
39. D'Acqui, L.P.; Pucci, A.; Janik, L.J. Soil properties prediction of western Mediterranean islands with similar climatic environments by means of mid-infrared diffuse reflectance spectroscopy. *Eur. J. Soil Sci.* **2010**, *61*, 865–876. [CrossRef]
40. Gitelson, A.A.; Schalles, J.F.; Hladik, C.M. Remote chlorophyll-a retrieval in turbid, productive estuaries: Chesapeake bay case study. *Remote Sens. Environ.* **2007**, *109*, 464–472. [CrossRef]
41. Huang, Y.; Jiang, D.; Zhuang, D.; Fu, J. Evaluation of hyperspectral indices for chlorophyll-*a* concentration estimation in Tangxun Lake (Wuhan, China). *Int. J. Environ. Res. Public Health* **2010**, *7*, 2437–2451. [CrossRef] [PubMed]
42. Gitelson, A.A. The peak near 700 nm on radiance spectra of algae and water: Relationships of its magnitude and position with chlorophyll concentration. *Int. J. Remote Sens.* **1992**, *13*, 3367–3373. [CrossRef]
43. Bennet, J.; Bogorad, L. Complementary chromatic adaptation in a filamentous blue-green alga. *J. Cell Biol.* **1973**, *58*, 419–435. [CrossRef]

44. Ma, R.H.; Ma, X.D.; Dai, J.F. Hyperspectral Feature Analysis of Chlorophyll a and Suspended Solids Using Field Measurements from Taihu Lake, Eastern China. *Hydrol. Sci. J.* **2007**, *52*, 808–824. [CrossRef]

45. Mittenzwey, K.H.; Breitwieser, S.; Penig, J.; Gitelson, A.A.; Dubovitzkii, G.; Garbusov, G.; Ullrich, S.; Vobach, V.; Müller, A. Fluorescence and reflectance for the in-situ determination of some quality parameters of surface waters. *Acta Hydrochim. Hydrobiol.* **1991**, *19*, 1–15. [CrossRef]

46. Song, K.; Li, L.; Tedesco, L.P.; Li, S.; Duan, H.; Liu, D.; Hall, B.E.; Du, J.; Li, Z.; Shi, K.; et al. Remote estimation of chlorophyll-a in turbid inland waters: Three-band model versus GA-PLS model. *Remote Sens. Environ.* **2013**, *136*, 342–357. [CrossRef]

47. Ryan, K.; Ali, K. Application of a partial least-squares regression model to retrieve chlorophyll-*a* concentrations in coastal waters using hyper-spectral data. *Ocean Sci. J.* **2016**, *51*, 209–221. [CrossRef]

48. Chen, D.; Cai, W.; Shao, X. Representative subset selection in modifiediterative predictor weighting (mIPW)-PLS models for parsimonious multivariate calibration. *Chemom. Intell. Lab. Syst.* **2007**, *87*, 312–318. [CrossRef]

49. Yacobi, Y.Z.; Moses, W.J.; Kaganovsky, S.; Sulimani, B.; Leavitt, B.C.; Gitelson, A.A. NIR-red reflectance-based algorithms for chlorophyll-a estimation in mesotrophic inland and coastal waters: Lake Kinneret case study. *Water Res.* **2011**, *45*, 2428–2436. [CrossRef] [PubMed]

50. Vasilkov, A.; Kopelevich, O. Reasons for the appearance of the maximum near 700 nm in the radiance spectrum emitted by the ocean layer. *Oceanology* **1982**, *22*, 697–701.

51. Gitelson, A.; Garbuzov, G.; Szilagyi, F.; Mittenzwey, K.; Karnieli, A.; Kaiser, A. Quantitative remote sensing methods for real-time monitoring of inland waters quality. *Int. J. Remote Sens.* **1993**, *14*, 1269–1295. [CrossRef]

52. Hu, Z.; Liu, H.; Zhu, L.; Lin, F. Quantitative inversion model of water chlorophyll-a based on spectral analysis. *Procedia Environ. Sci.* **2011**, *10*, 523–528. [CrossRef]

53. Thiemann, S.; Kaufman, H. Determination of chlorophyll content and tropic state of lakes using field spectrometer and IRS—IC satellite data in the Mecklenburg Lake Distract, Germany. *Rem. Sens. Environ.* **2000**, *73*, 227–235. [CrossRef]

54. Gons, H.J. Optical teledetection of chlorophyll a in turbid inland waters. *Environ. Sci. Technol.* **1999**, *33*, 1127–1132. [CrossRef]

remote sensing

MDPI

Article

Fluorescence-Based Approach to Estimate the Chlorophyll-A Concentration of a Phytoplankton Bloom in Ardley Cove (Antarctica)

Chen Zeng [1,2,*], Tao Zeng [3,4], Andrew M. Fischer [5] and Huiping Xu [6]

[1] School of Ocean and Earth Science, Tongji University, Shanghai 200092, China
[2] Department of Earth, Ocean and Atmospheric Sciences, University of British Columbia, Vancouver, BC V6T 1Z4, Canada
[3] National Satellite Ocean Application Service State Oceanic Administration, Beijing 100081, China; ztao10@mail.nsoas.org.cn
[4] Key Laboratory of Space Ocean Remote Sensing and Application State Oceanic Administration, Beijing 100081, China
[5] Institute for Marine and Antarctic Studies, University of Tasmania, Launceston, 7250, Australia; andy.fischer@utas.edu.au
[6] Institute of Deep-sea Science and Engineering, Chinese Academy of Sciences, Sanya 572000, China; xuhp@idsse.ac.cn
* Correspondence: 07zengchen@tongji.edu.cn; Tel.: + 1-778-885-0301

Academic Editors: Yunlin Zhang, Claudia Giardino, Linhai Li, Deepak R. Mishra and Prasad S. Thenkabail
Received: 23 September 2016; Accepted: 20 February 2017; Published: 25 February 2017

Abstract: A phytoplankton bloom occurred in Ardley Cove, King George Island in January 2016, during which maximum chlorophyll-a reached 9.87 mg/m^3. Records show that blooms have previously not occurred in this area prior to 2010 and the average chlorophyll-a concentration between 1991 and 2009 was less than 2 mg/m^3. Given the lack of in situ measurements and the poor performance of satellite algorithms in the Southern Ocean and Antarctic waters, we validate and assess several chlorophyll-a algorithms and apply an improved baseline fluorescence approach to examine this bloom event. In situ water properties including in vivo fluorescence, water leaving radiance, and solar irradiance were collected to evaluate satellite algorithms and characterize chlorophyll-a concentration, as well as dominant phytoplankton groups. The results validated the nFLH fluorescence baseline approach, resulting in a good agreement at this high latitude, high chlorophyll-a region with correlation at 59.46%. The dominant phytoplankton group within the bloom was micro-phytoplankton, occupying 79.58% of the total phytoplankton community. Increasing sea ice coverage and sea ice concentration are likely responsible for increasing phytoplankton blooms in the recent decade. Given the profound influence of climate change on sea-ice and phytoplankton dynamics in the region, it is imperative to develop accurate methods of estimating the spatial distribution and concentrations of the increasing occurrence of bloom events.

Keywords: chlorophyll-a estimation; fluorescence approach; King George Island; phytoplankton bloom

1. Introduction

Due to the extreme climate and the difficulties of conducting field research above 60° south, the Southern Ocean (SO), especially around Antarctica, lacks a systematic in situ sampling program of its peculiar bio-optics and micro-organism community structure [1–3]. As a result, satellite measurements from space still have large errors in estimating phytoplankton biomass [4,5] and global chlorophyll-a satellite algorithms typically underestimate chlorophyll-a in the Southern Ocean [6–11].

There are three main reasons for global algorithm underestimation. First is the difference in optical properties between global and SO waters. Compared with the global ocean, the SO has a narrower water leaving radiance in the green band [12]. This narrow gap leads to underestimation of chlorophyll-a in blue-green band ratio algorithms such as OC3 or OC4 [13,14]. Second, seasonal sea ice causes contaminated pixels, which underestimate chlorophyll-a in > 1.5 mg/m^3 and overestimates in < 1.5 mg/m^3 waters [15,16]. These are obvious patterns in the Arctic and are probably similar in Antarctica. Third, inappropriate atmospheric correction in the SO introduces error when converting top of atmosphere radiance into the water leaving radiance. The SO has unique aerosols and cloud coverage [17], and lacks a proper vertical atmospheric simulation to correct for aerosol influences [18].

In addition to band-ratio chlorophyll-a algorithms, state-of-the-art algorithms for global chlorophyll-a estimation include IOP (inherent optical property) bio-optical models (e.g., GSM01 [19] and QAA (quasi-analytical algorithm) [20]) and baseline algorithms, such as nFLH (normalized fluorescence line height). Bands setting at 667, 678, and 748 nm help MODIS become the only satellite to achieve the baseline nFLH approach, which estimates chlorophyll-a through fluorescence intensity from photosynthesis products [21].

$$F(\lambda_{em}) = E(\lambda_{ex})a^*(\lambda_{em})[Chla]\varphi_f Q_a^*(\lambda_{em}) \tag{1}$$

where, $F(\lambda_{em})$ is the fluorescence intensity (mol quanta m^{-3}·s^{-1}), $E(\lambda_{ex})$ is the incident intensity (mol quanta m^{-2}·s^{-1}), $a^*(\lambda_{ex})$ is are chlorophyll-a absorption coefficients (m^2·mg·Chla^{-1}) per chlorophyll unit (* average per [Chla]), $[Chla]$ is the chlorophyll-a concentration (mg·m^{-3}), φ_f is the fluorescence yield (mol quanta), and $Q_a^*(\lambda_{em})$ is the re-absorption coefficients in cells. From this equation, $a^*(\lambda_{ex})Q_a^*(\lambda_{em})$ reflects the phytoplankton composition features and $E(\lambda_{ex})\varphi_f$ is from the light acclimation mechanism (photosynthetic adjustment in response to light availability). Under stable phytoplankton community composition and light acclimation, chlorophyll-a will have a linear relationship with fluorescence intensity. The nFLH approach helps to avoid contamination of the chlorophyll-a signal by suspended sediments, detritus, and CDOM (colored dissolved organic matter) and typically produces more accurate results in case 2 coastal waters [22]. However, photosynthetic mechanisms are subject to NPQ (non-photochemical quenching) when phytoplankton encounters intense light [23]. Non-linearity then occurs between chlorophyll-a and fluorescence intensity, limiting the applications of nFLH in high-intensity light areas, such as midday direct solar radiation.

A phytoplankton bloom occurred in Ardley Cove near King George Island (KGI) in January 2016. Historical records showed that phytoplankton blooms had not previously occurred in this area prior to 2010 and the average of chlorophyll-a between 1991 and 2009 was less than 2 mg/m^3 [24]. An obvious phytoplankton bloom was reported until 2010, during which maximum chlorophyll-a reached 20 mg/m^3. In situ records show an increasing trend of phytoplankton biomass in this area, likely caused by increasing SST (sea surface temperature) related to global warming [25]. The aim of this study was to document an algal bloom in Ardley Cove through the validation and regionalization of the MODIS nFLH algorithm.

2. Materials and Methods

All the samples were collected on a Zodiac (an inflatable boat) in the Great Wall Cove and Ardley Cove near the China Great Wall Station on KGI between 6 and 27 January 2016 (Figure 1a,b). Clear sky, high solar evaluation (40°–50° in 10 am–2 pm), and light breeze (wind speed < 5m/s) were chosen as the threshold conditions during sampling to reduce the impact on above-water optical property retrieval.

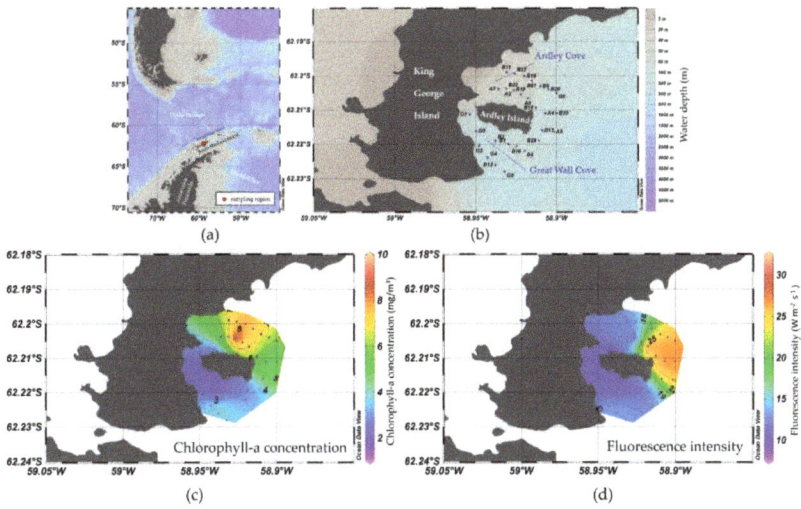

Figure 1. (**a**) Sampling area location. (**b**) Locations of in situ samples. (**c**) Chlorophyll-a spatial pattern interpolated from the in situ samples during January 2016. (**d**) In vivo fluorescence interpolated from the in situ samples during January 2016.

2.1. Apparent Optical Properties

We applied (45°, 135°) angles for above-water measurements recommended by NASA Ocean color protocol [26], to normalize water radiance from simultaneous sky radiance and to remove interference from environmental light conditions. A hand-held VNIR spectroradiometer (HH2, ASD Inc., Boulder, Colorado, USA), with a high spectral resolution of 1 nm, was used to measure water leaving radiance. When measuring, the Zodiac stopped its engine to prevent white capping. However, without power, the boat did not drift far from its original location. The black hull of the boat decreased the probability of reflected light from contaminating the spectrometer readings. Furthermore, being close to the water allowed measurements to be made from 20 cm above the surface of the water, qualifying the measurement as an 'at surface' reading. Every station had duplicate measurements and each duplicate had 15 samples for water, sky and standard plaque. Averaging of the 15 samples improved the signal-noise ratio. Upwelling radiance was converted into normalized water leaving reflectance with Equation (2),

$$R_{rs} = \left[S_{water} - r S_{sky} \right] \times \rho_p / \pi S_p \qquad (2)$$

where, S_{water}, S_{sky}, S_p are signals for water, sky, and standard plaque respectively, ρ_p is the reflectance rate for the standard plaque on whole bands (%, which is provided by factory calibration), and r is the sea-air interface reflectance rate (%). As a response to wind speed, r varies from 2.5%–2.7% (see details in Tang, et al. [27]). Solar irradiance has significant impacts on water upwelling radiance, while lesser influences on water reflectance (see Figures 4 and 5 in Mobley [28]). Since reflectance takes into account environmental factors from solar irradiance after normalization and all measurements were conducted at the same angles, they can be directly compared with each other. Water reflectance then went through a baseline correction to shift the infrared band to 0 (Rrs(λ) − Rrs(763)), and was smoothed with a five-point median filter. We also simultaneous collected sky iPAR (instant photosynthetically available radiation) from a cosine receptor setup on an ASD HH2 at every station.

2.2. In Situ Chlorophyll-a Concentration

Duplicate surface water samples (500 ml for each) were collected at each station (Figure 1b). Water samples were filtered under 50 kpa onto a GF/F filter. The filter was immersed into a flask with 10 ml 90% acetone and wrapped with aluminum foil. All steps were conducted under low light conditions to prevent chlorophyll-a decomposition. Extracted chlorophyll-a was stored for 24 h in a freezer. The flask was then placed in a centrifuge for 10 min at 4000 revolutions/min (TDL-60B, Anke Ins., Ninbo, China). Total chlorophyll-a was measured from its supernatant with three readings using a fluorometer (AquaFluor®Handheld Fluorometer and Turbidimeter, Turner Design, San Jose, CA, USA), with an excitation band of 430 nm and an emission band of 660 nm. Pheophytin concentration was measured again with the fluorometer after the supernatant had a chemical reaction with 10% HCl for 1 min. The chlorophyll-a concentration responsible for photosynthesis is total chlorophyll-a concentration minus the pheophytin concentration after converting the fluorescence intensity to the chlorophyll-a value from a chlorophyll-a: fluorescence curve. All duplicates show a mean deviation lower than 5.3%. Prior to chlorophyll-a evaluation, a spectrophotometer (HITACHI F-2700, Hitachi High-Technologies Corporation, Tokyo, Japan) was used to calibrate the various chlorophyll-a concentration values on the fluorometer using standard stock chlorophyll-a diluted gradient solutions.

2.3. In Vivo Fluorescence

A fluorometer (AquaFluor®Handheld Fluorometer and Turbidimeter, Turner Design, San Jose, CA, USA.) was also used to measure in vivo fluorescence at all stations. Surface water was directly placed in the measuring window to retain phytoplankton light acclimation and physiology information. All samples were measured 3 times. The in vivo fluorometer has an excitation band at 430 nm and emission band at 660 nm. This instrument directly transfers water fluorescence intensity into chlorophyll-a concentration from a preset linear equation based on factory calibration.

2.4. Satellite/In Situ Match-Ups

We limited satellite/in situ match-ups with three criteria. First, we used a mean 3×3 spatial windows on the satellite image. Second, the gap between in situ and satellite image was limited to less than four hours. Lastly, an AOT (aerosol optical thickness) index lower than 0.15 was chosen. All satellite data were L2 products from the Ocean Color Website collected by MODIS Aqua and Terra. Before converting to water leaving reflectance, the L2 products were subject to pixel-by-pixel atmospheric correction with the 6S model and kept at a spatial resolution of 1.1 km without any resampling. Matchups between in situ chlorophyll-a concentration and water optical properties (Rrs) were processed with global algorithms (Table 1) to estimate chlorophyll-a concentration accuracy and error around Ardley and Great Wall Cove.

To obtain regional coefficients for the nFLH algorithm, we applied the 'Leave One Out' cross-validation method [29]. For each iteration, a matching pair was left out and residual error and curve fitting coefficients were estimated. The 27 coefficients and residual errors were then averaged to obtain the unbiased estimation.

$$\text{RMSE} = \sqrt{\left(nFLH_{in\ situ} - nFLH_{est}\right)^2} \tag{3}$$

$$\text{LH} = c_1\ nlw_{678} + c_2\ nlw_{667} + c_3\ nlw_{748} \tag{4}$$

where, RMSE is the residual error, $nFLH_{in\ situ}$ is the fluorescence value from the field and $nFLH_{est}$ is the value estimated from the 26 matching pairs. Equation (4) is the fitting function and c_1, c_2, and c_3 are the fitting coefficients. The leave-one out approach helps to limit the problem of overfitting [29].

2.5. Phytoplankton Absorption Coefficent Spectrum

We applied bio-optical models to determine phytoplankton absorption coefficients from water leaving reflectance. There are two commonly used approaches in bio-optics, QAA [20] and GSM01 [19]. GSM01 retrieves simultaneous estimates for chlorophyll-a concentration, the absorption coefficients for dissolved, and detrital materials and particulate backscatter. Model parameters are then tuned through simulated annealing. The GSM01 model decreases residual errors from satellite and estimated Rrs through a multiple iteration approach (Maritorena, et al. [19]). The QAA (Lee, et al. [19]) analytically calculates coefficient values of total absorption and backscattering from remote sensing reflectance. In comparison to GSM01, the QAA approach has the benefit of not requiring any prior information about the spectral shape of $a_\Phi(\lambda)$ and thereby reduces potential errors and uncertainties with spectral models or inappropriate spectral shapes [19].

Considering the uniqueness of SO water optics [13,14], we intended to keep more original Rrs signals in phytoplankton absorption coefficients. Therefore, we applied QAA approach to invert the in situ Rrs and expanded the single band absorption into whole visible bands by introducing some of GSM01 equations. The QAA approach we applied here followed its V5 coefficients [30] and pure water absorption [31]. Details are shown in Appendix A.

Ciotti et al. [32] developed a micro-pico cell-size composition estimation approach from absorption coefficient spectrum. They used least-square fitting to gain abundance ($S_{\langle f \rangle}$) of micro and pico cells using whole bands.

$$\widehat{a_{\langle ph \rangle}}(\lambda) = \left[S_{\langle f \rangle} \cdot \overline{a_{\langle pico \rangle}}(\lambda) \right] + \left[(1 - S_{\langle f \rangle}) \cdot \overline{a_{\langle micro \rangle}}(\lambda) \right] \tag{5}$$

where $\widehat{a_{\langle ph \rangle}}(\lambda)$ are phytoplankton absorption coefficients, $S_{\langle f \rangle}$ is abundance of pico-phytoplankton, $\overline{a_{\langle pico \rangle}}(\lambda)$ and $\overline{a_{\langle micro \rangle}}(\lambda)$ are standard pico- and micro-absorption coefficients per unit from Table 3 in Ciotti, et al. [32]. This algorithm was used to estimate micro-phytoplankton fractions.

Table 1. Various models application on chlorophyll-a estimations in KGI phytoplankton bloom area.

Model Algorithms	Relationship (R) with In situ Chlorophyll-a	Relative Error (%)	Regions	References
chl = $10^{(0.573 - 2.259X + 0.203X2 - 1.300X3)}$ + 0.386; X = log((nlw 443 > nlw 460 > 520)/ nlw 545)	0.269214	0.367748	SO	Mitchell et.al. [33]
chl = 2.22X; chl < 1.5 mg/m³; X = log(nlw 440/ nlw 555)	0.191404	0.990063	WAP	Dierssen et.al. [7]
chl = $10^{(0.78 - 2.52X)}$; chl > 1.5 mg/m³; X = log(nlw 520/ nlw 555)	0.260399	0.632074	WAP	Dierssen et.al. [7]
chl = 0.45 + 0.53X; chl > 1.5 mg/m³; X = log(nlw 520/ nlw 555)	0.261184	1.160298	WAP	Dierssen et.al. [7]
chl = $10^{(0.641 - 2.058X - 0.442X2 - 1.140X3)}$; X = log(rrs490/rrs555)	0.28514	0.733532	WAP	Dierssen et.al. [7]
chl = $10^{(0.3914 + 1.0176X - 0.3114X2 + 0.0186X3 + 0.0610X4)}$; X = log(rrs490/rrs555)	0.28981	0.811026	WAP	Dierssen et.al. [7]

* nlw is water leaving radiance; WAP(West Antarctic Peninsula)

3. Results

3.1. Spatial Distribution of Water Optical Properties

Chlorophyll-a varied significantly between Ardley Cove and Great Wall Cove (Figure 1c). Ardley Cove had a large phytoplankton bloom, with a maximum of chlorophyll-a at 9.87 mg/m³. Relatively low chlorophyll-a concentrations occurred in the Great Wall Cove, reaching only 1.37 mg/m³. Simultaneous in vivo fluorescence did not follow the same pattern, with the maximum occurring on the east side of Ardley Island instead of the north side (Figure 1d).

KGI Rrs spectrum produced 4 different shapes, with varied slopes between 550–560 nm (Figure 2a). These shapes were driven by chlorophyll-a concentration, with chlorophyll-a concentration increasing the overall Rrs value across all wavelengths less than 600 nm. For red bands higher than 600 nm, 4 Rrs spectra decreased significantly due to pure water absorption [32].

Despite the different Rrs spectral shapes for the various chlorophyll-a concentrations, the phytoplankton absorption coefficients showed similar curves after removing non-phytoplankton absorptions and particle backscattering coefficients (Figure 2b). Two obvious peaks appear in the blue and red bands of phytoplankton absorption spectra. Higher chlorophyll-a concentration has higher absorption coefficients, and the 660 nm absorption band has a good agreement with chlorophyll-a concentration (52.6%, $p < 0.05$) (Figure 2c).

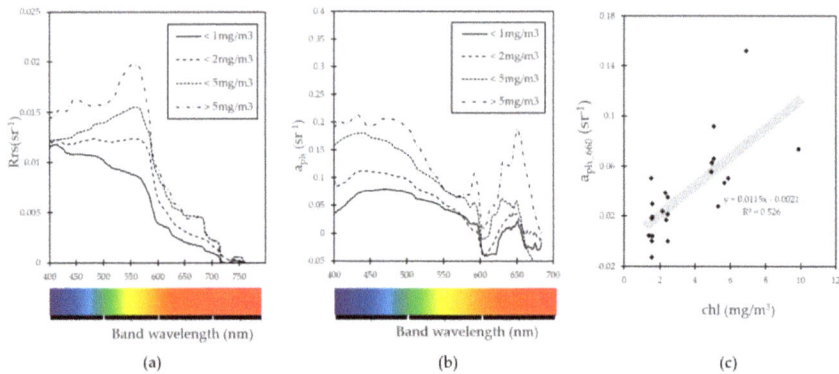

Figure 2. (**a**) Four selected above water Rrs spectra. (**b**) Four featured phytoplankton absorption coefficients derived from Figure 2a using QAA v5 presented in Appendix A. (**c**) Correlation between phytoplankton absorption coefficients and in situ chlorophyll-a concentration.

3.2. Algorithim Chlorophyll-a Estimation and In Situ/Satellite Match-Ups

Chlorophyll-a estimation from various algorithms were compared to in situ Rrs and residual errors were calculated. The results showed poor relationships with almost all previously developed algorithms (Table 1). All correlations were lower than 30%. An exception was the algorithm of Mitchell et.al [33]. Relative errors from estimated and in situ chlorophyll-a were greater than 60%. Although all these algorithms were built from the SO dataset, scarce samples in high latitude and high chlorophyll-a coastal water resulted in poor performance of chlorophyll-a estimation in KGI waters.

In addition, satellite-derived chlorophyll-a and fluorescence intensity from global empirical algorithms were evaluated. The comparison between in situ and satellite data showed a relatively good correlation with fluorescence intensity (55.35%) (Figure 3a). Direct band-ratio chlorophyll-a estimation showed a non-linear relationship with in situ chlorophyll-a with an 11.43% correlation (Figure 3b). Poor estimation was also apparent in the QAA bio-optical approach at a correlation of 1.62% (Figure 3b).

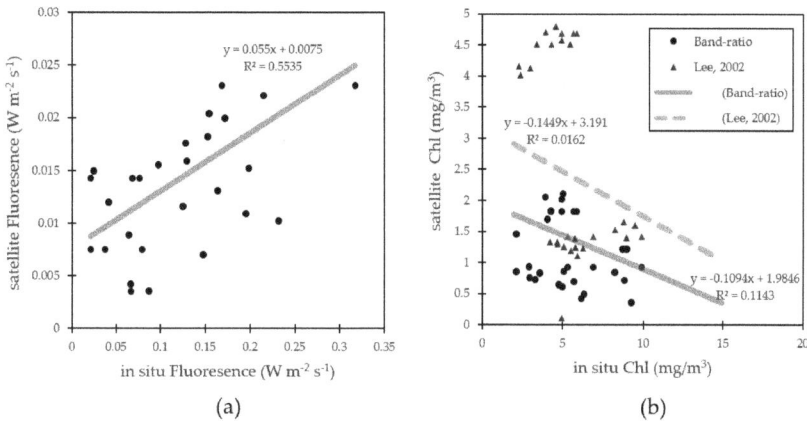

Figure 3. (**a**) Correlation between satellite derived fluorescence and in situ measured in vivo fluorescence (n = 28). The satellite fluorescence algorithm here is based on the global fluorescence coefficients from NASA ocean color group (https://oceancolor.gsfc.nasa.gov/atbd/nflh/). (**b**) Correlation between satellite and in situ chlorophyll-a concentrations. The satellite algorithm applied here is the OC3 global band-ratio (http://oceancolor.gsfc.nasa.gov/cms/atbd/chlor_a) and NASA IOP model derived absorption coefficient for phytoplankton (http://oceancolor.gsfc.nasa.gov/cms/atbd/giop).

3.3. Fluorescence Approach Estimation for Chlorophyll-a

Based on the relatively good linear relationship between in situ and satellite fluorescence measurements, we explored the feasibility of using the fluorescence approach for estimating chlorophyll-a in KGI. This issue of non-linearity between in vivo fluorescence and chlorophyll-a, caused by NPQ under intense solar irradiance, was considered. Our measurements also showed a decreasing exponential phase of fluorescence per chlorophyll-a: iPAR (Figure 4a). Increasing solar irradiance decreases the fluorescence of chlorophyll-a yield, illustrating the impact of NPQ.

Previous research [23] reported a turning point of NPQ when iPAR reaches 100 μ mol/sec. After that point, NPQ will grow slowly. Since our measurements were collected in the southern summer with long daytime and high solar irradiance, the in vivo fluorescence influence on chlorophyll-a decreased (Figure 4a). Their correlation was 40.64%. High chlorophyll-a concentration samples had larger underestimation and deviation from 1:1 line than low chlorophyll-a concentration measurements (Figure 4b).

For the relationship between global empirical fluorescence from satellite and in situ chlorophyll-a (Figure 4c), the global coefficients do not work well in estimating chlorophyll-a in Ardley Cove since the fluorescence reacts a lot on phytoplankton physiology and iron stress [34]. Therefore, we optimized the three coefficients in the nFLH algorithm from the 30 in vivo fluorescence samples for the KGI region to produce the following algorithm:

$$\text{nFLH} = \frac{13}{91}nlw_{678} + \frac{26}{91}nlw_{667} - \frac{53}{91}nlw_{748}, \ (r^2 = 65.15\%, \text{ rmse} = 0.056) \tag{6}$$

where at 678, 667, and 748 nm, respectively. This optimization improved fluorescence estimation by about 10% from global empirical coefficients according to Figure 3a. Then, NPQ (FLH/iPAR) and phytoplankton absorption package correction (FLH/$\overline{a_{ph}}$, where $\overline{a_{ph}}$ is absorption mean between 300–700 nm per chlorophyll-a unit following Babin et al. [35]) were applied to the fluorescence estimation. The final fluorescence: chlorophyll-a relationship increased from original 40.64% to 59.46% (Figure 4d). Those two corrections referenced Behrenfeld et al. [23] correction in their Figure 2b,d. We do not introduce a global satellite average for iPAR and a_{ph} to non-dimensionlization as they did in their correction. In our equation, the magnitude of in situ iPAR (10^3) and a_{ph} (10^{-3}) balanced out and produced agreeable results.

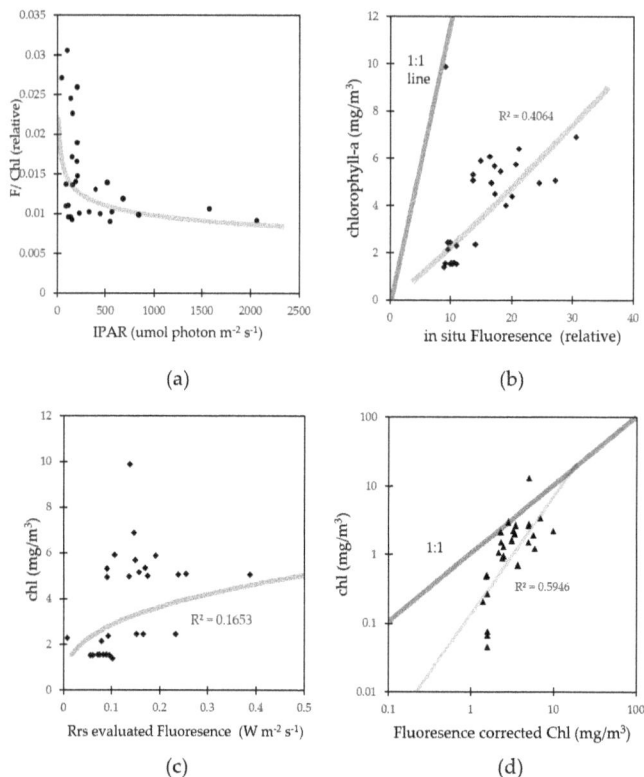

Figure 4. (a) Relationship between fluorescence per chlorophyll-a unit and iPAR. (b) Relationship between in situ chlorophyll-a and fluorescence intensity. (c) Relationship between in situ chlorophyll-a and in situ Rrs estimated fluorescence. (d) Relationship between in situ chlorophyll-a and in situ fluorescence estimated chlorophyll-a after regional coefficients improvement, NPQ correction and phytoplankton absorption package correction.

3.4. Nutrient, Light, and Phytoplankton Composition

Behrenfeld et al's. [23] algorithm was useful in providing phytoplankton physiological status for our samples. According to Behrenfeld et al. [23], the remaining fluorescence signal divided by chlorophyll-a after NPQ and phytoplankton absorption package corrections are defined as fluorescence quantum yield and reflects the factors limiting phytoplankton growth (e.g., iron, light, or nitrogen/phosphorus). NPQ correction removes the photo-protection impacts on fluorescence intensity, and phytoplankton absorption package correction decreases the gaps between various phytoplankton communities. Fluorescence quantum yield is then quantified to global phytoplankton growth limitation factors where > 1.4% stands for iron limitation and < 1.4% means light, nitrogen/phosphorus limitation (in Figure 4 from their publication). The quantum yield in our samples was < 0.7% indicating that light and nitrogen/phosphorus limitation is greater in KGI Great Wall Cove than Ardley Cove (Figure 5a).

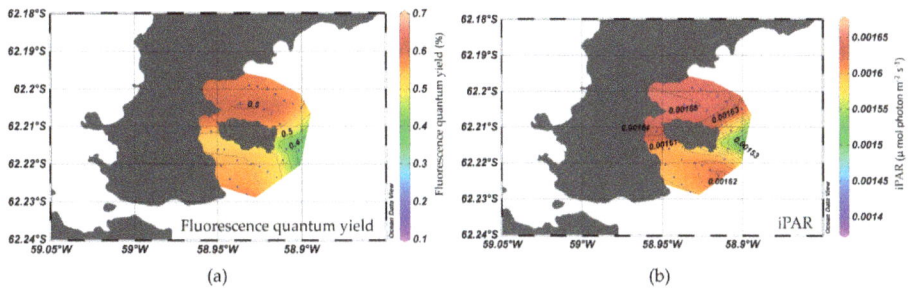

Figure 5. (**a**) Fluorescence quantum yield interpolated from in situ samples during January 2016, and (**b**) iPAR spatial pattern interpolated from in situ samples during January 2016.

The spatial pattern of iPAR (Figure 5b) is similar to fluorescence quantum yield and chlorophyll-a in Figure 1c., which showed a stronger positive link between solar irradiance and phytoplankton biomass.

Phytoplankton size, deduced from the phytoplankton absorption coefficients, showed that the dominant phytoplankton in the bloom area are micro-size algae (> 20µm) (Figure 6a). The abundance of micro-size algae occupied 79.58% of the entire bloom phytoplankton community in Ardley Cove during our study. The mean residual error for Sf (pico-phytoplankton fraction estimation) is 16.54%.

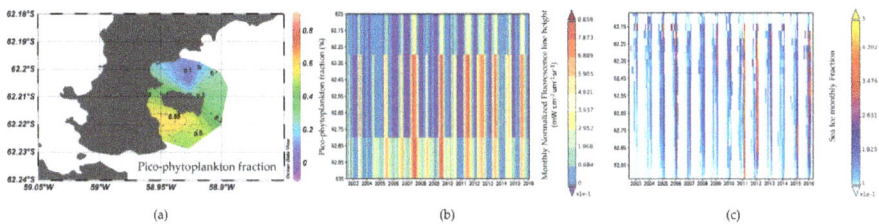

Figure 6. (**a**) Pico-phytoplankton fraction spatial pattern retrieved from the in situ spectrum algorithm (Equation (5)) during January 2016 (**b**) Hovmoller longitude-average plots of satellite records on fluorescence estimated chlorophyll-a from Giovanni (http://giovanni.gsfc.nasa.gov/giovanni/). (**c**) Hovmoller longitude-average plots of satellite records on sea ice concentration from Giovanni.

Additionally, satellite-estimated chlorophyll-a and sea ice in this area show an increasing trend after 2002 (Figure 6b,c). When we focus on the year with extended sea ice coverage and high sea ice concentration (2009, 2011, and 2015), fluorescence intensity also increased in the following southern summer (2010, 2012, and 2016). Those years with high phytoplankton concentrations included 2010 [7]. It is noted that the fluorescence data shown here did not go through NPQ and phytoplankton absorption package corrections, so Figure 6b illustrates the relative trend with in the most recent decade.

4. Discussion

4.1. Fluorescence Approach Estimation for Chlorophyll-a

Light energy is an important factor regulating phytoplankton growth in KGI though photo-inhibition. The spatial pattern of IPAR (Figure 5b), which showed good agreement with chlorophyll-a concentration (Figure 1c), clearly illustrated the influence of light limitation on phytoplankton growth. This physical condition prevents optimal light for phytoplankton growth and this along with intense wind and terrigenous particles, typically leads to low primary productivity

KGI [25]. From the perspective of this study, low light conditions ensured the reliability of the fluorescence approach and the nFLH algorithm.

Studies have shown improved application of the baseline fluorescence approach in case II waters over the global blue-green band-ratio algorithms [36]. The fluorescence approach also has higher effectiveness than the semi empirical algorithms like QAA (this study). Studies have also shown that inefficiencies in the SO bio-optics algorithms is generated from green band reflectance, leading to an obvious underestimation in chlorophyll-a when using global empirical coefficients [9]. In addition, the blue band is highly affected by CDOM, whose absorption coefficient was reported to occupy 70% of the non-water components in WAP (west of the Antarctica Peninsula) [37]. The absorption spectrum of CDOM has decreasing exponential influence, showing less influence on longer bands. Since the red band has longer wavelengths in the visible band, the selection of red bands will introduce less contamination in CDOM rich waters.

Conversely, this band selection has some issues. The red/NIR bands are more susceptible to atmospheric interference. Two recent programs, SOCTRATES [16] and WCRP [38] focused on the SO atmosphere, intend to improve the understanding of clouds, aerosols, air-sea exchanges over the SO. The improvement of SO atmospheric model will further meet the demands of input parameters for ocean color estimation in the near future.

SST-based corrections of fluorescence intensity have also been applied to SO waters. Browning et al. [39] found that extremely low SST in the SO has a strong connection with NPQ parameters and can be used to correct field chlorophyll-a fluorescence signals. They showed highly variable phytoplankton physiology features and communities under regional irradiance conditions and suggested more regional studies on NPQ capacity impacts. However, this correction needs an empirical relationship between SST and physiology coefficient B and was not applied in this study.

The byproduct of fluorescence quantum yield from the fluorescence approach also has a significant role in estimating primary productivity. State-of-the-art algorithms for primary productivity estimation are mainly based on chlorophyll-a absorption-fluorescence quantum yield equation or a photosynthesis-irradiance equation [40]. If we improve the estimation of fluorescence quantum yield from space, we will improve estimation for SO phytoplankton photosynthesis and carbon export. This will lead to a clearer understanding of the role of the SO in global climate change.

4.2. Factors on Phytoplankton Bloom in KGI

As an island hosting several international research stations, the ecological trend of the coastal waters surrounding KGI deserves attention. Historical records in KGI have shown that no phytoplankton blooms [6] occurred in these waters until 2010, during which maximum chlorophyll-a reached 20 mg/m^3. Research has assumed that the increasing phytoplankton growth was exacerbated by global warming [25]. This manuscript documented another phytoplankton bloom in 2016 and attempted to determine its potential cause from optics and satellite history records. Iron limitation was widely observed in SO phytoplankton growth, which leads to ferredoxin deficiency in photosystem I and decreases photosynthetic efficiency. Iron depletion is reported as a major restriction for SO primary productivity [41,42]. However, previous research in KGI found no obvious nutrient or iron deficiencies in phytoplankton growth [43]. The microphytoplankton dominant features in this bloom (Figure 6a) further confirms iron enrichment because micro-size cells favor the iron abundant water. Micro-phytoplankton maintains a high growth rate in iron rich water, and their large size prevents zooplankton grazing [44].

Diatoms found around the WAP are larger in size than those found in other regions [44,45]. The phytoplankton bloom reported in 2010 was dominated by micro-algae (e.g., *Porosira glacialis*, *Thalassiosira antarctica*, and *T. ritscheri* [25]). Phytoplankton tends to adapt their cellular physiology to optimize light harvesting and photoreception [44]. Generally, small size algae benefit from light limited conditions due to their high absorption-photosynthesis effectiveness and small cellular shading [46]. Micro-size algae have more self-protection under high light conditions because their large surface area will cause low absorption-photosynthesis effectiveness. Therefore, for the micro-phytoplankton dominant bloom in KGI, nutrient support and zooplankton grazing has more influence on phytoplankton community composition than light limitation.

Owing to extended sea ice coverage and high sea ice concentration (Figure 6b,c), KGI has more fresh water input following southern summer to form strong stratification and keep phytoplankton in surface waters. This shallow mixed layer increases the probability for phytoplankton to absorb light energy. In addition, sea ice melt brings nutrients for algal growth [46–49]. The increasing trend of sea ice increases nutrients input (e.g., iron or micronutrients), which will lead to phytoplankton blooms in spring and summer. In addition, phytoplankton tends to bloom in marginal sea ice regions in the coastal ocean of the SO [46–49]. Therefore, increasing sea ice extent and coverage in KGI is a likely trigger for increased phytoplankton blooms.

4.3. Errors for Fluorescence Approach Estimating Chlorophyll-a

Two factors are responsible for accurate satellite estimation of chlorophyll-a. These include the sea-air model for calculating chlorophyll-a from water leaving Rrs and the atmospheric radiation transfer model to transfer top-of-atmosphere radiance into water leaving Rrs. Atmospheric correction has a large probability of introducing errors, because its signal occupies 90% of the total signal received at the satellite [50]. Therefore, atmospheric correction deserves great attention, particularly in improving its accuracy. Until now, there has not been a proper high latitude atmospheric correction model for the southern hemisphere and high cloud coverage in the SO creates a challenge for accurate atmospheric correction [16,17,51–53].

Atmospheric correction requires several steps to remove the influence of the atmosphere, such as single scattering, multiple molecular scattering, aerosol/ozone, and other gas absorption [54]. The procedure usually removes cloud contaminated pixels by the threshold of the near infrared band albedo. However, this method does not work well on cloud edges. Current satellite algorithms provide estimates of cloud optical thickness [55], which can be used to quantify the impacts of clouds on ocean color data in the near future.

The ocean color signal reflected from water is generally the integrated light reflected from the light penetration depth [56,57]. With increasing depth, deep waters have less contribution to the water modulated reflectance signals. However, we still cannot ignore it in chlorophyll-a inversion algorithms. In the SO, the deep chlorophyll-a maximum increases the contribution from deeper waters. Research seldom applies chlorophyll-a profiles and integrated chlorophyll-a signals to compare in situ and satellite match-ups. Current reports have already shown the obvious underestimation from satellite algorithms [6–11]. If considering the deep chlorophyll-a maximum and chlorophyll-a profile, there is a larger underestimation in SO chlorophyll-a retrieval. Chlorophyll-a profile estimation should be considered in future research. In fact, due to wind mixing, a deep chlorophyll-a maximum is a common situation in the SO [58,59]. Sullivan et al. [60] built a model on polar phytoplankton growth and reduction. They found that there is a high probability of maintaining a subsurface chlorophyll-a maximum during the post-bloom period.

5. Conclusions

Decades of development in satellite ocean color has produced multiple algorithms to detect, among other things, chlorophyll-a, fluorescence intensity, particulate organic carbon, photosynthetic available radiation. Satellites provide almost every parameter needed for observation and understanding the global ocean. However, the uncertainty of those parameters varies between regions. Our research validated the fluorescence approach to estimate chlorophyll-a concentration in KGI, resulting in a good agreement (59.46%) at this high latitude, high chlorophyll-a region. The phytoplankton bloom in 2016 showed that the dominant phytoplankton is micro-phytoplankton, occupying 79.58% of the total phytoplankton community in Ardley Cove. Increasing sea ice coverage and sea ice concentration are possible reasons for the increasing occurrence phytoplankton blooms in the recent decade. Due to NPQ, the fluorescence approach does not accurately estimate chlorophyll-a in places that have intense solar irradiance. However, the fluorescence approach and red band selection has notable advantages in avoiding CDOM interference from blue bands and decreasing gaps from the peculiar bio-optics of SO green bands. Our future work intends to validate and extend the application of this algorithm to the entire SO. The photosynthesis mechanism revealed by the fluorescence approach will provide more information on SO primary productivity estimation and its role on global carbon cycling.

Acknowledgments: Our research work was supported by the Chinese Polar Environment Comprehensive Investigation & Assessment Programs (CHINARE2015-02-04 and CHINARE2016-02-04). Satellite Ocean Color data was provided by NASA Goddard Space Flight Center, Ocean Ecology Laboratory, Ocean Biology Processing Group. Analyses and visualizations used in this study were produced with the Giovanni online data system, developed and maintained by the NASA GES DISC. The support provided by China Scholarship Council (CSC) during a visit by Chen Zeng to UBC is acknowledged. We would also like to appreciate the hard work and time of the anonymous reviewers.

Author Contributions: Chen Zeng, Tao Zeng and Huiping Xu conceived and designed the experiments; Chen Zeng performed the experiments, analyzed the data and wrote the paper. Tao Zeng conducted the cross-validation for improving the regional nFLH algorithm. Andrew M. Fisher revised and polished the paper.

Conflicts of Interest: The authors declare no conflict of interest.

Appendix A

The QAA-v5 algorithm and expanded equations are listed as follows, which is a comprehensive application from QAA-v5 and GSM01.

Table A1. QAA-v algorithm and expanded rules combined from QAA-v5 and GSM01.

Symbol and Description	Equation and Process
r_{rs}, below-surface remote-sensing reflectance (sr^{-1}); R_{rs}, Above-surface remote-sensing reflectance (sr^{-1})	$r_{rs} = R_{rs}/(0.52 + 1.7R_{rs})$
u, ratio of backscattering coefficient to the sum of absorption and backscattering coefficients, $b_b/(a + b_b)$	$u(\lambda) = \frac{-g_0 + [g_0^2 + 4g_1 r_{rs}(\lambda)]^{1/2}}{2g_1}$, $g_0 = 0.089$, $g_1 = 0.125$
a_w, absorption coefficient of pure water (m^{-1}); 443, 490, 555, 667, band wavelength (nm)	$\chi = \log\left(\frac{r_{rs}(443) + r_{rs}(490)}{r_{rs}(555) + 5 \frac{r_{rs}(667)}{r_{rs}(490)} r_{rs}(667)}\right)$, $a(555) = a_w(555) + 10^{-1.146 - 1.366\chi - 0.469\chi^2}$
b_{bw}, backscattering coefficient of pure water (m^{-1}); b_{bp}, Backscattering coefficient of suspended particles (m^{-1})	$b_{bp}(555) = \frac{u(555)a(555)}{1 - u(555)} - b_{bw}(555)$
Y, spectral power of particle backscattering coefficient	$Y = 2.0\left\{1 - 1.2exp\left[-0.9\frac{r_{rs}(443)}{r_{rs}(555)}\right]\right\}$
λ, all band wavelength (nm)	$b_{bp}(\lambda) = b_{bp}(555)\left(\frac{555}{\lambda}\right)^Y$
a_g, absorption coefficient of gelbstoff and detritus (m^{-1}); S, Spectral slope for gelbstoff absorption coefficient	$\xi = \frac{a_g(410)}{a_g(440)} = \exp[S(443 - 411)]$, $S = 0.015 + \frac{0.002}{0.6 + r_{rs}(443)/r_{rs}(555)}$, $a_g(440) = \frac{[a(410) - \xi a(440)]}{\xi - \zeta} - \frac{[a_w(410) - \zeta a_w(440)]}{\xi - \zeta}$
η, spectral exponential coefficient for gelbstoff and detritus	$a_g(\lambda) = a_g(440) \exp[-\eta(\lambda - 440)], \eta = 0.015$
a_{ph}, absorption coefficient of phytoplankton	$a_{ph}(\lambda) = a(\lambda) - a_w(\lambda) - a_g(\lambda)$

References

1. Falkowski, P.G. The power of plankton. *Nature* **2012**. [CrossRef] [PubMed]
2. Pope, A.; Wagner, P.; Johnson, R.; Baeseman, J.; Newman, L. Community review of Southern Ocean satellite data needs. *Antarct. Sci.* **2016**. [CrossRef]
3. Babin, M.; Arrigo, K.; Bélanger, S.; Forge, M.-H. *Ocean Colour Remote Sensing in Polar Seas*; IOCCG Report Series, No. 16; International Ocean Colour Coordinating Group: Dartmouth, NS, Canada, 2015.
4. Reynolds, R.A.; Stramski, D.; Mitchell, B.G. A chlorophyll-a-dependent semi-analytical reflectance model derived from field measurements of absorption and backscattering coefficients within the Southern Ocean. *J. Geophys. Res.* **2001**, *106*, 7125–7138. [CrossRef]

5. Marrari, M.; Hu, C.; Daly, K. Validation of SeaWiFS chlorophyll-a a concentrations in the Southern Ocean: A revisit. *Remote Sens. Environ.* **2006**, *105*, 367–375. [CrossRef]

6. Moore, J.K.; Abbott, M.R. Phytoplankton chlorophyll-a distributions and primary production in the Southern Ocean. *J. Geophys. Res. Ocean.* **2000**, *105*, 28709–28722. [CrossRef]

7. Dierssen, H.M.; Smith, R.C. Bio-optical properties and remote sensing ocean color algorithms for Antarctic Peninsula waters. *J. Geophys. Res.* **2000**, *105*, 26301–26312. [CrossRef]

8. Gregg, W.W.; Casey, N.W. Global and regional evaluation of the SeaWiFS chlorophyll-a data set. *Remote Sens. Environ.* **2004**, *93*, 463–479. [CrossRef]

9. Kwok, R.; Comiso, J.C. Spatial patterns of variability in Antarctic surface temperature: Connections to the southern hemisphere annular mode and the southern oscillation. *Geophys. Res. Lett.* **2002**. [CrossRef]

10. Gabric, A.J.; Shephard, J.M.; Knight, J.M.; Jones, G.; Trevena, A.J. Correlations between the satellite-derived seasonal cycles of phytoplankton biomass and aerosol optical depth in the Southern Ocean: Evidence for the influence of sea ice. *Global Biogeochem. Cycles* **2005**, *19*, 1–10. [CrossRef]

11. Friedrichs, M.A.; Carr, M.E.; Barber, R.T.; Scardi, M.; Antoine, D.; Armstrong, R.A.; Asanuma, I.; Behrenfeld, M.J.; Buitenhuis, E.T.; Chai, F.; et al. Assessing the uncertainties of model estimates of primary productivity in the tropical Pacific Ocean. *J. Mar. Syst.* **2009**, *76*, 113–133. [CrossRef]

12. Zeng, C.; Xu, H.; Fischer, A. M. Chlorophyll-a estimation around the Antarctica peninsula using satellite algorithms: hints from field water leaving reflectance. *Sensors* **2016**. [CrossRef] [PubMed]

13. Szeto, M.; Werdell, P.J.; Moore, T.S.; Campbell, J.W. Are the world's oceans optically different? *J. Geophys Res.* **2011**. [CrossRef]

14. Arrigo, K.R.; van Dijken, G.L.; Bushinsky, S. Primary production in the Southern Ocean, 1997–2006. *J. Geophys. Res.* **2008**. [CrossRef]

15. Bélanger, S.; Ehn, J.K.; Babin, M. Impact of sea ice on the retrieval of water-leaving reflectance, chlorophyll-a a concentration and inherent optical properties from satellite ocean color data. *Remote Sens. Environ.* **2007**, *111*, 51–68. [CrossRef]

16. Wang, M.; Shi, W. Detection of ice and mixed ice–water pixels for MODIS ocean color data processing. *IEEE Trans. Geosci. Remote Sens.* **2009**, *47*, 2510–2518. [CrossRef]

17. McFarquhar, G.M.; Wood, R.; Bretherton, C.S.; Alexander, S.; Jakob, C.; Marchand, R.; Protat, A.; Quinn, P.; Siems, S.T.; Weller, R.A. The southern ocean clouds, radiation, aerosol transport experimental study (SOCRATES): An observational campaign for determining role of clouds, aerosolsand radiation in climate system. In Proceedings of the 2014 AGU Fall Meeting, San Francisco, CA, USA, 15–19 December 2014.

18. Kay, J.E.; Medeiros, B.; Hwang, Y.T.; Gettelman, A.; Perket, J.; Flanner, M.G. Processes controlling Southern Ocean shortwave climate feedbacks in CESM. *Geophys. Res. Lett.* **2014**, *41*, 616–622. [CrossRef]

19. Maritorena, S.; Siegel, D.A.; Peterson, A. Optimization of a semi-analytical ocean color model for global scale applications. *Appl. Opt.* **2002**, *41*, 2705–2714. [CrossRef] [PubMed]

20. Lee, Z.; Carder, K.L.; Arnone, R.A. Deriving inherent optical properties from water color: A multiband quasi-analytical algorithm for optically deep waters. *Appl. Opt.* **2002**, *41*, 5755–5772. [CrossRef] [PubMed]

21. Xing, X.; Claustre, H.; Blain, S.; D'Ortenzio, F.; Antoine, D.; Ras, J.; Guinet, C. Quenching correction for in vivo chlorophyll-a fluorescence acquired by autonomous platforms: A case study with instrumented elephant seals in the Kerguelen region (Southern Ocean). *Limnol. Oceanogr. Method.* **2012**, *10*, 483–495. [CrossRef]

22. Palmer, S.C.; Hunter, P.D.; Lankester, T.; Hubbard, S.; Spyrakos, E.; Tyler, A.N.; Presing, M.; Horvath, H.; Lamb, A.; Balzter, H.; et al. Validation of Envisat MERIS algorithms for chlorophyll-a retrieval in a large, turbid and optically-complex shallow lake. *Remote Sens. Environ.* **2015**, *157*, 158–169. [CrossRef]

23. Behrenfeld, M.J.; Westberry, T.K.; Boss, E.; O'Malley, R.T.; Siegel, D.A.; Wiggert, J.D.; Franz, B.A.; McClain, C.R.; Feldman, G.C.; Doney, S.C.; et al. Satellite-detected fluorescence reveals global physiology of ocean phytoplankton. *Biogeosciences* **2009**, *6*, 779–794. [CrossRef]

24. Schloss, I.R.; Abele, D.; Moreau, S.; Demers, S.; Bers, A.V.; González, O.; Ferreyra, G.A. Response of phytoplankton dynamics to 19-year (1991–2009) climate trends in Potter Cove (Antarctica). *J. Marine Syst.* **2012**, *92*, 53–66. [CrossRef]

25. Schloss, I.R.; Wasilowska, A.; Dumont, D.; Almandoz, G.O.; Hernando, M.P.; Michaud-Tremblay, C.A.; Saravia, L.; Rzepecki, M.; Monien, P.; Monien, D.; et al. On the phytoplankton bloom in coastal waters of southern King George Island (Antarctica) in January 2010: An exceptional feature? *Limnol. Oceanogr.* **2014**. [CrossRef]

26. Mueller, J.L.; Fargion, G.S.; McClain, C.R.; Pegau, S.; Zanefeld, J.R.V.; Mitchell, B.G.; Kahru, M.; Wieland, J.; Stramska, M. *Ocean Optics Protocols for Satellite Ocean Color Sensor Validation, Revision 4, Volume Iv: Radiometric Measurements and Data Analysis Protocols*; Goddard Space Flight Space Center: Greenbelt, MD, USA, 2003.

27. Tang, J.W.; Tian, G.L.; Wang, X.Y.; Wang, X.M.; Song, Q.J. The methods of water spectra measurement and analysis i: Above-water method. *J. Remote Sens. Beijing* **2004**, *8*, 37–44.

28. Mobley, C. Overview of Optical Oceanography. Available online: http://www.oceanopticsbook.info/view/overview_of_optical_oceanography/reflectances (accessed on 25 June 2015).

29. Kohavi, R. A study of cross-validation and bootstrap for accuracy estimation and model selection. In Proceedings of the 14th International Joint Conference on Artificial Intelligence, Montreal, QB, Canada, 20–25 August 1995.

30. Lee, Z.P.; Lubac, B.; Werdell, J.; Arnone, R. An update of the Quasi-Analytical Algorithm (QAA_v5). Available online: http://www.ioccg.org/groups/Software_OCA/QAA_v5.pdf (accessed on 28 July 2009).

31. Pope, R.M.; Fry, E.S. Absorption spectrum (380–700 nm) of pure water. II. Integrating cavity measurements. *Appl. Opt.* **1997**, *36*, 8710–8723. [CrossRef] [PubMed]

32. Ciotti, Á.; Lewis, M.R.; Cullen, J.J. Assessment of the relationships between dominant cell size in natural phytoplankton communities and the spectral shape of the absorption coefficient. *Limnol. Oceanogr.* **2002**, *2*, 404–417. [CrossRef]

33. Mitchell, B.G.; Kahru, M. Bio-optical algorithms for ADEOS-2 GLI. *J. Remote Sens. Soc. Jpn.* **2009**, *29*, 80–85.

34. Behrenfeld, M. J.; Halsey, K.; Milligan, A. J. Evolved physiological responses of phytoplankton to their integrated growth environment. *Phil. Trans. Royal Soc. B* **2008**, *363*, 2687–2703. [CrossRef] [PubMed]

35. Babin, M.; Morel, A.; Gentili, B. Remote sensing of sea surface sun-induced chlorophyll-a fluorescence: consequences of natural variations in the optical characteristics of phytoplankton and the quantum yield of chlorophyll-a a fluorescence. *Int. J. Remote Sens.* **1996**, *17*, 2417–2448. [CrossRef]

36. Gower, J. On the use of satellite-measured chlorophyll fluorescence for monitoring coastal waters. *Int. J. Remote Sens.* **2016**, *37*, 2077–2086. [CrossRef]

37. Ortega-Retuerta, E.; Frazer, T.K.; Duarte, C.M.; Ruiz-Halpern, S.; Tovar-Sánchez, A.; Arrieta López de Uralde, J.M.; Reche, I. Biogeneration of chromophoric dissolved organic matter by bacteria and krill in the Southern Ocean. *Limnol. Oceanogr.* **2009**, *54*, 1941–1950. [CrossRef]

38. Bony, S.; Stevens, B.; Frierson, D.M.; Jakob, C.; Kageyama, M.; Pincus, R.; Shepherd, T.G.; Sherwood, S.C.; Siebesma, A.P.; Sobel, A.H.; et al. Clouds, circulation and climate sensitivity. *Nat. Geosci.* **2015**, *8*, 261–268. [CrossRef]

39. Browning, T.J.; Bouman, H.A.; Moore, C.M. Satellite-detected fluorescence: Decoupling nonphotochemical quenching from iron stress signals in the South Atlantic and Southern Ocean. *Glob. Biogeochem. Cycles* **2014**, *28*, 510–524. [CrossRef]

40. Sathyendranath, S.; Platt, T. Spectral effects in bio-optical control on the ocean system. *Oceanologia.* **2007**, *49*, 5–39.

41. Boyd, P.W.; Arrigo, K.R.; Strzepek, R.; Dijken, G.L. Mapping phytoplankton iron utilization: Insights into Southern Ocean supply mechanisms. *J. Geophys. Res. Ocean.* **2012**, *117*, 304–315. [CrossRef]

42. Boyd, P.W.; Watson, A.J.; Law, C.S.; Abraham, E.R.; Trull, T.; Murdoch, R.; Bakker, D.C.; Bowie, A.R.; Buesseler, K.O.; Chang, H.; et al. A mesoscale phytoplankton bloom in the polar Southern Ocean stimulated by iron fertilization. *Nature* **2000**, *407*, 695–702. [CrossRef] [PubMed]

43. Hewes, C.D.; Reiss, C.S.; Holm-Hansen, O. A quantitative analysis of sources for summertime phytoplankton variability over 18 years in the South Shetland Islands (Antarctica) region. *Deep-Sea Res. Part I. Oceanogr. Res. Pap.* **2009**, *56*, 1230–1241. [CrossRef]

44. Tripathy, S.C.; Pavithran, S.; Sabu, P.; Naik, R.K.; Noronha, S.B.; Bhaskar, P.V.; Anilkumar, N. Is phytoplankton productivity in the Indian Ocean sector of Southern? *Curr. Sci.* **2014**, *107*, 1019–1026.

45. Brody, E.; Mitchell, B.G.; Holm-Hansen, O.; Vernet, M. Species-dependent variations of the absorption coefficient in the Gerlache Strait. *Antarct. J. USA* **1992**, *27*, 160–162.

46. Finkel, Z.V.; Beardall, J.; Flynn, K.J.; Quigg, A.; Rees, T.A.V.; Raven, J.A. Phytoplankton in a changing world: cell size and elemental stoichiometry. *J. Plankton Res.* **2009**, *32*, 119–137. [CrossRef]

47. Smith, W.O.; Nelson, D.M. Phytoplankton bloom produced by a receding ice edge in the Ross Sea: Spatial coherence with the density field. *Science* **1985**, *227*, 163–166. [CrossRef] [PubMed]

48. Boetius, A.; Albrecht, S.; Bakker, K.; Bienhold, C.; Felden, J.; Fernández-Méndez, M.; Hendricks, S.; Katlein, C.; Lalande, C.; Krumpen, T.; et al. Export of algal biomass from the melting Arctic sea ice. *Science* **2013**, *339*, 1430–1432. [CrossRef] [PubMed]

49. Arrigo, K.R.; Lowry, K.E.; van Dijken, G.L. Annual changes in sea ice and phytoplankton in polynyas of the Amundsen Sea, Antarctica. *Deep-Sea Res. Part II Top. Stud. Oceanogr.* **2012**, *71*, 5–15. [CrossRef]

50. Gordon, H.R. Removal of atmospheric effects from the satellite imagery of the oceans. *Appl. Opt.* **1978**, *17*, 1631–1636. [CrossRef] [PubMed]

51. Behrangi, A.; Stephens, G.; Adler, R.F.; Huffman, G.J.; Lambrigtsen, B.; Lebsock, M. An update on the oceanic precipitation rate and its zonal distribution in light of advanced observations from space. *J. Clim.* **2014**, *27*, 3957–3965. [CrossRef]

52. Haynes, J.M.; L'Ecuyer, T.S.; Stephens, G.L.; Miller, S.D.; Mitrescu, C.; Wood, N.B.; Tanelli, S. Rainfall retrieval over the ocean with spaceborne W-band radar. *J. Geophys. Res.* **2009**. [CrossRef]

53. Meskhidze, N.; Nenes, A. Phytoplankton and cloudiness in the Southern Ocean. *Science* **2006**, *314*, 1419–1423. [CrossRef] [PubMed]

54. Vermote, E.F.; Tanré, D.; Deuzé, J.L.; Herman, M.; Morcrette, J.J.; Kotchenova, S.Y. Second Simulation of A Satellite Signal in the Solar Spectrum-Vector (6SV), 6S User Guide Version 3. Available online: http://6s.ltdri.org/files/tutorial/6S_Manual_Part_1.pdf (accessed on 23 September 2016).

55. Marchand, R.; Ackerman, T.; Smyth, M.; Rossow, W.B. A review of cloud top height and optical depth histograms from MISR, ISCCP, and MODIS. *J. Geophys. Res.* **2010**. [CrossRef]

56. Soppa, M.A.; Dinter, T.; Taylor, B.B.; Bracher, A. Satellite derived euphotic depth in the Southern Ocean: Implications for primary production modelling. *Remote Sens. Environ.* **2013**, *137*, 198–211. [CrossRef]

57. Lee, Z.; Weidemann, A.; Kindle, J.; Arnone, R.; Carder, K.L.; Davis, C. Euphotic zone depth: Its derivation and implication to ocean-color remote sensing. *J. Geophys. Res.* **2007**. [CrossRef]

58. Schlitzer, R. Carbon export fluxes in the Southern Ocean: Results from inverse modeling and comparison with satellite-based estimates. *Deep-Sea Res. Part II Top. Stud. Oceanogr.* **2002**, *49*, 1623–1644. [CrossRef]

59. Uitz, J.; Claustre, H.; Griffiths, F.B.; Ras, J.; Garcia, N.; Sandroni, V. A phytoplankton class-specific primary production model applied to the Kerguelen Islands region (Southern Ocean). *Deep-Sea Res. Part I. Oceanogr. Res. Pap.* **2009**, *56*, 541–560. [CrossRef]

60. Sullivan, C.W.; McClain, C.R.; Comiso, J.C.; Smith, W.O. Phytoplankton standing crops within an Antarctic ice edge assessed by satellite remote sensing. *J. Geophys. Res. Ocean.* **1988**, *93*, 12487–12498. [CrossRef]

remote sensing

MDPI

Article

Seasonal and Interannual Variability of Satellite-Derived Chlorophyll-a (2000–2012) in the Bohai Sea, China

Hailong Zhang [1,2], Zhongfeng Qiu [1,2,*], Deyong Sun [1,2], Shengqiang Wang [1,2] and Yijun He [1,2]

[1] School of Marine Sciences, Nanjing University of Information Science & Technology, Nanjing 210044, China; zhanghl1205@163.com (H.Z.); sundeyong1984@163.com (D.S.); shengqiang.wang@nuist.edu.cn (S.W.); yjhe@nuist.edu.cn (Y.H.)

[2] Jiangsu Research Centre for Ocean Survey Technology, NUIST, Nanjing 210044, China

* Correspondence: zhongfeng.qiu@nuist.edu.cn; Tel.: +86-25-5869-5696

Academic Editors: Yunlin Zhang, Claudia Giardino, Linhai Li, Deepak R. Mishra and Prasad S. Thenkabail
Received: 28 February 2017; Accepted: 5 June 2017; Published: 10 June 2017

Abstract: Knowledge of the chlorophyll-a dynamics and their long-term changes is important for assessing marine ecosystems, especially for coastal waters. In this study, the spatial and temporal variability of sea surface chlorophyll-a concentration (Chl-a) in the Bohai Sea were investigated using 13-year (2000–2012) satellite-derived products from MODIS and SeaWiFS observations. Based on linear regression analysis, the results showed that the entire Bohai Sea experienced an increase in Chl-a on a long-term scale, with the largest increase in the central Bohai Sea and the smallest increase in the Bohai strait. Distinct seasonal patterns of Chl-a existed in different sub-regions of the Bohai Sea. A long-lasting Chl-a peak was observed from May to September in coastal waters (Liaodong bay, Qinhuangdao coast, and Bohai bay) and the central Bohai Sea, whereas Laizhou bay had relatively low Chl-a in early summer. In the Bohai strait, two pronounced Chl-a peaks occurred in March and September, but the lowest Chl-a was in summer. This pattern was quite different from those in other regions of the Bohai Sea. The water column condition (stratified or mixed) was likely an important physical factor that affects the seasonal pattern of Chl-a in the Bohai Sea. Meanwhile, increased human activity (e.g., river discharge) played a significant role in changing the Chl-a distribution in both coastal waters and the central Bohai Sea, especially in summer. The increasing trend of Chl-a in the Bohai Sea might be attributed to the increase in nutrient contents from riverine inputs. The Chl-a dynamics documented in this study provide basic knowledge for the future exploration of marine biogeochemical processes and ecosystem evolution in the Bohai Sea.

Keywords: satellite data; long-term changes; sub-regions; Bohai Sea

1. Introduction

Marine phytoplankton is a fundamental component of marine biogeochemical cycles and ecosystems, accounting for approximately 50% of global organic matter production [1,2]. It also influences the diversity of marine organisms and global climate processes [3,4]. Chlorophyll-a is widely used to indicate phytoplankton biomass [5], as it can generally reflect the situation of phytoplankton growth. Due to the limitation of field methods, chlorophyll-a concentrations (Chl-a) collected by field methods are usually insufficient for investigating the Chl-a dynamics.

Satellite ocean color observations can provide large spatial and temporal coverage, which is ideal for examining the spatial and temporal variability of Chl-a [6]. Empirical and semi-analytical Chl-a algorithms have been developed to infer information about Chl-a from space on both global and regional scales [7–10]. For instance, the Tassan-like algorithm [8] and OC4 algorithm [10] have been widely applied to satellite data. In addition, the OC4 algorithm was performed on SeaWiFS and MODIS data to derive the NASA standard products of Chl-a, which are provided at http://oceancolour.gsfc.nasa.gov/.

Recent efforts have been made to evaluate Chl-a in the global oceans using satellite-derived products. These studies clearly revealed that increased Chl-a occurred in many coastal waters (e.g., the eastern China seas), and the reason was generally attributed to the interaction between human activity and climate change [11–13]. For instance, Gong et al. [14] found that the rate of primary production in the subtropical East China Sea was regulated by seawater temperature during winter and early spring and nutrients during summer and autumn. Shi and Wang [15] studied the seasonal distribution of satellite-derived Chl-a, sea surface temperature (SST), and normalized water leaving radiance (nLw) spectra in the eastern China seas. These results showed that ocean color property variations were driven by the ocean stratification, sea surface thermodynamics, and river discharge, among other factors. Yamaguchi et al. [16] presented the seasonal and summer temporal variability of satellite-derived Chl-a in the Yellow and East China Seas from 1997 to 2007, and revealed that the inter-annual variation of Chl-a in summer was significantly influenced by Yangtze River discharge. He et al. [17] investigated the seasonal and inter-annual variability of phytoplankton blooms in the eastern China seas using satellite-derived Chl-a from 1998 to 2011. They reported that the doubling of the bloom intensity in the eastern China seas was mainly caused by an increase in nitrate and phosphate concentrations. Based on the 15-year (1997–2011) satellite Chl-a data derived using the OC4 algorithm, Liu et al. [18] analyzed the effects of bathymetry on seasonal and inter-annual patterns of Chl-a in a larger region including the Bohai and Yellow Seas, but did not examine the detailed Chl-a dynamics in sub-regions of the Bohai Sea. They also found that the correlation between Chl-a and SST was positive in coastal waters and negative in offshore waters.

Previous researches have commonly investigated the Chl-a dynamics in a large region including the Bohai Sea, such as the studies of He et al. [17] and Liu et al. [18]. However, studies specifically focusing on the Chl-a dynamics on large temporal scales in sub-regions of the Bohai Sea are limited. In addition, the possible factors of Chl-a variation in the Bohai Sea are not yet clear. Because of the highly variable environmental conditions (e.g., river discharge, circulations, and water masses), the Bohai Sea ecosystem is complex [19–21]. The phytoplankton growth needs nutrients and light [22]. The available light for photosynthesis depends on photosynthetically available radiation (PAR), extinction coefficient, and water clarity. Additionally, SST may influence the vertical structure of the water column, which would further change the nutrient supply and light conditions. Fortunately, both the SST and PAR parameters can be detected by satellite remote sensing technology, thus providing a large quantity of materials to explore the influences of these two factors on the Chl-a dynamics.

Because atmospheric correction models can be inaccurate due to the uncertainty of the aerosol and bio-optical algorithms in high suspended sediment areas (e.g., the Bohai Sea), it is difficult to obtain reliable Chl-a data from satellite ocean color data [23]. For instance, the OC4 standard algorithm performs well for Case-1 waters, but is not always valid for Case-2 waters [24]. Siswanto et al. [24] proposed an empirical local algorithm based on an extensive bio-optical dataset collected in the Yellow and East China Seas. In brief, they regionally tuned and combined the Tassan Chl-a and OC4v4 algorithms under low and high nLw_{555} ($2 \ mW \cdot cm^{-2} \cdot \mu m^{-1} \cdot sr^{-1}$) conditions, respectively. This combined Chl-a algorithm can improve the retrieval accuracy of Chl-a, especially for high suspended sediment area [25,26]. In addition, it was used as the standard algorithm in the Geostationary Ocean Color Imager Data Processing System (GDPS) [27–29].

We hypothesize that the distinct Chl-a patterns existed in different sub-regions of the Bohai Sea and may be influenced by separate mechanisms. Hence, in this study, we first assessed the performance of the Chl-a algorithm of Siswanto et al. [24] in the Bohai Sea by comparing satellite-derived Chl-a with in situ measurements. The main objectives of this study were: (1) to investigate the spatiotemporal variability and trend of Chl-a in the Bohai Sea during a 13-year period (2000–2012); (2) to analyze area differences in seasonal variations and long-term changes in Chl-a; and (3) to discuss the possible factors that affect the Chl-a dynamics and its trend.

2. Materials and Methods

2.1. Study Area

The Bohai Sea, located in northern China, is a shallow shelf sea with an average water depth of 18 m (with maximum depth of 80 m) and a total area of 77,000 km^2 [30] (Figure 1). It is connected to the Yellow Sea through the Bohai Strait. Numerous inland rivers flow into the Bohai Sea from Mainland China with a total annual runoff of 8.88 × 10^{10} m^3, nearly half of which comes from the Huanghe (Yellow River) [31]. The Bohai Sea is important as the main fishing ground and base of the marine fishery resources in northern China. Over the past several decades, the Bohai Sea has been influenced by human activity (e.g., agriculture, and industrial and domestic sewage) [32]. The Bohai Sea ecosystem has been gradually deteriorating due to red-tide events and eutrophication [33,34].

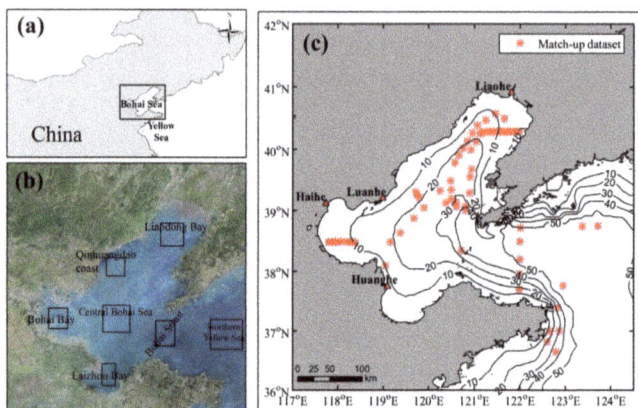

Figure 1. Location of the Bohai Sea (**a**) and the sampling locations of sub-regions, which are marked by squares (**b**), namely, Liaodong bay, Qinhuangdao coast, Bohai bay, Laizhou bay, central Bohai Sea, Bohai strait, and northern Yellow Sea. Locations of the match-up stations in the Bohai and northern Yellow Seas (**c**).

In this study, the Bohai Sea was subdivided into six sub-regions, following geographical regions (Figure 1b), including Liaodong bay (156 pixels), Qinhuangdao coast (144 pixels), Bohai bay (169 pixels), Laizhou bay (128 pixels), central Bohai Sea (256 pixels), and Bohai strait (165 pixels). The areas along the coastline of the Bohai Sea were excluded because of high levels of suspended sediment, which may increase the uncertainties of satellite-derived Chl-a [35,36]. Meanwhile, the northern Yellow Sea (400 pixels) was distinguished separately as a specific region. All clusters of pixels were isolated as regions of interest, and the data were averaged within each sub-region.

2.2. In Situ Chl-a Data

The in situ Chl-a data used in this study were collected during three cruises in June 2005, July 2011, and September 2012 in the Bohai Sea. Additionally, the Chl-a data collected during August 2015, June 2016, and December 2016 in the Bohai and northern Yellow Seas were also used in the algorithm validation, although the collecting time of these data was beyond that of our study period, because using the more data may give more reliable validation result. In total, the in situ dataset included 367 Chl-a samples.

For Chl-a analysis, seawater samples at the near surface (0–3 m) were collected using 12-liter Niskin bottles mounted on a CTD system. Water samples were filtered through 25-mm Whatman GF/F glass fiber filters under low vacuum pressure (<0.01 Mpa). After filtration, these samples were stored in liquid nitrogen until analysis in the laboratory. Prior to analysis, chlorophyllous pigments were extracted with *N,N*-dimethylformamide (DMF) for 24 h at 0 °C in the dark. Then, the florescence values of each sample, in fluorescent standard units (FSU), were measured three times using a Turner Design Fluorometer Model, and averaged these three measurements. Finally, the Chl-a was calculated from the corresponding florescence values based on the calibration curves.

2.3. Satellite Data

The daily remote sensing reflectance R_{rs} products of MODIS and SeaWiFS were acquired from the NASA ocean color website (http://oceancolour.gsfc.nasa.gov/). This dataset spanned 2000 to 2012 for a rectangular region (36–42°N and 117–124°E) that encompassed our study region. Additionally, the Level 3 monthly SST and PAR data with global coverage during our study period were obtained from the NASA ocean color website. These products were all cropped to the Bohai Sea. In addition, the bathymetric data were obtained from the ETOPO5 data (Earth Topography-5 Minute) at https://www.ngdc.noaa.gov/mgg/global/etopo5.html.

2.4. Chl-a Algorithm

In this study, we used the Chl-a algorithm of Siswanto et al. [24] to obtain satellite-derived Chl-a data. In regions with $nLw_{555} > 2$ mW·cm^{-2}·µm^{-1}·sr^{-1}, the regionally tuned Tassan-like algorithm was used:

$$log(Chl-a) = -0.166 - 2.518log_{10}^2(R) + 9.345log_{10}^2(R) \tag{1}$$

$$R = [(R_{rs443}/R_{rs555})(R_{rs412}/R_{rs490})]^{-0.463} \tag{2}$$

Under the low range of nLw_{555} (<2 mW·cm^{-2}·µm^{-1}·sr^{-1}), the regionally tuned OC4v4 algorithm was used:

$$log(Chl-a) = 0.248 - 2.703R + 1.695R^2 - 1.764R^3 + 1.092R^4 \tag{3}$$

$$R = log[\max(R_{rs443}/R_{rs555}, R_{rs490}/R_{rs555})] \tag{4}$$

where R is a function of spectra value and $R_{rs}(\lambda)$ is the remote sensing reflectance value at a given wavelength. Note that the invalid R_{rs} pixels were masked out based on the level 2 flagged pixels of the standard R_{rs} product. The daily Chl-a data were composed into monthly averages to match the SST and PAR datasets.

To assess the performance of the Chl-a retrieval algorithm, the coefficient of determination (R^2), root mean square error (RMSE), and mean absolute percentage error (MAPE) were calculated between satellite-derived Chl-a and these measured values as below:

$$RMSE = \frac{1}{n}\sqrt{\sum_{i=1}^{n}[(x_{i,\text{derived}} - x_{i,\text{field}})/x_{i,\text{field}}]^2} \tag{5}$$

$$MAPE = \frac{1}{n}\sum_{i=1}^{n}|(x_{i,\text{derived}} - x_{i,\text{field}})/x_{i,\text{field}}| \times 100\% \tag{6}$$

where n is the number of samples, and $x_{i,\text{derived}}$ and $x_{i,\text{field}}$ denote satellite-derived and in situ Chl-a data for the i-th sample, respectively.

2.5. Calculation of Trend and Information Flow

The linear trend model is commonly used in environmental and climate change research. The trend was obtained by: (1) subtracting the monthly climatological mean values from each corresponding month to remove the seasonal signal (producing monthly anomaly) [37]; and (2) calculating the linear trend using linear regression analysis. The line trend was the slope of the linear regression line for the time series of monthly anomaly, and its statistical significance was assessed with a statistical F-test. It is worth emphasizing that the linear trend in this study was assessed under a high percentage (>70%) of valid pixels to the total number of the monthly anomaly. If the percentage of valid pixels is too small (e.g., <30%), the trend statistical analysis may have large uncertainty due to insufficient valid data.

The correlation between different parameters is often detected using Pearson correlation analysis, but it is not designed to explore statistical dependencies between parameters [38]. In this study, we used a mathematical method based on the information flow (IF), which can quantitatively evaluate the cause and effect relation between time series [39]. The method was expressed as:

$$T_{2 \to 1} = (C_{11} \times C_{12} \times C_{2,d1} - C_{12}^2 \times C_{1,d1}) / (C_{11}^2 \times C_{22} - C_{11} \times C_{12}^2) \tag{7}$$

$$d1 = (X_{1,n+1} - X_{1,n}) / \Delta t \tag{8}$$

where $T_{2 \to 1}$ is the rate of information flowing from X_2 to X_1, C_{ij} is the sample covariance between X_i and X_j, and $C_{i,dj}$ is the covariance between X_i and d_j. If the $\mid T_{2 \to 1} \mid$ value is nonzero, there is causality between X_2 and X_1; if not, there is no causality between them. In this study, the Chl-a anomaly and SST (PAR) anomaly were symbolized by the subscripts "1"and "2" in Equation (7), respectively. The method could analyze the cause-effect relation and compare degrees of influence between different factors. It has been successfully applied to explain scientific problems in the real world. For instance, Liang [39] investigated the causal relation between the El Niño and Indian Ocean Dipole. Stips et al. [40] used this method to explore causality between different forcing components (e.g., anthropogenic, CO_2, aerosol, cloud, and solar) and annual global mean surface temperature anomalies since 1850.

3. Results

3.1. Validation of Satellite-Derived Chl-a in the Bohai Sea

We applied the Chl-a algorithm of Siswanto et al. [24] to satellite data to investigate the Chl-a dynamics in the Bohai Sea. Before the application, the performance of the Chl-a algorithm was assessed based on 69 pairs of in situ Chl-a and satellite $R_{rs}(\lambda)$ data (Figure 1c). This match-up dataset only consisted of satellite $R_{rs}(\lambda)$ with an overpass time window within 5 h before and after field data. To avoid the effects of outliers, the median $R_{rs}(\lambda)$ values for a 3×3 pixels window centered on the locations of the sampling stations were defined as satellite $R_{rs}(\lambda)$. As shown in Figure 2, satellite-derived Chl-a generally agreed well with in situ Chl-a, with R^2, RMSE and MAPE values of 0.53, 0.21 mg·m^{-3} and 38.38%, respectively. These results suggested that satellite-derived Chl-a in the Bohai Sea had high accuracy, which was considered generally acceptable in remote sensing research [24]. Therefore, we can further study the spatiotemporal variability of Chl-a in the Bohai Sea based on satellite-derived Chl-a.

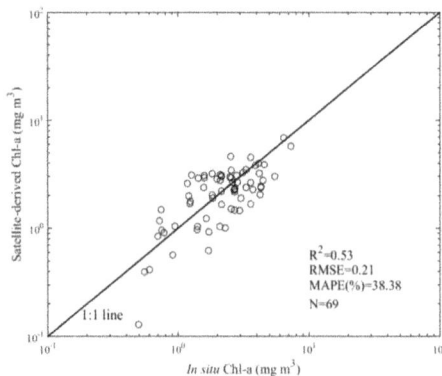

Figure 2. Comparison of satellite-derived Chl-a with in situ measured values.

3.2. Chl-a Spatial Distribution and Variability in the Bohai Sea

Using the monthly Chl-a data, the spatial and variability patterns of Chl-a were obtained by the temporal mean and standard deviation (SD) values during 2000–2012, respectively (Figure 3). In general, the Chl-a in the Bohai Sea showed much higher values than those in the northern Yellow Sea, and decreased gradually from coastal waters to offshore waters (Figure 3a). The highest Chl-a (>4.5 mg·m^{-3}) were in the Qinhuangdao coast, southern Laizhou bay, and northern Liaodong bay. In addition, the relatively high Chl-a values (2.7–4.5 mg·m^{-3}) were observed in coastal waters shallower than 20 m, whereas the relatively low values (<2.7 mg·m^{-3}) were in the central Bohai Sea and Bohai strait. As shown in Figure 3b, the highest variability (SD > 2.5 mg·m^{-3}) occurred in coastal area with <10 m isobaths. The higher variability (SD = 1.0–2.5 mg·m^{-3}) were distributed in coastal waters and the central Bohai Sea, whereas lower variability (SD < 1 mg·m^{-3}) appeared in the Bohai strait.

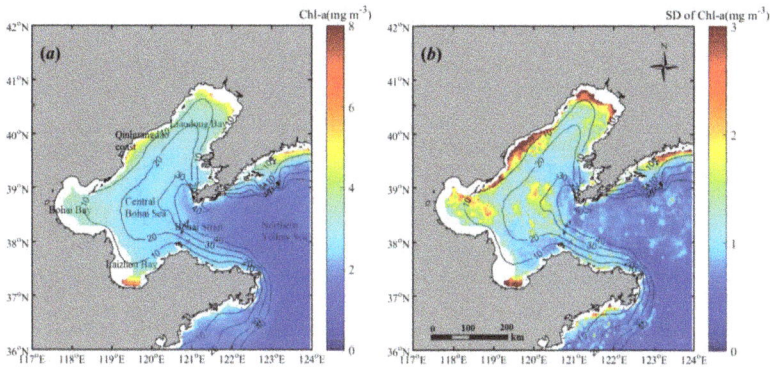

Figure 3. Distribution of the mean (**a**) and standard deviation (**b**) values of satellite-derived Chl-a during 2000–2012. The invalid pixels in the image are indicated by the white color.

3.3. Chl-a Seasonal Patterns in the Bohai Sea

The seasonal patterns of Chl-a in each month from 2000 to 2012, represented by climatological monthly images, are shown in Figure 4. The seasonal dynamics of Chl-a in the Bohai Sea resulted in a growth process from May to September and depletion from October to April. The Chl-a values during winter and early spring (December–April) (with most of the values below 2.5 mg·m^{-3}) were significantly lower than those from late spring to early autumn (May–September) (with most of the values above 3.5 mg·m^{-3}). Figure 4e–j shows that the Chl-a was relatively higher (>6 mg·m^{-3}) in coastal regions. The central Bohai Sea also had high Chl-a (>3 mg·m^{-3}) from June to September (Figure 4f–i).

3.4. Area Difference in Seasonal Variations of Chl-a

To gain more insight into seasonal variations of Chl-a over different locations, the Bohai Sea was divided into six sub-regions, as described in Section 2.1. The sampling area-averaged 13-year average of monthly Chl-a in the Bohai and northern Yellow Seas are shown in Figure 5. In the Liaodong bay, Qinhuangdao coast and Bohai bay, the seasonal patterns of Chl-a were characterized by a long-lasting Chl-a peak (>2.5 mg·m^{-3}) from May to September (Figure 5a–c). However, in these three areas, seasonal maxima of Chl-a appeared in June (4.7 ± 1.0 mg·m^{-3}), August (4.4 ± 1.7 mg·m^{-3}), and June (4.7 ± 1.7 mg·m^{-3}), respectively. In the central Bohai Sea (Figure 5d), the high Chl-a (>3 mg·m^{-3}) was observed from May, decreased from September, and then remained relatively low during winter. The maximum Chl-a dominated in July or August (3.5 ± 2.0 mg·m^{-3}). Two Chl-a maxima occurred in March (3.7 ± 1.4 mg·m^{-3}) and September (4.1 ± 1.2 mg·m^{-3}) in the Laizhou bay. However, the relatively low Chl-a was identified from April to June (Figure 5e), which was different from those in other coastal regions. Compared with other sub-regions, the Bohai strait had a distinct seasonal pattern of Chl-a with two Chl-a peaks in March (2.8 ± 0.7 mg·m^{-3}) and September (2.2 ± 0.7 mg·m^{-3}) and the lowest Chl-a in summer (Figure 5f). A similar seasonal pattern with the maximum in April (2.5 ± 0.6 mg·m^{-3}) was identified in the northern Yellow Sea, but no maximum occurred in August or September.

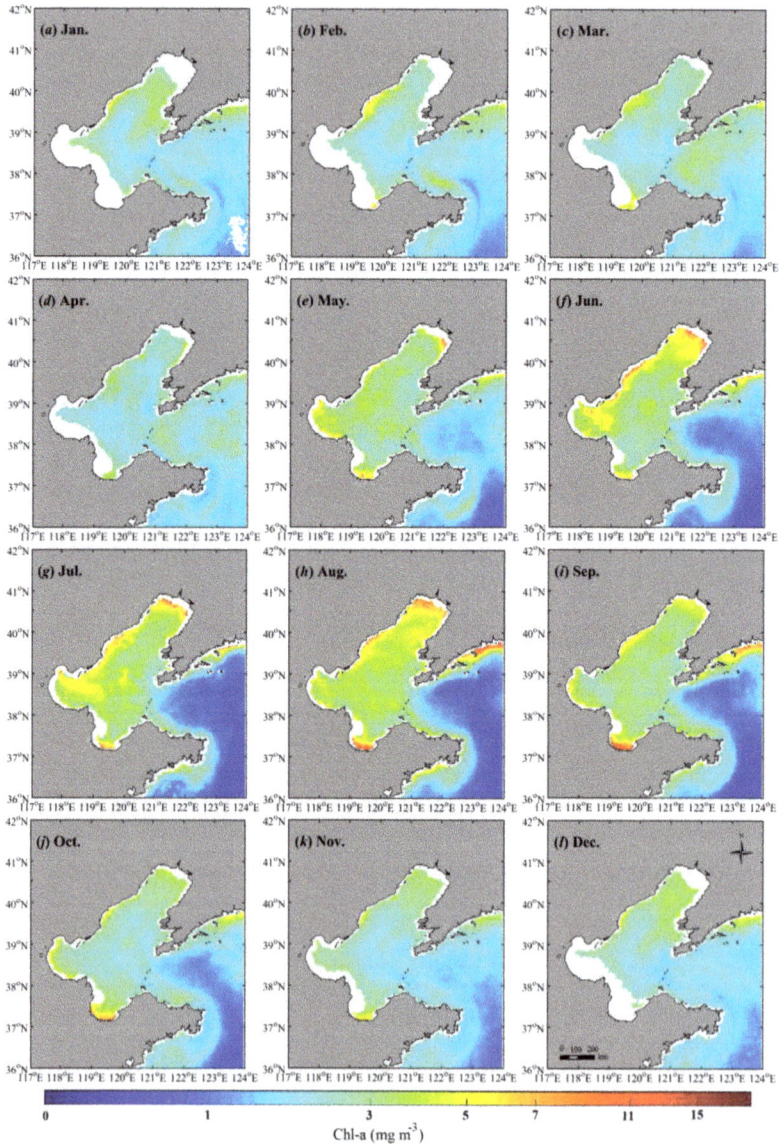

Figure 4. (a–l) Monthly climatological Chl-a images in the Bohai Sea during 2000–2012. The invalid pixels in the image are indicated by the white color.

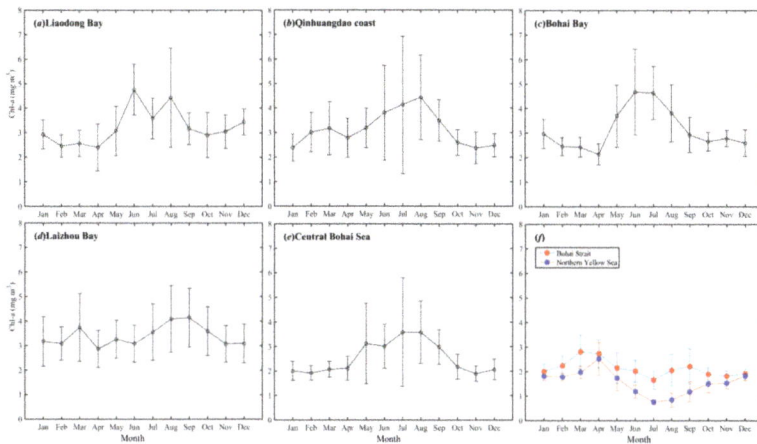

Figure 5. (a–f) Seasonal variation in 13-year averaged monthly Chl-a from January to December in seven sub-regions.

3.5. Chl-a Trend in the Bohai Sea

The long-term trend of Chl-a in the Bohai Sea from 2000 to 2012 is shown in Figure 6. In general, the Chl-a trend values in the Bohai Sea were higher compared with those in the northern Yellow Sea. The upward trends (>0.0018 mg·m^{-3}·month^{-1}) were detected in the entire Bohai Sea, and its pattern was heterogeneous. The larger increase in Chl-a (>0.0035 mg·m^{-3}·month^{-1}) prevailed over coastal waters, especially in northern Liaodong bay and Qinhuangdao coast. The central Bohai Sea also had large positive trends (0.0025–0.0038 mg·m^{-3}·month^{-1}). It is noted that some coastal regions in the image, such as the Laizhou bay and Bohai bay, showed invalid pixels (white color) because of small percentage ($<70\%$) of valid pixels.

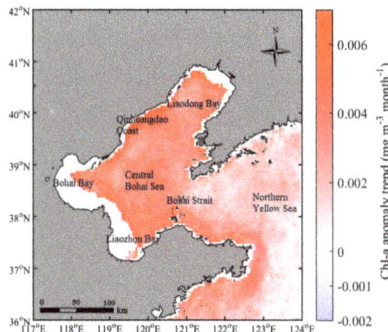

Figure 6. The long-term trend of monthly Chl-a anomaly from 2000 to 2012. The invalid pixels in the image are indicated by the white color.

In this study, spring, summer, autumn, and winter were defined as March to May, June to August, September to November, and December to February of the next year, respectively. The patterns of the Chl-a trend across four seasons are shown in Figure 7. Clear spatial and temporal variations of the Chl-a trend were observed in the Bohai Sea. In general, the Chl-a in the Bohai Sea displayed an increasing trend throughout the year. The Chl-a trend during summer and autumn showed higher values than those during winter and spring. At the temporal scale, the Chl-a trend was high in spring in the Bohai bay and Laizhou bay. In summer, the Chl-a trend was highest in most regions of the Bohai Sea, especially in the Bohai bay, Qinhuangdao coast, and central Bohai Sea. During autumn and winter, the Qinhuangdao coastal waters had high Chl-a trend.

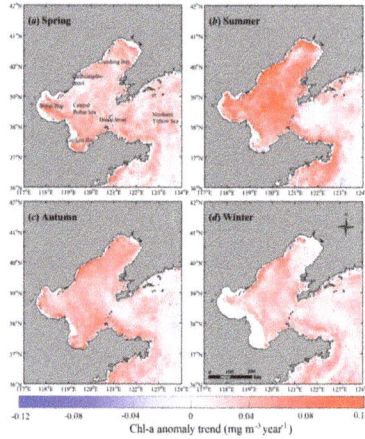

Figure 7. (a–d) The long-term trends of Chl-a anomaly in four seasons from 2000 to 2012.

The inter-annual variations of Chl-a during 2000–2012 displayed different patterns for the six sub-regions of the Bohai Sea (Figure 8). All the sub-regions had an increasing trend: Bohai strait (0.0018), central Bohai Sea (0.0032), Laizhou bay (0.003), Bohai bay (0.0027), Qinhuangdao coast (0.003), and Liaodong bay (0.0024). The largest increase in Chl-a was observed in the central Bohai Sea, whereas the smallest increase in Chl-a was in the Bohai strait.

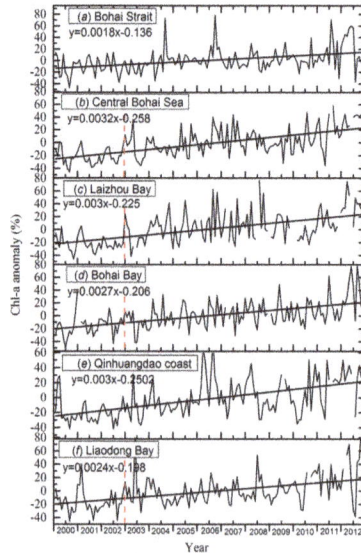

Figure 8. Linear trends of the Chl-a anomaly in the six sub-regions of the Bohai Sea. The black lines represent the linear trend, and the red lines represent the scratch line of the year 2003, as mentioned in Section 4.2.

3.6. The Causality between Chl-a Anomaly, SST Anomaly, and PAR Anomaly

To assess the causality between Chl-a and SST and PAR in the Bohai Sea, we calculated the information flow from the SST anomaly to the Chl-a anomaly (hereafter referred to as $IF_{SST \rightarrow Chl-a}$) and those from the PAR anomaly to the Chl-a anomaly (hereafter referred to as $IF_{PAR \rightarrow Chl-a}$) in four seasons using Equation (7) (Figure 9). Clearly, both the $IF_{SST \rightarrow Chl-a}$ and $IF_{PAR \rightarrow Chl-a}$ values were nonzero in the areas marked by red (Figure 9a,b), in the sense that phytoplankton growth could be affected by PAR and SST in spring. As shown in Figure 9c,d, the $IF_{PAR-Chl-a}$ values were higher than $IF_{SST \rightarrow Chl-a}$. The $IF_{PAR \rightarrow Chl-a}$ values in the Bohai Sea were above zero in summer, which indicated that PAR may be one of the factors affecting the growth of phytoplankton. During autumn, SST mainly showed significant IF in offshore waters (Figure 9e,f). $IF_{SST \rightarrow Chl-a}$ was close to zero in the Bohai Sea in winter (Figure 9g), thus essentially no causality could be identified here. The causality between PAR and Chl-a occurred in most regions of the Bohai Sea in winter (Figure 9h). These results, as shown in Figure 9, implied that the influences of environmental drivers (PAR and SST) on the Chl-a pattern were complex, which has been confirmed by previous studies [22,41,42]. At this stage, it should be stated that we investigated the causality between Chl-a and SST by mainly considering the indirect influences of SST on Chl-a. This is because the changes in SST may induce stratification or mixing of the water column, which further alter the light and nutrient conditions and thereby impact the phytoplankton growth.

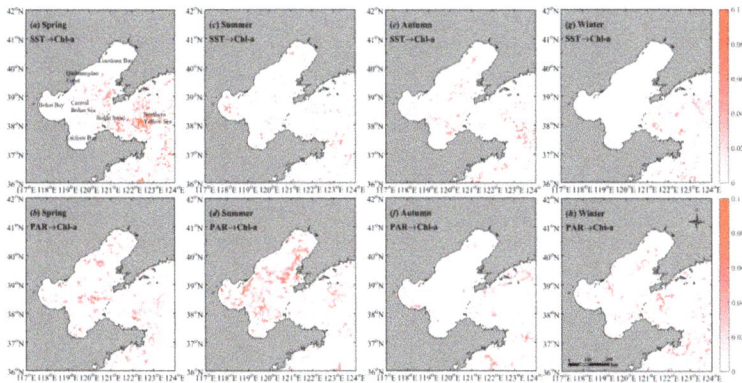

Figure 9. (**a–h**) The spatial distribution of information flow from the SST (PAR) anomaly to the Chl-a anomaly.

4. Discussion

4.1. The Chl-a Seasonal Patterns in the Bohai Sea

Our results on the seasonal patterns of Chl-a in the Bohai Sea, as shown in Figures 4 and 5, revealed that the Chl-a varied on both spatial and temporal scales. To better understand these results, we focus on the discussion of the related physical and chemical effects and human activity on seasonal scale, combined with the causality between Chl-a and environmental factors (SST and PAR), as below.

In spring, the information flow shown in Figure 8a,b indicated that SST and PAR can affect phytoplankton growth [43]. Increased SST and solar radiation gradually reduce the vertical mixing of the water column. In addition, weak wind stress can retain vertical mixing, which enhances the transportation of the nutrient-rich bottom water to the euphotic layer [44]. This allows phytoplankton to live longer in the upper euphotic layer and acquire sufficient nutrients and more PAR for phytoplankton growth. Thus, the spring bloom occurred in our study regions, especially in the Qinhuangdao coast, Laizhou bay, Liaodong bay, and Bohai strait (Figure 5).

During summer, the causality between Chl-a and PAR was observed in the Bohai Sea (Figure 9d). Surface warming and low wind stress would increase the stratification of the water column. Theoretically, the stronger stratification and less mixing not only provide a higher percentage of PAR that is available for photosynthesis, but also lead to high water clarity and thereby deepen the euphotic layer depths. These conditions can favor phytoplankton growth. However, light may not be a limiting factor for controlling phytoplankton growth during summer. In contrast, the nutrient supply is expected to be an important factor in different regions [42].

Therefore, the nutrient conditions in different sub-regions of the Bohai Sea are discussed below to help understand the area differences in seasonal variations of Chl-a during summer. In the Bohai strait, the surface layer of the water column is stratified, preventing nutrient-rich waters from the deeper layer entering the photic zone. Meanwhile, all nutrients are depleted. Thus, the growth of phytoplankton is restricted, and the Chl-a reaches a minimum in summer (Figure 5). In contrast to the pattern in the Bohai strait, a pronounced Chl-a peak from May to September was observed in coastal water bodies (Liaodong bay, Qinhuangdao coast, and Bohai bay) and the central Bohai Sea. It may be related to the nutrients added by river discharge. Because of freshwater discharge from inland rivers carrying abundant nutrients, the trophic level in coastal waters increases significantly, especially in summer [45,46]. This has also been confirmed by Tang et al. [47] who reported that most harmful algal blooms may be initiated by nutrients from river discharge. Thus, the increased nutrients may support higher Chl-a levels in coastal waters. A question is why the central Bohai Sea also had higher Chl-a in summer. This may be attributed to water exchange between coastal waters and offshore waters related to the Bohai Sea circulation (including the warm current extension, Liaodong coastal current, and southern Bohai coastal current) and wind-tide-thermohaline circulation [19,48,49]. The water-exchange can enhance coastal nutrient transporting to the central Bohai Sea, thereby promoting the phytoplankton growth. Therefore, during summer, the nutrient supply from river discharge might be a major controlling factor in the high Chl-a in coastal waters and the central Bohai Sea. In contrast, the seasonal pattern of Chl-a in the Laizhou bay showed the relatively low Chl-a in early summer (Figure 5). Liu et al. [18] also reported this phenomenon in the sea region near the Yellow River mouth. This could be related to the water storage of dams and reservoirs on the Yellow River. The decreased riverine inputs due to dams and reservoirs can reduce the nutrient load, and thus result in the limitation for phytoplankton growth. Gong et al. [50] and Jiao et al. [51] reported that the decreased nutrient load and primary productivity during summer were associated with freshwater discharge reduction caused by water storages. In the Laizhou bay, human activity (e.g., dams and reservoirs) might be the reason for the change in Chl-a in early summer.

In autumn, the causality between SST and Chl-a (Figure 9e) indicated that the change in SST may influence the phytoplankton growth. With a decreasing SST and stronger wind stress, the stratification is broken down, and the vertical mixing of the water column increases, which could provide the nutrient supply and a suitable environment for phytoplankton growth. In the Bohai Sea, seasonal water stratification appears in April and breaks down at the end of September [47]. Thus, the relatively high Chl-a was observed in the Bohai Sea, such as the Bohai strait, Laizhou bay, central Bohai Sea, and Qinhuangdao coast (Figure 5).

When winter comes, stratification disappears and vertical mixing of the water column becomes strong due to sea surface cooling and strong winds. A strong northerly monsoon wind from late November to March influences the Bohai Sea [52], which can increase the mixing in the water column. Nutrients are carried to the surface layer from underlying nutrient-rich waters, which could provide for the spring bloom in the next year [53]. However, the low temperature and instability of the water column make it difficult to support an optimal growth condition for phytoplankton. In addition, mixing of the water column may decrease water transparency and increase the extinction coefficient of the upper water, which could reduce the amount of light available to phytoplankton. These offer an explanation to help us understand the relatively low Chl-a in winter (Figure 5). Due to the lack of field nutrient data, currently, we can only give a general discussion on the influence of nutrients on

Chl-a. Further investigations focusing on this topic are still required in the future, when nutrient data become available.

4.2. The Increasing Trend of Chl-a in the Bohai Sea

The entire Bohai Sea exhibited an increasing trend of Chl-a from 2000 to 2012 (Figures 5 and 8). In particular, there were clear long-term increases in Chl-a since 2003 in the Laizhou bay, Bohai bay, Liaodong bay, Qinhuangdao coast, and central Bohai Sea (Figure 7b–f). There was a corresponding significant increase in the annual total runoff data of the three major rivers (Yellow River, Haihe River, and Liaohe River) since 2003, as shown in Figure 10a (date from Zhang et al. [54]). To further examine the relationship between the long-term changes in Chl-a and river runoff, we generated scatter plots to compare the annual Chl-a and annual total runoff for the six sub-regions of the Bohai Sea (Figure 10b). Although it is difficult to assess the effects of river discharge on the Chl-a trend in different sub-regions based only on 11-year time series data, we believe it is still useful to discuss their relationships. For these six sub-regions, the correlations between the annual Chl-a and annual total runoff were all positive, with high correlation coefficients ($R \geq 0.53$), indicating that the increasing trend of Chl-a in the Bohai Sea might be influenced by river discharge. The freshwater discharge from riverine inputs supplies large amounts of nutrients to the Bohai Sea, favoring the phytoplankton growth. Furthermore, the nutrients added by inland rivers has increased significantly over the past several decades, mainly due to the use of chemical fertilizers and industrial/domestic sewage discharge [55]. Similarly, Li et al. [56] reported that eutrophication in the Qinhuangdao coast was mainly affected by nutrients from river discharge. Additionally, the higher and the lowest correlation coefficients were in the central Bohai Sea and Bohai strait, respectively. This might offer an explanation for the different increases in Chl-a in the central Bohai Sea and Bohai strait (Figure 7). The coastal nutrient can be transposed to offshore waters by currents and wind-tide-thermohaline circulation [47,57]. Meanwhile, because of the limitation of the Bohai strait, the central Bohai Sea has a longer water exchange with the Yellow Sea, and thus, the long retention time can increase nutrient concentrations, which further promotes an increase in Chl-a [58]. In contrast, the high water-exchange ability in the Bohai strait leads to nutrients in a shorter retention time and limits the increase trend of Chl-a. However, it is noted that we only showed the relationship between the Chl-a and riverine inputs over a relatively short timescale (11 years). Longer time series data with high temporal resolution (e.g., monthly) are needed to reveal the influence of river discharge on the Chl-a trends in the future when more data become available.

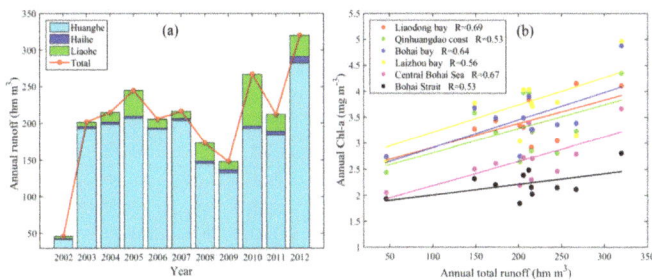

Figure 10. The annual runoff data of the three major rivers from 2002 to 2011 (data from Zhang et al. [54]) (**a**). The scatter plots between the annual Chl-a of six sub-regions and the annual total runoff from 2002 to 2012 (**b**).

5. Conclusions

This study investigated the Chl-a dynamics in the Bohai Sea using satellite-derived products. The seasonal patterns of Chl-a displayed a long-lasting summer peak (May–September) in the Liaodong bay, Qinhuangdao coast, Bohai bay, and the central Bohai Sea. The relatively low Chl-a in early summer occurred in the Laizhou bay. In the Bohai strait, the two seasonal peaks appeared in March and September, and the minimum Chl-a was observed in summer. These variations of Chl-a could be mainly explained by the vertical structure of the water column, climate conditions (e.g., SST), and human activity. Meanwhile, the inter-annual patterns of Chl-a from 2000 to 2012 showed an increasing trend in the entire Bohai Sea, particularly in the central Bohai Sea, which might be related to river discharge. To better understand the long-term changes in Chl-a and its mechanisms, further efforts should be dedicated to making more detailed materials available (e.g., nutrient, water quality, and wind).

Acknowledgments: This research was jointly supported by the National Natural Science Foundation of China (Nos. 41576172, 41506200, and 41276186), the National Key Research and Development Program of China (No. 2016YFC1400901), the Provincial Natural Science Foundation of Jiangsu in China (Nos. BK20151526, BK20150914, and BK20161532), the National Program on Global Change and Air-sea Interaction (No. GASI-03-03-01-01), the Public Science and Technology Research Funds Projects of Ocean (201005030), a project funded by "the Priority Academic Program Development of Jiangsu Higher Education Institutions (PAPD)", and the Research and Innovation Project for College Graduates of Jiangsu Province (1344051601032). Special thanks to Shaojie Sun in University of South Florida for data support.

Author Contributions: Hailong Zhang designed this study; Zhongfeng Qiu contributed to the data analyses and drafted the manuscript; Deyong Sun and Shengqiang Wang assisted with developing the research design and results interpretation; and Yijun He contributed to the interpretation of result.

Conflicts of Interest: The authors declare no conflict of interest.

References

1. Field, C.B.; Behrenfeld, M.J.; Randerson, J.T.; Falkowski, P. Primary Production of the Biosphere: Integrating Terrestrial and Oceanic Components. *Science.* **1998**, *281*, 237–240. [CrossRef] [PubMed]
2. Boyce, D.G.; Lewis, M.R.; Worm, B. Global phytoplankton decline over the past century. *Nature.* **2010**, *466*, 591–596. [CrossRef] [PubMed]
3. Roemmich, D.; Mcgowan, J. Climatic Warming and the Decline of Zooplankton in the California Current. *Science.* **1995**, *267*, 1324–1326. [CrossRef] [PubMed]
4. Chassot, E.; Bonhommeau, S.; Dulvy, N.K.; Watson, R.; Gascuel, D.; Pape, O.L. Global marine primary production constrains fisheries catches. *Ecol. Lett.* **2010**, *13*, 495–505. [CrossRef] [PubMed]
5. Henson, S.A.; Sarmiento, J.L.; Dunne, J.P.; Bopp, L.; Lima, I.; Doney, S.C.; John, J.; Beaulieu, C. Detection of anthropogenic climate change in satellite records of ocean chlorophyll and productivity. *Biogeosciences.* **2010**, *7*, 621–640. [CrossRef]
6. Sun, D.; Hu, C.; Qiu, Z.; Cannizzaro, J.P.; Barnes, B.B. Influence of a red band-based water classification approach on chlorophyll algorithms for optically complex estuaries. *Remote Sens. Environ.* **2014**, *155*, 289–302. [CrossRef]
7. Maritorena, S.; Siegel, D.A.; Peterson, A.R. Optimization of a semianalytical ocean color model for global-scale applications. *Appl. Opt.* **2002**, *41*, 2705–2714. [CrossRef] [PubMed]
8. Tassan, S. Local algorithms using SeaWiFS data for the retrieval of phytoplankton, pigments, suspended sediment, and yellow substance in coastal waters. *Appl. Opt.* **1994**, *33*, 2369–2378. [CrossRef] [PubMed]
9. Tang, J.; Wang, X.; Song, Q.; Li, T.; Chen, J.; Huang, H.; Ren, J. The statistic inversion algorithms of water constituents for the Huanghai Sea and the East China Sea. *Acta Oceanol. Sin.* **2004**, *23*, 617–626.
10. O'Reilly, J.E.; Maritorena, S.; Mitchell, B.G.; Siegel, D.A.; Carder, K.L.; Garver, S.A.; Kahru, M.; Mcclain, C. Ocean color chlorophyll algorithms for SeaWiFS. *J. Geophys. Res. Oceans* **1998**, *103*, 24937–24950. [CrossRef]
11. Harley, C.D.G.; Hughes, A.R.; Hultgren, K.M.; Miner, B.G.; Sorte, C.J.B.; Thornber, C.S.; Rodriguez, L.F.; Tomanek, L.; Williams, S.L. The impacts of climate change in coastal marine systems. *Ecol. Lett.* **2006**, *9*, 228–241. [CrossRef] [PubMed]
12. Smetacek, V.; Cloern, J.E. On Phytoplankton Trends. *Science.* **2008**, *319*, 1346–1348. [CrossRef] [PubMed]
13. Tan, J.; Cherkauer, K.A.; Chaubey, I.; Troy, C.D.; Essig, R. Water quality estimation of River plumes in Southern Lake Michigan using Hyperion. *J. Gt. Lakes Res.* **2016**, *42*, 524–535. [CrossRef]
14. Gong, G.-C.; Wen, Y.-H.; Wang, B.-W.; Liu, G.-J. Seasonal variation of chlorophyll a concentration, primary production and environmental conditions in the subtropical East China Sea. *Deep Sea Res. Part II Top. Stud. Oceanogr.* **2003**, *50*, 1219–1236. [CrossRef]

15. Shi, W.; Wang, M. Satellite views of the Bohai Sea, Yellow Sea, and East China Sea. *Prog. Oceanogr.* **2012**, *104*, 30–45. [CrossRef]
16. Yamaguchi, H.; Kim, H.C.; Son, Y.B.; Sang, W.K.; Okamura, K.; Kiyomoto, Y.; Ishizaka, J. Seasonal and summer interannual variations of SeaWiFS chlorophyll a in the Yellow Sea and East China Sea. *Prog. Oceanogr.* **2012**, *105*, 22–29. [CrossRef]
17. He, X.Q.; Bai, Y.; Pan, D.L.; Chen, C.-T.A.; Cheng, Q.; Wang, D.F.; Gong, F. Satellite views of seasonal and inter-annual variability of phytoplankton blooms in the eastern China seas over the past 14 years (1998–2011). *Biogeosci. Discuss.* **2013**, *10*, 111–155. [CrossRef]
18. Liu, D.; Wang, Y. Trends of satellite derived chlorophyll-a (1997–2011) in the Bohai and Yellow Seas, China: Effects of bathymetry on seasonal and inter-annual patterns. *Prog. Oceanogr.* **2013**, *116*, 154–166. [CrossRef]
19. Guan, B.X. Patterns and Structures of the Currents in Bohai, Huanghai and East China Seas. In *Oceanology of China Seas*; Springer: Dordrecht, The Netherlands, 1994.
20. Zhao, B.; Zhuang, G.; Cao, D.; Lei, F. Circulation, tidal residual currents and their effects on the sedimentations in the Bohai Sea. *Oceanol. Limnol. Sin.* **1995**, *26*, 466–473.
21. Zhang, Y.; He, X.; Gao, Y. Preliminary analysis on the modified water massea in the north Yellow Sea and the Bohai Sea. *Trans. Oceanol. Limnol.* **1983**, *2*, 19–26. (In Chinese)
22. Jackson, J.M.; Thomson, R.E.; Brown, L.N.; Willis, P.G.; Borstad, G.A. Satellite chlorophyll off the British Columbia Coast, 1997–2010. *J. Geophys. Res. Oceans* **2015**, *120*, 4709–4728. [CrossRef]
23. Jamet, C.; Loisel, H.; Kuchinke, C.P.; Ruddick, K.; Zibordi, G.; Feng, H. Comparison of three SeaWiFS atmospheric correction algorithms for turbid waters using AERONET-OC measurements. *Remote Sens. Environ.* **2011**, *115*, 1955–1965. [CrossRef]
24. Siswanto, E.; Tang, J.; Yamaguchi, H.; Ahn, Y.H.; Ishizaka, J.; Yoo, S.; Kim, S.W.; Kiyomoto, Y.; Yamada, K.; Chiang, C. Empirical ocean-color algorithms to retrieve chlorophyll-a, total suspended matter, and colored dissolved organic matter absorption coefficient in the Yellow and East China Seas. *J. Oceanogr.* **2011**, *67*, 627–650. [CrossRef]
25. Yamaguchi, H.; Ishizaka, J.; Siswanto, E.; Son, Y.B.; Yoo, S.; Kiyomoto, Y. Seasonal and spring interannual variations in satellite-observed chlorophyll-a in the Yellow and East China Seas: New datasets with reduced interference from high concentration of resuspended sediment. *Cont. Shelf Res.* **2013**, *59*, 1–9. [CrossRef]
26. Terauchi, G.; Tsujimoto, R.; Ishizaka, J.; Nakata, H. Preliminary assessment of eutrophication by remotely sensed chlorophyll-a in Toyama Bay, the Sea of Japan. *J. Oceanogr.* **2014**, *70*, 175–184. [CrossRef]
27. Ryu, J.H.; Han, H.J.; Cho, S.; Park, Y.J.; Ahn, Y.H. Overview of geostationary ocean color imager (GOCI) and GOCI data processing system (GDPS). *Ocean Sci. J.* **2012**, *47*, 223–233. [CrossRef]
28. Qiu, Z.; Zheng, L.; Zhou, Y.; Sun, D.; Wang, S.; Wu, W. Innovative GOCI algorithm to derive turbidity in highly turbid waters: A case study in the Zhejiang coastal area. *Opt. Express.* **2015**, *23*, A1179–A1193. [CrossRef] [PubMed]
29. Yuan, Y.; Qiu, Z.; Sun, D.; Wang, S.; Yue, X. Daytime sea fog retrieval based on GOCI data: A case study over the Yellow Sea. *Opt. Express.* **2016**, *24*, 787–801. [CrossRef] [PubMed]
30. Qiu, Z. A simple optical model to estimate suspended particulate matter in Yellow River Estuary. *Opt. Express* **2013**, *21*, 27891–27904. [CrossRef] [PubMed]
31. Zhou, H.; Zhang, Z.N.; Liu, X.S.; Tu, L.H.; Yu, Z.S. Changes in the shelf macrobenthic community over large temporal and spatial scales in the Bohai Sea, China. *J. Mar. Syst.* **2007**, *67*, 312–321. [CrossRef]
32. Liu, S.M.; Zhang, J.; Chen, H.T.; Zhang, G.S. Factors influencing nutrient dynamics in the eutrophic Jiaozhou Bay, North China. *Prog. Oceanogr.* **2005**, *66*, 66–85. [CrossRef]
33. Gao, X.; Zhou, F.; Chen, C.-T.A. Pollution status of the Bohai Sea: an overview of the environmental quality assessment related trace metals. *Environ. Int.* **2014**, *62*, 12–30. [CrossRef] [PubMed]
34. Peng, S. The nutrient, total petroleum hydrocarbon and heavy metal contents in the seawater of Bohai Bay, China: Temporal–spatial variations, sources, pollution statuses, and ecological risks. *Mar. Pollut. Bull.* **2015**, *95*, 445–451. [CrossRef] [PubMed]
35. Gong, G.-C. Absorption coefficients of colored dissolved organic matter in the surface waters of the East China Sea. *Terr. Atmos. Ocean. Sci.* **2004**, *15*, 75–88. [CrossRef]
36. Cheng, C.; Huang, H.; Liu, C.; Jiang, W. Challenges to the representation of suspended sediment transfer using a depth-averaged flux. *Earth Surf. Process. Landf.* **2016**, *41*, 1337–1357. [CrossRef]

37. Gregg, W.W.; Casey, N.W.; Mcclain, C.R. Recent trends in global ocean chlorophyll. *Geophys. Res. Lett.* **2005**, *32*, 259–280. [CrossRef]

38. Sies, H. A new parameter for sex education. *Nature* **1988**, 332. [CrossRef]

39. Liang, X.S. Unraveling the cause-effect relation between time series. *Phys. Rev. E.* **2014**, *90*, 052150. [CrossRef] [PubMed]

40. Stips, A.; Macias, D.; Coughlan, C.; Garcia-Gorriz, E.; San Liang, X. On the causal structure between CO_2 and global temperature. *Sci. Rep.* **2016**, *6*, 21691. [CrossRef] [PubMed]

41. Lin, C.; Su, J.; Xu, B.; Tang, Q. Long-term variation of temperature and salinity of the Bohai Sea and their influence on its ecosystem. *Prog. Oceanogr.* **2001**, *49*, 7–19. [CrossRef]

42. Waite, J.N.; Mueter, F.J. Spatial and temporal variability of chlorophyll-a concentrations in the coastal Gulf of Alaska, 1998–2011, using cloud-free reconstructions of SeaWiFS and MODIS-Aqua data. *Prog. Oceanogr.* **2013**, *116*, 179–192. [CrossRef]

43. Fennel, K. Convection and the timing of phytoplankton spring blooms in the western Baltic Sea. *Estuar. Coast. Shelf Sci.* **1999**, *49*, 113–128. [CrossRef]

44. Behrenfeld, M.J.; O'Malley, R.T.; Siegel, D.A.; McClain, C.R.; Sarmiento, J.L.; Feldman, G.C.; Milligan, A.J.; Falkowski, P.G.; Letelier, R.M.; Boss, E.S. Climate-driven trends in contemporary ocean productivity. *Nature* **2006**, *444*, 752–755. [CrossRef] [PubMed]

45. Lin, C.; Ning, X.; Su, J.; Lin, Y.; Xu, B. Environmental changes and the responses of the ecosystems of the Yellow Sea during 1976–2000. *J. Mar. Syst.* **2005**, *55*, 223–234. [CrossRef]

46. Ye, L.; Yujie, Z.; Shitao, P.; Qixing, Z.; Ma, L.Q. Temporal and spatial trends of total petroleum hydrocarbons in the seawater of Bohai Bay, China from 1996 to 2005. *Mar. Pollut. Bull.* **2010**, *60*, 238–243.

47. Tang, D.; Kawamura, H.; Oh, I.S.; Baker, J. Satellite evidence of harmful algal blooms and related oceanographic features in the Bohai Sea during autumn 1998. *Adv. Space Res.* **2006**, *37*, 681–689. [CrossRef]

48. Sündermann, J.; Feng, S. Analysis and modelling of the Bohai sea ecosystem—A joint German-Chinese study. *J. Mar. Syst.* **2004**, *44*, 127–140. [CrossRef]

49. Ning, X.; Lin, C.; Su, J.; Liu, C.; Hao, Q.; Le, F.; Tang, Q. Long-term environmental changes and the responses of the ecosystems in the Bohai Sea during 1960–1996. *Deep Sea Res. Part II Top. Stud. Oceanogr.* **2010**, *57*, 1079–1091. [CrossRef]

50. Gong, G.-C.; Chang, J.; Chiang, K.-P.; Hsiung, T.-M.; Hung, C.-C.; Duan, S.-W.; Codispoti, L.A. Reduction of primary production and changing of nutrient ratio in the East China Sea: Effect of the Three Gorges Dam? *Geophys. Res. Lett.* **2006**, *33*. [CrossRef]

51. Jiao, N.; Zhang, Y.; Zeng, Y.; Gardner, W.D.; Mishonov, A.V.; Richardson, M.J.; Hong, N.; Pan, D.; Yan, X.H.; Jo, Y.H. Ecological anomalies in the East China Sea: Impacts of the Three Gorges Dam? *Water Res.* **2007**, *41*, 1287–1293. [CrossRef] [PubMed]

52. Yuan, Y.; Su, J. Numerical modeling of the circulation in the East China Sea. *Ocean Hydrodyn. Jpn. East China Seas.* **1984**, *39*, 167–176.

53. Yamada, K.; Ishizaka, J.; Yoo, S.; Kim, H.C.; Chiba, S. Seasonal and interannual variability of sea surface chlorophyll a concentration in the Japan/East Sea (JES). *Prog. Oceanogr.* **2004**, *61*, 193–211. [CrossRef]

54. Zhang, M.; Dong, Q.; Cui, T.; Ding, J. Remote Sensing of Spatiotemporal Variation of Apparent Optical Properties in Bohai Sea. *IEEE J. Sel. Top. Appl. Earth Obs. Remote Sens.* **2015**, *8*, 1176–1184. [CrossRef]

55. Qu, H.J.; Kroeze, C. Past and future trends in nutrients export by rivers to the coastal waters of China. *Sci. Total Environ.* **2010**, *408*, 2075–2086. [CrossRef] [PubMed]

56. Li, Z.; Cui, L. Contaminative conditions of main rivers flowing into the sea and their effect on seashore of Qinhuangdao. *Ecol. Environ. Sci.* **2012**, *21*, 1285–1288. (In Chinese)

57. Hickox, R.; Belkin, I.; Cornillon, P.; Shan, Z. Climatology and seasonal variability of ocean fronts in the East China, Yellow and Bohai seas from satellite SST data. *Geophys. Res. Lett.* **2000**, *27*, 2945–2948. [CrossRef]

58. Wei, H.; Sun, J.; Moll, A.; Zhao, L. Phytoplankton dynamics in the Bohai Sea—Observations and modelling. *J. Mar. Syst.* **2004**, *44*, 233–251. [CrossRef]

remote sensing

MDPI

Article

Estimation of Water Quality Parameters in Lake Erie from MERIS Using Linear Mixed Effect Models

Kiana Zolfaghari * and Claude R. Duguay

Interdisciplinary Centre on Climate Change and Department of Geography and Environmental Management, University of Waterloo, Waterloo, ON N2L 3G1, Canada; crduguay@uwaterloo.ca
* Correspondence: kzolfagh@uwaterloo.ca; Tel.: +1-519-888-4026 (ext. 31322)

Academic Editors: Yunlin Zhang, Claudia Giardino, Linhai Li, Magaly Koch and Prasad S. Thenkabail
Received: 16 February 2016; Accepted: 30 May 2016; Published: 3 June 2016

Abstract: Linear Mixed Effect (LME) models are applied to the CoastColour atmospherically-corrected Medium Resolution Imaging Spectrometer (MERIS) reflectance, L2R full resolution product, to derive chlorophyll-a (chl-a) concentration and Secchi disk depth (SDD) in Lake Erie, which is considered as a Case II water (*i.e.*, turbid and productive). A LME model considers the correlation that exists in the field measurements which have been performed repeatedly in space and time. In this study, models are developed based on the relation between the logarithmic scale of the water quality parameters and band ratios: B07:665 nm to B09:708.75 nm for $\log_{10}$chl-a and B06:620 nm to B04:510 nm for $\log_{10}$SDD. Cross validation is performed on the models. The results show good performance of the models, with Root Mean Square Errors (RMSE) and Mean Bias Errors (MBE) of 0.31 and 0.018 for $\log_{10}$chl-a, and 0.19 and 0.006 for $\log_{10}$SDD, respectively. The models are then applied to a time series of MERIS images acquired over Lake Erie from 2004–2012 to investigate the spatial and temporal variations of the water quality parameters. Produced maps reveal distinct monthly patterns for different regions of Lake Erie that are in agreement with known biogeochemical properties of the lake. The Detroit River and Maumee River carry sediments and nutrients to the shallow western basin. Hence, the shallow western basin of Lake Erie experiences the most intense algal blooms and the highest turbidity compared to the other sections of the lake. Maumee Bay, Sandusky Bay, Rondeau Bay and Long Point Bay are estimated to have prolonged intense algal bloom.

Keywords: Lake Erie; MERIS; CoastColour; linear mixed effect model; chlorophyll-a; Secchi disk depth

1. Introduction

Lake Erie, a turbid and regionally eutrophic lake, is the most southern and shallowest of the Laurentian Great Lakes. Total suspended matters (TSM) are a major contributor to the lake's low water clarity [1]. The problem of excess nutrients and resulting algal blooms are also threatening the ecosystem of the lake and the economic activities of the surrounding regions. The ecological state of Lake Erie significantly affects its role as a natural, social, and economic resource, considering that the lake is as an essential drinking water source that also offers many opportunities for recreational activities, fisheries and tourism. As a result, the Lakewide Management Plan was signed in 1972 to restore and maintain the ecological health of the lake [2]. Ongoing efforts to support this plan require high-resolution measurements of the water quality parameters on a variety of spatial and temporal scales. Conventional field-based measurements of these parameters can be expensive and they are often sparse in either space or time, or both. Remote sensing has the potential to infer the lake bio-optical/water quality parameters, overcoming these concerns.

The emerging water radiance measured by remote sensing instruments depends on the water itself and its constituents. The water constituents interact with light and modify the incoming and outgoing radiation at various wavelengths. Therefore, remote sensing measurements of water leaving radiance can be related to the composition and concentration of water constituents. In Case I waters, chl-a concentration (phytoplankton population) and its co-varying particles are dominating the optical properties, which is the case in nearly all open ocean waters. However, optically complex inland waters and coastal waters are referred to as Case II waters, where chl-a alone is a poor predictor of light attenuation, and variations in TSM and colored dissolved organic matter (CDOM) are also important in light scattering or absorption, and therefore the water leaving radiance [3].

The delivery of data on water color has been explored using satellite sensors such as Landsat Thematic Mapper (TM)/Enhanced Thematic Mapper (ETM+) due to their relatively high spatial resolution of 30 m [4–6]. However, the shortcomings of Landsat in other capacities, such as its relatively low temporal resolution (*i.e.*, 16 days), spectral and radiometric sensitivity, make the use of MODIS (Moderate-resolution Imaging Spectroradiometer) and MERIS (Medium Resolution Imaging Spectrometer) data more attractive for water quality monitoring [7–9]. Images from these satellite sensors compensate for the limitations of Landsat at the expense of a lower spatial resolution (*ca.* 250–500 m). MERIS was originally designed for water quality monitoring applications. Therefore, compared to MODIS, it has a more suitable spectral resolution in the red and near-infrared (NIR) to derive the secondary chlorophyll-a (chl-a) absorption maximum [10]. This is essential for Case II waters, as is the case for Lake Erie, where chl-a is not the predominant color-producing agent (CPA) and multiple non-covarying CPAs may also confound the reflectance signal, particularly at shorter wavelengths. Hence, the traditional and empirical blue/green band ratio algorithms, used for Case I waters, result in large uncertainties in Case II waters due to the limited ability to distinguish signals coming from the independent water constituents.

As a consequence of the ambiguities related to the shorter wavelengths, several authors have investigated the applicability of red and NIR wavelengths for estimating chl-a concentration in turbid optically complex waters to aim for a minimal sensitivity to other water-coloring parameters. The effect of CDOM can certainly be neglected at these wavelengths [1]. Red-NIR band ratio algorithms have been found to work well in Lake Chagan (chl-a concentration: 6.4 to 58.21 mg·m^{-3}) [11], as well as Curonian Lagoon (chl-a concentration: 44.1 to 85.3 mg·m^{-3}) [12] and Zeekoevlei Lake (chl-a concentration: 61 to 247.4 mg·m^{-3}) [13] that have chl-a concentration ranges typical of mesotrophic lakes and eutrophic Lake Erie [1]. Band ratio algorithms developed to derive Secchi disk depth (SDD) variations make use of bands in the visible range of the spectrum. Two multiple linear regression models have been developed separately, based on blue and red bands of Landsat TM and MODIS, to predict the logarithm of SDD in Poyang Lake National Nature Reserve in China [14]. A linear regression model has also been proposed based on the logarithmic transformation of MERIS band ratio (490 nm to 620 nm) to estimate the natural logarithm of SDD in the Baltic Sea [15]. However, the correlated errors resulting from repeated measurements in space and time are not considered in the regression models developed in these studies. Multiple measurements per variable will result in non-independency, which violates the assumptions of regression methods. The Linear Mixed Effect (LME) model [16] approach developed herein is appropriate for cases where observations are collected in time and/or space for the same parameter, and therefore represent clustered or dependent data.

The applicability of LME models is tested in this study to estimate chl-a concentration and SDD from the CoastColour (CC) atmospherically corrected MERIS reflectance product [17] in support of water quality monitoring in Lake Erie. Although *in situ* measurements remain the most accurate solution for water quality monitoring programs, satellite remote sensing can be added for routine and synoptic measurements [18]. Chl-a is widely measured as an indicator of eutrophication and primary production. SDD is another environmental descriptor that is indicative of water clarity. It also provides a highly relevant measure of the extent of the euphotic layer where primary production is possible [19,20]. Therefore, both parameters are of interest in this study. This work aimed to derive chl-a and SDD by applying LME models on MERIS data. Also temporal and spatial variations of these parameters were examined.

2. Materials and Methods

2.1. Study Site

Lake Erie (42°11′N, 81°15′W; Figure 1) is the smallest (by volume), the shallowest, and the warmest of the Laurentian Great Lakes [21]. It is a monomictic lake (with occasional dimictic years) covering an area of 25,700 km^2, with average and maximum depths of 19 m and 64 m, respectively [22]. The lake is naturally divided into three basins of different depths: the shallow western basin, the central basin, and the deep eastern basin (Table 1). The basins are separated approximately based on the Lake Erie Islands (~82°49′W) and the Long Point-Erie Ridge (~80°25′W) (Figure 1) [1]. River discharge into Lake Erie originates mostly from the St. Clair River and Lake St. Clair through the Detroit River. Other smaller rivers and streams in the territory of Lake Erie also contribute to water inflow into the lake. Lake Erie drains into Lake Ontario through the Niagara River and shipping canals [2,23].

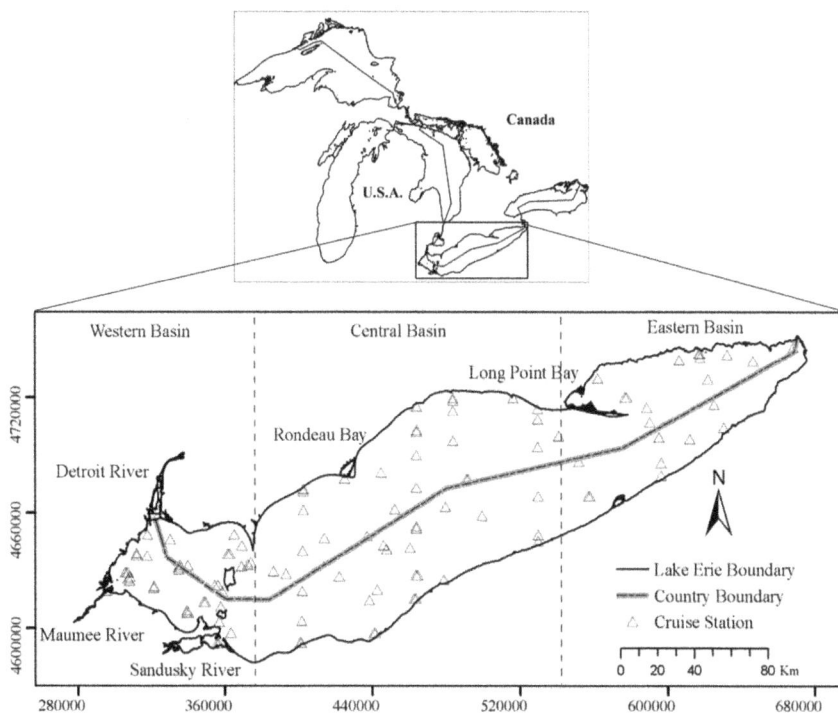

Figure 1. Location of Lake Erie and its boundary (Canada and US). *In situ* sampling stations from cruises that took place in September 2004, May, July, and September 2005, May and June 2008, July and September 2011, and February 2012 are illustrated by empty triangles.

Table 1. Lake Erie Basins Information (source: [23]).

Lake Erie	Mean Depth (m)	Maximum Depth (m)
West Basin	7.4	19
Central Basin	18.3	25
East Basin	25	64

Lake Erie is exposed to greater stress than any other of the Great Lakes due to agricultural practices and urbanization in its surroundings. Chemically enriched runoff from agricultural lands in the basin flows into the lake. In addition, the lake receives the most effluents from wastewater treatment works [2,23]. Lake Erie has experienced substantial eutrophication over the past half century due to excess phosphorus loads from point and nonpoint sources producing algal blooms [21]. In general, phosphorus concentration in Lake Erie decreases from west to east and from near-shore to the offshore [24].

2.2. Field Measurements of Water Quality Parameters

Sample collection in Lake Erie was conducted on board of the Canadian Coast Guard ship *Limnos* during September 2004, May, July, and September 2005, May and June 2008, July and September 2011, and February 2012. A total of 89 distributed stations were visited to provide measurements of a wide range of optical properties as well as concentrations of the main CPAs in different locations of the lake (Figure 1).

Composite water samples were collected at all stations, during 2004 to 2012, from the surface mixed layer of the lake using Niskin bottles. The samples were filtered through a Whatman GF/F fiber filter (0.7 μm) in the field.

The filtered samples are then frozen and sent to the laboratory (the National Laboratory for Environmental Testing (NLET)) for extraction of the CPAs concentrations including chl-a. The chl-a measurement method is based on the trichromatic spectrophotometry following fixation using a 90% acetone solution and centrifugation. Absorption of the residue at specified wavelengths of 663 nm, 645 nm, and 630 nm are determined. Chl-a values are calculated using SCOR/UNESCO equations in the analytical range of 0.1–100 mg·m^{-3}. The following trichromatic Equation 1 is recommended by SCOR/UNESCO to measure chl-a concentrations:

$$chl\_a = 11.64\,e_{663} - 2.16\,e_{645} + 0.10\,e_{630} \tag{1}$$

where chl-a is in μg·cm^{-3}; and e$_{663}$, e$_{645}$, e$_{630}$ are the absorbances (cm^{-1}) of light path at 663, 645, 630 nm after subtracting the 750 nm reading [25]. The reported chl-a concentration also contains phaeopigments, which are degradation products of chl-a: phaeophytin and pheophorbide [26].

Secchi disk measurement is a worldwide accepted procedure to estimate water clarity in water bodies [27]. Light propagation and Secchi disk readings decreases exponentially due to light attenuation phenomenon [28]. SDD is regularly conducted during the Lake Erie cruises. Chl-a and SDD measurement methods follow the Ocean Optics Protocols for Satellite Ocean Color Sensor Validation [29,30].

2.3. Satellite Data and Processing

Launched by the European Space Agency (ESA) on 1 March 2002, the MERIS sensor was one of the instruments operating on the Envisat polar-orbiting satellite platform. Contact was lost with Envisat on 8 April 2012, which marked the end of the mission. MERIS was primarily dedicated to ocean color studies. MERIS was a push-broom imaging spectrometer that could measure the solar radiation reflected from the Earth's surface in a high spectral and radiometric resolution (15 spectral bands across the range 390 nm to 1040 nm) with a dual spatial resolution (300 and 1200 m). MERIS scanned the Earth with global coverage every 2–3 days.

In this study, CC L2R (Version 2) MERIS reflectance full resolution images with full or partial coverage of Lake Erie between September 2004 and February 2012 were acquired through the Calvalus on-demand processing portal. The CC MERIS Level 2R product is generated using an atmospheric correction algorithm applied to the Level 1P product, which is a refined top of atmosphere radiance product with improved geolocation, calibration, equalization, smile correction, in addition to precise coastline and additional pixel characterization information (e.g., cloud, snow). The atmospheric correction procedure is based on two processors implemented in the Basic ERS & ENVISAT (A)ATSR MERIS (BEAM) software (Version 5.0): the Case II Regional (C2R) lake processor and also glint correction processor [17]. A detailed description of C2R can be found in Doerffer and Schiller (2007) [31].

The MERIS images were selected to be within a 2-day time window of *in situ* water quality measurements for the study period (2004–2012). This criterion was set to maximize the number of possible satellite and *in situ* measurements match-ups; at the same time reducing the effect of time heterogeneity of water quality parameters, assuming that these parameters would not change significantly in this time frame. Atmospherically corrected MERIS L2R reflectance values were extracted from pixels covering the geographic location of the stations. A valid pixel expression was defined that excluded all pixels with properties listed in Table 2. Spatial averaging of pixels surrounding the station could be a technical solution to increase the number of resulting match-ups, when the considered pixel is excluded due to flags [32]. However, the horizontal spatial heterogeneity of parameters over the lakes prevents the averaging analysis.

Table 2. Flags of excluded pixels.

Level 1	Level 1P	Level 2
Glint_risk	Land	AOT560_OOR (Aerosol optical thickness at 550 nm out of the training range)
Suspect	Cloud	TOA_OOR (Top of atmosphere reflectance in band 13 out of the training range)
Land_ocean	Cloud_ambigious	TOSA_OOR (Top of standard atmosphere reflectance in band 13 out of the training range)
Bright	Cloud_buffer	Solzen (Large solar zenith angle)
Coastline	Cloud_shadow	
Invalid	Snow_ice	
	MixedPixel	

2.4. Water Quality Parameters Algorithms

Semi-empirical algorithms are based on the regression between individual bands or band ratios, and the dependent variables, which are chl-a and SDD in this study. Different combinations can be considered based on the 15 MERIS spectral bands. Lakes with various optical and biological properties can produce different levels of correlation with these band combinations. This makes the use of semi-empirical algorithms a robust approach that can work on the lake of interest and within the time period that data samples were collected. The best band combinations were determined from the highest calculated Pearson correlation coefficients (R) against the logarithmic scale of *in situ* measurements of water quality parameters. It has long been known that the chl-a concentration distribution in the ocean is lognormal [33]. Also, the logarithmic function linearizes the relationship of *in situ* observations to band ratios and makes the distribution more symmetric (normal) (see Section 3.1). Considering the optical complexity of Lake Erie (see Section 3.1), red and NIR bands are required to derive chl-a concentration. Therefore, the selected band ratio for estimating chl-a was chosen among MERIS bands centered at B05:560 nm, B06:620 nm, B07:665 nm, B008:681.25 nm, B09:708.75 nm, B10:753.75 nm, B12:778.75 nm, and B13:865 nm. The band ratio for deriving SDD was selected among the visible bands: B01:412.5 nm, B02:442.5 nm, B03:490 nm, B04:510 nm, B05:560 nm, B06:620 nm, B07:665 nm, and B08:681.25 nm.

Sampling-wise, the same *in situ* measurements are repeated in time (month) over Lake Erie during the study period. Multiple measurements per variable in space or time will generally result in correlated errors and clustered data, which violate the assumptions of regression methods. Accordingly, random effect of time has to be added to the error term of the general regression models to account for measurements being made in clusters of time. The effect of time on the variation of water quality parameters can be considered to differ on a monthly basis. Therefore, *in situ* data collected repeatedly over the same month are considered to be correlated and this is the reason LME is used in this study, and month is selected as the random effect. Also, the measurements for different stations are inter-dependent. Different locations can affect each other's measurements, depending on their distance. Therefore, there is spatial dependency in *in situ* observations. To consider both random and fixed effects in the regression, a LME model approach was selected to handle the repeated measurements and also the spatial autocorrelation of *in situ* observations.

Separate LME models were developed between the logarithmic scale of *in situ* chl-a and SDD with selected individual bands or band ratios of MERIS atmospherically corrected reflectance. The models were then used to predict chl-a and SDD over Lake Erie at different times. A LME model allows the prediction to be made at the outermost level (level = 0: predictions only based on fixed effects, as it would be in a standard regression model), and the innermost level (Level ≠ 0: predictions based on estimated random and fixed effects).

2.5. Accuracy Assessment

Cross validation was performed to assess the accuracy of derived chl-a and SDD estimates. Ten rounds were repeated by splitting the *in situ* measurements into training (70%) and testing (30%) datasets. The model performance indicators were reported as the average over the iterations. The mean bias error (MBE) and the root mean square error (RMSE) were used as the model performance indicators to describe chl-a and SDD retrieval accuracies. The MBE, and RMSE statistics are defined as follows:

$$RMSE = \sqrt{mean[(x_{pr_i} - x_{obs_i})^2]} \tag{2}$$

$$MBE = mean(x_{pr_i} - x_{obs_i}) \tag{3}$$

where x_{pr_i} is the predicted value, and x_{obs_i} is the observed value of the quantity which is measured in the field. Predicted and observed values should have the same scales. Therefore, the calculation of RMSE and MBE in logarithmic scale were performed on log-transformed *in situ* observations (algorithms' output were already in the logarithmic scale). The RMSE is a comprehensive metric as it combines the mean and variance of the error distribution into a single term [34]. MBE also reveals the systematic errors [18].

Statistical analyses including: (1) finding the best band ratios using the MERIS atmospherically corrected reflectance, based on correlation with *in situ* observations; (2) development of LME models (regression method) based on selected band ratios and *in situ* observations; and (3) cross validation were performed in the R programming language (Version 3.2.1) [35].

3. Results

3.1. Lake Erie Optical Properties

Descriptive statistics of various bio-optical parameters measured in Lake Erie over the 2004–2012 period are summarized in Table 3.

Table 3. Descriptive statistics of *in situ* measurements for Lake Erie (2004–2012). N is the number of times samples were collected at stations. St. dev. is standard deviation. Chl-a is in mg·m^{-3} and total suspended matters (TSM) is in g·m^{-3}, a$_{CDOM}$ in m^{-1}, and Secchi disk depth (SDD) in m.

	N	Min	Max	Mean	St. dev.
Chl-a	190	0.20	70.10	4.27	7.82
TSM	190	0.18	50.50	5.75	8.00
a$_{CDOM}$	160	0.04	2.36	0.31	0.33
SDD	117	0.20	11.00	3.69	2.68

Figure 2 shows the contribution of different water constituents to water clarity in Lake Erie by investigating correlations between the concentrations of chl-a and TSM, and also absorption of CDOM in 440 nm (a$_{CDOM}$(440)) with the measured SDD. The graphs reveal that the concentrations of chl-a, TSM, and a$_{CDOM}$(440) are correlated with SDD over the period of measurements. TSM and to a lesser extent CDOM are important contributors to water clarity observed in Lake Erie with coefficient of determination (R^2) values of 0.67 and 0.54, respectively. There is a low correlation between *in situ* measured chl-a and TSM demonstrating that these CPAs are non-covarying and independent.

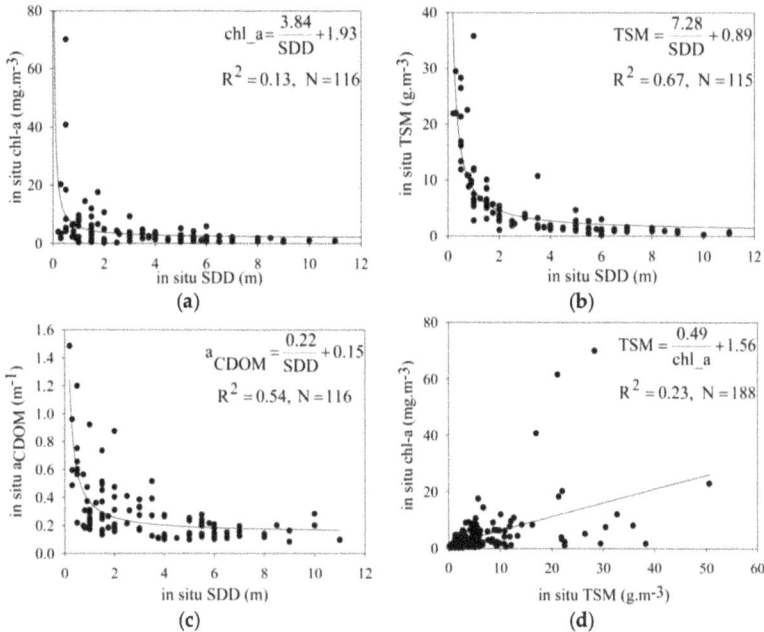

Figure 2. Relationships between *in situ* SDD and three bio-optical parameters of the water: chl-a (**a**); TSM (**b**); and $a_{CDOM}(440)$ (**c**); Relationship between *in situ* chl-a and *in situ* TSM is also shown (**d**).

The relative contribution of TSM compared to CDOM can be reduced in microtidal estuaries, or depending on the bathymetry of the lake [36]. In shallow Lake Erie, re-suspension of bottom sediments leads to higher water turbidity. Also, the Detroit and Maumee rivers contribute large sediment loads into the western basin of the lake. Kemp *et al.* (1997) identified the regions off Long Point and the mouths of these two rivers as the points of highest sedimentation rates in the lake [37]. The study of Binding *et al.* (2012) identified these zones as the highest turbid areas in the lake, confirming that TSM plays a major role in optically complex Lake Erie [1].

Based on the results presented in Table 3 and in Figure 2, Lake Erie can be classified as a typical Case II water system where other water constituents play a major and independent role in its low water clarity, besides chl-a concentration. Therefore, red and NIR reflectances are consistently the most reliable and expedient remote sensing variables in predictive algorithms for chl-a concentrations assessments in Lake Erie [11]. Figure 3 shows the probability density function (PDF) of *in situ* chl-a and SDD in Lake Erie. The PDFs for both variables demonstrate a lognormal distribution. Therefore, logarithmic transformations of chl-a concentration and SDD are used to develop the regressions.

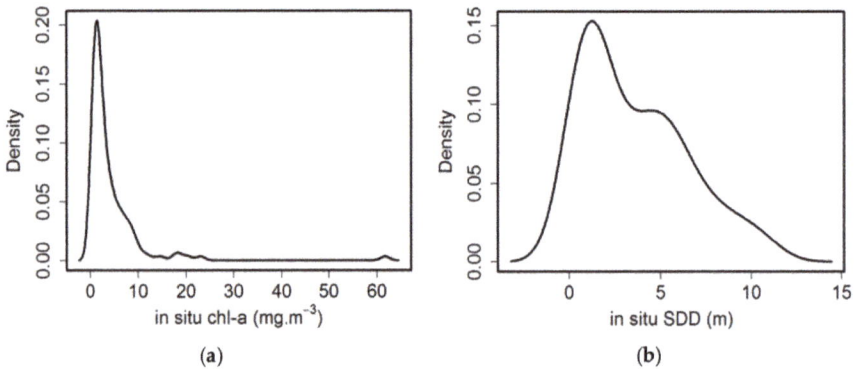

Figure 3. Probability density function of *in situ* chl-a (**a**) and SDD (**b**) in Lake Erie.

3.2. Linear Mixed Effect Models Calibration

The natural variation of the water quality parameter being measured determines the required period of concurrency between satellite overpasses and *in situ* observations [38]. In this study, satellite images were selected in a 2-day time window of *in situ* data collection. Using this criterion resulted in 16 MERIS CC L2R images being available for analysis. Applying defined flags produced 117 (60) pairs of atmospherically corrected reflectance and *in situ* chl-a (SDD) observations.

Pearson correlation (R) coefficients were calculated from MERIS bands or band ratios against the logarithmic scale of chl-a concentration and SDD measurements to select the best band or band ratios for the regression analysis of chl-a and SDD, separately. Figure 4 shows the range of correlation coefficients between the parameters of interest (chl-a and SDD) in logarithmic scale and both individual bands and band ratios of atmospherically corrected reflectance. The ratio of B07:665 nm to B09:708.75 nm has the highest correlation with chl-a concentration (R = −0.68) and the ratio of B13:865 nm to B10:753.75 nm the weakest correlation (R = −0.14). The highest and lowest correlation coefficients between SDD measurements and individual spectral bands or the ratio of them are observed for the band ratio of B06:620 nm to B04:510 nm (R = −0.90) and ratio of B04:510 nm to B02:442 nm (R = −0.16), respectively. From this analysis, band ratio B07:665 nm to B09:708.75 nm and band ratio of B06:620 nm to B04:510 nm were selected to investigate their predictive capability in estimating chl-a concentration and SDD, respectively, using LME regression models.

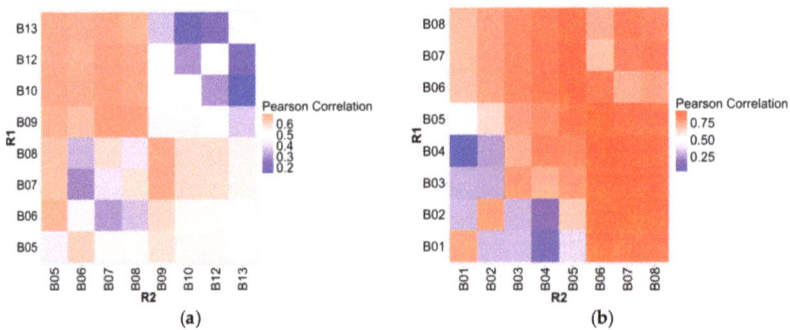

Figure 4. Correlation coefficients between MERIS atmospherically corrected reflectance ratio and *in situ* chl-a (**a**) and between MERIS atmospherically corrected reflectance ratio and *in situ* SDD (**b**) R1 and R2 represent nominator and denominator, respectively. Values along the diagonal line from lower left to top right indicate correlation with reflectance of a single wavelength.

Resulting scatterplots of selected band ratios against *in situ* data are shown in Figure 5 for chl-a concentration and SDD measurements.

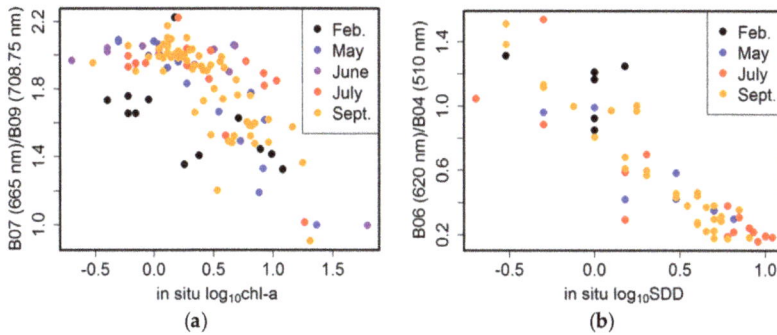

Figure 5. Scatter plots between selected MERIS atmospherically corrected reflectance ratio and *in situ* chl-a (**a**) and between selected MERIS atmospherically corrected reflectance ratio and *in situ* SDD (**b**).

The LME models were developed from the relationships between selected band ratios and the logarithmic scale of chl-a and SDD. The month of *in situ* data collection represented the random effect due to repeated measurements in time. Random slopes and intercepts were considered for each group of measurements in a single month. Goodness of fit (r-squared; R^2) for the outermost and innermost levels of prediction was 0.49 and 0.56 for chl-a and 0.78 and 0.83 for SDD, respectively. Therefore, the innermost level of predictions was used to predict chl-a concentration and SDD, with different values of slope and intercept for each month. The model developed for chl-a has average values of -1.16 (standard deviation: 3×10^{-5}, standard error: 0.1) and 2.44 (standard deviation: 0.12, standard error: 0.2) for slope and intercept, and the model developed to predict SDD has average values of -1.04 (standard deviation: 0.09, standard error: 0.08) and 0.99 (standard deviation: 0.05, standard error: 0.08) for slope and intercept. The estimated slope and intercept values for both models are significant ($p < 0.005$). The derived models are summarized using only the fixed effects as follows:

$$\log_{10}(chl\_a) = \frac{B07\ (665\ nm)}{B09\ (708.75\ nm)} \times (-1.16) + 2.44 \tag{4}$$

$$\log_{10}(SDD) = \frac{B06\ (620\ nm)}{B04\ (510\ nm)} \times (-1.04) + 0.99 \tag{5}$$

3.3. Evaluation of Linear Mixed Effect Models

The model performance indicators were derived for both models. Chl-a concentration is estimated with RMSE and MBE values of 0.31, and 0.018, respectively, in a logarithmic scale ($N = 117$; in the actual scale of chl-a: RMSE = 2.48 mg·m^{-3}, MBE = -0.58 mg·m^{-3}). SDD is predicted with RMSE value of 0.19 and MBE value equal to 0.006 in a logarithmic scale ($N = 60$; in the actual scale of SDD: RMSE = 1.40 m, MBE = -0.25 m). Comparisons between the measured and predicted $\log_{10}$chl-a and $\log_{10}$SDD using the LME models show that the values are in close agreement with paired observations, mostly evenly distributed along the 1:1 line (Figure 6). The chl-a model is, however, not sensitive enough to detect changes in low concentrations (below 0.1 in logarithmic scale, *ca.* 1 mg·m^{-3}) and, as a result, the predicted values are not showing the variations corresponding to the small amount of *in situ* chl-a concentration measurements.

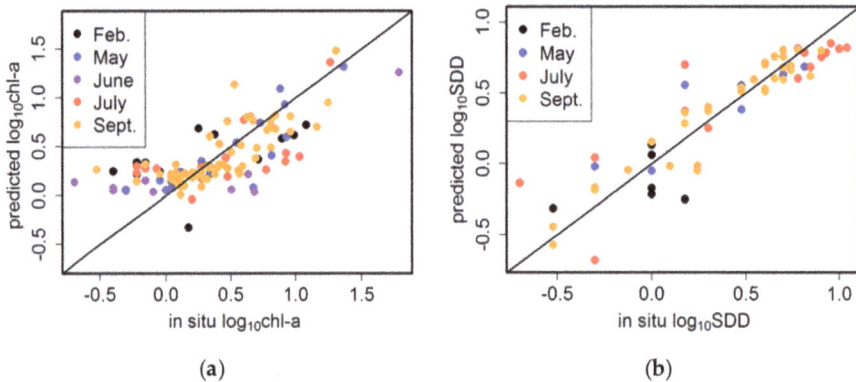

Figure 6. Comparison between MERIS estimates of chl-a (**a**) and SDD (**b**) using LME models and *in situ* measurements for Lake Erie. The solid diagonal line is the 1:1 line.

3.4. Spatial and Temporal Variations of Chl-a and SDD

The average chl-a concentration and SDD for each month between 2005 and 2011 are shown in Figures 7a and 8a. *In situ* data in 2004 were only collected in September, and the values were not estimated for the months before September. Also, the Envisat satellite stopped operating in April 2012, hence there were no full year time series estimated for 2004 and 2012. These years were therefore disregarded in the time series analysis below. The statistics related to the number of available pixels for each month is included in Figure 7a, which are the same for SDD measurements.

The three basins of Lake Erie are characterized by distinct physical, chemical, biological, and optical properties. The highest chl-a concentrations and turbidity are experienced in different times of the year for each basin. The western basin always experiences a more intense algal bloom compared to the two other basins, with the most and least concentrations in September $(6.62 \pm 4.67$ mg·m$^{-3})$–October $(4.83 \pm 3.67$ mg·m$^{-3})$ and June $(2.39 \pm 3.68$ mg·m$^{-3})$, respectively. Lake Erie's central basin experiences spring bloom in April $(2.08 \pm 0.67$ mg·m$^{-3})$ and a more intense bloom in fall (October: 2.80 ± 0.79 mg·m$^{-3})$. The eastern basin shows the least chl-a concentrations of the three basins, and its highest algal intensity occurs in summer (August, 1.78 ± 1.27 mg·m$^{-3})$. Some more specific areas of the lake are affected by prolonged intense algal bloom, including Maumee Bay, Sandusky Bay, Rondeau Bay and Long Point Bay. The highest SDD values for the full lake are estimated in July $(5.38 \pm 1.16$ m$)$, whereas the lowest SDD estimates are observed in March $(2.44 \pm 1.20$ m$)$ and October $(2.52 \pm 1.13$ m$)$. There is also a north-south gradient noticeable in both chl-a concentration and SDD in western Lake Erie.

Standard deviations of chl-a (Figure 7b) and SDD (Figure 8b) were also calculated for each month (March to October) to show variations of $\log_{10}$chl-a and $\log_{10}$SDD from the average values. Figures 7 and 8 show that the greatest variability occurs in the western basin for both $\log_{10}$Chl-a and $\log_{10}$SDD, with its largest variability in March. The least variations in chl-a concentration and SDD patterns occur in the offshore areas and eastern basin.

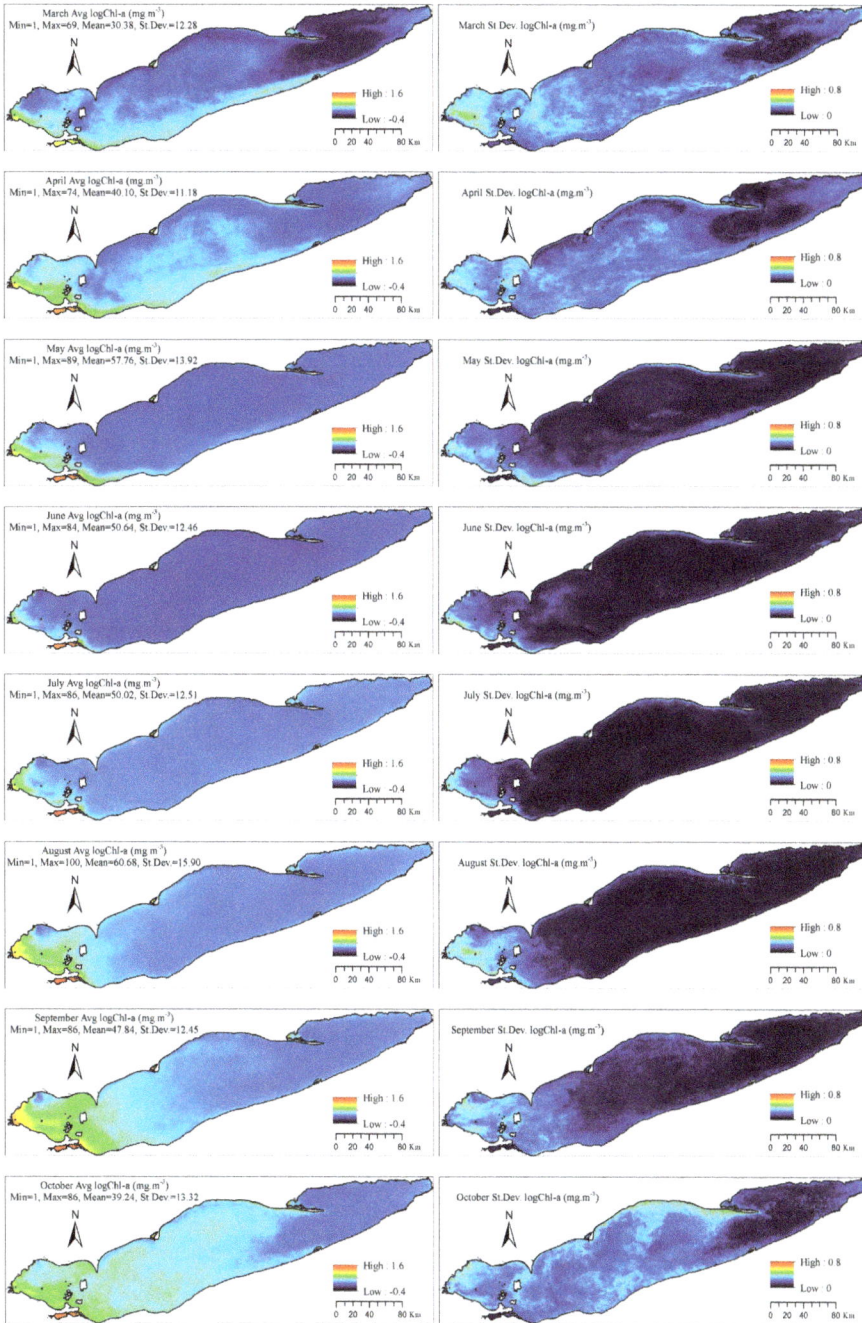

Figure 7. Chl-a average (Avg, (**a**)) and standard deviation (St.Dev., (**b**)) in $\log_{10}$ scale from March to October for the study period (2005–2011). The statistics shown on the left figures are related to the number of available pixels for each month.

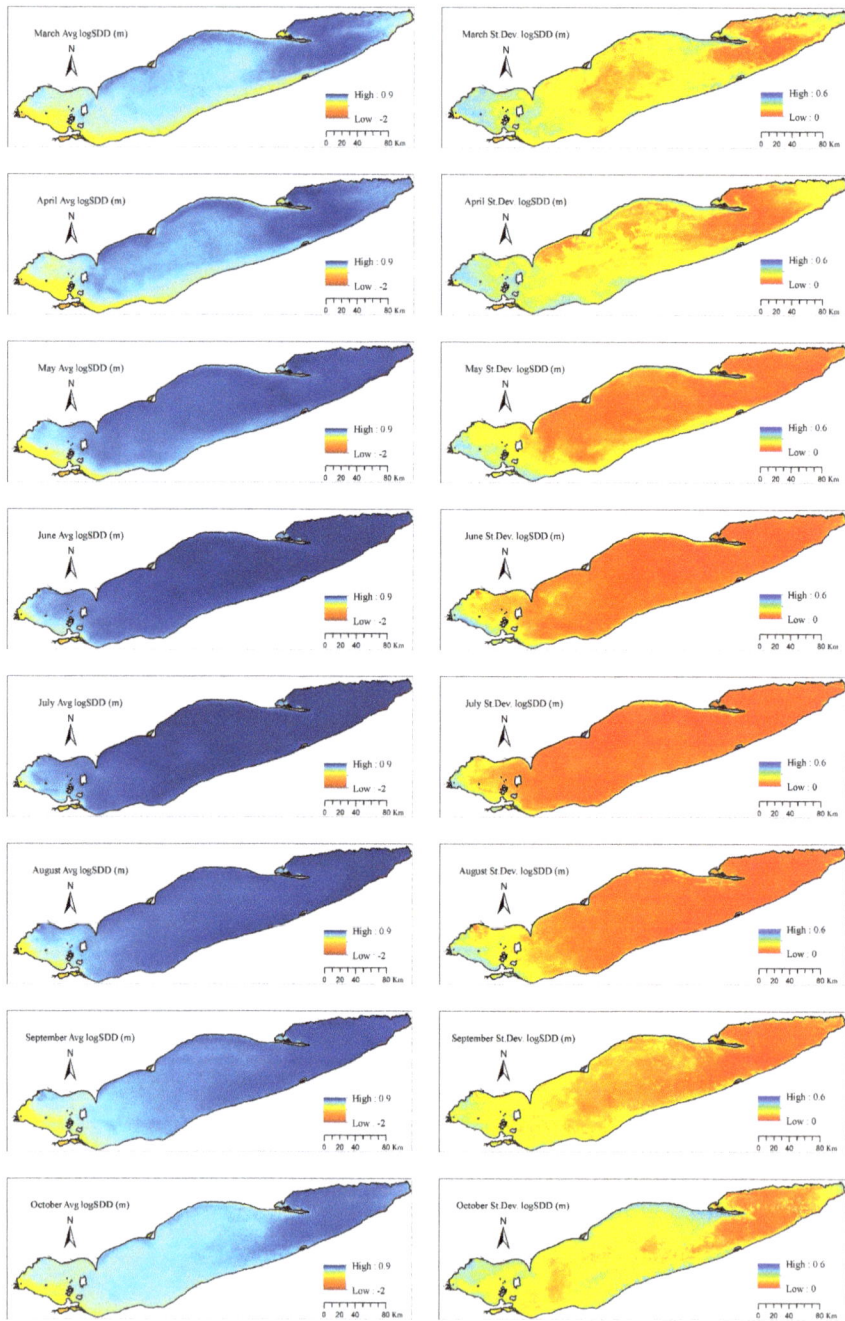

Figure 8. SDD average (Avg, (**a**)) and standard deviation (St. Dev., (**b**)) in $\log_{10}$ scale from March to October for the study period (2005–2011). The number of available pixels for each month is shown in Figure 7a.

4. Discussion

4.1. Linear Mixed Effect Model Results

Our study agrees with previous studies to the effect that Lake Erie is to be considered as a Case II water. O'Donnell *et al.* (2010) employed modern instrumentation to measure IOPs and AOPs in the western basin of Lake Erie. The study also concluded that the characterization of IOPs and AOPs supports the fact that the western basin of Lake Erie is an optically complex Case II system. Therefore, in such case, red and NIR band ratios are the most reliable predictors in regression algorithms to estimate chl-a concentration in Lake Erie. Results from the LME models calibration show that the band ratio of B07:665 nm to B09:708.75 nm is highly negatively correlated with the variations of chl-a concentration in Lake Erie. Gitelson *et al.* (2007) applied a two-band model, as the special case of a conceptual three-band model [39], to the turbid (Case II) waters of Chesapeake Bay to estimate chl-a concentration [40]. The tuning process found the ratio of 720 nm to 670 nm as the optimal spectral band ratio, with the maximal R^2 of 0.79 in a positive correlation. Water samples collected from Chesapeake Bay contained widely variable chl-a concentrations (9 to 77.4 mg·m^{-3}), when SDD ranged from 0.28 to 1.5 m. Duan *et al.* (2010) also found the band ratio of 710/670 nm to be positively correlated with chl-a concentration in eutrophic Lake Chagan with $R^2 = 0.70$. Chl-a concentration in this lake was between 6.40 and 58.21 mg·m^{-3} and SDD rarely exceeded 0.50 m [11]. Simis *et al.* (2005, 2007) correlated the absorption of chl-a to MERIS band ratio of 708.75 to 665 nm for turbid, cyanobacteria-dominated lakes in the Netherlands and Spain. Hicks *et al.* (2013) reported that the logarithmic scale of SDD measurements and logarithmic scale of Landsat 7 ETM+ band ratio of B01(0.450–0.515 nm)/B03(0.630–0.690 nm) were positively correlated with a high correlation (R = 0.82). This study was conducted for shallow lakes (ranging from 1.8 to 8.7 m depth) in the Waikato region in New Zealand, with SDD *in situ* observations varying between 0.005 and 3.78 m [41]. In our study, the highest correlation between MERIS band ratios and SDD variations in Lake Erie was estimated for the band ratio of B06:620 nm to B04:510 nm.

The selected band ratios were used to develop two separate LME models to estimate chl-a concentration and SDD in Lake Erie. The models were evaluated using the testing data in a cross validation approach. Results showed that LME model was in a high agreement with the chl-a *in situ* observations with RMSE and MBE values of 0.31, and 0.018, respectively, in a logarithmic scale. The overestimation of chl-a concentration derived from the LME model can be attributed to the contribution of TSM in the red/NIR region of the spectrum that is not necessarily correlated with chl-a concentration. In turbid waters such as Lake Erie, the signals measured in the red and NIR regions can no longer be attributed to the chl-a concentration absorption and fluorescence and water alone, while TSM can also confound the signal [1]. Although, formation of blooms in a thick surface layer can dominate the reflectance and eliminate much of the contribution of TSM to reflectance [42]. The LME model is particularly overestimating while not showing sensitivity to chl-a values lower than 0.1 in logarithmic scale (see Figure 6a). Binding *et al.* (2013) assessed the sensitivity of maximum chlorophyll index (MCI; measures a peak in red/NIR region near 708 nm relative to a baseline which is drawn between two suitable wavelengths) to mineral sediments. The modeling results in this study suggested that the sensitivity of MCI to mineral turbidity particularly increases at low chlorophyll concentrations when mineral sediments can contribute to reflectance and lead to substantial increase in the resulted MCI [42]. This study derived a strong linear relationship between *in situ* MCI and chl-a concentration in Lake Erie with R^2 value of 0.70, suggesting a minimal contamination of the MCI signal from sediments under intense surface algal blooms [42]. A blue/green band ratio algorithm was tested for western basin Lake Erie in Ali *et al.* (2014) and resulted in R^2 value of 0.46. For chl-a concentrations below 6 mg·m^{-3}, a closer 1:1 relationship with the *in situ* measurements was derived [43]. However, Witter *et al.* (2009) found systematic overestimation of low chl-a concentration in the western basin of Lake Erie applying regionally calibrated quadratic algorithms (employing blue/green bands as the predictor) on SeaWiFS imagery. Therefore, the difficulties associated with estimation of chl-a using blue/green band ratio algorithms in turbid, optically complex waters is demonstrated. In this region of spectrum, CDOM confounds the signals as well as TSM and chl-a concentrations [44].

Moore *et al.* (2014) applied a blending approach, to manage the selection between two band ratio algorithms in the blue/green and red/NIR regions, based on the optical water type classification of Lake Erie. RMSE and MBE values were 0.32, and 0.023 in logarithmic chl-a units [18]. Sá *et al.* (2015) evaluated CC chl-a products including: OC4, NN, and merged products, for the Western Iberian coast. The uncertainty estimation analysis was presented on the logarithmic scale of chl-a (0.249 < RMSE < 0.278, 0.139 < MBE < 0.200; for 3-hour time intervals) [45]. The derived LME model for SDD estimation resulted in RMSE and MBE values of 0.19, and 0.006, in logarithmic units. Wu *et al.* (2008) estimated SDD in Poyang Lake in China from two multiple regression models. The models were developed using spectral bands of Landsat TM and MODIS, separately. In both models the blue and red bands were used in the regression. The logarithmic scale of SDD was predicted with RMSE values of 0.20 and 0.37 for the models, respectively [14]. Results from our study indicate that the LME models can be used to derive the bio-optical quantities; the models provide accuracies comparable to that of other studies. A good agreement between the selected band ratios (B07/B09 for chl-a and B06/B04 for SDD) of atmospherically corrected CC L2R MERIS data and *in situ* measurements of chl-a and SDD in logarithmic scale were derived for Lake Erie for the 2004–2012 study period.

4.2. Interpretation of Spatial and Temporal Variations in Chl-a and SDD

Monthly maps of chl-a concentration and SDD for Lake Erie (Figures 7 and 8) show that Maumee Bay, Sandusky Bay, Rondeau Bay and Long Point Bay have persistent intense algal blooms. These specific areas are known to experience cyanobacteria blooms due to constant nutrient enrichment [1]. Maumee River drains a large watershed which is dominated by agricultural fields, and also is a tributary of the largest storm runoff within the Lake Erie basin [46,47]. There was also a north-south gradient in western basin for chl-a concentration and SDD estimations. This gradient can be explained by inflows from the Detroit River. The Detroit River is a major source of flows from the upper Great Lakes into Lake Erie, which carries contaminated sediments and nutrients from a highly urbanized and industrialized watershed into western Lake Erie [1,48]. However, the comparatively clearer water that is carried through this river from the upper Great Lakes can create the north-south gradient in Lake Erie [1]. Also, Dolan (1993) reported that municipal phosphorus loads from US sources have a higher magnitude compared to the Canadian ones during the period 1986–1990 [49]. Therefore, if the same trend of phosphorus loads in those years occurs during the time period of this study, the observed differences between north and south near-shore algal productivity can be enlightened [1].

Re-suspension, shoreline erosion and loading from different sources such as rivers are among the most important factors influencing SDD estimates. Wind, as the primary source of kinetic energy, affects the sediment redistribution in the water column in Lake Erie [50]. The high-energy and short-lived winter storms are a characteristic of Lake Erie wave climate that interrupts a long period of relative calm weather [47]. These strong storms usually occur before the lake freezes (in October, November, and December) and also in spring after ice break-up (March and April) [1]. However, it should be noted that the depth of the lake directly affects the amount of kinetic energy generated by wind. In other words, the re-suspension of sediment loads generated by wind in the shallower areas can be more pronounced than in the deeper areas of Erie. Comparing SDD estimates (Figure 8) with lake depths in Figure 9, it can clearly be seen that the deeper areas are relatively clearer, while the shallow areas are more turbid. The maximum depth of the western basin is only 11 m [51]. Hence, being the shallowest area, the western basin is the most vulnerable to physical processes such as re-suspension. Therefore, re-suspension of TSM can result in a prolonged constant turbidity in West Erie basin. Rivers and streams can supply suspended matters to the lake; and result in SDD reduction. The Detroit River (1.6 million tons/year) and the Maumee River (1.2–1.3 tons/year) have the major role in loading fine-grained sediments into Lake Erie [52]. Rainstorms can even strengthen the contribution of river discharges to load sediments in the lakes [53].

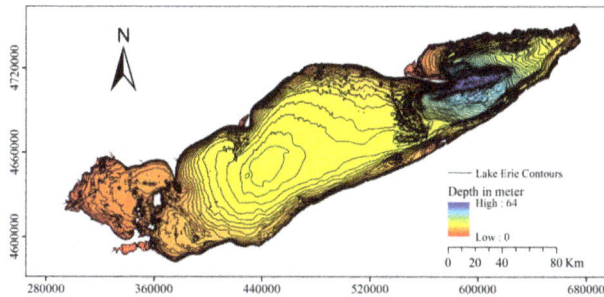

Figure 9. Bathymetry of Lake Erie (source: NOAA).

The largest variations in chl-a concentration (Figure 7) and SDD (Figure 8) occur in the western basin in March. This area of the lake is the estuary of the Detroit River and close to Maumee Bay and Sandusky Bay. Precipitation and runoff during this time of year, after the ice break-up period on the lake, cause more variations in nutrient availability and water column re-suspension effect on algal biomass and lake turbidity. The offshore areas and eastern basin have the least variations in chl-a concentration and SDD patterns. These lake sections appear to experience low fluctuations in the availability of required resources for algal bloom such as nutrients. Also, eastern basin of Lake Erie is the deepest with an average depth of 24 m (max depth = 64 m). Physical processes such as re-suspension have the least effect on the turbidity and its variations in the deep parts of the lake, as opposed to the shallow western basin.

Meteorological forcings can also have an impact on the magnitude and timing of blooms. In general, a temperature increase leads to higher rates of photosynthesis and therefore to a greater phytoplankton growth rate under adequate resource supplies such as nutrients and light. Light-limited photosynthesis rate is insensitive to temperature, whereas a light-saturated one increases with temperature [54]. The resource availability of light and nutrients can be accompanied by vertical mixing. Therefore, the seasonal cycles of stratification and wind-induced vertical mixing are the key variables that condition the growth rate of phytoplankton in the water column [54]. Stratification results in a nutrient-depleted condition at the water surface, when the upward flux of nutrients from the deep water layers is suppressed. Also, the overall impact of windiness decreases light availability in the lower depth due to re-suspension of sediments [54]. As a result, the balance found between meteorological forcings, which sometimes can have opposite effects, is one of the driving factors determining the bloom condition. Phytoplankton production is a complex function and can be controlled by resources dynamics, species composition, and predator–prey interactions in the ecosystem [54].

4.3. Limitations and Uncertainties of the Applied Linear Mixed Effect Model on MERIS

The influence of other existing particulates in a Case II water, such as CDOM and TSM, will be significantly decreased employing the chosen wavelengths to develop the chl-a LME model, as opposed to empirical blue-to-green band ratio algorithms. The absorption of CDOM is greatest in the blue region and certainly decreases exponentially with increasing wavelength, being near negligible in the NIR for the majority of the Great Lakes waters [1]. The wavelengths (665 and 708.75 nm) have a minimal sensitivity to other CPAs, but the absorption and scattering of suspended matters can still interfere within the chl-a algorithm selected wavebands. Increasing sediment loads result in the reflectance peak to move from blue to green to red in turbid waters [55]. Therefore, the semi-empirical models need to be tuned for each water body of interest characterized by different optical properties, in order to obtain the optimized wavelengths that can discriminate algal from suspended matters, and result in improved retrieval accuracy. Binding *et al.* (2012) presented a method to discriminate algal from particulate matters. The method simultaneously extract algal and suspended matters for Lake Erie from red and NIR bands of MODIS-Aqua sensor. The study resulted in estimated concentrations in close agreement with *in situ* observations with RMSE and R^2 values of 2.21 mg·m^{-3} and 0.95 for chl-a and 1.04 g·m^{-3} and 0.91 for TSM, respectively [1].

In addition, one has to consider that there is a relatively higher level of errors in computing remote sensing reflectance at longer wavelengths. Water absorption in red-NIR is strong and produces less remote sensing reflectance and, accordingly, a lower signal-to-noise ratio. This error is even higher in the case of clear waters where there is a low concentration of CDOM and TSM to produce remote sensing reflectance [1]. In Lake Erie, however, the contribution of suspended and dissolved matters in remote sensing reflectance is high enough to allow the use of the proposed wavelengths from this study and produce a strong agreement between the modeled and observed values.

Atmospheric corrections are a critical step over water bodies, since the radiance signal emerging from the water column is much less than that of land. Atmospheric corrections become even more challenging in highly turbid inland and coastal waters where the 'black pixel' assumption of negligible water-leaving radiance in the near-infrared (NIR) is no longer valid due to scattering from suspended matters [50]. Thus, typical atmospheric corrections fail and other schemes based on radiative transfer models or other approaches are required [13]. The accuracy of atmospheric correction algorithms used in different models is very important to evaluate the satellite-derived water quality products. However, in a band ratio algorithm with bands near each other, atmospheric effects are normalized [13].

In the northern part of the western basin of Lake Erie, benthic algae can be seen at the surface when the water is clear enough. Consequently, in some remote sensing methods, benthic algae can contribute to the remote sensing reflectance [1]. In the present study, however, there is no need to distinguish benthic algae from surface algae. The rapid in-water attenuation of the wavelengths selected in this study for chl-a model means that the remote sensing reflectance in these wavelengths originates mostly from the upper 30 cm of water column in the lake (depends on the diffuse attenuation coefficient) [50]. Therefore, there is no contribution of reflectance from algae at the bottom of the lake or subsurface. The estimated chl-a concentration is attributed to the surface or near surface algae even in the shallow areas or sections of the lake with clear water. The *in situ* samples to measure chl-a concentration in Lake Erie were collected from the surface mixed layer. There is a constant relationship between chl-a concentration at the surface and the one averaged over the mixed layer, as Lake Erie is shallow and exposed to strong wind-driven mixing to create a mainly mixed water column condition [50].

In situ data are required for algorithm evaluation purposes and also for parameterizing the LME models. The water quality parameters measured in the field can change at a scale smaller than that of the satellite image pixel resolution (300 m for MERIS), especially in Lake Erie due to different river inputs and wind effect. Thus, multiple measurements around stations are necessary to consider spatial heterogeneity. Also, the time lapse between satellite overpasses and *in situ* data collection may characterize a large change in the water quality parameter magnitude. The extent of these variations depends on the particular condition in the water body and defines the time window to be considered between satellite and *in situ* measurements. A time window of 3 h between satellite overpass and *in situ* data collection is recommended for open ocean waters [56]. However 2-day time window was selected for inland waters of Lake Erie to increase the number of matchups for validation and training purposes. There are also some uncertainties associated with *in situ* data collections. Over- or underestimation of chl-a concentration measurements in the field is inevitable when the collected samples contain all of the pigments, due to spectral absorption overlaps [57–60]. SDD measurements are subjective and may vary depending on the operator's ability. In shallow water bodies, disk contrast disappears at a shorter depth due to bottom reflections. Also, the disk can reach the bottom of the shallow parts of the lake without disappearing [28], which is not the case in Lake Erie as the depth measured in the survey was always larger than SDD [1].

Although MERIS is no longer active, the upcoming Sentinel-3a and b satellite missions of ESA, which will each carry the OLCI (Ocean and Land Colour Instrument) sensor (heritage of MERIS), will mark a new era in the measurement of lake water quality parameters from space. OLCI has an optimized design to minimize sun-glint and will provide 21 spectral bands compared to the 15 bands available from MERIS. Therefore the band ratio selection is between a larger numbers of bands that are improved with regards to radiometric correction.

5. Conclusions

This paper presented and assessed a remote sensing approach that utilizes spectral bands in the red and NIR portions of the spectrum to estimate chl-a concentration, and visible bands to determine SDD from MERIS images obtained over Lake Erie for the 2004–2012 period. LME models were developed based on the selected bands and *in situ* measurements. This method presents advantages over the traditional regression models that are only based on fixed effects, and that do not consider the correlation that stems from repeated measurements in space and time. Also, the LME models for chl-a and SDD are semi-empirical models that, unlike the semi-analytical models, do not require detailed knowledge of the IOPs of the CPAs in water.

Despite the limitations of remote sensing methods, they can still be considered as providing a complementary approach for the estimation of parameters related to water optical properties for many lakes over large areas and with frequent temporal coverage. Measurements at an acceptable frequency are required in order to discern potential water quality problems associated with the lake. *In situ* measurements of water quality parameters at sufficient temporal and spatial resolutions are, on the other hand, problematic due to field logistics and extended periods without sampling as a result of changes in funding priorities by agencies. Remote sensing has the potential to infer the lake bio-optical/water quality parameters overcoming these concerns.

Acknowledgments: The authors would like to thank Caren Binding (Environment Canada) for providing the optical *in situ* data of Lake Erie. Financial assistance was provided through a Discovery Grant from the Natural Sciences and Engineering Research Council of Canada (NSERC) to Claude Duguay. We are grateful to the anonymous reviewers for their constructive comments that helped improve our manuscript.

Conflicts of Interest: The authors declare no conflict of interest.

Abbreviations

The following abbreviations are used in this manuscript:

BEAM	Basic ERS & ENVISAT (A)ATSR MERIS
CC	CoastColour
CDOM	Colored Dissolved Organic Matters
Chl-a	Chlorophyll-a
CPA	Color-Producing Agent
ESA	European Space Agency
IOP	Inherent Optical Property
LME	Linear Mixed Effect
MBE	Mean Bias Error
MCI	Maximum Chlorophyll Index
MERIS	Medium Resolution Imaging Spectrometer
MODIS	Moderate-resolution Imaging Spectrometer
NLET	National Laboratory for Environmental Testing
OLCI	Ocean and Land Colour Instrument
RMSE	Root Mean Square Error
SDD	Secchi Disk Depth
TSM	Total Suspended Matters

References

1. Binding, C.E.; Greenberg, T.A.; Bukata, R.P. An analysis of MODIS-derived algal and mineral turbidity in Lake Erie. *J. Gt. Lakes Res.* **2012**, *38*, 107–116. [CrossRef]
2. Daher, S. *Lake Erie LAMP Status Report*; Environment Canada: Burlington, ON, Canada, 1999.
3. Morel, A.; Prieur, L. Analysis of variations in ocean color. *Limnol. Oceanogr.* **1977**, *22*, 709–722. [CrossRef]
4. Zhao, D.; Cai, Y.; Jiang, H.; Xu, D.; Zhang, W.; An, S. Estimation of water clarity in Taihu Lake and surrounding rivers using Landsat imagery. *Adv. Water Resour.* **2011**, *34*, 165–173. [CrossRef]
5. McCullough, I.M.; Loftin, C.S.; Sader, S.A. Combining lake and watershed characteristics with Landsat TM data for remote estimation of regional lake clarity. *Remote Sens. Environ.* **2012**, *123*, 109–115. [CrossRef]
6. Tebbs, E.J.; Remedios, J.J.; Harper, D.M. Remote sensing of chlorophyll-a as a measure of cyanobacterial biomass in Lake Bogoria, a hypertrophic, saline–alkaline, flamingo lake, using Landsat ETM+. *Remote Sens. Environ.* **2013**, *135*, 92–106. [CrossRef]
7. Binding, C.E.; Greenberg, T.A.; Bukata, R.P. Time series analysis of algal blooms in Lake of the Woods using the MERIS maximum chlorophyll index. *J. Plankton Res.* **2011**, *33*, 1847–1852. [CrossRef]
8. McCullough, I.M.; Loftin, C.S.; Sader, S.A. High-frequency remote monitoring of large lakes with MODIS 500 m imagery. *Remote Sens. Environ.* **2012**, *124*, 234–241. [CrossRef]
9. Saulquin, B.; Hamdi, A.; Gohin, F.; Populus, J.; Mangin, A.; D'Andon, O.F. Estimation of the diffuse attenuation coefficient k_{dPAR} using MERIS and application to seabed habitat mapping. *Remote Sens. Environ.* **2013**, *128*, 224–233. [CrossRef]
10. Odermatt, D.; Pomati, F.; Pitarch, J.; Carpenter, J.; Kawka, M.; Schaepman, M.; Wüest, A. MERIS observations of phytoplankton blooms in a stratified eutrophic lake. *Remote Sens. Environ.* **2012**, *126*, 232–239. [CrossRef]
11. Duan, H.; Ma, R.; Xu, J.; Zhang, Y.; Zhang, B. Comparison of different semi-empirical algorithms to estimate chlorophyll-a concentration in inland lake water. *Environ. Monit. Assess.* **2010**, *170*, 231–244. [CrossRef] [PubMed]
12. Bresciani, M.; Giardino, C. Retrospective analysis of spatial and temporal variability of chlorophyll-a in the Curonian Lagoon. *J. Coast. Conserv.* **2012**, *16*, 511–519. [CrossRef]
13. Matthews, M.W.; Bernard, S.; Winter, K. Remote sensing of cyanobacteria-dominant algal blooms and water quality parameters in Zeekoevlei, a small hypertrophic lake, using MERIS. *Remote Sens. Environ.* **2010**, *114*, 2070–2087. [CrossRef]
14. Wu, G.; Leeuw, J.D.; Skidmore, A.K.; Prins, H.H.T.; Liu, Y. Comparison of MODIS and Landsat TM5 images for mapping tempo–spatial dynamics of secchi disk depths in Poyang Lake national nature reserve, China. *Int. J. Remote Sens.* **2008**, *29*, 2183–2198. [CrossRef]
15. Kratzer, S.; Brockmann, C.; Moore, G. Using MERIS full resolution data to monitor coastal waters—A case study from Himmerfjärden, a fjord-like bay in the Northwestern Baltic Sea. *Remote Sens. Environ.* **2008**, *112*, 2284–2300. [CrossRef]
16. Pinheiro, J.; Bates, D.; DebRoy, S.; Sarkar, D. Linear and Nonlinear Mixed Effects Models. Available online ftp: //ftp.uni-bayreuth.de/pub/math/statlib/R/CRAN/doc/packages/nlme.pdf (accessed on 16 May 2016).
17. Ruescas, A.; Brockmann, C.; Stelzer, K.; Tilstone, G.H.; Beltrán-Abaunza, J.M. *DUE CoastColour Final Report, Version 1*; Brockmann Consult: Geesthacht, Germany, 2014.
18. Moore, T.S.; Dowell, M.D.; Bradt, S.; Ruiz-Verdu, A. An optical water type framework for selecting and blending retrievals from bio-optical algorithms in lakes and coastal waters. *Remote Sens. Environ.* **2014**, *143*, 97–111. [CrossRef] [PubMed]
19. Kratzer, S.; Håkansson, B.; Sahlin, C. Assessing secchi and photic zone depth in the Baltic Sea from satellite data. *AMBIO* **2003**, *32*, 577–585. [CrossRef] [PubMed]
20. Fleming-Lehtinen, V.; Laamanen, M. Long-term changes in secchi depth and the role of phytoplankton in explaining light attenuation in the Baltic Sea. *Estuar. Coast. Shelf Sci.* **2012**, *102–103*, 1–10. [CrossRef]
21. Michalak, A.M.; Anderson, E.J.; Beletsky, D.; Boland, S.; Bosch, N.S.; Bridgeman, T.B.; Chaffin, J.D.; Cho, K.; Confesor, R.; Daloglu, I.; *et al.* Record-setting algal bloom in Lake Erie caused by agricultural and meteorological trends consistent with expected future conditions. *Proc. Natl. Acad. Sci. USA* **2013**, *110*, 6448–6452. [CrossRef] [PubMed]
22. Bootsma, H.A.; Hecky, R.E. A comparative introduction to the biology and limnology of the African Great Lakes. *J. Gt. Lakes Res.* **2003**, *29*, 3–18. [CrossRef]

23. Lake Erie LaMP Work Group. *Lake Erie Lakewide Action and Management Plans (LAMPs)*; US Environmental Protection Agency: Chicago, IL, USA, 2000.

24. International Joint Commission Canada and United States. *Lake Erie Ecosystems Priority, Scientific Findings and Policy: Recommendations to Reduce Nutrient Loadings and Harmful Algal Blooms*; International Joint Commission: Washington, DC, USA, 2013.

25. UNESCO. *Determination of Photosynthetic Pigments in Sea-Water*; UNESCO Monographs on Oceanographic Methodology: Paris, France, 1966.

26. Environment Canada. *Manual of Analytical Methods*; Environmental Conservation Service—ECD; Canadian Communications Group: Toronto, ON, Canada, 1997.

27. Effler, S. Secchi disk transparency and turbidity. *J. Environ. Eng.* **1988**, *114*, 1436–1447. [CrossRef]

28. Civera, J.I.; Miró, N.L.; Breijo, E.G.; Peris, R.M. Secchi depth and water quality control: Measurement of sunlight extinction. In *Mediterranean Sea: Ecosystems, Economic Importance and Environmental Threats*; Hughes, T.B., Ed.; Nova Science: New York, NY, USA, 2013; pp. 91–114.

29. Mueller, J.L.; Bidigare, R.R.; Trees, C.; Balch, W.M.; Dore, J.; Drapeau, D.T.; Karl, D.; Heukelem, L.V.; Perl, J. Biogeochemical and bio-optical measurements and data analysis protocols. *Ocean Optics Protocols for Satellite Ocean Color Sensor Validation*; Mueller, J.L., Fargion, G.S., Mcclain, C.R., Eds.; National Aeronautical and Space Administration: Washington, DC, USA, 2003.

30. Pegau, S.; Zaneveld, J.R.V.; Mitchell, B.G.; Mueller, J.L.; Kahru, M.; Wieland, J.; Stramska, M. Inherent optical properties: Instruments, characterizations, field measurements and data analysis protocols. *Ocean Optics Protocols for Satellite Ocean Color Sensor Validation*; Mueller, J.L., Fargion, G.S., McClain, C.R., Eds.; National Aeronautical and Space Administration: Washington, DC, USA, 2002.

31. Doerffer, R.; Schiller, H. The MERIS Case 2 water algorithm. *Int. J. Remote Sens.* **2007**, *28*, 517–535. [CrossRef]

32. Heim, B.; Abramova, E.; Doerffer, R.; Günther, F.; Hölemann, J.; Kraberg, A.; Lantuit, H.; Loginova, A.; Martynov, F.; Overduin, P.P.; *et al.* Ocean colour remote sensing in the Southern Laptev Sea: Evaluation and applications. *Biogeosciences* **2014**, *11*, 4191–4210. [CrossRef]

33. Campbell, J.W. The lognormal distribution as a model for bio-optical variability in the sea. *J. Geophys. Res.* **1995**, *100*, 13237–13254. [CrossRef]

34. Szeto, M.; Werdell, P.J.; Moore, T.S.; Campbell, J.W. Are the world's oceans optically different? *J. Geophys. Res.* **2011**, *116*. [CrossRef]

35. R Development Core Team. *R: A Language and Environment for Statistical Computing*; R Foundation for Statistical Computing: Vienna, Austria, 2015.

36. Branco, A.B.; Kremer, J.N. The relative importance of chlorophyll and colored dissolved organic matter (CDOM) to the prediction of the diffuse attenuation coefficient in shallow estuaries. *Estuaries* **2005**, *28*, 643–652. [CrossRef]

37. Kemp, A.L.W.; MacInnis, G.A.; Harper, N.S. Sedimentation rates and a revised sediment budget for Lake Erie. *Gt. Lakes Res.* **1977**, *3*, 221–233. [CrossRef]

38. Shi, K.; Zhang, Y.; Liu, X.; Wang, M.; Qin, B. Remote sensing of diffuse attenuation coefficient of photosynthetically active radiation in Lake Taihu using MERIS data. *Remote Sens. Environ.* **2014**, *140*, 365–377. [CrossRef]

39. Dall'Olmo, G.; Gitelson, A.A. Effect of bio-optical parameter variability and uncertainties in reflectance measurements on the remote estimation of chlorophyll-a concentration in turbid productive waters: Modeling results. *Appl. Opt.* **2006**, *44*, 412–422. [CrossRef]

40. Gitelson, A.; Schalles, J.; Hladik, C. Remote chlorophyll-a retrieval in turbid, productive estuaries: Chesapeake Bay case study. *Remote Sens. Environ.* **2007**, *109*, 464–472. [CrossRef]

41. Hicks, B.J.; Stichbury, G.A.; Brabyn, L.K.; Allan, M.G.; Ashraf, S. Hindcasting water clarity from Landsat satellite images of unmonitored shallow lakes in the Waikato region, New Zealand. *Environ. Monit. Assess.* **2013**, *185*, 7245–7261. [CrossRef] [PubMed]

42. Binding, C.E.; Greenberg, T.A.; Bukata, R.P. The MERIS maximum chlorophyll index; its merits and limitations for inland water algal bloom monitoring. *J. Gt. Lakes Res.* **2013**, *39*, 100–107. [CrossRef]

43. Ali, K.A.; Witter, D.; Ortiz, J.D. Application of empirical and semi-analytical algorithms to MERIS data for estimating chlorophyll-a in Case 2 waters of Lake Erie. *Environ. Earth Sci.* **2014**, *71*, 4209–4220. [CrossRef]

44. Witter, D.L.; Ortiz, J.D.; Palm, S.; Heath, R.T.; Budd, J.W. Assessing the application of SeaWiFS ocean color algorithms to Lake Erie. *J. Gt. Lakes Res.* **2009**, *35*, 361–370. [CrossRef]

45. Sá, C.; D'Alimonte, D.; Brito, A.C.; Kajiyama, T.; Mendes, C.R.; Vitorino, J.; Oliveira, P.B.; da Silva, J.C.B.; Brotas, V. Validation of standard and alternative satellite ocean-color chlorophyll products off Western Iberia. *Remote Sens. Environ.* **2015**, *168*, 403–419. [CrossRef]

46. Bolsenga, S.J.; Herdendorf, C.E. *Lake Erie and Lake St. Clair Handbook*; Wayne State University Press: Detroit, MI, USA, 1993.

47. Morang, A.; Mohr, M.C.; Forgette, C.M. Longshore sediment movement and supply along the U.S. Shoreline of Lake Erie. *J. Coast. Res.* **2011**, *27*, 619–635. [CrossRef]

48. Marvin, C.H.; Charlton, M.N.; Reiner, E.J.; Kolic, T.; MacPherson, K.; Stern, G.A.; Braekevelt, E.; Estenik, J.F.; Thiessen, L.; Painter, S. Surficial sediment contamination in Lakes Erie and Ontario: A comparative analysis. *J. Gt. Lakes Res.* **2002**, *28*, 437–450. [CrossRef]

49. Dolan, D.M. Point source loadings of phosphorus to Lake Erie: 1986–1990. *J. Gt. Lakes Res.* **1993**, *19*, 212–223. [CrossRef]

50. Binding, C.E.; Jerome, J.H.; Bukata, R.P.; Booty, W.G. Suspended particulate matter in Lake Erie derived from MODIS aquatic colour imagery. *Int. J. Remote Sens.* **2010**, *31*, 5239–5255. [CrossRef]

51. Ortiz, J.D.; Witter, D.L.; Ali, K.A.; Fela, N.; Duff, M.; Mills, L. Evaluating multiple colour-producing agents in Case II waters from Lake Erie. *Int. J. Remote Sens.* **2013**, *34*, 8854–8880. [CrossRef]

52. Carter, C.H. *Sediment–Load Measurements along the United States Shore of Lake Erie*; Ohio Division of Geological Survey: Columbus, OH, USA, 1977.

53. Zhang, Y.; Shi, K.; Zhou, Y.; Liu, X.; Qin, B. Monitoring the river plume induced by heavy rainfall events in large, shallow, Lake Taihu using MODIS 250 imagery. *Remote Sens. Environ.* **2016**, *173*, 109–121. [CrossRef]

54. Winder, M.; Sommer, U. Phytoplankton response to a changing climate. *Hydrobiologia* **2012**, *698*, 5–16. [CrossRef]

55. Bukata, R.P.; Jerome, J.H.; Kondratyev, A.S.; Pozdnyakov, D.V. *Optical Properties and Remote Sensing of Inland and Coastal Waters*; CRC Press: Boca Raton, FL, USA, 1995.

56. Wardell, P.J.; Bailey, S.W. An improved *in-situ* bio-optical data set for ocean color algorithm developement and satellite data product validation. *Remote Sens. Environ.* **2005**, *98*, 122–140. [CrossRef]

57. Arar, J.E. *Determination of Chlorophylls-a and b and Identification of Other Pigments of Interest in Marine and Freshwater Algae Using High Performance Liquid Chromatography with Visible Wavelength Detection*; EPA: Cincinnati, OH, USA, 1997; pp. 1–20.

58. Arar, J.E. *Determination of Chlorophylls-a, b, c 1c and Pheopigments in Marine and Freshwater Algae by Visible Spectrophotometry*; EPA: Cincinnati, OH, USA, 1997; pp. 1–26.

59. Arar, J.E.; Collins, G.B. *Determination of Chlorophyll-a and Pheophytin a in Marine and Freshwater Algae by Fluorescence*; EPA: Cincinnati, OH, USA, 1997; pp. 1–22.

60. DosSantos, A.C.A.; Calijuri, M.C.; Moraes, E.M.; Adorno, M.A.T.; Falco, P.B.; Carvalho, D.P.; Deberdt, G.L.B.; Benassi, S.F. Comparison of three methods for chlorophyll determination: Spectrophotometry and fluorimetry in samples containing pigment mixtures and spectrophotometry in samples with separate pigments through High Performance Liquid Chromatography. *Acta Limnol. Bras.* **2003**, *15*, 7–18.

remote sensing

MDPI

Article

Remote Sensing of Particle Cross-Sectional Area in the Bohai Sea and Yellow Sea: Algorithm Development and Application Implications

Shengqiang Wang [1,2], Yu Huan [1], Zhongfeng Qiu [1,2,*], Deyong Sun [1,2], Hailong Zhang [1], Lufei Zheng [1] and Cong Xiao [1]

[1]   School of Marine Sciences, Nanjing University of Information Science & Technology (NUIST), Nanjing 210044, Jiangsu, China; shengqiang.wang@nuist.edu.cn (S.W.); huanyu0624@foxmail.com (Y.H.); sundeyong1984@163.com (D.S.); zhanghailong.1205@163.com (H.Z.); lufei.z@foxmail.com (L.Z.); rafaxiaocong@163.com (C.X.)

[2]   Jiangsu Research Center for Ocean Survey Technology, NUIST, Nanjing 210044, Jiangsu, China

*   Correspondence: zhongfeng.qiu@nuist.edu.cn; Tel.: +86-25-5869-5692

Academic Editors: Claudia Giardino, Linhai Li, Yunlin Zhang, Xiaofeng Li and Prasad S. Thenkabail
Received: 26 July 2016; Accepted: 8 October 2016; Published: 22 October 2016

Abstract: Suspended particles in waters play an important role in determination of optical properties and ocean color remote sensing. To link suspended particles to their optical properties and thereby remote sensing reflectance ($R_{rs}(\lambda)$), cross-sectional area is a key factor. Till now, there is still a lack of methodologies for derivation of the particle cross-sectional area concentration (AC) from satellite measurements, which consequently limits potential applications of AC. In this study, we investigated the relationship between AC and $R_{rs}(\lambda)$ based on field measurements in the Bohai Sea (BS) and Yellow Sea (YS). Our analysis confirmed the strong dependence of $R_{rs}(\lambda)$ on AC and that such dependence is stronger than on mass concentration. Subsequently, a remote sensing algorithm that uses the slope of $R_{rs}(\lambda)$ between 490 and 555 nm was developed for retrieval of AC from satellite measurements of the Geostationary Ocean Color Imager (GOCI). In situ evaluations show that the algorithm displays good performance for deriving AC and is robust to uncertainties in $R_{rs}(\lambda)$. When the algorithm was applied to satellite data, it performed well, with a coefficient of determination of 0.700, a root mean squared error of 2.126 m^{-1} and a mean absolute percentage error of 40.7%, and it yielded generally reasonable spatial and temporal distributions of AC in the BS and YS. The satellite-derived AC using our algorithm may offer useful information for modeling the inherent optical properties of suspended particles, deriving the water transparency, estimating the particle composition and possibly improving particle mass concentration estimations in future.

Keywords: particle cross-sectional area; remote sensing; retrieval model; the Bohai Sea and Yellow Sea; Geostationary Ocean Color Imager (GOCI)

1. Introduction

Suspended particles are an important type of matter in coastal and oceanic waters because of their significant roles in marine physical and biogeochemical processes [1,2]. More specifically, sunlight entering the near surface of the sea is attenuated by suspended particles due to absorption and scattering. This process significantly impacts the depth of penetration of sunlight into the sea, which regulates marine primary production [3,4]. At the same time, certain fractions of sunlight are scattered backward by suspended particles to emerge from the sea surface. This backscattered light is the essential foundation of ocean color remote sensing for monitoring of suspended particles, water transparency, turbidity, etc., using satellite measurements [5]. Thus, knowledge of the properties of suspended particles related to light attenuation is fundamentally important for improved understanding of water radiative transfer, aquatic photosynthesis processes and ocean color algorithms.

Theoretically, when a photon enters the sea, the probability of the photon interacting with a particle depends on its cross-sectional area [6,7]. The optical properties (e.g., light attenuation, scattering and backscattering coefficients) are therefore directly related to the particle cross-sectional area concentration (AC) [7,8]. Taking the backscattering coefficient $b_{bp}(\lambda)$ as an example, based on the assumption of a sphere for particle, the relationship between $b_{bp}(\lambda)$ and AC is expressed as follows [9].

$$b_{bp}(\lambda) = Q_{bbe}(\lambda)AC \tag{1}$$

where $Q_{bbe}(\lambda)$ indicates the backscattering efficiency. Though AC is known to be important for understanding optical properties of particles, field measurement of AC has been a technical difficulty in the past. Therefore, mass concentration of total suspended matters (TSM) which can be easily measured using filtration of water samples in the field is often used by researchers to understand optical properties [4,6]. However, note that there is no exact theoretical basis between TSM and optical properties. The observed relationship between TSM and optical properties is actually based on the correlation of TSM with AC, but such correlation is not stable due to additional impacts of the size distribution and density of particles [7,8].

In the past decade, field instruments such as digital cameras and laser diffraction (e.g., laser in situ scattering and transmissometry device, LISST) have gradually made it easier to measure the AC of suspended particles [10–12]. For instance, volume concentrations of suspended particles in a given size bin can be derived through inversion of the near-forward scattering measured by a LISST instrument. Based on information on the distribution of particle volume with size, the AC of suspended particles can be calculated by assuming a spherical shape and material homogeneity for the particles. Due to the available measurements of AC, increasing numbers of laboratory and field studies have been conducted in recent years to investigate the impacts of AC on the optical properties of particles under various conditions (e.g., flocculation and phytoplankton bloom) [8,9,11–15]. These studies have clearly showed that the optical properties of suspended particles strongly depend on AC, and this dependence is stronger than TSM. For instance, Hatcher et al. [11] observed that for suspension of aggregates (>10 μm), although a drop in TSM was noted and caused poor correlation between $b_{bp}(\lambda)$ and TSM, the linear relationship between $b_{bp}(\lambda)$ and AC still performed well. In two typical shallow and semi-enclosed seas, namely, the Bohai Sea (BS) and the Yellow Sea (YS), Wang et al. [15] also found that AC is strongly related to both attenuation coefficient and $b_{bp}(\lambda)$ (coefficient of determination $R^2 > 0.90$). Meanwhile, LISST measurements of volume concentrations of suspended particles in a given size also enable determining more biogeochemical information about particles, such as particle size distribution and density; and thus to better understand optical properties of particles [9,15,16].

As the improvement in knowledge of controlling factors (e.g., AC, particle size distribution and density) of optical properties of particles, some studies have attempted to derive more plentiful biogeochemical properties of particles from satellite observations in addition to TSM, such as particle size distribution [17]. For AC, several studies have also examined its relationship with remote sensing reflectance ($R_{rs}(\lambda)$), and compared with that between $R_{rs}(\lambda)$ and TSM [7,8,14,18]. For instance, during a laboratory experiment, Bale et al. [18] observed that when the size of the suspended particles varied from large to fine, $R_{rs}(\lambda)$ increased 10-fold despite the constant value of TSM, while such increases in $R_{rs}(\lambda)$ were strongly correlated with AC, showing a R^2 value of 0.897. Meanwhile, based on field observations, Mikkelsen [8] showed that reflectance spectra are highly dependent on AC rather than TSM. Similarly, using in situ measurements in shelf seas and estuaries, Bowers & Braithwaite [7] claimed that reflectance is more closely linked to the AC of suspended mineral particles (mainly in flocculation conditions) than they are to TSM, with R^2 values of 0.82 and 0.53 with AC and TSM, respectively. At this stage, we must note that although previous studies have examined the relationship between $R_{rs}(\lambda)$ with AC, and provided clear evidence indicating a closer relationship between $R_{rs}(\lambda)$ with AC than TSM. There is still a lack of methodologies for successful derivation of AC from satellite measurements. This gap consequently limits potential satellite applications of AC, such as modeling of the inherent optical properties of particles, detecting water turbidity, and retrieval of particle composition, among others.

In this study, we mainly focus on the development of the AC estimation model. Based on an in situ dataset composed of AC of particles derived from LISST measurements and the corresponding $R_{rs}(\lambda)$ collected during three cruises in the BS and YS, we firstly investigated the relationships between AC and $R_{rs}(\lambda)$ and evaluated the capability of several spectral indicators of $R_{rs}(\lambda)$ for derivation of AC. Consequently, a remote sensing algorithm was proposed for mapping AC in the BS and YS from ocean color data, and potential satellite applications of AC were further discussed.

2. Materials and Methods

2.1. Study Area and Sampling

The data used in this study were collected during three cruises in May 2014, November 2014 and August 2015 in the BS and the YS aboard the R/V *Dongfanghong* 2 (Figure 1). The BS is known as a shallow (average depth of 18 m) and highly productive inner sea of China [19]. A notably large amount of mineral-rich particles carried by the Yellow River flows into the BS. Additionally, in the past several decades, the BS has been strongly impacted by human activities because of the rapid proliferation of surrounding industries, agriculture, aquaculture and domestic sewage [19]. The YS is one of the largest marginal seas in the western North Pacific, with an area of 380,000 km^2 and an average water depth of 44 m. The YS is semi-enclosed by the contiguous lands of China in the west and Korea in the east and is connected with the BS through the Bohai Strait. Because of freshwater discharge from inland rivers carrying abundant nutrients and sediments and water mixing driven by winds and tides, water properties in the YS are regionally and seasonally varied over a wide dynamic range [20,21]. In addition, the YS is influenced by industrial pollution, agricultural runoff, and domestic sewage [21].

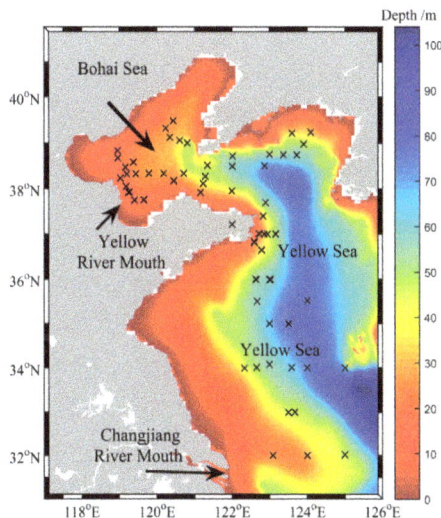

Figure 1. Locations of sampling stations in the Bohai Sea and the Yellow Sea during three cruises in 2014 and 2015. Colors indicate water depth.

At each sampling station, a profiling package was used to measure various water properties. This package includes a Sequoia Scientific LISST-100X (type C) for determining the particle size distribution (PSD), a Seabird SBE911P conductivity-temperature-depth (CTD) profiler for measuring the hydrological characteristics of the water column, and other optical instruments (i.e., a WET labs AC-S and a HOBI Labs Hydroscat-6) used to determine the inherent optical properties of water. The profiling package was immersed in surface waters for several minutes to equilibrate the sensor temperature and the seawater temperature, and was subsequently slowly (0.2 m·s^{-1}) lowered from the surface to 2–3 m above the bottom to measure a vertical profile. We used only downcast measurements for data analysis to avoid perturbations of the package on the water column. At the daytime stations, a Satlantic Hyper-Profiler II radiometer was used to derive $R_{rs}(\lambda)$. Similarly, this instrument was also stabilized at the surface water for several minutes to reduce the influence of temperature differences between sensor and water. This process also allowed the instrument to drift further away from the ship to avoid the ship's shadow. Using a hand-controlled cable, we lowered the instrument from the surface in vertical free-fall mode until it reached the euphotic layer, which is defined as the depth with 1% surface photosynthetically active radiation [22]. The downcast measurements were used in determining of $R_{rs}(\lambda)$. In addition, surface water samples for measurement of TSM concentrations were simultaneously collected using 12-liter Niskin bottles mounted on a CTD/rosette system. In total, 86 surface samples with good quality were obtained that incorporated all measurements from the profiling package, Hyper-Profiler II and TSM concentrations.

2.2. In Situ Data Measurements

The cross-sectional area concentration AC values were calculated from field measurements of the volume scattering function using a LISST-100X Type-C particle size analyzer (Sequioa Scientific Inc., Bellevue, WA, USA). In brief, this instrument records the light scattering of particles at a wavelength of 670 nm at 32 logarithmically spaced scattering angles in the near-forward direction [10]. The volume concentrations with mean diameters of 32 particle size bins (range: 2.5–500 μm) are determined through inversion of the angular forward-scattering pattern based on Mie theory calculations [23]. By assuming the particles as spherical, the cross-sectional area concentration of particles of each size class was calculated as

$$AC_i = \frac{3}{2D_i} VC_i, \tag{2}$$

where AC_i, D_i and VC_i denote the cross-sectional area concentration, mean diameter and volume concentration of particles in size bin i, respectively. As reported in previous studies [9,24], measurements from the LISST instrument usually show significant instability at the smallest and largest size ranges. These instabilities are possibly caused by the presence of particles that are smaller and larger than the measured size range and that of the smallest size ranges might also be associated with stray light [25]. Therefore, the data from the first and the last size bins, which respectively correspond to the smallest and the largest size ranges, were excluded from our analysis. The total cross-sectional area concentration AC was subsequently obtained as

$$AC = \sum_{i=2}^{31} AC_i, \tag{3}$$

where i ranges from 2 to 31 with corresponding values of 3.2 μm $\leq D_i \leq$ 390 μm.

Remote sensing reflectance $R_{rs}(\lambda)$ values were determined using measurements from the Hyper-Profiler II (Satlantic Inc., Halifax, NS, Canada). This instrument includes three inter-calibrated radiometers, i.e., a deck sensor that measures the above-water surface downwelling irradiance $(E_d(\lambda, 0^+))$ and two underwater sensors that respectively measure the vertical profiles of downwelling irradiance $(E_d(\lambda, z))$ and upwelling radiance $(L_u(\lambda, z))$. These measured spectra range from 349 to 804 nm with a mean bandwidth of approximately 3.3 nm. The radiometric measurements were processed to perform calibration, data filtering, binning and interpolation using the manufacturer-provided software Prosoft 7.7.16 [26]. Data with tilt angles >5° or/and downward velocity >0.5 m·s^{-1} were excluded during the processing, and $R_{rs}(\lambda)$ was obtained as

$$R_{rs}(\lambda) = \frac{L_w(\lambda)}{E_d(\lambda, 0^+)}, \tag{4}$$

where $L_w(\lambda)$ is the water-leaving radiance determined from the profile of $L_u(\lambda, z)$ at the upper layer, as described by Rudorff et al. [26].

In this study, we intended to investigate the utility of ocean color measurements for retrieval of AC. Recently, the Geostationary Ocean Color Imager (GOCI), the world's first geostationary ocean color satellite sensor, has become increasingly popular due to its high temporal resolution. GOCI observes the Northeast Asian region 8 times within a day from 8:15 to 15:15 local time (temporal resolution of 1 h). In the visible region, GOCI contains 6 bands with central wavelengths of 412, 443, 490, 555, 660 and 680 nm. Therefore, in this study, we resampled the in situ hyper-spectral $R_{rs}(\lambda)$ to the GOCI bands based on the spectral response function.

TSM concentrations were determined gravimetrically by filtration of 0.5–2 L of seawater onto 47-mm Whatman GF/F glass fiber filters [27]. These filters were pre-weighed on a balance with an accuracy of 0.01 mg and stored in a desiccator before the cruise. Water samples were filtered onto the pre-weighed filters under low vacuum pressure (<0.01 MPa). After filtration, the filters were rinsed three times using 50 mL MilliQ water to remove immersed salt and immediately frozen at $-20\ ^\circ$C before drying at 105° for 4 h in the laboratory. The filters were reweighed on a balance to obtain the TSM concentrations until the difference between the last two measurements was within 0.02 mg. The TSM concentrations were corrected for the sea salt plus water of hydration retention using the method proposed by Stavn et al. [28]. The salinity data used for the correction was from the CTD measurements.

2.3. Satellite Data

Level 1B products of top-of-atmosphere radiance of GOCI (with 8 images per day) in 2015 and during our cruise period in 2014 were obtained from the Korea Ocean Satellite Center (KOSC). These products were cropped to the BS and the YS and processed to Level 2 data for $R_{rs}(\lambda)$ using the GOCI Data Processing Software (GDPS, version 1.3) set to the default parameters and standard atmospheric correction [29]. Subsequently, to derive the monthly variations of AC in the BS and the YS, we produced monthly $R_{rs}(\lambda)$ in 2015 from the hourly Level 2 $R_{rs}(\lambda)$ using the combined time-series tool of GDPS.

2.4. Accuracy Assessment

The performance of the AC estimation model proposed in this study was assessed with comparison between the $R_{rs}(\lambda)$-retrieved AC and those measured values [30]. The coefficient of determination R^2 calculated in the log-log space was used to show how well the derived AC agreed with the measured values. Two other indicators, the root mean squared error (RMSE) and the mean absolute percentage error (MAPE), were also used to demonstrate the scatter between the measured and estimated values, which were calculated as follows [31]:

$$\text{RMSE} = \sqrt{\frac{1}{N}\sum_{i=1}^{N}\left[y_{est,i} - y_{mea,i}\right]^2} \tag{5}$$

$$\text{MAPE} = \frac{1}{N}\sum_{i=1}^{N}\left|\frac{y_{est,i} - y_{mea,i}}{y_{mea,i}}\right| \times 100\% \tag{6}$$

where $y_{est,i}$ and $y_{mea,i}$ represent the estimated and measured AC for the ith sample, respectively, and N is the total number of samples.

3. Results

3.1. Data Distributions

The particle concentrations showed wide dynamic ranges and significant variability (Table 1). The TSM varied from 0.53 to 43.88 $g \cdot m^{-3}$ with a mean value of 7.60 $g \cdot m^{-3}$ (standard variation of 7.25 $g \cdot m^{-3}$). The variation coefficient of TSM was 95.4%. Compared with TSM, the AC showed more significant variation with a range from 0.12 to 13.00 m^{-1}. The values for the mean, standard variation and variation coefficient of AC were 1.74 m^{-1}, 2.14 m^{-1} and 123.0%, respectively. In general, the AC was log-normally distributed, as shown in Figure 2.

Correspondingly, the remote sensing reflectance $R_{rs}(\lambda)$ values of all samples showed large variations in both magnitude and spectral shape, although the spectral shapes generally appeared similar (Table 1 and Figure 3). The spectral variability in magnitude and shape of $R_{rs}(\lambda)$ were generally in accord with those reported in previous studies in coastal regions [32,33]. For the majority of samples, the $R_{rs}(\lambda)$ spectra displayed peaks between 550 and 600 nm, whereas the spectral peaks of some samples mainly collected from offshore waters of the Bohai Strait and YS during late summer and autumn were shifted to around 490 nm. The in situ hyper-spectral $R_{rs}(\lambda)$ were resampled to the GOCI 6 bands according their spectral response functions. The statistics of synthetic GOCI $R_{rs}(\lambda)$ at 6 bands were summarized in Table 1.

Table 1. Statistics of concentrations of total suspended matters (TSM) and cross-sectional area (AC) and remote sensing reflectance $R_{rs}(\lambda)$ from all stations.

Variable	Band (nm)	Min	Max	Mean	S.D.	CV (%)
TSM ($g \cdot m^{-3}$)		0.53	43.88	7.60	7.25	95.4
AC (m^{-1})		0.12	13.00	1.74	2.14	123.0
$R_{rs}(\lambda)$ (sr^{-1})	412	0.0006	0.0111	0.0042	0.0030	70.7
	443	0.0008	0.0152	0.0060	0.0045	75.2
	490	0.0012	0.0217	0.0086	0.0066	76.0
	555	0.0010	0.0333	0.0110	0.0094	85.7
	660	0.0001	0.0239	0.0045	0.0057	126.1
	680	0.0001	0.0220	0.0040	0.0051	127.8

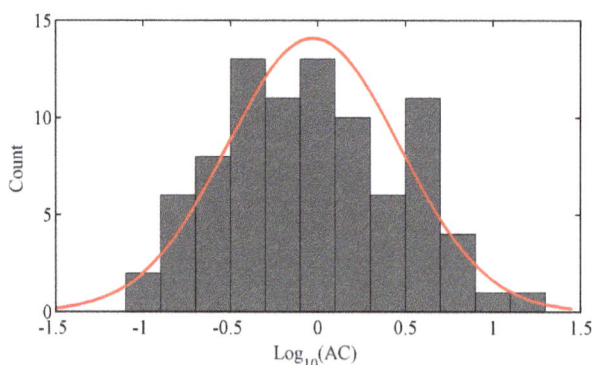

Figure 2. Frequency distribution of cross-sectional area concentration (AC) of particles. The red line represents a log-normally distributed fitting curve.

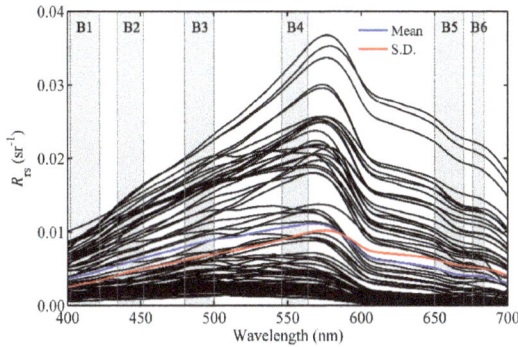

Figure 3. Remote sensing reflectance $R_{rs}(\lambda)$ from in situ measurements. The values of mean and standard deviation (S.D.) are indicated using blue and red lines, respectively. GOCI 6 channels are indicated using gray bars.

3.2. Development and Performance of AC Retrieval Model

Before the model development, we analyzed the relationship between $R_{rs}(\lambda)$ and AC, and compared with those between $R_{rs}(\lambda)$ and TSM. Considering the log-normal distribution of AC (Figure 2), the correlations of log-transformed AC ($\log_{10}(AC)$) with $R_{rs}(\lambda)$ were analyzed for each band of GOCI. The highest correlation between $\log_{10}(AC)$ and $R_{rs}(\lambda)$ was found at 555 nm with a value of 0.85 (Figure 4). The second highest correlations occurred at 490 and 660 nm with values of 0.77 and 0.77, respectively. We also analyzed the correlation between $\log_{10}(TSM)$ and $R_{rs}(\lambda)$. Although the variation pattern of the correlations along the wavelength between $\log_{10}(TSM)$ and $R_{rs}(\lambda)$ was similar to that between $\log_{10}(AC)$ and $R_{rs}(\lambda)$, the values of the correlations between $\log_{10}(TSM)$ and $R_{rs}(\lambda)$ were much lower, with the highest correlation of only 0.62 at 555 nm. The causes of these differences are discussed later in Section 4.2.

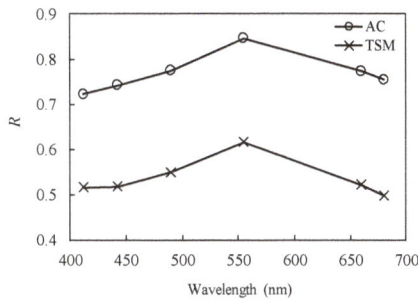

Figure 4. Correlation coefficient (R) of $\log_{10}(AC)$ and $\log_{10}(TSM)$ with remote sensing reflectance $R_{rs}(\lambda)$ at six bands of GOCI.

Empirical models have been widely used to derive water quality parameters [34,35] because of their simple and easy implementation. Considering the log-normal distribution and the above correlation analysis, we designed an empirical model for retrieval of AC from $R_{rs}(\lambda)$, the general form of which can be expressed as

$$\log_{10}(AC) = k_1 X^2 + k_2 X + k_0 \tag{7}$$

where X represents the spectral indicator constructed from $R_{rs}(\lambda)$, and k_0, k_1, and k_2 are model parameters that can be derived by regression analysis. This general form of the retrieval model for AC is similar to that used for TSM in previous studies, and single band, band ratios and band arithmetic calculations are often used in the spectral indicator X [36–38]. In this study, we designed four forms of X to derive the best model for estimating AC. Regarding the four forms, X_1 and X_2 represent the single band and band ratio, respectively; X_3 denotes the difference of $R_{rs}(\lambda)$ at two bands; and X_4 represents the arithmetic calculation of $R_{rs}(\lambda)$, which was referred from the spectral form used for estimation of TSM [34] (Table 2).

For each spectral indicator X, its ability for estimating AC was examined for all possible combinations from GOCI 6 $R_{rs}(\lambda)$ bands. The best band combinations that displayed the highest R^2 and the lowest RMSE and MAPE for estimation of AC are shown in Table 2. Among the four types of spectral indicator, X_3 indicating the difference of $R_{rs}(\lambda)$ between 490 and 555 nm showed the best performance, and the performance of X_1 representing the single band of $R_{rs}(\lambda)$ at 555 nm was the worst. The X_2 and X_4 showed slightly poorer ability than X_3. Therefore, we recommend X_3 as the best spectral indicator for estimation of AC from $R_{rs}(\lambda)$ and decided to use it in our model. The fitting plots between $\log_{10}(AC)$ and X_3 are shown in Figure 5. It can be observed that most of the samples cluster around the fitting curve with an R^2 value of 0.843. Comparisons between the measured AC and estimated values for the same dataset also showed good agreement with RMSE and MAPE values of 1.145 m^{-1} and 38.9%, respectively (Table 2).

Note that the model fitting of this study was based on the log-transformed AC data, while the fitting can also be conducted directly based on untransformed data. We have compared the two manners for the model fitting, the results indicated that the model based on log-transformed AC data generally performed better than the model based on untransformed data (data not shown), probably because that log-transformation may reduce uncertainty of outliers to some extent. Thus, the log-transformed AC data are recommended to use for model calibration.

Table 2. Summary of the performance of different spectral indicators X for retrieval of the cross-sectional area. X_1 to X_4 indicate four forms of X.

X	General Form	Best Band (nm)	R^2	RMSE	MAPE (%)
X_1	$R_{rs}(\lambda_1)$	$\lambda_1 = 555$	0.755	1.084	50.3
X_2	$R_{rs}(\lambda_1)/R_{rs}(\lambda_2)$	$\lambda_1 = 660, \lambda_2 = 412$	0.807	1.095	44.2
X_3	$R_{rs}(\lambda_1) - R_{rs}(\lambda_2)$	$\lambda_1 = 555, \lambda_2 = 490$	0.843	1.145	38.9
X_4	$\dfrac{R_{rs}(\lambda_1)+R_{rs}(\lambda_3)}{(R_{rs}(\lambda_2)/R_{rs}(\lambda_1))^{12}}$	$\lambda_1 = 555, \lambda_2 = 490, \lambda_3 = 412$	0.816	1.167	44.3

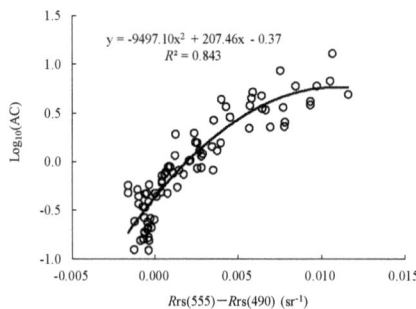

Figure 5. Scatter plots of $\log_{10}(AC)$ versus the difference between $R_{rs}(555)$ and $R_{rs}(490)$ of GOCI. Solid black line indicates the fitted curve.

To further evaluate the performance and test the robustness of the AC estimation model, we conducted a cross-validation [39]. In brief, one sample was selected from the full dataset, which was used in model validation. The remaining samples were used in model calibration, and the calibrated model was adapted to derive the AC for the remaining one sample. This procedure was repeated until all of the samples of the full dataset were selected as the validation sample. The estimated AC values obtained from the above steps were independent of each other and formed a new validation dataset. Subsequently, comparisons were performed between these estimated AC and the measured values, and R^2, RMSE and MAPE were calculated. As shown in Figure 6, a strong linear relationship was found between the estimated AC and measured values in logarithmic space, with values of R^2, RMSE and MAPE of 0.783, 1.744 m^{-1} and 48.4%, respectively.

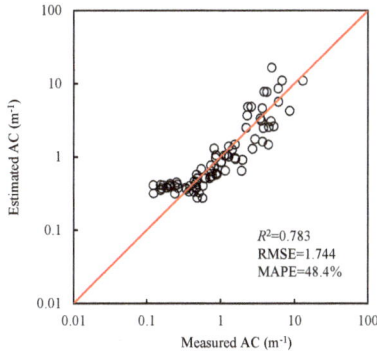

Figure 6. Scatter plots of estimated AC versus measured AC during the cross-validation. The solid red line indicates the 1:1 line.

3.3. Model Sensitivity Analysis

The sensitivities of the developed models to the uncertainties in $R_{rs}(\lambda)$ were examined. We added random errors with a standard deviation of m and average value of 0 to the spectral indicator X_3 (i.e., the difference of $R_{rs}(\lambda)$ between 490 or 555 nm) of all samples. This process was repeated 1000 times; thus an approximately normal distribution was produced in the added random errors. The mean of the relative errors in the estimated AC compared with those derived from $R_{rs}(\lambda)$ without addition of errors was subsequently calculated by averaging the errors across the 1000 repetitions. Considering the uncertainty in X_3 may reach around 40% for GOCI satellite measurements as shown in Figure 8 later, we varied the standard deviation m of the errors from 5% to 45% with an interval of 5% and conducted the above procedure for each value of m.

As shown in Figure 7, it was found that the relative errors in the estimations of AC increased with increases in the random errors added to spectral indicator X_3, and the increase rate gradually became large. Note that although the relative errors in the estimations of AC gradually became large with the increases of the uncertainty in X_3, when 45% random errors were added to X_3, the relative errors yielded by the model was 20.1%. These results indicate that the AC estimation model is robust for input errors.

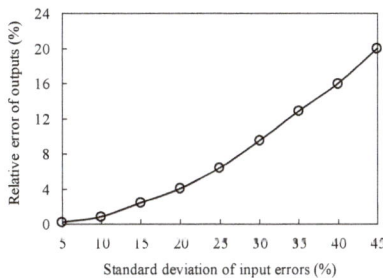

Figure 7. Relative errors in estimation of AC when random errors with standard deviation of 5%, 10%, 15%, 20%, 25%, 30%, 35%, 40% and 45% were respectively added to the spectral indicator X_3.

3.4. Model Application to Satellite Data

We applied the developed AC retrieval model to the GOCI satellite data to preliminarily investigate the spatial and temporal variations of AC in the BS and the YS. Before the application, the satellite derived $R_{rs}(\lambda)$ and the performance of the AC retrieval model for satellite $R_{rs}(\lambda)$ was validated based on a dataset containing 28 pairs of in situ measurements and GOCI satellite data. This match-up dataset only included satellite $R_{rs}(\lambda)$ with an overpass time within 3 h before and after field measurements [30,40,41]. To avoid the influences of outliers, the median $R_{rs}(\lambda)$ values from a 3 × 3 pixel window centered on the locations of the field sampling stations were used as GOCI satellite $R_{rs}(\lambda)$. In general, satellite derived $R_{rs}(490)$ and $R_{rs}(555)$ of the most samples showed good agreement with in situ measured values, while large errors were also observed for some samples (marked by crosses) (Figure 8a,b). However, note that the model of this study uses the spectral indicator X_3 (i.e., the difference between $R_{rs}(555)$ and $R_{rs}(490)$) rather than the single band. We found that although these samples showed large errors in satellite $R_{rs}(\lambda)$, X_3 of them were consistent with in situ measured values (Figure 8c). Subsequently, these 28 satellite $R_{rs}(\lambda)$ were used to derive AC by the model which was calibrated based on the remaining subset ($N = 58$) of the full 86 in situ measurements (independent of validation dataset). As shown in Figure 8d, the satellite-derived AC generally agreed well with field measured AC, with values of R^2, RMSE and MAPE of 0.700, 2.126 m^{-1} and 40.7%, respectively. These results suggest that our model has good performance for satellite application of GOCI data.

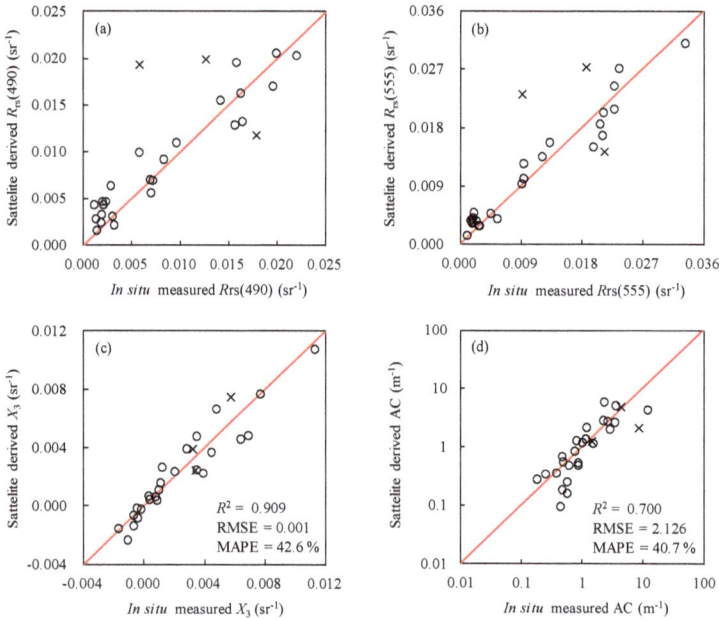

Figure 8. Comparisons of GOCI satellite derived $R_{rs}(490)$ (**a**); $R_{rs}(555)$ (**b**); spectral indicator X_3 (i.e., the difference between $R_{rs}(555)$ and $R_{rs}(490)$) (**c**) and AC (**d**) with in situ measured values. The red lines indicate the 1:1 line.

The distributions of the AC estimated from composited GOCI monthly $R_{rs}(\lambda)$ were shown in Figures 9 and 10. Clear spatial and temporal variations of AC were observed in the BS and the YS (Figure 9). In general, the AC of coastal shallow waters showed much higher values than those in offshore waters, while variations of the AC in coastal regions were lower compared with those in offshore waters (Figure 10). The high AC in the Yellow River mouth and the Changjiang River mouth and their adjacent regions might be related to sediment loads from these rivers, and the high AC in other coastal regions is possibly associated with re-suspension of particles driven by tidal mixing and/or wind stress [42,43]. For offshore waters, the AC in the BS is higher than in the YS. At the temporal scale, the AC was the lowest in summer in most regions of the BS and the YS, whereas in winter, the AC showed the highest values, especially in December. During spring and autumn, high values of AC were observed in offshore waters. The lowest and the highest AC during summer and winter might be related to water stratification and strong mixing, respectively, and the high AC in spring and autumn in offshore waters is possibly caused by algae bloom [20,44]. Understanding the controlling factors of the spatial and temporal variability in AC is beyond the scope of this study, but this topic will be further investigated. However, as mentioned above, re-suspension of particles, sediment loads from rivers and water mixing are possible causes. Meanwhile, we note that some coastal regions in images show invalid pixels (white color). These invalid pixels are mainly related to the effects of unsuccessful atmospheric correction. In this study, we used the standard atmospheric correction method to process GOCI data, by which turbid waters are often masked. Atmospheric correction in coastal regions has been a difficult point for ocean color remote sensing for a long time. Further efforts focusing on this issue are still required.

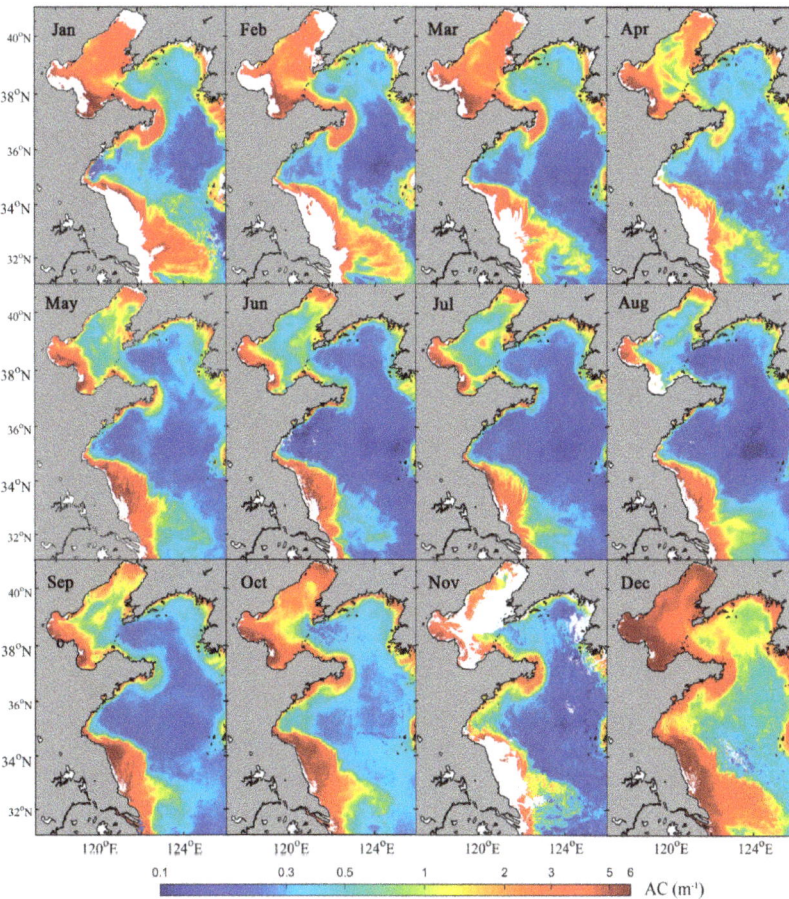

Figure 9. Distributions of the AC derived from GOCI satellite measurements in 2015.

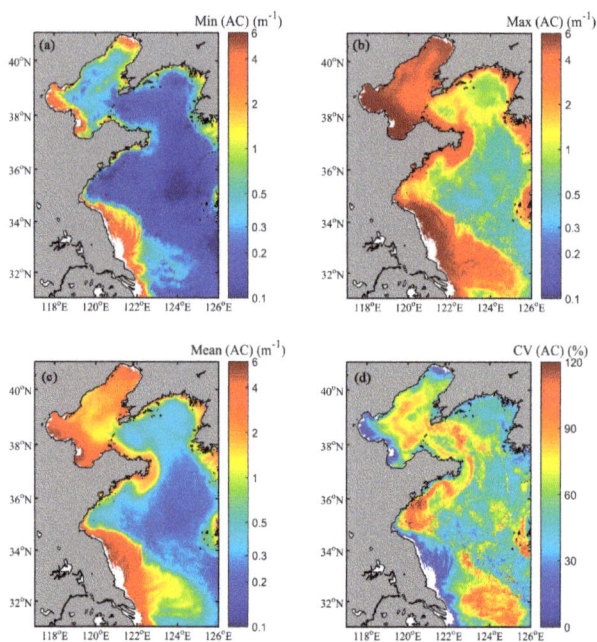

Figure 10. Distributions of the minimum (**a**); maximum (**b**); mean (**c**) and coefficient of variation (**d**) values of the AC derived from monthly GOCI satellite data in 2015.

4. Discussion

4.1. Rationality and Limitation of AC Retrieval Model

Suspended matters significantly impact the underwater light field, algae photosynthesis and the ocean color signal through light absorption and scattering processes [3,4]. Theoretical, experimental and in situ studies have proved that the inherent optical properties of suspended particles are more sensitive to AC than TSM [6,11,12,15] because of the direct dependences of the chance of a photon interacting with a particle on cross-sectional area rather than mass [7]. Consequently, remote sensing reflectance $R_{rs}(\lambda)$ as a function of light absorption and backscattering is also more strongly related to AC than TSM [7,8,18]. In this study, as expected, our analysis showed that although the AC varied over a large range, the correlations between AC and $R_{rs}(\lambda)$ at six bands of GOCI were strong and were approximately 1.4 times higher than those between $R_{rs}(\lambda)$ and TSM (Figure 4). The strong relationship between AC and the inherent optical properties, thus $R_{rs}(\lambda)$, offers a solid optical basis for deriving AC from $R_{rs}(\lambda)$.

The retrieval model of AC was empirically determined by comparing the performances of four types of spectral indicators constructed from all possible GOCI satellite band combinations. The difference between $R_{rs}(\lambda)$ at 490 and 555 nm, which in fact represents the slope of $R_{rs}(\lambda)$ between these two bands, is recommended as the best spectral indicator for estimation of AC (Table 2 and Figure 5). This observation is consistent with our correlation analysis, which indicated that these two bands showed the highest correlations with the AC (Figure 4). The model sensitivity analysis suggests that our model is robust for uncertainties in $R_{rs}(\lambda)$ (Figure 7). In general, the AC retrieval model is simple and can be easily implemented in other regions as well as in other ocean color sensors, such as the Sea-viewing Wide Field-of-view Sensor (SeaWiFS), the Moderate Resolution Imaging Spectroradiometer (MODIS) and the Medium Resolution Imaging Spectrometer (MERIS) because bands similar to 490 and 550 nm are commonly set for most of the ocean color sensors. This was confirmed when we derived a retrieval model of AC for MODIS based on a synthetic dataset of

MODIS $R_{rs}(\lambda)$ using the same approach as that for GOCI. We found that for MODIS, the difference between $R_{rs}(555)$ and $R_{rs}(488)$ was also the best for deriving the AC, and the model performances were comparable to those of the model for GOCI, with R^2, RMSE and MAPE values of 0.842, 1.146 m^{-1} and 38.9%, respectively (Figure 11). However, due to the lack of sufficient match-ups of MODIS satellite data and in situ measurements, currently it is difficult to evaluate the performance of the algorithm for real MODIS satellite data.

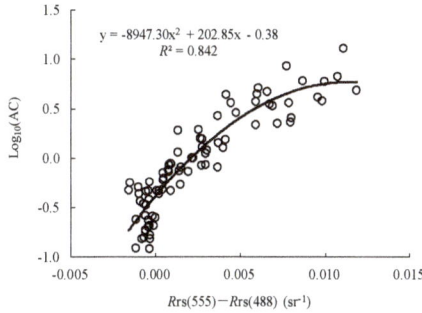

Figure 11. Scatter plots of $\log_{10}$(AC) versus the difference between $R_{rs}(555)$ and $R_{rs}(488)$ of MODIS. Solid black line indicates the fitted curve.

It was noted that the data used in this study were collected mainly from coastal regions, and a subset even originated from turbid waters, showing mean values of TSM and AC of 7.60 g·m^{-3} and 1.74 m^{-1}, respectively, and the highest values reached 43.88 g·m^{-3} and 13.00 m^{-1}, respectively (Table 1). These results imply that the model developed in this study might be mostly suitable for coastal regions. We verified this speculation by analyzing the distribution of the percentage errors (PE) in estimations of AC. As shown in Figure 12, when the AC was larger than 0.20 m^{-1}, our model performed well, showing that the PE were uniformly distributed around the two sides of the 0% line with a mean absolute value of 30.9%. At a notably low AC range (AC < 0.20 m^{-1}), the samples showed significant estimation errors, with PE >50%, and the PE values of these samples were positive, indicating overestimations in AC. However, it was noted that the samples were few ($N = 11$). A future hybrid model applicable for both coastal and clear waters, and particularly for clear waters, might further improve the accuracy in estimations of AC, but it is difficult to conduct such an analysis at this stage because of limited samples from clear waters in our dataset. In addition, bio-optical properties of particles in coastal regions are complex and dynamic due to changes in particle assemblages [45]. Although the algorithm of this study performed generally well in the BS and YS, its applicability in additional coastal regions must be validated in further investigations.

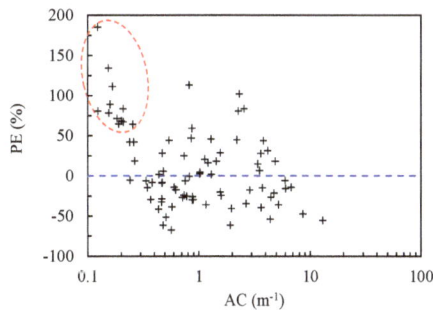

Figure 12. Distributions of percentage error (PE) in estimations of AC using the model developed in this study. Blue dash indicates the PE of 0%. Samples with large PE are circled.

The AC estimation model proposed in this study uses the difference between $R_{rs}(550)$ and $R_{rs}(490)$. It is known that $R_{rs}(\lambda)$ is regulated by the absorption of pure seawater, particles and colored dissolved organic matter (CDOM), and the backscattering of pure seawater and particles [46]. Usually, absorption and backscattering coefficients of pure seawater are treated as constant values [46]. In this sense, changes in the absorption of CDOM may impact the AC estimation model, since although the absorption of CDOM at 490 and 555 nm is usually low, it should not be ignored. Due to the lack of in situ measurements of CDOM, we could not exactly examine to what extent CDOM would impact the AC estimation model. The in situ data used in this study collected from large regions covering river-influenced coastal waters and offshore waters, in which CDOM might vary greatly, while the AC model performed generally well. This observation implies that the AC estimation model may be generally robust to CDOM in the BS and YS, while further investigations are required based on available CDOM measurements.

It should be noted that the AC data used in this study were determined from LISST measurements. Due to the inherent limitation of LISST instrument, the AC data only account for information of particles within the size range of 3.2–390 μm. Risović [47] showed that sub-micron particles may have significant contribution to backscattering signal, so that the particles measured by LISST may not appropriately explain changes in $b_{bp}(\lambda)$ and $R_{rs}(\lambda)$ as well. However, in our previous studies conducted in the BS and YS [15], we found that more than 90% of variability in $b_{bp}(\lambda)$ could be explained by the AC of particles measured by LISST. Therefore, although LISST measurements only effectively cover a size range of 3.2–390 μm, we believe in our study region, the AC values derived from LISST measurements can effectively explain variations in optical properties. In addition, to describe the relationship between AC and optical properties, particles are usually assumed to be spherical. This assumption is useful and has been widely used in studies of bio-optical properties of particles [9,13,16,48], while we have to admit that this assumption may be not necessarily valid in natural environments, which may bring uncertainty in the relationship between AC and optical properties described in this study. Therefore, regarding the above two issues, more advanced field instruments that can cover larger particle size ranges and reveal more adequate particle shapes should be developed in future that will facilitate development of a more accurate AC estimation model.

4.2. Implications of Remote Sensing Applications of AC

The cross-sectional area of suspended particles is a key factor to link the suspended particles with their optical properties. Satellite derived AC using the model proposed in this study provide a new property of suspended particles from space, which may improve our understanding of particle biogeochemical properties and the role of particles in determination of water optical properties. First of all, AC is closely related to the inherent optical properties (i.e., attenuation, scattering and backscattering coefficients) as shown in Equation (1) for the backscattering coefficient as an example, and the relationships are often observed to be strong [11,12,15]. This observation indicates that the inherent optical properties can be modeled from AC with knowledge of the corresponding efficiency factors. Additionally, because the chances of photons interacting with particles directly depend on the cross-sectional area, the light conditions underwater might also be estimated based on the information about AC. In our dataset, strong correlation between AC and transparency ($R^2 = 0.87$) in log-log space was observed, implying that AC retrieved from satellite observations using our model might have great potential for mapping transparency and photosynthetically active radiation, which is important for phytoplankton photosynthesis.

Inherent optical properties are linked with AC by efficiency factors. Many field investigations have observed clear differences in attenuation and backscattering efficiencies among different particle compositions (e.g., phytoplankton and mineral particles) [9,15,48–51]. For instance, Vaillancourt et al. [50] showed that $Q_{bbe}(\lambda)$ levels at 620 nm for cultured 28 phytoplankton species varied from 0.001 to 0.068 with a mean of 0.011; and Peng and Effler [51] found that $Q_{bbe}(\lambda)$ levels at 650 nm of mineral particles ranged from 0.046 to 0.062 with a median value of 0.051. Particularly,

Bowers et al. [48] observed a strong positive relationship between $Q_{bbe}(665)$ and particle composition (the ratio of minerals to total suspended matters) ($R^2 = 0.62$), implying that $Q_{bbe}(665)$ might be used to detect particle composition. Similarly, in the BS and the YS, Wang et al. [15] found that $Q_{bbe}(640)$ clearly classified particles into two types respectively dominated by organic or relatively large particles and small mineral particles. As shown in this study, AC can be successfully derived from satellite measurements using the model developed in this work. At the same time, values of $b_{bp}(\lambda)$ can also be retrieved from satellite $R_{rs}(\lambda)$ using an inherent optical properties algorithm [52,53]. Consequently, $Q_{bbe}(\lambda)$ can be easily calculated as the ratio of $b_{bp}(\lambda)$ to AC. When satellite $Q_{bbe}(\lambda)$ data become available, particle composition or particle classification might be mapped based on the findings of previous studies [15,48], and specific studies focusing on this topic are expected in future.

Although TSM has been widely used to quantify the amount of particles, and various remote sensing algorithms have been developed to map TSM from satellite measurements [27,54–57], the correlation of the inherent optical properties with TSM is weaker compared with AC. This may cause an unstable relationship between TSM and $R_{rs}(\lambda)$, and thereby bring large errors in TSM estimations which could reach a factor of 10 if their values are derived from satellite measurements [8,18,58]. The reason for the large uncertainties is attributed to the fact that $R_{rs}(\lambda)$ reflects the information not only for TSM but also for particle size distribution and density which might also vary by up to several orders of magnitude from location to location due to changes in particle composition and flocculation [13,59]. However, because of a more direct relationship between AC and $R_{rs}(\lambda)$, retrieval of AC from $R_{rs}(\lambda)$ suffers fewer influences from changes in particle size distribution and density. The relationship between TSM and AC can be approximately expressed as [7,9].

$$\text{TSM} = \frac{2AC}{3\rho_a D_A} \tag{8}$$

This implies that satellite derived AC might offer an alternative approach for mapping of TSM. However, at this stage, it is important to note that estimation of TSM from AC also requires additional information on particle size and density, which is beyond of the scope of this study. Meanwhile, it is still needed to further evaluate whether this approach can improve the estimation of TSM.

5. Conclusions

The cross-sectional area of suspended particles is of fundamental importance for understanding water optical properties. Based on in situ measurements conducted in the Bohai Sea (BS) and the Yellow Sea (YS), this study proposed an algorithm for deriving concentration of particle cross-sectional area (AC) from GOCI satellite measurements. The designed algorithm uses the slope of the remote sensing reflectance $R_{rs}(\lambda)$ between 490 and 555 nm as an input. Both in situ and satellite validations indicated that the algorithm shows good performances, with the R^2 values of 0.783 and 0.700, RMSE values of 1.744 and 2.126 m^{-1} and MAPE values of 48.3% and 40.7%, respectively. This result indicates the great potential of the algorithm for mapping of AC from satellite measurements, which was confirmed by the generally reasonable patterns of the temporal and spatial distributions of AC derived using the new algorithm in the BS and the YS. Overall, this study provides a technological basis for estimation of AC from satellite measurements in the BS and the YS, and it may be easily implemented in other regions and for other ocean color sensors, while further investigations are required. The values of AC derived from satellite measurements might offer new potential applications of ocean color, such as modeling of the inherent optical properties of particulates, deriving water transparency, estimating particle composition and possibly improvement in the concentration estimations of total suspended particles, topics that will be specifically studied in future work.

Acknowledgments. Special thanks to Shuguo Chen and other members of the group led by Tinglu Zhang in Ocean University of China, other members of the group led by Zhongfeng Qiu in Nanjing University of Information Science & Technology and the captains, officers, and crews of R/V *Dongfanghong* 2 for their significant contributions on field sampling and measurements. We also would like to thank the anonymous reviewers for their constructive comments towards improving this manuscript. This work is jointly supported by the National Natural Science Foundation of China (NSFC) (41506200, 41276186, 41576172), the Natural Science Foundation of Jiangsu Province (BK20150914, BK20151526), the National Key Research and Development Program of China (2016YFC1400901, 2016YFC1400904), the Natural Science Foundation of the Jiangsu Higher Education Institutions of China (15KJB170015), the Priority Academic Program Development of Jiangsu Higher Education Institutions (PAPD), the Public Science and Technology Research Funds Projects of Ocean (201005030), the Startup Foundation for Introducing Talent of NUIST (2015r007), the National Programme on Global Change and Air-sea Interaction (GASI-03-03-01-01), the Priority Academic Program Development of Jiangsu Higher Education Institutions (PAPD), the Public Science and Technology Research Funds Projects of Ocean (201005030), and sponsored by Qing Lan Project.

Author Contributions: Shengqiang Wang designed this study and the algorithm; Yu Huan contributed to the data analyses and drafted the manuscript; Zhongfeng Qiu, Deyong Sun and Hailong Zhang assisted with developing the research design and results interpretation; Lufei Zheng and Cong Xiao contributed to the interpretation of results and manuscript revision.

Conflicts of Interest: The authors declare no conflict of interest.

References

1. Doxaran, D.; Froidefond, J.-M.; Castaing, P.; Babin, M. Dynamics of the turbidity maximum zone in a macrotidal estuary (the Gironde, France): Observations from field and MODIS satellite data. *Estuar. Coast. Shelf Sci.* **2009**, *81*, 321–332. [CrossRef]

2. Anderson, R.F.; Hayes, C.T. Characterizing marine particles and their impact on biogeochemical cycles in the geotraces program. *Prog. Oceanogr.* **2015**, *133*, 1–5. [CrossRef]

3. Morel, A. Optics of marine particles and marine optics. In *Particle Analysis in Oceanography*; Demers, S., Ed.; Springer: Berlin, Germany, 1991; pp. 141–188.

4. Babin, M.; Morel, A.; Fournier-Sicre, V.; Fell, F.; Stramski, D. Light scattering properties of marine particles in coastal and open ocean waters as related to the particle mass concentration. *Limnol. Oceanogr.* **2003**, *48*, 843–859. [CrossRef]

5. Gordon, H.R.; Morel, A.Y. *Remote Assessment of Ocean Color for Interpretation of Satellite Visible Imagery: A Review*; Springer: Berlin, Germany, 1983; Volume 4.

6. Mobley, C.D. *Light and Water: Radiative Transfer in Natural Waters*; Academic Press: Cambridge, MA, USA, 1994.

7. Bowers, D.; Braithwaite, K. Evidence that satellites sense the cross-sectional area of suspended particles in shelf seas and estuaries better than their mass. *Geo-Mar. Lett.* **2012**, *32*, 165–171. [CrossRef]

8. Mikkelsen, O.A. Variation in the projected surface area of suspended particles: Implications for remote sensing assessment of TSM. *Remote Sens. Environ.* **2002**, *79*, 23–29. [CrossRef]

9. Neukermans, G.; Loisel, H.; Mériaux, X.; Astoreca, R.; McKee, D. In situ variability of mass-specific beam attenuation and backscattering of marine particles with respect to particle size, density, and composition. *Limnol. Oceanogr.* **2012**, *57*, 124–144. [CrossRef]

10. Agrawal, Y.; Pottsmith, H. Instruments for particle size and settling velocity observations in sediment transport. *Mar. Geol.* **2000**, *168*, 89–114. [CrossRef]

11. Hatcher, A.; Hill, P.; Grant, J. Optical backscatter of marine flocs. *J. Sea Res.* **2001**, *46*, 1–12. [CrossRef]

12. Flory, E.; Hill, P.; Milligan, T.; Grant, J. The relationship between floc area and backscatter during a spring phytoplankton bloom. *Deep Sea Res. Part I Oceanogr. Res. Pap.* **2004**, *51*, 213–223. [CrossRef]

13. Boss, E.; Slade, W.; Hill, P. Effect of particulate aggregation in aquatic environments on the beam attenuation and its utility as a proxy for particulate mass. *Opt. Express* **2009**, *17*, 9408–9420. [CrossRef] [PubMed]

14. Bowers, D.; Braithwaite, K.; Nimmo-Smith, W.; Graham, G. The optical efficiency of flocs in shelf seas and estuaries. *Estuar. Coast. Shelf Sci.* **2011**, *91*, 341–350. [CrossRef]

15. Wang, S.; Qiu, Z.; Sun, D.; Shen, X.; Zhang, H. Light beam attenuation and backscattering properties of particles in the Bohai Sea and Yellow Sea with relation to biogeochemical properties. *J. Geophys. Res. Oceans* **2016**, *121*, 3955–3969. [CrossRef]

16. Bowers, D.; Braithwaite, K.; Nimmo-Smith, W.; Graham, G. Light scattering by particles suspended in the sea: The role of particle size and density. *Cont. Shelf Res.* **2009**, *29*, 1748–1755. [CrossRef]

17. Kostadinov, T.S.; Siegel, D.A.; Maritorena, S. Retrieval of the particle size distribution from satellite ocean color observations. *J. Geophys. Res.* **2009**, *114*, C09015. [CrossRef]

18. Bale, A.; Tocher, M.; Weaver, R.; Hudson, S.; Aiken, J. Laboratory measurements of the spectral properties of estuarine suspended particles. *Neth. J. Aquat. Ecol.* **1994**, *28*, 237–244. [CrossRef]

19. Wei, H.; Sun, J.; Moll, A.; Zhao, L. Phytoplankton dynamics in the Bohai Sea—Observations and modelling. *J. Mar. Syst.* **2004**, *44*, 233–251. [CrossRef]

20. Chen, C.-T.A. Chemical and physical fronts in the Bohai, Yellow and East China Seas. *J. Mar. Syst.* **2009**, *78*, 394–410. [CrossRef]

21. Zhang, M.; Tang, J.; Song, Q.; Dong, Q. Backscattering ratio variation and its implications for studying particle composition: A case study in Yellow and East China Seas. *J. Geophys. Res. Oceans* **2010**, *115*. [CrossRef]

22. Qiu, Z.; Wu, T.; Su, Y. Retrieval of diffuse attenuation coefficient in the china seas from surface reflectance. *Opt. Express* **2013**, *21*, 15287–15297. [CrossRef] [PubMed]

23. Agrawal, Y.; Whitmire, A.; Mikkelsen, O.A.; Pottsmith, H. Light scattering by random shaped particles and consequences on measuring suspended sediments by laser diffraction. *J. Geophys. Res. Oceans* **2008**, *113*. [CrossRef]

24. Traykovski, P.; Latter, R.J.; Irish, J.D. A laboratory evaluation of the laser in situ scattering and transmissometery instrument using natural sediments. *Mar. Geol.* **1999**, *159*, 355–367. [CrossRef]

25. Reynolds, R.; Stramski, D.; Wright, V.; Woźniak, S. Measurements and characterization of particle size distributions in coastal waters. *J. Geophys. Res. Oceans* **2010**, *115*. [CrossRef]

26. De Moraes Rudorff, N.; Frouin, R.; Kampel, M.; Goyens, C.; Meriaux, X.; Schieber, B.; Mitchell, B.G. Ocean-color radiometry across the southern atlantic and southeastern pacific: Accuracy and remote sensing implications. *Remote Sens. Environ.* **2014**, *149*, 13–32. [CrossRef]

27. Miller, R.L.; McKee, B.A. Using MODIS Terra 250 m imagery to map concentrations of total suspended matter in coastal waters. *Remote Sens. Environ.* **2004**, *93*, 259–266. [CrossRef]

28. Stavn, R.H.; Rick, H.J.; Falster, A.V. Correcting the errors from variable sea salt retention and water of hydration in loss on ignition analysis: Implications for studies of estuarine and coastal waters. *Estuar. Coast. Shelf Sci.* **2009**, *81*, 575–582. [CrossRef]

29. Wang, M.; Gordon, H.R. A simple, moderately accurate, atmospheric correction algorithm for SeaWiFS. *Remote Sens. Environ.* **1994**, *50*, 231–239. [CrossRef]

30. Hyde, K.J.; O'Reilly, J.E.; Oviatt, C.A. Validation of SeaWiFS chlorophyll a in Massachusetts Bay. *Cont. Shelf Res.* **2007**, *27*, 1677–1691. [CrossRef]

31. Antoine, D.; d'Ortenzio, F.; Hooker, S.B.; Bécu, G.; Gentili, B.; Tailliez, D.; Scott, A.J. Assessment of uncertainty in the ocean reflectance determined by three satellite ocean color sensors (MERIS, SeaWiFS and MODIS-A) at an offshore site in the mediterranean sea (BOUSSOLE project). *J. Geophys. Res. Oceans* **2008**, *113*. [CrossRef]

32. He, X.; Bai, Y.; Pan, D.; Huang, N.; Dong, X.; Chen, J.; Chen, C.-T.A.; Cui, Q. Using geostationary satellite ocean color data to map the diurnal dynamics of suspended particulate matter in coastal waters. *Remote Sens. Environ.* **2013**, *133*, 225–239. [CrossRef]

33. Sun, D.; Hu, C.; Qiu, Z.; Cannizzaro, J.P.; Barnes, B.B. Influence of a red band-based water classification approach on chlorophyll algorithms for optically complex estuaries. *Remote Sens. Environ.* **2014**, *155*, 289–302. [CrossRef]

34. Tassan, S. Local algorithms using seawifs data for the retrieval of phytoplankton, pigments, suspended sediment, and yellow substance in coastal waters. *Appl. Opt.* **1994**, *33*, 2369–2378. [CrossRef] [PubMed]

35. O'Reilly, J.E.; Maritorena, S.; Mitchell, B.G.; Siegel, D.A.; Carder, K.L.; Garver, S.A.; Kahru, M.; McClain, C. Ocean color chlorophyll algorithms for SeaWiFS. *J. Geophys. Res. Oceans* **1998**, *103*, 24937–24953. [CrossRef]

36. Doxaran, D.; Froidefond, J.-M.; Lavender, S.; Castaing, P. Spectral signature of highly turbid waters: Application with SPOT data to quantify suspended particulate matter concentrations. *Remote Sens. Environ.* **2002**, *81*, 149–161. [CrossRef]

37. Mao, Z.; Chen, J.; Pan, D.; Tao, B.; Zhu, Q. A regional remote sensing algorithm for total suspended matter in the East China Sea. *Remote Sens. Environ.* **2012**, *124*, 819–831. [CrossRef]

38. Siswanto, E.; Tang, J.; Yamaguchi, H.; Ahn, Y.-H.; Ishizaka, J.; Yoo, S.; Kim, S.-W.; Kiyomoto, Y.; Yamada, K.; Chiang, C. Empirical ocean-color algorithms to retrieve chlorophyll-a, total suspended matter, and colored dissolved organic matter absorption coefficient in the Yellow and East China Seas. *J. Oceanogr.* **2011**, *67*, 627–650. [CrossRef]

39. Kohavi, R. A study of cross-validation and bootstrap for accuracy estimation and model selection. In Proceedings of the 14th International Joint Conference on Artificial Intelligence, Montreal, QC, Canada, 20–25 August 1995; Volume 14, pp. 1137–1145.

40. Bailey, S.W.; Werdell, P.J. A multi-sensor approach for the on-orbit validation of ocean color satellite data products. *Remote Sens. Environ.* **2006**, *102*, 12–23. [CrossRef]

41. Cui, T.; Zhang, J.; Groom, S.; Sun, L.; Smyth, T.; Sathyendranath, S. Validation of MERIS ocean-color products in the Bohai Sea: A case study for turbid coastal waters. *Remote Sens. Environ.* **2010**, *114*, 2326–2336. [CrossRef]

42. Yang, S.Y.; Jung, H.S.; Lim, D.I.; Li, C.X. A review on the provenance discrimination of sediments in the Yellow Sea. *Earth-Sci. Rev.* **2003**, *63*, 93–120. [CrossRef]

43. Jiang, W.; Pohlmann, T.; Sun, J.; Starke, A. SPM transport in the Bohai Sea: Field experiments and numerical modelling. *J. Mar. Syst.* **2004**, *44*, 175–188. [CrossRef]

44. Xu, Y.; Ishizaka, J.; Yamaguchi, H.; Siswanto, E.; Wang, S. Relationships of interannual variability in SST and phytoplankton blooms with giant jellyfish (*Nemopilema nomurai*) outbreaks in the Yellow Sea and East China Sea. *J. Oceanogr.* **2013**, *69*, 511–526. [CrossRef]

45. Braithwaite, K.; Bowers, D.; Smith, W.N.; Graham, G.; Agrawal, Y.; Mikkelsen, O. Observations of particle density and scattering in the Tamar Estuary. *Mar. Geol.* **2010**, *277*, 1–10. [CrossRef]

46. Gordon, H.R.; Brown, O.B.; Evans, R.H.; Brown, J.W.; Smith, R.C.; Baker, K.S.; Clark, D.K. A semianalytic radiance model of ocean color. *J. Geophys. Res. Atmos.* **1988**, *93*, 10909–10924. [CrossRef]

47. Risović, D. Effect of suspended particulate-size distribution on the backscattering ratio in the remote sensing of seawater. *Appl. Opt.* **2002**, *41*, 7092–7101. [CrossRef] [PubMed]

48. Bowers, D.; Hill, P.; Braithwaite, K. The effect of particulate organic content on the remote sensing of marine suspended sediments. *Remote Sens. Environ.* **2014**, *144*, 172–178. [CrossRef]

49. Ahn, Y.-H.; Bricaud, A.; Morel, A. Light backscattering efficiency and related properties of some phytoplankters. *Deep Sea Res. Part I Oceanogr. Res. Pap.* **1992**, *39*, 1835–1855. [CrossRef]

50. Vaillancourt, R.D.; Brown, C.W.; Guillard, R.R.; Balch, W.M. Light backscattering properties of marine phytoplankton: Relationships to cell size, chemical composition and taxonomy. *J. Plankton Res.* **2004**, *26*, 191–212. [CrossRef]

51. Peng, F.; Effler, S.W. Characterizations of individual suspended mineral particles in Western Lake Erie: Implications for light scattering and water clarity. *J. Gt. Lakes Res.* **2010**, *36*, 686–698. [CrossRef]

52. Lee, Z.; Carder, K.L.; Arnone, R.A. Deriving inherent optical properties from water color: A multiband quasi-analytical algorithm for optically deep waters. *Appl. Opt.* **2002**, *41*, 5755–5772. [CrossRef] [PubMed]

53. Smyth, T.J.; Moore, G.F.; Hirata, T.; Aiken, J. Semianalytical model for the derivation of ocean color inherent optical properties: Description, implementation, and performance assessment. *Appl. Opt.* **2006**, *45*, 8116–8131. [CrossRef] [PubMed]

54. Binding, C.; Bowers, D.; Mitchelson-Jacob, E. Estimating suspended sediment concentrations from ocean colour measurements in moderately turbid waters; the impact of variable particle scattering properties. *Remote Sens. Environ.* **2005**, *94*, 373–383. [CrossRef]

55. Qiu, Z. A simple optical model to estimate suspended particulate matter in Yellow River Estuary. *Opt. Express* **2013**, *21*, 27891–27904. [CrossRef] [PubMed]

56. Chen, J.; Cui, T.; Qiu, Z.; Lin, C. A three-band semi-analytical model for deriving total suspended sediment concentration from HJ-1A/CCD data in turbid coastal waters. *ISPRS J. Photogramm.* **2014**, *93*, 1–13. [CrossRef]

57. Kong, J.-L.; Sun, X.-M.; Wong, D.W.; Chen, Y.; Yang, J.; Yan, Y.; Wang, L.-X. A semi-analytical model for remote sensing retrieval of suspended sediment concentration in the gulf of Bohai, China. *Remote Sens.* **2015**, *7*, 5373–5397. [CrossRef]

58. Baker, E.T.; Lavelle, J.W. The effect of particle size on the light attenuation coefficient of natural suspensions. *J. Geophys. Res.* **1984**, *89*, 8197–8203. [CrossRef]

59. Mikkelsen, O.; Pejrup, M. In situ particle size spectra and density of particle aggregates in a dredging plume. *Mar. Geol.* **2000**, *170*, 443–459. [CrossRef]

![remote sensing logo] *remote sensing*

MDPI

Article

An Empirical Ocean Colour Algorithm for Estimating the Contribution of Coloured Dissolved Organic Matter in North-Central Western Adriatic Sea

Alessandra Campanelli [1], Simone Pascucci [2,*], Mattia Betti [1], Federica Grilli [1], Mauro Marini [1], Stefano Pignatti [2] and Stefano Guicciardi [1]

[1]   CNR-ISMAR Institute of Marine Sciences, National Research Council, L.go Fiera della Pesca, 2, 60125 Ancona, Italy; a.campanelli@ismar.cnr.it (A.C.); m.betti@an.ismar.cnr.it (M.B.); f.grilli@ismar.cnr.it (F.G.); m.marini@ismar.cnr.it (M.M.); stefano.guicciardi@an.ismar.cnr.it (S.G.)
[2]   CNR-IMAA Institute of Methodologies for Environmental Analysis, National Research Council, C.da S. Loja, 85050 Tito Scalo (PZ), Italy; stefano.pignatti@imaa.cnr.it
*   Correspondence: simone.pascucci@imaa.cnr.it; Tel.: +39-06-4993-4685

Academic Editors: Yunlin Zhang, Claudia Giardino, Linhai Li, Deepak R. Mishra and Prasad S. Thenkabail
Received: 7 December 2016; Accepted: 9 February 2017; Published: 21 February 2017

Abstract: The performance of empirical band ratio models were evaluated for the estimation of Coloured Dissolved Organic Matter (CDOM) using MODIS ocean colour sensor images and data collected on the North-Central Western Adriatic Sea (Mediterranean Sea). Relationships between in situ measurements (2013–2016) of CDOM absorption coefficients at 355 nm ($a_{CDOM}355$) with several MODIS satellite band ratios were evaluated on a test data set. The prediction capability of the different linear models was assessed on a validation data set. Based on some statistical diagnostic parameters (R^2, APD and RMSE), the best MODIS band ratio performance in retrieving CDOM was obtained by a simple linear model of the transformed dependent variable using the remote sensing reflectance band ratio $R_{rs}(667)/R_{rs}(488)$ as the only independent variable. The best-retrieved CDOM algorithm provides very good results for the complex coastal area along the North-Central Western Adriatic Sea where the Po River outflow is the main driving force in CDOM and nutrient circulation, which in winter mostly remains confined to a coastal boundary layer, whereas in summer it spreads to the open sea as well.

Keywords: Western Adriatic Sea; CDOM; MODIS; remote sensing

---

1. Introduction

Ocean colour imagery collected over the past decades from many satellites has the potential for providing synoptic retrievals of bio-optical properties in the world's oceans [1,2]. Unfortunately, the optical complexity of coastal waters affects remote sensing products and their accuracy. Thus, the pressing need of validating large-scale remote sensing coastal products makes attractive the availability of cross-site consistent in situ observations performed applying comprehensively assessed and standardized measurement protocols. Improving retrievals of water borne constituents in coastal systems requires the optimization of algorithms using regional in situ measurements of optical and biogeochemical water properties.

In this context, the Coloured Dissolved Organic Matter (CDOM) represents the optically active fraction of DOM in natural waters and plays various roles in physical and biogeochemical processes. CDOM absorption is characterized by an exponential absorption decrease from ultraviolet (UV) to visible wavelengths. It can dominate the inherent light absorption at the blue wavelengths in surface waters of the coastal seas (20%–70% at 440 nm; [3,4]), pelagic seas and oceans (>50% at 440 nm; [5]), hence competing with phytoplankton for photosynthetically active radiation [6,7]. In many coastal areas, CDOM absorption is several times that of chlorophyll and confounds the retrieval of the latter from ocean colour satellite observations due to overlapping absorbance spectra at the blue wavelengths [8–10]. Furthermore, CDOM has proven to be a useful tracer not only for carbon but also as a proxy for mixing in a wide variety of environments [11–13].

Primary sources of CDOM are rivers and groundwater near coastlines, which carry CDOM primarily from soils, but coastal waters, can contain plankton-derived CDOM produced in rivers and estuaries, as well as anthropogenic compounds from runoff, sewage discharge and other effluents [14,15]. However, biological processes such as phytoplankton growth, zooplankton grazing and microbial activity can also contribute to marine-derived CDOM in continental margins and pelagic ocean [16–19]. On the other hand, photobleaching is the dominant process for CDOM removal from natural waters [20], while microbial decomposition is of less importance [21,22]. The balance between sources and sinks controls the patterns of CDOM distribution.

Historically, many ocean colour algorithms have been developed to retrieve biogeochemical properties (e.g., chlorophyll-a, total suspended matter and CDOM) based on direct empirical relationship with the remote sensing reflectance, $R_{rs}(\lambda)$, or ratios of R_{rs} at various wavelengths. Many of these models estimate the absorption coefficient of CDOM (a_{CDOM}) and detrital (non-pigmented) particles as a single parameter (a_{CDM}), because CDOM and detritus have similar spectral responses in the visible spectrum [23–33]. The retrieval of a_{CDM} by these algorithms yields reasonable results in open ocean regions. However, such algorithms typically do not work well in coastal waters due to the optical complexity of turbid coastal waters (high levels of CDOM, coloured detrital particles, and phytoplankton; [34]).

During the last years, many authors [32,35,36] have attempted to separate a_{CDOM} from a_{CDM} and to derive the CDOM spectral slope in various spectral regions [37–39]. Other authors [34,40] described how retrievals of bio-optical properties with a spectral references band near 640 nm showed significant improvement over the standard 555 nm band because total light absorption near 640 nm is primarily greatest for seawater itself rather than for other constituents in the water column. Anyway, the primary limitation to rigorous validation is the lack of sufficient data of coincident field measurements and satellite observations that are independent from the data used to develop the algorithms.

Satellite ocean colour optical sensors currently operating are: (a) SEVIRI (Spinning Enhanced Visible and Infrared Imager) onboard the Meteosat Second Generation (MSG-3; 2005-present) geostationary platforms with a Ground Sampling Distance (GSD) of 1–4 km; (b) Moderate Resolution Imaging Spectroradiometer on board the polar-orbiting Terra/Aqua satellites (MODIS-A; 2002–present) and the Visible Infrared Imager/Radiometer Suite (VIIRS; 2011–present) on the NPOESS Preparatory Project (NPP) satellite mission, with a GSD of 250–1000 m; (c) the Operational Land Imager (OLI; 2013–present) on the polar-orbiting Landsat-8 satellite with a GSD of 30 m; and (d) Sentinel-2A (launched on 23 June 2015) and Sentinel 3A (launched on 16 February 2016 and available from 17 November 2016) polar-orbiting multispectral high-resolution imaging ESA missions (Copernicus Services) with a GSD of 10–300 m. The data recorded by these satellite optical sensors cover the range of temporal (15 min to 15 days) and spatial (10 m to 4000 m) resolutions that are currently free available for the monitoring of sea surface dynamics.

In this study, we evaluate and analyse the performance of the MODIS-A-derived CDOM estimations using the Copernicus Marine Environment Monitoring Service CMEMS Remote sensing reflectance (R_{rs} at wavelengths of 412, 443, 488, 531, 547 and 667 nm) for the North-Central Western Adriatic Sea (NCWAS) in Italy, which is a complex system influenced by the Po River, one of the largest Mediterranean rivers [41]. Ocean colour satellite retrievals of a_{CDOM} could have many potential applications and in this work, some regional CDOM empirical algorithms for the NCWAS were tested and validated on MODIS-Aqua-measured $R_{rs}(\lambda)$ and a_{CDOM} in situ data were collected from 2013 to 2016. After validation, the optimal model was used to produce a_{CDOM} maps of the NCWAS to describe his distribution throughout the basin. The results of the work are drawn from field measurements supported by multiple research projects in order to collect sufficient coincident satellite data with field observations.

2. Materials and Methods

2.1. Study Site

The North Adriatic Sea (Figure 1) is the northernmost part of the Mediterranean Sea, extending as far North as 45°47'N. The area is strongly influenced by river floods [42–45], which affect both the circulation through buoyancy input and the ecosystem by introducing a large amount of matters [46–48]. The Po River provides the major buoyancy flux with an annual mean freshwater discharge rate of about 1500 m^3·s^{-1} [41,49]. On the annual basis, the Po River alone carries 28% of the total Adriatic Sea runoff (5500−5700 m^3·s^{-1}), 45% comes from the eastern coast (about half of which from Albania), 19% from the northern coast and 8% from the entire western coast [41,50]. The southward coastal flow, i.e., the Western Adriatic Current (WAC [50–53]), is driven by the Po River buoyancy flux (low-salinity waters) and northeastern Bora winds that characterize this region during the winter months. Bora winds cause elevated sea surface height along the western coasts, producing the downwelling and transport of coastal dense waters toward the open sea [54].

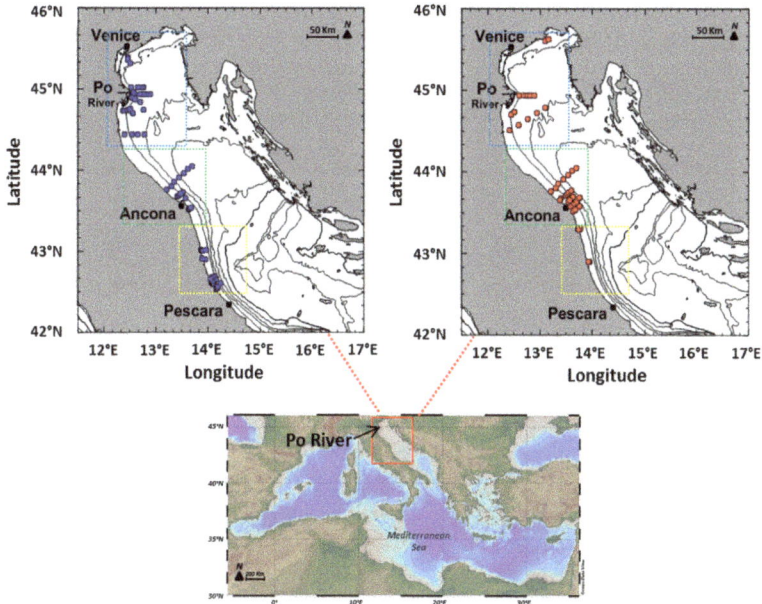

Figure 1. Study area. In the upper pictures, the blue dotted box indicates Zone A, the green dotted box Zone B and the yellow dotted box Zone C (see text). In the top left picture, stations pertaining to the test data set are indicated by blue dots, in the top right picture the red dots indicate stations pertaining to the validation data set. Many stations were sampled more than once. In the lower picture, the red box indicates the location of the study area in the Mediterranean basin.

The Po River freshwater may flow southward along the shelf strengthening the Western Adriatic Current (WAC) or cross the shelf extending toward Istria in the central basin [55]. In particular, during periods of weak stratification, in absence of wind forcing, vortices constrain fresh Po River waters to a southward flow along the Italian shelf [56]. In contrast, in periods of stratification, particularly in spring and summer, the Po River fresh water plume spreads across the basin to the Istrian coast to form a front that divides the northern basin [57,58]. The extent of Po riverine flow and wind forcing, therefore, modulate fresh water penetration in the North Adriatic [59].

Eutrophication, which periodically occurs along the Italian coast of the Adriatic Sea [60–63], is mainly caused by the discharge of organic and inorganic matter from the Po River and detrital coming from other rivers flowing into the NCWAS and contributes to make this area a complex optical system due to simultaneous presence of CDOM, coloured detrital particles, and phytoplankton [34]. Therefore, in this work, only CDOM-based algorithms (a_{CDM}-based algorithms works better for other study sites where detrital particles concentrations are not so high) were considered, as CDOM is the variable of main interest for the study area; moreover, detrital particles and phytoplankton were not measured as extensively as CDOM.

2.2. Field Measurements of Seawater

As part of numerous national and international projects, several oceanographic cruises in NCWAS were conducted between November 2013 and July 2016. One hundred seventy-two water samples, corresponding to clear-sky days during which MODIS-A satellite has acquired images over the selected study area, were collected (Figure 1; Table 1). For each of these water samples, CDOM and salinity were measured at the sea surface level (in situ CDOM).

Table 1. List of field data from cruises for the absorption coefficient of Coloured Dissolved Organic Matter (a_{CDOM}) and salinity (S).

Region	Cruise	Dates	No. of Measurements
			a_{CDOM}, S
Zone A	Ritmare A	29 November 2013	6
Zone B	ANOC14_1	13 February 2014	4
Zone B	ANOC14_2	21 Marh 2014	5
Zone A	Balmas 1	25 March 2014	3
Zone B	Balmas 2	24 May 2014	4
Zone B	Balmas 3	21–22 August 2014	3
Zone C	Ecosee/a 1	26 August 2014	15
Zone A	ANOC_4	22 September 2014	7
Zone C	Ecosee/a 2	7 October 2014	6
Zone B	Ecosee/a 3	17 October 2014	4
Zone B	Balmas 4	29 October 2014	3
Zone A	Ritmare B	11–12 December 2014	18
Zone B	ANOC15_1	11 February 2015	6
Zone B	ANOC15_2	14 March 2015	4
Zone B	ANOC15_2	6 May 2015	5
Zone B	Ecosee/a 4	22 July 2015	11
Zone B	ANOC15_4	18 September 2015	7
Zone B	Escavo AN	10 November 2015	10
Zone A	Solemon2015	29 November 2015	2
Zone B	ANOC16_1	11 February 2016	7
Zone B	ANOC16_2	18–19 March 2016	15
Zone A	Ritmare C	28–29 April 2016	7
Zone B	ANOC16_4	18 June 2016	6
Zone A	Medias2016	23 June 2016	6
Zone C	Post-Ecosee/a	15 July 2016	8

Salinity (S) of the surface waters was measured using a SeaBird Electronics 911-plus CTD (Conductivity Temperature Depth). The CTD data were processed according to UNESCO standards [64].

Seawater samples for analysis of CDOM absorbance spectra were filtered through sterile Whatman GD/X 0.2 μm filters, under low pressure. The filters were washed with 100 mL of sample before its collection in order to avoid any CDOM contamination. Samples were: (a) filtered immediately after the sampling in order to avoid alteration due to microbial activity; (b) collected in amber glass bottles, previously acid-soaked (10% HCl); (c) rinsed with Milli-Q water; and (d) rinsed three times with the sample before its collection. The filtered water was kept at 4 °C until the optical measurements were performed, at the latest within four weeks.

CDOM absorbance was measured throughout a dual beam UV-VIS spectrophotometer (SHIMADZU 2600 Series), using a 10 cm path length cells with ultraviolet oxidized Milli-Q water as the blank and reference [65]. Instrument scan settings were as follows: 250–750 nm wavelength scan range, fast scan speed; 1 nm sampling interval; 0.5 nm slit width. The quartz cuvettes for blanks and samples were acid-soaked for one hour and then rinsed with Milli-Q water and sample aliquots. The absorbance (A) was converted into absorption coefficients ($a_{CDOM}\lambda$) by Equation (1):

$$a_{CDOM}\lambda = 2.303 \times A_\lambda / L \tag{1}$$

where A_λ is the absorbance at wavelength λ and L is the path length expressed in meters. Before determining $a_{CDOM}\lambda$, absorbance data were corrected by subtracting the mean absorbance from 680 to 690 nm from each wavelength. The absorbance measurements ranged from 250 to 450 nm representing wavelengths where the signal was sufficiently high for a reliable estimation of $a_{CDOM}\lambda$ [48,66]. Specifically, the absorption coefficients at 355 nm ($a_{CDOM}355$) were calculated because this wavelength shows the maximum excitation for humic-like substances [66], i.e., those of terrestrial origin. Since the NCWAS is an area impacted by high river discharges we chose this absorption coefficient for the retrieval algorithms on the assumption that most of CDOM is of terrestrial origin.

2.3. Satellite Data and Processing

MODIS-Aqua data over the study areas in the NCWAS were acquired from CMEMS online catalogue [67]. MODIS is an ocean colour sensor aboard the polar-orbiting Aqua (MODIS-A) and Terra (MODIS-T) satellite platforms with 36 spectral bands and three spatial resolutions of 250 m, 500 m and 1 km. For the Mediterranean Sea, R_{rs} and diffuse attenuation coefficient of light at 490 nm is operationally produced and uploaded on the CMEMS catalogue by the Group for Satellite Oceanography (GOS-ISAC) of the Italian National Research Council, in Rome, for near real time data from MODIS-A and NPP-VIIRS sensors [1,68,69].

$R_{rs}(\lambda)$ is the fundamental quantity to be derived from ocean colour sensors and it is defined as the ratio of upwelling radiance and downwelling irradiance at any wavelength that can be expressed as the ratio of normalized water leaving radiance and the extra-terrestrial solar irradiance. The spectral remote-sensing reflectance R_{rs} is defined as [70]:

$$R_{rs}(\theta, \phi, \lambda) = \frac{L_w(\text{in air, } \theta, \phi, \lambda)}{E_d(\text{in air,} \lambda)} \tag{2}$$

where the depth argument of "in air" indicates that R_{rs} is calculated using the water-leaving radiance L_w and E_d in the air, just above the water surface. The R_{rs} is a measure of how much of the downwelling radiance that is incident onto the water surface in any direction is eventually returned through the surface into a small solid angle $\Delta\Omega$ centered on a particular direction (θ, ϕ). However, R_{rs} is usually computed for nadir-viewing directions only.

In this study, nominal MODIS-A six R_{rs} bands (412, 443, 488, 531, 547 and 667 nm) were used for CDOM retrieval, with a spatial resolution of 1.1 km and acquired in the same dates of the field campaign period 2013–2016 (temporal window between satellite overpass and the time of field sampling = ±12 h). Figure 2a shows an example of R_{rs} MODIS imagery (R: 547 nm; G: 488 nm; B: 443 nm) resized on the Adriatic Sea in Italy centred on the Ancona coastal area (red box in Figure 2a). MODIS R_{rs} imagery was acquired on 22 July 2015, i.e., on the same date of the eleven measurement stations performed in these coastal areas (white dots in Figure 2b; Ecosee/a 4 Cruise). In Figure 2c the graph shows the R_{rs} spectra relative to the eleven coastal measurement stations. As expected, R_{rs} spectra relative to stations 3, 10 and 11 show higher values in the blue band at 443 nm. This is also evident from the MODIS R_{rs}

RGB imagery in Figure 2 that shows both the CDOM gradient along the coastal area occurring in this date of acquisition (22 July 2015) and the higher values in the blue band at 443 nm for the stations more distant from the coast (i.e., 3, 10 and 11).

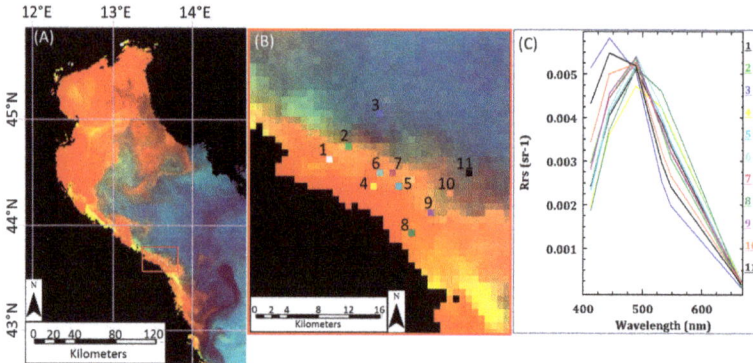

Figure 2. (**A**) Overview of MODIS R_{rs} imagery (R: 547 nm; G: 488 nm; B: 443 nm) with the R_{rs} spectra relative to the oceanographic cruises stations acquired in the Ancona coastal area (Italy) on 22 July 2015 shown in inset (**B**) as coloured dots (with the relative station number). Eastings and Northings are in geographic coordinates (latitude/longitude). (**C**) R_{rs} spectra of the eleven measurement stations shown in (**B**) and represented with the same colour and number of the dots.

2.4. CDOM Models

Due to the lack of coincident in situ measurements and satellite observations, previous attempts to retrieve bio-information from satellite data in this complex optical region found better results on the retrieval of chlorophyll and suspended matter rather than of CDOM [71–74]. Having a consistent number of in situ and satellite data, the main aim of this work is on the definition of empirical models useful to effectively retrieve $a_{CDOM}355$ values using the optical satellite data acquired on the northwestern Adriatic coastal sea waters. Simple linear algorithms were chosen among several algorithms available in literature for CDOM retrieval, as they proved to be nearly effective with respect to the more complex models in CDOM retrieval [39]. On the other hand, empirical algorithms are known to be valid only regionally as they are particularly sensitive to changes in the specific composition of water constituents when boundary conditions are changed [75]. For the fitting and evaluation of our empirical models, the collected data were divided into two data sets: data collected from November 2013 to May 2015 (test data set, 93 observations) and data collected from July 2015 to July 2016 (validation data set, 79 observations). The initial data set was split according to two different time intervals in order to assess the prediction capability of the models. The spatial coverage of the two data sets was about the same, Figure 1. Even if the spatial coverage could be considered not very extensive, it must be kept in mind that this limitation was due to the necessity to have corresponding data from in situ and satellite measurements. The mathematical models were fitted on the test data set and the prediction performance evaluated on the validation data set. In all cases, a least squares analysis was carried out in order to identify the best relationship between the dependent variable, the CDOM absorption coefficient $a_{CDOM}355$, and the independent variables. As independent variables, the following MODIS R_{rs} band ratios were considered: (a) $R_{rs}(547)/R_{rs}(412)$; (b) $R_{rs}(547)/R_{rs}(488)$; (c) $R_{rs}(667)/R_{rs}(412)$; (d) $R_{rs}(667)/R_{rs}(488)$; (e) $R_{rs}(667)/R_{rs}(443)$; and (f) $R_{rs}(531)/R_{rs}(412)$. The inverse band ratios (a) and (b), i.e., $R_{rs}(412)/R_{rs}(547)$ and $R_{rs}(488)/R_{rs}(547)$, were used to define empirical algorithms in coastal waters by many authors [23,37,39,76]. However, the reference wavelength at 547 nm is affected by particles light scattering [77] that in coastal waters interferes with the CDOM retrieval. Since the NCWAS is strongly influenced by freshwaters/terrestrial input, a wavelength in the red or near infrared (667 nm) was added as numerator of the band ratios (c) and (d) in order to reduce the effects of detritus particles, as suggested by other authors [40,78,79]. At first, a simple multilinear model was tested by ordinary least square (OLS) using both the dependent and the independent variables transformed according to the log function [39]. This first analysis provided a quite low value of the adjusted coefficient of determination (adjusted R^2 = 0.5671). Therefore, we started using a simple multilinear model with untransformed variables and, as shown in the Results Section, proceeded according to model check and diagnostics. Once obtained some final models, their prediction capability was assessed against the validation data set.

The accuracy of model prediction was evaluated according to the following statistical indicators: (a) the adjusted coefficient of determination (adjusted R^2) between the in situ validation values $a_{CDOM}355_{is}$'s and the calculated values $a_{CDOM}355_c$'s; (b) the mean and standard deviation of the absolute per cent difference (APD) (APD = $100 \times |a_{CDOM}355_{is} - a_{CDOM}355_c| / a_{CDOM}355_{is}$); and (c) the root mean square error (RMSE) between $a_{CDOM}355_{is}$'s and $a_{CDOM}355_c$'s. All the statistical analyses were performed using the open source statistical software R ver. 3.3.1 [80] at a significance level of 0.05.

Finally, the retrieved optimal model was used to construct CDOM time maps for this geographical area to further evaluate the applicability of the $a_{CDOM}355$ algorithm. These maps were then compared with the corresponding Po river discharges, which as aforementioned has a great influence in CDOM and nutrient circulation in the northwestern Adriatic Sea coastal areas.

3. Results

3.1. Field Distribution of CDOM

The surface distribution of salinity and CDOM ($a_{CDOM}355$) are summarized in Table 2. Because of the complexity of the basin, the NCWSA was divided in three sub-areas: the northern part (Zone A), directly influenced from Po River runoff, and the central (Zone B) and southern parts (Zone C) both influenced by minor river discharges and indirectly by Po River runoff carried out along the northwestern Adriatic coastal areas by the WAC.

Table 2. Mean and standard deviation (s.d.) of in situ $a_{CDOM}355$ and salinity (S) measurements as a function of the area, see Figure 1. The last column shows the number of observations (*n*).

Region	$a_{CDOM}355$ (m^{-1})		S		*n*
	Mean	s.d.	Mean	s.d.	
Zone A	0.81	0.49	31.75	5.25	42
Zone B	0.38	0.30	36.27	2.26	99
Zone C	0.38	0.16	35.16	0.52	29

River discharges along the Western Adriatic Sea coastal areas enrich seawater in CDOM, especially in Zone A (Figure 1). A strong linear relationship between $a_{CDOM}355$ and salinity confirms this assumption (R^2 = 0.80; Figure 3). Salinity is a useful parameter commonly used to characterize the main terrestrial origin of CDOM, i.e., riverine CDOM, along the NCWAS.

The NCWAS is characterized by a decrease of CDOM from north to south and from coast to off-shore coupled with a salinity increase as expected. Throughout the study period a wide range of CDOM and salinity (S) are observed (S ranges from 18 to 38.7 and $a_{CDOM}355$ from 0.08 to 2.18 m^{-1}), as shown by the high standard deviations (Table 2). The high spatial and temporal variability of the data is however useful for satellite observation in order to define a properly CDOM retrieval algorithm.

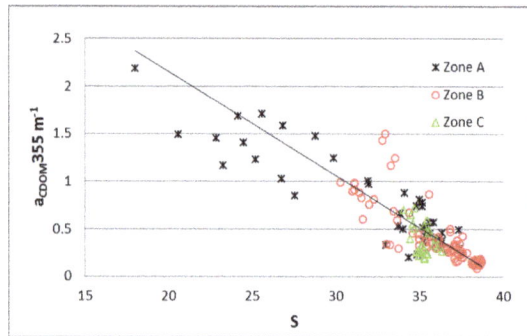

Figure 3. Relationship between $a_{CDOM}355$ and salinity (S) from in situ observations. The symbols represents the three areas of the Western Adriatic Sea, see text. The solid line is a regression line.

3.2. Model Definition

3.2.1. Simple OLS Linear Regression

Before starting the model tuning, a simple correlation analysis was carried out among the independent variables, i.e., the band ratios, to avoid the problem of collinearity. It turned out that the band ratio $R_{rs}(667)/R_{rs}(443)$ was nearly perfectly correlated with the band ratio $R_{rs}(667)/R_{rs}(488)$, Pearson correlation coefficient = 0.991, and the band ratio $R_{rs}(531)/R_{rs}(412)$ was nearly perfectly correlated with the band ratio $R_{rs}(547)/R_{rs}(412)$, Pearson correlation coefficient = 0.996. Since the aim of this work is the development of a good prediction algorithm for the dependent variable [81], the band ratios $R_{rs}(667)/R_{rs}(443)$ and $R_{rs}(531)/R_{rs}(412)$ were dropped in the following analysis. According to [78], the band ratio $R_{rs}(667)/R_{rs}(488)$ was maintained because riverine CDOM still shows significant absorption at 488 nm, whereas chlorophyll does not. The band ratio $R_{rs}(547)/R_{rs}(412)$ was maintained in order to compare our results to previous works [23,37].

At the beginning, a simple multilinear model with untransformed variables was chosen as starting algorithm:

$$a_{CDOM}355 = b_1 + b_2 \times R_{rs}(547)/R_{rs}(412) + b_3 \times R_{rs}(667)/R_{rs}(412) + \\ b_4 \times R_{rs}(667)/R_{rs}(488) + b_5 \times R_{rs}(547)/R_{rs}(488), \tag{3}$$

where the b_i's are the parameters to be estimated using satellite R_{rs} band ratios and the in situ $a_{CDOM}355$ values (test data set, 93 observations collected from November 2013 to May 2015). The results of the OLS regression for this model on the test data set are briefly summarized in Table 3. The adjusted coefficient of determination R^2 of this model was 0.8502.

Table 3. Results of the OLS regression of $a_{CDOM}355$ algorithm (3) on the test data set.

Parameter	Parameter Estimate	Standard Error	*t*-Value	*p*
b_1	0.22966	0.08290	2.770	0.006831 *
b_2	0.01374	0.03707	0.371	0.711856
b_3	0.26675	0.24283	1.098	0.274996
b_4	1.63251	0.47735	3.420	0.000951 *
b_5	−0.18052	0.15756	−1.146	0.255012

* Significant at $p = 0.05$.

From Table 3, only the intercept and the band ratio $R_{rs}(667)/R_{rs}(488)$ can be considered significant for the model. Therefore, the following simplified algorithm (Model 1) was evaluated:

$$\text{Model 1: } a_{CDOM}355 = b_1 + b_2 \times R_{rs}(667)/R_{rs}(488) \text{ by OLS.} \tag{4}$$

The regression analysis of Model 1 provided the results presented in Table 4. The Model 1 fit is shown in Figure 4 along with the experimental data. The adjusted coefficient of determination R^2 was 0.8433. This high value of adjusted R^2 is comforting to pursue further the simple linear model.

Table 4. Results of the OLS regression of Model 1 on the test data set.

Parameter (Model 1)	Parameter Estimate	Standard Error	*t*-Value	*p*
b_1	0.1165	0.0226	5.157	1.46×10^{-6} *
b_2	1.9089	0.0857	22.273	$<2 \times 10^{-16}$ *

* Significant at $p = 0.05$.

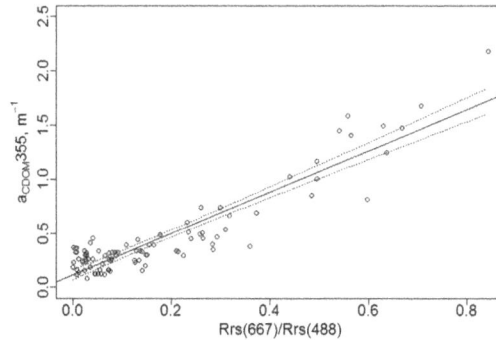

Figure 4. Plot of the in situ $a_{CDOM}355$ values versus the in situ band ratio $R_{rs}(667)/R_{rs}(488)$ of the test data set. The solid and dashed lines are the OLS fitted Model 1 and the 95% confidence interval, respectively.

The diagnostic check of Model 1 was carried out according to the normal distribution of the residuals and the constancy of the variance (i.e., the homoscedasticy). The Shapiro–Wilk normality test on the model residuals gave a P value of 0.7093. However, as shown in Figure 5, the plot of the residuals against the fitted values displays a variance which increases with the increment of the fitted value (i.e., the heteroscedasticity). Heteroscedasticity was also confirmed by the studentized Breusch–Pagan test [82], which gave a p-value of 1.091×10^{-6}. Besides heteroscedasticity, the residuals trend is slightly upward concave, Figure 5. This additional feature is discussed in Section 3.2.4.

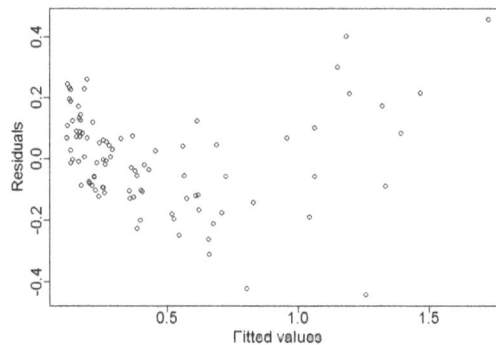

Figure 5. Plot of the residuals of Model 1 versus fitted values. Note the spread increase of the residual values with the increase of the fitted values.

Even if the OLS estimates of the parameters and the coefficient of determination (R^2) are not biased by heteroscedasticity, this violation affects the confidence intervals of OLS parameters [82]. Two different approaches can be adopted to overcome heteroscedasticity. The first approach is a dependent variable transformation according to the Cox-Box analysis [83]. The second approach is to carry out a generalized least square (GLS) regression with a specific variance structure [84]. Both approaches are adopted in the following.

3.2.2. Dependent Variable Transformation

The dependent variable transformation was carried out according to the Box-Cox transformation, i.e., $y' = y^{\alpha}$, where y is the original dependent variable, y' the transformed variable and α the transformation exponent. The plot in Figure 6, as produced by the R package MASS [85], suggests a square root transformation [83], i.e., $y' = y^{0.5}$, with the following Model 2 to be tested:

$$\text{Model 2: } a_{\text{CDOM}}355^{0.5} = b_1 + b_2 \times R_{rs}(667)/R_{rs}(488) \text{ by OLS} \tag{5}$$

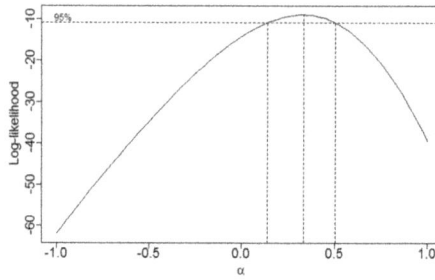

Figure 6. Plot of the log-likelihood function as a function of λ for the variable transformation $y' = y^{\alpha}$.

Table 5 reports the regression results obtained for Model 2 on the test data set. The model fit is shown in Figure 7 along with the experimental data. The adjusted coefficient of determination was 0.8369.

Table 5. Results of the OLS regression of Model 2 on the test data set.

Parameter (Model 2)	Parameter Estimate	Standard Error	*t*-Value	*p*
b_1	0.42387	0.01403	30.21	$<2 \times 10^{-16}$ *
b_2	1.15761	0.05321	21.75	$<2 \times 10^{-16}$ *

* Significant at $p = 0.05$.

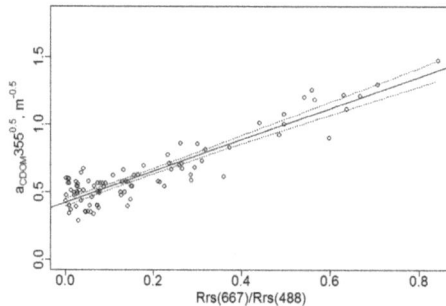

Figure 7. Plot of the in situ $a_{\text{CDOM}}355^{0.5}$ values versus the in situ band ratio $R_{rs}(667)/R_{rs}(488)$ of the test data set. The solid and dashed lines are the OLS fitted Model 2 and the 95% confidence interval, respectively.

Both the plots of the residuals (not shown) and the studentized Breusch–Pagan test ($p = 0.7571$) indicate that the problem of heteroscedasticity was solved.

3.2.3. Generalized Least Squares (GLS)

Another way of tackling the problem of heteroscedasticity is to perform a generalized least squares (GLS) regression analysis. In this case, the dependent and independent variables are untransformed, but the variance of the dependent variable is let to vary instead of being considered constant as in the OLS. Model 3 is, therefore, given by

$$\text{Model 3: } a_{CDOM}355 = b_1 + b_2 \times R_{rs}(667)/R_{rs}(488) \tag{6}$$

with the fitting parameters b_i's to be estimated by GLS. The fitting results of this model, which were carried out by using the R package "nlme" with a variance structure of the type "varPower()" [86], are shown in Table 6. Figure 8 shows the experimental data and the model fit. The adjusted coefficient of determination was 0.6325. Comparing the values of intercept b_1 and slope b_2 of Model 1 (Table 4) and Model 2 (Table 6), we can see that they are statistically different. This is unexpected, since heteroscedasticity should not affect the estimate of the regression parameters. Evidently, heteroscedasticity is not the only issue that emerges from the analysis of the residuals plot. In the next section, this problem is analysed more deeply.

Table 6. Results of the GLS regression Model 3 on the test data.

Parameter (Model 3)	Parameter Estimate	Standard Error	t-Value	p
b_1	0.18325	0.01552	11.81	$<2 \times 10^{-16}$ *
b_2	1.48088	0.11732	12.62	$<2 \times 10^{-16}$ *

* Significant at $p = 0.05$.

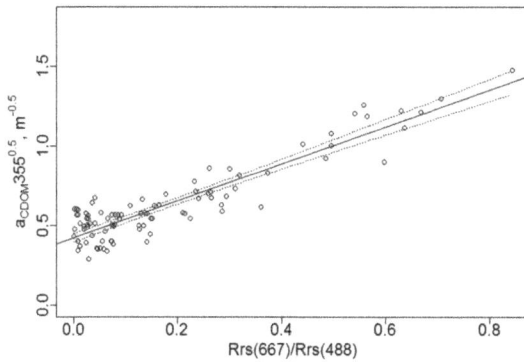

Figure 8. Plot of the in situ $a_{CDOM}355$ values versus the in situ band ratio $R_{rs}(667)/R_{rs}(488)$ of the test data set. The solid and dashed lines are the GLS fitted Model 3 and the 95% confidence interval, respectively.

3.2.4. Curvature of the Residuals

As shown in Figure 5, a slight curved trend is present in the residuals of Model 1. This curvature usually implies the presence of higher order terms in the model [87]. In this section, besides the possibility of introducing terms of higher order, also the interaction between independent variables was considered. The detailed steps of this kind of analysis are described in Crawley [88]. According to this procedure, preliminarily to the regression analysis, two explorative investigations should be carried out: (1) a regression tree to see if the model has a complex structure, i.e., there are interaction terms between the independent variables; and (2) a generalized additive model (GAM) analysis to see if high order terms are necessary. Once ascertained the general model, a repeated OLS regression is performed cancelling, each time, the non-significant terms starting from the highest interaction terms. When all the surviving terms are significant, the regression procedure stops. However, if heteroscedasticity is present in the final model diagnostic, the regression procedure must be repeated from the beginning after the transformation of the dependent variable. In our case, the regression tree and the GAM analysis both indicated that we had to start with a model comprising all the main effects, i.e., all the independent band ratios, all the interaction terms and a quadratic term for the band ratio $R_{rs}(667)/R_{rs}(488)$. At the end of the regression steps, the studentized Breusch–Pagan test indicated that the dependent variable had to be transformed, also in this case as $y' = y^{0.5}$. Repeating all the regression steps, the final model was the one with the dependent variable transformed as $y' = y^{0.5}$ and with the following lengthy form:

$$\text{Model 4: } a_{CDOM}355^{0.5} = b_1 \times R_{rs}(547)/R_{rs}(412) + b_2 \times R_{rs}(667)/R_{rs}(412) + b_3 \times$$
$$R_{rs}(667)/R_{rs}(488) + b_4 \times R_{rs}(547)/R_{rs}(488) + b_5 \times [R_{rs}(547)/R_{rs}(412)]^2 + b_6 \times$$
$$[R_{rs}(667)/R_{rs}(412)]^2 + b_7 \times [Rrs(667)/Rrs(488)]^2 + b_8 \times [Rrs(547)/Rrs(488)]^2 + b_9 \times \qquad (7)$$
$$Rrs(547)/Rrs(412) \times Rrs(667)/Rrs(412) + b_{10} \times Rrs(547)/Rrs(412) \times Rrs(667)/Rrs(488) +$$
$$b_{11} \times Rrs(667)/Rrs(412) \times Rrs(667)/Rrs(488)$$

The adjusted coefficient of determination R^2 was 0.9873. The results of the regression analysis carried out by OLS are summarized in Table 7.

Table 7. Results of the OLS regression of Model 4 on the test data set.

Parameter (Model 4)	Parameter Estimate	Standard Error	*t*-Value	*p*
b_1	−0.39669	0.07517	−5.277	1.05×10^{-6} *
b_2	2.18707	0.74484	2.936	0.004309 *
b_3	−5.50891	1.12927	−4.878	5.18×10^{-6} *
b_4	1.70642	0.12586	13.55	$<2 \times 10^{-16}$ *
b_5	0.06978	0.01424	4.901	4.74×10^{-6} *
b_6	3.18145	1.13802	2.796	0.006450 *
b_7	13.1729	3.34276	3.941	0.000170 *
b_8	−0.74416	0.10134	−7.343	1.37×10^{-10} *
b_9	−1.15771	0.27683	−4.182	7.20×10^{-5} *
b_{10}	3.41274	0.61429	5.556	3.34×10^{-7} *
b_{11}	−13.29309	3.88291	−3.423	0.000967 *

* Significant at $p = 0.05$.

3.2.5. Model Validation

In order to validate the four models tested so far, the calculated values $a_{CDOM}355_c$ of each model were compared to the in situ values of the validation data $a_{CDOM}355_{is}$ (validation data set, 79 observations collected from 2015 to 2016). The calculated $a_{CDOM}355_c$ were obtained by inserting the appropriate values of the band ratios of the validation data set into the equations of the four fitted models. The accuracy of the four models was then assessed by comparing the calculated values against the in situ values of the validation data set according to following statistical indicators: (a) the adjusted coefficient of determination (adjusted R^2) between $a_{CDOM}355_{is}$ and $a_{CDOM}355_c$; (b) the mean and standard deviation of the APD (APD = $100 \times |a_{CDOM}355_{is} - a_{CDOM}355_c|/a_{CDOM}355_{is}$); and (c) the RMSE between $a_{CDOM}355_{is}$ and $a_{CDOM}355_c$. Table 8 summarizes the results of the validation. For comparison, we have considered three additional empirical models developed by other authors. When a band of the additional models was not coincident with any of the bands considered for this work, the closest band among those at hand was chosen.

Table 8. Comparison results of the prediction ability of the four models on the validation data set. Three additional models taken from bibliography are also included.

Algorithm	Adjusted R^2	APD	RMSE (m^{-1})
Model 1	0.8426	33 (28) *	0.1421
Model 2	0.8322	31 (24)	0.1482
Model 3	0.8426	35 (31)	0.1496
Model 4	0.5528	72 (65)	0.3344
Mannino et al. [37]	0.5594	28 (23)	0.2580
Del Castillo and Miller [78]	0.0098	74 (71)	0.3541
D'Sa and Miller [76]	0.4122	39 (30)	0.2922

* Mean value (standard deviation).

The second column of Table 8 indicates a very good correlation between the calculated and the in situ values for the first three models, with adjusted R^2 values in the range of 0.832–0.843. The adjusted R^2 value for Model 4 was far below in spite of its highest fitting of the test data. The good agreement for the first three models must be truly appreciated considering that the validation data were collected more than one year after the test data. According to the APD values, Model 2 has the best prediction accuracy, while Model 4 the lowest one. The best value, 31 for Model 2, is comparable with APD's found by other authors [39,71] even if Mannino et al. obtained values as low as 18–20 [37]. On the absolute scale, the prediction accuracy was almost comparable for the first three models, as indicated by the RMSE values. With the highest RMSE value, the low prediction capability of Model 4 was confirmed. The performances of the three additional models were quite low and comparable to Model 4, in some cases even worse.

3.3. Application of Satellite-Derived CDOM

In order to verify the ability of Model 2 (i.e., the one with the best prediction accuracy) to capture the dynamic range of CDOM for the test areas, maps of satellite $a_{CDOM}355$ were produced for the studied region. The selected days to produce the maps were chosen on the basis of daily flow rate of the Po River [89] and the availability of cloud free MODIS-A R_{rs} imagery in order to show the distribution of riverine CDOM along the NCWAS. The processed MODIS-A satellite R_{rs} images, as shown as examples in Figure 9, indicate that the proposed Model 2 algorithm can be confidently applied to generate consistent MODIS-A maps of $a_{CDOM}355$. The satellite retrieved CDOM distribution fits quite well based on the field measurements and the processes that affect the distribution of biochemical properties along the NCWAS.

Furthermore, the field measurements as well as optical satellite data demonstrate a decrease in $a_{CDOM}355$ from North (Po River mouth) towards South and from nearshore to offshore. Increasing and decreasing of CDOM due to differences in Po River flow rate is a quite evident process (as expected) along the Western coast of the Adriatic Sea.

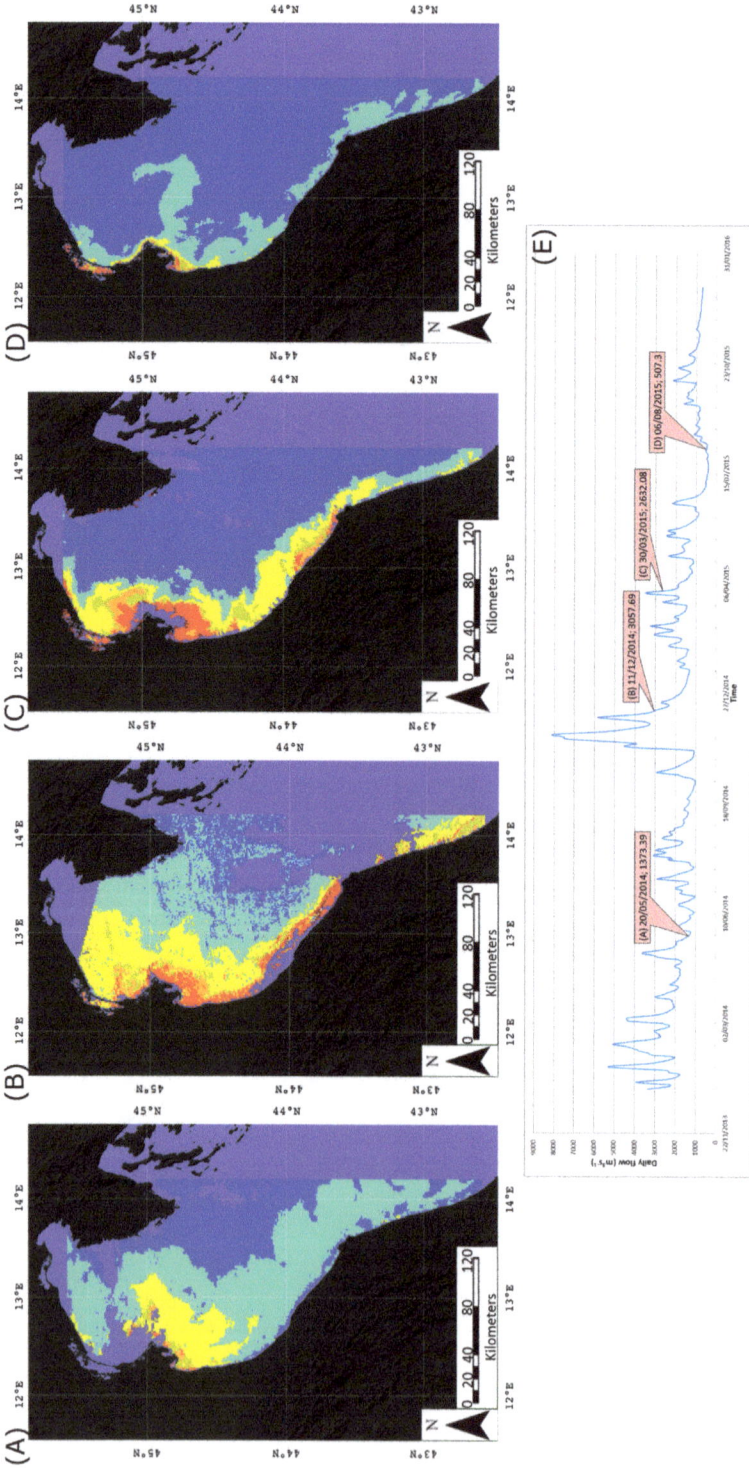

Figure 9. Satellite-derived a$_{CDOM}$355 maps obtained using MODIS-A R_{rs} computed using the Model 2 equation (a$_{CDOM}$355$^{0.5}$ = b_1 + b_2 × R_{rs}(667)/ R_{rs}(488)) for the days: (**A**) 20 May 2014; (**B**) 11 December 2014; (**C**) 30 March 2015; and (**D**) 6 August 2015. The calculated maps matches pretty well the flow rates of the Po River (**E**). CDOM maps are overlaid (only for visualization purposes) on ASTER GDEM World Wide Elevation data (1 arc-second of spatial resolution) where dark green colour depicts land and the blue-violet colour the sea (in our case, no-data values due to the clouds).

4. Discussion

According to the above model analysis, the band ratio $R_{rs}(667)/R_{rs}(488)$ was the most significant independent variable for the retrieval of $a_{CDOM}355$ in the study area of NCWAS, which is characterized by high river discharges. The results confirm that the performance of empirical algorithms in a complex basin, i.e., a basin strongly influenced by freshwater/terrestrial input, can be significantly improved by selecting at least one band with a wavelength greater than 600 nm [34,40,78]. The additional spectral bands in the red or near infrared could be also helpful in better accounting for detritus particles because they are less sensible to the constituents in the seawater column so that the $R_{rs}(667)$ can be assumed to be mainly invariant as not influenced by detritus particles. On the sea surface at the wavelength of 667 nm the total light absorption is highest for seawater rather than for other constituents in the seawater column so that the $R_{rs}(667)$ can be assumed to be constant [34]. Moreover, riverine CDOM shows a higher absorption at 488 nm than the chlorophyll [40]. On the other hand, the $R_{rs}(547)$ nm is mostly driven by particles light scattering that in coastal water could reduce the uncertainty of CDOM retrieval when used as a reference point in band ratios [77]. Furthermore, the NCWAS is well described by many authors to be impacted by particles coming from many rivers [60–62,90]. Therefore, we applied the ratio $R_{rs}(667)/R_{rs}(488)$ that is more sensitive to changes in CDOM rather than in chlorophyll because, in low salinity waters as in the NCWAS, most of the light attenuation is controlled by CDOM [14,91]. The high river discharges on the NCWAS should therefore imply the use of a band ratio normally used in retrieving algorithms for river-plume waters. In this work, only simple linear models were considered. However, as shown in the Section 3, they proved to be quite effective for modelling the available data. Of the final four algorithms, only Model 2 and Model 4 fulfil the fundamental requirements for a correct linear regression. Considering the prediction capability, from our analysis Model 2 can be indicated as the best algorithm in order to retrieve $a_{CDOM}355$ in NCWAS starting from the band ratio $R_{rs}(667)/R_{rs}(488)$. The predictions of this model in data collected more than one year after the model fitting showed a very good comparison with the in situ measured data with an adjusted R^2 value of 0.8322, Table 8. Just for comparison, similar validation analyses for coastal sea waters gave R^2 values of: 0.37 in the same geographical area for $a_{CDOM}412$ [71], 0.61 for $a_{CDOM}443$ in the western Canada coastal waters [92], 0.62 for $a_{CDOM}412$ in the Gulf of Mexico [91], 0.70 for $a_{CDOM}442$ in the North Sea and Western English Channel [32], 0.83 $a_{CDOM}420$ in a Finnish lake [93] and 0.96 for $a_{CDOM}355$, $a_{CDOM}412$ and $a_{CDOM}443$ in coastal waters in U.S. Middle Atlantic Bight [37]. To better illustrate the agreement between calculated and measured values, in Figure 10 a graphical comparison between the in situ validation data and the Model 2 calculated data is shown. Figure 10 indicates that the prediction capability of the model seems to be rather good. It is however fair to add that the model performances could be overestimated due to the size of the validation data set, which is not very large, and the limited range of $a_{CDOM}355$ values, which roughly spans only one order of magnitude.

Plots of satellite-derived distributions of $a_{CDOM}355$ are spatially and temporally consistent with changes in Po River discharges (Figure 9) and the NCWAS. Since terrestrial DOM is the main source of CDOM to estuaries and coastal seas, $a_{CDOM}355$ values should generally decrease from the estuary to off-shore. The CDOM maps of Figure 9, moreover, show that the minimum CDOM concentration appears in 2015 only after a long period of reduced Po River flows (daily flow less than 500 $m^3 \cdot s^{-1}$), nevertheless determining the distribution of the sediments along the western Adriatic coast. The increase in surface $a_{CDOM}355$ corresponds to high river discharge.

As an example, Figures 9b and 11 show how an exceptional Po River discharge, happened in autumn 2014 (mean discharge 10 November–10 December 2014: 5028.2 $m^3 \cdot s^{-1}$) spreads high concentrations of riverine CDOM over the NCWAS compared with other periods, reaching the Istrian coast [44]. In particular, Figure 11 shows the CDOM map obtained by applying Model 2 to the R_{rs} Landsat 8 (30 m/pixel) bands closest to the $R_{rs}(667)$ and $R_{rs}(488)$ MODIS bands. In this case, we exceptionally tested the Model 2 equation on the Landsat R_{rs} imagery, which was besides the severe cloud cover affecting the study area the only available for the day of the maximum Po River discharge between 2014 and 2016 (i.e., 19 November 2014; see plot of Figure 9e). Landsat retrieved CDOM map, because of the higher GSD, moreover, correctly depicts the detrital contribution of all the minor rivers' discharges occurring in the North and West Adriatic coasts, as well as the Venice Lagoon sub-basins. However, it is important to remark that MODIS R_{rs} are recursively generated as standard product [68,69], while Landsat R_{rs} were retrieved by applying the following commonly used R_{rs} equation using as input Landsat radiance data.

$$R_{rs}(\lambda) = L_{tot}(\lambda) - L_{surf}(\lambda)/E_d(\lambda) = L_w(\lambda)/E_d(\lambda) \qquad (8)$$

where $L_{tot}(\lambda) = L_0(\lambda) - L\uparrow(\lambda)$ is the total spectral radiance signal above the water surface; $L_{surf}(\lambda)$ is the contribution to the radiance due to the reflection of the water surface; $L_w(\lambda)$ is the water leaving spectral radiance; and $E_d(\lambda) = E(\lambda)\cos\theta_z\tau_\downarrow(\lambda)$ is the incident irradiance above the water surface. Once the surface reflection is removed, the R_{rs} can be considered Lambertian for short variation of the viewing angle around the zenith. In order to estimate and then remove the signal reflected by the water surface (L_{surf}) we used the Hydrolight 4.3 software that is a radiative transfer numerical model that computes radiance distributions and derived quantities for water bodies [94].

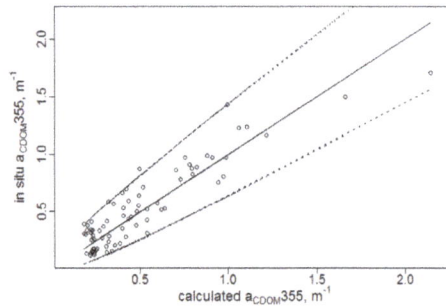

Figure 10. Plot of the in situ $a_{CDOM}355$ values versus the $a_{CDOM}355$ values calculated according to Model 2. All data are from the validation data set. The solid line corresponds to the one-to-one relationship and the dashed lines are the 95% prediction intervals of Model 2.

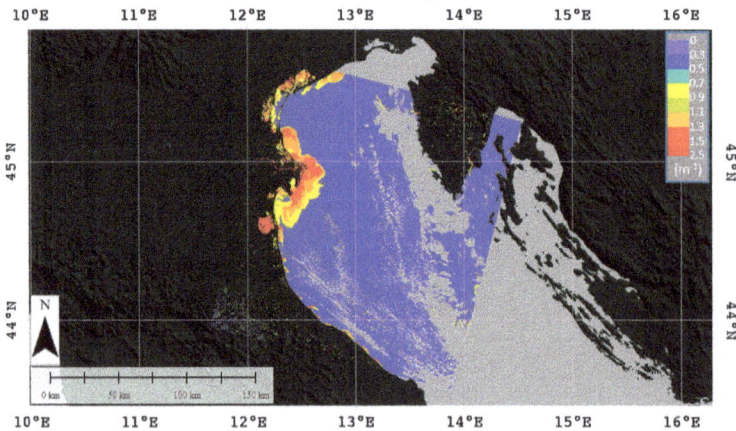

Figure 11. CDOM (m^{-1}) map retrieved by applying Model 2 equation to the Landsat 8 R_{rs} acquired on the NCWAS on 19 November 2016. CDOM map is overlaid (only for visualization purposes) on ASTER GDEM World Wide Elelvation data (1 arc-second of spatial resolution) where dark green colour depicts land and the light grey colour the sea (in our case no-data values due to the clouds).

Landsat atmospheric correction and R_{rs} retrieval are anyhow out of the scope of this study and the use of Landsat data is not an important step for consistency validation of the CDOM algorithm here proposed and validated only for MODIS—A R_{rs} on the NCWAS coastal areas. In addition, the central wavelengths of the two bands applied for the a$_{CDOM}$ algorithm (667 nm and 488 nm) from Landsat-8 and MODIS are close, while their bandwidths and SNRs (signal to noise ratio) are different.

The decrease in surface a$_{CDOM}$355 towards Southern Adriatic Sea and off-shore during spring 2014 and summer 2015 is consistent with photobleaching due to high solar radiation (both periods; [10,20]) as well as low river discharge (summer period).

5. Conclusions and Future Work

Based on the in situ data collected from several cruises in the North-Central Western Adriatic Sea (Mediterranean Sea) and on the corresponding MODIS ocean colour data, empirical models for the estimation of Coloured Dissolved Organic Matter (CDOM) using sensor data were tested. Data were split into two data sets: a test data set for model fitting (data collected from 2013 to 2015) and a validation data set (data collected from 2015 to 2016) for model assessment. The model set up was based on some statistical indicators: adjusted R^2, absolute percentage difference (APD) and root mean square error (RMSE).

Following statistical check and diagnostic, four final simple linear models were obtained. Three of them gave very good results in terms of comparison between the calculated and the in situ measured values of the validation data set (adjusted R^2 in the range 0.83–0.84). On the base of APD and RMSE, the best model resulted to be a simple linear model (Model 2) with the dependent variable transformed according to $y' = y^{0.5}$ and the band ratio $R_{rs}(667)/R_{rs}(488)$ as the only independent variable. The use of the $R_{rs}(667)$ instead of the $R_{rs}(547)$ as a reference point in band ratio seems to reduce the uncertainty of CDOM retrieval in the NCWAS characterized by high light particles scattering. However, in this view, future work will be dedicated to demonstrate that retrievals are effective even for high particulate concentrations, e.g., by showing their efficacy even for high $R_{rs}(547)$ conditions and for particularly high river outflows.

The validation methods and processing of MODIS-A data show that Model 2 provides a good retrieval for a$_{CDOM}$355 and it is consistent with the spatial and temporal distribution of freshwater discharges along the Western Adriatic Sea.

The spatial and temporal limitations associated with in situ and remote sensing data has introduced some uncertainties. Anyway, the retrieved algorithm (Model 2) is very promising for CDOM mapping in this complex basin and should be considered when applying ocean colour remote sensing data to quantify other ocean constituents (e.g., chlorophyll).

MODIS-A data provision started on 2002 and its quality in terms of bands and Signal to Noise Ratio is going down, therefore, future work will be focused on developing new CDOM algorithms for the recently launched multispectral optical satellite sensors (e.g., Landsat 8 and Sentinel 2 and 3 with a GSD of 30 m) including also VIIRS observations and next generation hyperspectral sensor (e.g., EnMAP and PRISMA with a GSD of 30 m and the VNIR VENµS Superspectral Camera with a GSD of 10 m) and their validation strategy using novel in situ measurements.

Acknowledgments: The research activities were conducted thanks to the funding projects RITMARE (Ricerca ITaliana per il MARE financed by the Italian University and Research Ministry, 2012–2016), ECOSEE/A (DG-MARE Guardian of the Sea co-financed by the European Union) and 'Smart Basilicata' (Contract No. 6386-3, 20 July 2016) that was approved by the Italian Ministry of Education, University and Research (Notice MIUR n. 84/Ric 2012, PON 2007–2013 of 2 March 2012) and funded with the Cohesion Fund 2007–2013 of the Basilicata Regional authority. We also thank the crews of Research Vessels G. Dallaporta, Tecnopesca II, and Eco 1 for their help in sampling activities during the cruises. We also thanks Raffaele D'Adamo to make the laboratories of CNR-ISMAR of Lesina available for CDOM analysis. Additionally, we would like to thank three anonymous reviewers for their valuable comments, which helped to improve the quality of this paper.

Author Contributions: A. Campanelli and S. Pascucci developed the idea for the study and its design. A. Campanelli, S. Pascucci, F. Grilli and M. Betti performed the in situ measurements, laboratories'analysis, data processing and the construction and validation of the dataset. S. Guicciardi was responsible for the statistical analysis of the empirical modelling and its presentation. A. Campanelli, S. Pascucci, S. Guicciardi, S. Pignatti and M. Marini shared in the analysis and development of the discussion. A. Campanelli, S. Pascucci and S. Guicciardi finalized the manuscript. All authors read and approved the manuscript.

Conflicts of Interest: The authors declare no conflicts of interest.

References

1. Volpe, G.; Santoleri, R.; Vellucci, V.; Ribera d'Alcala, M.; Marullo, S.; D'Ortenzio, F. The colour of the Mediterranean Sea: Global versus regional bio-optical algorithms evaluation and implication for satellite chlorophyll estimates. *Remote Sens. Environ.* **2007**, *107*, 625–638. [CrossRef]
2. McClain, C.R. A decade of satellite ocean color observation. *Annu. Rev. Mar. Sci.* **2009**, *1*, 19–42. [CrossRef] [PubMed]
3. Del Vecchio, R.; Subramaniam, A. Influence of the Amazon River on the surface optical properties of the western tropical North Atlantic Ocean. *J. Geophys. Res.* **2004**, *109*, C11001. [CrossRef]
4. Pan, X.; Mannino, A.; Russ, M.E.; Hooker, S.B. Remote sensing of the absorption coefficients and chlorophyll a concentration in the U.S. southern Middle Atlantic Bight from SeaWiFS and MODIS-Aqua. *J. Geophys. Res.* **2008**, *113*, C11022. [CrossRef]
5. Siegel, D.A.; Maritorena, S.; Nelson, N.B.; Hansell, D.A.; Lorenzi-Kayser, M. Global distribution and dynamics of colored dissolved and detrital organic materials. *J. Geophys. Res.* **2002**, *107*, C123228. [CrossRef]
6. Keith, D.J.; Yoder, J.A.; Freeman, S.A. Spatial and temporal distribution of coloured dissolved organic matter (CDOM) in Narragansett Bay, Rhode Island: Implication for phytoplankton in coastal waters. *Estuar. Coast. Shelf Sci.* **2002**, *55*, 705–717. [CrossRef]
7. McKee, D.; Cunningham, A.; Jones, K.J. Optical and hydrographic consequences of freshwater run-off during spring phytoplankton growth in a Scottish fjord. *J. Plankton Res.* **2002**, *24*, 1163–1171. [CrossRef]
8. Bricaud, A.; Morel, A.; Prieur, L. Absorption by dissolved organic matter of the sea (yellow substance) in the UV and visible domains. *Limnol. Oceanogr.* **1981**, *26*, 43–53. [CrossRef]
9. Nelson, J.R.; Guarda, S. Particulate and dissolved spectral absorption on the continental shelf of the southeastern United States. *J. Geophys. Res.* **1995**, *100*, 8715–8732. [CrossRef]
10. Coble, P. Marine Optical Biogeochemistry: The Chemistry of Ocean Color. *Chem. Rev.* **2007**, *107*, 402–418. [CrossRef] [PubMed]
11. Swan, C.M.; Siegel, D.A.; Nelson, N.B.; Carlson, C.A.; Nasir, E. Biogeochemical and hydrographic controls on chromophoric dissolved organic matter distribution in the Pacific Ocean. *Deep-Sea Res. I* **2009**, *56*, 2175–2192. [CrossRef]
12. Nelson, N.B.; Siegel, D.A.; Carlson, C.A.; Swan, C.M. Tracing global biogeochemical cycles and meridional overturning circulation using chromophoric dissolved organic matter. *Geophys. Res Lett.* **2010**, *37*. [CrossRef]
13. Stedmon, C.A.; Osburn, C.L.; Gragh, T. Tracing water mass mixing in the Baltic–North Sea transition zone using the optical properties of coloured dissolved organic matter. *Estuar. Coast. Shelf Sci.* **2010**, *87*, 156–162. [CrossRef]
14. Del Castillo, C.E.; Coble, P.G.; Morell, J.M.; López, J.M.; Corredor, J.E. Analysis of the optical properties of the Orinoco River plume by absorption and fluorescence spectroscopy. *Mar. Chem.* **1999**, *66*, 35–51. [CrossRef]
15. Del Vecchio, R.; Blough, N.V. Spatial and seasonal distribution of chromophoric dissolved organic matter (CDOM) and dissolved organic carbon (DOC) in the Middle Atlantic Bight. *Mar. Chem.* **2004**, *89*, 169–187. [CrossRef]
16. Rochelle-Newall, E.J.; Fisher, T.R. Chromophoric dissolved organic matter and dissolved organic carbon in Chesapeake Bay. *Mar. Chem.* **2002**, *77*, 23–41. [CrossRef]
17. Steinberg, D.K.; Nelson, N.; Carlson, C.A.; Prusak, A.C. Production of chromophoric dissolved organic matter (CDOM) in the open ocean by zooplankton and the colonial cyanobacterium *Trichodesmium* spp. *Mar. Ecol. Prog. Ser.* **2004**, *267*, 45–56. [CrossRef]
18. Andrew, A.A.; Del Vecchio, R.; Subramaniam, A.; Blough, N.V. Chromophoric dissolved organic matter (CDOM) in the equatorial Atlantic Ocean: Optical properties and their relation to CDOM structure and source. *Mar. Chem.* **2013**, *148*, 33–43. [CrossRef]
19. Nelson, N.B.; Siegel, D.A. The global distribution and dynamics of chromophoric dissolved organic matter. *Annu. Rev. Mar. Sci.* **2013**, *5*, 447–476. [CrossRef] [PubMed]

20. Mopper, K.; Keiber, D.J. Photochemistry and the cycling of carbon, sulfur, nitrogen and phosphorus. In *Biogeochemistry of Marine Dissolved Organic Matter*; Hansel, D.A., Carlson, C.A., Eds.; Academic Press: San Diego, CA, USA, 2002; pp. 455–507.

21. Moran, M.A.; Sheldon, W.M.; Zepp, R.G. Carbon loss and optical property changes during long-term photochemical and biological degradation of estuarine dissolved organic matter. *Limnol. Oceangr.* **2000**, *45*, 1254–1264. [CrossRef]

22. Boyd, T.J.; Osburn, C.L. Changes in CDOM fluorescence from allochthonous and autochthonous sources during tidal mixing and bacterial degradation in two coastal estuaries. *Mar. Chem.* **2004**, *89*, 189–210. [CrossRef]

23. Carder, K.L.; Chen, F.R.; Lee, Z.P.; Hawes, S.K.; Kamykowski, D. Semianalytic moderate-resolution imaging spectrometer algorithms for chlorophyll and absorption with bio-optical domains based on nitrate-depletion temperatures. *J. Geophys. Res.* **1999**, *104*, 5403–5421. [CrossRef]

24. Hoge, F.E.; Wright, C.W.; Lyon, P.E.; Swift, R.N.; Yungel, J.K. Inherent optical properties imagery of the western North Atlantic Ocean: Horizontal spatial variability of the upper mixed layer. *J. Geophys. Res.* **2001**, *106*, 31129–31140. [CrossRef]

25. Lee, Z.P.; Carder, K.L.; Arnone, R.A. Deriving inherent optical properties from water color: A multiband quasi-analytical algorithm for optically deep waters. *Appl. Opt.* **2002**, *41*, 5755–5772. [CrossRef] [PubMed]

26. Maritorena, S.; Siegel, D.A.; Peterson, A.R. Optimization of a semianalytical ocean color model for global-scale applications. *Appl. Opt.* **2002**, *41*, 2705–2714. [CrossRef] [PubMed]

27. Siegel, D.A.; Maritorena, S.; Nelson, N.B.; Behrenfeld, M.J. Independence and interdependencies among global ocean color properties: Reassessing the bio-optical assumption. *J. Geophys. Res.* **2005**, *110*, C07011. [CrossRef]

28. Siegel, D.A.; Maritorena, S.; Nelson, N.B.; Behrenfeld, M.J.; McClain, C.R. Colored dissolved organic matter and its influence on the satellite-based characterization of the ocean biosphere. *Geophys. Res. Lett.* **2005**, *32*, L20605. [CrossRef]

29. Doerffer, R.; Schiller, H. The MERIS case 2 water algorithm. *Int. J. Remote Sens.* **2007**, *28*, 517–535. [CrossRef]

30. Lee, Z.P.; Arnone, R.A.; Hu, C.; Werdell, P.J.; Lubac, B. Uncertainties of optical parameters and their propagations in an analytical ocean color inversion algorithm. *Appl. Opt.* **2010**, *49*, 369–381. [CrossRef] [PubMed]

31. Bricaud, A.; Ciotti, A.M.; Gentili, B. Spatial-temporal variations in phytoplankton size and colored detrital matter absorption at global and regional scales, as derived from twelve years of SeaWiFS data (1998–2009). *Glob. Biogeochem. Cycles* **2012**, *26*, GB1010. [CrossRef]

32. Tilstone, G.H.; Peters, S.W.M.; Van Der Woerd, H.J.; Eleveld, M.A.; Ruddick, K.; Schönfeld, W.; Krasemann, H.; Martinez-Vicente, V.; Blondeau-Patissier, D.; Röttgers, R.; et al. Variability in specific-absorption properties and their use in a semianalytical ocean colour algorithm for MERIS in north sea and western English channel coastal waters. *Remote Sens. Environ.* **2012**, *118*, 320–338. [CrossRef]

33. Werdell, P.J.; Franz, B.A.; Bailey, S.W.; Feldman, G.C.; Boss, E.; Brando, V.E.; Dowell, M.; Hirata, T.; Lavender, S.J.; Lee, Z.; et al. Generalized ocean color inversion model for retrieving marine inherent optical properties. *Appl. Opt.* **2013**, *52*, 2019–2037. [CrossRef] [PubMed]

34. Aurin, D.A.; Dierssen, H.M. Advantages and limitations of ocean color remote sensing in CDOM-dominated, mineral-rich coastal and estuarine waters. *Remote Sens. Environ.* **2012**, *125*, 181–197. [CrossRef]

35. Dong, Q.; Shang, S.; Lee, Z. An algorithm to retrieve absorption coefficient of chromophoric dissolved organic matter from ocean color. *Remote Sens. Environ.* **2013**, *128*, 259–267. [CrossRef]

36. Matsuoka, A.; Hooker, S.B.; Bricaud, A.; Gentili, B.; Babin, M. Estimating absorption coefficients of colored dissolved organic matter (CDOM) using a semi-analytical algorithm for southern Beaufort Seawaters: Application to deriving concentrations of dissolved organic carbon from space. *Biogeosciences* **2013**, *10*, 917–927. [CrossRef]

37. Mannino, A.; Russ, M.E.; Hooker, S.B. Algorithm development and validation for satellite-derived distributions of DOC and CDOM in the U.S. Middle Atlantic Bight. *J. Geophys. Res.* **2008**, *113*, C07051. [CrossRef]

38. Fichot, C.G.; Kaiser, K.; Hooker, S.B.; Amon, R.M.W.; Babin, M.; Belanger, S.; Walker, S.A.; Benner, R. Pan-Arctic distributions of continental runoff in the Arctic Ocean. *Sci. Rep.* **2013**, *3*, 1053. [CrossRef] [PubMed]

39. Mannino, A.; Novak, M.G.; Hooker, S.B.; Hyde, K.; Aurin, D. Algorithm development and validation of CDOM properties for estuarine and continental shelf waters along the northeastern U.S. coast. *Remote Sens. Environ.* **2014**, *152*, 576–602. [CrossRef]

40. Zhu, W.; Yu, Q.; Tian, Y.Q.; Becker, B.L.; Zheng, T.; Carrick, H. An assessment of remote sensing for colored dissolved organic matter in complex freshwater environments. *Remote Sens. Environ.* **2014**, *140*, 766–778. [CrossRef]

41. Raicich, F. On the fresh water balance of the Adriatic coast. *J. Mar. Syst.* **1996**, *9*, 305–319. [CrossRef]

42. Marini, M.; Fornasiero, P.; Artegiani, A. Variations of Hydrochemical Features in the Coastal Waters of Monte Conero: 1982–1990. *Mar. Ecol.* **2002**, *23*, 258–271. [CrossRef]

43. Campanelli, A.; Fornasiero, P.; Marini, M. Physical and Chemical characterization of water column in the Piceno coastal area (Adriatic Sea). *Fresenius Environ. Bull.* **2004**, *13*, 430–435.

44. Campanelli, A.; Grilli, F.; Paschini, E.; Marini, M. The influence of an exceptional Po River flood on the physical and chemical oceanographic properties of the Adriatic Sea. *Dyn. Atmos. Oceans* **2011**, *52*, 284–297. [CrossRef]

45. Giani, M.; Djakovac, T.; Degobbis, D.; Cozzi, S.; Solidoro, C.; Umani, S.F. Recent changes in the marine ecosystems of the northern Adriatic Sea. *Estuar. Coast. Shelf Sci.* **2012**, *115*, 1–13. [CrossRef]

46. Degobbis, D.; Precali, R.; Ivančić, I.; Smodlaka, N.; Fuks, D.; Kveder, S. Long-term changes in the northern Adriatic ecosystem related to anthropogenic eutrophication. *J. Environ. Pollut.* **2000**, *13*, 495–533. [CrossRef]

47. Marini, M.; Jones, B.H.; Campanelli, A.; Grilli, F.; Lee, C.M. Seasonal variability and Po River plume influence on biochemical properties along western Adriatic coast. *J. Geophys. Res.* **2008**, *113*, C05S90. [CrossRef]

48. Berto, D.; Giani, M.; Savelli, F.; Centanni, E.; Ferrari, C.R.; Pavoni, B. Winter to spring variations of chromophoric dissolved organic matter in a temperate estuary (Po River, northern Adriatic Sea). *Mar. Environ. Res.* **2010**, *70*, 73–81. [CrossRef] [PubMed]

49. Cozzi, S.; Giani, M. River water and nutrient discharges in the Northern Adriatic Sea: Current importance and long term changes. *Cont. Shelf Res.* **2011**, *31*, 1881–1893. [CrossRef]

50. Poulain, P.M.; Raicich, F. Forcings. In *Physical Oceanography of the Adriatic Sea*; Cushman-Roisin, B., Gacic, M., Poulain, P.M., Artegiani, A., Eds.; Kluwer Academic Publishers: Dordrecht, The Netherlands, 2001; pp. 45–65.

51. Zore-Armanda, M.; Gačić, M. Effects of Bora on the circulation in the North Adriatic. *Ann. Geophys.* **1987**, *5B*, 93–102.

52. Artegiani, A.; Bregant, D.; Paschini, E.; Pinardi, N.; Raicich, F.; Russo, A. The Adriatic Sea general circulation. Part I. Air-sea interactions and water mass structure. *J. Phys. Oceanogr.* **1997**, *27*, 1492–1514. [CrossRef]

53. Artegiani, A.; Bregant, D.; Paschini, E.; Pinardi, N.; Raicich, F.; Russo, A. The Adriatic Sea general circulation. Part II: Baroclinic Circulation Structure. *J. Phys. Oceanogr.* **1997**, *27*, 1515–1532.

54. Boldrin, A.; Carniel, S.; Giani, M.; Marini, M.; Bernardi Aubry, F.; Campanelli, A.; Grilli, F.; Russo, A. Effects of bora wind on physical and biogeochemical properties of stratified waters in the northern Adriatic. *J. Geophys. Res.* **2009**, *114*, C08S92. [CrossRef]

55. Poulain, P.M.; Cushman-Roisin, B. Circulation. In *Physical Oceanography of the Adriatic Sea*; Cushman-Roisin, B., Gačič, M., Poulain, P.M., Artegiani, A., Eds.; Kluwier Academic Publisher: Dordrecht, The Netherlands, 2001; pp. 67–109.

56. Orlić, M.; Gačič, M.; La Violette, P.E. The currents and circulation of the Adriatic Sea. *Oceanol. Acta* **1992**, *15*, 109–124.

57. Kourafalou, V.H. Process studies on the Po River plume, north Adriatic Sea. *J. Geophys. Res.* **1999**, *104*, 29963–29985. [CrossRef]

58. Kourafalou, V.H. River plume development in semi-enclosed Mediterranean regions: North Adriatic Sea and Northwestern Aegean Sea. *J. Mar. Syst.* **2001**, *30*, 181–205. [CrossRef]

59. Jeffries, M.A.; Lee, C.M. A climatology of the northern Adriatic Sea's response to Bora and river forcing. *J. Geophys. Res.* **2007**, *112*, 1–18. [CrossRef]

60. Sangiorgi, F.; Donders, T.H. Reconstructing 150 years of eutrophication in the north-western Adriatic Sea (Italy) using dinoflagellate cysts, pollen and spores. *Estuar. Coast. Shelf Sci.* **2004**, *60*, 69–79. [CrossRef]

61. Socal, G.; Acri, F.; Bastianini, M.; Bernardi Aubry, F.; Bianchi, F.; Cassin, D.; Coppola, J.; De Lazzari, A.; Bandelj, V.; Cossarini, G.; et al. Hydrological and biogeochemical features of the Northern Adriatic Sea in the period 2003–2006. *Mar. Ecol.* **2008**, *29*, 449–468. [CrossRef]

62. Marini, M.; Campanelli, A.; Sanxhaku, M.; Kljajić, Z.; Betti, M.; Grilli, F. Late spring characterization of different coastal areas of the Adriatic Sea. *Acta Adriat.* **2015**, *56*, 27–46.

63. Specchiulli, A.; Bignami, F.; Marini, M.; Fabbrocini, A.; Scirocco, T.; Campanelli, A.; Penna, P.; Santucci, A.; D'Adamo, R. The role of forcing agents on biogeochemical variability along the southwestern Adriatic coast: The Gulf of Manfredonia case study. *Estuar. Coast. Shelf Sci.* **2016**, *183*, 136–149. [CrossRef]

64. United Nations Educational, Scientific, and Cultural Organization (UNESCO). *The Acquisition, Calibration and Analysis of CTD Data*; A Report of SCOR WG 51; United Nations Educational, Scientific, and Cultural Organization: Paris, France, 1988; pp. 1–59.

65. Mitchell, B.G.; Kahru, M.; Wieland, J.; Stramska, M. Determination of spectral absorption coefficient of particles, dissolved material and phytoplankton for discrete water samples. In *Ocean Optics Protocols for Satellite Ocean Colour Sensor Validation*; NASA/TM-2003-211621/Rev4-Volume IV; Fargion, G.S., Mueller, J.L., McClain, C.R., Eds.; NASA Goddard Space Flight Center: Greenbelt, MD, USA, 2003; pp. 39–64.

66. Vignudelli, S.; Santinelli, C.; Murru, E.; Nannicini, L.; Seritti, A. Distributions of dissolved organic carbon (DOC) and chromophoric dissolved organic matter (CDOM) in coastal of the northern Tyrrhenian Sea (Italy). *Estuar. Coast. Shelf Sci.* **2004**, *60*, 133–149. [CrossRef]

67. Copernicus Marine Environment Monitoring Service (CMEMS) Online Catalogue. Available online: http://marine.copernicus.eu/services-portfolio/access-to-products/ (accessed on 10 October 2016).

68. Volpe, G.; Colella, S.; Forneris, V.; Tronconi, C.; Santoleri, R. The Mediterranean Ocean Colour Observing System—System development and product validation. *Ocean Sci.* **2012**, *8*, 869–883. [CrossRef]

69. Santoleri, R.; Volpe, G.; Marullo, S.; Buongiorno Nardelli, B. Open Waters Optical Remote Sensing of the Mediterranean Sea. In *Remote Sensing of the European Seas*; Barale, V., Gade, M., Eds.; Springer: Berlin/Heidelberg, Germany, 2008; pp. 103–116.

70. Mobley, C.D. Estimation of the remote-sensing reflectance from above-surface measurements. *Appl. Opt.* **1999**, *38*, 7442–7455. [CrossRef] [PubMed]

71. D'Alimonte, D.; Zibordi, G.; Berthon, J.-B.; Canuti, E.; Kajiyama, T. Performance and applicability of bio-optical algorithms in different European seas. *Remote Sens. Environ.* **2012**, *124*, 402–412. [CrossRef]

72. Braga, F.; Giardino, C.; Bassani, C.; Matta, E.; Candiani, G.; Strombeck, N.; Adamo, M.; Bresciani, M. Assessing water quality in the northern Adriatic Sea from HICO (TM) data. *Remote Sens. Lett.* **2013**, *4*, 1028–1037. [CrossRef]

73. Kajiyama, T.; D'Alimonte, D.; Zibordi, G. Regional algorithms for European Seas: A case study based on MERIS data. *IEEE Geosci. Remote Sens.* **2013**, *10*, 283–287. [CrossRef]

74. Brando, V.E.; Braga, F.; Zaggia, L.; Giardino, C.; Bresciani, M.; Matta, E.; Bellafiore, D.; Ferrarin, C.; Maicu, F.; Benetazzo, A.; et al. High-resolution satellite turbidity and sea surface temperature observations of river plume interactions during a significant flood event. *Ocean Sci.* **2015**, *11*, 909–920. [CrossRef]

75. Sathyendranath, S. *Remote Sensing of Ocean Colour in Coastal, and Other Optically-Complex Waters*; Report Number 3; International Ocean-Colour Coordinating Group (IOCCG): Dartmouth, NS, Canada, 2003; pp. 1–140.

76. D'Sa, E.J.; Miller, R.L. Bio-optical properties in waters influenced by the Mississippi River during low flow conditions. *Remote Sens. Environ.* **2003**, *84*, 538–549. [CrossRef]

77. Belanger, S.; Babin, M.; Larouche, P. An empirical ocean color algorithm for estimating the contribution of chromophoric dissolved organic matter to total light absorption in optically complex waters. *J. Geophys. Res.* **2008**, *113*, C04027. [CrossRef]

78. Del Castillo, C.E.; Miller, R.L. On the use of ocean color remote sensing to measure the transport of dissolved organic carbon by the Mississippi River Plume. *Remote Sens. Environ.* **2008**, *112*, 836–844. [CrossRef]

79. Tiwari, S.P.; Shanmugam, P. An optical model for the remote sensing of coloured dissolved organic matter in coastal/ocean waters. *Estuar. Coast. Shelf Sci.* **2011**, *93*, 396–402. [CrossRef]

80. R Core Team. *R: A Language and Environment for Statistical Computing*; R Foundation for Statistical Computing: Vienna, Austria, 2016; Available online: https://www.R-project.org/ (accessed on 22 June 2016).

81. Graybill, F.A.; Iyer, H.K. *Regression Analysis: Concepts and Applications*; Duxbuty Press: Belmont, TN, USA, 1994; p. 3978.

82. Wooldridge, J.M. Heteroscedasticity. In *Introductory Econometrics*, 6th ed.; Cengage Learning: Boston, MA, USA, 2016; pp. 240–252.

83. Draper, N.R.; Smith, H. *Applied Regression Analysis*, 3rd ed.; John Wiley and Sons: New York, NY, USA, 1998; pp. 280–282.

84. Zuur, A.F.; Ieno, E.N.; Walker, N.J.; Saveliev, A.A.; Smith, G.M. *Mixed Effects Models and Extensions in Ecology with R*; Springer: New York, NY, USA, 2009; pp. 74–75.

85. Venables, W.N.; Ripley, B.D. *Modern Applied Statistics with S*, 4th ed.; Springer: New York, NY, USA, 2002; p. 171.

86. Pinheiro, J.; Bates, D.; DebRoy, S.; Sarkar, D.; R Core Team. NLME: Linear and Nonlinear Mixed Effects Models. R Package Version 3.1-128. 2016. Available online: http://CRAN.R-project.org/package=nlme (accessed on 18 October 2016).

87. Christensen, R. *Plane Answers to Complex Questions: The Theory of Linear Models*, 4th ed.; Springer: New York, NY, USA, 2011; pp. 368–370.

88. Crawley, M.J. *Statistics: An Introduction Using R*, 2nd ed.; John Wiley and Sons: Chichester, UK, 2015; pp. 193–202.

89. Arpa Emilia Romagna-Idro-Meteo-Clima. Available online: https://www.arpae.it/sim/ (accessed on 16 January 2016).

90. Harris, C.K.; Sherwood, C.R.; Signell, R.P.; Bever, A.J.; Warner, J.C. Sediment dispersal in the northwestern Adriatic Sea. *J. Geophys. Res.* **2008**, *113*, C11S03. [CrossRef]

91. D'Sa, E.J.; Miller, R.L.; Del Castillo, C. Bio-optical properties and ocean color algorithms for coastal waters influenced by the Mississippi River during a cold front. *Appl. Opt.* **2006**, *45*, 7410–7428. [CrossRef] [PubMed]

92. Komick, N.M.; Costa, M.P.F.; Gower, J. Bio-optical algorithm evaluation for MODIS for western Canada coastal waters: An exploratory approach using in situ reflectance. *Remote Sens. Environ.* **2009**, *113*, 794–804. [CrossRef]

93. Kutsera, T.; Piersona, D.C.; Kalliob, K.Y.; Reinarta, A.; Sobek, S. Mapping lake CDOM by satellite remote sensing. *Remote Sens. Environ.* **2005**, *94*, 535–540. [CrossRef]

94. Mobley, C.D.; Sundman, L.K. *HydroLight 5.2-EcoLight 5.2 Technical Documentation (2013)*; Sequoia Scientific, Inc.: Bellevue, WA, USA, 2013; Available online: http://www.oceanopticsbook.info/view/references/publications (accessed on 15 October 2016).

remote sensing

MDPI

Article

Turbidity in Apalachicola Bay, Florida from Landsat 5 TM and Field Data: Seasonal Patterns and Response to Extreme Events

Ishan D. Joshi [1], Eurico J. D'Sa [1,*], Christopher L. Osburn [2] and Thomas S. Bianchi [3]

[1] Department of Oceanography and Coastal Sciences, Louisiana State University, Baton Rouge, LA 70803, USA; ijoshi1@lsu.edu

[2] Department of Marine, Earth and Atmospheric Sciences, North Carolina State University, Raleigh, NC 27695, USA; closburn@ncsu.edu

[3] Department of Geological Sciences, University of Florida, Gainesville, FL 32611, USA; tbianchi@ufl.edu

* Correspondence: ejdsa@lsu.edu; Tel.: +1-225-578-0212

Academic Editors: Yunlin Zhang, Claudia Giardino, Linhai Li, Deepak R. Mishra and Prasad S. Thenkabail
Received: 8 February 2017; Accepted: 9 April 2017; Published: 13 April 2017

Abstract: Synoptic monitoring of estuaries, some of the most bio-diverse and productive environments on Earth, is essential to study small-scale water dynamics and its role on spatiotemporal variation in water quality important to indigenous marine species and surrounding human settlements. We present a detailed study of turbidity, an optical index of water quality, in Apalachicola Bay, Florida (USA) using historical in situ measurements and Landsat 5 TM data archive acquired from 2004 to 2011. Data mining techniques such as time-series decomposition, principal component analysis, and classification tree-based models were utilized to decipher time-series for examining variations in physical forcings, and their effects on diurnal and seasonal variability in turbidity in Apalachicola Bay. Statistical analysis showed that the bay is highly dynamic in nature, both diurnally and seasonally, and its water quality (e.g., turbidity) is largely driven by interactions of different physical forcings such as river discharge, wind speed, tides, and precipitation. River discharge and wind speed are the most influential forcings on the eastern side of river mouth, whereas all physical forcings were relatively important to the western side close to the major inlet, the West Pass. A bootstrap-optimized and atmospheric-corrected single-band empirical relationship (Turbidity (NTU) $= 6568.23 \times$ (Reflectance $(Band\ 3))^{1.95}$; $R^2 = 0.77 \pm 0.06$, range $= 0.50$–0.91, N $= 50$) is proposed with seasonal thresholds for its application in various seasons. The validation of this relationship yielded $R^2 = 0.70 \pm 0.15$ (range $= -0.96$–0.97; N $= 38$; RMSE $= 7.78 \pm 2.59$ NTU; Bias (%) $= -8.70 \pm 11.48$). Complex interactions of physical forcings and their effects on water dynamics have been discussed in detail using Landsat 5 TM-based turbidity maps during major events between 2004 and 2011. Promising results of the single-band turbidity algorithm with Landsat 8 OLI imagery suggest its potential for long-term monitoring of water turbidity in a shallow water estuary such as Apalachicola Bay.

Keywords: Apalachicola Bay; Landsat; turbidity; bootstrapping; classification tree; PCA; turbidity maps

---

1. Introduction

Estuaries are economically important and bio-diverse zones located between terrestrial and marine environments. Highly dynamic physicochemical conditions, such as nutrient richness, steep light and salinity gradients, and highly variable water column stratification conditions, well-mixed water column, make these transition zones some of the most productive regions in the coastal ocean [1]. Apalachicola Bay, Florida (USA), is such an estuary that has experienced significant urbanization and population-stress due to its economic-importance, thus necessitating a regular and synoptic monitoring of the bay's water quality [2,3]. Turbidity is an optical index of water quality that directly impacts light availability to the photosynthetic organisms (e.g., phytoplankton and submerged vegetation), and in many cases can be directly linked to the amounts of suspended particulate matter (SPM) [4–7]. High turbidity is commonly responsible for obscuring the vision of marine nekton, damaging early-stage larval development, affecting prey-predator interactions, and causing significant mortality of commercially-important species (e.g., oysters) [8,9]. Spatiotemporal monitoring of water turbidity can further provide useful information about SPM and its impact on marine biota [4,10,11], the distribution and transport of SPM-coupled pollutants [12], and on depositional/erosional events [13–15] in coastal and inland environments.

Conventional turbidity monitoring requires an extensive network of in situ observations that can be demanding in both time and cost. In addition, traditional in situ methods are limited by poor spatial and temporal coverage to resolve estuarine-scale processes. Alternatively, continuous measurement strategies (e.g., point locations with data-loggers) may resolve temporal variations in water turbidity at specified locations, yet fail to provide synoptic representations of water dynamics and bay-shelf exchanges. Turbidity is a measure of incoming light attenuation, mainly due to particle scattering [16], while, water-leaving radiance carries information about optically-active water constituents [17]. Hence, a successful linkage between these optical parameters is a key tool in obtaining synoptic views of estuarine-scale turbidity distribution—using remote sensing platforms (e.g., satellites). Numerous studies have reported effective ways to utilize satellites in monitoring turbidity and suspended particulate matter in inland and coastal waters [18–24]. However, it is important to remember that water turbidity is closely linked to the optical characteristics of particles (e.g., particle size, shape, and composition), rather than just mass-specific properties such as particle weight and concentration [25]. Hence, the direct comparisons of particles in suspension and turbidity are still critically needed in these dynamic environments where particle properties change rapidly.

In the past few decades, advances in satellite technology have opened new horizons for achieving efficient monitoring of water quality in coastal and inland waters [19,21,26–28]. The ocean color satellite sensors such as MODIS (Aqua and Terra), OCM-2, VIIRS, and GOCI provide temporal coverage of hours to days with coarse spatial resolution. Although the temporal coverage of these sensors may aid the study of short time-scale processes (e.g., hours to days), their coarser spatial resolution limits applications to estuarine-scale water dynamics and turbidity distribution. In contrast, Landsat 5 TM sensor provides fine resolution of ~30 m to study small estuaries (e.g., Apalachicola Bay), but is limited by a longer revisit time of 16 days and poor signal-to-noise ratio compared to its successors (e.g., ETM+ and OLI) and the ocean color satellites.

In this paper, we investigate turbidity climatology over an 8-year period (2004–2011) in Apalachicola Bay. The aim of this study was to examine seasonally-varying water dynamics and its effect on the bay's turbidity distribution, and therefore, importance was given to the spatial resolution rather than temporal coverage of a satellite sensor. Using historical Landsat 5 TM imagery in conjunction with in situ point measurements, we propose a single-band empirical relationship for monitoring turbidity in Apalachicola Bay, Florida (USA). The performance of Landsat 5 TM—turbidity relationship was also evaluated on the recently launched Landsat 8 OLI imagery. A time-series of turbidity was then analyzed using advanced statistical techniques to investigate the effects of meteorological, hydrological, and astronomical forcing on the water turbidity in Apalachicola Bay. Using the Landsat-based turbidity maps and in situ observations, we examined spatial variability in the bay's turbidity and water dynamics seasonally and during major hydrological and meteorological events, such as the Apalachicola River floods, passages of the cold fronts, and the hurricane landfall during our study period. Water quality monitoring strategy presented in this study supports the US Environmental Protection Agency's (EPA) water quality standards under the Clean Water Act to protect human health and the environment.

2. Materials and Methods

2.1. Study Area

Apalachicola Bay is an elongated shallow (average depth = ~3 m) estuarine system located in the Florida's Panhandle that covers an area of ~540 km² (Figure 1) [29]. Apalachicola River, the largest river in Florida, is the main source of freshwater and nutrients in the bay [30]. The bay is bounded by four barrier islands (St. Vincent, St. George, St. Little George, and Dog), and exchanges water through three natural passes (Indian, West, and East) and one man-made pass (Sikes Cut). Apalachicola Bay is one the most productive estuarine systems in North America. For example, it is well-known for its oyster harvest that supplies ~90% of the oysters in Florida, and accounts for ~10% of nationwide oyster production [31]. However, environmental stresses such as salt-water intrusion [32], tropical storms [33], Deep Water Horizon oil spill [34], and droughts/floods [35,36] have negatively affected the bay's commercial oyster industry. Historical sediment records showed a decrease in nutrient levels in Apalachicola Bay possibly due to the reduction in river discharge and rising sea-level [37]. The bay is located in a transition zone, where diurnal tides of the western Gulf change to semi-diurnal tides towards the Florida's Panhandle [38,39]. It also experiences relatively shorter periods of strong winds during extreme weather events, such as cold fronts, storms, and hurricanes that can have large effects on the bay's water quality [40,41].

Figure 1. Apalachicola Bay, USA (white star in inset). In situ turbidity is observed near DB (Dry Bar) and CP (Cat Point) stations. White star represents a meteorological station maintained by Apalachicola National Estuarine Research Reserve (ANERR). Arrows represent various natural and man-made connections between the bay and the Gulf of Mexico.

2.2. Data Sources

A list of in situ and satellite measurements with their sources and purpose in this analysis is given in Table 1. Meteorological and hydrological data, such as wind speed, wind direction, river discharge, tidal height, and rainfall, were requested from various state and federal agencies. Water quality measurements, such as turbidity and salinity, were collected from the *Apalachicola National Estuarine Research Reserve* (ANERR)-maintained YSI-6600 series sondes (YSI Inc., Yellow Springs, OH, USA) positioned ~0.3 m above the bottom at two locations, Cat Point (CP) and Dry Bar (DB) (Figure 1). Three sets of clear-sky Landsat imagery with no sun-glint artifact were requested from *USGS Landsat data archive* that include 19 images of Landsat TM, ETM+ and OLI sensors for validating ENVI-FLAASH atmospheric correction, 57 images of Landsat TM sensor for developing turbidity algorithm and spatial analysis, and 17 images of Landsat OLI sensor for evaluating the performance of proposed algorithm on Landsat 8 OLI imagery.

Table 1. A list of data that are used in 8-year turbidity analysis (2004 to 2011) in Apalachicola Bay.

Data	N (Duration)	Location	Source	Purpose
Field Measurements				
Turbidity (NTU)	2758 (CP) 2745 (DB) (2004–2011)	Cat point (CP) (29.702°N, −84.875°W) Dry Bar (DB) (29.675°N, −85.058°W)	Apalachicola National Estuarine Research Reserve (ANERR)	Temporal and statistical analysis of turbidity, Satellite-based turbidity maps
Wind speed (ms^{-1}) and Wind direction	2842 (2004–2011)	East Bay (EB) (29.791°N, −84.883°W)	ANERR	Effects on turbidity
Tidal height (m)	2922 (2004–2011)	ID-8728690 (29.435°N, −84.90°W)	NOAA Tides and Currents	Effects on turbidity
River discharge (m^{3}s^{-1})	2922 (2004–2011)	ID-02359170 (near Sumatra, Florida)	USGS Water Data for the Nation	Effects on turbidity
Rainfall (mm)	2439 (2004–2011)	ID-080211 (Apalachicola Airport)	Florida Climate Center	Effects on turbidity
Salinity	2922 (2004–2011)	Cat point (CP) (29.702°N, −84.875°W) Dry Bar (DB) (29.675°N, −85.058°W)	Apalachicola National Estuarine Research Reserve (ANERR)	Effects on turbidity
Remote Sensing Measurements				
Landsat TM, ETM+ & OLI	19 clear-sky images (2011–2014)	Path 22, Row 40	Landsat Data Archive (USGS)	Validation of ENVI-FLAASH atmospheric correction
Radiance L_w (mWcm^{-2}μm^{-1}sr^{-1}) and AOT	19 match-ups (2011–2014)	CSI-6 (28.867°N, −90.483°W)	AERONET-OC (WAVCIS)	Validation of ENVI-FLAASH atmospheric correction
Landsat 5 TM images	57 images with clear-sky conditions (2004–2011)	(Path 18/19, Row 39/40)	Landsat Data Archive (USGS)	Landsat based turbidity maps, Analysis of spatiotemporal changes in turbidity
Landsat 8 OLI images	17 images with clear-sky conditions (2014–2016)	(Path 18/19, Row 39/40)	Landsat Data Archive (USGS)	Performance evaluation of turbidity algorithm on Landsat 8 OLI

2.3. Methods

2.3.1. Overview

The main aim of this study was to examine the effects of physical forcings (e.g., Apalachicola River, winds, precipitation, and tides) on water turbidity, and to investigate the possible use of Landsat sensors (e.g., Landsat 5 TM) to get synoptic views of estuarine water dynamics and turbidity distribution in Apalachicola Bay. Regular monitoring of water quality parameters using space-borne sensors generally requires two necessary steps: (1) to achieve reasonable estimates of water-leaving radiance by removing atmospheric contributions from a signal received at the top-of-atmosphere (TOA); and (2) to obtain a robust relationship between water quality (e.g., turbidity) and satellite-based optical parameters (e.g., remote sensing reflectance—*Rrs* and surface reflectance—ρ) [42]. Remote sensing over water is quite challenging mainly due to low contributions of water-leaving light, and strong influence of atmosphere (~90%) to light sensed at the TOA. Consequently, it is necessary to remove atmospheric signal prior to the usage of satellite data for ocean color applications [43]. Numerous methods have been proposed for the removal of the atmospheric contribution to get better estimation of water-leaving radiance [43–48]. The complexity associated with these schemes makes them time-intensive and unsuitable for day-to-day monitoring purposes. Freely available tools (e.g., SeaDAS (OBPG, NASA)) incorporate many of these schemes for satellite data processing, but these tools do not support the older Landsat 5 TM sensor data. This study used a commercially available ENVI-FLAASH® (Fast Line of sight Atmospheric Analysis of Spectral Hypercubes) atmospheric correction module to correct clear-sky Landsat 5 TM imagery acquired for the analysis. However, the validation of FLAASH results (e.g., surface reflectance) was a necessary step before algorithm development and analysis. Furthermore, this study mainly focused on the historical data, and in situ matchups for validation analysis were not available over the study area. To obtain confidence on FLAASH's performance in moderately turbid coastal waters, the atmospheric-correction module was tested first near the only available AERONET-OC site in the Gulf of Mexico. Subsequently, Landsat 5 TM—turbidity relationships were analyzed for different seasons, and a robust general relationship was developed using statistical techniques for converting atmospheric-corrected Landsat imagery into the turbidity maps in Apalachicola Bay. A schematic diagram (Figure 2) illustrates the pathways to generate turbidity maps in the bay wherein, AERONET-FLAASH comparison is represented by Path 1 (dashed lines), while satellite and in situ data quality check/processing are shown by Path 2 and Path 3, respectively.

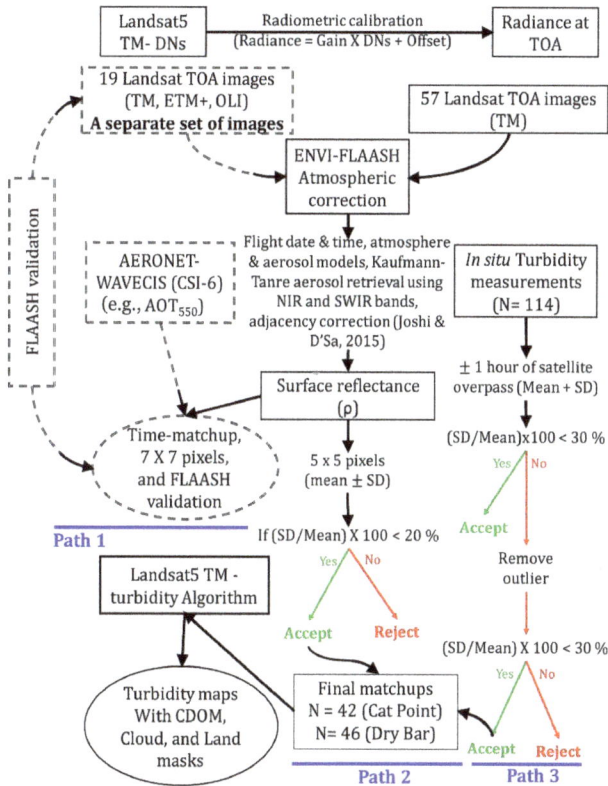

Figure 2. A schematic diagram of processing pathways for developing an empirical relationship between in situ turbidity and atmospheric-corrected Landsat 5 TM red band (Band 3). Dashed lines represent processing units for AERONET—FLAASH validation on a separate set of 19 clear-sky images. DNs = Digital numbers, TM = Thematic mapper, ETM+ = Enhanced thematic mapper, OLI = Operational land imager, TOA = Top-of-the-atmosphere, and CDOM = Chromophoric Dissolved Organic Matter.

2.3.2. Landsat Image Processing

Raw at-sensor radiance images were corrected for the atmosphere using the ENVI-FLAASH® module (Exelis Visual Information Solutions, 2015). Radiometric-calibration was applied first to all of the raw images to convert at-sensor DNs (digital numbers) to at-sensor radiance values (L_{TOA}) at each spectral band,

$$L_n = C_{0n} + C_{1n} \times DN_n \tag{1}$$

where C_0, is the offset and C_1 the gain coefficients of Landsat 5 TM for the corresponding band n. Surface-leaving radiance (L_u) conveys valuable information about the in-water constituents, but it contributes only about 10% to the L_{TOA}. Major part of L_{TOA} is contributed by the atmosphere (aerosols and gases), specular reflection of directly transmitted sunlight from the water surface (sun glint), direct reflected radiance from the water surface (sky light) and radiance reflected from whitecaps. The FLAASH module facilitates automated atmospheric correction with MODTRAN4 code over water using NIR (Band 4) and SWIR (Band 5) channels. The FLAASH processing methodology [49] is briefly described next. At sensor radiance is given by,

$$L_{TOA} = L_{atm} + \left(\frac{A \times \rho}{1 - \rho_x \times S} \right) + \left(\frac{B \times \rho_x}{1 - \rho_x \times S} \right) \tag{2}$$

where, ρ = surface reflectance of target pixels, ρ_x = spatially averaged surface reflectance of surrounding region, A and B = coefficients that depend on atmospheric and geometric conditions, S = the spherical albedo of the atmosphere, L_{TOA} = radiance that is received at the top-of-the-atmosphere (TOA), L_{atm} = radiance that is back-scattered by the atmosphere (aerosols, gases, and water molecules) to the sensor's field of view. In Equation (2), L_{TOA} includes radiance contributions from mainly 3 paths, (1) path radiance which is light scattered by the atmosphere (first term); (2) surface radiance which includes water-leaving radiance (L_w), part of sky light (L_{sky1}), and part of sun-glint (L_{glint1}) from that reflected from target pixels (second term); and (3) surface radiance that is diffusely transmitted into sensor, giving rise to "adjacency effect" (third term). It also includes part of skylight (L_{sky2}) and part of sun-glint (L_{glint2}) from the surrounding pixels.

In processing, FLAASH first utilizes MODTRAN4 radiative transfer model to calculate A, B, S and L_{atm} by using manually provided image-based information such as viewing and solar angles (date and time of image acquisition), the mean surface elevation of the measurement (~705 km for Landsat 5 TM), model atmosphere (Mid-Latitude summer or Tropical based on in situ air temperature and month of image acquisition), and aerosol model (maritime-aerosol model). FLAASH includes a method for retrieving the aerosol amount and estimating a scene average visibility using an automated dark-pixel reflectance ratio method [50]. A dark-pixel is obtained by using an infrared wavelength (usually 2.1 μm; Landsat 5 TM Band 7) at which reflectance retrieval is generally insensitive to visibility. FLAASH then uses the Landsat bands 4 (NIR) and 7 (SWIR) with a nominal reflectance ratio (1 or 0.81) and a band 4 reflectance cutoff of 0.03. This assumes that the source of water reflectance in the infrared is spectrally flat or foam (white caps) with reflectance ratio 1 representing spectrally flat assumption whereas 0.81 is better suited for foam conditions (Moore, Voss, and Gordon, 2000). The visibility estimate is then combined with MODTRAN4 aerosol representation to describe the atmosphere. The values of A, B, S and L_{atm} are also strongly dependent on the water vapor column amount, which is generally not well-known and may vary across the scene. Landsat sensor does not have the appropriate bands to perform water retrieval, so it is determined according to the standard water column amount for the selected atmospheric model which is then multiplied by an optional water column multiplier (e.g., Mid-Latitude summer—2.08 gcm^{-2} and Tropical—4.11 gcm^{-2} in our case).

Next, spatial averaging is implied by convolving radiance image of L_{TOA} (Equation (2)) to find $L_{equivalent}$ (at sensor) by using upward diffused transmission spatial point spread function that describes the relative contributions to the pixel radiance from points on the ground at different distances from the direct line of sight. ρ_x is then obtained by neglecting contribution of surrounding pixels,

$$L_{equivalent} \approx L_{atm} + \left(\frac{(A+B) \times \rho_x}{1 - \rho_x \times S} \right) \tag{3}$$

Once, L_{atm}, ρ_x, A, B, and S are available, Equation (2) is solved for ρ (surface reflectance). It is important to note that adjacency correction partially removes sun-glint and skylight that is contributed by neighboring pixels; however, their exact values are unknown as they are removed during adjacency correction. Therefore, water surface reflectance (ρ) may have some effects of residual skylight and sun-glint. FLAASH-based image processing was used as described in Joshi and D'Sa (2015) [51]. After applying suitable quality-check criteria to remove the effects of outliers, average values of 5×5 and 7×7 pixels were calculated from the atmospherically-corrected Landsat images for turbidity algorithm and FLAASH—AERONET analysis, respectively, at each sampling site. We have used SeaDAS 7.2 (OBPG, NASA) to generate turbidity maps for further analysis (Figure 2).

2.3.3. FLAASH vs. AERONET

A separate analysis was performed to examine the validity of the FLAASH's atmospheric correction. We used a different set of Landsat images (e.g., TM, ETM+, and OLI) collected over the AERONET WAVCIS CSI-6 site located to the south of Terrebonne Bay close to the Mississippi River delta (Figure 3). A validation analysis at CSI-6 can apply to Apalachicola Bay because of the similarity in turbidity conditions of the bay and the relatively low influence from the Mississippi-Atchafalaya river system at CSI-6. Nineteen clear-sky Landsat images (100% clear-sky image with no sun glint) were acquired over the AERONET station between 2011 and 2014. They included five Landsat 5 TM images, eight Landsat 7 ETM+ images and six Landsat 8 OLI images.

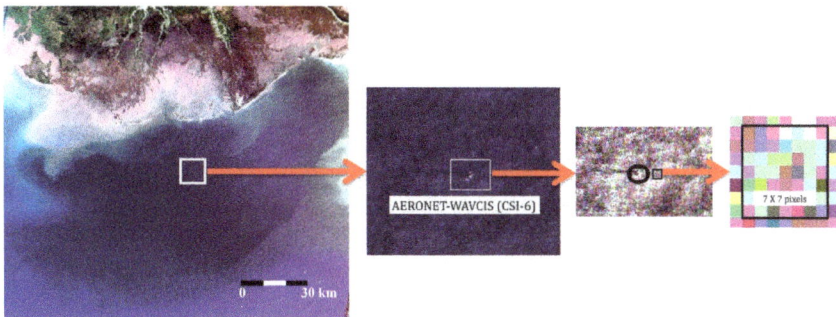

Figure 3. AERONET WAVCIS (CSI-6) station, Gulf of Mexico (white box). A 7×7 box was chosen to the right of AERONET site rather than centered at the site to avoid noise from the station itself (black box).

The validity of FLAASH-corrected satellite observations was evaluated using two approaches, namely, (1) comparing the aerosol optical thickness (AOT); and (2) relating water-leaving radiance (L_w) or normalized water-leaving reflectance ($[\rho_w]_{normalized}$) obtained from the in situ and satellite measurements. The first approach used AOT obtained from in situ AERONET measurements and Landsat images. In situ AOT acquired just before and after the satellite overpass were averaged to keep comparison closer to satellite measurement. Since the in situ AERONET measurements do not provide AOT (550 nm), these were obtained by the interpolation of nearby wavelengths. Furthermore, FLAASH utilizes MODTRAN4 code (which is not freely available) to calculate AOT (550 nm) using image data and historical climatology, but it does not provide AOT as an output in atmospheric-correction procedure. However, FLAASH does provide horizontal visibility which can be approximately converted to AOT using *Koschmieder's equation* [52],

$$Visibility \ (km) = \frac{3.912}{AOT_{550}}$$ (4)

The second approach compared normalized water-leaving reflectance obtained from in situ measurements and atmospheric-corrected Landsat imagery. AERONET uses water-leaving radiance L_w, ozone optical thickness (τ_o), Rayleigh optical thickness (τ_r), and aerosol optical thickness (τ_a) to obtain normalized water-leaving radiance ($[L_w]_{normalized}$). $[L_w]_{normalized}$ is then converted to $[\rho_w]_{normalized}$ by using the following equation,

$$[\rho_w]_{normalized} = \frac{\pi}{F_0} \times [L_w]_{normalized} \quad (\text{AERONET} - \text{OC}) \tag{5}$$

where, F_0 = Extraterrestrial solar irradiance ($\text{mWcm}^{-2}\mu\text{m}^{-1}$) [53]. Satellite-based water-leaving reflectance (ρ_w) was obtained from FLAASH-corrected imagery with 7×7 box centered near to the AERONET site (Figure 3), which is then converted to $[\rho_w]_{normalized}$ [54],

$$[\rho_w]_{normalized} = \rho_w \times \exp\left[\left(\frac{\tau_r}{2} + \tau_o + M \times \tau_a\right) \times \frac{1}{\cos\theta}\right] \quad (\text{Landsat 5 TM}) \tag{6}$$

where, θ = solar zenith angle, M is related to scattering phase function of aerosol.

2.3.4. Statistical Analysis

Time-series decomposition, principal component analysis (PCA), and classification-tree based models were used to examine relative importance of meteorological (e.g., wind speed and wind direction), hydrological (e.g., river discharge and salinity), and astronomical (e.g., tidal height) factors on the historical trends of turbidity in Apalachicola Bay. Time-series decomposition was applied using a *loess*-based seasonal-trend decomposition procedure (STL) [55]. The STL decomposes time-series into seasonal, yearly trend, and remainder components to extract useful information about seasonal, inter-annual, and random variations from the original time-series ("STL" package in R). PCA was used to show the complexity of Apalachicola Bay; this is a powerful dimension reduction technique that reduces a large number of variables into a few orthogonally separated principal components. Principal component is a linear transformation of the variables into a lower dimensional space that retains the maximum amount of information about the original variables. We used *R statistical analysis software* with "*prcomp*" package for the PCA analysis. Likewise, the relationship between turbidity and physical forcings, for example, could be strongly non-linear and involve complex interactions that could not be explained by commonly-used statistical modeling approaches. Classification (categorical dependent variable) and regression (numerical dependent variable) trees are the modern statistical techniques for exploring and modeling such a complexity in data [56], and have been widely used in a variety of fields such as agriculture, coastal environment [51], and freshwater and marine ecology [57,58]. Trees explain the variation of a single dependent variable corresponding to one or more explanatory variables by splitting data recursively based on the most influential independent variable. At each node, model splits observations into two mutually-exclusive groups based on a threshold value of the most influential explanatory variable by keeping each group as homogenous as possible with a minimum residual sum of squares. We used *R statistical analysis software* with "CART" package for the tree-based analysis.

It is complex to derive a general relationship for the diversity of sub-habitats in estuarine systems with a small sample size. Furthermore, the regression relationships between dependent and independent variables changes completely if a different set of training data is drawn from the sample every time. Therefore, uncertainty analysis is necessary for the reasonable estimates of regression coefficients and a robust general relationship for the study area. A bootstrapping approach was used with 5000 simulations to get reasonable estimates of regression coefficients for the Landsat 5 TM—turbidity relationship, and to obtain the validation statistics (e.g., Bias and root mean squared error (RMSE)) [59,60]. The bootstrapping is a machine-learning technique that provides reasonable inferential statistics for population with random and repeated sample selection, especially when the sample size is limited. In regression analysis, algorithm development and validation require the data to be divided in training and testing sets (e.g., training data = 50; testing data = 38 total data = 88). In processing, the first bootstrap-run picks up a training set from the given sample to develop a relationship between dependent and independent variables. Then, it validates this relationship on remaining data (e.g., testing set), and provides statistical inferences for the prediction accuracy. Next, the second bootstrap runs on the new set of randomly selected training and testing data sets, and provides statistical parameters and inferences of regression and validation (e.g., slope, intercept, R^2, Bias, and RMSE). If bootstrapping procedure is run for N times on the given sample size, it can provide good estimates of regression coefficients and validation accuracy for the population. Furthermore, the means of regression coefficients (e.g., slope and intercept) can be used as the reasonable estimates for developing a general regression model if their distributions are not highly skewed. In this analysis, 5000 bootstrap simulations (training data = 50 and testing data = 38) provided population distributions of regression estimates, inferential statistics, and validation accuracy for the Landsat 5 TM—turbidity relationship in Apalachicola Bay.

3. Results and Discussion

3.1. Satellite Assessment of Water Turbidity in Apalachicola Bay

3.1.1. Assessing the Performance of ENVI-FLAASH Based Atmospheric Correction

The FLAASH derived AOT_{550} showed a good correlation (R^2 = 0.96, RMSE = 0.011, N = 19) to in situ AOT_{550} that indicated a reasonable retrieval of aerosol properties by MODTRAN4 code (Figure 4a). However, the uncertainty increased towards low visibility (or high AOT) conditions. The Koschmieder's equation (Equation (4)) assumes the equality in vertical and horizontal extinction coefficients with no effects of vertical extinction profile on the AOT. However, vertical extinction coefficient may become larger than horizontal extinction coefficient in many cases (e.g., high aerosol concentration in the upper atmospheric layers, and temperature inversions). Furthermore, the conversion of AOT from horizontal visibility may cause large RMSE especially for the visibility <30 km [52]. FLAASH-corrected $[\rho_w]_{normalized}$ and in situ $[\rho_w]_{normalized}$ also showed a reasonable agreement between satellite and above-water measurements (R^2 = 0.86; N = 24) (Figure 4b).

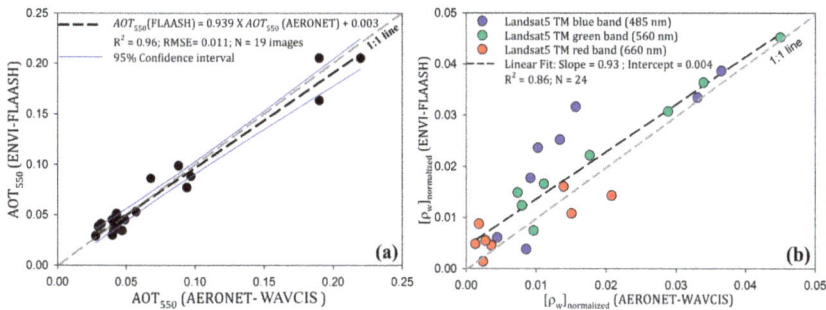

Figure 4. An assessment of FLAASH-based atmospheric correction using in situ (**a**) aerosol optical thickness (AOT_{550}), and (**b**) normalized water-leaving reflectance ($[\rho_w]_{normalized}$) observed at AERONET-WAVCIS site in the northern Gulf of Mexico.

3.1.2. Band Selection

In principle, water turbidity is mainly a measure of the light scattered (back or side scattering of white or NIR light based on instrument design) from particles in suspension [16]. Therefore, it can be related to water-leaving light (e.g., Rrs and ρ) recorded by satellite sensor by assuming, (1) water column is homogenously mixed; and (2) surface water is spatially homogenous around the study area. Generally, estuaries, such as Apalachicola Bay, are highly productive due to their nutrient richness and light availability for the primary production [61]. Furthermore, they receive considerable amounts of sediments and dissolved organic matter (e.g., CDOM) from the river, its tributaries, and adjacent wetlands [62]. Also, physical forcings such as winds and tides make these shallow water systems optically complex due to water column mixing and bottom sediment re-suspension. Therefore, band selection for the turbidity algorithm is difficult especially in the estuarine environment where larger concentrations of major optical constituents do not co-vary.

Water turbidity can be mainly related to the presence of reflective particulate matter in a water column (e.g., non-algal particulate matter (NAP) and detritus). In natural waters, back-scattering spectra generally show a broad peak in blue-green region (e.g., 500–600 nm) that decreases towards blue and red regions possibly due to strong pigment absorption [63,64]. Additionally, the presence of CDOM further downgrades the water-leaving radiance at the blue and green wavelengths especially in CDOM-rich coastal waters such as Apalachicola Bay [65,66]. Therefore, these wavelengths are generally avoided for the remote sensing of suspended particles and turbidity especially in optically complex estuarine and inland waters. The CDOM and NAP absorptions decrease toward the red and NIR wavelength, and particle back-scattering becomes a major contributor to the surface reflectance except in chlorophyll-a absorption region (~670–680 nm). Several studies have demonstrated the successful application of red and NIR wavelengths to monitor water quality in coastal and estuarine environments [19–21,67–69]. Landsat 5 TM NIR band is broad (760–900 nm, band width = ~140 nm), and water absorption increases about an order of 4 from the red (~660 nm) to the NIR wavelength (~727.5 nm) [70]. Therefore, NIR band could be useful only in highly-turbid waters where particle back-scattering prevails over the water absorption, but it may not provide useful information about water clarity in low to moderately turbid waters [71]. This could be one of the reasons that NIR band showed relatively poor correlation ($R^2 = 0.55$) with turbidity in Apalachicola Bay (Figure S1). On the other hand, red band could be useful as it has relatively narrower (630–690 nm) bandwidth with the dominance of back-scattering in the larger portion of bandwidth compared to the chlorophyll-a absorption, thus we have used surface reflectance (ρ) of Band 3 (red band) which was obtained from FLAASH-corrected Landsat imagery for algorithm development.

3.1.3. Landsat 5 TM-Turbidity Algorithm and Validation

Landsat 5 TM—turbidity relationships were evaluated separately for different seasons. Despite the influence of the chlorophyll absorption peak, the correlation between in situ turbidity and Landsat 5 TM reflectance (Band 3: 630–690 nm) was reasonably good in the spring ($R^2 = 0.79$, N = 22), the fall ($R^2 = 0.89$, N = 19), the winter ($R^2 = 0.76$, N = 27) except in the summer ($R^2 = 0.50$, N = 21) (Figure S2). The temporal variability in in situ measurements ($\pm$ 1 h of satellite overpass) and pixel-to-pixel variability in surface reflectance (5 $\times$ 5 pixels centered close to in situ station) were relatively smaller in the spring, and the winter likely due to energetic mixing of water column by river discharge, and winds along with sustained-influence of these forces during the time of measurements. Both Cat Point and Dry Bar stations are located on the hard bottom substrate (with no vegetation) of productive oyster bars [72]. The large uncertainties in reflectance measurements in relatively clearer water scenarios indicated the possible contamination due to the variability in bottom reflectance at each pixel [73].

The bootstrap approach yielded reasonable statistics for turbidity algorithm ($R^2 = 0.77 \pm 0.06$, range = 0.50–0.91, N = 50), and its validation ($R^2 = 0.70 \pm 0.15$, range = -0.96 – 0.97, N = 38; RMSE = 7.78 $\pm$ 2.59 NTU) (Table S1; Figure S3). The regression coefficients for the mean RMSE (= 7.78 NTU) of simulations were considered as the reasonable estimates of slope and intercept for the following power-law relationship between turbidity and Landsat 5 TM red band (Figure 5a),

$$Turbidity\ (NTU) = 6568.23 \times (Reflectance\ (Band\ 3))^{1.95} \tag{7}$$

3.1.4. An Optimization of the General Turbidity Algorithm Based on Seasonal Thresholds

Because the slope and range of the general turbidity algorithm are different than individual seasonal Landsat 5 TM—turbidity relationships, the usage of a general algorithm outside the ranges of individual relationships can cause erroneous estimates of turbidity. Therefore, using reflectance thresholds corresponding to different seasons further optimized the algorithm. The lower threshold (surface reflectance = 0.01), was kept the same for all the seasons whereas upper threshold, was selected where modeled seasonal relationships deviated significantly from the general turbidity algorithm as shown in Figure 5b (e.g., $\rho = 0.11$ for the spring, $\rho = 0.08$ for the winter, $\rho = 0.07$ for the fall, and $\rho = 0.04$ for the summer). Furthermore, any signature of strong bottom reflectance and skylight will result in very high reflectance at the surface. If signature is too high to cross the upper limits, the pixel will be masked out from the image and thus, erroneous pixels can be avoided in the analysis with the proposed thresholds.

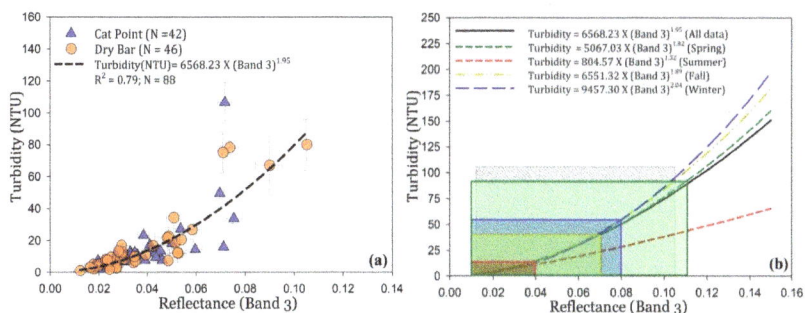

Figure 5. (a) Turbidity algorithm using bootstrap estimations; (b) optimization of turbidity algorithm based on seasonal thresholds. Grey box represents data bounds for the general turbidity algorithm (black line), whereas colored boxes represents threshold bounds for different seasons (e.g., green for spring ($\rho_{min} = 0.01$ and $\rho_{max} = 0.11$), blue for winter ($\rho_{min} = 0.01$ and $\rho_{max} = 0.08$), yellow for fall ($\rho_{min} = 0.01$ and $\rho_{max} = 0.07$), and red for summer ($\rho_{min} = 0.01$ and $\rho_{max} = 0.04$).

3.2. Temporal Variations in Physical Forcings in Apalachicola Bay

3.2.1. Time-Series Analysis

Apalachicola Bay generally experienced elevated river flows in late winter and spring that gradually reduced to minimal flows towards the warm seasons (Figure 6a). Furthermore, a right-skewed frequency distribution of discharge ($\sim$558.21 $\pm$ 499.10 m^{3}s^{-1}) suggested that the bay usually experiences long periods of low-flow conditions and strong episodes of elevated river flow (Table S2; Figure S4). High river discharge were normally observed during the Apalachicola River flood events (e.g., March–April, 2005; 2008; 2009), and during passages of strong cold-frontal systems (e.g., January-February, 2010) (also see blue bars in Figure 6a). However, heavy precipitation and associated run-off in the lower ACF basin (Apalachicola-Chattahoochee-Flint) can also occasionally affected the freshwater inflow to the bay. Seasonally-decomposed precipitation revealed frequent occurrences of rainfall events in the summer and fall seasons (Figure 6b; Figure S5). The anomalously high river discharge and precipitation observed during winter and spring may have also been related to the weak and moderate El Niño events in 2004–2005 and 2009–2010, respectively [74,75].

Apalachicola Bay was relatively saline near to the freshwater sources in warm seasons, and was relatively fresher in winter and spring indicating seasonal-influence on the bay's water quality (Figure 6c). Mean-daily salinity showed similar trends at Cat Point (CP) (21.69 $\pm$ 7.78) and Dry Bar (DB) (21.99 $\pm$ 7.45) stations, but they varied considerably in many instances possibly due to complex interactions of river, wind, precipitation, and tidal forcings during the study period (also see red ellipses in Figure 6c). The bay experiences a micro-tidal environment with tidal height of 0.28 m (± 0.06 m) above the mean lower low water datum (MLLW = 1.307 m). Diurnal tidal-range was relatively larger in spring and summer than in the fall and winter (Figure 6d). Furthermore, mean-daily tidal height showed a bimodal distribution with relatively higher tides in the summer and fall seasons, and relatively lower tides in winter. Although wind speed can vary significantly at shorter time intervals (e.g., minutes), the bay experienced low to moderate mean-daily wind speed (2.67 ± 1.06 ms^{-1}) from 2004 to 2011 (Table S2). Wind vectors also showed yearly trends of strong northerly winds in winter and early spring, and relatively milder southerly winds in the summer and fall (Figure 6e).

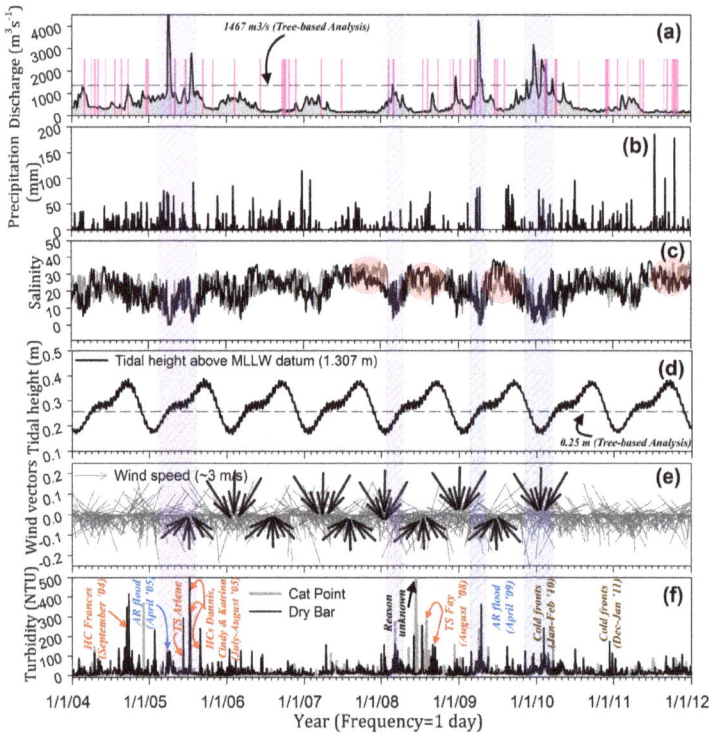

Figure 6. Time-series of (**a**) Apalachicola River discharge (m^{3}s^{-1}); (**b**) precipitation (mm); (**c**) salinity, (**d**) tidal height (m); (**e**) wind speed (ms^{-1}) and wind direction (°N); and (**f**) water turbidity (NTU) from 2004 to 2011. The gray-dashed line represents Cat Point (CP) station and black-solid line represents Dry Bar (DB) station in figures (**c,f**). Blue boxes illustrate the effects of high river flow conditions on salinity and water turbidity, e.g., salinity decreases and turbidity increases as river discharge increases. Red ellipses indicate the difference in salinity at two stations during the low-flow conditions. Black arrows illustrate wind direction during cold (downward) and warm (upward) seasons in general. Red bars in (**a**) represent Landsat 5 TM clear sky match-ups for turbidity algorithm and validation analysis during the study period. HC = Hurricane, TS = Tropical storm, AR = Apalachicola River.

A time-series of mean-daily turbidity showed irregular variations that could be attributed to complex interactions between the Apalachicola River plume, daily tidal cycles, precipitation, and erratic behavior of winds across the bay (Figure 6f). Mean turbidity at DB was relatively higher than at CP likely due to shallower depth and proximity to the major inlet at West Pass (Figure 1; Table S2). Large turbidity peaks were well-correlated with high-flow conditions in the Apalachicola River (blue boxes in Figure 6) and tropical storms [14,41]. A seasonally-decomposed turbidity suggested the difference in trends at two stations, e.g., Cat Point experienced turbid waters, whereas Dry Bay station remained less turbid in the summer (Figures S6 and S7). Despite the proximity to major river mouth, the observed differences in turbidity ranges could also be associated with complex mechanisms of plume dynamics in the bay (Table S2).

3.2.2. Principal Component Analysis

In Figure 7, each point represents physical state of the bay based on different physical variables, e.g., salinity, tidal height, river discharge etc. It is difficult to summarize the bay's "conditions" with this many variables because some of them remain stable for most of the days (e.g., air temperature, while others vary more frequently during different days (e.g., wind speed). PCA generated new characteristics (e.g., principal components) with maximum possible variance in data for easier explanation (Figure 7). The PCA results showed that ~50% of the information (variance) contained in the data are retained by the first two principal components, which indicated that the variance in data can be explained by more than two principal components and in more than two dimension spaces. Principal component-1 (PC1) explained ~26% variance, and generally described the data based on seasonality (e.g., days in the winter and spring to the right) and in summer and fall to the left (Figure 7). Therefore, blue/green points in figure are the days of winter and spring when the bay experienced elevated river discharge, higher air pressures, more turbid waters at Dry Bar than at Cat Point, cold air temperatures, and became relatively fresher as compared to the summer and fall. Furthermore, a positive association of PC1 to the river discharge (~49%) and negative association with salinity (at Cat Point (~40%) and Dry Bar (~46%)) and tidal height (~40%), supported the time-series results which showed that the bay generally experienced moderate river discharge, low tidal heights, and low salinity environments in winter and spring, and vice versa.

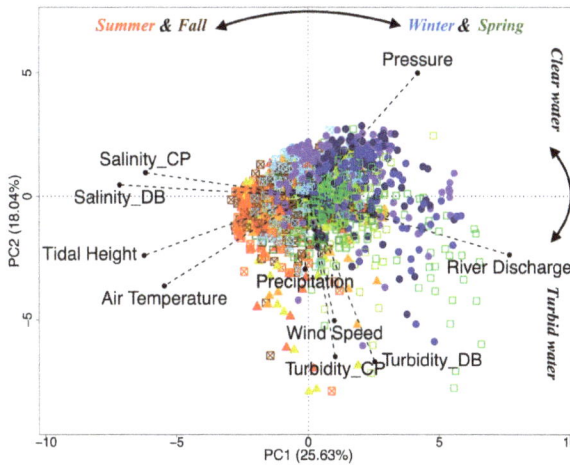

Figure 7. Principal Component Analysis (N = 2377 days) on meteorological (wind speed, wind direction, and air temperature), hydrological (river discharge, precipitation, turbidity, and salinity), and astronomical (tidal height) variables in Apalachicola Bay. A color scheme and symbols are used to illustrate seasonality, e.g., blue color represent the winter (November to February), green square represent the spring (March to May), red triangle represents the summer (June to August), brown crossed-square boxes represent the fall (September & October). The shades of group color indicate months in each group, e.g., November—sky blue, January—Navy blue, July—orange, August—red. Turbidity_DB = Turbidity at Dry Bar, Turbidity_CP = Turbidity at Cat Point, Salinity_DB = Salinity at Dry Bar, and Salinity_CP = Salinity at Cat Point.

Conversely, principal component-2 (PC2) explained data generally based on water turbidity, whereby, less turbid days had positive scores on PC2 and more turbid days had negative scores on PC2. The PC2 was strongly related to wind speed and water turbidity at both locations indicating the importance of winds on diurnal variability in the bay's turbidity. Loadings of river discharge were located closer to Dry Bar turbidity (Turbidity_DB), suggesting stronger influence of river discharge in the western part of bay (e.g., DB) than the eastern part (e.g., CP)—except during periods of strong westerly winds that could veer the river plume towards St. George Sound (Figure 1). Similarly, tides could affect more the turbidity at the CP compared to the DB station. As expected, high pressure and low temperatures were generally observed in the winter. Moreover, in many instances we observed positive correlations between high barometric pressures and river discharge, possibly due to wind-driven river flow during frontal passages. However, these interactions of frontal winds and river discharge, and their effects on turbidity, were not apparent in the PCA results. Overall, a small difference in first two principal components (25.63% and 18.04%, respectively) to explain variance in data indicated that various physical parameters strongly interact with each other, making Apalachicola Bay a complex estuarine system. Classification tree-based models were analyzed to further examine the relative importance of wind, tides, river discharge, and precipitation on the bay's turbidity in the eastern and western regions of the bay.

3.2.3. Tree-Based Classification Models

Tree-based classification models showed relative importance of major physical forcings on the mean-daily turbidity at Cat Point (CP) and Dry Bar (DB) locations in Apalachicola Bay from 2004 to 2011. The mean turbidity at CP was 11 NTU (n = 2170 days). Similar to PCA, tidal height was the important factor at Cat Point that was used to classify data in two groups based on tidal influence, e.g., Node 2 (Tide < 0.25 m, turbidity = 8.3 NTU, n = 653 days), and Node 3 (Tide > 0.25 m, turbidity = 12 NTU, n = 1517 days) (Figure 8a). A tidal height threshold (0.25 m) separated data based on seasonal amplitudes, e.g., low tidal-height in winter and early spring, whereas high tidal-height in late spring, summer, and fall (Figure 6d). River discharge was a major contributor to the turbidity at Node 2. High-river flow conditions (RD > 1467 m^3s^{-1}) showed the largest increase in turbidity (Node 5, mean = 24 NTU, n = 33 days) that is likely to represent flood days (e.g., March–April, 2009; also see Figure 6a), and elevated river flow during frontal passages in winter months (e.g., December and January–2010; also see Figure 6a). Apalachicola River discharge is usually low in late spring, summer, and fall (Figures 6a and 7), however the periods of moderate winds can likely lead to increase in water turbidity through water-column mixing or by veering of the river plume water towards the eastern part of the bay (Node 7, mean turbidity = 15 NTU, n = 682 days). Variable importance analysis showed that Apalachicola River discharge was the most influential forcing (~43%) associated with the largest variation in water turbidity, followed by wind speed (~32%) and tidal height (~19%). In contrast, precipitation had a least influence (~6%) on variability in turbidity to the eastern region of the bay (e.g., Cat Point).

At Dry Bar station, wind speed was the primary factor that explained variability in turbidity during the study period (Figure 8b). Mean water turbidity was generally ~50% higher during the strong winds (Node 3, Mean turbidity = 18, wind speed > 3 ms^{-1}, n = 507) than weak winds (Node 2, Mean turbidity = 12 NTU, wind speed < 3 ms^{-1}, n = 1453 days). In addition, periods of strong winds during high river flow and precipitation events showed significant increases in water turbidity (~142% and ~84% at Node-7 and Node-13, respectively) in comparison to the weak winds (Node 2) at the Dry Bar station. Variable importance analysis showed that wind speed (~36%) was the most important factor for daily turbidity variations that was followed by river discharge (~27%), precipitation (~23%), and tidal height (~14%) towards the western region of bay (e.g., Dry Bar).

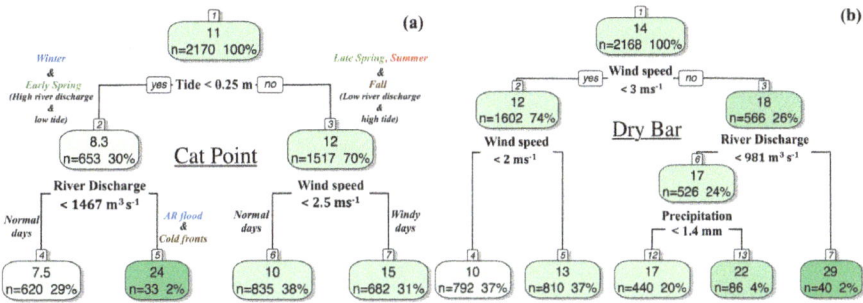

Figure 8. Classification tree model for turbidity at (**a**) Cat Point; and (**b**) Dry Bar (right) in Apalachicola Bay from 2004 to 2011. Wind speed in ms^{-1}, Apalachicola River discharge in m^3s^{-1}, tidal height in meter, and precipitation in millimeter.

3.2.4. Seasonal Turbidity Patterns in Apalachicola Bay

Apalachicola Bay's turbidity ranges from 1 to 70 NTU during normal days [29]. According to US EPA's (Environmental Protection Agency) water quality standards under the Clean Water Act (CWA), the bay's water is classified as CLASS II water which is suitable for shellfish propagation and harvesting. Although turbidity classification of low, moderate and high levels varies with locations and importance of a water body, a turbidity reading of <10 NTU indicates fairly clean water for most of the lakes and rivers. Furthermore, the EPA has set a turbidity criterion of <29 NTU above background (5–10 NTU) as normal condition in the bay. Often however values can increase beyond 100 NTU especially during extreme events such as cold fronts, floods, and tropical storms. Based on this information, we classified turbidity levels as low (<10 NTU), moderate (10–50 NTU), and high (>50 NTU) in Apalachicola Bay.

Cross-year (2004–2011) seasonally mean turbidity maps showed distinct variations in spatial patterns of turbidity in Apalachicola Bay. Moderate turbidity (>10 NTU) was observed in central and east bays in the spring and the winter (Figure 9a,d). However, spatial extent of moderately turbid water was usually widespread and uniform in the spring indicating a strong influence of river discharge on the bay's water turbidity in the wet season. In contrast, highly turbid water, mainly observed on shallow regions of the bay (e.g., EB, VC, and near barrier islands), was likely due to wind-supported sediment resuspension and river discharge in winter. Apalachicola Bay generally experiences low turbidity (<10 NTU) during low flow conditions (e.g., Figure 9b,c). Frequent periods of precipitation and run-off activities could be responsible for showing well-distributed turbid waters close to terrestrial sources as observed in summer. Nonetheless, the effects of bottom reflectance on the seasonal maps cannot be ignored (e.g., VS, CP, and DB) during low flow conditions (e.g., summer and fall). Bottom reflectance can be very pronounced in the fall when the bay lacks adequate supply of particulate material—via freshwater sources (e.g., rivers and run-off). However, short periods of wind-induced sediment resuspension could have resulted in moderate turbidity in central bay in the fall. St. George Sound (GS) remained relatively clear in all seasons possibly due to lack of freshwater sources and relatively deeper water column [72]. Seasonal turbidity maps also indicated that Apalachicola Bay can be generally classified as a low to moderately turbid estuary.

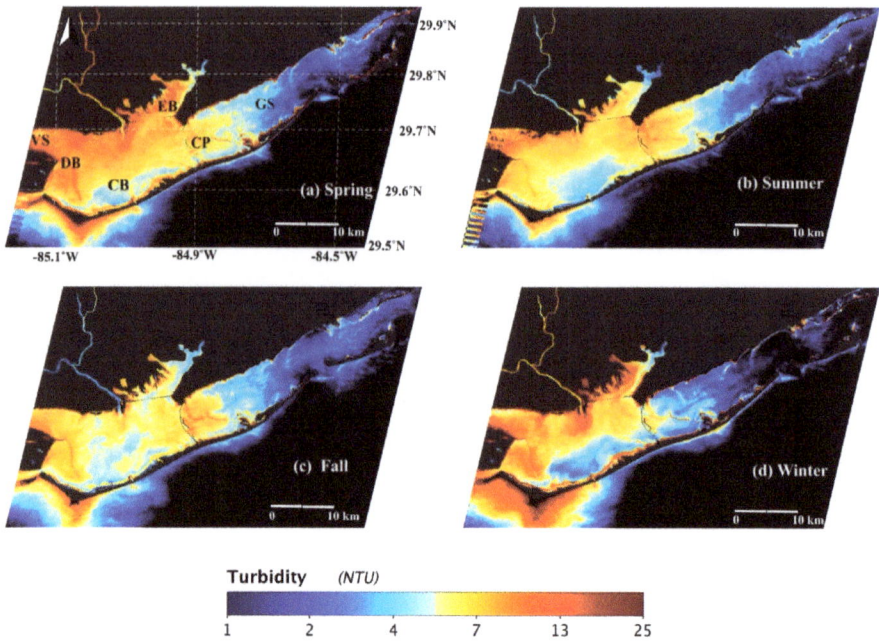

Turbidity *(NTU)*

Figure 9. Mean seasonal turbidity maps in Apalachicola Bay. (**a**) Spring (March, April, and May)—12 images; (**b**) summer (June, July, and August)—12 images; (**c**) fall (September and October)—13 images, and (**d**) winter (November, December, January, and February)—12 images. Data outside the seasonal reflectance and turbidity thresholds (as in Section 3.1.4) have been masked with black color. Clouds mask ($\rho_{band-7} > 0.011$) is applied similar to Wang and Shi (2006) [76]. VS = St. Vincent Sound, DB = Dry Bar, CB = Central Bay, EB = East Bay, CP = Cat Point, and GS = St. George Sound.

3.3. Turbidity Maps during Extreme Events in Apalachicola Bay

The general turbidity algorithm was used with applications of season-dependent thresholds to convert atmospheric-corrected satellite images to the turbidity maps in Apalachicola Bay. Three extreme case scenarios were examined to investigate the effects of major physical forcings on the distribution of turbidity and water dynamics in the bay.

3.3.1. Apalachicola River Flood Conditions (4 April 2005 and 15 April 2009)

Landsat 5 TM images were obtained during two extreme case scenarios when (1) a peak river discharge (~4580 m^3s^{-1}) was observed in the strongest Apalachicola River flood event (April-2005) in the 8-year study period; and (2) the diminishing phase of river discharge (~1910 m^3s^{-1}) during the strong flood event in 2009 (also see Figure 6a). On 4 April 2005, Apalachicola River discharge was ~8 times higher than the mean discharge (~558 m^3s^{-1}) of study period (Figure 10a). Tidal height indicated flooding phase approaching to slack water, and northwesterly winds were relatively weaker (~2 ms^{-1}) to play significant role in bay's water dynamics (Figure 10c). Turbidity map showed highly turbid water mass (>50 NTU) in central bay directed towards the east, and gradually diluted by the saline Gulf water entering through the East Pass (Figure 10e). Apalachicola Bay experienced longer period of ebb tidal phase just few hours before the image acquisition. Although the tidal range was small, a combination of ebb tidal currents and high river discharge could have drawn significant amount of water in the central bay that later could have been trapped inside the bay due to shorter transition period from ebb to successive flood tidal phase. Flooding-supported intrusion of less turbid saline waters through the West and Indian passes is clearly evident in the maps. Furthermore, St. George Sound is relatively deeper than St. Vincent Sound [72]. Therefore, west to east surface gradient and increasing tidal height could have caused the net eastward movement of turbid fresh water plume during the image acquisition. Interestingly, distinct dark colored water enters from the marshes, Tate's hell long-leaf pine forest (East Bay), and the Carrabelle River plume to the bay (See Figure 1). This CDOM-rich water has been masked out in the turbidity map.

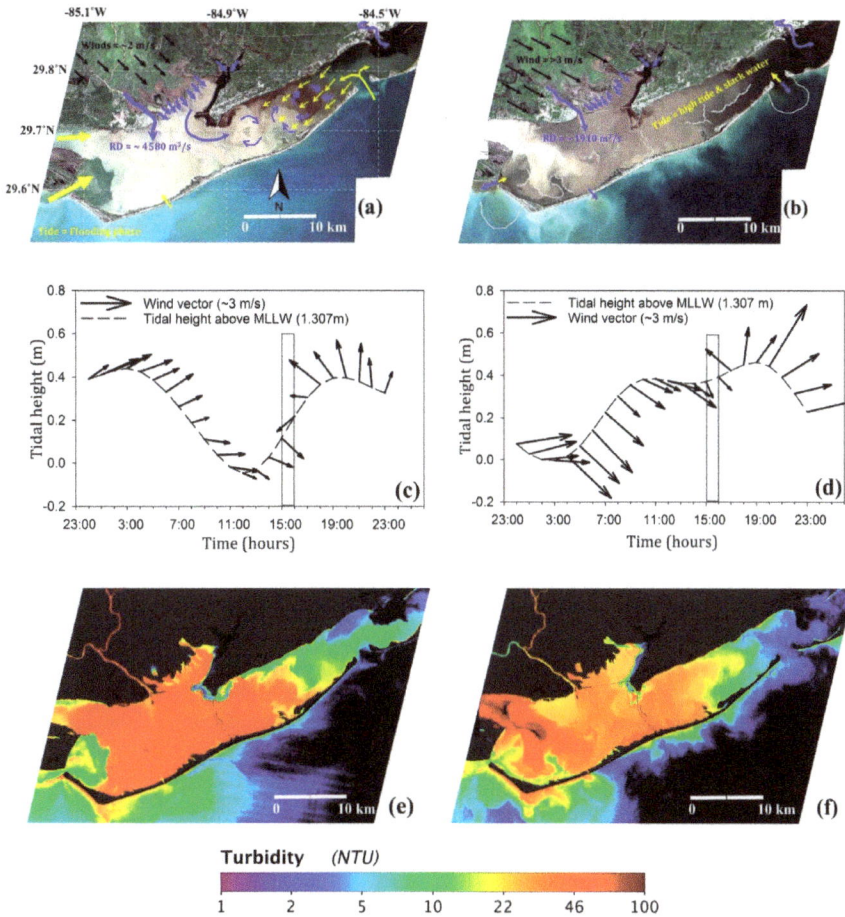

Figure 10. (a–f) A true color image (**top**), tidal-height and wind time-series (**center**), and turbidity map (**bottom**) for the Apalachicola River flood condition on 4 April 2005 (**left panel**) and on 15 April 2009 (**right panel**).

Second case represents the influence of wind speed and river discharge during low energy (slack water) and high tidal period (Figure 10b). The bay experienced sustained periods of moderate northwesterly winds (>3 ms^{-1}) during the flooding phase before image acquisition (Figure 10d). Hence, a distinct feature of turbid water mass in St. Vincent Sound could be an outcome of wind induced sediment re-suspension rather than the river supplied sediments possibly due to the relatively shallower depths of this region. Furthermore, tidal forcing was weak during the image acquisition, and hence moderate exchange of turbid river plume could have occurred via the West Pass and

Sikes Cut despite high tides in the bay (Figure 10f). Moreover, high tidal phase could have caused a two-layered water column at the inlets with relatively low-density river water exiting the bay over underlying denser saline water. The eastward transport, mixing, and dilution of sediment rich water were similar as observed in the previous case.

3.3.2. Passages of Strong Cold Fronts over Apalachicola Bay (12 January 2010 and 13 February 2010)

In winter and early spring, Apalachicola Bay experiences frequent passages of southward propagating high-pressure systems characterized by cold air temperatures and strong winds. This scenario is represented by two cases, (1) a remnant of a strong frontal system that passed on 9 January 2010; and (2) a passage of strong frontal system on 13 February 2010. On 12 January 2010, a frontal remnant passed over the bay in the morning, and moderate northwesterly winds (>2.5 ms^{-1}) sustained throughout the day. Furthermore, the bay experienced a severe *"cold stun"* event (7–14 January 2010) when water temperature fell below 10 °C affecting marine life for several days [77]. Apalachicola River discharge was about 1.5 times higher than the mean discharge of study period (Figure 11a). There were no records of precipitation over the bay within a time span of 3 days before image acquisition. Tidal height distribution showed flooding phase, however, to be relatively weaker during this time of a year as discussed in the temporal analysis (Figures 7d and 11c). The elevated river discharge in absence of precipitation indicated influence of sustained winds that could have drawn off the river water and increased water turbidity in the bay due to mixing and re-suspension. Furthermore, northwesterly winds could have directed turbid plume water towards the barrier islands from where it could have been further diverted to the east towards St. George Sound and to the west towards central bay. Although water level in the bay was relatively high at the end of flood tidal phase, wind supported salt-water intrusion and shallower depths of the western region could have blocked the plume from escaping freely through the West Pass. This hypothetical view is further supported by turbidity patterns in the map that could have occurred due to strong interactions between turbid fresh water mass and a relatively clear marine water mass in central bay (Figure 11e). The net water movement could have occurred towards relatively deeper St. George Sound. The effect of flooding phase is clearly evident as intrusion of less turbid and saline water through the East Pass and Sikes Cut.

On 13 February 2010, a strong frontal system passed over Apalachicola Bay. The river discharge was 4 times higher than the mean discharge in the bay (Figure 11b). Apalachicola Bay experienced anomalously higher precipitation since the beginning of the year 2010 with ~16 mm of rain on the day of image acquisition. During winter and early spring, the anomalously high rainfall events in the southwest USA have been previously associated to the ENSO [74,75]. The year 2009–2010 experienced a moderate *El Niño* event that can be linked to the change in precipitation and river discharge pattern in the Apalachicola Basin. Also, the bay experienced moderate northerly winds that prevailed much longer before image acquisition (Figure 11d). Therefore, river plume dynamics could have been similar to the previous case. However, turbid water could have to escape through the West Pass, the East Pass, and Sikes Cut despite the flooding phase possibly due to unique combination of wind and river forcing. The highest turbidity was generally observed over the shallow regions of the bay, e.g., oyster bars, possibly due to sediment re-suspension, although the turbidity was masked because it exceeded the reflectance thresholds generally observed during winter (Figure 11f).

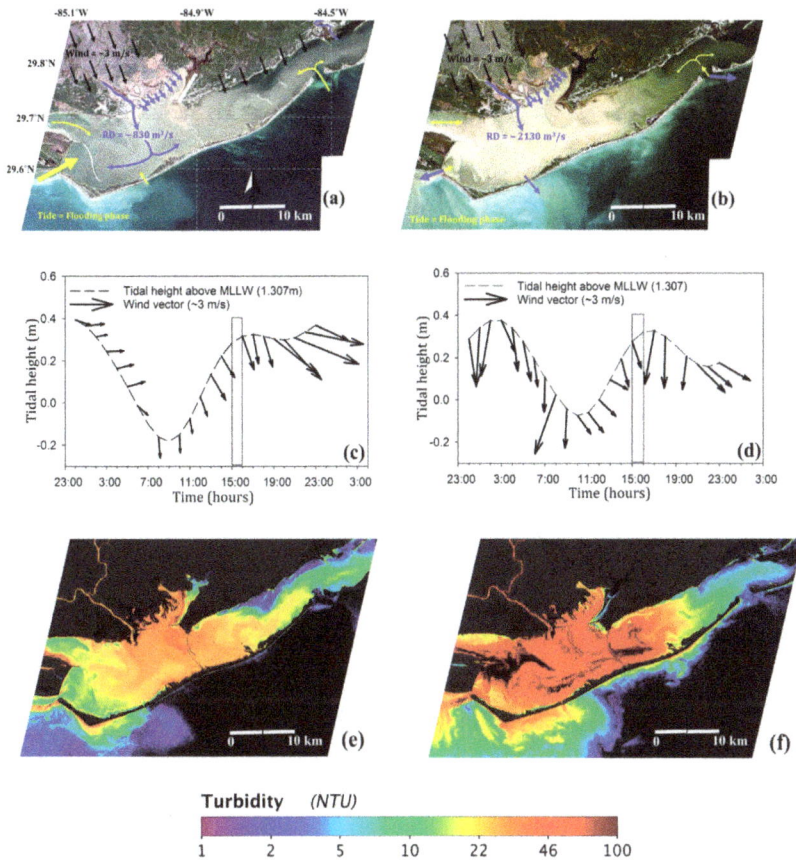

Figure 11. (**a–f**) A true color image (**top**), tidal-height and wind time-series (**center**), and turbidity map (**bottom**) during the cold-front passages on 12 January 2010 (**left panel**) and on 13 February 2010 (**right panel**).

3.3.3. Low Flow Conditions in Apalachicola Bay

This scenario represents two cases, (1) the effect of a tropical storm during low flow condition on 8 September 2004; and (2) a strong frontal passage during low flow conditions on 22 November 2008. The first case shows turbidity map two days after the Hurricane Frances that made landfall to the East of Apalachicola Bay on 6 September 2004, when river discharge was ~50% lower than the average discharge (~558 m^3s^{-1}) of the study period (Figure 12a). As expected, the bay experienced sustained periods of southwesterly winds—as generally observed to the southwest of tropical storms. Tidal height showed the beginning of ebb tidal phase during the image acquisition (Figure 12c). However, winds could have been more dominant than the tidal forcing near the West and Indian passes, which could have veered the turbid river plume towards East Bay and St. George Sound (Figure 12e). Overall, the well-mixed nature of the bay indicated the dominance of storm-induced southerly winds during low flow conditions. Chen et al. (2009) used MODIS high-resolution imagery to generate total suspended solids (TSS) maps, and studied the effects of Hurricane Frances in Apalachicola Bay. They observed similar patterns of total suspended solids and southwesterly winds after the passage of Hurricane Frances.

In the second case, a strong frontal system passed perpendicular to the Apalachicola River. The strongest winds (>6 ms^{-1}) were observed among all the cases, however river discharge was ~50% lower than the mean discharge of ~558 m^{3}s^{-1} (Figure 12b). In contrast to the northwesterly winds, perpendicular forcing of the northeasterly winds could have confined the river water to the river channel. Interestingly, river discharge was higher before and after the cold front, whereas it decreased during the frontal passage. Despite the ebb tidal phase, wind could have supported the water intrusion through the East Pass, and net water movement from the East to West (Figure 12d). Turbidity was the highest in the shallower parts of central bay (e.g., Dry Bar) possibly due to wind-induced sediment re-suspension and sediment transport from St. George Sound and East Bay (Figure 12f).

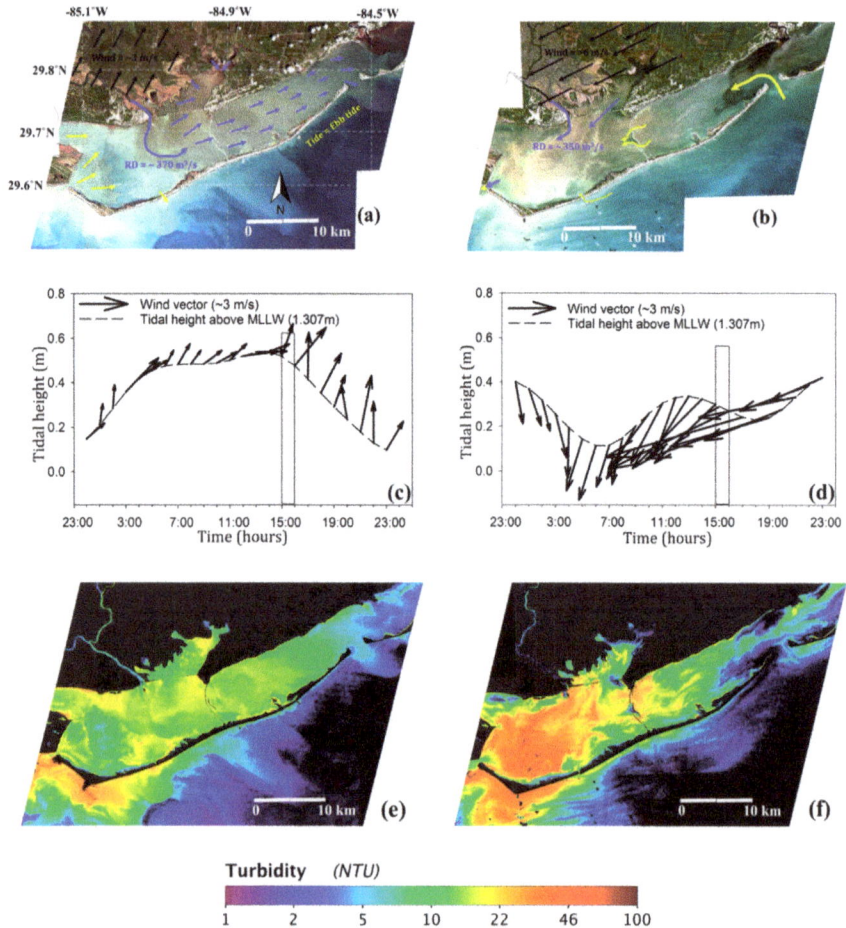

Figure 12. (a–f) A true color image (**top**), tidal-height and wind time-series (**center**), and turbidity map (**bottom**) after a hurricane passage on 8 September 2004 (**left panel**) and during frontal passage in low flow condition on 22 November 2008 (**right panel**).

4. Assessing Potential Applicability of Turbidity Algorithm to Landsat 8 OLI

Landsat 8 OLI, a successor of Landsat 5 TM and Landsat 7 ETM+ sensors, contains spectral bands similar to its predecessors except additional channels in deep blue (430–450 nm) and infrared (1360–1380 nm) spectral regions that further improve its capability in monitoring water resources and cirrus cloud detection. In addition, the signal-to-noise ratio and radiometric resolution (12 bits) of OLI were higher than Landsat 5 TM (8 bits). The OLI red band has a better spectral resolution (640–670 nm, bandwidth = 30 nm) than Landsat 5 TM (630–690 nm, bandwidth = 60 nm). Furthermore, the red band is located just outside chlorophyll-a absorption peak which might reduce the chlorophyll contamination to the particulate back-scattered signal and hence, Landsat 5 TM—turbidity relationship could be applicable to Landsat 8 OLI. A set of FLAASH-corrected 17 clear-sky Landsat 8 OLI images (Table S3) were converted to turbidity using the proposed single band empirical relationship. The matchup comparison between modeled and in situ turbidity showed promising results along with turbidity maps as illustrated in Figure 13. Turbidity maps represent images for 3 seasons: winter (18 February 2015) (Figure 13b), spring (8 March 2016) (Figure 13c), and summer (7 July 2016) (Figure 13d). In spring, river discharge was approximately two and three folds higher (~1050 m^3/s) than in winter and summer, respectively. However, wind speed and direction appears to be major controlling factors of the bay's water dynamics and turbidity distribution despite elevated river discharge in the spring and the winter. Interestingly, these individual images are also matching with seasonally averaged turbidity maps (Figure 6), further supporting the applicability of the Landsat 5 TM-based turbidity algorithm to Landsat 8 OLI, and its potential for long-term monitoring of water turbidity in a shallow water estuary such as Apalachicola Bay.

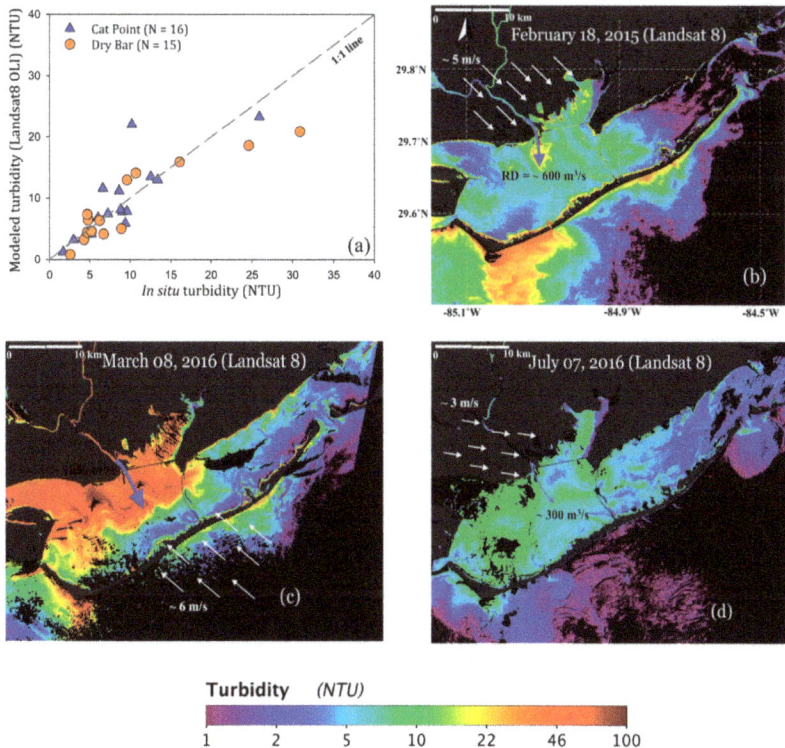

Figure 13. (**a**) Performance of Landsat 5 TM-based red band empirical relationship on Landsat 8 OLI images (R^2 = 0.80, N = 31); (**b–d**) Turbidity maps derived using Landsat 8 OLI imagery (18 February 2015, 08 March 2016, and 7 July 2016, respectively) and the general turbidity algorithm. Reflectance values beyond seasonal thresholds and clouds are masked in the maps.

5. Conclusions

Apalachicola Bay is a bar-built estuary located to the west of Florida's panhandle and is well-known for its commercial oyster yields. We investigated turbidity climatology (2004–2011) to understand roles of meteorological, hydrological and astronomical factors on the bay's water dynamics and turbidity distribution. Together, time-series, PCA, and classification tree-based analysis conveyed important information that the bay is highly dynamic in nature, both diurnally and seasonally, and its water quality (e.g., turbidity) is largely driven by interactions of different physical forcings, such as river discharge, tides, winds, and precipitation associated run-off, except during the extreme conditions when one or more of these factors become the dominant regulator of the bay's water dynamics (e.g., river flood, cold fronts, and tropical storms). The following conclusions can be derived from the temporal analysis of physical forcings in Apalachicola Bay:

(1) Apalachicola Bay, especially the western and eastern sides of river mouth, experiences higher turbidity in the spring and the winter mainly due to elevated supply of sediments by the Apalachicola River and wind-induced sediment re-suspension and mixing of the shallow water column. In addition, periods of strong winds, mainly associated with the passage of low-pressure systems (e.g., storms and hurricanes), can result in higher flow velocity and increased water turbidity for a few days under low flow conditions in summer and fall [14,41].

(2) Apalachicola River discharge and wind speed were the major influential forcing at Cat Point (the eastern region) station, but tidal height could be the common factor for the variations in diurnal turbidity especially in the St. George Sound possibly due to the strong influence of semi-diurnal tides and its remoteness to the fresh water sources [38,39]. In contrast, all the forcings were relatively important at Dry Bar (the western region) due to its shallower depths and proximity to major inlets and land.

(3) The study of water dynamics and turbidity distribution is highly complex due to the multiple inlets and strong interactions of major physical forcings in Apalachicola Bay. The water dynamics is mainly controlled by the interactions of three forces: river discharge, tide, and winds. River plume generally flow from north to south with tendency to escape the bay through the closest inlet, the West Pass. In contrast, tides and winds either control the free flow of river plume either by blocking/diverting it in the opposite direction or by assisting the river mass to escape through the tidal inlets.

Landsat sensors (e.g., TM, ETM+, and OLI) are mainly designed for land application; however they can be utilized to investigate estuarine water dynamics and turbidity distribution. We have used ENVI's FLAASH-based atmospheric correction, as it is a crucial step in ocean color remote sensing. However, the validation of FLAASH performance was necessary before using the atmospheric-corrected Landsat reflectance in the algorithm development. FLAASH-AERONET comparison yielded promising results to provide enough confidence for using FLAASH-derived surface reflectance in low to moderately turbid waters in Apalachicola Bay. Although this comparison contains many errors mainly due to the effects of skylight, it gave a preliminary indication of reasonable performance of the FLAASH. We have explored historical archives of Landsat 5 TM and in situ turbidity measurements to develop a simple single band (Band 3: 630–690 nm) Landsat 5 TM—turbidity relationship in Apalachicola Bay. Bootstrap-based uncertainty analysis overall yielded a reasonable performance of a turbidity algorithm with realistic possibility of training and testing data sets (N = 5000 simulations). The bootstrap results provided reasonable estimates of a constant and power law exponent for a turbidity algorithm for the Apalachicola Bay. We also tested seasonal dependency of the Landsat 5 TM—turbidity relationship and found that the individual seasonal relationship deviates significantly from the general turbidity algorithm. Hence, we restricted the use of the turbidity algorithm to different seasons by applying seasonal thresholds. Seasonally-mean turbidity maps showed distinct patterns of turbidity in Apalachicola Bay with moderate to highly turbid waters in spring and winter, in contrast to low to moderately turbid waters in summer and fall. Shallow regions of the bay (e.g., East Bay, St. Vincent Sound, Cat Point Bar, and Dry Bar) remain turbid throughout the year while relatively deeper parts (e.g., Central Bay and St. George sound) of the bay remain clearer—except during strong winds and high fresh water inputs.

We presented synoptic views of three extreme case scenarios to examine water dynamics and turbidity distribution in Apalachicola Bay. Turbidity maps generally showed good agreement with the meteorological, hydrological, and astronomical forcings in the bay. The following results can be derived from these extreme case scenarios:

(1) Apalachicola River discharge is the major controlling factor on turbidity during the high-flow conditions. Winds may affect turbidity, especially in shallow regions of the bay, through sediment re-suspension and mixing, but only if it prevails over longer periods. Tidal-influence mainly depends on the flood and ebb tidal phases, except during the slack waters when tidal forcing is weak. The flood tides increase the tidal height by introducing saline water in the bay, whereas ebb tides decrease the tidal height by drawing water out of the bay. Turbidity maps indicated that the synchronization of ebb tidal flow and river discharge during high river flow conditions may have stronger effects on bay's water turbidity than the combination of river discharge and flood tides – which can cause dilution of the bay's water.

(2) Apalachicola Bay experiences strong northerly winds during the passage of cold-frontal systems. The sustained winds can increase water turbidity by drawing off the river water and by re-suspending sediments in shallow regions of the bay. Similarly, anomalously high precipitation and associated run-off can further increase water turbidity in the bay. Westerly and easterly components of northerly winds can divert river plume waters to different regions of the bay. Likewise, flood and ebb tidal phases affect the sediment transport and its exchange to the shelf. Overall, Apalachicola Bay experiences extremely turbid environments when two forcings, winds and river flow, are in phase.

(3) Wind becomes the dominant factor controlling water dynamics during low flow conditions. The bay becomes well-mixed and relatively saline after the passage of tropical storms and hurricanes. However, the bay's turbidity depends on the strength of winds. Furthermore, if strong winds move perpendicular to the river channel, it can reduce the freshwater input to the bay by confining river water within the river channel.

Turbidity maps showed reasonable performance of a turbidity algorithm in Apalachicola Bay; however several issues may have affected the performance of the algorithm that still needs to be addressed. These include the following: (1) contamination of signal by the bottom reflectance; (2) interference due to high concentrations of CDOM; (3) instrumentation error in in situ measurements; (4) sensor calibration; (5) contributions from skylight and sun-glint on the surface reflectance products; (6) performance of the atmospheric correction module and; (7) simplistic assumptions of water column homogeneity. The aim of this study was to investigate the turbidity climatology in Apalachicola Bay and thus, historic Landsat 5 TM data were used in the analysis. The performance of single-band empirical relationships showed promising results on more advanced instruments (e.g., Landsat 8 OLI) and its applicability should be further explored in other coastal regions using much robust ocean color parameters (e.g., remote sensing reflectance (Rrs)). Although the proposed algorithm has been shown to work well in Apalachicola Bay, it could be subject to change in band combination, coefficients, and/or regression function, if it is used in other coastal regions. Nonetheless, the promising performance of the empirical Landsat 5 TM—turbidity relationship supports the use of Landsat data in monitoring water quality in estuarine environments.

Supplementary Materials: The following are available online at www.mdpi.com/2072-4292/9/4/367/s1, Figure S1: Landsat 5 TM turbidity algorithm with NIR band (Band 4), Figure S2: Turbidity-Landsat 5 TM relationships for different seasons. Blue and orange symbols represent measurements at Cat Point and Dry Bar stations, Figure S3: Bootstrap simulation (N = 5000) results, (a) intercept or log (constant) (Mean $\pm$ SD = 8.32 $\pm$ 0.50, range = 6.2–10.16), (b) slope or power law exponent (Mean $\pm$ SD = 1.76 $\pm$ 0.14, range = 1.16–2.35), c) R2 for turbidity-Landsat 5 TM algorithm, (d) R2 for algorithm validation, (e) RMSE for algorithm validation, and (f) bias for algorithm validation., Figure S4: Histogram of mean-daily Apalachicola River discharge from 2004 to 2011, Figure S5: Decomposition of precipitation time-series (2004–2011) into, (a) daily (N = 2439 days), (b) seasonal, (c) yearly trend, (d) random or remainder component (=daily time-series – (seasonal component + yearly trend)), Figure S6: Time-series of turbidity was decomposed into seasonal, trend, and random components for recognizing the underlying pattern during the 8-years of study period at Cat Point station, Figure S7: The turbidity time-series at Dry Bar was decomposed after removing outliers, Table S1: Bootstrapping statistics for the turbidity algorithm and its validation, Table S2: Sample size, mean, standard deviation, and range of mean-daily river discharge, precipitation, salinity, tidal height, wind speed, water depth, and turbidity from 2004 to 2011 in Apalachicola Bay, Table S3: Turbidity matchups at Cat Point and Dry Bar stations used to evaluate applicability of the Landsat 5 TM-based turbidity algorithm to Landsat 8 OLI images (17 images).

Acknowledgments: Authors acknowledge NASA funding (Project No. NNX14A043G). We would like to thank NOAA NERRS Centralized Data Management Office for the turbidity and meteorological data. We are also thankful to Alan Weidermann, Bill Gibson and Robert Arnone for the AERONET-Ocean color data and USGS for Landsat imagery.

Author Contributions: I.J. and E.D. conceived and designed the research; I.J. analyzed the data and drafted the work and all authors critically revised the paper for intellectual content.

Conflicts of Interest: The authors declare no conflict of interest.

References

1. Cloern, J.E.; Foster, S.; Kleckner, A. Phytoplankton primary production in the world's estuarine-coastal ecosystems. *Biogeosciences* **2014**, *11*, 2477–2501.
2. Chelsea Nagy, R.; Graeme Lockaby, B.; Kalin, L.; Anderson, C. Effects of urbanization on stream hydrology and water quality: The Florida Gulf Coast. *Hydrol. Process.* **2012**, *26*, 2019–2030.
3. Giosan, L.; Syvitski, J.; Constantinescu, S.; Day, J. Climate change: Protect the world's deltas. *Nature* **2014**, *516*, 31–33. [CrossRef] [PubMed]
4. Fabricius, K.E. Effects of terrestrial runoff on the ecology of corals and coral reefs: Review and synthesis. *Mar. Pollut. Bull.* **2005**, *50*, 125–146. [PubMed]
5. Kenworthy, W.J.; Fonseca, M.S. Light requirements of seagrasses *Halodule* wrightii and *Syringodium* filiforme derived from the relationship between diffuse light attenuation and maximum depth distribution. *Estuaries* **1996**, *19*, 740–750.
6. Pedersen, T.M.; Gallegos, C.L.; Nielsen, S.L. Influence of near-bottom re-suspended sediment on benthic light availability. *Estuar. Coast. Shelf Sci.* **2012**, *106*, 93–101. [CrossRef]
7. Thrush, S.F.; Hewitt, J.E.; Cummings, V.J.; Ellis, J.I.; Hatton, C.; Lohrer, A.; Norkko, A. Muddy Waters: Elevating Sediment Input to Coastal and Estuarine Habitats. *Front. Ecol. Environ.* **2004**, *2*, 299–306.
8. Ryan, P.A. Environmental effects of sediment on New Zealand streams: A review. *N. Z. J. Mar. Freshw. Res.* **1991**, *25*, 207–221.
9. Wang, H.; Huang, W.; Harwell, M.A.; Edmiston, L.; Johnson, E.; Hsieh, P.; Milla, K.; Christensen, J.; Stewart, J.; Liu, X. Modeling oyster growth rate by coupling oyster population and hydrodynamic models for Apalachicola Bay, Florida, USA. *Ecol. Model.* **2008**, *211*, 77–89.
10. Robertson, M.J.; Scruton, D.A.; Clarke, K.D. Seasonal effects of suspended sediment on the behavior of juvenile Atlantic salmon. *Trans. Am. Fish. Soc.* **2007**, *136*, 822–828.
11. Pollock, F.J.; Lamb, J.B.; Field, S.N.; Heron, S.F.; Schaffelke, B.; Shedrawi, G.; Bourne, D.G.; Willis, B.L. Sediment and turbidity associated with offshore dredging increase coral disease prevalence on nearby reefs. *PLoS ONE* **2014**, *9*, e102498. [CrossRef] [PubMed]
12. Rügner, H.; Schwientek, M.; Beckingham, B.; Kuch, B.; Grathwohl, P. Turbidity as a proxy for total suspended solids (TSS) and particle facilitated pollutant transport in catchments. *Environ. Earth Sci.* **2013**, *69*, 373–380. [CrossRef]
13. Stubblefield, A.P.; Reuter, J.E.; Goldman, C.R. Sediment budget for subalpine watersheds, Lake Tahoe, California, USA. *Catena* **2009**, *76*, 163–172. [CrossRef]
14. Chen, S.; Huang, W.; Wang, H.; Li, D. Remote sensing assessment of sediment re-suspension during Hurricane Frances in Apalachicola Bay, USA. *Remote Sens. Environ.* **2009**, *113*, 2670–2681. [CrossRef]
15. D'Sa, E.J.; Ko, D.S. Short-term Influences on Suspended Particulate Matter Distribution in the Northern Gulf of Mexico: Satellite and Model Observations. *Sensors* **2008**, *8*, 4249–4264. [CrossRef] [PubMed]
16. Kemker, C. Turbidity, Total Suspended Solids and Water Clarity. In *Fundamentals of Environmental Measurements*; Fondriest Environmental, Inc.: Fairborn, OH, USA, 2014; Available online: http://www.fondriest.com/environmental-measurements/parameters/water-quality/turbidity-total-suspended-solids-water-clarity/ (accessed on 30 August 2016).

17. Mobley, C.; Boss, E.; Roesler, C. Ocean Optics Web Book. 2010. Available online: http://www. oceanopticsbook.info (accessed on15 March 2016).

18. Dogliotti, A.; Ruddick, K.; Nechad, B.; Doxaran, D.; Knaeps, E. A single algorithm to retrieve turbidity from remotely-sensed data in all coastal and estuarine waters. *Remote Sens. Environ.* **2015**, *156*, 157–168. [CrossRef]

19. Chen, Z.; Hu, C.; Muller-Karger, F. Monitoring turbidity in Tampa Bay using MODIS/Aqua 250-m imagery. *Remote Sens. Environ.* **2007**, *109*, 207–220. [CrossRef]

20. Choi, J.K.; Park, Y.J.; Ahn, J.H.; Lim, H.S.; Eom, J.; Ryu, J.H. GOCI, the world's first geostationary ocean color observation satellite, for the monitoring of temporal variability in coastal water turbidity. *J. Geophys. Res. Oceans* **2012**, *117*. [CrossRef]

21. Miller, R.L.; McKee, B.A. Using MODIS Terra 250 m imagery to map concentrations of total suspended matter in coastal waters. *Remote Sens. Environ.* **2004**, *93*, 259–266. [CrossRef]

22. Minella, J.P.; Merten, G.H.; Reichert, J.M.; Clarke, R.T. Estimating suspended sediment concentrations from turbidity measurements and the calibration problem. *Hydrol. Process.* **2008**, *22*, 1819–1830. [CrossRef]

23. Onderka, M.; Pekárová, P. Retrieval of suspended particulate matter concentrations in the Danube River from Landsat ETM data. *Sci. Total Environ.* **2008**, *397*, 238–243. [CrossRef] [PubMed]

24. Zhang, M.; Tang, J.; Dong, Q.; Song, Q.; Ding, J. Retrieval of total suspended matter concentration in the Yellow and East China Seas from MODIS imagery. *Remote Sens. Environ.* **2010**, *114*, 392–403. [CrossRef]

25. Gippel, C.J. Potential of turbidity monitoring for measuring the transport of suspended solids in streams. *Hydrol. Process.* **1995**, *9*, 83–97. [CrossRef]

26. D'Sa, E.J.; Miller, R.L.; McKee, B.A. Suspended particulate matter dynamics in coastal waters from ocean color: Application to the Northern Gulf of Mexico. *Geophys. Res. Lett.* **2007**, *34*, L23611. [CrossRef]

27. Palmer, S.C.J.; Kutser, T.; Hunter, P.D. Remote sensing of inland waters: Challenges, progress and future directions. *Remote Sens. Environ.* **2015**, *157*, 1–8. [CrossRef]

28. D'Sa, E.J.; Roberts, H.H.; Allahdadi, M.N. Suspended particulate matter dynamics along the Louisiana-Texas coast from satellite observation. In Proceedings of the Coastal Sediments 2011, Miami, FL, USA, 2–6 May 2011; Rosati, J.D., Wang, P., Roberts, T.M., Eds.; pp. 2390–2402.

29. Edmiston, H. *A River Meets the Bay—A Characterization of the Apalachicola River and Bay System. Apalachicola National Estuarine Research Reserve*; Florida Department of Environmental Protection: Tallahassee, FL, USA, 2008; pp. 1–200.

30. Leitman, H.M.; Sohm, J.E.; Franklin, M.A. *Wetland Hydrology and Tree Distribution of the Apalachicola River Flood Plain, Florida*; US Geological Survey: Reston, VA, USA, 1984.

31. Whitfield, W.; Beaumariage, D.S. Shellfish management in Apalachicola Bay: Past, Present, Future. In Proceedings of the Conference on the Apalachicola Drainage System, Gainesville, FL, USA, 23–24 April 1977; Florida Department of Natural Resources Marine Research Laboratory: Gainesville, FL, USA, 1977; pp. 130–140.

32. Havens, K.; Allen, M.; Camp, E.; Irani, T.; Lindsey, A.; Morris, J.; Kane, A.; Kimbro, D.; Otwell, S.; Pine, B. *Apalachicola Bay Oyster Situation Report*; Florida Sea Grant College Program, Technical Publication TP-200; Florida Sea Grant: Gainesville, FL, USA, 2013; Available online: https://www.flseagrant.org/news/2013/04/apalachicola-oyster-report/ (accessed on 20 May 2016).

33. Edmiston, H.L.; Fahrny, S.A.; Lamb, M.S.; Levi, L.K.; Wanat, J.M.; Avant, J.S.; Wren, K.; Selly, N.C. Tropical Storm and Hurricane Impacts on a Gulf Coast Estuary: Apalachicola Bay, Florida. *J. Coast. Res.* **2008**, 38–49. [CrossRef]

34. Grattan, L.M.; Roberts, S.; Mahan, W.T., Jr.; McLaughlin, P.K.; Otwell, W.S.; Morris, J.G., Jr. The Early Psychological Impacts of the Deepwater Horizon Oil Spill on Florida and Alabama Communities. *Environ. Health Perspect.* **2011**, *119*, 838–843. [CrossRef] [PubMed]

35. Petes, L.E.; Brown, A.J.; Knight, C.R. Impacts of upstream drought and water withdrawals on the health and survival of downstream estuarine oyster populations. *Ecol. Evol.* **2012**, *2*, 1712–1724. [CrossRef] [PubMed]

36. Livingston, R.J. *Climate Change and Coastal Ecosystems: Long-Term Effects of Climate and Nutrient Loading on Trophic Organization*; CRC Press: Boca Raton, FL, USA, 2014.

37. Surratt, D.; Cherrier, J.; Robinson, L.; Cable, J. Chronology of sediment nutrient geochemistry in Apalachicola Bay, Florida (U.S.A). *J. Coast. Res.* **2008**, *24*, 660–671. [CrossRef]

38. Koch, M.; Sun, H. Tidal and Non-Tidal Characteristics of water Levels and Flow in the Apalachicola Bay, Florida. *WIT Trans. Built Environ.* **1970**, *43*, 357–366. [CrossRef]

39. Huang, W.; Sun, H.; Nnaji, S.; Jones, W. Tidal Hydrodynamics in a Multiple-Inlet Estuary: Apalachicola Bay, Florida. *J. Coast. Res.* **2002**, 674–684.
40. Huang, W.; Chen, S.; Yang, X. Remote sensing for water quality monitoring in Apalachicola Bay, USA. In *Advances in Earth Observation of Global Change*; Chuvieco, E., Li, J., Yang, X., Eds.; Springer: Dordrecht, The Netherlands, 2010; pp. 69–78.
41. Liu, X.; Huang, W. Modeling sediment resuspension and transport induced by storm wind in Apalachicola Bay, USA. *Environ. Model. Softw.* **2009**, *24*, 1302–1313. [CrossRef]
42. Hu, C.; Chen, Z.; Clayton, T.D.; Swarzenski, P.; Brock, J.C.; Muller–Karger, F.E. Assessment of estuarine water-quality indicators using MODIS medium-resolution bands: Initial results from Tampa Bay, FL. *Remote Sens. Environ.* **2004**, *93*, 423–441. [CrossRef]
43. Gordon, H.R.; Wang, M. Retrieval of water-leaving radiance and aerosol optical thickness over the oceans with SeaWiFS: A preliminary algorithm. *Appl. Opt.* **1994**, *33*, 443–452. [CrossRef] [PubMed]
44. Chavez, P.S. Image- based atmospheric corrections-revisited and improved. *Photogramm. Eng. Remote Sens.* **1996**, *62*, 1025–1035.
45. Hu, C.; Carder, K.L.; Muller-Karger, F.E. Atmospheric correction of SeaWiFS imagery over turbid coastal waters: A practical method. *Remote Sens. Environ.* **2000**, *74*, 195–206. [CrossRef]
46. Ruddick, K.G.; Ovidio, F.; Rijkeboer, M. Atmospheric correction of SeaWiFs imagery for turbid coastal and inland waters. *Appl. Opt.* **2000**, *39*, 897–912. [CrossRef] [PubMed]
47. Siegel, D.A.; Wang, M.; Maritorena, S.; Robinson, W. Atmospheric Correction of Satellite Ocean Color Imagery: The Black Pixel Assumption. *Appl. Opt.* **2000**, *39*, 3582–3591. [CrossRef] [PubMed]
48. Wang, M.; Shi, W. The NIR-SWIR combined atmospheric correction approach for modis ocean color data processing. *Opt. Express* **2007**, *15*, 15722–15733. [CrossRef] [PubMed]
49. Adler-Golden, S.; Berk, A.; Bernstein, L.; Richtsmeier, S.; Acharya, P.; Matthew, M.; Anderson, G.; Allred, C.; Jeong, L.; Chetwynd, J. FLAASH, a MODTRAN4 atmospheric correction package for hyperspectral data retrievals and simulations. In Proceedings of the 7th Ann. JPL Airborne Earth Science Workshop, Pasadena, CA, USA, 12–16 January 1998; pp. 9–14.
50. Kaufman, Y.; Wald, A.; Remer, L.; Gao, B.-C.; R-R, L.; Flynn, L. The MODIS 2.1-μm channel-correlation with visible reflectance for use in remote sensing of aerosol. *IEEE Trans. Geosci. Remote Sens.* **1997**, *35*, 1286–1298. [CrossRef]
51. Joshi, I.; D'Sa, E.J. Seasonal Variation of Colored Dissolved Organic Matter in Barataria Bay, Louisiana, Using Combined Landsat and Field Data. *Remote Sens.* **2015**, *7*, 12478–12502. [CrossRef]
52. Wilson, R.; Milton, E.; Nield, J. Are visibility-derived aot estimates suitable for parameterizing satellite data atmospheric correction algorithms? *Int. J. Remote Sens.* **2015**, *36*, 1675–1688. [CrossRef]
53. Thuillier, G.; Hersé, M.; Foujols, T.; Peetermans, W.; Gillotay, D.; Simon, P.; Mandel, H. The solar spectral irradiance from 200 to 2400 nm as measured by the SOLSPEC spectrometer from the ATLAS and EURECA missions. *Sol. Phys.* **2003**, *214*, 1–22. [CrossRef]
54. Gordon, H.R. Atmospheric correction of ocean color imagery in the Earth Observing System era. *J. Geophys. Res. Atmos.* **1997**, *102*, 17081–17106.
55. Cleveland, R.B.; Cleveland, W.S.; McRae, J.E.; Terpenning, I. Stl: A Seasonal-Trend Decomposition Procedure Based on Loess. *J. Off. Stat.* **1990**, *6*, 3–73.
56. Breiman, L.; Friedman, J.; Stone, C.J.; Olshen, R.A. *Classification and Regression Trees*; CRC Press: Boca Raton, FL, USA, 1984.
57. De'ath, G.; Fabricius, K.E. Classification and Regression Trees: A Powerful Yet Simple Technique for Ecological Data Analysis. *Ecology* **2000**, *81*, 3178–3192. [CrossRef]
58. Zhang, Y.; Liu, X.; Qin, B.; Shi, K.; Deng, J.; Zhou, Y. Aquatic vegetation in response to increased eutrophication and degraded light climate in Eastern Lake Taihu: Implications for lake ecological restoration. *Sci. Rep.* **2016**, *6*. [CrossRef] [PubMed]
59. Efron, B.; Tibshirani, R.J. *An Introduction to the Bootstrap*; CRC Press: Boca Raton, FL, USA, 1994.
60. Montgomery, D.C.; Peck, E.A.; Vining, G.G. *Introduction to Linear Regression Analysis*; John Wiley & Sons: Somerset, NJ, USA, 2015.
61. Mortazavi, B.; Iverson, R.L.; Landing, W.M.; Lewis, F.G.; Huang, W. Control of phytoplankton production and biomass in a river-dominated estuary: Apalachicola Bay, Florida, USA. *Mar. Ecol. Prog. Ser.* **2000**, *198*, 19–31. [CrossRef]

62. Del Castillo, C.E.; Gilbes, F.; Coble, P.G.; Müller-Karger, F.E. On the dispersal of riverine colored dissolved organic matter over the West Florida Shelf. *Limnol. Oceanogr.* **2000**, *45*, 1425–1432. [CrossRef]

63. Gordon, H.R.; Lewis, M.R.; McLean, S.D.; Twardowski, M.S.; Freeman, S.A.; Voss, K.J.; Boynton, G.C. Spectra of particulate backscattering in natural waters. *Opt. Express* **2009**, *17*, 16192–16208. [CrossRef] [PubMed]

64. Slade, W.H.; Boss, E. Spectral attenuation and backscattering as indicators of average particle size. *Appl. Opt.* **2015**, *54*, 7264–7277. [CrossRef] [PubMed]

65. Fan, C.; Warner, R.A. Characterization of water reflectance spectra variability: Implications for hyperspectral remote sensing in estuarine waters. *Mar. Sci.* **2014**, *4*, 1–9.

66. Joshi, I.D.; D'Sa, E.J.; Osburn, C.L.; Bianchi, T.S.; Ko, D.S.; Oviedo-Vargas, D.; Arellano, A.R.; Ward, N.D. Assessing chromophoric dissolved organic matter (CDOM) distribution, stocks, and fluxes in Apalachicola Bay using combined field, VIIRS ocean color, and model observations. *Remote Sens. Environ.* **2017**, *191*, 359–372. [CrossRef]

67. Doxaran, D.; Froidefond, J.-M.; Castaing, P.; Babin, M. Dynamics of the turbidity maximum zone in a macrotidal estuary (the Gironde, France): Observations from field and MODIS satellite data. *Estuar. Coast. Shelf Sci.* **2009**, *81*, 321–332. [CrossRef]

68. Wang, H.; Hladik, C.M.; Huang, W.; Milla, K.; Edmiston, L.; Harwell, M.; Schalles, J.F. Detecting the spatial and temporal variability of chlorophyll-a concentration and total suspended solids in Apalachicola Bay, Florida using MODIS imagery. *Int. J. Remote Sens.* **2010**, *31*, 439–453. [CrossRef]

69. Bustamante, J.; Pacios, F.; Díaz-Delgado, R.; Aragonés, D. Predictive models of turbidity and water depth in the Doñana marshes using Landsat TM and ETM+ images. *J. Environ. Manag.* **2009**, *90*, 2219–2225. [CrossRef] [PubMed]

70. Pope, R.M.; Fry, E.S. Absorption spectrum (380–700 nm) of pure water. II. Integrating cavity measurements. *Appl. Opt.* **1997**, *36*, 8710–8723. [CrossRef] [PubMed]

71. Novoa, S.; Doxaran, D.; Ody, A.; Vanhellemont, Q.; Lafon, V.; Lubac, B.; Gernez, P. Atmospheric corrections and multi-conditional algorithm for multi-sensor remote sensing of suspended particulate matter in low-to-high turbidity levels coastal waters. *Remote Sens.* **2017**, *9*, 61. [CrossRef]

72. Twichell, D.; Edmiston, L.; Andrews, B.; Stevenson, W.; Donoghue, J.; Poore, R.; Osterman, L. Geologic controls on the recent evolution of oyster reefs in Apalachicola Bay and St. George Sound, Florida. *Estuar. Coast. Shelf Sci.* **2010**, *88*, 385–394. [CrossRef]

73. Reichstetter, M.; Fearns, P.R.; Weeks, S.J.; McKinna, L.I.; Roelfsema, C.; Furnas, M. Bottom Reflectance in Ocean Color Satellite Remote Sensing for Coral Reef Environments. *Remote Sens.* **2015**, *7*, 16756–16777. [CrossRef]

74. Schmidt, N.; Lipp, E.; Rose, J.; Luther, M. ENSO influences on Seasonal Rainfall and River Discharge in Florida. *J. Clim.* **2001**, *14*, 615–628. [CrossRef]

75. Sittel, M.C. *Marginal Probabilities of the Extremes of ENSO Events for Temperature and Precipitation in the Southeastern United States*; Center for Ocean-Atmosphere Prediction Studies, Florida State University: Tallahassee, FL, USA, 1994.

76. Wang, M.; Shi, W. Cloud Masking for Ocean Color Data Processing in the Coastal Regions. *IEEE Trans. Gesci. Remote Sens.* **2006**, *44*, 3196–3205. [CrossRef]

77. Roberts, K.; Collins, J.; Paxton, C.H.; Hardy, R.; Downs, J. Weather patterns associated with green turtle hypothermic stunning events in St. Joseph Bay and Mosquito Lagoon, Florida. *Phys. Geogr.* **2014**, *35*, 134–150. [CrossRef]

remote sensing

MDPI

Article

Temporal and Spatial Dynamics of Phytoplankton Primary Production in Lake Taihu Derived from MODIS Data

Yubing Deng [1,2], **Yunlin Zhang** [1,*], **Deping Li** [2], **Kun Shi** [1] and **Yibo Zhang** [1]

[1] Taihu Lake Laboratory Ecosystem Research Station, State Key Laboratory of Lake Science and Environment, Nanjing Institute of Geography and Limnology, Chinese Academy of Sciences, Nanjing 210008, China; 18229888004@163.com (Y.D.); kshi@niglas.ac.cn (K.S.); hbxgzyb@126.com (Y.Z.)

[2] College of Resources & Environment, GIS Research Center, Hunan Normal University, Changsha 410081, China; lideping106@aliyun.com

* Correspondence: ylzhang@niglas.ac.cn; Tel.: +86-25-8688-2198; Fax: +86-25-5771-4759

Academic Editors: Linhai Li, Claudia Giardino, Deepak R. Mishra, Xiaofeng Li and Prasad S. Thenkabail
Received: 24 November 2016; Accepted: 20 February 2017; Published: 24 February 2017

Abstract: We investigated the long-term variations in primary production in Lake Taihu using Moderate Resolution Imaging Spectroradiometer (MODIS) data, based on the Vertically Generalized Production Model (VGPM). We firstly test the applicability of VGPM in Lake Taihu by comparing the results between the model-derived and the in situ results, and the results showed that a strong significant correlation (R^2 = 0.753, p < 0.001, n = 63). Then, VGPM was used to map temporal-spatial distributions of primary production in Lake Taihu. The annual mean daily primary production of Lake Taihu from 2003 to 2013 was 1094.06 ± 720.74 mg·C·m^{-2}·d^{-1}. Long-term primary production maps estimated from the MODIS data demonstrated marked temporal and spatial variations. Spatially, the primary production in bays, especially in Zhushan Bay and Meiliang Bay, was consistently higher than that in the open area of Lake Taihu, which was caused by chlorophyll-a concentrations resulting from high nutrient concentrations. Temporally, the seasonal variation of primary production from 2003 to 2013 was: summer > autumn > spring > winter, with significantly higher primary production found in summer and autumn than in winter (p < 0.005, t-test), primarily caused by seasonal variations in water temperature. On a monthly scale, the primary production exerts a clear character of bimodality, increasing from January to May, decreasing in June or July, and finally reaching its highest value during August or September. Wind is another important factor that could affect the spatial variations of primary production in the large, eutrophic and shallow Lake Taihu.

Keywords: primary production; VGPM model; Lake Taihu; carbon cycle; remote sensing estimation

---

1. Introduction

Phytoplankton is at the bottom of the food chain, creating fresh organic matter from inorganic nutrients, carbon dioxide and energy from sunlight, which strongly influence nutrient concentrations and support higher trophic levels such as zooplankton and filter feeder. When either system is employed, phytoplankton creates organic matter from inorganic compounds and carbon dioxide, which is called phytoplankton primary production [1,2]. The rate of phytoplankton primary production is a fundamental property of aquatic system and measurements of primary production are critical to our understanding of the carbon cycle [3]. Estimation of the phytoplankton primary production is also an important topic in fisheries resource management and global change [4].

One of the major goals of modern biological oceanography is to acquire a better understanding of primary production in various oceanic provinces, with a special emphasis on marine carbon cycling and climate change on regional to global scales [5]. As an important part of the global carbon cycle, estimating the primary production in lakes is also the major goals of modern limnology, and important for us to understand the regional ecological environment [4]. Actually, lake primary production, which is commonly regarded as the photosynthetic capacity of phytoplankton for the unit volume, is an important parameter to describe ecosystem and environmental characteristics of lake [6].

Traditionally, the primary production of an aquatic environment was obtained from cruise samples. While in situ experiments provided accurate estimates of primary production in small volumes of water, they may not be easily extrapolated to lake-wide estimates [7–9]. Moreover, these experiments provide an integrated measure of production that is dependent on many variables (e.g., phytoplankton biomass, light, temperature, etc.), thus limiting their predictive value. Early ways to estimate primary production for the great lakes may be biased because of deficiencies in traditional collection and incubation techniques [10–15].

Satellite data have been widely used to derive several lake biogeochemical parameters, such as, total suspended matter, chlorophyll a (Chl*a*), and diffuse attenuation coefficient of photosynthetically active radiation (K_d(PAR)). Satellite remote sensing has many advantages in estimation of lake biogeochemical parameters. The repeated coverage by remote sensing enables the detection of the temporal and spatial variation, which has proven beneficial for rapidly estimating lake biogeochemical parameters [16]. Additionally, satellite remote sensing is more practical and economical than the other monitoring methods and also can be easily integrated into geographic information [17–19]. However, remote sensing data cannot directly provide information on primary production without support of models [20]. Vertically Generalized Productivity Model (VGPM) is the most widely used to estimate primary production due to its minimal input parameters, which has been validated by thousands of in situ measured data over several orders of magnitude. Among these input parameters, Chl*a* and euphotic depth are could be derived from remote sensing data, providing a possibility for estimating primary production using remote sensing data. Many studies have successfully estimated phytoplankton primary production by combining remote sensing data to VGPM for open ocean [21–23]. However, due to difficulty in retrieval of Chl*a* and euphotic depth in complexly optical waters, there have been few applications of the VGPM to lakes [24–26].

As the third-largest freshwater lake in China, Lake Taihu has a water surface area of 2338 km^2, and a mean depth of 1.9 m [24]. This large, shallow lake is located in one of the world's most heavily populated regions, which has experienced rapid economic development in recent years. Lake Taihu serves as a key drinking water resource for the approximately 10 million local residents residing in several large nearby cities, such as Shanghai, Suzhou, Wuxi and Huzhou [27,28]; it also has additional economic functions including tourism, fisheries and shipping [29]. Besides, Lake Taihu Basin also contributes 11.6% of China's gross domestic product, despite accounting for only 0.38% of China's total area. With economic development, a lot of wastewater, sewage and polluted water surrounding urban and rural areas were discharged to Lake Taihu, resulting in phytoplankton blooms and then triggering a drinking water crisis in 2007 [30]. Phytoplankton are commonly the most important primary producer in lake ecosystems, strongly influencing nutrient concentrations and supporting higher trophic levels such as zooplankton and filter feeders. Phytoplankton primary production is an important indicator of water quality and algal biomass. Therefore, describing the temporal and spatial variations in phytoplankton assemblages and primary production is crucial for understanding the development of eutrophication and ecosystem dynamics in Lake Taihu [31].

Biogeochemical parameters in Lake Taihu are always characterized by highly temporal-spatial dynamics, indicating that estimates of primary production using traditional sampling methods may lead to large uncertainty when we investigate primary production dynamics in the whole lake for a long time period.

In recent years, we have calibrated and validated Chl*a* and euphotic depth estimation model in Lake Taihu, making it possible for us to estimate the phytoplankton primary production based on the VGPM model. Therefore, the aims of ours study were to (1) assess the accuracy and feasibility of VGPM in Lake Taihu using in situ measurement data; (2) estimate the phytoplankton primary production of Lake Taihu from 2003 to 2013 using MODIS-derived data; (3) analyze the temporal and spatial variations in primary production in Lake Taihu and discuss the potential affecting factors.

2. Materials and Methods

2.1. Study Area

Lake Taihu is a crucial water source for the Yangtze River Delta, located between 30°56'N–31°33'N and 119°54'E–120°36'E (Figure 1) [30]. In this article, Lake Taihu can be divided into six areas based on shoreline geometry, environmental factors, and human activities: Meiliang Bay, Zhushan Bay, Gonghu Bay, Xukou Bay, East Lake Taihu and the open area (Figure 1) [32,33]. It should be noted that the East Lake Taihu, Gonghu Bay and Xukou Bay were outside of the scope of the study area, defined as the non-region-of-interest (NROI) for this research, because the retrieved Chl*a* were unusually high in these regions. In fact, these regions were marcophytic-dominated rather than phytoplankton-dominated [34], and the abundant submerged plants resulted in the deviation of retrieved Chl*a* by MODIS data because of similar spectral characteristics with algae.

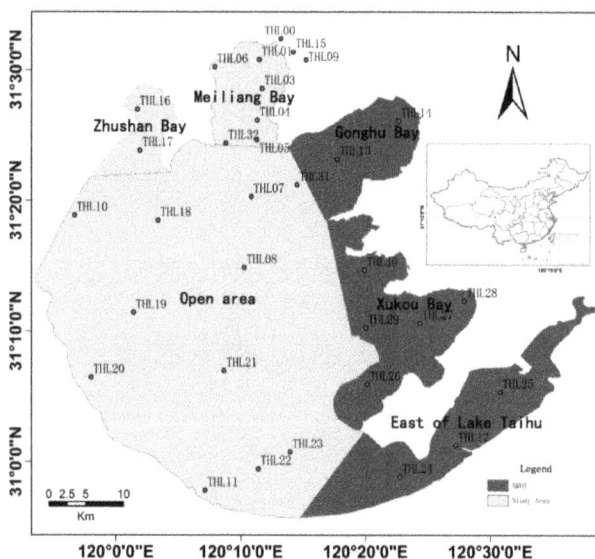

Figure 1. Maps showing location of Lake Taihu and the sampling sites. The in situ primary production were measured at two sites, THL04 (31.44583°N, 120.18883°E) and THL08 (31.24816°N, 120.10062°E), from 2005 to 2013. The in situ nutrient concentration in 2013 including total nitrogen and total phosphorus were collected from 32 sites THL00-THL32 (except THL02).

2.2. Field Data

A total of 63 samples were used to validate the VGPM model and a total of 20 samples were used to analyze the correlation between phytoplankton primary production and nutrients. 63 phytoplankton primary production samples were collected between 2005 and 2013 from the two sites (THL04 and THL08). Rates of phytoplankton primary production and respiration were measured using the light–dark bottle incubation method. Within the dark bottle, only respiration can occur; and within the light bottle, both respiration and photosynthesis can occur. From the difference in O_2 within the light and dark bottles relative to the initial O_2, the rate of primary production can be calculated [35–37].

The nutrient concentration data in 2013 were collected from 32 sites THL00-THL32 (except THL02) (Figure 1).

2.3. Image Data Description and Processing

The MODIS-Aqua data have been freely available since 2002 and has a maximum spatial resolution of 250 m (band 1 and band 2) and a very short revisit interval (1 image/day). MODIS-Aqua L-0 data from January 2003 to December 2013 (more than 4000 images) were downloaded from NASA's Goddard Space Flight Center (GSFC, http://oceancolor.gsfc.nasa.gov/). Due to the clouds, cloud shadows, or thick aerosols, not all downloaded images were used in this study. We selected 1086 cloud-free images of Lake Taihu from January 2003 to December 2013. These images were processed to Level-1 (calibrated spectral radiance) using the software package SeaDAS (version 6.0).

2.4. Method Description

After investigating the variability observed in primary production by assembling a dataset of 11,283 ^{14}C-based measurements of daily carbon fixation collected at 1698 oceanographic stations in both open ocean and coastal waters, Behrenfeld and Falkowski discovered a consistent trend in the vertical distribution of primary production and observed that the VGPM well accounted for the trend in normalized primary production estimated from depth integrated primary production [14]. VGPM model has been validated by thousands of in situ measured data points on a large scale and in different water areas. Therefore, it has been used widely to accurately estimate primary production.

The equation of VGPM is given below:

$$PP_{eu} = P_{opt}^B \times D_{irr} \times \int_{z=0}^{Zeu} \frac{(1 - e^{\frac{-E_z}{Emax}})e^{(\beta_d * E_d)}}{(1 - e^{\frac{-Eopt}{Emax}})e^{(\beta_d * E_{opt})}} \times C_z \times d_z \tag{1}$$

where PP_{eu} represents primary production from sea surface to the euphotic depth (mg·C·m^{-2}·d^{-1}.), P_{opt}^B is the maximum optimal rate of carbon fixation of water mass, D_{irr} is the illumination period, Zeu is the euphotic depth, E_z represents Photosynthetically Active Radiation(PAR) that is photosynthetic effective radiation at depth Z (mol quanta/m^2), E_{max} represents the maximum photosynthetic effective radiation, E_{opt} is the PAR where P_{opt}^B is located and β_d is the initial slope of curve P-I chlorophyll concentration at depth Z.

After contrasting the results calculated by the VGPM model and ship-measured data, Behrenfeld and Falkowski considered that the model describes 79% of temporal-spatial variety of primary production ($n = 10857$). After simplification, the model can be presented as:

$$PP_{eu} = 0.66125 \times P_{opt}^B \times \frac{E_0}{E_0 + 4.1} \times Z_{eu} \times C_{opt} \times D_{irr} \tag{2}$$

where P_{opt}^B is the maximum rate of carbon fixation within a water column (mg C/(mg Chl·h)), E_0 is PAR of the surface photosynthetic effective radiation of the lake, Z_{eu} is the euphotic depth, C_{opt} is Chl$a$ concentration where P_{opt}^B is located, which can be replaced by remote sensed surface Chl$a$ concentration, and D_{irr} is day length in decimal hours.

The model parameters can be acquired through the following methods:

(1) Calculation of P_{opt}^B is based on the experienced relationship between P_{opt}^B and the lake surface temperature (LST, °C) obtained with field investigation data provided by Behrenfeld and Falkowski.

Therefore, P_{opt}^{B} can be commonly considered as a function of the lake surface temperature, which was described by Behrenfeld and Falkowski [12,14] as:

$$P_{opt}^{B} = \begin{cases} 1.13 & (T \leq -1.0\,^{\circ}\text{C}) \\ 4.00 & (T \geq 28.5\,^{\circ}\text{C}) \\ P_{opt}^{B\prime} & (-1.0 < T < 28.5\,^{\circ}\text{C}) \end{cases} \tag{3}$$

where

$$P_{opt}^{B\prime} = 1.2956 + 2.749 \times 10^{-1}T + 6.17 \times 10^{-2}T^2 - 2.05 \times 10^{-2}T^3 + 2.462 \times 10^{-3}T^4$$
$$-1.348 \times 10^{-4}T^5 + 3.4132 \times 10^{-6}T^6 - 3.27 \times 10^{-8}T^7 \tag{4}$$

The LST data from Lake Taihu can be calculated from MODIS data. Previous studies have showed that MODIS-derived LST and in situ water temperatures in Lake Taihu were significantly correlated, with a coefficient of determination higher than 0.96 and a root mean square error between 1.2 °C and 1.8 °C [38]. Then, the P_{opt}^{B} can be derived from MODIS-derived LST using the Equations (3) and (4).

(2) Chl*a* can also be acquired from the MODIS data based on the model which was developed by Shi et al using Rayleigh-corrected reflectance as follows [31]:

$$\text{Chl}a = -1454.3 \times \text{Index}_{\text{MODIS}} + 69.35 \tag{5}$$

where $\text{Index}_{\text{MODIS}} = (\text{EXP}(R_{rc}645) - \text{EXP}(R_{rc}859))/(\text{EXP}(R_{rc}645) + \text{EXP}(R_{rc}859))$. A significant linear correlation was found between in situ and estimated Chl*a* with the normalized spectral index ($R^2 = 0.72$, $p < 0.005$; *t*-test). The relative error of the model for the validation dataset ranged from 0.4% to 64.5% with a mean absolute percent error of 27.1% (*RMSE* = 15.01 µg/L). The good agreement showed that the Chl*a* can be used to estimate the primary production of Lake Taihu.

(3) The euphotic depth (Z_{eu}) in this study was calculated using the relationship between the euphotic depth and the PAR diffuse attenuation coefficient ($K_d(\text{PAR})$) as follows:

$$Z_{eu}(\text{PAR}) = \frac{4.605}{K_d(\text{PAR})} \tag{6}$$

In this study, the $K_d(\text{PAR})$ can be derived using a estimation model developed for Lake Taihu using MODIS-Aqua remote sensing reflectance at the 645 nm band ($R_{rs}(645)$) (Unpublished data) as follows.

$$K_d(\text{PAR}) = 1.56 \times \exp[44.603 \times R_{rs}(645)] \tag{7}$$

This model has been validated with independent random samples, and indicated a good performance for Taihu $K_d(\text{PAR})$ estimation ($R^2 = 0.71$; $p < 0.001$, *t*-test; RMSE = 1.06 m^{-1}; MAPE = 25%). It is noted that there will be such a situation arose in which estimation euphotic zone depth exceeded bottom depth and if so, we use the bottom depth to replace the estimation euphotic zone depth [39].

(4) D_{irr} can be calculated based on the time and position of the water column. In this study, northern, southern and central regions of the Lake Taihu were selected to calculate the day length of each day.

(5) E_0 data from Lake Taihu can be calculated according to the following equation from Zhang and Qin [40]:

$$E_0 = (0.2909 + 0.0764 \times \lg E^*) \times Q \tag{8}$$

where E_0 is monthly mean total daily PAR and E^* is the vapor pressure corrected by air pressure,

$$E^* = P_0 \times E/P \tag{9}$$

where P_0 and P are the air pressure at sea level and at the monitoring station, respectively, E is the monthly mean vapor pressure and Q is the monthly mean daily global solar radiation. Q can be calculated according to the percentage of sunshine as follows [41]:

$$Q = Q_0 \times (0.1351 + 0.5707 \times s_1) \tag{10}$$

where Q is the monthly mean total daily global solar radiation, Q_0 is the monthly mean total daily extra-terrestrial radiation outside the Earth's atmosphere and s_1 is the percentage of actual sunshine expressed relative to the duration of possible sunshine. This model has been validated with independent random samples, and indicated a good performance for Taihu E_0 estimation ($R^2 = 0.83$, $p < 0.001$; t-test)

With the VGPM model and the Chl*a* data, lake surface photosynthetic available radiation, day length, optimal rate of daily carbon fixation and the euphotic depth, daily primary production in Lake Taihu from 2003 to 2013 has been calculated using the Equation (2).

2.5. Statistical Analysis and Accuracy Assessment

Statistical analysis, including calculation of the maximum, minimum, mean, median values and linear and non-linear regressions were performed using the Statistical Program for Social Sciences (SPSS) 17.0 software. The significance level was reported to be significant ($p \leq 0.05$) or not significant ($p > 0.05$). The fitting determination coefficient (R^2) was calculated through linear regression using theOriginPro8.5 software.

The accuracy of the phytoplankton primary production estimation was assessed for the two validation datasets by comparing the estimation with the measured primary production values. This comparison was quantified by means of the percent difference (MNB) between the estimated primary production (PP_{esti}) and analytically measured phytoplankton primary production (PP_{meas}):

$$MNB = \left| \frac{PP_{esti,i} - PP_{meas,i}}{PP_{esti,i}} \right| \times 100\% \tag{11}$$

Systematic and randomized errors were characterized by the root mean square error (RMSE), and Mean absolute percentage error (MAPE).These metrics were defined as follows:

$$RMSE = \sqrt{\frac{1}{N} \sum_{i=1}^{N} (PP_{esti,i} - PP_{meas,i})^2} \tag{12}$$

$$MAPE = \frac{1}{N} \times \sum_{i=1}^{N} \frac{|Y_{measured,i} - Y_{estimated,i}|}{Y_{measured,i}} \times 100\% \tag{13}$$

3. Results

3.1. Validation of VGPM

In theory, the estimated phytoplankton primary production through satellite data and in situ measured phytoplankton primary production should be consistent within a limited period. We set the criterion for matching the satellite data and the in situ observations to ≤ 12 h (the time interval between the in situ measurements and the corresponding estimated data), according to the hourly variations of the process being measured, to minimize the effects of the temporal difference between the in situ and estimated phytoplankton primary production. Among all collected samples, there were 63 pairs of in situ data and estimated phytoplankton primary production that met the matching criterion. The accuracy of the VGPM was then evaluated through the comparison of the estimated phytoplankton primary production, and in situ phytoplankton primary production is shown in Figure 2. The results show that these values were in good agreement, with a highly significant linear relationship

(R^2 = 0.753, RMSE = 384.68 mg·C·m^{-2}·d^{-1}, MAPE = 40.5%). For all data sets, the minimum and mean of MNB were 1.3% and 41.7%, respectively, between estimated phytoplankton primary production and in situ measurements. The good performance of the VGPM is encouraging and clearly demonstrates that the VGPM could be used for remotely sensed estimations of phytoplankton primary production in the turbid waters of Lake Taihu.

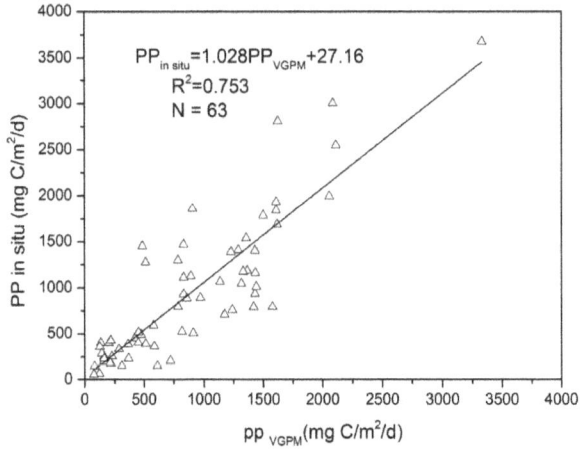

Figure 2. Relationship between measured and estimated primary production by VGPM.

3.2. Temporal-Spatial Distribution of Phytoplankton Primary Production

The daily average data were considered on a monthly, seasonal and annual basis, with the seasons defined as follows: winter, December–February; spring, March–May; summer, June–August; and autumn, September–November [23].

3.2.1. Annual Variation of Phytoplankton Primary Production

The annual means of the MODIS-derived phytoplankton primary production for Lake Taihu are presented in Figure 3, demonstrating inter-annual changes in the phytoplankton primary production from 2003 to 2013. The mean phytoplankton primary production of entire Lake Taihu ranged from 952.84 ± 759.36 to 1259.39 ± 859.71 mg·C·m^{-2}·d^{-1}. The mean daily phytoplankton primary production in Lake Taihu (Open area, Meiliang Bay, Zhushan Bay) from 2003 to 2013 is 1094.06 ± 720.74 C/m^2/d, which is far higher than the average of case I water and some case II water, such as Lake Michigan and Lake Huron [42,43]. In detail, the phytoplankton primary production of entire Lake Taihu in 2007 (1259.39 ± 859.71 mg·C·m^{-2}·d^{-1}) and 2008 (1215.48 ± 866.74 mg·C·m^{-2}·d^{-1}) was higher than that of 2005 (977.39 ± 634.76 mg·C·m^{-2}·d^{-1}) and 2013 (952.85 ± 759.36 mg·C·m^{-2}·d^{-1}), revealing that Lake Taihu experienced moderate spatial and inter-annual variations in phytoplankton primary production. It is worth noting that the highest phytoplankton primary production value was recorded in 2007. This was probably due to abnormal weak cold air in farther northern part of the region, which caused high temperature during the winter and spring of 2007, and the abnormal lack of precipitation may have jointly contributed to the monthly variations in 2007 [44]. The phytoplankton primary production in Lake Taihu experienced three markedly different variations from 2003 to 2013: (1) decreased from 2003 to 2005; (2) increased sharply from 2005 to 2007; and (3) decreased from 2007 to 2013.

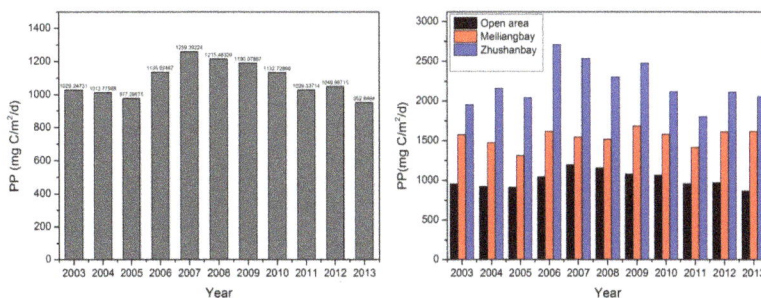

Figure 3. Annual variations of phytoplankton primary production in Lake Taihu from 2003 to 2013.

3.2.2. Seasonal Variation of Phytoplankton Primary Production

The time series of seasonal averages of MODIS-derived phytoplankton primary production values from 2003 to 2013 were constructed for the three regions (open area, Meiliang Bay, Zhushan Bay) by spatially and temporally averaging all valid pixels over the lake in each region. Overall, phytoplankton primary production exhibited typical seasonal variability over Lake Taihu, with significantly higher values in summer (1529.65 ± 782.77 mg·C·m^{-2}·d^{-1}) and autumn (1364.75 ± 697.83 mg·C·m^{-2}·d^{-1}) than in spring (921.47 ± 536.79 mg·C·m^{-2}·d^{-1} or winter (378.66 ± 206.81 mg·C·m^{-2}·d^{-1}) (t-test, $p < 0.001$) (Figures 4 and 5).

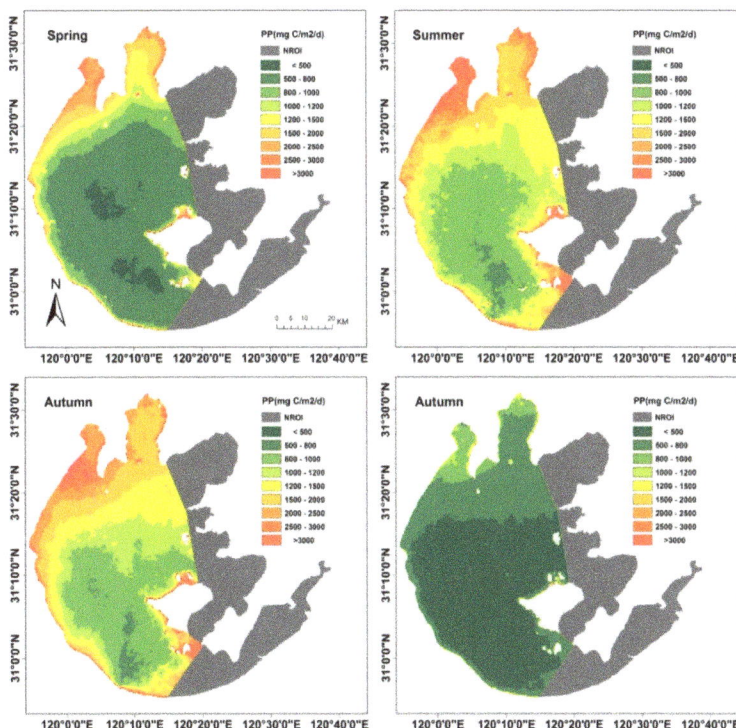

Figure 4. Maps of the MODIS-derived mean daily phytoplankton primary production for the four seasons in Lake Taihu generated using the VGPM.

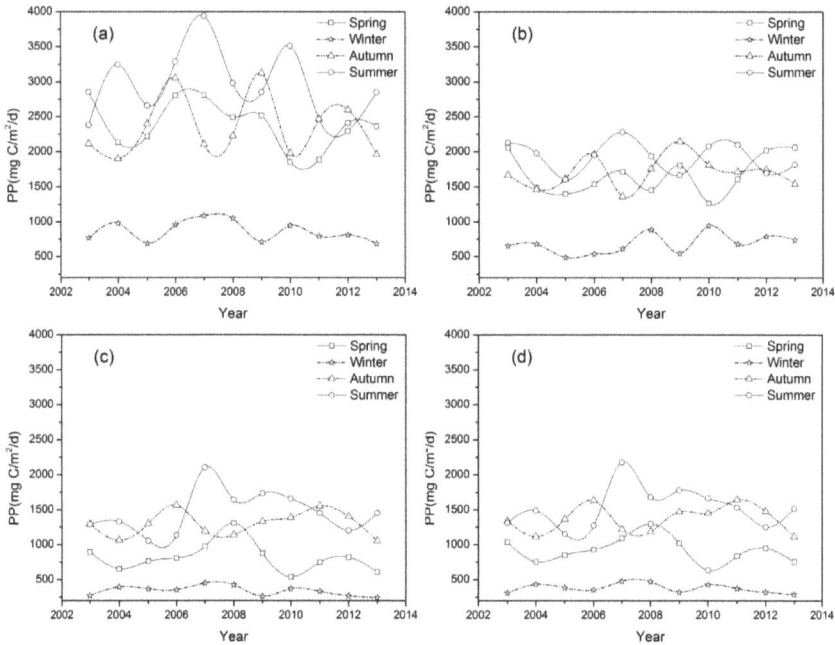

Figure 5. Seasonal variation of phytoplankton primary production in each region based on VGPM estimation from 2003 to 2013 in (**a**) Zhanshan Bay; (**b**) Meiliang Bay; (**c**) the open area and (**d**) the entire lake.

3.2.3. Monthly Variation of Phytoplankton Primary Production

Maps of monthly phytoplankton primary production derived from the VGPM are shown in Figures 6 and 7, demonstrating the inter-monthly changes in the phytoplankton primary production from 2003 to 2013. The highest monthly mean daily phytoplankton primary production in Lake Taihu during the study interval from 2003 to 2013 was measured in August, with an average value of 1675.49 ± 710.36 mg·C·m^{-2}·d^{-1}, nearly 7 times higher than the lowest value of 296.01 ± 130.83 mg·C·m^{-2}·d^{-1} in January. Some interesting characteristics demonstrated that there is a clear bimodality in character in the monthly variability in Lake Taihu. Specifically, the monthly distribution of phytoplankton primary production increased from January to May, then decreased in June or July, and finally reached highest value in August or September (Figure 7).

3.2.4. Spatial Variation of Phytoplankton Primary Production

The MODIS-derived phytoplankton primary production data from 2003 to 2013 were averaged to calculate the regional phytoplankton primary production distribution for Lake Taihu. It can be concluded that the decrease of phytoplankton primary production in Lake Taihu from the bay of the lake to the open area is in accordance with the basic distribution principle of the spatial variability (Figures 4 and 6). The area around the estuary such as Meiliang Bay and Zhushan Bay are the regions where the maximum value of phytoplankton primary production was recorded. In the district of the Meiliang Bay and Zhushan Bay, the annual mean daily phytoplankton primary production was 2205.47 ± 1397.98 mg·C·m^{-2}·d^{-1}, 1551.39 ± 851.36 mg·C·m^{-2}·d^{-1}, respectively, which was nearly 2 times and 1.5 times higher than the lowest value of 1019.25 ± 721.23 mg·C·m^{-2}·d^{-1} that was recorded in open area (Table 1).This is in agreement with previous observations of productivity in Lake Taihu [24,26]. However, there was no significant spatial variation in the monthly mean daily phytoplankton primary production in winter, especially in January and February, possibly due to the degradation of algae in the whole lake (Figure 6).

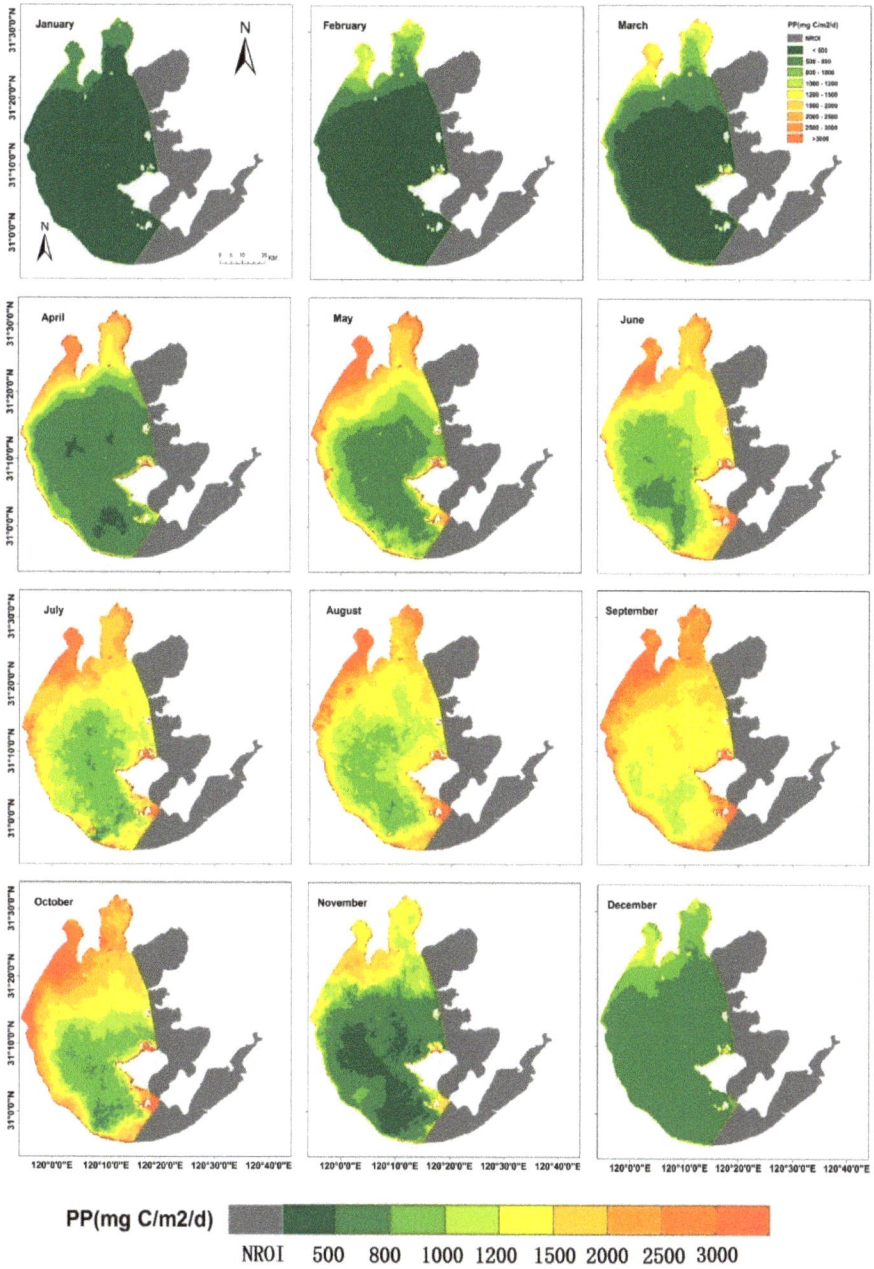

Figure 6. Monthly mean daily phytoplankton primary production of Lake Taihu from January to December.

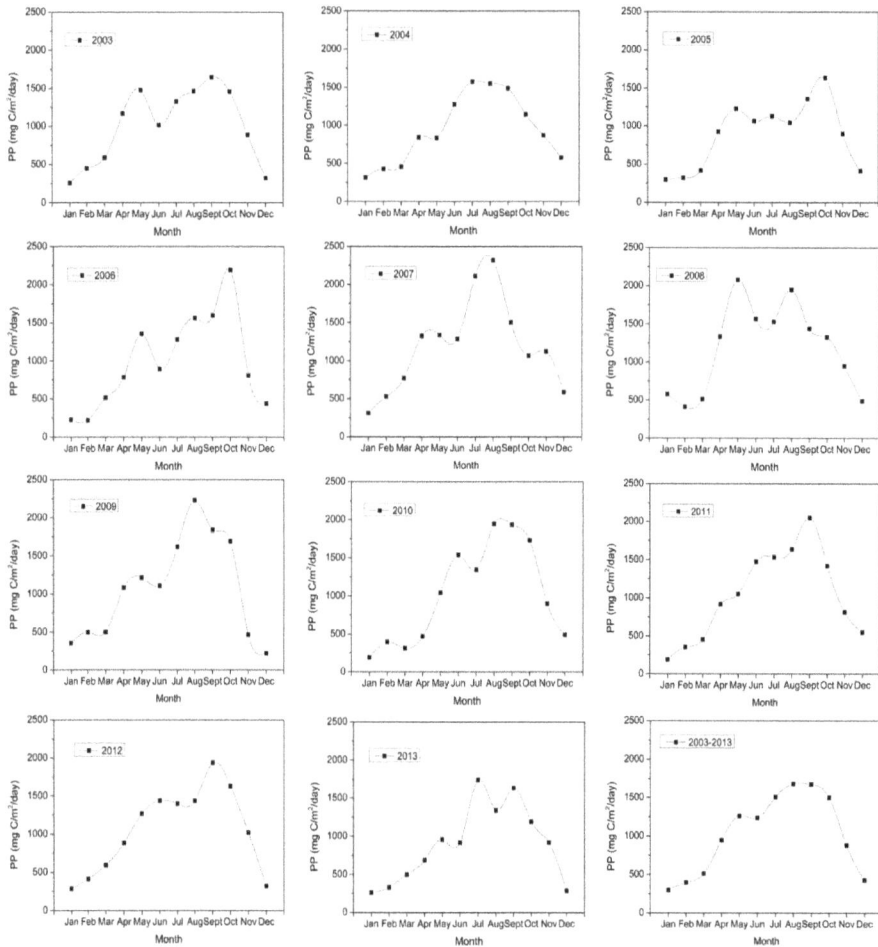

Figure 7. Time series of monthly phytoplankton primary production in Lake Taihu from 2003 to 2013.

Table 1. Phytoplankton primary production variations of different seasons in different regions $(mg \cdot C \cdot m^{-2} \cdot d^{-1})$.

Lake Taihu	Spring	Summer	Autumn	Winter	Mean
Open area	815.52	1458.07	1296.31	339.72	1019.25
Meiliang Bay	1673.32	1931.07	1707.68	685.55	1551.39
Zhushan Bay	2394.08	2951.64	2356.00	860.39	2205.47
Whole lake	921.47	1529.65	1364.75	378.66	1094.61

4. Discussion

4.1. Correlation between the Primary Production with Chla and Nutrient Concentrations

Previous studies have suggested that phytoplankton biomass is the most important factor affecting the temporal variations of phytoplankton primary production [45–52]. The linear relationship between Chla and phytoplankton primary production demonstrates that the Chla is strongly positively correlated with phytoplankton primary production in Lake Taihu (Figure 8). Deng reported that

Chl*a* was low in winter and early spring, and increased in late spring to a maximum in late summer and early autumn; the algae bloom began in May or June, and lasted until October. This finding is consistent with the recorded temporal variations in phytoplankton primary production in Lake Taihu [53]. The spatial variation in phytoplankton primary production was also consistent with the variations in Chl*a*; the highest productivity was observed near Zhushan Bay, where Chl*a* was also the highest among all the study areas (Figure 9).

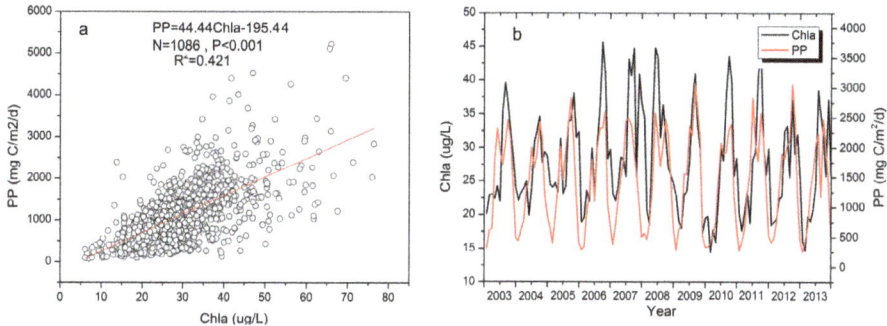

Figure 8. (**a**) Relationship between Chl*a* and MODIS derived primary productivity of Lake Taihu from 2003–2013 (**b**) Monthly mean MODIS derived primary productivity and Chl*a* of Lake Taihu from 2003–2013.

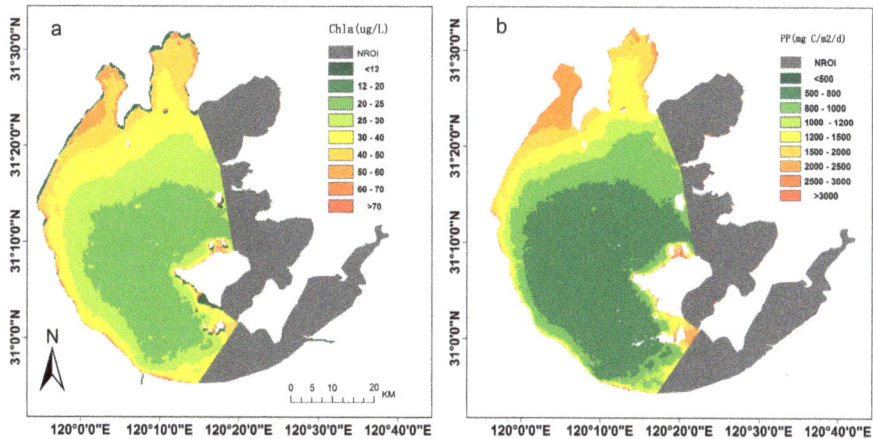

Figure 9. (**a**) Chl*a* distribution in Lake Taihu averaged from all Chl*a* estimates from the MODIS data gathered from 2003–2013 and (**b**) spatial PP distribution in Lake Taihu based on all PP estimates from the MODIS data gathered from 2003–2013.

The phenomenon in which both higher phytoplankton primary production and higher Chl*a* appeared in the bay is related to nutrient distribution. The nutrients mainly affect phytoplankton biomass to control the level of the phytoplankton primary production of phytoplankton. In fact, previous studies have showed that total nutrients, especially the total phosphorus, play a key role in determining phytoplankton in Lake Taihu [25,54,55]. Higher total nitrogen and total phosphorus concentrations at ZhuShan Bay and MeilLiang Bay resulted in a marked increase in Chl*a*. As a highly eutrophic lake, high concentrations of nutrients in Lake Taihu provide good conditions for phytoplankton to grow. The high phytoplankton biomass keeps phytoplankton primary production at a very high level, and the spatial distribution of nutrients can affect the spatial variation in phytoplankton primary production of phytoplankton through affecting the Chl*a*. Some studies have shown that the spatial distribution of nutrient concentrations in Lake Taihu was consistent with the variations in Chl*a* [56,57]. Based on the above analysis, it can be found that the influence of phytoplankton primary production of phytoplankton in Lake Taihu by nutrients is highly significant, and the concentration and spatial distribution of nutrients are major factors responsible for the level and spatial variations in phytoplankton primary production in Lake Taihu. Additionally, the temporal variations of phytoplankton primary production in Lake Taihu appear to show a correlation with the concentrations of nutrients, especially the total phosphorus concentration (Figure 10). Therefore, one of the reasons for the temporal-spatial variations of phytoplankton primary production in Lake Taihu is the temporal-spatial variation in Chl*a*, originated by different nutrient concentrations.

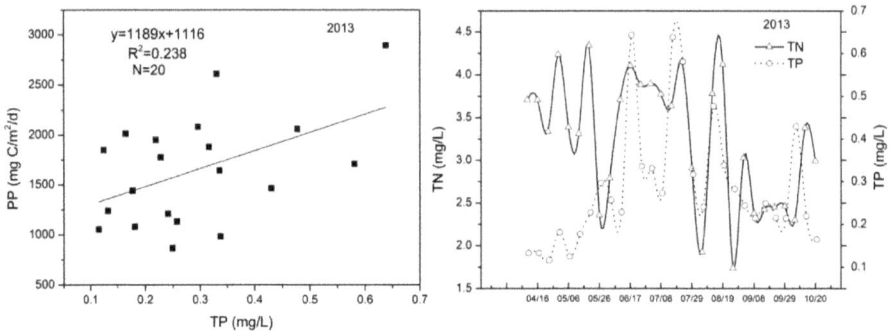

Figure 10. Time series of total nitrogen and total phosphorusin 2013, and the correlation between phytoplankton primary production and total phosphorus.

4.2. Correlation between the Primary Production and Lake Water Temperature

Although influenced by nutrient concentrations, the temporal variations of phytoplankton primary production were also likely affected by water temperature. Chl*a* and phytoplankton primary production peaks appeared during different months (Figure 8b), suggesting that the water temperature was likely to be responsible for this difference. The P_{opt}^B is potentially the most important variable in the estimation of phytoplankton primary production, and one of the principal approaches for estimating P_{opt}^B is to define predictive relationships between P_{opt}^B and lake surface temperature by a high-order polynomial. The suitable water temperature for phytoplankton growth was approximately 20 °C [34]; a temperature that is too high or too low will suppress the phytoplankton photosynthesis rate, therefore maximal P_{opt}^B values occur from18–23 °C (Figure 11). In January, the water temperature was the lowest, which corresponded to a very low value for P_{opt}^B, greatly inhibiting phytoplankton primary production. As temperatures increased, the phytoplankton primary production dropped briefly due to the value of inhibition of P_{opt}^B result from the high water temperature of 23.6 °C in June and 26.6 °C in July. In August and September, although the temperature remained high, the phytoplankton primary production began to gradually rise to its maximum due to the explosive growth of the phytoplankton biomass. In November and December, as with the decrease of phytoplankton biomass, the photoperiod and P_{opt}^B, phytoplankton primary production began to drop quickly. Based on the above analysis, the water temperature in Lake Taihu seem to influence the temporal distribution by affecting the maximum optimal rate of carbon fixation of water mass. Additionally, the mutual movement between phytoplankton biomass and P_{opt}^B was the intrinsic c motivation of the temporal variation mechanism of phytoplankton primary production in Lake Taihu.

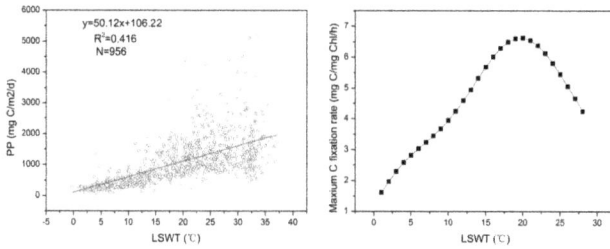

Figure 11. Correlations between water temperature and maximum carbon fixation rate and phytoplankton primary production.

4.3. The Effects of Wind-Driven Sediment Resuspension on the Spatial Distribution of Phytoplankton Primary Production

Wind-induced sediment resuspension can affect phytoplankton primary production by influencing euphotic depth and nutrient availability. To investigate the effect of wind-driven sediment resuspension on the spatial distribution of phytoplankton primary production in Lake Taihu, we performed a correlated analysis between the MODIS-derived phytoplankton primary production and K_d(PAR). The results (Figure 12) showed the existence of a negative linear correlation between the phytoplankton primary production and the K_d(PAR) in the open area, indicating that the process of primary production in the open area was easily disturbed by wind-derived waves. This result is in agreement with our finding that lower phytoplankton primary production was found in the open area compared to the other regions of Lake Taihu. The dominant wind directions in Lake Taihu were found to be NNW and ESE, resulting in longer wind fetches for northern and western parts of Lake Taihu. Longer wind fetches yield stronger wind forces and thus more sediment resuspension in the northern and western regions of Lake Taihu. In the littoral zones and bays, such as Zhushan Bay and Meiliang Bay, the wind fetches are shorter than in the open area. Therefore, wind-driven sediment resuspension, expressed using disturbance index, could explain the spatial variations in the phytoplankton primary production in Lake Taihu.

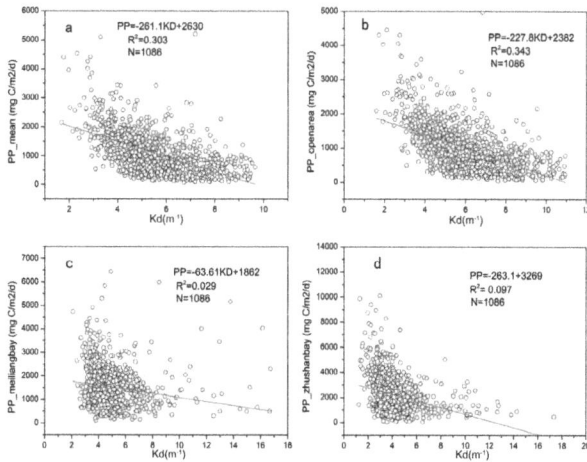

Figure 12. Correlations between PAR diffuse attenuation coefficient and phytoplankton primary production for the regions of (**a**) the entire Lake Taihu; (**b**) the open area; (**c**) Meiliang Bay and (**d**) Zhushan Bay.

5. Conclusions

The efficient and accurate estimation of the distribution and variation of phytoplankton primary production is important for ecosystem and water resource management of lake. In this study, we addressed one technical challenge and understanding of the long-term primary production distribution variations in Lake Taihu, from 2003 to 2013. The technical challenge is that the VGPM model is a good model for phytoplankton primary production estimation in Lake Taihu ($R^2 = 0.753$, RMSE = 384.68 mg·C·m^{-2}·d^{-1}). It holds considerable promise and could be useful for primary production estimation in large lakes.

The annual mean daily phytoplankton primary production was 1094.06±720.74 mg·C·m^{-2}·d^{-1} in Lake Taihu from 2003 to 2013. The spatial pattern in daily mean phytoplankton primary production was similar to that of Chl*a*, decreasing from bay to open area. The phytoplankton primary production exhibited typical seasonal variability over the entire Lake Taihu, and there is a clear bimodality characteristic of phytoplankton primary production in the monthly variability of Lake Taihu.

Temporal-spatial variations in phytoplankton primary production appeared to be mainly driven by Chl*a* variations, and the spatial distribution of nutrients affected the spatial variation of phytoplankton primary production of through affecting Chl*a*. High nutrient concentrations determined the high Chl*a* and then resulted in the high phytoplankton primary production of Lake Taihu. Other contributing factors include water temperature, surface PAR, photoperiod and euphotic depth responsible for the spatial variation of the phytoplankton primary production. In addition, as a large, shallow lake, the wind-driven sediment resuspension plays an important role in spatial distribution of phytoplankton primary production, especially in the open area.

Acknowledgments: This study was jointly supported by the National Natural Science Foundation of China (Grants 41271355 and 41230744), the CAS/SAFEA International Partnership Program for Creative Research Teams (KZZD-EW-TZ-08), International Science & Technolog Cooperation Program of China (2014DFG91780), and the Taihu Lake Laboratory Ecosystem Research Station (TLLER). The authors would like to thank the anonymous reviewers for their useful and constructive comments.

Author Contributions: Y.D. wrote the main manuscript text. Y.Z., K.S., Y.Z. and D.L. designed and conducted the experiments, collected and analyzed the data. Y.Z., K.S., Y.Z contributed to analyzing the data, writing and editing the manuscript. All authors reviewed the manuscript.

Conflicts of Interest: The authors declare no conflict of interest.

References

1. Lakshmi, E.; Pratap, D.; Nagamani, P.V.; Rao, K.H.; Latha, T.P.; Choudhury, S.B. Time Series Analysis of Primary production along the East Coast of India Using Oceansat-2 Ocean Colour Monitor (O cm). *Int. Arch. Photogramm. Remote Sens. Spat. Inf. Sci.* **2014**, *40*, 1049–1053. [CrossRef]
2. Deines, A.M.; Bunnell, D.B.; Rogers, M.W. A review of the global relationship among freshwater fish, autotrophic activity, and regional climate. *Rev. Fish Biol. Fish.* **2015**, *25*, 323–336. [CrossRef]
3. Behrenfeld, M.J.; Boss, E.; Siegel, D.A.; Shea, D.M. Carbon-based ocean productivity and phytoplankton physiology from space. *Glob. Biogeochem. Cycles* **2005**, *19*, 177–202. [CrossRef]
4. Wetzel, R.G. Limnology: Lake and River Ecosystem. *EOS Trans. Am. Geophys. Union* **2001**, *21*, 1–9.
5. Sterner, R.W. In situ-measured primary production in Lake Superior. *J. Great Lakes Res.* **2010**, *36*, 139–149. [CrossRef]
6. Smith, R.C.; Prezelin, B.B.; Bidigare, R.R.; Baker, K.S. Bio-optical modeling of photosynthetic production. *Limnol. Oceanogr.* **1989**, *34*, 1524–1544. [CrossRef]
7. Li, G.; Ping, G.; Fang, W.; Qiang, L. Estimation of ocean primary production and its spatio-temporal variation mechanism for East China Sea based on VGPM model. *J. Geogr. Sci.* **2004**, *14*, 32–40. [CrossRef]
8. Fahnenstiel, G.L.; Sayers, M.J.; Shuchman, R.A.; Yousef, F.; Pothoven, S.A. Lake-wide phytoplankton production and abundance in the Upper Great Lakes: 2010–2013. *J. Great Lakes Res.* **2016**, *42*, 619–629. [CrossRef]

9. Morel, A.; Berthon, J.F. Surface pigments, algal biomass profiles, and potential production of the euphotic layer: Relationships reinvestigated in view of remote sensing applications. *Limnol. Oceanogr.* **1989**, *34*, 1545–1562. [CrossRef]

10. Fahnenstiel, G.; Pothoven, S.; Vanderploeg, H.; Klarer, D.; Nalepa, T.; Scavia, D. Recent Changes in Primary Production and Phytoplankton in the Offshore Region of Southeastern Lake Michigan. *J. Great Lakes Res.* **2010**, *36*, 20–29. [CrossRef]

11. Eppley, R.W.; Stewart, E.; Abbott, M.R.; Heyman, U. Estimating ocean primary production from satellite chlorophyll—Introduction to regional differences and statistics for the southern California bight. *J. Plankton Res.* **1985**, *7*, 227–234. [CrossRef]

12. Behrenfeld, M.J.; Falkowski, P.G. A consumer's guide to phytoplankton primary production models. *Limnol. Oceanogr.* **1997**, *42*, 1479–1491. [CrossRef]

13. Carr, M.E.; Friedrichs, M.A.M.; Schmeltz, M.; Aita, M.N.; Antoine, D.; Arrigo, K.R.; Asanuma, I.; Aumont, O.; Barber, R.; Behrenfeld, M. A comparison of global estimates of marine primary production from ocean color. *Deep Sea Res. Part II Top. Stud. Oceanogr.* **2006**, *53*, 741–770. [CrossRef]

14. Behrenfeld, M.J.; Falkowski, P.G. Photosynthetic rates derived from satellite-based chlorophyll concentration. *Limnol. Oceanogr.* **1997**, *42*, 1–20. [CrossRef]

15. Tripathy, S.C.; Ishizaka, J.; Siswanto, E.; Shibata, T.; Mino, Y. Modification of the vertically generalized production model for the turbid waters of Ariake bay, southwestern Japan. *Estuar. Coast Shelf* **2012**, *97*, 66–77. [CrossRef]

16. Kameda, T.; Ishizaka, J.; Murakami, H. Two-phytoplankton community model of primary production for ocean color satellite data. *Hyperspectr. Remote Sens. Ocean.* **2000**, *4154*, 159–165.

17. Ye, H.B.; Chen, C.Q.; Sun, Z.H.; Tang, S.L.; Song, X.Y.; Yang, C.Y.; Tian, L.Q.; Liu, F.F. Estimation of the primary production in pearl river estuary using MODIS data. *Estuar. Coast Shelf* **2015**, *38*, 506–518. [CrossRef]

18. Vollenweider, R.A.; Munawar, M.; Stadelmann, P. A comparative Review of Phytoplankton and Primary Production in the Laurentian Great Lakes. *J. Fish. Res. Board Can.* **2011**, *31*, 739–762. [CrossRef]

19. Kemili, P.; Putri, M.R. Estimation of primary production in Banda Sea using the vertical distribution model. *AIP Conf. Proc.* **2014**, *1589*, 389–393.

20. Field, C.B.; Behrenfeld, M.J.; Randerson, J.T.; Falkowski, P. Primary production of the biosphere: Integrating terrestrial and oceanic components. *Science* **1998**, *281*, 237–240. [CrossRef] [PubMed]

21. Kahru, M.; Mitchell, B.G. Influence of the El Niño—La Niña cycle, on satellite-derived primary production, in the California Current. *Investig. Mar.* **2002**, *29*, 27-1–27-4. [CrossRef]

22. Mcclain, C.R.; Christian, J.R.; Signorini, S.R.; Lewis, M.R.; Asanuma, I.; Turk, D.; Dupouy-Douchement, C. Satellite ocean-color observations of the tropical pacific ocean. *Deep Sea Res. Part II Top. Stud. Oceanogr.* **2002**, *49*, 2533–2560. [CrossRef]

23. Gregg, W.W.; Conkright, M.E.; Ginoux, P.; O'Reilly, J.E.; Casey, N.W. Ocean primary production and climate: Global decadal changes. *Geophys. Res. Lett.* **2003**, *30*, 157–168. [CrossRef]

24. Zhang, Y.L.; Qin, B.Q.; Liu, M.L. Temporal—Spatial variations of chlorophyll a and primary production in meiliang bay, Lake Taihu, China from 1995 to 2003. *J. Phytoplankton Res.* **2007**, *29*, 707–719. [CrossRef]

25. Zhang, Y.; Feng, S.; Ronghua, M.A.; Liu, M.A.; Qin, B. Spatial pattern of euphotic depth and estimation of phytoplankton primary production in Lake Taihu in autumn 2004. *J. Lake Sci.* **2008**, *20*, 380–388.

26. Yin, Y.; Zhang, Y.; Shi, Z.; Liu, X. Estimation of spatial and seasonal changes in phytoplankton primary production in meiliang bay, Lake Taihu, based on the vertically generalized production model and MODIS data. *Acta Ecol. Sin.* **2012**, *32*, 3528–3537. [CrossRef]

27. Cheng-Feng, L.E.; Yun-Mei, L.I.; Zha, Y.; Sun, D.Y.; Heng, L.U. Simulation of backscattering properties of Taihu Lake. *Adv. Water Sci.* **2009**, *20*, 707–713.

28. Le, C.F.; Li, Y.M.; Zha, Y.; Sun, D.; Huang, C.; Lu, H. A four-band semi-analytical model for estimating chlorophyll a in highly turbid lakes: The case of Taihu Lake, China. *Remote Sens. Environ.* **2009**, *113*, 1175–1182. [CrossRef]

29. Qin, B.; Xu, P.; Wu, Q.; Luo, L.; Zhang, Y. Environmental issues of Lake Taihu, China. *Hydrobiologia* **2007**, *581*, 3–14. [CrossRef]

30. Le, C.F.; Li, Y.M.; Zha, Y.; Sun, D.Y. Specific absorption coefficient and the phytoplankton package effect in Lake Taihu, China. *Hydrobiologia* **2009**, *619*, 27–37. [CrossRef]

31. Shi, K.; Zhang, Y.; Zhou, Y.; Liu, X.; Zhu, G.; Qin, B.; Gao, G. Long-term MODIS observations of cyanobacterial dynamics in Lake Taihu: Responses to nutrient enrichment and meteorological factors. *Sci. Rep.* 2017. [CrossRef] [PubMed]

32. Qin, B.; Zhu, G.; Gao, G.; Zhang, Y.; Wei, L.; Paerl, H.W.; Carmichael, W.W. A drinking water crisis in Lake Taihu, China: Linkage to climatic variability and lake management. *Environ. Manag.* **2010**, *45*, 105–112. [CrossRef] [PubMed]

33. Zhang, Y.; Shi, K.; Liu, X.; Zhou, Y.; Qin, B. Lake topography and wind waves determining seasonal-spatial dynamics of total suspended matter in turbid Lake Taihu, China: Assessment using long-term high-resolution MERIS data. *PLoS ONE* **2014**, *9*, e98055. [CrossRef] [PubMed]

34. Bai, X.; Xiaohong amp, G.U.; Yang, L. Analyses on water quality and its protection in east Lake Taihu. *J. Lake Sci.* **2006**, *18*, 91–96.

35. Kauer, T.; Kutser, T.; Arst, H.; Danckaert, T.; Nõges, T. Modelling primary production in shallow well mixed lakes based on MERIS satellite data. *Remote Sens. Environ.* **2015**, *163*, 53–261. [CrossRef]

36. Fahnenstiel, G.L.; Carrick, H.J. Primary production in lakes Huron and Michigan: In vitro and in situ comparisons. *J. Plankton Res.* **1988**, *10*, 1273–1283. [CrossRef]

37. Gaarder, T.; Gran, H.H. Investigations of the production of plankton in the Oslo Fjord. *J. Conseil—Conseil Permanent Int. l'Explor. Mer.* **1927**, *42*, 1–48.

38. Liu, G.; Ou, W.; Zhang, Y.; Wu, T.; Zhu, G.; Shi, K.; Qin, B. Validating and mapping surface water temperatures in Lake Taihu: Results from MODIS land surface temperature products. *IEEE J. Sel. Top. Appl. Earth Obs. Remote Sens.* **2015**, *8*, 1–15. [CrossRef]

39. Shi, K.; Zhang, Y.; Liu, X.; Wang, M.; Qin, B. Remote sensing of diffuse attenuation coefficient of photosynthetically active radiation in Lake Taihu using MERIS data. *Remote Sens. Environ.* **2014**, *140*, 365–377. [CrossRef]

40. Zhang, Y.; Qin, B. The basic characteristic and climatological calculation of the photosythetically available radiation in Taihu region. *Acta Energiae Solaris Sinica* **2002**, *21*, 118–123.

41. Zhang, Y.; Sciences, C.A.O. Nanjing. Climatological calculation and characteristic analysis of global radiation over Wuxi region. *Q. J. Appl. Meteorol.* **2003**, *14*, 339–347.

42. Shuchman, R.A.; Sayers, M.; Fahnenstiel, G.L.; Leshkevich, G. A model for determining satellite-derived primary production estimates for Lake Michigan. *J. Great Lakes Res.* **2013**, *39*, 46–54. [CrossRef]

43. Warner, D.M.; Lesht, B.M. Relative importance of phosphorus, invasive mussels and climate for patterns in chlorophyll a and primary production in Lakes Michigan and Huron. *Freshw. Biol.* **2015**, *60*, 1029–1043. [CrossRef]

44. Asano, T.; Li, G.M.; Hirai, Y.; Kim, D.C.; Ito, T. Eutrophication of lake taihu in China and post-response to the plague of algal bloom in 2007. *E-Journal GEO* **2011**, *5*, 138–153. [CrossRef]

45. Hu, C.; Zhongping, L.; Ma, R.; Yu, K.; Li, D.; Shang, S. Moderate resolution imaging spectroradiometer (MODIS) observations of cyanobacteria blooms in Taihu Lake, China. *Prehosp. Disaster Med.* **2010**, *29*, 303–306. [CrossRef]

46. Wang, M.; Shi, W.; Tang, J. Water property monitoring and assessment for China's inland Lake Taihu from MODIS-Aqua measurements. *Remote Sens. Environ.* **2011**, *115*, 841–854. [CrossRef]

47. Ma, W.; Chai, F.; Xiu, P.; Xue, H.; Tian, J. Modeling the long-term variability of phytoplankton functional groups and primary production in the South China Sea. *J. Oceanogr.* **2013**, *69*, 527–544. [CrossRef]

48. Pei, G.; Hong, X. Primary production of benthic algae community in the Donghu Lake. *J. South-Central Univ. Natly (Nat. Sci. Ed.)* **2010**, *29*, 28–31.

49. Bergamino, N.; Horion, S.; Stenuite, S.; Cornet, Y.; Loiselle, S.; Plisnier, P.D.; Descy, J.P. Spatio-temporal dynamics of phytoplankton and primary production in lake Tanganyika using a MODIS based bio-optical time series. *Remote Sens. Environ.* **2010**, *114*, 772–780. [CrossRef]

50. Ardyna, M.; Babin, M.; Gosselin, M.; Devred, E. Parameterization of vertical chlorophyll a in the Arctic ocean: Impact of the subsurface chlorophyll maximum on regional, seasonal and annual primary production estimates. *Biogeosciences* **2013**, *10*, 4383–4404. [CrossRef]

51. Tan, S.; Shi, G.; Tan, S.; Shi, G. Satellite-derived primary production and its spatial and temporal variability in the China seas. *J. Geogr. Sci.* **2006**, *16*, 447–457. [CrossRef]

52. Kameda, T.; Ishizaka, J. Size-fractionated primary production estimated by a Two-phytoplankton community model applicable to ocean color remote sensing. *J. Oceanogr.* **2005**, *61*, 663–672. [CrossRef]

53. Deng, J.; Qin, B.; Paerl, H.W.; Zhang, Y.; Ma, J.; Chen, Y. Earlier and warmer springs increase cyanobacterial (*microcystis* spp.) blooms in subtropical Lake Taihu, China. *Freshw. Biol.* **2014**, *59*, 1076–1085. [CrossRef]
54. Deng, J.; Qin, B.; Paerl, H.W.; Zhang, Y.; Wu, P.; Ma, J.; Chen, Y. Effects of nutrients, temperature and their interactions on spring phytoplankton community succession in Lake Taihu, China. *PLoS ONE* **2014**, *9*, e113960. [CrossRef] [PubMed]
55. Xu, S.; Wang, Y.; Huang, B.; Wei, Z.B.; Miao, A.J.; Yang, L.Y. Nitrogen and phosphorus limitation of phytoplankton growth in different areas of Lake Taihu, China. *J. Freshw. Ecol.* **2014**, *30*, 113–127. [CrossRef]
56. Yuan, H.Z.; Shen, J.; Liu, E.F.; Wang, J.J.; Meng, X.H. Space distribution characteristics and diversity analysis of phosphorus from overlying water and surface sediments in Taihu Lake. *J. Environ. Sci.* **2010**, *31*, 954–960.
57. Zhao, X. Temporal and spatial distribution of physicochemical characteristics and nutrients in sediments of Lake Taihu. *J. Lake Sci.* **2007**, *19*, 698–704.

remote sensing

MDPI

Article

A MODIS-Based Novel Method to Distinguish Surface Cyanobacterial Scums and Aquatic Macrophytes in Lake Taihu

Qichun Liang [1,2], Yuchao Zhang [1,3,*], Ronghua Ma [1,3], Steven Loiselle [4], Jing Li [1,2] and Minqi Hu [1,2]

1   Key Laboratory of Watershed Geographic Sciences, Nanjing Institute of Geography and Limnology, Chinese Academy of Sciences, Nanjing 210008, China; lqc_niglas@163.com (Q.L.); rhma@niglas.ac.cn (R.M.); crystalleegis@163.com (J.L.); huminqi16@mails.ucas.ac.cn (M.H.)
2   University of Chinese Academy of Sciences, Beijing 100049, China
3   Jiangsu Collaborative Innovation Center of Regional Modern Agriculture & Environmental Protection, Huaiyin Normal University, Huai'an 223300, China
4   Dipartimento di Biotecnologie, Chimica e Farmacia, University of Siena, CSGI, Via Aldo Moro 2, Siena 53100, Italy; loiselle@unisi.it
*   Correspondence: yczhang@niglas.ac.cn; Tel.: +86-25-8688-2165

Academic Editors: Yunlin Zhang, Claudia Giardino, Linhai Li, Deepak R. Mishra, Richard Gloaguen and Prasad S. Thenkabail
Received: 30 August 2016; Accepted: 26 January 2017; Published: 6 February 2017

Abstract: Satellite remote sensing can be an effective alternative for mapping cyanobacterial scums and aquatic macrophyte distribution over large areas compared with traditional ship's site-specific samplings. However, similar optical spectra characteristics between aquatic macrophytes and cyanobacterial scums in red and near infrared (NIR) wavebands create a barrier to their discrimination when they co-occur. We developed a new cyanobacteria and macrophytes index (CMI) based on a blue, a green, and a shortwave infrared band to separate waters with cyanobacterial scums from those dominated by aquatic macrophytes, and a turbid water index (TWI) to avoid interference from high turbid waters typical of shallow lakes. Combining CMI, TWI, and the floating algae index (FAI), we used a novel classification approach to discriminate lake water, cyanobacteria blooms, submerged macrophytes, and emergent/floating macrophytes using MODIS imagery in the large shallow and eutrophic Lake Taihu (China). Thresholds for CMI, TWI, and FAI were determined by statistical analysis for a 2010–2016 MODIS Aqua time series. We validated the accuracy of our approach by in situ reflectance spectra, field investigations and high spatial resolution HJ-CCD data. The overall classification accuracy was 86% in total, and the user's accuracy was 88%, 79%, 85%, and 93% for submerged macrophytes, emergent/floating macrophytes, cyanobacterial scums and lake water, respectively. The estimated aquatic macrophyte distributions gave consistent results with that based on HJ-CCD data. This new approach allows for the coincident determination of the distributions of cyanobacteria blooms and aquatic macrophytes in eutrophic shallow lakes. We also discuss the utility of the approach with respect to masking clouds, black waters, and atmospheric effects, and its mixed-pixel effects.

Keywords: Cyanobacteria and macrophytes index; Floating algae index; Cyanobacteria blooms; aquatic macrophytes; Lake Taihu; MODIS

1. Introduction

Cyanobacteria dominated blooms are ubiquitous in many freshwater ecosystems affected by human activities, increasing in frequency and distribution since the 1940s [1,2]. These blooms have multiple adverse impacts on local environments and economy, with a classic example being the extensive and long-lasting bloom in Lake Taihu in 2007 [3].

Aquatic macrophytes support critical ecological services in most shallow lakes by providing the habitat for a diverse and economically important faunal community, sequestering carbon and nutrients, as well as stabilizing sediment and shorelines [4–6]. Many studies indicate that eutrophication may lead to a reduction in macrophyte coverage as competition for available light and nutrients is impacted by an increase in phytoplankton, suspended detritus and periphyton [7]. Macrophyte coverage can indicate catchment scale and lake scale impacts of climate change on storm frequency, lake temperature, and sediment inflow [8]. The distributional changes in the coverage of aquatic vegetation and cyanobacterial scums can be considered an indicator of lake ecosystem conditions. The availability of accurate long-term information of the distribution of cyanobacterial scums and aquatic macrophytes can be a fundamental tool to lake management.

In reality, in situ measurements are often inappropriate to deal with the complex temporal and spatial dynamics of cyanobacterial scums, due to rapid vertical migration [9] and very fast replication [10], while the study of aquatic macrophytes requires a seasonal approach. In recent decades, remote sensing became a fundamental tool to explore the spatial and temporal behavior of aquatic ecosystems. Many studies have successfully used optical remote sensing to map algal blooms [11–13] and aquatic macrophytes [14–16] in coastal, lacustrine, and lagoon environments.

Remote sensing of cyanobacterial scums includes indirect approaches using water quality parameters such as transparency and direct approaches based on photosynthetic pigments concentrations (chlorophyll-a and phycocyanobilin) and spectrum shape characteristics (Table 1) [17]. These chlorophyll-a algorithms become problematic in the event of surface scums, where thick foams and complex chemical constituents are present [18], and have led to the development of new qualitative approaches. Algorithms using near infra-red (NIR) bands have shown particular promise, based on the red-edge effect to vegetation of blooms [18] including single band [19–22], band ratio [23–29] and band difference [30,31], and some land vegetation indices such as Normalized Difference Vegetation Index [25,32–35] and Enhanced Vegetation Index [25,36]. NIR bands are also the basis for spectral shape methods, which use a computational equivalent to the second derivative [37], including Fluorescent Line Height [38], Maximum Chlorophyll Index [39], Spectral Shape index [40–42], Cyanobacterial Index [37,43], Maximum Peak-Height algorithm [44,45], Floating Algae Index [46–50], and a Classification And Regression Tree [51].

Methods for identifying aquatic macrophytes by remote sensing have also undergone important developments (Table 2), including unsupervised classifier [52–57], supervised classifier [58,59], classification trees [60–67] and remote sensing combined with ancillary data [68,69]. Ratios of band reflectances or band transformations (e.g., normalized difference index) based on the spectral characteristics of macrophytes in the visible and NIR wavelengths can be used for macrophyte detection and classification.

Due to their similar spectral characteristics, especially in red and NIR wavelengths, it is often difficult to distinguish cyanobacterial scums and aquatic macrophytes by remote sensing. According to Gao's [70] and Rogers and Kearney's [71] results that short-wave infrared (SWIR) bands were sensitive to vegetation liquid water, Oyama [67] used SWIR bands of Landsat TM for distinguishing cyanobacterial scums and aquatic macrophytes, with the floating algae index (FAI) and normalized difference water index (NDWI) in Japanese lakes. However, extremely thick cyanobacterial surface scums in Lake Taihu with relatively sparse aquatic macrophytes present a more complex condition. Furthermore, SWIR is sensitive to conditions of elevated turbidity, a common characteristic of most shallow lakes. The approach combined vegetation presence frequency (VPF) and FAI, suggested by Liu [68], suffered from similar challenges from highly turbid or highly absorbing "black" waters. Otherwise, this VPF method could not be utilized for routine monitoring because it is one of the retrospective evaluation approaches.

Compared to Landsat TM/ETM+ , MERIS, and MODIS imagery provide advantages for large lakes due to their short overpass period and good spatial resolution [72,73]. MERIS, accessible until April 2012, provided important insights into the concentrations of optically active substances in large lakes [45,74]. MODIS (1999–present for Terra, 2002–present for Aqua) provides frequent (daily) and synoptic global observations and is equipped with several medium-resolution bands ("sharpening" bands designed for land use), and is also the prototype for VIIRS (the Visible Infrared Imager/Radiometer Suite) [75,76], allowing MODIS algorithms to inform algorithm development for VIIRS.

In the present study, we develop and validate a new approach to distinguish lake water, cyanobacterial scums and different aquatic macrophytes in complex aquatic and atmospheric optical conditions and mixed-pixel effects.

Table 1. Remote sensing identification methods on algal blooms in lakes and coastal sea with satellite data.

	Algorithm Form	Satellite Data	Study Area	References
Single band	B_{NIR}	MODIS TM	Lake Taihu	[19]
	B_{RED}	CZCS AVHHR MODIS VIIRS	Northeast coast of the Atlantic Baltic Sea	[20,21]
Band ratio	B_{NIR}/B_{RED} B_{NIR}/B_{GREEN} B_{GREEN}/B_{BLUE}	CZCS MODIS GOCI	Yellow Sea East China Sea Lake Taihu Southeastern Mediterranean Black Sea Northwest European continental shelf	[23–25,27–29]
	B_{NIR}/B_{RED}	AVHRR	Northwest European continental shelf Lake Pontchartrain	[26,27]
Band difference	$B_{NIR} - B_{RED}$	AVHRR MODIS	Western shore of Canada Paracas Bay, Peru	[30,31]
NDVI	$(B_{NIR} - B_{RED})/(B_{NIR} + B_{RED})$	AVHRR MODIS TM/ETM+ GOCI	the Baltic Sea Yellow Sea East China Sea Lake Taihu Lake Dianchi	[25,32–35]
EVI	$G \times (B_{NIR} - B_{RED})/(B_{NIR} + C_1 \times B_{RED} - C_2 \times B_{BLUE} + C_3)$	MODIS GOCI	Yellow Sea East China Sea	[25,46]
Spectrum shape	FLH MCI SS MPH	MERIS MODIS	Lake Erie Baltic Sea Lake Taihu Lake Victoria Lake Michigan	[38–45]
	FAI	MODIS TM/ETM+	Lake Taihu Lake Chaohu Yellow Sea East China Sea West Florida Shelf	[46–50]
	CART	MODIS	Lakes in southern Quebec	[51]

NDVI: Normalized Difference Vegetation Index; EVI: Enhanced Vegetation Index; FLH: Fluorescent Line Height; MCI: Maximum Chlorophyll Index; SS: Spectral Shape index; MPH: Maximum Peak-Height algorithm; FAI: Floating Algae Index; CART: Classification and Regression Tree.

Table 2. Aquatic macrophytes mapping approaches by satellite data.

Approaches	Satellite Data	Study Area	References
Unsupervised classifier	Landsat-1 TM/ETM+ IRS-1B LISS-II Quickbird	North Dakota California's Central Valley Great Bay, New Hampshire Grand Teton National Park, USA Lake Mogan Chwaka Bay	[52–57]
Supervised classifier	TM/ETM+	Lower Mekong Basin Yakima River	[58,59]
Classification trees	TM/ETM+ SPOT HJ-CCD	DelawareWater Gap National Recreation Area Gallatin Valley of Southwest Montana, USA Yellowstone National Park Camargue or Rhône delta Lake Taihu Japanese lakes	[60–67]
Classification trees with ancillary data	MODIS	Lake Taihu	[68,69]

2. Study Area and Data

2.1. Lake Taihu

Lake Taihu, the third largest freshwater lake in China, is located in the Yangtze River Delta (latitude 30°55'40''–31°32'58''N; longitude 119°52'32''–120°36'10''E, Figure 1). The lake has a surface area of 2338 km^2, a maximum and average depth of 2.6 and 1.9 m, respectively, with a mean water residence time of approximately 309 days [77]. Due to the rapid economic growth through urbanization and industrialization of Lake Taihu basin, increasing eutrophication and recurrent cyanobacterial scums (Microcystis) have occurred in recent years, posing a significant threat to millions of people relying on the lake for drinking water supply [78]. Harmful cyanobacterial scums regularly occur in the northern Meiliang, Zhushan, and Gonghu bays and on the western lakeshore [19]. The eastern and southern lake areas are characterized by extensive areas of macrophytes, where turbidity is low and cyanobacterial scums are rare [68,79]. Thus many researchers always divided Lake Taihu into cyanobacteria-dominated zone and a macrophytes-dominated zone [64–66]. On the basis of havitat, aquatic macrophytes can always be subdivided as emergent, floating-leaved, and submerged macrophytes [80]. Fourteen dominant species of aquatic macrophytes were identified in historical surveys in Lake Taihu (Table 3), including two emergent species, three floating species and nine submerged species [81]. In recent field investigations between 2013–2016, submerged macrophytes (i.e., Potamogeton crispus) have re-appeared in areas where they have been absent for more than a decade, Zhushan and Meiliang Bays.

2.2. Field Data

Two hundred and sixty six in situ investigations were carried out in July and August of 2013, May of 2014, June of 2015, and May of 2016 (Figure 1), covering the whole lake. Of the investigations 82.7% were located in the bays, especially in the eastern lake, always covered with aquatic macrophytes or cyanobacterial scums. The name and classification (emergent, floating, and submerged macrophytes) of aquatic vegetation, GPS and photographs of each site were recorded, and the percentage coverage of the aquatic macrophytes were estimated by eye from the boat during field investigations. Otherwise, 37 in situ water leaving reflectance spectra $R_{rs}(\lambda)$ were measured in September of 2014 and April of 2015.

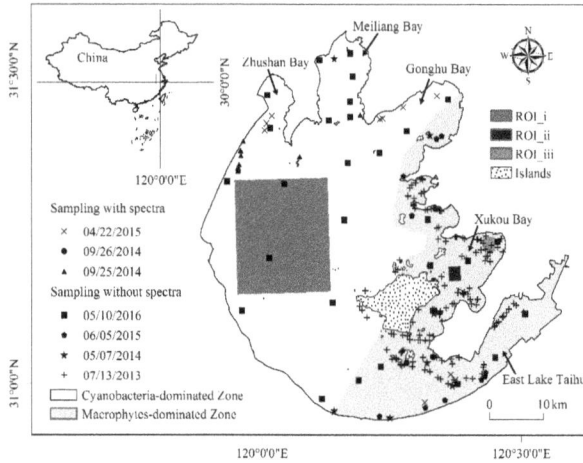

Figure 1. The locations of field investigation samples collected from 2013 to 2016.

Table 3. Main dominant aquatic macrophytes in Lake Taihu due to field investigation.

Type	Havitat	Dominant Macrophyte	Max Height
Emerged macrophytes	Frequently growing above the waterline of lakes and wetlands with only their roots located in wet or damp soils	*Phragmites australis*	2–5 m
		Zizania latifolia	1.6–2 m
Floating-leaved macrophytes	Having roots located into sediment and stems to lift the leaves floating above the water surface	*Nymphoides peltatum;* *Nymphoides indica;* *Trapa maximowiczii;*	Above the water surface
Submerged macrophytes	Being usually but not always rooted, and putting their whole body under the water except flowers	*Potamogeton maackianus;* *Potamogeton malaianus;* *Ceratophyllum demersum;* *Hydrilla verticillata;* *Myriophyllum spicatum;* *Elodea nuttalli;* *Potamogeton crispus*	Under the water surface
		Vallisneria natans	–1.2 m
		Chara	–0.5 m

We measured the $R_{rs}(\lambda)$ for two emergent macrophytes (*Zizania caduciflora; Phragmites australis*), one floating macrophytes (*Nymphoides peltatum*) and two submerged macrophytes (*Potamogeton crispus; Potamogeton maackianus*). In situ $R_{rs}(\lambda)$ were measured with a FieldSpec FR spectroradiometer (Analytical Spectral Devices, Boulder, CO) with the wavelength range from 350 to 2500 nm following the NASA Ocean Optics protocols [82]. As recommended by Mobley [83], a viewing geometry with an azimuth of 135° and zenith of 40°, was used to avoid water surface reflection from direct sun. Each water spectrum was sampled 90° azimuth with respect to the sun and with a viewing angle of 45°. The detector integration time was either 136 ms or 272 ms. A separate dark reading was obtained each time the integration time was changed. The measurement sequence, repeated five times for each measurement, began with a measurement of a standard 25 cm × 25 cm plaque (25% reflectivity), water and sky radiances (each preceded by a dark offset reading). The water surface reflectance factor ρ was assumed to be 0.028 but clearly depended on sky conditions, wind speed, and solar zenith angle [83]. The viewing direction of 40° and 135° from the sun was considered a reasonable compromise. This method has been used successfully in sea and inland lakes studies in conditions where wind speeds less than 5 m/s [83–86]. The viewing direction was set by adjusting the instrument angle to minimize the effects of sun glint and non-uniform sky radiance while also avoiding instrument shading. Measurements were made from a location that minimized shading, reflections from superstructure, ship's wake, associated foam patches and whitecaps, and specular reflection of sunlight.

2.3. Satellite Image Data Processing

MODIS Level-0 data collected by Aqua (2010–2016) covering the study region were obtained from the NASA Goddard Space Flight Center through its Ocean Biology Processing Group (OBPG) [87]. We used five bands from 469 nm to 1240 nm designed for land and atmosphere use, with a lower dynamic range than the ocean bands, and therefore unlikely to saturate in the highly turbid conditions of this shallow lake. The ground resolution of the 645 and 859 nm band is 250 m, and 500 m for the bands at 469, 555, and 1240 nm.

MODIS data were processed using SeaDAS (version 7.0). First, the data were converted to calibrated radiance (Level-1B). A key challenge remains in Case II waters, to perform a full atmospheric correction where significant errors in visible range occur for turbid waters with classic atmospheric correction methods, such as Gordon and Clark [88], Ruddick, Ovidio, and Rijkeboer [89], Wang and Shi [90], and Bailey, Franz, and Werdell [91] etc. Therefore, to avoid this problem, a partial atmospheric correction to correct for the gaseous absorption (mainly by ozone) and Rayleigh (molecular) scattering effects was applied to the Level-1B data, resulting in Rayleigh corrected reflectance (Rrc, dimensionless):

$$R_{rc}(\lambda) = \rho_t(\lambda) - \rho_r(\lambda) = \rho_a(\lambda) + \pi \times t(\lambda) \times t_0(\lambda) \times R_{rs}(\lambda) \tag{1}$$

where ρ_t is the top of atmosphere (TOA) reflectance after adjustment of the atmospheric (gas) absorption, ρ_r is the reflectance due to Rayleigh scattering, ρ_a is the reflectance due to aerosol scattering and aerosol-Rayleigh interactions, t and t_0 are diffuse transmittance from the image pixel to the satellite and from the sun to the image pixel, respectively. Note that ρ_a, t, and t_0 are functions of aerosol type, aerosol optical thickness, and solar/viewing geometry. The above formulation assumes negligible contributions from whitecaps and sun glint.

The Rrc data were mapped to a cylindrical equidistant projection for further analysis. First, the Rrc data at 645 nm, 555 nm, and 469 nm were used to compose the Red-Green-Blue true color images to screen for clouds and sun glint. After visual inspection, a total of 176 data granules between 2010 and 2015 (Table 4) were found to contain no cloud cover and sun glint, therefore suitable for thresholds determination and method development.

Table 4. Temporal distribution of MODIS Aqua imageries used in this study.

	2010	2011	2012	2013	2014	2015	2016	Total
January	0	0	2	0	5	5	5	17
February	0	0	0	0	0	5	13	18
March	1	0	2	4	10	2	5	24
April	0	2	1	7	0	3	4	17
May	4	1	5	6	4	2	3	25
June	1	1	0	0	0	2	2	6
July	1	3	3	2	1	2	2	14
August	4	0	0	5	0	5		14
September	2	4	1	1	0	2		10
October	1	2	5	3	6	5		22
November	1	0	3	7	4	1		16
December	6	0	0	2	14	5		27
Total	21	13	22	37	44	39	34	210

To further validate the results, we used the aquatic macrophyte distributions done by Luo [66] using HJ-CCD images. HJ-1A and HJ-1B satellites were launched by the China Center for Resources Satellite Data and Application (CRESDA) on September 2008, have a high revisit time (2 days) and wavebands (B1: 430–520; B2: 520–600; B3: 630–690; B4: 760–900 nm) that are appropriate for vegetation mapping. According to Luo's article [66], radiometric calibrations were made first using coefficients provided with the image (e.g., gains and offsets), then the atmospherically corrected images were geometrically corrected against a historical Landsat TM image with geometric accuracy of <0.5 pixel, and lastly the FLAASH module was applied for atmospheric correction. All of the lake except Gonghu Bay and the eastern lake was masked in Luo's research.

3. Methods

3.1. Spectral Features of Lake Water, Cyanobacterial Scums, and Aquatic Macrophytes

Spectra for clear lake water had the maximum reflectance at the wavelengths between 550–580 nm (Figure 2). Reflectance decreased sharply on both sides of the maximum, with reflectance at 500 nm higher than at 650 nm [92]. The reflectance spectra of turbid waters were typical, with a low reflectance in blue range and high reflectance in green range, owing to the absorption by dissolved organic matter and tripton as well as backscattering by particulate matter [93]. Reflectance in the red region (600–700 nm) had two minima around 620 nm and 675 nm, associated to phycocyanobilin and chlorophyll-a absorption [94,95]. A distinct peak around 700 nm shifted from 690 nm at low chlorophyll-a concentrations to 715 nm at high chlorophyll-a concentrations, which resulted from both high backscattering and minimum absorption by all optically active constituents, including pure water [96]. For low chlorophyll-a and high suspended sediments, this peak was reduced. The scattering by all particulate matter controlled the variations of NIR reflectance [93].

The reflectance spectra of cyanobacterial scums not only had the typical chlorophyll-a and phycocyanobilin absorption at 442, 665, and 620 nm respectively, but also an increase in wavelengths ranging from 700 and 1800 nm with chlorophyll-a concentrations compared to that at visible and longer than 1900 nm range [67]. This is similar to reflectance spectra of aquatic macrophytes.

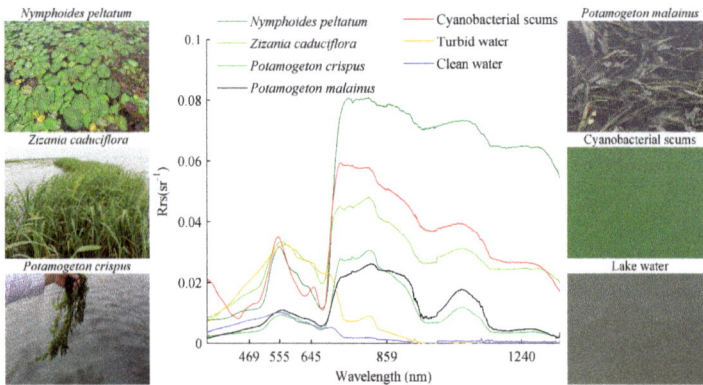

Figure 2. The in situ reflectance spectra and corresponding photographs of water, turbid water, cyanobacterial scums, and some typical aquatic macrophytes observed in Lake Taihu. Spectra were determined by the average spectra of each aquatic macrophytes.

For aquatic macrophytes, high reflectance values at the NIR region are mainly due to the cellular structure in the leaves [97]. A small absorption around 970 nm has been associated to water content [14], and is similar to that observed in the reflectance spectra of cyanobacterial scums [67]. All aquatic macrophytes investigated in Lake Taihu have similar reflectance spectra with pigment concentration and cellular structure responsible for observable differences [16]. Submerged macrophytes have lower reflectance values across the whole range due to the water absorption [16].

3.2. A New Classification Method for MODIS

3.2.1. CMI, FAI, and TWI

Considering the similar NIR reflectance spectra between cyanobacterial scums and aquatic macrophytes, indices based on other wavelengths were necessary to distinguish them. Compared to aquatic macrophytes, the Rrc spectra of cyanobacterial scums have a distinguishable peak at 555 nm and an obvious minimum at 469 nm (Figure 3). We note that the Rrc data used removed Rayleigh scattering effects but not aerosol or other atmospheric effects. The use of the 1240 nm band to construct a baseline with 469 nm removed additional impacts from atmospheric effects. This suggests that the difference between Rrc(555) and baseline between 469 nm and 1240 nm, can be used to distinguish cyanobacterial scums and aquatic macrophytes. The resulting Cyanobacteria and Macrophytes Index (CMI) was defined as:

$$CMI = R_{rc,GREEN} - R_{rc,BLUE} - [R_{rc,SWIR} - R_{rc,BLUE}] \times (\lambda_{GREEN} - \lambda_{BLUE})/(\lambda_{SWIR} - \lambda_{BLUE}) \quad (2)$$

following a baseline subtraction similar to the FAI index [46] with clearly different band combinations.

FAI, widely used for detecting floating algae in lakes [47,50] was expanded in this study to identify different types of aquatic macrophytes as well as floating algae. FAI was originally developed using MODIS red, NIR, and SWIR bands,

$$FAI = R_{rc,NIR} - R_{rc,RED} - [R_{rc,SWIR} - R_{rc,RED}] \times (\lambda_{NIR} - \lambda_{RED})/(\lambda_{SWIR} - \lambda_{RED}) \quad (3)$$

When very high suspended sediments are present, they can dominate the optical signal, leading both CMI and FAI to identify areas of high turbidity as cyanobacterial scums. To avoid this, we created a Turbid Water Index (TWI), following algorithm approaches for suspended solids in Case II waters [98]:

$$TWI = R_{rc,RED} - R_{rc,SWIR} \quad (4)$$

For MODIS, these are λ_{BLUE} = 469 nm, λ_{GREEN} = 555 nm, λ_{RED} = 645 nm, and λ_{SWIR} = 1240 nm.

3.2.2. Classification Tree

The relationship between CMI, FAI, and NDWI [67] from typical lake waters (Figure 3) showed $CMI_{cyano} > CMI_{water} > CMI_{float} > CMI_{sub}$. While CMI of floating macrophytes was much lower than that of cyanobacterial scums, respective FAIs overlapped significantly. Likewise, the FAI of the open lake water and clean water with submerged macrophytes in Xukou Bay were similar. Due to Oyama's research, NDWI showed a much better capability to distinguish cyanobacterial scums and aquatic macrophytes using Landsat TM/ETM+ [67]. Figure 3b indicated that NDWI of cyanobacterial scums, aquatic macrophytes, and lake water mixed together based on MODIS data. The impact of high concentrations of suspended solids on water reflectance at green, red, and NIR range was evident, as was its impact on CMI and FAI in turbid waters.

A classification decision tree based on these relationships was constructed with five steps (Figure 4). Step 1: identify and remove pixels of turbid water by TWI; Step 2: chose the CMI and FAI thresholds due to the pixel location (cyanobacteria-dominated or macrophytes-dominated zone); Step 3: identify which is existing in the pixel, cyanobacterial scum or aquatic macrophytes by CMI threshold; Step 4: if cyanobacterial scum exists, identify lake waters using FAI = −0.004; Step 5: if aquatic macrophytes exists, distinguish between floating and emergent/submerged macrophytes using FAI threshold.

Three regions of interest (ROI_i, ROI_ii, and ROI_iii) respectively in open lake and Xukou Bay of Lake Taihu were used to determine threshold values from MODIS Rrc data directly (Figure 5). It should be noted that the values from in situ reflectance measurements could not be used for threshold development directly due to the complex optical conditions of the waterbody and impossibility to apply a comparative full atmospheric correction.

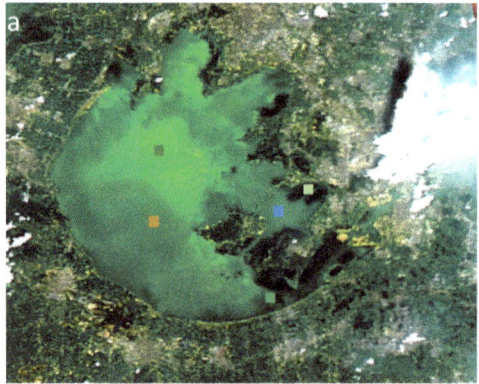

(a)

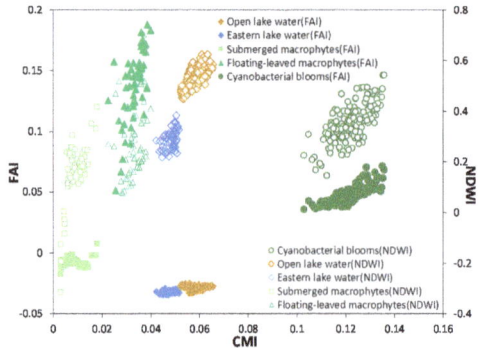

(b)

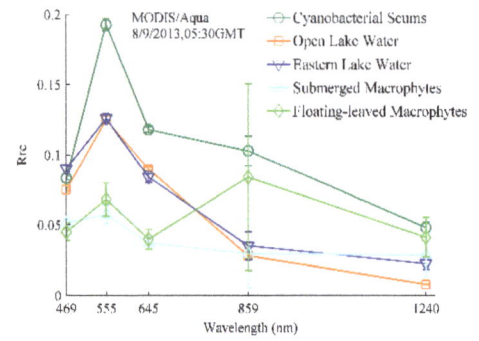

(c)

Figure 3. RGB image (**a**), the relationship of Cyanobacteria and Macrophytes Index (CMI), Floating Algae Index (FAI), and Normalized Difference Water Index (NDWI) (**b**) and the Rayleigh-corrected MODIS spectral reflectance (Rrc(λ)) (**c**) of lake areas with different optical conditions on 8 September 2013 GMT.

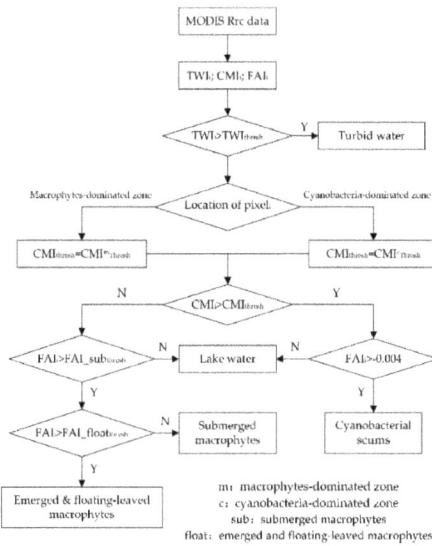

Figure 4. Classification decision tree to distinguish cyanobacterial scums and aquatic macrophytes in Lake Taihu.

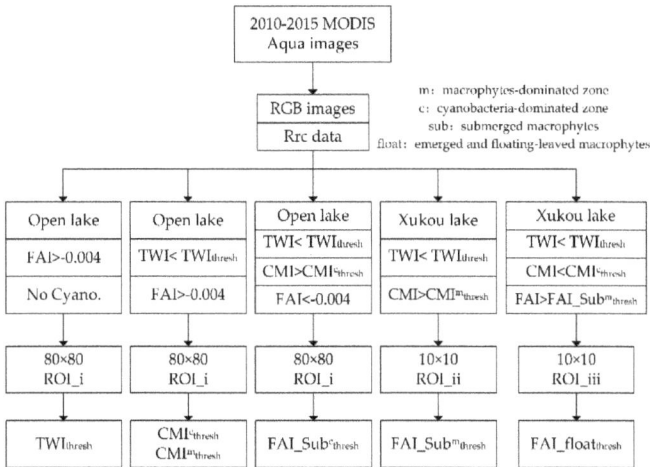

Figure 5. The flow chart for determining TWI, CMI, and FAI thresholds. ROI i was 80×80 pixels while ROI_ii and ROI_iii were 10×10 pixels, see Figure 1 for location.

3.2.3. Accuracy Assessment and Validation

Three approaches were used to validate our new classification tree: first, in situ remote sensing reflectance spectra were used to compare CMI to FAI and NDWI; second, user's accuracy, overall accuracy [99] and normalized accuracy [100] were calculated by comparing the satellite-derived distribution to the ground-truth results during 2013 and 2016; thirdly, aquatic macrophytes distributions derived from MODIS imageries were compared to those from 30 m spatial-resolution HJ-CCD imageries.

User's accuracy ($p_u(i)$) is a measure of the commission error associated with a class and is derived from the number of pixels correctly allocated to a class relative to the total number of pixels predicted to belong to that class in the accuracy assessment,

$$p_u(i) = \frac{p_c(i)}{p_t(i)} \tag{5}$$

where $p_c(i)$ is the number of correctly classified pixels of type i, and $p_t(i)$ is the number of total pixels of type i based on the ground-truth measurements.

The overall accuracy (p_o) was defined as the percentage of samples that were classified correctly and calculated using the following equation,

$$p_o = \frac{\sum_{i=1}^{n} p_c(i)}{p_t} \tag{6}$$

where n is the total number of types that were classified, $p_c(i)$ is the number of correctly classified pixels of type i, and p_t is the total number of pixels in the validation data set.

An iterative proportional fitting procedure which forces each row and column in the matrix to sum to one was used to normalized error metrics [100,101]. This process then changes the cell values along the major diagonal of the matrix (correct classifications) and therefore a normalized overall accuracy can be computed for each matrix by summing the major diagonal and dividing by the total of the entire matrix. The normalized accuracy is considered as a better representation of accuracy than the overall accuracy computed from the original matrix because it contains information about the off-diagonal cell values [100].

3.3. Classification Method for HJ-CDD Data

A spectral indices $(B_4 - \sum_{j=1}^{3} B_j)/(B_4 + \sum_{j=1}^{3} B_j)$ (SF$_1$), the second principal components (SF$_2$) and the difference between the greenness index and the brightness index from the tasseled-cap transform (SF$_3$) were used to construct Luo's classification tree using HJ-CDD data [66]. First, SF$_1$ was utilized to detect emergent mocrophytes, then floating-leaved macrophytes were identified by SF$_2$, and SF$_3$ discriminated submerged macrophytes and open water lastly. The thresholds of SFs in the classification tree were determined by corresponding field investigation data.

3.4. Analysis Methods

3.4.1. Atmospheric Effects

Considering MODIS Rayleigh corrected reflectance used in our research, in order to estimate atmospheric effects on CMI, we simulated the influence of the different atmospheric conditions on CMI using in situ reflectance measurements and an atmospheric radiance transfer model in the absence of whitecaps and sun glint (Equation 1). Aerosol optical thickness (0–1 at intervals of 0.2), six aerosol types (r30, 50, 70, 75, 80, and 85) and six typical solar/viewing geometry of Lake Taihu (oc1: solz $= 40°$, senz $= 20°$, phi $= 40°$; oc2: solz $= 40°$, senz $= 20°$, phi $= -120°$; oc3: solz $= 40°$, senz $= 40°$, phi $= 40°$; oc4: solz $= 40°$, senz $= 40°$, phi $= -120°$; oc5: solz $= 60°$, senz $= 20°$, phi $= -120°$; oc6: solz $= 60°$, senz $= 20°$, phi $= 40°$) were extracted from SeaDAS LUTs.

3.4.2. Mixed Pixels Effects

The best spatial resolution of MODIS data is 250 m, which is much rougher than that of Landsat TM/ETM+ and HJ-CCD. In order to assess mixed pixels effects on thresholds of CMI and FAI, a MODIS image with both of the cyanobacterial scums and all the kinds of macrophytes should be chosen in the following analysis. We therefore took the MODIS image on 9 August 2013 for an example to investigate the influence of mixed pixels on the thresholds of CMI and FAI by varying the fraction of cyanobacteria blooms, submerged macrophytes or emergent/floating macrophytes from 0% to 100% at intervals of 10%. However, first, the 5 × 5 pixels covering with only one of lake water, cyanobacteria blooms, submerged macrophytes or emergent/floating macrophytes were detected based on in situ investigations in 2013. Also, the average upper 25% values of CMI and FAI were determined as the pure pixel thresholds for cyanobacterial scums or emergent/floating macrophytes. The average lower 25% values of CMI and FAI were decided as the pure pixel threshold for submerged macrophytes. Then the CMI and FAI means of 5 × 5 lake water pixels were used as the pure lake water thresholds. A simple linear model was chosen with two endpoints: one was pure lake water, and the other was pure cyanobacterial scums, submerged macrophytes or emerged/floating-leaved macrophytes.

4. Results

4.1. Thresholds

4.1.1. TWI Threshold Determination for Turbid Waters

Considering the location and timing of elevated turbidity in Lake Taihu, a region of interest (ROI_i) with 80 × 80 pixels in the open lake was chosen and 76 images from December to February were visually checked for cyanobacterial scums and aquatic macrophytes. All pixels, with FAI > −0.004, were put together to determine TWI thresholds (Figure 6a). Based on the near normal distribution with a mean (μ) and standard deviation (σ) of 0.1326 and 0.0126, the TWI threshold for turbid waters was 0.107 ($\mu - 2\sigma$).

4.1.2. CMI Threshold Determination for Distinguishing Aquatic Macrophytes and Cyanobacterial Scums

Based on the existed FAI threshold for cyanobacterial scums in Lake Taihu, FAI > −0.004 [47], the CMI threshold was based on 176 images from ROI_i, which satisfied TWI ≤ 0.107 and FAI > −0.004 at the same time. The CMI distribution of cyanobacterial scums pixels was also in a normal fashion with μ = 0.0455 and σ = 0.0085 (Figure 6b). The priori knowledge on Lake Taihu shows that the macrophytes-dominated zones are located in the east sections of the lake [65]. In order to minimize the false positive, we used two different CMI thresholds for detecting cyanobacterial scums in cyanobacteria-dominated zone and macrophytes-dominated zone (Figure 1). For cyanobacteria-dominated zones, we chose 0.0285($\mu - 2\sigma$) as the CMI$^a_{thresh}$, and for non-cyanobacteria-dominated zones we chose a more strict 0.0455(μ) as CMI$^g_{thresh}$.

4.1.3. FAI Threshold Determination for Detection Different Types of Aquatic Macrophytes

The published FAI threshold of −0.004 was used to distinguish cyanobacterial scum and water [47]. As FAI can also discriminate different types of aquatic macrophytes (Figure 3), and considering the small area of emergent macrophytes and coarse spatial resolution of MODIS, we combined emergent and floating macrophytes in one group. It should be noted that water quality in the macrophytes-dominated zone was much better than that in the cyanobacteria-dominated zone [79]. We decided to use two thresholds FAI_sub$^a_{thresh}$ and FAI_sub$^g_{thresh}$ in these two zones to discriminated waters and aquatic macrophytes respectively. FAI_sub$^a_{thresh}$ was based on ROI_i (Figure 6c) where TWI < 0.107, CMI < 0.0285 and FAI < −0.004. Similarly, FAI_sub$^g_{thresh}$ was determined from ROI_ii (Figure 6d), where TWI < 0.107, CMI < 0.0455 and FAI < −0.004. FAI_sub$^a_{thresh}$ and FAI_sub$^g_{thresh}$ were −0.0122 and −0.011 ($\mu - 2\sigma$) respectively. In the same way, ROI_iii in Xukou Bay was used to determine the FAI threshold of 0.05 for submerged and emergent/floating macrophytes, interestingly the same FAI value used by Oyama with Landsat TM/ETM [67].

All of the thresholds have been summarized in Table 5.

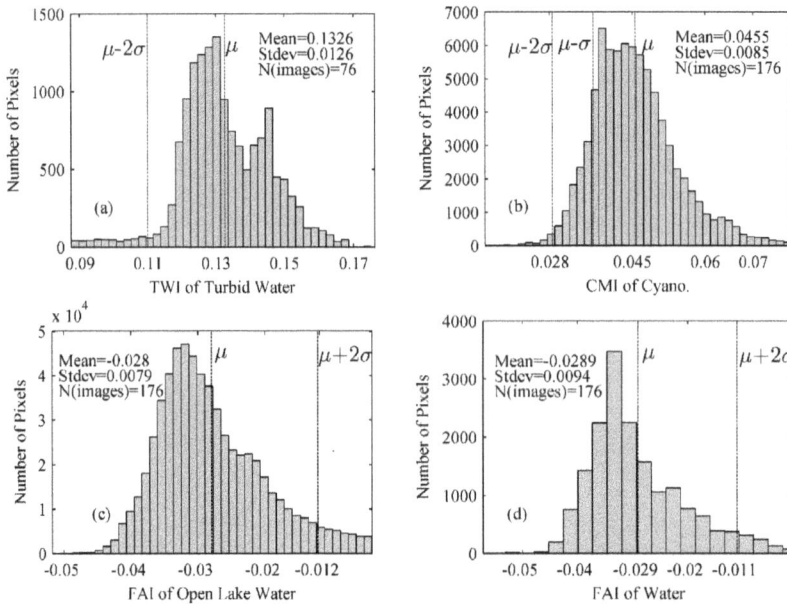

Figure 6. Distribution of TWI (**a**), CMI (**b**) and FAI (**c,d**) for images to distinguish thresholds for turbid water, cyanobacterial scums, submerged macrophytes, and floating macrophytes.

Table 5. The thresholds of TWI, CMI, and FAI used in the new classification method.

Index	Usage	Cyanobacteria-Dominated Zone	Macrophytes Dominated Zone
TWI	To detect high turbid water	0.107	
CMI$_{thresh}$	To distinguish lake water with cyanobacterial scums or with macrophytes	0.0285	0.0455
FAI_cyano	To detect cyanobacterial scums	−0.004	−0.004
FAI_sub$_{thresh}$	To detect submerged macrophytes	−0.0122	−0.011
FAI_float$_{thresh}$	To detect emerged and floating-leaved macrophytes	0.05	0.05

4.2. Validation by in Situ RRS Measurements

We first validated the applicability of CMI, FAI, and NDWI to distinguish lake water, cyanobacterial scums, submerged macrophytes, and emergent/floating macrophytes using in situ reflectance spectra of 2014 and 2015. All of these indices were calculated by in situ measured reflectance in simulated MODIS bands (Figure 7).

FAI was effective at separating lake water and others, but less useful for distinguishing cyanobacterial scums and aquatic macrophytes (Figure 7b). In contrast, CMI showed significant differences between the cyanobacterial scums and aquatic macrophytes, but was less useful for detecting open lake waters (Figure 7a). The modelled MODIS band 2 (841–876 nm) and band 6 (1628–1652 nm) from in situ spectra were chosen to estimate NDWI, which also showed a good separation between cyanobacterial scums and floating-leaved macrophytes from Figure 7c as that achieved using Landsat TM/ETM+ [67]. However, if MODIS Rrc data were used, NDWI was less useful due to Figure 3b. This may be the result of very thick cyanobacterial scums, relative disperse and sparse aquatic macrophytes, and turbid waters as well as the spectral differences between MODIS and Landsat TM. This validation by in situ Rrs indicated the improved effectiveness of combining CMI and FAI.

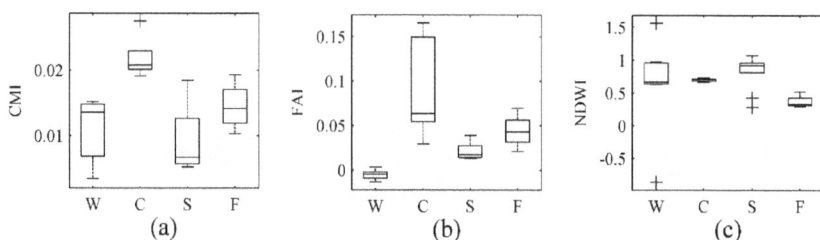

Figure 7. Box-and–whisker plots of CMI (**a**), FAI (**b**) and NDWI (**c**) for lake water, cyanobacterial scums, macrophytes based on in situ reflectance spectra of Lake Taihu. The solid line in the box represents the median value. The upper and lower fences represent the 1st and 3rd quartiles (Q1 and Q3), respectively. The lower and upper whiskers were calculated from (Q1 − 1.5 × IQR) and (Q3 + 1.5 × IQR), respectively, where IQR is the inter quartile range represented by the width of the box (i.e., Q3 − Q1). The data above or under the whisker were defined as the outliers and are shown as cross. (W: Lake water; C: Cyanobacterial scums; S: Submerged macrophytes; F: Emergent and floating macrophytes).

4.3. Validation by Field Investigation

To verify the stability and accuracy of the new method, we validated the predicted results with synchronous ground-truth investigations in Lake Taihu (Table 6). From 2013–2016, the user's accuracy of the submerged macrophytes, emergent and floating macrophytes, cyanobacterial scums and lake water was 83%–89%, 79%–83%, 75%–92% and, 92%–100% respectively, and the overall classification accuracy each year was 87%, 92%, 80%, and 86%, respectively. Based on the off-diagonal matrix values, the normalized accuracy was lower than the overall classification accuracy, which was 86.2%, 62.6%, 56.5%, and 81.0% for 2013–2016 respectively. In total, the overall classification accuracy and normalized accuracy was almost similar, 86.0% and 86.8% respectively. The classification accuracy was influenced by: the number and spatial distribution of ground-truth data [68] and the spatial resolution of MODIS where mixed pixels effects could lower the overall accuracy.

Table 6. Accuracy assessment of classification results from 2013 to 2016.

| Year | Measured | Predicted | | | | User's Accuracy | Overall Accuracy | Normalized Accuracy |
		S	E & F	C	W			
2013	S	71	5		4	89%	87%	86.2%
	E & F	7	26			79%		
	W	3			34	92%		
2014	S	11	1		1	85%	92%	62.6%
	C			11	1	92%		
	W				3	100%		
2015	S	10			2	83%	80%	56.5%
	E & F	1	5			83%		
	C	1		3		75%		
2016	S	7			1	88%	86%	81.0%
	C			3	1	75%		
	W	1			16	94%		
Total	S	99	6		8	88%	86%	86.8%
	E & F	8	31			79%		
	C	1		17	2	85%		
	W	4			53	93%		

S: Submerged macrophytes; E & F: Emergent & floating macrophytes; C: Cyanobacterial scums; W: Water.

4.4. Validation by HJ-CCD Data

High-spatial resolution HJ-CCD data were used to validate spatial distributions of water, cyanobacterial scums and different aquatic macrophytes for 9 and 16 August, and 16 and 26 September

2013 (Figure 8). Aquatic vegetation spatial distributions on 16 August and 26 September 2013 were provided only in the eastern part of Lake Taihu by Luo's CT method [66] using HJ-CCD data. The two datasets showed consistent spatial distributions of aquatic macrophytes in the macrophytes-dominated areas: in the south shore of Gonghu Bay, there existed many submerged macrophytes and sparse floating macrophytes; submerged macrophytes dominated Xukou Bay; floating macrophytes were mainly located in the East lake especially in summer. However, emergent macrophytes distributed dispersedly close to the shores were more difficult for MODIS data to identify.

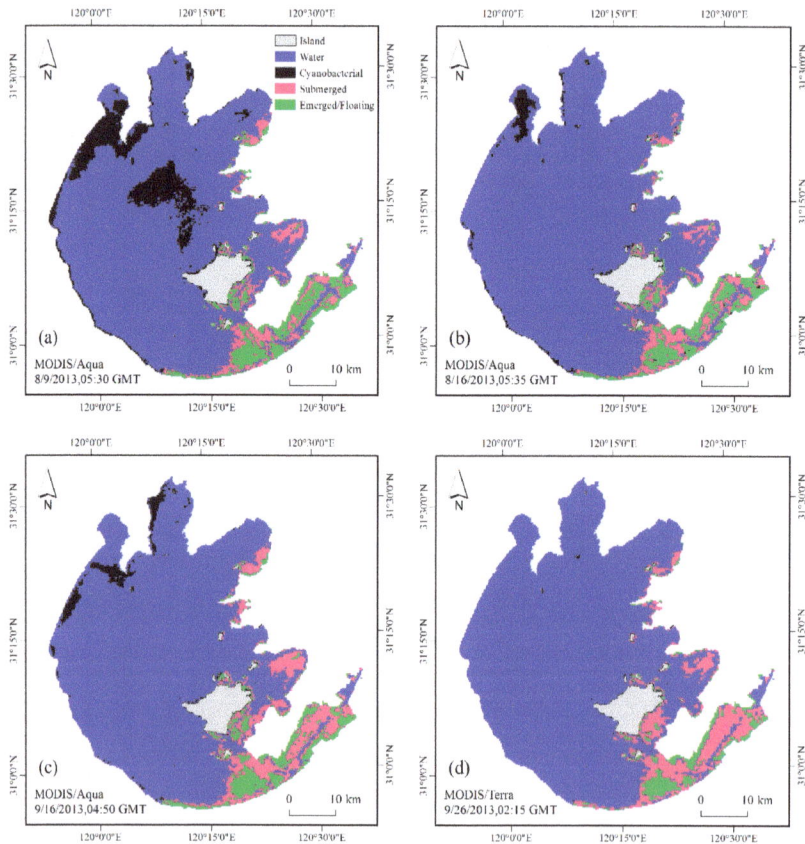

Figure 8. Spatial distributions of lake water, cyanobacterial scums and different aquatic macrophytes in Lake Taihu on 9 (**a**) and 16 (**b**) August, and 16 (**c**) and 26 (**d**) September 2013.

The expected temporal consistency was met, as submerged and emergent/floating macrophytes in the macrophytes-dominated zone were 174.5 and 178.3 km^2 on 16 August, and 255.7 and 105.6 km^2 on 26 September, respectively and their location was consistent (Figure 8). Cyanobacterial scums showed a much larger variation, changing from 68.4 to 0.75 km^2 over the same period.

5. Discussion

5.1. Data Quality versus Data Quantity

Time-series and trend analyses for aquatic vegetation, both macro and micro, in inland water bodies have been difficult to examine using satellite ocean color data due to challenging atmospheric conditions (e.g., clouds, thick aerosols, and sun glint) [72]. This complicates the separation of in-water reflectance from the top of atmospheric reflectance. The use of Rrc(869) > 0.027 to screen cloud pixels and low quality data significantly reduced data availability for Lake Taihu even under cloud-free and glint-free conditions [72]. Not only thick aerosols but also cyanobacteria blooms and aquatic macrophytes can increase reflectance in the near infra-red and shortwave infra-red wavelengths. For turbid waters, Wang and Shi recommended a cloud threshold of Rrc(1240) = 0.0235 and Rrc(1640) = 0.0215 [102], while Hu suggested Rrc(1640) = 0.03 [47]. None of these are suitable for the complex atmospheric, aquatic, and ecological conditions of Lake Taihu, where thick aerosols, sun glint, cyanobacteria blooms, aquatic macrophytes, and elevated turbidity all co-exist. In order to avoid false positive cloud masking from cyanobacteria blooms and aquatic macrophytes, we utilized Qi's Rrc(555) > 0.25 [72] and Rrc(1240) > 0.10 as constraint conditions for cloud masking. This resulted in >90% valid data coverage per image for every climatological month (Figure 9). In comparison, the mean valid data coverage for the Rrc(1240) < 0.0235 threshold was 35%–74%.

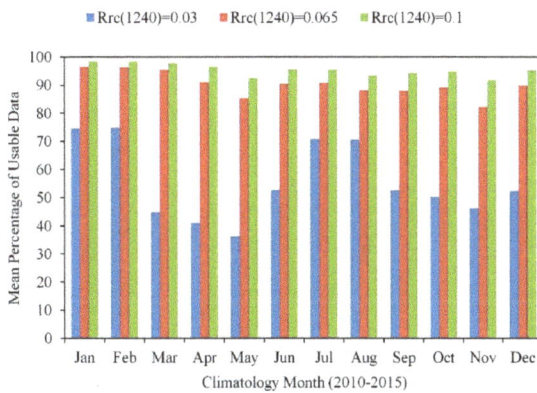

Figure 9. Mean percentage of usable data per image during each climatological month, based on different Rrc(1240) thresholds.

5.2. Use in Highly Turbid Waters

For shallow and turbid Lake Taihu, re-suspended solids have a major impact on visible and near infra-red reflectance spectra [103]. A comparison of reflectance spectra from cyanobacteria blooms, typical aquatic macrophytes, and turbid waters with different suspended solid (SS) concentrations shows that CMIs increased at higher suspended solid concentrations (Figure 10), impacting the detection of cyanobacteria blooms and emergent/floating macrophytes. Figure 11 is an example showing that the turbid water index, TWI, avoided these misclassifications, even under high wind and high turbidity conditions. MODIS RGB image (Figure 11a) showed lake water in the open lake was yellow and very turbid especially close to the west (W) and south (S) lakeshores, correspondingly CMI(W: 0.040 ± 0.003; S: 0.043 ± 0.003) and FAI (W: -0.006 ± 0.008; S: 0.0001 ± 0.008) of high turbid water were much higher than the other regions (Figure 11b,c). Therefore, high turbid water could be misclassified to cyanobacterial scums in cyanobacteria-dominated zone and submerged macrophytes in macrophyte dominated zones (Figure 11d). (W: 0.141 ± 0.014; S: 0.147 ± 0.010) As our new method considered the disturbance from turbid water via TWI (Figure 11f), original false positives for cyanobacteria blooms were rejected.

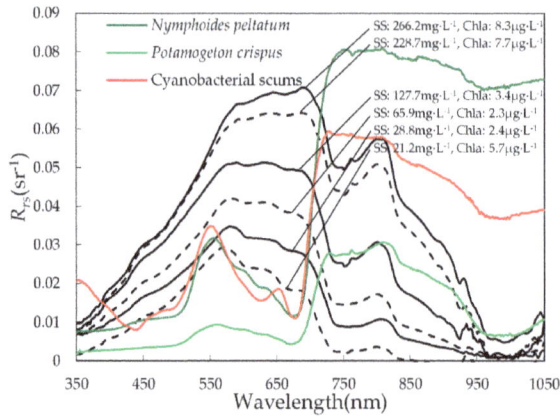

Figure 10. A comparison of in situ water leaving reflectance (Rrs) spectra for different aquatic macrophytes under different concentrations of suspended solids and cyanobacterial scums in Lake Taihu.

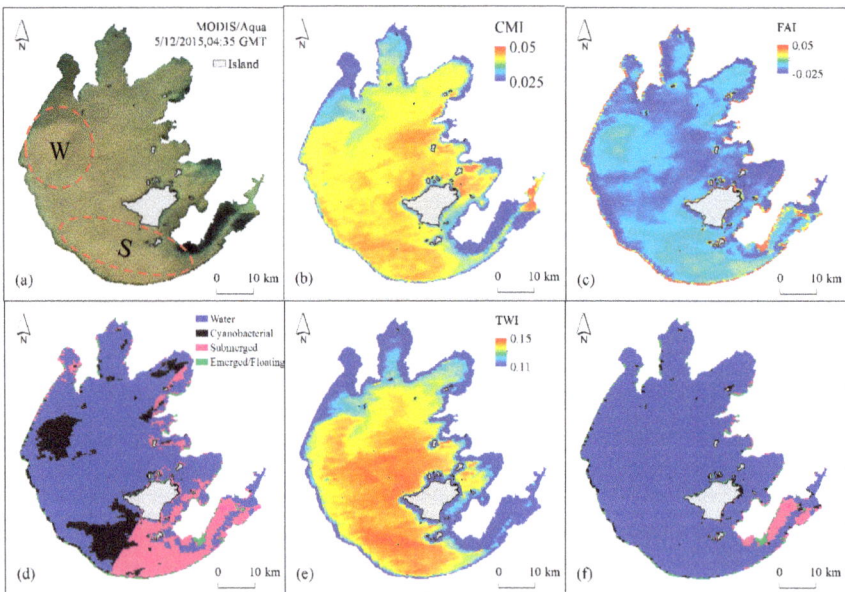

Figure 11. Spatial distributions of RGB (**a**), CMI (**b**), FAI (**c**) and TWI (**e**) based on MODIS Aqua image of 12 May 2015; and spatial distribution of lake water, cyanobacterial scums, and macrophytes in Lake Taihu before (**d**) and after (**f**) using TWI.

5.3. Impact of Black Waters

Black waters are another typical environmental phenomenon in eutrophic lakes, often associating to lake areas with cyanobacteria blooms and macrophyte stands [104]. They are characterized by high absorption due to elevated concentrations of colored dissolved organic matter (CDOM) [104,105]. In Lake Taihu, further studies on the 2007 algal bloom event indicated that the intrusion of a black water bloom in the main water intake of Wuxi City was the key cause of the water source crisis [106].

Studies show that Zhanshan Bay, Meiliang Bay, and Gonghu Bay were the main sensitive area to black waters in Lake Taihu [104,107,108], all areas of high-frequency cyanobacterial scums [47]. Attempts to use remote sensing to detect black waters have found that interference by cyanobacteria blooms and submerged macrophytes can be significant [105]. Field observations indicate that black waters in Lake Taihu have limited spatial and temporal scope, usually persisting for less than five days. They occur in areas of submerged macrophytes (i.e., *Potamogeton crispus*) from March to May [109]. By integrating the static location and growth period of submerged macrophytes, the possible interference of black waters on the remote sensing of macrophyte and bloom was achieved.

5.4. Atmospheric Effects

The impact of aerosol thickness, aerosol type, and solar/viewing geometry on classification indices was variable. The CMI value was supposed as 0.04 at AOT(555) = 0, aerosol type = r30, solz = 40°, senz = 20°, and phi = 40°. CMI was least sensitive to aerosol thickness, as only 12.1% changed when AOT(555) varied from 0 to 1 (Figure 12a). Although the effects of aerosol thickness in shorter wavelengths are stronger than those in longer wavelengths [110], the baseline subtraction of CMI, like FAI, helped reduce the effects from different aerosol thicknesses. CMI changed <0.5% and <15.2% with different aerosol types and solar/viewing geometry, respectively (Figure 12b,c). FAI has been shown to be less sensitive to changes in aerosol and observing conditions [46].

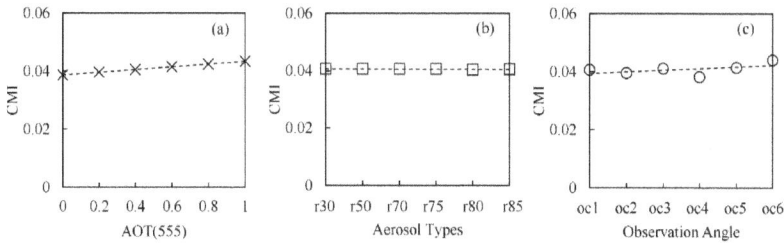

Figure 12. CMI value variation with changing atmospheric conditions: optical thickness (**a**), aerosol type (**b**) and solar/viewing geometry (**c**), based on model simulations (oc1: solz = 40°, senz = 20°, phi = 40°; oc2: solz = 40°, senz = 20°, phi = -120°; oc3: solz = 40°, senz = 40°, phi = 40°; oc4: solz = 40°, senz = 40°, phi = −120°; oc5: solz = 60°, senz = 20°, phi = −120°; oc6: solz = 60°, senz = 20°, phi = 40°).

5.5. Mixed Pixels Effects

Differences between the classification and ground-truth observations (Table 4) are likely to have resulted from scale effects. Mixed MODIS pixels containing both open lake water and cyanobacteria blooms or aquatic macrophytes were validated in field observations. Figure 13a shows that CMI values of different fractions of cyanobacteria blooms were above the CMI thresholds of 0.0455 in macrophytes-dominated zones and 0.0285 in cyanobacteria-dominated zones. This confirmed that the CMI thresholds detected cyanobacteria blooms correctly regardless of where cyanobacterial scums occurred. The CMI threshold of 0.0455 allowed for the identification of aquatic macrophytes in macrophytes-dominated zones (Figure 13b), not only emergent/floating macrophytes but also submerged macrophytes. In cyanobacteria-dominated zone, however, the CMI threshold of 0.0285 misclassified emergent/floating macrophytes as cyanobacteria blooms, and similarly submerged macrophytes were detected as cyanobacteria when submerged macrophytes coverage was lower than 33.3%. Usually, few emergent/floating macrophytes grow in cyanobacteria-dominated zones of Lake Taihu [111]. Due to this, we analyzed the influence of the CMI threshold of 0.0285 on the detection of the pixel mixing with cyanobacteria blooms and submerged macrophytes (Figure 13c). Submerged macrophytes were classified correctly only when their fraction was more than 79.2%. In macrophyte-dominated zones, the CMI threshold of 0.0455 distinguished cyanobacteria blooms and aquatic macrophytes perfectly; but in cyanobacteria-dominated zones, the CMI threshold of 0.0285 was more appropriate for bloom detection.

The FAI threshold of −0.011 correctly detected submerged macrophytes from lake water when their fractions were larger than 47.0% (Figure 14), and when the fraction of emergent/floating macrophytes was more than 23.8%, the mixed pixel could be classified as emergent/floating macrophytes due to the FAI threshold of 0.05.

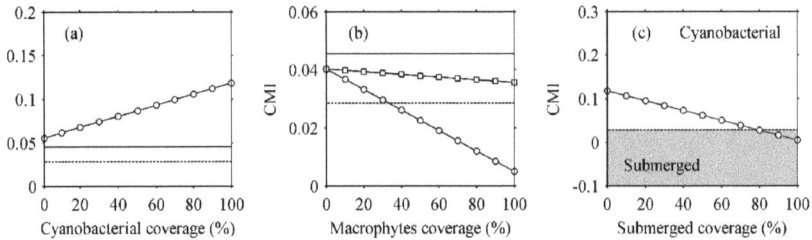

Figure 13. (a–c) CMI variation in different lake water conditions, dominated by cyanobacterial scums and aquatic macrophytes. The dotted line represents the CMI threshold between cyanobacterial scums and different aquatic macrophytes in cyanobacteria-dominated zone, and the full line represents the CMI threshold between cyanobacterial scums and different aquatic macrophytes in macrophyte-dominated zones. The gray areas represent the area recognized as submerged macrophytes by the threshold.

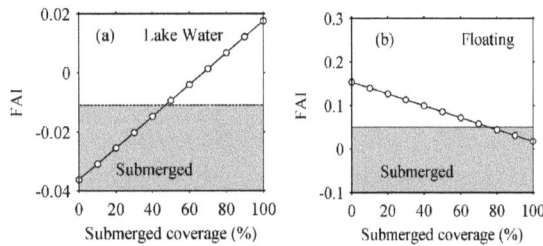

Figure 14. (a,b) FAI variation with different composition of lake water and aquatic macrophytes. The dotted line represents the FAI threshold between lake water and submerged macrophytes, and the full line represents the FAI threshold between submerged macrophytes and emergent/floating macrophytes. The gray areas represent the area recognized as submerged macrophytes by the threshold.

6. Conclusions

In this study, a new method to examine the spatial distribution of cyanobacterial scums and macrophyte vegetation dominated areas of optically complex shallow lakes was developed. The approach was validated by in situ reflectance spectra, field investigations and high spatial resolution HJ-CCD data. The results show an overall classification accuracy of 86% and a normalized classification accuracy of 86.8% with a specific accuracy of 88%, 79%, 85%, and 93% for submerged macrophytes, emergent/floating macrophytes, cyanobacterial scums, and lake water, respectively. The possibility to separate open lake water, areas dominated by cyanobacterial scums and those with different aquatic macrophytes opens new possibilities to explore the impacts of large scale changes in lake and catchment management.

The MODIS indices were first developed; TWI, CMI, and FAI were tested in the highly complex atmospheric and aquatic optical conditions of a large highly eutrophic and shallow lake in southeast China. The classification method addressed first areas of high turbidity (TWI), then separated pixels into lake waters with cyanobacterial scums or aquatic macrophytes (CMI). Lastly, FAI thresholds were used to separate cyanobacterial scums and lake water, and to classify the remaining areas into submerged macrophytes, emergent/floating macrophytes or lake water.

The new method provided high accuracy, low sensitivity to changing atmospheric conditions, high background reflectance (turbidity) or absorption (black waters), and mixed-pixels effects. The overall approach to combining specific indices in a decision tree to identify and separate lake areas with dominant optical properties can easily be extended to other lake ecosystems. The indices applied in the present study are those most appropriate to shallow eutrophic lakes, therefore most directly applicable to these more complex waterbodies. In deeper or less impacted lakes, the use of turbidity indices (TWI) is not appropriate, and more specific indices related to the lake- specific dominant optical components (e.g., CDOM) should be used. Furthermore, additional testing would be required for lakes with significantly different types of macrophytes or where macrophytes and algae are both present. Further validation to test the applicability of the new method using other satellite sensors would extend the use of this approach.

Acknowledgments: This work was supported by Key Program of the National Natural Science Foundation of China (Grant No. 41431176 & No. 41671371), the National Natural Science Foundation of China (Grant No. 41471287), National Key Technology Research and Development Program of the Ministry of Science and Technology of China during the "12th Five-Year Plan" (Grant No. 2015BAD13B06) and the ESA-MOST (China) Dragon 4 Cooperation Program project 32442. Field data were provided by Scientific Data Sharing Platform for Lake and Watershed, Nanjing Institute of Geography and Limnology, Chinese Academy of Sciences. The Rrc data of MODIS images were processed and provided by Chuanmin Hu and his optical oceanography laboratory, University of South Florida. We also express our gratitude to Minwei Zhang for his support on atmospheric transfer simulation.

Author Contributions: Y.Z. and Q.L. developed the idea for the study and its design. Q.L., J.L. and M.H. were responsible for the construction and validation of the dataset. Q.L., Y.Z., R.M. and S.L. shared in the analysis and development of the discussion. Q.L., Y.Z. and S.L. finalized the manuscript. All authors read and approved the manuscript.

Conflicts of Interest: The authors declare no conflict of interest.

References

1. Huisman, J.; Matthijs, H.C.P.; Visser, P.M.E. *Harmful Cyanobacteria*; Springer: Dordrecht, The Netherlands, 2005.
2. Granéli, E.; Turner, J.T.E. *Ecology of Harmful Algae*; Springer: Berlin/Heidelberg, Germany, 2006.
3. Guo, L. Doing battle with the green monster of Taihu Lake. *Science* **2007**, *317*. [CrossRef] [PubMed]
4. Duarte, C.M.; Middelburg, J.J.; Caraco, N. Major role of marine vegetation on the oceanic carbon cycle. *Biogeosciences* **2006**, *2*, 1–8. [CrossRef]
5. Orth, R.J.; Carruthers, T.J.B.; Dennison, W.C.; Duarte, C.M.; Fourqurean, J.W.; Heck, K.L.; Hughes, A.R.; Kendrick, G.A.; Kenworthy, W.J.; Olyarnik, S.; et al. A global crisis for seagrass ecosystems. *BioScience* **2006**, *56*, 987–996. [CrossRef]
6. Carr, J.; D'Odorico, P.; McGlathery, K.; Wiberg, P. Stability and bistability of seagrass ecosystems in shallow coastal lagoons: Role of feedbacks with sediment resuspension and light attenuation. *J. Geophys. Res.* **2010**, *115*. [CrossRef]
7. Kolada, A. The use of aquatic vegetation in lake assessment: Testing the sensitivity of macrophyte metrics to anthropogenic pressures and water quality. *Hydrobiologia* **2010**, *656*, 133–147. [CrossRef]
8. Scheffer, M.; van Nes, E.H. Shallow lakes theory revisited: Various alternative regimes driven by climate, nutrients, depth and lake size. *Hydrobiologia* **2007**, *584*, 455–466. [CrossRef]
9. Xue, K.; Zhang, Y.; Duan, H.; Ma, R.; Loiselle, S.; Zhang, M. A remote sensing approach to estimate vertical profile classes of phytoplankton in a eutrophic lake. *Remote Sens.* **2015**, *7*, 14403–14427. [CrossRef]
10. Walsby, A.E. Gas vesicles. *Microbiol. Rev.* **1994**, *58*, 94–144. [CrossRef] [PubMed]
11. Matthews, M.W. A current review of empirical procedures of remote sensing in inland and near-coastal transitional waters. *Int. J. Remote Sens.* **2011**, *32*, 6855–6899. [CrossRef]
12. Odermatt, D.; Gitelson, A.; Brando, V.E.; Schaepman, M. Review of constituent retrieval in optically deep and complex waters from satellite imagery. *Remote Sens. Environ.* **2012**, *118*, 116–126. [CrossRef]

13. Patissier, D.B.; Gower, J.F.R.; Dekker, A.G.; Phinn, S.R.; Brando, V.E. A review of ocean color remote sensing methods and statistical techniques for the detection, mapping and analysis of phytoplankton blooms in coastal and open oceans. *Prog. Oceanogr.* **2014**, *123*, 123–144. [CrossRef]

14. Peñuelas, J.; Gamon, J.A.; Griffin, K.L.; Field, C.B. Assessing community type, plant biomass, pigment composition, and photosynthetic efficiency of aquatic vegetation from spectral reflectance. *Remote Sens. Environ.* **1993**, *46*, 110–118. [CrossRef]

15. Gitelson, A.A.; Kaufman, Y.J.; Stark, R.; Rundquist, D.C. Novel algorithms for remote estimation of vegetation fraction. *Remote Sens. Environ.* **2002**, *80*, 76–87. [CrossRef]

16. Silva, T.S.; Costa, M.P.; Melack, J.M.; Novo, E.M. Remote sensing of aquatic vegetation: Theory and applications. *Environ. Monit. Assess.* **2008**, *140*, 131–145. [CrossRef] [PubMed]

17. Olmanson, L.G.; Bauer, M.E.; Brezonik, P.L. A 20-year Landsat water clarity census of Minnesota's 10,000 lakes. *Remote Sens. Environ.* **2008**, *112*, 4086–4097. [CrossRef]

18. Bresciani, M.; Adamo, M.; De Carolis, G.; Matta, E.; Pasquariello, G.; Vaičiūtė, D.; Giardino, C. Monitoring blooms and surface accumulation of cyanobacteria in the curonian lagoon by combining MERIS and ASAR data. *Remote Sens. Environ.* **2014**, *146*, 124–135. [CrossRef]

19. Duan, H.; Ma, R.; Xu, X.; Kong, F.; Zhang, S.; Kong, W.; Hao, J.; Shang, L. Two-decade reconstruction of algal blooms in China's lake Taihu. *Environ. Sci. Technol.* **2009**, *43*, 3522–3528. [CrossRef] [PubMed]

20. Kahru, M.; Savchuk, O.P.; Elmgren, R. Satellite measurements of cyanobacterial bloom frequency in the Baltic Sea: Interannual and spatial variability. *Mar. Ecol. Prog. Ser.* **2007**, *343*, 15–23. [CrossRef]

21. Kahru, M.; Elmgren, R. Multidecadal time series of satellite-detected accumulations of cyanobacteria in the Baltic Sea. *Biogeosciences* **2014**, *11*, 3619–3633. [CrossRef]

22. Groom, S.B.; Holligan, P.M. Remote sensing of coccolithophore blooms. *Adv. Space Res.* **1987**, *7*, 73–78. [CrossRef]

23. Duan, H.; Zhang, S.; Zhang, Y. Cyanobacteria bloom monitoring with remote sensing in Lake Taihu. *J. Lake Sci.* **2008**, *20*, 145–152. (In Chinese)

24. Ma, R.; Kong, F.; Duan, H.; Zhang, S.; Kong, W.; Hao, J. Spatiotemporal distribution of cyanobacterial scums based on satellite imageries in Lake Taihu, China. *J. Lake Sci.* **2008**, *20*, 687–694. (In Chinese)

25. Son, Y.B.; Min, J.E.; Ryu, J.H. Detecting massive green algae (*Ulva prolifera*) blooms in the Yellow Sea and East China Sea using geostationary ocean color imager (GOCI) data. *Ocean Sci. J.* **2012**, *47*, 359–375. [CrossRef]

26. Stumpf, R.P.; Tomlinson, M.C. Remote sensing of harmful algal blooms. *Remote Sens. Coast. Aquat. Environ.* **2005**, *7*, 277–296.

27. Holligan, P.M.; Viollier, M.; Harbour, D.S.; Camus, P.; Philippe, M.C. Satellite and ship studies of coccolithophore production along a continental shelf edge. *Nature* **1983**, *304*, 339–342. [CrossRef]

28. Gitelson, A.; Karnieli, A.; Goldman, N.; Yacobi, Y.Z.; Mayo, M. Chlorophyll estimation in the southeastern Mediterranean using CZCS images: Adaption of an algorithm and its validation. *J. Mar. Syst.* **1996**, *9*, 283–290. [CrossRef]

29. Kopelevich, O.V.; Sheberstov, S.V.; Yunev, O.; Basturk, O.; Finenko, Z.Z.; Nikonov, S.; Vedernikov, V.I. Surface chlorophyll in the Black Sea over 1978–1986 derived from satellite and in situ data. *J. Mar. Syst.* **2002**, *36*, 145–160. [CrossRef]

30. Gower, J.F.R. Red tide monitoring using AVHRR HRPT imagery from a local receiver. *Remote Sens. Environ.* **1994**, *48*, 309–318. [CrossRef]

31. Kahru, M.; Mitchell, B.G.; Diaz, A.; Miura, M. MODIS detects a devastating algal bloom in Paracas Bay, Peru. *EOS Trans. Am. Geophys. Union* **2004**, *85*, 465–472. [CrossRef]

32. Rouse, J.W.; Haas, R.H.; Schell, J.A.; Deering, D.W. Monitoring vegetation systems in the Great Plains with ERTS-1. In *3rd Earth Resources Technology Satellite Symposium*; NASA: Washington, DC, USA, 1973; pp. 309–317.

33. Hu, C.; He, M. Origin and offshore extent of floating algae in Olympic sailing area. *EOS Trans. Am. Geophys. Union* **2008**, *89*, 302–303. [CrossRef]

34. Garcia, R.A.; Fearns, P.; Keesing, J.K.; Liu, D. Quantification of floating macroalgae blooms using the scaled algae index. *J. Geophys. Res. Oceans* **2013**, *118*, 26–42. [CrossRef]

35. Prangsma, G.J.; Roozekrans, J.N. Using NOAA AVHRR imagery in assessing water quality parameters. *Int. J. Remote Sens.* **1989**, *10*, 811–818. [CrossRef]

36. Huete, A.; Justice, C.; Leeuwen, W.V. *MODIS Vegetation Index (MOD13) Algorithm Theoretical Basis Document (Ver 3.0)*; NASA: Washington, DC, USA, 1999.

37. Wynne, T.T.; Stumpf, R.P.; Tomlinson, M.C.; Dybleb, J. Characterizing a cyanobacterial bloom in western Lake Erie using satellite imagery and meteorological data. *Limnol. Oceanogr.* **2010**, *55*, 2025–2036. [CrossRef]

38. Gower, J.F.R.; Doerffer, R.; Borstad, G.A. Interpretation of the 685nm peak in water-leaving radiance spectra in terms of fluorescence, absorption and scattering, and its observation by MERIS. *Int. J. Remote Sens.* **1999**, *20*, 1771–1786. [CrossRef]

39. Gower, J.; King, S.; Borstad, G.; Brown, L. Detection of intense plankton blooms using the 709 nm band of the MERIS imaging spectrometer. *Int. J. Remote Sens.* **2005**, *26*, 2005–2012. [CrossRef]

40. Wynne, T.T.; Stumpf, R.P.; Tomlinson, M.C.; Warner, R.A.; Tester, P.A.; Dyble, J.; Fahnenstiel, G.L. Relating spectral shape to cyanobacterial scums in the Laurentian Great Lakes. *Int. J. Remote Sens.* **2008**, *29*, 3665–3672. [CrossRef]

41. Wynne, T.T.; Stumpf, R.P.; Briggs, T.O. Comparing MODIS and MERIS spectral shapes for cyanobacterial bloom detection. *Int. J. Remote Sens.* **2013**, *34*, 6668–6678. [CrossRef]

42. Wynne, T.T.; Stumpf, R.P.; Tomlinson, M.C.; Fahnenstiel, G.L.; Dyble, J.; Schwab, D.J.; Joshi, S.J. Evolution of a cyanobacterial bloom forecast system in western lake Erie: Development and initial evaluation. *J. Great Lakes Res.* **2013**, *39*, 90–99. [CrossRef]

43. Stumpf, R.P.; Wynne, T.T.; Baker, D.B.; Fahnenstiel, G.L. Interannual variability of cyanobacterial scums in lake Erie. *PLoS ONE* **2012**, *7*. [CrossRef] [PubMed]

44. Matthews, M.W.; Bernard, S.; Robertson, L. An algorithm for detecting trophic status (chlorophyll-a), cyanobacterial-dominance, surface scums and floating vegetation in inland and coastal waters. *Remote Sens. Environ.* **2012**, *124*, 637–652. [CrossRef]

45. Matthews, M.W.; Odermatt, D. Improved algorithm for routine monitoring of cyanobacteria and eutrophication in inland and near-coastal waters. *Remote Sens. Environ.* **2015**, *156*, 374–382. [CrossRef]

46. Hu, C. A novel ocean color index to detect floating algae in the global oceans. *Remote Sens. Environ.* **2009**, *113*, 2118–2129. [CrossRef]

47. Hu, C.H.; Lee, Z.L.; Ma, R.M.; Yu, K.; Li, D. Moderate resolution imaging spectroradiometer (MODIS) observations of cyanobacteria blooms in Taihu Lake, China. *J. Geophys. Res.* **2010**, *115*. [CrossRef]

48. Hu, C.; Cannizzaro, J.; Carder, K.L.; Karger, F.E.M.; Hardya, R. Remote detection of *Trichodesmium* blooms in optically complex coastal waters: Examples with MODIS full-spectral data. *Remote Sens. Environ.* **2010**, *114*, 2048–2058. [CrossRef]

49. Huang, C.; Li, Y.; Yang, H.; Sun, D.; Yu, Z.; Zhang, Z.; Chen, X.; Xu, L. Detection of algal bloom and factors influencing its formation in Taihu lake from 2000 to 2011 by MODIS. *Environ. Earth Sci.* **2014**, *71*, 3705–3714. [CrossRef]

50. Zhang, Y.; Ma, R.; Zhang, M.; Duan, H.; Loiselle, S.; Xu, J. Fourteen-year record (2000–2013) of the spatial and temporal dynamics of floating algae blooms in lake Chaohu, observed from time series of MODIS images. *Remote Sens.* **2015**, *7*, 10523–10542. [CrossRef]

51. Alem, A.E.; Chokmani, K.; Laurion, I.; Adlouni, S.E. An adaptive model to monitor chlorophyll-a in inland waters in southern Quebec using downscaled MODIS imagery. *Remote Sens.* **2014**, *6*, 6446–6471. [CrossRef]

52. Work, E.A.; Gilmer, D.S. Utilization of satellite data for inventorying prairie ponds and potholes. *Photogramm. Eng. Remote Sens.* **1976**, *42*, 685–694.

53. Kempka, R.G.; Kollasch, R.P.; Koeln, G.T. Ducks unlimited: Using GIS to preserve the pacific flyway's wetland resource. *GIS World* **1992**, *5*, 46–52.

54. Jakubauskas, M.; Kindscher, K.; Debinski, D. Multitemporal characterization and mapping of montane sagebrush communities using Indian IRS LISS-II imagery. *Geocarto Int.* **1998**, *13*, 65–74. [CrossRef]

55. Macleod, R.D.; Congalton, R.G. A quantitative comparison of change-detection algorithms for monitoring eelgrass from remotely sensed data. *Photogramm. Eng. Remote Sens.* **1998**, *64*, 207–216.

56. Gullström, M.; Lundén, B.; Bodin, M.; Kangwe, J.; Öhman, M.C.; Mtolera, M.S.P.; Björk, M. Assessment of changes in the seagrass-dominated submerged vegetation of tropical Chwaka bay (Zanzibar) using satellite remote sensing. *Estuar. Coast. Shelf Sci.* **2006**, *67*, 399–408. [CrossRef]

57. Dogan, O.K.; Akyurek, Z.; Beklioglu, M. Identification and mapping of submerged plants in a shallow lake using Quickbird satellite data. *J. Environ. Manag.* **2009**, *90*, 2138–2143. [CrossRef] [PubMed]

58. Hewitt, M.J. Synoptic inventory of riparian ecosystems: The utility of Landsat thematic mapper data. *For. Ecol. Manag.* **1990**, *33–34*, 605–620. [CrossRef]

59. MacAlister, C.; Mahaxay, M. Mapping wetlands in the lower Mekong Basin for wetland resource and conservation management using Landsat ETM images and field survey data. *J. Environ. Manag.* **2009**, *90*, 2130–2137. [CrossRef] [PubMed]

60. De Colstoun, E.B. National park vegetation mapping using multitemporal Landsat 7 data and a decision tree classifier. *Remote Sens. Environ.* **2003**, *85*, 316–327. [CrossRef]

61. Baker, C.; Lawrence, R.; Montagne, C.; Patten, D. Mapping wetlands and riparian areas using Landsat ETM+ imagery and decision-tree-based models. *Wetlands* **2006**, *26*, 465–474. [CrossRef]

62. Wright, C.; Gallant, A. Improved wetland remote sensing in Yellowstone national park using classification trees to combine tm imagery and ancillary environmental data. *Remote Sens. Environ.* **2007**, *107*, 582–605. [CrossRef]

63. Davranche, A.; Lefebvre, G.; Poulin, B. Wetland monitoring using classification trees and SPOT-5 seasonal time series. *Remote Sens. Environ.* **2010**, *114*, 552–562. [CrossRef]

64. Zhao, D.; Jiang, H.; Yang, T.; Cai, Y.; Xu, D.; An, S. Remote sensing of aquatic vegetation distribution in Taihu lake using an improved classification tree with modified thresholds. *J. Environ. Manag.* **2012**, *95*, 98–107. [CrossRef] [PubMed]

65. Zhao, D.; Lv, M.; Jiang, H.; Cai, Y.; Xu, D.; An, S. Spatio-temporal variability of aquatic vegetation in Taihu lake over the past 30 years. *PLoS ONE* **2013**, *8*. [CrossRef] [PubMed]

66. Luo, J.; Ma, R.; Duan, H.; Hu, W.; Zhu, J.; Huang, W.; Lin, C. A new method for modifying thresholds in the classification of tree models for mapping aquatic vegetation in Taihu Lake with satellite images. *Remote Sens.* **2014**, *6*, 7442–7462. [CrossRef]

67. Oyama, Y.; Matsushita, B.; Fukushima, T. Distinguishing surface cyanobacterial scums and aquatic macrophytes using Landsat/TM and ETM+ shortwave infrared bands. *Remote Sens. Environ.* **2015**, *157*, 35–47. [CrossRef]

68. Liu, X.; Zhang, Y.; Shi, K.; Zhou, Y.; Tang, X.; Zhu, G.; Qin, B. Mapping aquatic vegetation in a large, shallow eutrophic lake: A frequency-based approach using multiple years of MODIS data. *Remote Sens.* **2015**, *7*, 10295–10320. [CrossRef]

69. Zhang, Y.; Liu, X.; Qin, B.; Shi, K.; Deng, J.; Zhou, Y. Aquatic vegetation in response to increased eutrophication and degraded light climate in eastern lake Taihu: Implications for lake ecological restoration. *Sci. Rep.* **2016**, *6*. [CrossRef] [PubMed]

70. Gao, B.C. NDWI—A normalized difference water index for remote sensing of vegetation liquid water from space. *Remote Sens. Environ.* **1996**, *58*, 257–266. [CrossRef]

71. Rogers, A.; Kearney, M. Reducing signature variability in unmixing coastal marsh thematic mapper scenes using spectral indices. *Int. J. Remote Sens.* **2004**, *25*, 2317–2335. [CrossRef]

72. Qi, L.; Hu, C.; Duan, H.; Cannizzaro, J.; Ma, R. A novel meris algorithm to derive cyanobacterial phycocyanin pigment concentrations in a eutrophic lake: Theoretical basis and practical considerations. *Remote Sens. Environ.* **2014**, *154*, 298–317. [CrossRef]

73. Qi, L.; Hu, C.; Duan, H.; Barnes, B.; Ma, R. An EOF-based algorithm to estimate chlorophyll a concentrations in Taihu Lake from MODIS land-band measurements: Implications for near real-time applications and forecasting models. *Remote Sens.* **2014**, *6*, 10694–10715. [CrossRef]

74. Palmer, S.C.J.; Hunter, P.D.; Lankester, T.; Hubbard, S.; Spyrakos, E.; Tyler, A.N.; Présing, M.; Horváth, H.; Lamb, A.; Balzter, H.; et al. Validation of ENVISAT Meris algorithms for chlorophyll retrieval in a large, turbid and optically-complex shallow lake. *Remote Sens. Environ.* **2015**, *157*, 158–169. [CrossRef]

75. Hu, C.; Chen, Z.; Clayton, T.D.; Swarzenski, P.; Brock, J.C.; Muller-Karger, F.E. Assessment of estuarine water-quality indicators using MODIS medium-resolution bands: Initial results from Tampa bay, FL. *Remote Sens. Environ.* **2004**, *93*, 423–441. [CrossRef]

76. Le, C.; Hu, C.; English, D.; Cannizzaro, J.; Chen, Z.; Feng, L.; Boler, R.; Kovach, C. Towards a long-term chlorophyll-a data record in a turbid estuary using MODIS observations. *Prog. Oceanogr.* **2013**, *109*, 90–103. [CrossRef]

77. Qin, B.Q.; Hu, W.P.; Chen, W.M. *Process and Mechanism of Environmental Changes of the Taihu Lake*; Science Press: Beijing, China, 2004.

78. Qin, B.; Xu, P.; Wu, Q.; Luo, L.; Zhang, Y. Environmental issues of lake Taihu, China. *Hydrobiologia* **2007**, *581*, 3–14. [CrossRef]

79. Ma, R.; Duan, H.; Gu, X.; Zhang, S. Detecting aquatic vegetation changes in Taihu lake, China using multi-temporal satellite imagery. *Sensors* **2008**, *8*, 3988–4005. [CrossRef] [PubMed]

80. Kalff, J. *Limnology: Inland Water Ecosystems*; Prentice Hall: Upper Saddle River, NJ, USA, 2002.

81. Lei, Z. Study on Aquatic Macrophyte Vegetations and Their Environment Effects in Taihu Lake. Ph.D. Thesis, Jinan University, Jinan, China, 2006.

82. Mueller, J.L.; Fargion, G.S.; McClain, C.R. *Ocean Optics Protocols for Satellite Ocean Color Sensor Validation, Revision 4. Volume VI: Special Topics in Ocean Optics Protocols and Appendices*; NASA Goddard Space Flight Center: Greenbelt, MD, USA, 2003.

83. Mobley, C.D. Estimation of the remote-sensing reflectance from above-surface measurements. *Appl. Opt.* **1999**, *38*, 7442–7455. [CrossRef] [PubMed]

84. Ahn, Y.H.; Shanmugam, P. Detecting the red tide algal blooms from satellite ocean color observations in optically complex Northeast-Asia coastal waters. *Remote Sens. Environ.* **2006**, *103*, 419–437. [CrossRef]

85. Dev, P.J.; Shanmugam, P. A new theory and its application to remove the effect of surface-reflected light in above-surface radiance data from clear and turbid waters. *J. Quant. Spectrosc. Radiat. Transf.* **2014**, *142*, 75–92. [CrossRef]

86. Sun, D.; Hu, C.; Qiu, Z.; Shi, K. Estimating phycocyanin pigment concentration in productive inland waters using Landsat measurements: A case study in lake Dianchi. *Opt. Exp.* **2015**, *23*, 3055–3074. [CrossRef] [PubMed]

87. NASA's OceanColor Web. Available online: http://oceancolor.gsfc.nasa.gov (accessed on 16 December 2016).

88. Gordon, H.R.; Clark, D.K. Clear water radiances for atmospheric correction of coastal zone color scanner imagery. *Appl. Opt.* **1981**, *20*, 4175–4180. [CrossRef] [PubMed]

89. Ruddick, K.G.; Ovidio, F.; Rijkeboer, M. Atmospheric correction of SeaWIFS imagery for turbid coastal and inland waters. *Appl. Opt.* **2000**, *39*, 897–912. [CrossRef] [PubMed]

90. Wang, M.; Shi, W. The NIR-SWIR combined atmospheric correction approach for MODIS ocean color data processing. *Opt. Exp.* **2007**, *15*, 15722–15733. [CrossRef]

91. Bailey, S.W.; Franz, B.A.; Werdell, P.J. Estimation of near-infrared water-leaving reflectance for satellite ocean color data processing. *Opt. Express* **2010**, *18*, 7521–7527. [CrossRef] [PubMed]

92. Reinart, A.; Herlevi, A.; Arst, H.; Sipelgas, L. Preliminary optical classification of lakes and coastal waters in Estonia and South Finland. *J. Sea Res.* **2003**, *49*, 357–366. [CrossRef]

93. Gitelson, A.A.; Dall'Olmo, G.; Moses, W.M.; Rundquist, D.C.; Barrow, T.; Fisher, T.R.; Gurlin, D.; Holz, J. A simple semi-analytical model for remote estimation of chlorophyll-a in turbidwaters: Validation. *Remote Sens. Environ.* **2008**, *112*, 3582–3593. [CrossRef]

94. Bricaud, A.; Roesler, C.; Zaneveld, J.R.V. In situ methods for measuring the inherent optical properties of ocean waters. *Limnol. Oceanogr.* **1995**, *40*, 393–410. [CrossRef]

95. Dekker, A.G.; Malthus, T.J.; Goddijn, L.M. Monitoring cyanobacteria in eutrophic waters using airborne imaging spectroscopy and multispectral remote sensing systems. In Proceedings of the 6th Australasian Remote Sensing Conference, Wellington, NZ, USA, 2–6 November 1992; pp. 204–214.

96. Le, C.; Li, Y.; Zha, Y.; Sun, D.; Huang, C.; Lu, H. A four-band semi-analytical model for estimating chlorophyll a in highly turbid lakes: The case of Taihu Lake, China. *Remote Sens. Environ.* **2009**, *113*, 1175–1182. [CrossRef]

97. Hoffer, R.M. Biological and physical considerations in applying computer-aided analysis techniques to remote sensor data. In *Remote Sensing: The Quantitative Approach*; Swain, P.H., Davis, S.M., Eds.; McGraw-Hill: New York, NY, USA, 1978; pp. 227–289.

98. Feng, L.; Hu, C.; Chen, X.; Cai, X.; Tian, L.; Chen, L. Human induced turbidity changes in Poyang Lake between 2000 and 2010: Observations from MODIS. *J. Geophys. Res.* **2012**, *117*. [CrossRef]

99. Foody, G.M. Local characterization of thematic classification accuracy through spatially constrained confusion matrices. *Int. J. Remote Sens.* **2005**, *26*, 1217–1228. [CrossRef]

100. Congalton, R.G. A review of assessing the accuracy of classifications of remotely sensed data. *Remote Sens. Environ.* **1991**, *37*, 35–46. [CrossRef]

101. New Hampshire View Web. Available online: http://www.nhview.unh.edu/accuracyprograms.html (accessed on 16 December 2016).

102. Wang, M.; Shi, W. Cloud masking for ocean color data processing in the coastal regions. *IEEE Trans. Geosci. Remote Sens.* **2006**, *44*, 3196–3205. [CrossRef]

103. Ma, R.; Duan, H.; Liu, Q.; Loiselle, S.A. Approximate bottom contribution to remote sensing reflectance in Taihu Lake, China. *J. Great Lakes Res.* **2011**, *37*, 18–25. [CrossRef]

104. Duan, H.; Ma, R.; Loiselle, S.A.; Shen, Q.; Yin, H.; Zhang, Y. Optical characterization of black water blooms in eutrophic waters. *Sci. Total Environ.* **2014**, *482–483*, 174–183. [CrossRef] [PubMed]

105. Zhao, J.; Hu, C.; Lapointe, B.; Melo, N.; Johns, E.; Smith, R. Satellite-observed black water events off southwest florida: Implications for coral reef health in the Florida keys national marine sanctuary. *Remote Sens.* **2013**, *5*, 415–431. [CrossRef]

106. Yang, M.; Yu, J.W.; Li, Z.L.; Guo, Z.H.; Burch, M.; Lin, T.F. Taihu lake not to blame for Wuxi's woes. *Science* **2008**, *319*. [CrossRef] [PubMed]

107. Lu, G.; Ma, Q.; Zhang, J. Analysis of black water aggregation in Taihu Lake. *Water Sci. Eng.* **2011**, *4*, 374–385.

108. Lei, Z.; Bing, Z.; Junsheng, L.; Qian, S.; Fangfang, Z.; Ganlin, W. A study on retrieval algorithm of black water aggregation in Taihu Lake based on HJ-1 satellite images. In *IOP Conference Series: Earth and Environmental Science*; IOP Publishing: Bristol, UK, 2014.

109. Mi, W.J.; Zhu, D.W.; Zhou, Y.Y.; Zhou, H.D.; Yang, T.W.; Hamilton, D.P. Influence of *Potamogeton crispus* growth on nutrients in the sediment and water of lake Tangxunhu. *Hydrobiologia* **2007**, *603*, 139–146. [CrossRef]

110. Antoine, D.; Morel, A. Relative importance of multiple scattering by air molecules and aerosols in forming the atmospheric path radiance in the visible and near-infrared parts of the spectrum. *Appl. Opt.* **1998**, *37*, 2245–2259. [CrossRef] [PubMed]

111. Zhao, D.; Jiang, H.; Cai, Y.; An, S. Artificial regulation of water level and its effect on aquatic macrophyte distribution in Taihu lake. *PLoS ONE* **2012**, *7*. [CrossRef] [PubMed]

![remote sensing logo] *remote sensing*

MDPI

Article

Spatio-Temporal Change of Lake Water Extent in Wuhan Urban Agglomeration Based on Landsat Images from 1987 to 2015

Yue Deng [1,2], Weiguo Jiang [1,2,*], Zhenghong Tang [3], Jiahong Li [4,*], Jinxia Lv [1,2], Zheng Chen [1,2] and Kai Jia [1,2]

[1]   Academy of Disaster Reduction and Emergency Management, Beijing Normal University, Beijing 100875, China; dengyue@mail.bnu.edu.cn (Y.D.); 201621480076@mail.bnu.edu.cn (J.L.); lubyn@126.com (Z.C.); sokee_studio@yeah.net (K.J.)
[2]   Key Laboratory of Environmental Change and Natural Disaster, Beijing Normal University, Beijing 100875, China
[3]   Community and Regional Planning Program, University of Nebraska-Lincoln, Lincoln, NE 68588, USA; ztang2@unl.edu
[4]   National Remote Sensing Center of China, MOST, Beijing 100036, China
*   Correspondence: jiangweiguo@bnu.edu.cn (W.J.); lijiahong@nrscc.gov.cn (J.L.)

Academic Editors: Yunlin Zhang, Claudia Giardino, Linhai Li, Xiaofeng Li and Prasad S. Thenkabail
Received: 22 November 2016; Accepted: 14 March 2017; Published: 15 March 2017

Abstract: Urban lakes play an important role in urban development and environmental protection for the Wuhan urban agglomeration. Under the impacts of urbanization and climate change, understanding urban lake-water extent dynamics is significant. However, few studies on the lake-water extent changes for the Wuhan urban agglomeration exist. This research employed 1375 seasonally continuous Landsat TM/ETM+/OLI data scenes to evaluate the lake-water extent changes from 1987 to 2015. The random forest model was used to extract water bodies based on eleven feature variables, including six remote-sensing spectral bands and five spectral indices. An accuracy assessment yielded a mean classification accuracy of 93.11%, with a standard deviation of 2.26%. The calculated results revealed the following: (1) The average maximum lake-water area of the Wuhan urban agglomeration was 2262.17 km^2 from 1987 to 2002, and it decreased to 2020.78 km^2 from 2005 to 2015, with a loss of 241.39 km^2 (10.67%). (2) The lake-water areas of loss of Wuhan, Huanggang, Xianning, and Xiaogan cities, were 114.83 km^2, 44.40 km^2, 45.39 km^2, and 31.18 km^2, respectively, with percentages of loss of 14.30%, 11.83%, 13.16%, and 23.05%, respectively. (3) The lake-water areas in the Wuhan urban agglomeration were 226.29 km^2, 322.71 km^2, 460.35 km^2, 400.79 km^2, 535.51 km^2, and 635.42 km^2 under water inundation frequencies of 5%–10%, 10%–20%, 20%–40%, 40%–60%, 60%–80%, and 80%–100%, respectively. The Wuhan urban agglomeration was approved as the pilot area for national comprehensive reform, for promoting resource-saving and environmentally friendly developments. This study could be used as guidance for lake protection and water resource management.

Keywords: urban lake water; water inundation frequency; Landsat; random forest; Wuhan urban agglomeration; climate change

---

1. Introduction

Urban lakes provide numerous ecosystem services that are closely related to human well being in both the present and future. These ecosystem services include agricultural production, fishery resources, flood mitigation, water storage, and entertainment locations [1,2]. All of these ecosystem services are affected by fluctuations in the lake-water extent over time and space [3]. Because the urban lake-water extent has been dramatically impacted due to urbanization, understanding the spatiotemporal patterns of lake-water extent dynamics is necessary for sustainable urban development.

The Wuhan urban agglomeration—consisting of Wuhan, Huangshi, Ezhou, Xiaogan, Huanggang, Xianning, Xiantao, Qianjiang, and Tianmen—is one of five national urban agglomerations and has extensive lake-water extent, due to its wet climate. For example, Wuhan is nicknamed "Water City" (Jiang Cheng). Over the past three decades, the Wuhan urban agglomeration has experienced a rapid expansion and has faced increasing challenges in protecting the lake water, both qualitatively and quantitatively, especially for Wuhan city [4–9]. However, few studies have focused on the whole urban agglomeration. Moreover, these studies adopted imaging data for several specific years. Obviously, mapping lake-water extent dynamics is unreasonable without considering the ephemerality in the lake-water extent. A clear understanding of the spatio-temporal change of the lake-water extent in the Wuhan urban agglomeration is thus important.

Remote-sensing data have been widely used for mapping the lake-water extent over time and space [10–17], with Landsat image data being one of the most common data types for monitoring and analyzing long-term lake-water extent changes, due to their high spatial resolution (30–60 m) and long data record [11–16].

Two major methods exist for extracting the water extent, based on remote-sensing data. One approach is the use of water indices, such as the Modified Normalized Difference Water Index (MNDWI) [18], the Normalized Difference Water Index (NDWI) [19], and the Automated Water Extraction Index (AWEI) [20]. However, an ideal single threshold for water indices, in order to distinguish between water bodies and non-water bodies, is difficult to determine because the spectral signature of water varies in space and time [13]. Automatic threshold methods have been adopted for determining the optimal value, like Otsu [21,22]. Some comparative analyses between these water indices have been conducted [23,24], but the best water indices cannot be determined. The other method is classification based on a combination of spectral bands and other variables (e.g., water indices, shape features) [25]. These classification methods include Support Vector Machines (SVM), Maximum Likelihood (ML) and Decision Tree (DT), et al. [26,27]. It is important to note that a best classification method doesn't exist for all cases. For long-term seasonally continuous water mapping in a large area, DTs are more popular, primarily for their ease of application [26]. However, the DT method performs less well than algorithms such as SVM and neural networks. But, the performance of Random Forest (RF), an improved implementation of DT, is superior to a traditional DT [26,27]. Therefore, some studies employed the RF to classify the continuous remote images [13,25,28]. For example, Tulbure et al. adopted RF to produce continuous surface water time series at a subcontinental scale in an Australian semi-arid region, the Murray-Darling Basin [13].

With an increasing amount of available remote sensing data, the mapping of seasonally continuous lake water is possible. This study employed a method that utilizes 1375 seasonally continuous Landsat Thematic Mapper (TM)/Enhanced Thematic Mapper Plus (ETM+)/ Operational Land Imager (OLI) data scenes from 1987 to 2015, to map the long-term lake-water extent in the Wuhan urban agglomeration. This study is the first to evaluate the lake-water extent dynamics in space and time in the Wuhan urban agglomeration, based on seasonally continuous Landsat images. The specific objectives are: (1) to obtain the long-term seasonally continuous lake water extent; (2) to map the water inundation frequency to explore the spatial distribution of the lake-water extent; (3) to analyze the temporal variation of the lake-water extent; and (4) to explore the impacts of urbanization and climate change on the lake-water extent. This study could be used as guidance for lake protection and water resource management for the Wuhan urban agglomeration.

2. Materials

2.1. Study Area

The Wuhan urban agglomeration, located in the eastern part of the Hubei province, China, covers an area of 580,151.9 km^2 (Figure 1). It consists of Wuhan, Huangshi, Ezhou, Xiaogan, Huanggang, Xianning, Xiantao, Qianjiang, and Tianmen and was approved by the Chinese Government in 2008. As one of five national urban agglomerations, the Wuhan urban agglomeration plays a critical role in central China. On the one hand, the Wuhan urban agglomeration has witnessed rapid urban expansion. For example, the built-up area in Wuhan city increased by 10.9% from 1990–2000 and increased by 154.5% from 2000–2013 [7]. On the other hand, rapid urban expansion has encroached upon lake water bodies. Lake wetlands and marsh wetlands decreased by 18.71% and 50.3%, respectively, from 1987–2005.

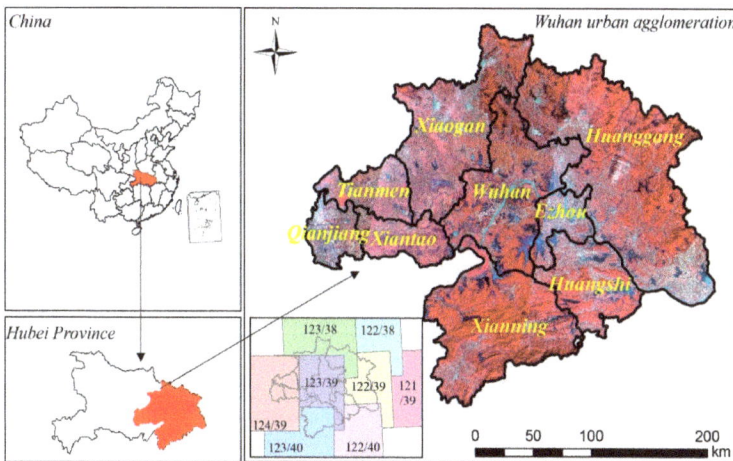

Figure 1. The location of the study area.

2.2. Study Data

2.2.1. Landsat Time Series

We used 1987–2015 data from the Landsat 4–5 TM, Landsat 7 Enhanced ETM+, and Landsat-8 OLI. We downloaded level 1 terrain-corrected (L1T) images with a cloud cover assessment threshold of ≤40% for the study period, from USGS/EROS (http://landsat.usgs.gov/).

The Wuhan urban agglomeration is located within eight scenes (Path/Row: 121/39, 122/38, 122/39, 122/40, 123/38, 123/39, 123/40, and 124/39), and Table 1 lists the number of the Landsat TM, ETM+, and OLI scenes processed per path/row. A total of 1375 scenes were used, including 1019 scenes from the TM from 1987–2011, 180 scenes from the ETM+ from 1999–2003, and 176 scenes from the OLI from 2013–2015. After mid–late 2003, we did not use the scenes from the ETM+ with the Scan-Line-Corrector-Off (SLC-Off) problem, because this problem creates data gaps in the ETM+ imagery, thereby leading to missing data. Since 2013, the Landsat 8 OLI datasets were utilized because Landsat 5 was terminated in 2011. The temporal distribution of the adopted Landsat images for the study area is shown in Figure 2. Although these images are unevenly distributed in time, they can generally be used for the long-term mapping of surface water. Overall, the available time series were relatively adequate for our study.

Table 1. Number of Landsat TM, ETM, and OLI scenes processed per path/row.

Sensor	Path/Row								
	121/39	**122/38**	**122/39**	**122/40**	**123/38**	**123/39**	**123/40**	**124/39**	**Total**
TM (1987–2011)	122	130	141	124	139	129	115	119	1019
ETM (1999–2003)	26	23	23	19	24	22	22	21	180
OLI (2013–2015)	25	22	20	25	26	23	21	14	176
Total	173	175	184	168	189	175	158	154	1375

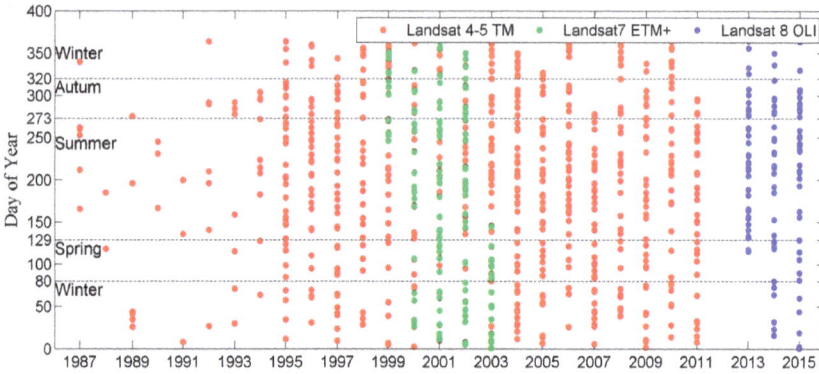

Figure 2. Temporal distribution (day of year) of Landsat images used in this study.

2.2.2. Landuse, Precipitation and DEM data

The landuse data acquired in 1995 and 2009 for the Wuhan urban agglomeration was collected, to analyze the impacts of urbanization on the lake area. The data were collected from the LUCC database of China. Moreover, to explore the correlation between the lake-water extent and precipitation, daily precipitation levels from eight stations around the Wuhan urban agglomeration from 1987–2015 were collected from the China Meteorological Data Sharing System. Then, the average precipitations of the eight stations were calculated. Moreover, the Advanced Spaceborne Thermal Emission and Reflection Radiometer Global Digital Elevation Model (ASTE GDEM) data with a 30-m spatial resolution were collected from the Geospatial Data Cloud (http://www.gscloud.cn), to aid in lake-water extraction.

3. Methods

3.1. Flowchart

The flowchart of extracting the lake-water extent in this study is displayed in Figure 3. It mainly consisted of five parts: the collection of feature variables, collection of training samples, image classification based on the random forest model, rule filter, and accuracy assessment. The details are described in the following sections.

3.2. Collection of Feature Variables

First, to diminish the negative influence from the images acquired over different locations and times, as well as the solar irradiance, etc. [13,25], we converted digital numbers (DN) to Top of Atmosphere (TOA) reflectance, according to previous research [29]. Bands 1–5 and 7 from the TM/ETM+, and bands 2–7 from the OLI, were processed. Second, MNDWI [18], NDWI [19], EVI [30], NDVI [31], and NDMI [32] were calculated, based the TOA reflectance. These spectral indices are common for water extraction [11,33]. Finally, a total of six bands and five spectral indexes were used in the model.

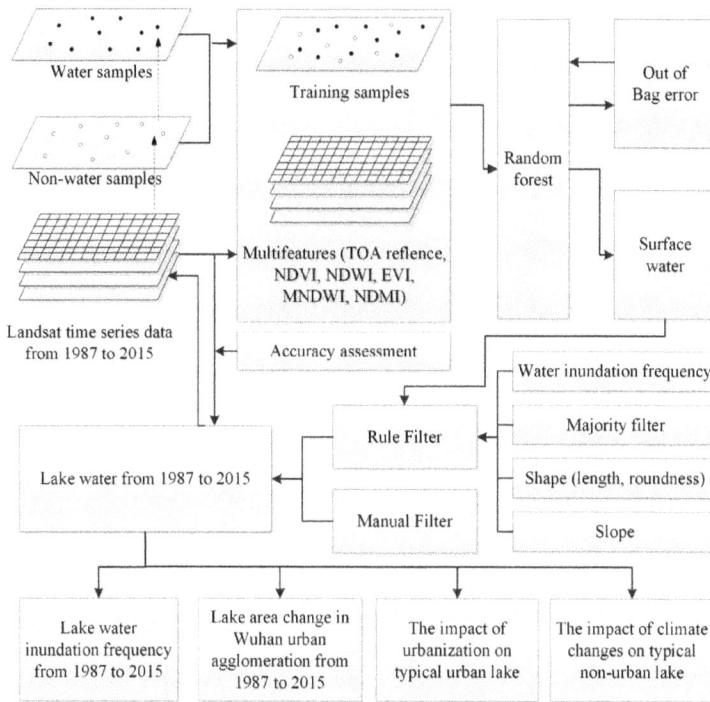

Figure 3. Flowchart of the study.

3.3. Collection of Training Samples

We collected training samples through the visual interpretation of the Landsat TM, ETM+, and OLI images. In addition, the five calculated spectral indices were also used to improve the confidence. The selected training samples were acquired from different regions, years, seasons, and land-cover types, to reduce possible biases [13,28]. We reclassified these training samples into water pixels (1; i.e., reservoirs, rivers, lakes) and non-water pixels (0; i.e., building land, vegetation, cropland, bare land, cloud). The research used 567,320 pixels (46,041 water pixels and 521,279 non-water pixels) from the TM, 235,663 pixels (63,272 water pixels and 172,391 non-water pixels) from the ETM+, and 243,421 pixels (54,300 water pixels and 189,121 non-water pixels) from the OLI (Table 2). Large amounts of training samples can more comprehensively represent the spectral signature of both water and non-water pixels, contributing to a higher model performance.

Table 2. Collected sample size of different sensor types.

Samples	TM	ETM	OLI
water bodies	46,041	63,272	54,300
non-water bodies	521,279	172,391	189,121
Total	567,320	235,663	243,421

3.4. Image Classification Based on Random Forest Model

In this study, RF was chosen and implemented for two reasons. RF does not overfit when the number of trees increases and does not need any additional feature selection or model performance test method, since a random selection of features and test methods is built into it [34–36]. Given these advantages, RF has been extensively used to characterize remotely sensed datasets across space and time [3,13,28]. We also employed RF as our classifier. The built-in "Out-of-Bag" (OOB) accuracy, which is unbiased and can be used to substitute the cross-validation or independent test datasets [34], was used to assess the performance of water/non-water pixel classification. In the random forest model, the number of classification trees (T) and the number of features (m) at each node for spilling, is critical to the results [36]. This study employed an RF classifier with 5, 10, 15, 20, 25, 30, 35, 40, 45, and 50 classification trees and chose the optimal item based on the OOB accuracy, using water and non-water samples as the input data. Our sensitivity tests indicate that 20 classification trees are best because, beyond that, the classification gains little performance improvement. For the number of features, previous studies suggested m to be the root of the total number of features [35,37]. Hence, we set the number of features at four. RF was supported by the scikit-learn package 0.17.

3.5. Extracting Lakes Based on Shape Features and Water Inundation Frequency

Our random forest model is not able to distinguish lakes and rivers, and it classified some artificial ponds and paddy fields as water bodies. Given that these features have particular shape features, such as length and roundness (a shape measure that compares the area of the polygon to the square of the maximum diameter of the polygon), the study further adopted rule-based feature extraction for extracted lakes. First, the study composed a maximum water area by stacking all of the classification results. Then, the study built two shape feature rules (length >10,000 m and roundness <0.15) to identify rivers, as well as some artificial ponds and paddy fields. In addition, the study distinguished shadow and water bodies by setting the slope threshold (slope < 15°). These thresholds were selected as the most suitable ones after a trial-and-error process. Next, to diminish the effect of "salt and pepper noise" in the classification process, the study calculated the water inundation frequency (WIF, the number of times a pixel was flagged as a water body divided by the number of cloud-free observations per pixel, expressed as 0% to 100%). Those areas with a lower WIF (~5%) were regarded as noise and non-water bodies, although this process may also remove all traces of the low-frequency surface water that characterizes flooding and ephemeral water bodies. On the other hand, a spatial filtering with a 3×3 window was also employed, to reduce the effect of "salt and pepper" based on a "majority vote" rule. After that, remaining non-lake pixels are very little, and they are almost paddy fields, which can easily be distinguished from lakes. Therefore, we manually removed these pixels and obtained the composited maximum lake areas. The composited maximum lake extent was used as a mask to extract the lakes in each image.

3.6. Accuracy Assessment of Extracted Lakes

Considering that most lakes are located in the tile (path/row 123/39), we selected eight images for this tile and outlined the lake-water extent as being the true lake extent using visual interpretation. The eight selected images consisted of four TM images from 5 December 1995, 13 February 2004, 22 July 2004, and 9 September 2009; two ETM images from 11 January 2000 and 22 July 2000; and two OLI images from 17 September 2013 and 26 November 2015, accounting for four or five year intervals, season changes, and sensor influences (Table 3). Moreover, these eight images have a good quality due to small influence of clouds. Then, we calculated the percentage of overlapped lake-water extent between the extracted lake-water extent and the true lake-water extent.

4. Results

4.1. Determining Optimal Classification Trees

Figure 4 shows our sensitivity test results regarding the optimal classification trees, based on the OOB error. Overall, the OOB errors are very low, ranging from 0.005 to 0.04. When the number of classification trees is less than 20, the TM RF obtains a lower OOB error, followed by the OLI and ETM+ RF. Their OOB errors decrease quickly as the number of classification trees increases and reach a stable point ($T = 20$). From then on, increasing the number of classification trees has no significant impact on the OOB error. Therefore, the study set the optimal classification trees at 20 for the TM, ETM+, and OLI RF, when the number of features (m) was four.

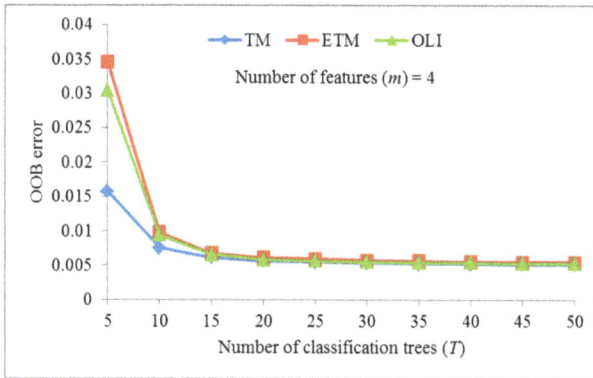

Figure 4. Impacts of different numbers of trees in the Random Forest based on OOB error assessment.

4.2. Accuracy of the Extracted Lakes

The accuracies of the eight images were 93.17%, 95.14%, 96.04%, 91.71%, 95.53%, 89.68%, 90.17%, and 93.41% (Table 3). The mean value and standard deviation of accuracy were 93.11% and 2.26%, respectively. The results showed that lake-water extraction based on the random forest model is feasible.

Table 3. Accuracy of extracted lakes based on the random forest model.

Sensor Type	Date	Extracted Lake Area (km^2)	True Lake Area (km^2)	Accuracy (%)
TM	1995/12/5	1003.64	1077.24	93.17
ETM	2000/1/11	618.19	649.76	95.14
ETM	2000/7/22	658.91	686.08	96.04
TM	2004/2/13	828.95	903.92	91.71
TM	2004/7/22	789.64	826.57	95.53
TM	2009/9/9	760.88	848.40	89.68
OLI	2013/9/17	810.32	898.67	90.17
OLI	2015/11/26	899.70	963.17	93.41

4.3. Water Inundation Frequency from 1987 to 2015

According to continuous Landsat time series, we calculated the WIF and divided it into six grades: 5%–10%, 10%–20%, 20%–40%, 40%–60%, 60%–80%, and 80%–100% (Figure 5). The highest WIF occurred over core areas of permanent lakes with a WIF over 80%. In contrast, the edge areas of lakes experienced a decreasing WIF. Table 4 shows the statistics of five WIF grades across the entire study area. The area of water bodies is 226.29 km^2, 322.71 km^2, 460.35 km^2, 400.79 km^2, 535.51 km^2, and 635.42 km^2 under WIFs of 5%–10%, 10%–20%, 20%–40%, 40%–60%, 60%–80%, and 80%–100%, respectively.

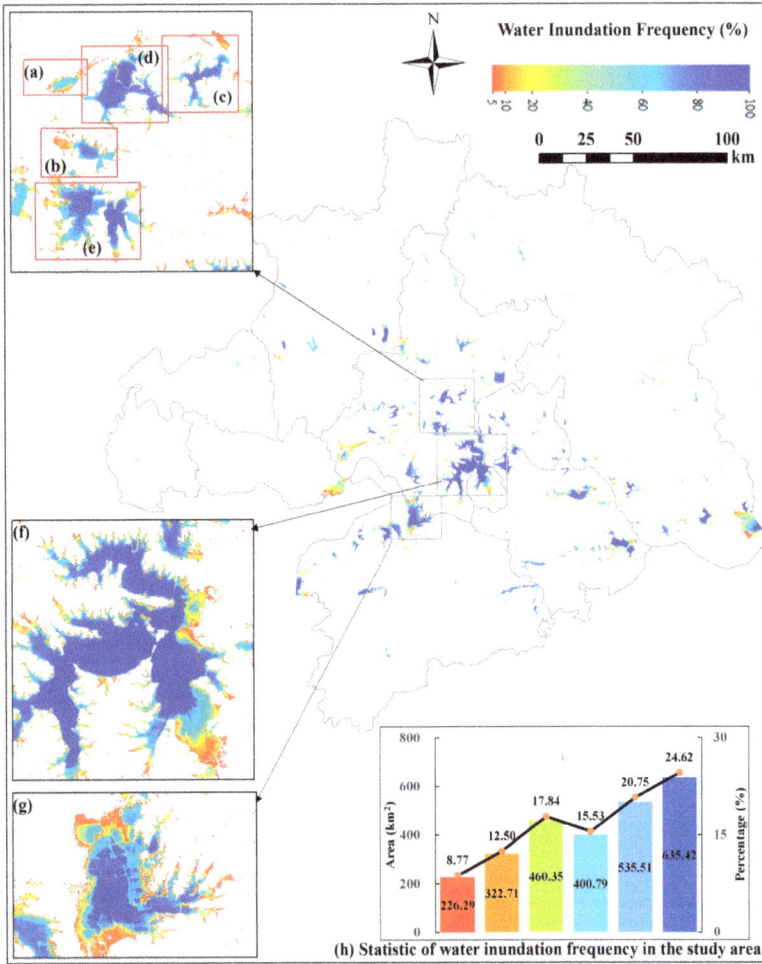

Figure 5. Water inundation frequency from 1987~2015. (**a**)–(**g**) represents Shahu, Nanhu, Yanxihu, Donghu, Tangxunhu, Liangzihu, and Futouhu, respectively. (**h**) shows the lake-water area at a water inundation frequency of 5%–10%, 10%–20%, 20%–40%, 40%–60%, 60%–80%, and 80%–100% from left to right, respectively.

4.4. Lake Water Extent Changes from 1987~2015 in Different Cities

We further explored the annual maximum lake-water extent across the entire study area from 1987–2015 (Figure 6). Considering that the earlier images are much smaller in number and cannot cover the entire study area, we composited two multi-year maximum lake areas from 1987–1990 and from 1991–1994. The lake in the study area experienced an apparent decrease from 1987–2015. From 1987–2002, the lake area shows a fluctuating tendency. However, it experienced a continuous decrease from 2003–2007, followed by a fluctuating tendency from 2008–2015, but the lake area did not return to the previous condition. The calculated results revealed that the average maximum lake area was 2276.03 km^2 from 1995–1999, and it decreased to 1966.94 km^2 from 2011–2015, with a total loss of 309.09 km^2 and a percentage loss of 13.58%.

Among the eight cities, Wuhan experienced the most serious lake-water area loss from 1987–2015. From 1995–1999, the average lake-water area was 803.01 km², and the area decreased to 688.18 km² from 2011–2015, with a total loss of 114.83 km² or 14.30%. Huanggang, Xianning, and Xiaogan also experienced relatively serious losses, totaling an area of loss of 44.40 km², 45.39 km², and 31.18 km², and a percentage loss of 11.83%, 13.16%, and 23.05%, respectively.

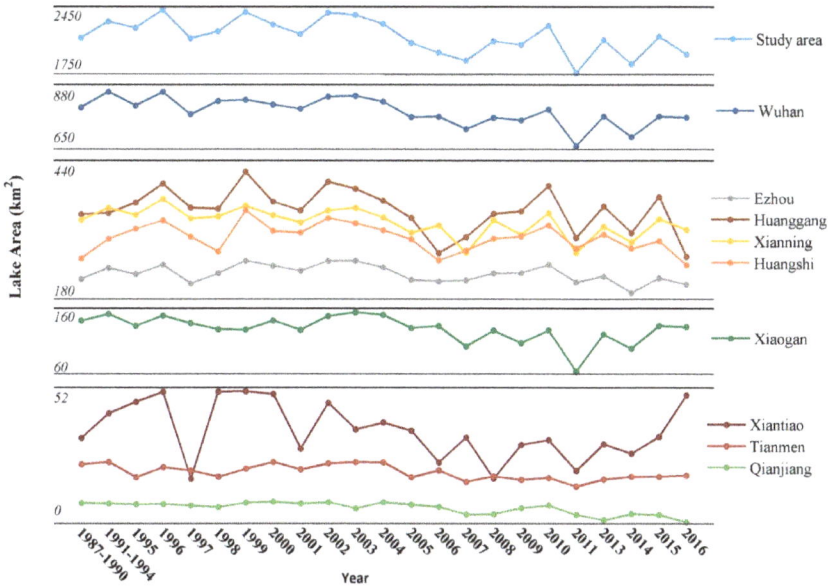

Figure 6. Lake water extent changes from 1987~2015.

4.5. Decreasing Lake Water Extent Due to Urbanization in Urban Areas

From 1995 to 2009, about 193.85 km² of lake bodies have been changed into agricultural land (paddy fields), and about 16.60 km² of lake bodies have been changed into building land in the Wuhan urban agglomeration, which indicates that a rapid urbanization over the past several years is an important reason for the loss of lake areas, especially for Wuhan [7,8]. Therefore, we selected five typical urban lakes (Tangxunhu, Donghu, Yanxihu, Nanhu, and Shahu) in Wuhan to explore the phenomena (Figures 7 and 8, Table 4). These five urban lakes are majorly encroached by building land.

Table 4. Lake-water extent changes of typical urban lakes.

Lake Name	Lake Area (km²)				Loss Rate (km²/year)		
	1995	2000	2008	2015	1995–2000	2000–2008	2008–2015
TangXunHu	44.52	46.40	43.70	41.98	−0.38	0.34	0.25
DongHu	33.62	33.82	32.24	30.53	−0.04	0.20	0.24
YanXiHu	12.59	12.53	12.25	12.01	0.01	0.04	0.03
NanHu	10.68	10.18	8.03	7.34	0.10	0.27	0.10
ShaHu	4.72	4.05	3.06	2.48	0.13	0.12	0.08

Note: *Loss rate* = $(Area_i - Area_j)/(j - i)$. $Area_i$ represents lake area in the previous year i, $Area_j$ represents lake area in the later year j. Therefore, the positive value indicates the decrease while the negative value indicates the increase.

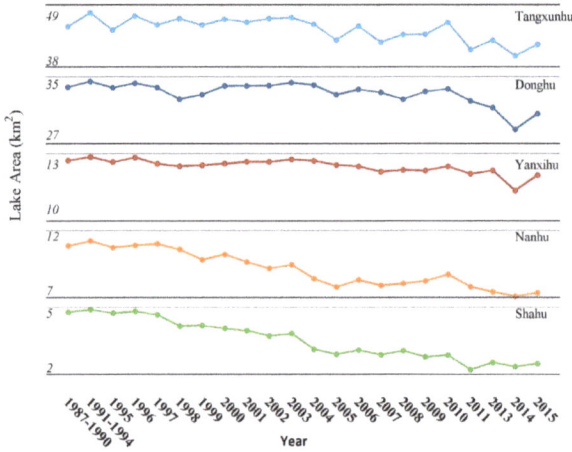

Figure 7. Lake water extent changes from 1987~2015.

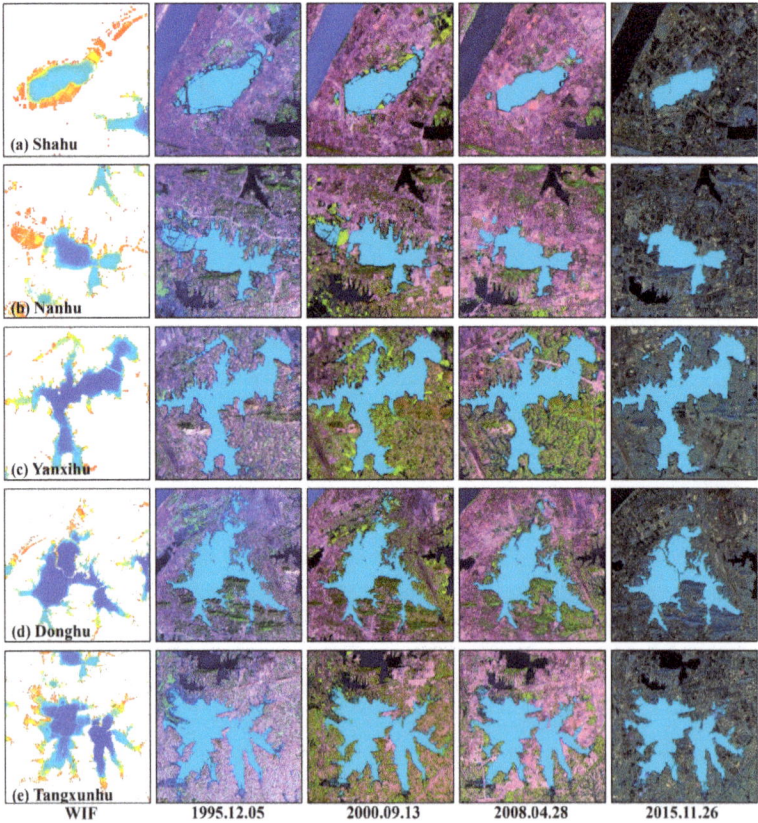

Figure 8. Lake water area changes from 1987~2015. First column represents the water inundation frequency (WIF) from 1987 to 2015 for these five lakes. The other four columns represent the composited maximum lake-water extents of these five lakes at different years: 1995, 2000, 2008, and 2015.

The study observed that Shahu and Nanhu have been declining from 1987–2015. The total areas of loss of Shahu and Nanhu were 2.23 km^2 and 3.35 km^2, respectively, from 1995–2015, representing 47.35% and 31.31% of the lake-water areas in 1995. Yanxihu showed a relatively stable tendency before 2003 and then experienced a rapid declining tendency from 2013. The total area of loss of Yanxihu was 0.58 km^2 from 1995–2015, accounting for 4.61% of the lake-water area in 1995. Donghu and Tangxunhu displayed a fluctuating tendency before 2003. From 2003–2010, the two cities experienced a slight declining tendency, followed by a rapid declining tendency after 2010. The total loss areas in Donghu and Tangxunhu were 3.10 km^2 and 2.54 km^2, respectively, from 1995–2015, accounting for 9.21% and 5.71% of the lake-water areas in 1995.

4.6. Lake Water Extent Changes Due to Climate Changes in Non-Urban Areas

Precipitation variation plays a critical role in lake areas. This study chose Liangzihu and Futouhu as non-urban lakes where urbanization has had little impact, to calculate the Pearson's correlation coefficient between lake areas and the annual maximum precipitation from 1987–2015. In addition, the correlation of the entire study area was also calculated.

These two non-urban lakes showed a significant correlation (*p*-values < 0.05). For Futouhu and Liangzihu, their *r*-values in area were 0.6771 and 0.6558, respectively. Overall, the lake-water areas of the entire study area showed a significant correlation with precipitation. It is evident from Figure 9 that much rainfall in the study area in 1996, 1998, 2002, and 2010 resulted in floods, which increased the lake-water area (Figure 10), while little rainfall in the study area in 1997, 2001, 2006, and 2011 decreased the lake-water area (Figure 11), because of droughts. From 1999–2001 and from 2002–2006, rainfall in the study area continuously decreased, resulting in a continuous decrease in the lake-water area, whereas, from 2007–2010, rainfall in the study area increased, resulting in an increase in the lake-water area. Moreover, for Futouhu, we found that relatively regular enclosure nets are distributed in Lake Futohu from Figure 5, indicating the influence of human activities. Therefore, the stable and increased extent of Lake Futouhu from 2004 to 2006, while the precipitation was decreasing, is possible.

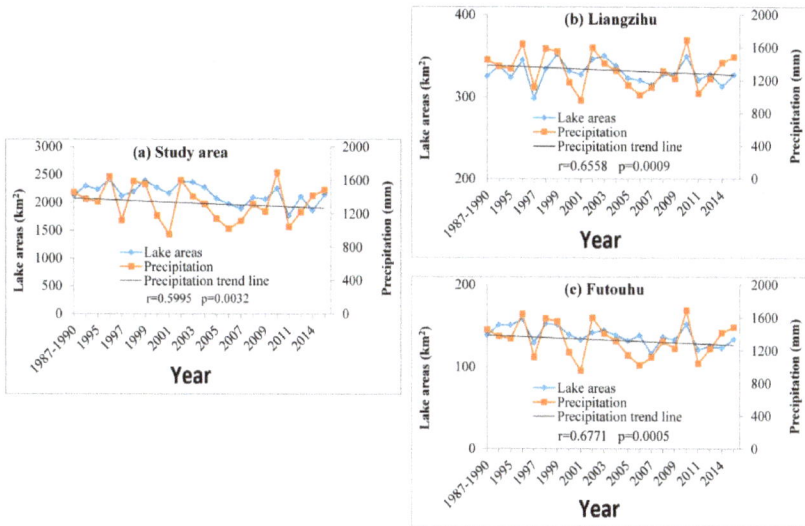

Figure 9. Correlation between annual maximum lake areas and annual precipitation from 1987 to 2015.

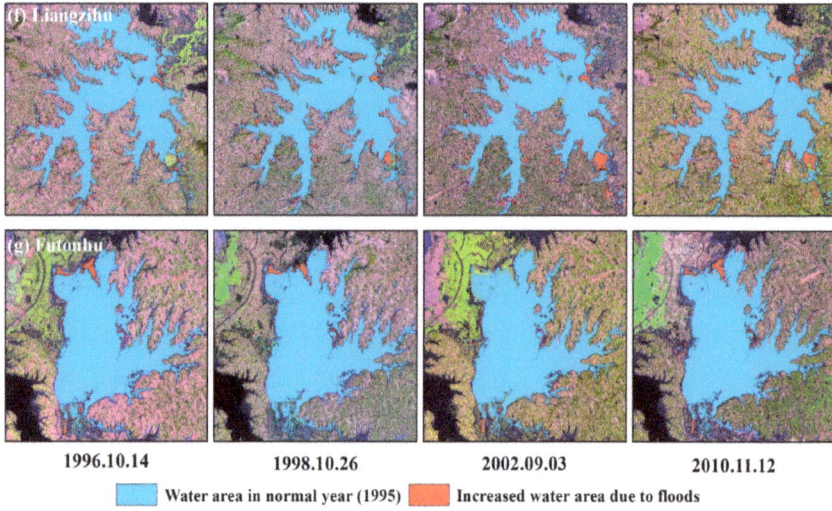

Figure 10. Increased water area due to floods in four periods: 1996, 1998, 2002, and 2010. The lake water area refers to the composited maximum area in the same year.

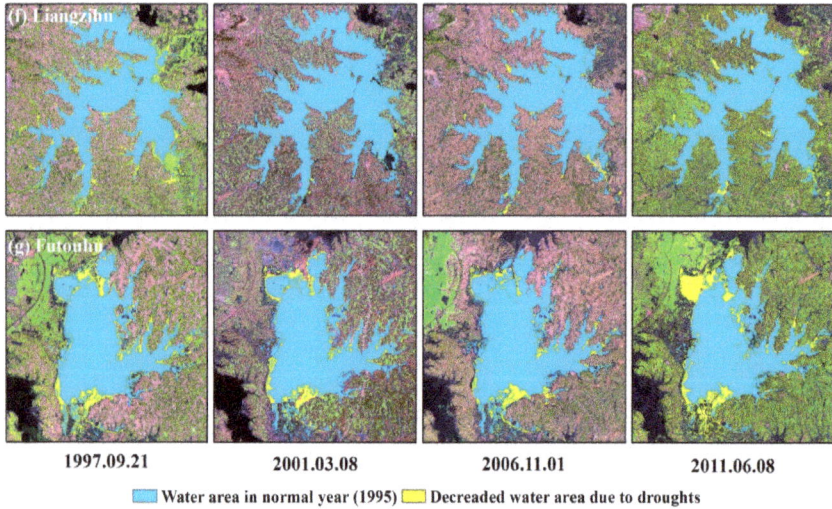

Figure 11. Decreased water area due to droughts in four periods: 1997, 2001, 2006, and 2011. The lake water areas refer to the composited maximum area in the same year.

5. Discussions

5.1. Long-Term Lake Water Extent Dynamics from Landsat Time Series Data

This study applied all available Landsat time series data from 1987–2015 to generate a comprehensive historical record of lake-water extent dynamics (1987–2015) at a high spatial resolution (30 m), for the entire Wuhan urban agglomeration. The study justified using the random forest model based on the multi-feature variables to extract water bodies from massive remote-sensing data. Tulbure [13] also applied the same method to produce continuous surface water–time series at the subcontinental scale in an Australian semi-arid region, the Murray-Darling Basin, with an overall classification accuracy of 99.9%. Given that the water-body types and land-use types in the Wuhan urban agglomeration are more complicated than they are in the Murray-Darling Basin, the mean value (93.11%) and standard deviation (2.26%) for this study are reasonable. In addition, this study did not select the same feature variables as those in Tulbure's study. For example, in Tulbure's study, AWEI [20] was used for the TM and ETM+ images. However, in this study, the use of AWEI is not feasible for the OLI images because the SWIR band is different from the TM band.

5.2. Decreasing Lake Water Extent Dynamics Under Urbanization Process and Climate Changes

From 1987–2015, the lake-water extent in the Wuhan urban agglomeration suffered from a substantial decrease, for two reasons. One import factor is urbanization. Due to rapid urban development in the Wuhan urban agglomeration over the past several years, the lake was converted into built-up areas, causing a substantial decrease in the lake-water extent (Figure 8). Chen et al. [7] revealed that the population is the main driving factor causing the morphological change of an urban lake in Donghu, Wuhan. Increasing numbers of people flowed into the urban area. Thus, increasing numbers of urban lands were needed, causing land-use changes. However, the transformation did not occur simultaneously. For example, Shahu is close to the city center, so it suffered a decrease in the lake-water extent at the earliest point in time. As the urban area expanded outward, lakes such as Tangxunhu, that were once far from the city's center, began to experience damage due to the conversion into built-up areas. In 2008, the Wuhan urban agglomeration was approved as the national comprehensive reform pilot area for promoting resource-saving and environmentally friendly developments. Increasing attention has been placed on environmental and ecological issues, and lake protection has been contained in development planning. Therefore, the lake's loss rate from 2008–2015 was lower than that from 1987–2015, for Tangxunhu, Yanxihu, Nanhu, and Shahu (Table 4). Another more important factor is climate change, especially the variation in precipitation. The total lake-water extent is strongly correlated with annual precipitation (Figure 9). For example, in 2011, the lake area of the Wuhan urban agglomeration reached the minimum value. At the same time, annual precipitation also reached the relative minimum value (Figure 9). In addition, this study found that annual precipitation showed a decreasing trend, which may explain the decrease in lake area from the perspective of climate change (Figure 9).

5.3. Uncertainties and Prospects

Some uncertainties exist in this study's mapping of long-term lake-water extent. One uncertainty is due to cloud cover. Cloud cover not only reduces the number of clear observations, but also causes cloud shadow that is difficult to distinguish from water bodies. The assumption that clouds cannot persist at the same place [38] could diminish the adverse effect by removing low-frequency water bodies. Moreover, atmospheric effects play a very critical role in datasets across time and space. However, due to the high cost of atmospheric correction in time and process, the study didn't pursue atmospheric correction. However, TOA reflectance was used as the feature variable to correct for the difference of the sun–earth distance, exoatmospheric solar irradiance, and solar geometry between images [25]. Future study will focus on the impacts of urbanization on the lake landscape and the resulting adverse influences, based on long-term lake mapping.

6. Conclusions

As sustainable development is vigorously advocated by the government, lakes will play an increasing role in providing urban ecosystem services for the Wuhan urban agglomeration. Therefore, a clear understanding of lake-water extent dynamics is necessary. This research revealed the lake-water extent dynamics in the Wuhan urban agglomeration, based on seasonally continuous Landsat TM/ETM+/OLI data from 1987–2015 and the random forest model. Overall, this study provides the following conclusions:

(1) This study justified the use of the random forest model based on the multi-feature variables to extract water bodies from massive remote-sensing data. Because this method is relatively simple, feasible, and highly accurate, it is a promising method for identifying water bodies at a large scale, such as nationally.

(2) The lake-water areas of the Wuhan urban agglomeration were 226.29 km^2, 322.71 km^2, 460.35 km^2, 400.79 km^2, 535.51 km^2, and 635.42 km^2 under water inundation frequencies of 5%–10%, 10%–20%, 20%–40%, 40%–60%, 60%–80%, and 80%–100%, respectively.

(3) The lake-water area of the Wuhan urban agglomeration experienced a substantial decrease. The area was 2276.03 km^2 from 1995–1999, and it decreased to 1966.94 km^2 from 2011–2015. In particular, the lake-water areas of loss of Wuhan, Huanggang, Xianning, and Xiaogan were 114.83 km^2, 44.40 km^2, 45.39 km^2, and 31.18 km^2, with loss percentages of 14.30%, 11.83%, 13.16%, and 23.05%, respectively.

Acknowledgments: This work was supported by funds from the National Natural Science Foundation of China (41571077) and the Fundamental Research Funds for the Central Universities.

Author Contributions: Yue Deng, Weiguo Jiang, and Jiahong Li conceived and designed the study. Yue Deng and Weiguo Jiang performed the experiments and wrote the paper. Jiahong Li and Zheng-Hong Tang served as scientific advisors. Jinxia Lv, Zheng Chen, and Kai Jai helped perform the experiments and write the paper. Jiahong Li. Yue Deng, Weiguo Jiang, Jiahong Li, Jinxia Lv, Zheng Chen, and Kai Jai reviewed and edited the manuscript. All authors read and approved the manuscript.

Conflicts of Interest: The authors declare no conflict of interest.

References

1. Gibbs, J.P. Wetland loss and biodiversity conservation. *Conserv. Biol.* **2000**, *14*, 314–317. [CrossRef]
2. Gleason, R.; Laubhan, M.; Euliss, N. USGS Professional Paper 1745: Ecosystem Services Derived from Wetland Conservation Practices in the United States Prairie Pothole Region with an Emphasis on the U.S. Department of Agriculture Conservation Reserve and Wetlands Reserve Programs. Available online: http://pubs.usgs.gov/pp/1745/ (accessed on 16 October 2016).
3. Gabrielsen, C.G.; Murphy, M.A.; Evans, J.S. Using a multiscale, probabilistic approach to identify spatial-temporal wetland gradients. *Remote Sens. Environ.* **2016**, *184*, 522–538. [CrossRef]
4. Yang, B.; Ke, X. Analysis on urban lake change during rapid urbanization using a synergistic approach: A case study of Wuhan, China. *Phys. Chem. Earth Parts ABC* **2015**, *89–90*, 127–135. [CrossRef]
5. Wang, X.; Ning, L.; Yu, J.; Xiao, R.; Li, T. Changes of urban wetland landscape pattern and impacts of urbanization on wetland in Wuhan City. *Chin. Geogr. Sci.* **2008**, *18*, 47–53. [CrossRef]
6. Xu, K.; Kong, C.; Liu, G.; Wu, C.; Deng, H.; Zhang, Y.; Zhuang, Q. Changes of urban wetlands in Wuhan, China, from 1987 to 2005. *Prog. Phys. Geogr.* **2010**, *34*, 207–220.
7. Chen, K.; Wang, X.; Li, D.; Li, Z. Driving force of the morphological change of the urban lake ecosystem: A case study of Wuhan, 1990–2013. *Ecol. Model.* **2015**, *318*, 204–209. [CrossRef]
8. Du, N.; Ottens, H.; Sliuzas, R. Spatial impact of urban expansion on surface water bodies—A case study of Wuhan, China. *Landsc. Urban Plan.* **2010**, *94*, 175–185. [CrossRef]
9. Xu, H.; Bai, Y. Evaluation of urban lake evolution using Google Earth Engine—A case study of Wuhan, China. In Proceedings of the 2015 Fourth International Conference on Agro-Geoinformatics (Agro-geoinformatics), Istanbul, Turkey, 20–24 July 2015; pp. 322–325.
10. Elmi, O.; Tourian, M.; Sneeuw, N. Dynamic river masks from multi-temporal satellite imagery: An automatic algorithm using graph cuts optimization. *Remote Sens.* **2016**, *8*, 1005. [CrossRef]
11. Li, L.; Xia, H.; Li, Z.; Zhang, Z. Temporal-spatial evolution analysis of lake size-distribution in the middle and lower Yangtze River Basin using Landsat imagery data. *Remote Sens.* **2015**, *7*, 10364–10384. [CrossRef]
12. Halabisky, M.; Moskal, L.M.; Gillespie, A.; Hannam, M. Reconstructing semi-arid wetland surface water dynamics through spectral mixture analysis of a time series of Landsat satellite images (1984–2011). *Remote Sens. Environ.* **2016**, *177*, 171–183. [CrossRef]

13. Tulbure, M.G.; Broich, M.; Stehman, S.V.; Kommareddy, A. Surface water extent dynamics from three decades of seasonally continuous Landsat time series at subcontinental scale in a semi-arid region. *Remote Sens. Environ.* **2016**, *178*, 142–157. [CrossRef]

14. Mueller, N.; Lewis, A.; Roberts, D.; Ring, S.; Melrose, R.; Sixsmith, J.; Lymburner, L.; McIntyre, A.; Tan, P.; Curnow, S.; Ip, A. Water observations from space: Mapping surface water from 25 years of Landsat imagery across Australia. *Remote Sens. Environ.* **2016**, *174*, 341–352. [CrossRef]

15. Han, X.; Chen, X.; Feng, L. Four decades of winter wetland changes in Poyang Lake based on Landsat observations between 1973 and 2013. *Remote Sens. Environ.* **2015**, *156*, 426–437. [CrossRef]

16. Carroll, M.; Wooten, M.; DiMiceli, C.; Sohlberg, R.; Kelly, M. Quantifying surface water dynamics at 30 meter spatial resolution in the North American high northern latitudes 1991–2011. *Remote Sens.* **2016**, *8*, 622. [CrossRef]

17. Wu, G.; Liu, Y. Mapping Dynamics of inundation patterns of two largest river-connected lakes in China: A comparative study. *Remote Sens.* **2016**, *8*, 560. [CrossRef]

18. Xu, H. Modification of Normalised Difference Water Index (NDWI) to enhance open water features in remotely sensed imagery. *Int. J. Remote Sens.* **2006**, *27*, 3025–3033. [CrossRef]

19. Mcfeeters, S.K. The use of the Normalized Difference Water Index (NDWI) in the delineation of open water features. *Int. J. Remote Sens.* **1996**, *17*, 1425–1432. [CrossRef]

20. Feyisa, G.L.; Meilby, H.; Fensholt, R.; Proud, S.R. Automated water extraction index: A new technique for surface water mapping using Landsat imagery. *Remote Sens. Environ.* **2014**, *140*, 23–35. [CrossRef]

21. Otsu, N. A Threshold selection method from gray-level histograms. *IEEE Trans. Syst. Man Cybern.* **1979**, *9*, 62–66. [CrossRef]

22. Li, W.; Du, Z.; Ling, F.; Zhou, D.; Wang, H.; Gui, Y.; Sun, B.; Zhang, X. A Comparison of land surface water mapping using the normalized difference water index from TM, ETM+ and ALI. *Remote Sens.* **2013**, *5*, 5530–5549. [CrossRef]

23. Rokni, K.; Ahmad, A.; Selamat, A.; Hazini, S. Water feature extraction and change detection using multitemporal Landsat imagery. *Remote Sens.* **2014**, *6*, 4173–4189. [CrossRef]

24. Yang, Y.; Liu, Y.; Zhou, M.; Zhang, S.; Zhan, W.; Sun, C.; Duan, Y. Landsat 8 OLI image based terrestrial water extraction from heterogeneous backgrounds using a reflectance homogenization approach. *Remote Sens. Environ.* **2015**, *171*, 14–32. [CrossRef]

25. Tulbure, M.G.; Broich, M. Spatiotemporal dynamic of surface water bodies using Landsat time-series data from 1999 to 2011. *ISPRS J. Photogramm. Remote Sens.* **2013**, *79*, 44–52. [CrossRef]

26. Gómez, C.; White, J.C.; Wulder, M.A. Optical remotely sensed time series data for land cover classification: A review. *ISPRS J. Photogramm. Remote Sens.* **2016**, *116*, 55–72. [CrossRef]

27. Khatami, R.; Mountrakis, G.; Stehman, S.V. A meta-analysis of remote sensing research on supervised pixel-based land-cover image classification processes: General guidelines for practitioners and future research. *Remote Sens. Environ.* **2016**, *177*, 89–100. [CrossRef]

28. Li, X.; Gong, P.; Liang, L. A 30-year (1984–2013) record of annual urban dynamics of Beijing City derived from Landsat data. *Remote Sens. Environ.* **2015**, *166*, 78–90. [CrossRef]

29. Chander, G.; Markham, B.L.; Helder, D.L. Summary of current radiometric calibration coefficients for Landsat MSS, TM, ETM+, and EO-1 ALI sensors. *Remote Sens. Environ.* **2009**, *113*, 893–903. [CrossRef]

30. Huete, A.; Didan, K.; Miura, T.; Rodriguez, E.P.; Gao, X.; Ferreira, L.G. Overview of the radiometric and biophysical performance of the MODIS vegetation indices. *Remote Sens. Environ.* **2002**, *83*, 195–213. [CrossRef]

31. Tucker, C.J. Red and photographic infrared linear combinations for monitoring vegetation. *Remote Sens. Environ.* **1979**, *8*, 127–150. [CrossRef]

32. Gao, B. NDWI—A normalized difference water index for remote sensing of vegetation liquid water from space. *Remote Sens. Environ.* **1996**, *58*, 257–266. [CrossRef]

33. Tang, Z.; Li, Y.; Gu, Y.; Jiang, W.; Xue, Y.; Hu, Q.; LaGrange, T.; Bishop, A.; Drahota, J.; Li, R. Assessing Nebraska playa wetland inundation status during 1985–2015 using Landsat data and Google Earth Engine. *Environ. Monit. Assess.* **2016**, *188*, 654. [CrossRef] [PubMed]

34. Breiman, L. Random Forests. *Mach. Learn.* **2001**, *45*, 5–32. [CrossRef]

35. Gislason, P.O.; Benediktsson, J.A.; Sveinsson, J.R. Random Forests for Land Cover Classification. *Pattern Recogn. Lett.* **2006**, *27*, 294–300. [CrossRef]

36. Yu, X.; Hyyppä, J.; Vastaranta, M.; Holopainen, M.; Viitala, R. Predicting individual tree attributes from airborne laser point clouds based on the random forests technique. *ISPRS J. Photogramm. Remote Sens.* **2011**, *66*, 28–37. [CrossRef]

37. Stumpf, A.; Kerle, N. Object-oriented mapping of landslides using random forests. *Remote Sens. Environ.* **2011**, *115*, 2564–2577. [CrossRef]

38. Zhu, Z.; Woodcock, C.E. Automated cloud, cloud shadow, and snow detection in multitemporal Landsat data: An algorithm designed specifically for monitoring land cover change. *Remote Sens. Environ.* **2014**, *152*, 217–234. [CrossRef]

MDPI AG

St. Alban-Anlage 66

4052 Basel, Switzerland

Tel. +41 61 683 77 34

Fax +41 61 302 89 18

http://www.mdpi.com

Remote Sensing Editorial Office

E-mail: remotesensing@mdpi.com

http://www.mdpi.com/journal/remotesensing